PROCEEDINGS OF THE 2ND INTERNATIONAL CONFERENCE ON TEACHER EDUCATION AND PROFESSIONAL DEVELOPMENT (INCOTEPD 2017), 21–22 OCTOBER 2017, YOGYAKARTA, INDONESIA

Character Education for 21st Century Global Citizens

Editors

Endah Retnowati, Anik Ghufron, Marzuki, Kasiyan,
Adi Cilik Pierawan & Ashadi
Universitas Negeri Yogyakarta, Indonesia

Routledge
Taylor & Francis Group

LONDON AND NEW YORK

Published 2019 by Routledge
2 Park Square, Milton Park, Abingdon, Oxon OX14 4RN
605 Third Avenue, New York, NY 10017

First issued in paperback 2020

Routledge is an imprint of the Taylor & Francis Group, an informa business

Typeset by V Publishing Solutions Pvt Ltd., Chennai, India

ISBN 13: 978-0-367-73320-9 (pbk)
ISBN 13: 978-1-138-09922-7 (hbk)

Published by: CRC Press/Balkema
            Schipholweg 107C, 2316 XC Leiden, The Netherlands
            e-mail: Pub.NL@taylorandfrancis.com
            www.crcpress.com – www.taylorandfrancis.com

Table of contents

Teacher professional development for enhancing character education

*Curriculum syllabus lesson plan learning materials development
for integrated values education*

Developing learning activities tasks strategies for character education

Preface

The 2nd International Conference on Teacher Education and Professional Development was held in Yogyakarta (Indonesia) on 21–22 October 2017. The conference is an annual event, conducted by Universitas Negeri Yogyakarta's Institute of Educational Development and Quality Assurance. Similar to the previous conference, this conference received enthusiastic response from scholars and practitioners particulcarly interested in character education. Participants from Australia, Japan, Malaysia, the Netherlands, and many cities in Indonesia attended this year's conference.

Exploring the theme "Character Education for 21st Century Global Citizens", the committee has invited Prof. dr. Ali Ghufron Mukti, M.Sc., Ph.D. (General Director of Higher Education and Human Resources from the Ministry of Research, Technology, and Higher Education of the Republic of Indonesia) as a keynote speaker. Moreover, the committee has also invited Prof. Azyumardi Azra, Ph.D., Prof. Dr. Wiel Veugelers, Asst. Prof. Dr. Betania Kartika Muflih, Emeritus Prof. Dr. Terry Lovat, Prof. Dr. Kerry John Kennedy and Prof. Suyanto, Ph.D as invited speakers. Participants presented their papers, which are categorized under subthemes: 1) Values for 21st century global citizens, 2) Preparing teachers for integrative values education, 3) Teacher professional development for enhanced character education, 4) Curriculum/syllabus/lesson plan/learning materials development for integrated values education, 5) Developing learning activities/tasks/strategies for character education, 6) Assessing student's character development (values acquisition assessment), 7) Creating/managing conducive school culture to character education, and 8) Parents and public involvement in character education.

There were approximately 232 submissions from various countries to the conference. The committee selected 127 papers to be presented in this year's conference. The scientific committee has reviewed 117 papers that are qualified for publication. After a careful consideration, there are 83 papers (covering sub-themes 1 to 7) included in the proceeding of the conference that is published by CRC Press/Balkema and submitted for indexation to Thomson Reuters/Scopus.

Marzuki, *Universitas Negeri Yogyakarta, Indonesia*
Endah Retnowati, *Universitas Negeri Yogyakarta, Indonesia*
Anik Ghufron, *Universitas Negeri Yogyakarta, Indonesia*
Kasiyan, *Universitas Negeri Yogyakarta, Indonesia*
Adi Cilik Pierawan, *Universitas Negeri Yogyakarta, Indonesia*
Ashadi, *Universitas Negeri Yogyakarta, Indonesia*

Acknowledgement

Prof. Monica J. Taylor, Ph.D., *University of London, UK*
Prof. Lesley Harbon, *University of Technology Sydney, Australia*
Dr. Alfredo Bautista, *National Institute of Education, Singapore*
Heidi Layne, Ph.D., *University of Helsinki, Finland*
Prof. Dr. Ng Shun Wing, *Hong Kong Institute of Education, Hong Kong*
Dr. Derek Patton, *Melbourne Graduate School of Education, Australia*
Dorothea Wilhelmina Hancock, Ph.D., *Queensland University of Technology, Australia*
Prof. Micha De Winter, Ph.D., *Uthrecht University*
Prof. Anne Burns, BA (Hons), Diploma in Adult TESOL, PhD, Med, *UNSW, Australia*
Prof. Azyumardi Azra, Ph.D., *Universitas Islam Negeri Syarif Hidayatullah, Jakarta, Indonesia*
Prof. Wiel Veugelers, *University of Humanistic Studies in Utrecht*
Ass. Prof. Dr. Betania Kartika, M.A., *International Islamic University, Malaysia*
Emiritus Prof. Terry Lovat, Ph.D., *University of New Castle, Australia*
Prof. Kerry Kennedy., *Education University of Hong Kong, Hong Kong*
Prof. Suyanto, Ph.D., *Universitas Negeri Yogyakarta, Indonesia*
Prof. Dr. Anik Ghufron, M.Pd., *Universitas Negeri Yogyakarta, Indonesia*
Prof. Dr. Suwarna, M.Pd., *Universitas Negeri Yogyakarta, Indonesia*
Prof. Sukirno, Ph.D., *Universitas Negeri Yogyakarta, Indonesia*
Prof. AK Projosantosa, Ph.D., *Universitas Negeri Yogyakarta, Indonesia*
Prof. Suwarsih Madya, Ph.D., *Universitas Negeri Yogyakarta, Indonesia*
Prof. Darmiyati Zuchdi, Ed.D., *Universitas Negeri Yogyakarta, Indonesia*
Prof. Dr. Sri Atun, *Universitas Negeri Yogyakarta, Indonesia*
Prof. Sugirin, Ph.D., *Universitas Negeri Yogyakarta, Indonesia*
Prof. Dr. Sunaryo Kartadinata, *Universitas Pendidikan Indonesia, Bandung, Indonesia*
Dr. Marzuki, *Universitas Negeri Yogyakarta, Indonesia*
Endah Retnowati, Ph.D., *Universitas Negeri Yogyakarta, Indonesia*
Dr. Kasiyan, *Universitas Negeri Yogyakarta, Indonesia*
Adi Cilik Pierawan, Ph.D., *Universitas Negeri Yogyakarta, Indonesia*
Ashadi, Ed.D., *Universitas Negeri Yogyakarta, Indonesia*

Organizing committees

PATRON

Prof. Dr. Sutrisna Wibawa, M.Pd.

ADVISOR

Prof. Dr. Anik Ghufron, M.Pd.

CONFERENCE CHAIR

Dr. Marzuki, M.Ag.

COMMITTEES

Kasiyan
Endah Retnowati
Joko Priyana
Ade Gafar Abdullah
Suwarna
Adi Cilik Pirewan
Samsuri
Sudiyatno
Sukirno
AK Projosantosa
Suwarsih Madya
Sugirin
Sunaryo Kartadinata
Ashadi
Heri Retnawati
Retna Hidayah
Nunik Sugesti
Sri Handayani
Mutaqin
Rukiyati
Joko Sudomo
Rosita E. Kusmaryanti
Budi Sulistiya
Supoyo
Dani Hendra Kristiawan
Binar Winantaka
Arpiaka Harani Pornawan
M. Zayin Adib
Rifqi Nur Setiawan
Marita Bahriani

Values for 21st century global citizens

Character Education for 21st Century Global Citizens – Retnowati et al. (Eds)
© 2019 Taylor & Francis Group, London, ISBN 978-1-138-09922-7

Moral and citizenship education in 21st century: The role of parents and the community

W. Veugelers
University of Humanistic Studies, Utrecht, The Netherlands

ABSTRACT: Individuals, communities, and societies can have different ideas about moral values and citizenship education: it can be focused on adaptation, individualisation, or on social justice and democracy. There can be differences between goals, practices and learning effects. Also in the concept of global citizenship education (gce) there are different orientations: on open gce with openness for other cultures, a moral gce that supports human development and concern for the world, and a social-political gce that addresses unequal power relations and is oriented to social justice. We will present data of a small comparative study of how Dutch and Indonesian students of teacher education for moral education and citizenship education think about global citizenship education. These studies show that it is important to recognise that people can think different about moral values and citizenship, that education should stimulate reflection and dialogue about what are relevant values and citizenship education, and create practices that challenge students. What can be the role of parents and communities in it? How can a more inclusive education be stimulated?

1   MORAL VALUES, CITIZENSHIP, AND CHARACTER

Education helps students in developing their identity and their preparation for participating in society. Different concepts are used for this kind of education: f.e. moral education, character education and citizenship education. Interesting is that both in theory and in practice there is a strong and increasing relationship between the personal and the societal, and the moral and the political. We think these links are fruitful to analyse theory, policy, and practice of education oriented to personal, social, societal, and political development.

In this article we use in particular the concepts moral values and citizenship education. Moral values express what people find important in life, for themselves and to live with others. Citizenship education is the way education helps students in developing their identity and their participation in society. In recent decades the concept of citizenship has been what we call broaded: from national, to regional, and global (Veugelers, 2011b). And deepened: from the political, to the social, and to the cultural. Citizenship is not anymore only formulated at the level of the political, but also as a social-cultural concept, that refers to identity, even sometimes on very concrete levels of preferred behaviour and character traits. We prefer the concept moral values above character, because moral values refer more to the ideas behind behaviour, to the purpose of life. Character is a concept that refers more to behaviour and express less explicitly the ideas embedded in it.

1.1   *Different levels of the curriculum*

To speak and think about citizenship education it is important to look at what kind of pedagogical aims are involved, what curriculum material, what kind of teaching methods in schools, and what learning activities students perform, and last but not least what students learn. It is necessary to distinguish the different levels of the curriculum, in our case of citizenship education.

Goodlad (1989) made an interesting division in levels of the curriculum:

- The *ideal* level. The general pedagogical goals, as formulated in speeches and documents.
- The *formal* level. The regulations and formal descriptions of the curriculum. The guidelines teachers have to follow.
- The *interpreted* level. This is how a teacher interpreted the formal curriculum. Each teacher makes always an own interpretation of the curriculum. Add examples and other topics, uses other methods, and sometimes skip content. Educational systems differ in the amount of freedom they give to the teachers to make their interpretation. But even in a very tight system there is always some space to make an own interpretation.
- The *operationalized* level. This is about what teachers exactly do in their classroom. This practice can differ what they intend to do. The practice is the reality researchers can observe.
- The *experienced* level. This is what students experience of the curriculum. Sometimes they miss a part of the content or they re-interpretate it. This level is really about what a student experience. Different student can have different experiences.
- The *effected* level. Van den Akker (1992) added this level to focus more on the outcomes of the curriculum. What students really learn of it. In particular in the moral domain the distinction between experiences and effects is very relevant. For example, students can experience the value of care when helping other people. How this experience changes their attitude, how they value care is the outcome of another process. It is the result of giving a personal meaning to the experience and reflecting on the values you have.

For thinking, speaking, teaching and researching moral education it is essential to be aware if these different levels of the curriculum. They seem to be quite top-down, but bottom-up processes interact with these top-down processes (Veugelers, 2004). A good example of using the different levels is the phd research of Bartels on Philosophy for Children in which he inquired the different levels of the introduction of this programme in the Netherlands as part citizenship education (Bartels, Onstenk & Veugelers, 2015).

1.2 *Different moral values*

People can think differently about moral values and about aims and practices of citizenship education. These differences can be at the level of persons, but also of communities and within nations, and between nations. These moral values and their aims can change over times, in particular their concrete articulations.

When people are speaking about moral values they often suggest that their moral values are the best and the only ones people should support. They often don't present them as their perspective, their view, but as THE moral values. Also, many people say that there are universal values: moral values that count for everybody. The example mostly given in the Western world is human rights. These values should count for everybody. However, even human rights are moral values that are formulated and supported by certain people, they are not universal. Human rights were formulated in the United Nations in a particular period in history, just after the second world war and influenced by the idea to avoid war between countries, human destruction like the holocaust, and try to balance better the power of states and the rights of individuals.

In the United Nations they succeeded in getting a general agreement on the formulation of the human rights, but only after an intensive debate between a liberal focus on individual rights and a more social oriented focus on social and economic rights. The first was strongly a western voice, the second more an eastern perspective. The final formulation of the human rights was more an individual than a social perspective, and expressing clearly the political power relations at that time in history.

Personally, I think human rights are important, but the social perspective should be more included in formulations of human rights. Why not update them each 10 years. Personally, and this is my humanist perspective, I really want to stress the value of democracy. Not only as voting system, but as what Dewey (1916) called 'a way of life': a way of organising society, politics and interpersonal relationships by deliberation, concensus-building, respect for minorities and for freedom of speech. So people can differ about moral values, but they have to live together and organise their life and society. How can education contribute to this?

2 DIFFERENT EDUCATIONAL GOALS AND TYPES OF CITIZENSHIP

Educational systems, schools within a system, and teachers within a school, they all can have different educational goals. In several research projects (with both quantitative and qualitative instruments), we asked teachers, students, and parents which educational goals they find important. Statistical analyses show three clusters of educational goals: discipline, autonomy, and social involvement (Leenders, Veugelers, & De Kat, 2008a; 2008b). (See for more conceptual explorations of these clusters Veugelers, 2007; 2017).

> *Discipline*, for example, has to do with listening and behaving well. These are goals that are emphasized especially in the educational movement that is called 'character education' (Lickona, 1991). It is about promoting good behavior and following norms. In socialization research, like in the work of the sociologist Emile Durkheim (1923), disciplining is considered an educational task: education teaches you how you should behave.
>
> *Autonomy* refers to setting pedagogical goals as personal empowerment and formulating your own opinion. These goals are central to the moral development tradition of Kohlberg (Power, Higgins & Kohlberg, 1989; Zizek, Garz & Nowak, 2015) but also in the structural sociology of Giddens (1990), with the emphasis on 'agency'. Autonomy can be defined as the experience of freedom, and giving meaning to your own life. In the Western world and in modernity, autonomy-development of people is considered very important.
>
> The third cluster, *social involvement*, shows a broad spectrum of social goals: from an instrumental coexistence, a social-psychological empathy, to a social justice-based solidarity and combating inequality in society. Under the social spectrum, different scientific orientations can be found: the justice approach of Rawls and Kohlberg, the concept of care of Noddings (2002), and empowerment of the Brazilian pedagogue Freire (1985). Social involvement can vary greatly in political orientation.
>
> Our research, with both quantitative and qualitative instruments, shows that these three clusters discipline, autonomy and social involvement are important educational goals for teachers, students, and parents.

2.1 *Types of citizenship*

Further analyzing our data (with person-centred factor analyses) we can construct three types of citizenship, which express different orientations:

> The first type is *adaptive* citizenship. This type scores high on discipline and social involvement—socially involved not in a political sense, but in a moral commitment to each other, especially your own community. For autonomy, though, the scores are not so high for the adaptive type.
>
> The second type, *individualized* citizenship, scores high on autonomy and fairly high on discipline but relatively low on social involvement. This type has a strong focus on personal development and freedom, not on the social.
>
> The third type, *critical-democratic* citizenship, scores high on social involvement and on autonomy. On discipline this type scores low. We call this type critical-democratic because of its focus on the social and on society, with a critical engagement that leaves room for individual autonomy and personal articulation.

In a survey of Dutch teachers in secondary education, with a representative sample, we could conclude that 53% of teachers are pursuing a critical-democratic citizenship, 29% an adaptive type, and 18% an individualizing type. This variety is not the same on the different levels of education: in pre-university secondary education we see more support for the individualized type and in the pre-vocational education for the adaptive type. A reproduction of social class power relationships becomes visible in these citizenship orientations (Leenders, Veugelers, & De Kat, 2008a).

2.2 *Types of citizenship and civic education practice*

These three types of citizenship each correspond to a specific practical operationalization of citizenship education with a specific methodology and a focus on certain goals:

> *Adaptive*: much transmission of values, in particular adaptive values, and attention to standards and norms. Teacher-directed education and students seating in rows. Values are embedded in the hidden curriculum.

> *Individualized*: great attention to development of independence of students, and to learning critical thinking. Students work a lot individually. Own choice of values.

> *Critical-democratic*: focus on learning to live together and to appreciate diversity, and on active student participation in dialogues. Cooperative and inquiry-oriented learning is practiced often. Attention for social values and critical-reflection on values.

Of course, the types of citizenship and the corresponding practical classroom interpretations are ideal-typical constructions. In the views of people and in educational practice we find many hybrid forms with a combination of these types of citizenship and citizenship education. But these three types of citizenship and citizenship education demonstrate that citizenship is not a matter of bad or good citizenship and that different orientations in the political nature of citizenship are possible. It also shows that schools and teachers can make choices in their educational goals and in their practice of citizenship education.

3   DIFFERENCES BETWEEN GOALS, PRACTICES, AND EXPERIENCES OF CITIZENSHIP EDUCATION

Until now we talked about the goals that teachers want to pursue. Do teachers realize these goals in practice? Teachers say that they are often unable to realize these objectives entirely. This is particularly true of the goals of autonomy and social involvement. It is striking that in the Netherlands teachers, as well as parents, indicate that discipline in education still receives relatively a lot of attention and is also fairly well developed in students. They realize that it is much more difficult to develop good autonomy, where students take real responsibility for their own actions and deliberate on alternatives in a grounded manner. The social orientation, and especially to realizing the attitude in it, gets much less attention in educational practice and is also more difficult to achieve (Veugelers, 2011; 2017). This is the difference between the ideal curriculum and the interpreted curriculum and the operationalized curriculum, the practice.

In traditional education the disciplinary mode gets attention. In more modern ways of teaching and in more child-centered pedagogical perspectives the individual is more central. This individual orientation and identity development is further strengthened in a specific manner by the competition and selection that is strongly embedded in many educational systems. Students have to compete with each other and are made responsible for their own educational success. We see this individual educative competitive orientation in the Western world and the Netherlands is a good example of it, but also in Asian countries like Korea, Singapore and Taiwan.

The social seems to be less intertwined in educational systems. We see some more social orientations in Scandinavian countries; countries with a strong social democratic political orientation (Green & Janmaat, 2012), in some Latin America countries as results of strong social movements (Teodoro & Guilherme, 2014; Veugelers, De Groot, & Llovomate, 2017). And in some Asian countries as part of a more collective culture (Wing On Lee, 2014; Sim, 2015).

International comparative studies like the International Civic and Citizenship Education Study (2010) show how adolescents think and act in the area of citizenship. In particular, many youngsters support democracy and individual freedom on an abstract level. But these studies also show that in many Western countries the social involvement of youngsters is not very strong. For example in Northwest European countries like the Netherlands, Great Britain, and Belgium, youngsters indicate a lack of interest in being involved in politics or the common good; however they do express certain political opinions such as restricting rights and support to immigrants. In our own research with the three types of citizenship, we find among youngsters a strong focus on autonomy, and a social involvement which is more psychological and focused on their own communities rather than global and social-justice oriented (Leenders, Veugelers & De Kat, 2008b; 2012).

4 TEACHING COMMON VALUES DEMOCRACY AND TOLERANCE IN EUROPE

In a recent study we inquired how democracy and tolerance are taught in Europe. Democracy and tolerance are considered as crucial values in the European Union. On request of the Parliament of the European Union we inquired how in all 28 EU Member States the policy is about teaching democracy and tolerance in secondary education (Veugelers, De Groot & Stolk, 2017).

The European Union has always stressed the relevance of the values of democracy and tolerance for Europe as a community as well as for its Member States. This research focused on the policy of teaching the common values of democracy and tolerance in secondary schools, and how this policy is implemented in practice. Further, it covers how teachers, local communities and NGOs influence the teaching of common values. Data on the formal education policies of all 28 European Union Member States has been collected by national academic experts in the field of moral education and citizenship education. They analyzed curriculum documents and research reports and answered a questionnaire developed by the research team. In the second phase of the study in-depth curriculum studies have been performed in 12 Member States by the same experts. They interviewed policy makers, teachers and NGO's representatives.

In the conceptual framework of the study we distinguish three components of democracy: participation, democratic politics, and democratic society. Participation focuses on voting, democratic politics on deliberation, and democratic society on democracy as way of life in society, on social justice and on freedom of speech. Also on tolerance we distinguish three components: interpersonal relations, tolerance towards different social and cultural groups, and an inclusive society. Further, a distinction was made between a national and an international orientation.

A review of existing relevant studies shows some evidence that the value development of students is stimulated by a whole school approach that incorporates the teaching of values in four ways (CDP, ICCS, 2009):

– A specific value-oriented subject;
– Integration of values into related subjects;
– Cross-curricular activities and establishing links with the community;
– More dialogical methodologies of teaching and learning

And organizing schools in two ways:

– A democratic school culture involving student participation
– An inclusive education bringing together different groups of students and teachers.

In this research, we investigated if these elements are part of the education policies of the EU Member States, and if schools and teachers can realize them in practice.

4.1 *Conclusions policy*

1. Greater attention to the teaching of values, including democracy and tolerance, is evident in the education policies of all EU Member States.
2. Though Teaching Common Values (TCV) is fairly important in half of the EU Member States, compared to other topics and subject areas, attention given to TCV is still lacking.

3. Analysis of the practice of TCV in 12 EU Member States shows that there are only a few Member States where the different components of teaching for democracy and tolerance receive systematic attention across schools.
4. TCV is often not strongly implemented in education policy in terms of concrete curriculum instruments and supporting measures. This results in practices that do not always give real attention to TCV. Moreover, the EU Member States differ in the extent to which they steer TCV policy.
5. In several EU Member States, there is a strong tendency to separate students into different groups based on different learning capacities. This reduces possibilities to learn about social and cultural differences. A second element that limits diversity amongst students is the prevalence of private or religious schools.

4.2 Conclusions curriculum

1. In most EU Member States, there is a focus on political participation. However, attention should also be given to democracy as a process of deliberation and consensus-building, and to the creation of a democratic society that is just and inclusive and values freedom of speech and equality. Strong education practices that relate to all these different components of democracy are scarce.
2. Tolerance is mostly addressed in education at the interpersonal level and to a larger extent at the level of cultural groups, but very little at the level of an inclusive society.
3. While national orientation gets abundant attention in education policy, attention given to the international dimension is not very strong, although it is growing. Teaching about own nations is often susceptible to an uncritical approach.

4.3 Recommendations policy

How attention to teaching democracy and tolerance can be improved?
1. Both the EU and each EU Member State has to take responsibility to support democracy and tolerance as common societal values and to support the sustainability of such a society. An intensive dialogue in society on what constitute the common values and the role of education in promoting them is an expression of a lively democracy, and is a challenge for tolerance. EU Member States and the EU should support such dialogues.
2. Education policy steering should target aims, guidelines for content and subjects, as well as activities. Further, education policy should challenge schools to use their relative autonomy to demonstrate their own vision and practice of TCV.
3. The EU can challenge its Member States to develop their own educational vision on Teaching Common Values like democracy and tolerance, stimulate the development of innovative practices, promote teacher and student exchange to help them experience different political and educational practices, and stimulate comparative research.

4.4 Recommendations curriculum

1. Greater attention should be given in education policy and practice to all three components of democracy. TCV also has to address all three elements of value development, namely knowledge, skills and a democratic attitude.
2. Besides tolerance, concepts with more positive attitudes such as appreciation, pluralism, and respectful engagement should be used. All three levels of tolerance (interpersonal relations, social and cultural groups, inclusive society) need more attention in education policy and practice.
3. Learning democracy and tolerance can be strengthened by social and cultural diversity in schools and classrooms. Education policy should stimulate diversity in education (amongst both students and teachers).
4. Each country has to find a good balance in education between national and international orientation, so as to strengthen democracy and tolerance both nationally and internationally and address both levels in a critical way.

To summarize: there is not only a gap between policy and practice, there is also a gap within policy: between aims and concrete measures. Teaching values can be stronger integrated in policy and in practice, with using more different elements of a whole school approach. In particular more dialogical teaching methods, a more democratic school culture, and more diversity within schools are necessary. Regarding the content both democracy and tolerance can be more inclusive and social justice oriented. At all levels of policy, the curriculum and schools progress can be made.

5 GLOBAL CITIZENSHIP: LINKING THE MORAL AND THE POLITICAL

In the last part of the article we will give another example on how people can differ about citizenship and citizenship education. Now we shift the focus to global citizenship.

Politicians, researchers, and practitioners often use the concept global citizenship. Many people get a warm feeling when they hear this word. However when people start talking about it, it becomes clear that people can have really different ideas about what the concept global citizenship means.

In a study we explore different meanings and practices of global citizenship and global citizenship education (Veugelers, 2011). We analysed the literature and we had interviews with teachers about their concepts and practices. Based on the literature review we distinguished three types of global citizenship:

- An open global citizenship with a focus on knowledge about different cultures and an openness for other cultures (f.e. Soros).
- A moral global citizenship education that supports human development, humanity and a concern about the global world (Nussbaum, Hansen).
- A social-political global citizenship that address also unequal power relations and is oriented to social justice and political change (Freire, Mouffe, post-colonialism studies).

Our interviews with Dutch teachers showed that they are reluctant to pay attention in education to a social-political global citizenship; they prefer a more moral concept of global citizenship. They are reluctant in making education political. This research shows clearly that people, also teachers, can think differently about global citizenship and global citizenship education and creates differences in the curriculum.

The concept of globalisation itself can be criticised more, in particular on its contradictory effects: favouring the mobility and capital accumulation of social, cultural and financial elites, and giving up the protection by the nation state of the working class who faces a spiral of job possibilities and wages to the bottom. A second criticism is the favouring of Western social and cultural capital, and (re-) colonising life styles, discourses and cultures in the non-Western world.

5.1 Research on global citizenship

We continue working on inquiring global citizenship and developed based on above three types of global citizenship a questionnaire. For each type we formulated three items.

Open global citizenship

- I find it important to get knowledge about different cultures.
- I am open to new cultural experiences.
- I want to meet people from different parts of the world.

Moral global citizenship

- I feel responsible for our global world and for humanity.
- I want to work on enlarging the opportunities of all human being.
- I appreciate cultural differences between people.

Table 1.   Goals of global citizenship in the Netherlands and Indonesia.

	NED (n = 30)	INDO (n = 52)
Open global citizenship		
1. Important knowledge about different cultures	4.1	4.3
2. Open to new cultural experiences	4.2	4.0
3. Meet people from different parts of the world	3.5	4.1
Total	11.8	12.4
Moral global citizenship		
4. Responsible for our global world and humanity	4.6	4.7
5. Enlarging the opportunities of all human being	3.9	4.2
6. Appreciate cultural differences between people	4.2	4.3
Total	12.7	13.2
Social-political global citizenship		
7. Social + political power relations global world	4.0	4.3
8. More equal social + political global world	3.7	4.2
9. Support underprivileged people to gain power	3.5	3.6
Total	11.2	12.1

Social-political global citizenship

- I am aware of social and political power relations in the global world.
- I want to strive for more equal social and political in the global world.
- I support the struggle of underprivileged people to gain more power.

On a 5-point scale we ask how important these goals are. In an exploratory study we used this questionnaire in the Netherlands and Indonesia with student-teachers in moral and citizenship education. In the Netherlands we included students of the education department (moral and citizenship education) of the University of Humanistic Studies (30 students) and in Indonesia students of the department Citizenship Education Yogyakarta State University (52 students). We asked them how important are these goals according to you for secondary education schools?

Maybe you would expect that the Dutch students are more oriented to global citizenship education than the Indonesian. It are however the Indonesian students who score often higher. There are not so many differences for an open and a moral global citizenship, but there are really differences on the social-political global level. Indonesian students are far more political and social-justice oriented than Dutch students. This study shows again that people can differ in their ideas about citizenship and that these differences need to be addressed in research, political debates, and in education itself.

5.2   Desirable educational strategy

What does this all means for teachers?
    We end with formulating some guidelines for teachers' practice:

1. Giving voice to everybody.
2. Showing diversity and different perspectives.
3. Showing that the critical political is one of the alternatives.
4. Stimulating openness/dialogue and valuing diversity.
5. Focusing on moral values embedded in social and political power relations.
6. Care about the world, autonomy, humanity and democracy.
7. Showing own values, I hope social justice.

REFERENCES

Apple, M.A., & Beane, J.A. (Eds.) (1995). *Democratic schools*. Alexandria, VA: ASCD.

Barber, B. (2003). *Strong democracy*. Berkeley, CA: University of California Press.

Bron, J., & Veugelers, W. (2014). Why we need to involve our students in curriculum design: Five arguments for student voice. *Curriculum and Teaching Dialogue, 16* (1), 125–139.

De Groot, I. (2013). *Adolescents' democratic engagement*. (Thesis). Utrecht, Netherlands: University of Humanistic Studies.

Dewey, J. (1923). *Democracy and education*. New York, NY: Macmillan.

Durkheim, E. (1971[1923]). *Moral education*. New York, NY: Free Press.

Freire, P. (1985). *The politics of education*: Culture, power and liberation. South Hadley, MA: Bergin & Garvey.

Giddens, A. (1990). *The consequences of modernity*. Cambridge, UK: Polity.

Haste, H. (2004). Constructing the citizen. *Political Psychology*, 25 (3), 413–440.

International Civic and Citizenship Education Study (2010). *ICCS 2009 International Report*. Amsterdam: IEA.

Kennedy, K., Lee, W.O., & Grossman, D.L. (Eds.). (2010). *Citizenship pedagogies in Asia and the Pacific*. Dordrecht, Netherlands: Springer.

Leeman. Y. (2006). Teaching in ethnically diverse schools: teachers' professionalism. *European Journal of Teacher Education*, 29 (3), 341–356.

Leenders, H., Veugelers, W., & De Kat, E. (2008a). Teachers' views on citizenship in secondary education in the Netherlands. *Cambridge Journal of Education*, 38 (2), 155–170.

Leenders, H., Veugelers, W., & De Kat, E. (2008b). Moral education and citizenship education at pre-university schools. In F.K. Oser & W. Veugelers (Eds.), *Getting involved: Global citizenship development and sources of moral values* (pp. 57–74). Rotter dam, Netherlands: Sense Publishers.

Lickona, T. (1991). *Educating for character*. New York, NY: Bantam Books.

Mouffe, C. (2005). *On the political*. London, UK: Routledge.

Noddings, N. (2002). *Educating moral people*. New York, NY: Teachers College Press.

Nussbaum, M.C. (1997). *Cultivating humanity*. Cambridge, MA: Harvard University Press.

Parker, W. (2004). *Teaching democracy*. New York, NY: Teachers College Press.

Power, F.C., Higgins, A., & Kohlberg, L. (1989). *Lawrence Kohlberg's approach to moral education*. New York, NY: Columbia University Press.

Putnam, R.D. (2000). *Bowling alone*. New York, NY: Simon and Schuster.

Schuitema, J., & Veugelers, W. (2011). Multicultural contacts in education. *Educational Studies*, 37 (1), 101–114.

Sim, J. B-Y. (2011). Social studies and citizenship for participation in Singapore: How one state seeks to influence its citizens. *Oxford Review of Education*, 37 (6), 743–761.

Teodoro, T. & Guilherme, M. (Eds.) (2014). *European and Latin American higher education between mirrors: Conceptual frameworks and politics of equity and social cohesion*. Rotterdam, Netherlands: Sense Publishers.

Veugelers, W. (2007). Creating critical-democratic citizenship education: Empowering humanity and democracy in Dutch education. *Compare*, 37 (1), 105–119.

Veugelers, W. (Ed.) (2011a). *Education and humanism*. Rotterdam, Netherlands: Sense Publishers.

Veugelers, W. (2011b). *The moral and the political in global citizenship education*: Appreciating differences in education. Globalisation, Societies and Education, 9 (3–4), 473–485.

Veugelers, W., De Groot, I, & Stolk, V. (2017). *Research for Cult Committee—Teaching Common Values in Europe*. Brussels: European Parliament, Policy Department for Structural and Cohesion Policy.

Westheimer, J., & Kahne, J. (2004). What kind of citizen? The politics of educating for democracy. *American Educational Research Journal*, 41 (2), 237–269.

Zizek, B., Garz, D., & Nowak, E. (Eds.) (2015). *Kohlberg revisited*. Rotterdam, Netherlands: Sense Publishers.

How to control hate speech and hoaxes: Character language for character citizens

J. Jumanto
Universitas Dian Nuswantoro, Indonesia

ABSTRACT: This research explores hate speech and hoaxes. It proposes a character language formulation to control them. This research paper is aimed at identifying the principles and values of character language which are expected to be effective in character education for the 21st century global citizens. This qualitative research employs two methods: grounded theory and autoethnography. The data obtained from theory reviews of hate speech and hoaxes through interpretive techniques are then further analyzed using a coding technique. The coding technique is used to find the main characteristics of hate speech and hoaxes. The principles and values of character language are applied as a means of controlling them, by reducing their textual transactions or avoiding them completely in communication. The control is elaborated into two aspects: character language to control hate speech, and character language to control hoaxes. This paper advocates the theories of *face works, probabilities in language use, and elaborate types of hearer in the aspects of power and solidarity*. This study empirically promotes harmony among speakers of languages, thus enhancing character citizen encouragement in the world, especially in cross-cultural verbal and non-verbal communication.

1 INTRODUCTION

Using language in the real-life world as well as in the virtual world is the right of a speaker. However, that right should be adjusted to that of others, or of a particular hearer, or even of a particular hearer's group. Every speaker or every hearer has the right to speak or to hear in the society, and this right should be appreciated. This concept of want or will has been pragmatically termed as *face* (Goffman, 1959), and should be considered carefully in interactions or communication. In the development of this face works theory, there have emerged the so-called interpersonal face and social face. The face works are then regarded as one of the key factors in harmonious interactions or communication, as a speaker is usually obliged to maintain the interpersonal as well as the social face.

The wild phenomenon emerging today, threatening the face of a particular hearer or of a particular hearer's group, is the attack in the form of hate speech and hoaxes. Both are usually launched in public spaces, real or virtual, making the society at large panic or become restless in uncomfortable situations. Hate speech and hoaxes, like viruses in the real world, are now becoming viral in the virtual world, often or even usually uncontrolled or beyond control.

Creating a control for hate speech and hoaxes is not easy. Sometimes, one has the power to control or not to create a hate speech or hoax. However, probably due to their lack of good morality and empathy, some people deliberately launch a hate speech or hoaxes to satisfy their lust or evil desire and to mislead others. This is the first group of creators of hate speech and hoaxes. On the other hand, some do not know when their posts of hate speech and hoaxes are harming others. In this sense, they then probably make a mistake honestly or by accident. This latter group of creators is the target audience of this research, while the former is in the hands of ethical, morale, and religions guardians, thus unfortunately beyond the scope of this study. It is the honestly mistaken audience, or the naive or innocent citizens, who should learn best from what the author is proposing in this research paper, regarding character language to control hate speech and hoaxes.

2 RESEARCH METHODOLOGY

This qualitative research employs two methods: grounded theory and autoethnography. In the social sciences, grounded theory is a systematic methodology for constructing theory through data analysis (Martin & Turner, 1986; Faggiolani, 2011). It is based or grounded, and also developed from observations or data (Ralph et al., 2014). Meanwhile, autoethnography is used by an author for self-reflection and writing, in the exploration of personal experience and its connection to wider cultural, political, and social meanings and understandings (Ellis, 2004; Maréchal, 2010). This research is based on data in the form of theory reviews on hate speech and hoaxes, which are then verified by the researcher by using his own working experience, hence being autoethnographic.

The researcher analyses the data on theory reviews through interpretive techniques, and the data obtained is then further analyzed through a coding technique, which consists of three steps: open coding, axial coding, and selective coding (Strauss & Corbin, 1990; Holloway, 1997; Bohm, 2004; Saladana, 2012). The data is separated and conceptualized in the open coding. The researcher then reunites the separated data in the axial coding to find the major categories. Finally, the researcher discovers the main phenomena in the selective coding. Here, the researcher then relates the major categories of hate speech and hoaxes to the principles and values of character language, in other words, character language to control hate speech and hoax.

3 ISSUES ON HATE SPEECH AND HOAXES

3.1 *Hate speech*

Hate speech attacks a person or group on the basis of attributes such as gender, ethnic origin, religion, race, disability, or sexual orientation (Nockleby, 2000). According to Rosenfeld (2017), hate speech is designed to promote hatred on the basis of race, religion, ethnicity or national origin, and it poses vexing and complex problems for contemporary constitutional rights to freedom of expression. As a part of hate crime, hate speech is generally understood to mean a crime against a victim specifically chosen due to their race, religion, gender, national origin, disability, or sexual orientation (Roleff, 2001).

To control hate speech is not easy, even if there is a real law passed on this matter. Banks explains that far-right and extremist websites, and hate-based activity in cyberspace is increasing, and that the anonymity and mobility from the Internet makes harassment and expressions of hate beyond the control of traditional law enforcement (Banks, 2010). In the marketplace of ideas, hate speech has been for over 15 years a subject of hot debate at the height of the culture wars (Lee, 2017). Hate speech is also designed to threaten certain groups publicly, and acts as a means for offline hate groups to share ideology and spread propaganda to recruit new converts and advocate violence (Cohen-Almagor, 2014).

Efforts to restrict hate speech in the media and instructional settings continue, despite the lack of a convincing need to do so (Jay, 2009) and due to the difficulty in outlawing the speech of this kind (Baker, 2009). To help address hate speech in print media, radio, television, and new technologies, as well as public speaking, the United States Holocaust Memorial Museum (2009) convened a seminar to explore contemporary case studies as well as the international law governing this area. To restrict hate speech, Amnesty International (2012) has campaigned for non-discrimination, with its principle that all human rights are universal, indivisible and interdependent, and interrelated. Hate speech is also very carefully taken into account by the European Court of Human Rights (2017). It suggests and covers aspects, which, among others, are ethnic and racial hatred, apology of violence and incitement to hostility, denigrating national identity, incitement to religious intolerance, and the insult of state officials.

3.2 Hoaxes

Hoax (noun) is a deliberately fabricated falsehood made to masquerade as the truth (MacDougall, 1968), while to hoax (verb) is to trick into believing or accepting as genuine something that is false and often preposterous. Hoaxes are part of deceptive news, which may be misleading or even harmful, especially when they are disconnected from their original sources and contexts (Chen et al., 2015).

The spread of hoaxes is accelerated by the Internet, as well as by other advances in telecommunications, and is part of a media frenzy. Hoaxes depend on emotions and unfounded beliefs, rather than scientific data and logical thinking (Chang & Gershwin, 2005). Hoaxes are for instance present as fake reviews on product websites and about manipulative statements of celebrities and politicians (Li et al., 2012; Gupta et al., 2013).

There are three reasons for why hoaxes happen. The first reason is misinformation; the text is conveyed in the honest but mistaken belief that the relayed incorrect facts are true (Kumar et al., 2016). The second is disinformation; false facts are conceived in order to deliberately deceive or betray an audience (Hernon, 1995; Fallis, 2014). The third is called *bullshit*, which is to convey a certain impression of one's self (Frankfurt, 1986).

Today, almost everybody can use the Internet to hoax or fake the news. Social medias are open, access to them belongs to users in the world, and they have a potential to address millions of users and possible future customers (Krombholz et al., 2012. Social media content can be relayed among users with no significant third-party filtering, fact-checking, or editorial judgment, so an individual user with no track record or reputation can in some cases reach as many readers as can Fox News, CNN, or *The New York Times* (Allcott & Gentzkow, 2017). Part of hoaxing is the use of fake profiles by malicious users to present themselves as fictitious or real persons (Krombholz et al., 2012).

Hoax or fake news can easily spread, reach, and influence the greater world audience in a relatively short time. Within minutes or hours, a hoax from a lone tweet or badly sourced report can be a story repeated by dozens of news websites, generating thousands of shares, and may gradually find its *true quality*, as repetition by a critical mass has a powerful effect on belief; it then becomes true for readers simply by virtue of its ubiquity (Silverman, 2017). Probably one of the best cases is a case study of faulty or fraudulent research relating to DDT. The chemical compound has saved more human lives than any other in history, but was banned by order of one man, the head of the US Environmental Protection Agency (EPA), the worldwide effect of which has been millions of preventable deaths (Edwards, 2004).

3.3 Characteristics of hate speech and hoax

Based on the accounts of the hate speech and hoaxes above, their characteristics can be identified as follows:

1. Hate speech is verbal, while hoaxes can be verbal or non-verbal;
2. Both hate speech and hoaxes are spread publicly by malicious users to reach a worldwide audience for an attacking or deceptive purpose;
3. Hate speech and hoaxes attack or deceive others on the basis of dangerous topics or false facts, such as gender, ethnic origin, religion, or manipulative statements about celebrities, politicians, and fake profiles;
4. Repeatedly shared hate speech and hoaxes powerfully affect belief, to the extent that they then become true due to their virtual omnipresence or ubiquity.

4 CHARACTER LANGUAGE: ITS PRINCIPLES AND VALUES

The term *character language* has been proposed by Jumanto (2011, 2012, 2014a, b), and has earned a definition that it is a language which is able to function as a means of communication (ability), has qualities which makes the language is different from the others (quality), and is effective in a correct formality (validity). Character language proposition has made use of theories of face (Goffman, 1959), interpersonal face (e.g. Leech, 1983; Arndt & Janney, 1985; Brown & Levinson, 1987), social face (e.g. Ide, 1989; Gu, 1990), and the assertion that language use is a matter of probabilities (Jumanto, 2014a, b). Character language is used to bring politeness to a particular hearer to bring politeness (i.e. everything good that has been uttered as well as acted by the speaker to the hearer within a particular context), to maintain their interpersonal face as well as their social face (Jumanto, 2014a). This particular hearer is elaborated, whether or not they are close to a speaker, based on the theory of types of hearer in the aspects of power and solidarity (Brown & Gilman, 1968). Implied from the definition, face works (i.e. interpersonal face and social face) are also involved in character language. The most recent contribution on this matter is elaboration of character language in Indonesian language use, hence Distant Indonesian Language (DIL) and Close Indonesian Language (CIL) (Jumanto, 2017). The elaboration has been theoretically critically interpreted from the concepts of negative and positive face (Goffman, 1959), negative and positive politeness strategies (Brown & Levinson, 1987), respect and solidarity politeness (Renkema, 1993), and politeness and camaraderie (Jumanto, 2014a). Based on the concepts, distant language and close language have then come into being (Jumanto, 2016).

4.1 *Principles of character language*

A principle is a comprehensive and fundamental law, doctrine, or assumption, or a belief that is accepted as a reason for acting or thinking in a particular way. The principles of character language accounted for here are thus as follows:

1. Character language, comprising of distant language and close language, is elaborated in line with face works, interpersonal face and social face, and with types of hearer in the aspects of power and solidarity;
2. Character language elaborates language use as a matter of probabilities and is a formulation for maintenance of politeness (i.e. either distant language to superiors or close language to close people);
3. Distant language employs formal, indirect, non-literal utterances, with relatively safe and common topics, while close language employs informal, direct, literal utterances, with any topics including those that are touchy and dangerous.

4.2 *Values of character language*

A value is relative worth, utility, or importance or a belief about what is right and wrong and what is important in life. Hence, the values of character language based on the principles above are as follows:

1. Character language is useful for a speaker in maintaining interpersonal face as well as social face in interactions with a particular hearer or a particular hearer's group;
2. Character language is important to equip a speaker with knowledge and skills of language use concerning politeness, impoliteness, rude situations, and awkward situations;
3. Character language provides a speaker with potential working knowledge and hints of distant language and close language, and enables them to confirm (respect) politeness to superiors, to instill solidarity (politeness) to close people, and to avoid impoliteness (rude situations or awkward situations).

4.3 *Practices of character language*

4.3.1 *Politeness: Distant language to superiors*
Distant language with formal, indirect, non-literal utterances, and with safe and common topics is used for politeness to superiors, for example:

1. I thank you very much, Mr. John Smith. (= formal; topic: gratitude);

16

2. The day is not very sunny today. (= indirect; topic: weather);
3. Excuse me, may I go to the restroom, please? (= non-literal; topic: permission).

4.3.2 *Politeness: Close language to close people*
Close language with informal, direct, literal utterances, and with any topics is used for politeness to close people, for example:

1. Damn! Where are you fucking around this week? Or getting lost? (= informal; topic: swearing);
2. That celebrity's gotten pregnant after so many blind dates. (= direct; topic: gossiping);
3. Where's the bathroom? (= literal; topic: permission).

4.3.3 *Impoliteness (Rude situations): Close language to superiors*
Use of close language to superiors may lead to rude situations, for example:

1. Thanks again, John! (= informal; topic: gratitude) (?);
2. You are a noted journalist, Ms. Turner. Are you married? (= direct; topic: status of marriage) (?);
3. How old are you, Mr. Smith? I just met you today. (= literal; topic: age) (?).

4.3.4 *Impoliteness (Awkward situations): Distant language to close people*
Use of distant language to close people may lead to awkward situations, for example:

1. I apologize for not letting you know soon about this, Mr. Allan Willis. (= formal; topic: apology) (?);
2. I think that it is better if we talk about this matter later, Sir. (= indirect; topic: problem) (?);
3. Excuse me, may I go to the restroom, please? (= non-literal; topic: permission) (?).

4.3.5 *Mixed code to close people*
A mixed code of distant language and close language may happen due to confusion in factors of power and solidarity on the part of a hearer (i.e. whether a close person has power, or a superior is close to a speaker). However, as the term suggests, the mixed code refers to informality, hence solidarity instills and politeness is maintained, for example:

1. Damn. This is very good! Where did you buy this porn mag? (informal + formal; topic: pornography);
2. I think that it is better if we talk about this matter later, John. (= indirect + informal; topic: problem);
3. That celebrity is a spinster. What d'you say? (= formal + non-literal; topic: gossiping).

5 CHARACTER LANGUAGE TO CONTROL HATE SPEECH AND HOAXES

5.1 *Character language to control hate speech*

Based on the characteristics of hate speech and hoax, and the principles and values of character language, hate speech can be controlled as follows:

1. We should bear in mind that public space is open, with interpersonal face as well as social face we have to deal with, and, therefore, we should use distant language, instead of close language, for maintenance of politeness and avoidance of rude situations. Formality is part of distant language, and formality and politeness are regarded as equivalent (Sifianou, 2013);
2. We had better not post something of hate in public space, as when we do, we are threatening not only a particular hearer's face but also that particular hearer's ethnic group's face;
3. We should not use close language in public space, as we never know what is going to happen to people that we are not close to, and as we are not alone with a particular hearer, but others are watching;

4. Probably fed up with intolerable hate we have gotten, we should try to find the best way of saying something or creating utterances: formal is better than informal, indirect is better than direct, and non-literal is better than literal.

5.2 *Character language to control hoax*

Based on the characteristics of hate speech and hoaxes, and the principles and values of character language, hoaxes can be controlled as follows:

1. We should not post or forward a text which we are lacking or without knowledge of, we should not immediately believe in any text forwarded to us, and should check a text out to make sure the news is true;
2. We had better not post a text with a touchy or dangerous topic, and should not create a hoax, as it threatens the social face of the public;
3. We should control our desire carefully and responsibly for sharing texts with others, e
4. specially those texts with dangerous topics;
5. We should support expanded media and an information literacy program to make people more aware of the need to be responsible, tolerant, and other needs regarded to communications (White, 2017).

6 CONCLUSION

Character language can be elaborated for informing as well as equipping *children of tomorrow* with distant language and close language, so the citizens can get a control of their language, verbal and non-verbal, and not to produce hate speech and hoaxes. Verbally, the citizens have a character in their language: distant language (formal, indirect, non-literal utterances) for superiors, and close language (informal, direct, literal utterances) for close people. With topics of character language (i.e. common and safe topics for superiors and any topics for close people) the citizens have a control not to threaten interpersonal face or social face of superiors or close people. In line with character language and public space, in real-life or in the virtual world, it is recommended that distant language is used instead of close language. In this way, the form as well as the topics of distant language help a speaker to control their language verbally and/or non-verbally against hate speech and hoaxes. With distant language, a speaker is aware of the presence of interpersonal face as well as social face in public, so harmony in verbal interactions and communication is preserved. In this sense, producing hate speech is not done or avoided at all, as the awareness controls, and creating hoaxes is not very necessary, as it will harm others.

REFERENCES

Allcott, H. & Gentzkow, M. (2017). Social media and fake news in the 2016 election. *Journal of Economic Perspectives, 3*(2), 211–236. doi:10.1257/jep.31.2.211.

Amnesty International. (2012). *Amnesty International: Written contribution to the thematic discussion on Racist Hate Speech and Freedom of Opinion and Expression, 28 August 2012.* London, UK: United Nations Committee on Elimination of Racial Discrimination. Index:

Arndt, H. & Janney, R.W. (1985). Politeness revisited: Cross-Modal supportive strategies. *International Review of Applied Linguistics in Language Teaching, 23*(4), 281–300.

Baker, C.E. (2009). Hate speech and incitement to violence. *Paper Presented of Workshop Series.* Columbia University School of Law.

Banks, J. (2010). Regulating hate speech online. *International Review of Law, Computers and Technology, 24*(3), 233–239.

Bohm, A. (2004). Theoretical coding: Text analysis in grounded theory. In U. Flick, E. Kardorff & I. Steinke (Eds.), *A Companion to Qualitative Research* (pp. 270–275). London: SAGE Publications.

Brown, R. & Gilman, A. (1968). The pronouns of power and solidarity. In J.A. Fishman (Ed.), *Readings in the Sociology of Language* (pp. 252–275). The Hague, Netherlands: Mouton & Co. N.V. Publishers.

Brown, P. & Levinson, S.C. (1987). *Politeness: Some universals in language usage.* New York, NY: Cambridge University Press.

Chang, C. & Gershwin, M.E. (2005). Mold hysteria: Origin of the hoax. *Clinical & Developmental Immunology, 12*(2), 151–158. doi:10.1080/17402520500131409.

Chen, Y., Conroy, N.J. & Rubin, V.L. (2015). News in an online world: The need for an 'automatic crap detector'. *Proceedings of the Association for Information Science and Technology Annual Meeting.* November 6–10.

Cohen-Almagor, R. (2014). Countering hate on the internet. *Annual Review of Law and Ethics, 22*(2014), 431–443.

Davis, V. (2009). Influencing positive change: The vital behaviors to turn schools toward success. *Teacher Librarian, 37*(2), 8.

Edwards, J.G. (2004). DDT: A case study in scientific fraud. *Journal of American Physicians and Surgeons, 9*(3), 83–88.

Ellis, C. (2004). *The ethnographic I: A methodological novel about auto ethnography.* Walnut Creek, CA: Alta Mira Press.

European Court of Human Rights. (2017). *Information note 205: Case-law of the European Court of Human Rights.* European: Council of Europe.

Faggiolani, C. (2011). Perceived identity: Applying grounded theory in libraries. *Italian Journal of Library, Archives and Information Science, 2*(1), 1–33. doi:10.4403/jlis.it-4592.

Fallis, D. (2014). A functional analysis of disinformation. *IConference 2014 Proceeding* (pp. 621–627). Berlin, March 4–7.

Frankfurt, H. (1986). On bullshit. *Raritan Quarterly Review, 6*(2), 81–100.

Goffman, E. (1959). *The presentation of self in everyday life.* New York, NY: Anchor Books.

Gu, Y. (1990). Politeness phenomena in modern Chinese. *Journal of Pragmatics, 14*(2), 237–257.

Gupta, A., Lamba, H., Kumaraguru, P. & Joshi, A. (2013). Faking sandy: Characterizing and identifying fake images on Twitter during hurricane Sandy. In *Proceedings of the 22nd International Conference on World Wide Web.* Rio de Janeiro, Brazil, May 13–17.

Hernon, P. (1995). Disinformation and misinformation through the internet: Findings of an exploratory study. *Government Information Quarterly, 12*(2), 133–139.

Holloway, I. (1997). *Basic concepts for qualitative research.* Oxford, UK: Blackwell Science Ltd.

Ide, S. (1989). Formal forms and discernment: Two neglected aspects of universals of linguistic politeness. *Multilingua, 8*(2–3), 223–248.

Jay, T. (2009). Do offensive words harm people? *Psychology, Public Policy, and Law, 15*(2), 81–101. doi:10.1037/a0015646.

Jumanto, J. (2011). Pragmatics and character language building. *The 58th TEFLIN International Conference on Language Teaching and Character Building: 329–340.* Semarang, Indonesia.

Jumanto, J. (2012). Teaching a character BIPA (Indonesian for non-native speakers). *The 2012 KIP-BIPA VIII-ASILE International Conference: 1–20.* Salatiga, Indonesia.

Jumanto, J. (2014a). Politeness and camaraderie: How types of form matter in Indonesian context. *Proceeding: The Second International Conference on Education and Language (2nd ICEL): II-335–350.* Universitas Bandar Lampung, Indonesia.

Jumanto, J. (2014b). Towards a character language: A probability in language use. *Open Journal of Modern Linguistics, 4*(2), 333–349. doi:10.4236/ojml.2014.42027.

Jumanto, J. (2016). Distant language, close language, and impoliteness in the Indonesian context. *Journal of Global Research in Education and Social Science, 8*(3), 131–137.

Jumanto, J. (2017). Educating the Indonesian language: A proposed verbal social project for the national harmony. *Advances in Social Sciences, Education and Humanities Research, 66,* 215–221. doi:10.2991/yicemap–17.2017.36.

Krombholz, K., Merkl, D. & Weippl, E. (2012). Fake identities in social media: A case study on the sustainability of the Facebook business model. *Journal of Service Science Research, 4*(2), 175–212. doi:10.1007/s12927-012-0008-z.

Kumar, S. West, R. & Leskovec, J. (2016). Disinformation on the web: Impact, characteristics, and detection of Wikipedia hoaxes. *Proceedings of the 25th International Conference on World Wide Web* (pp. 591–602). 11–15 April, Montreal, Quebec, Canada.

Lee, S.P. (2010). Hate speech in the marketplace of ideas. In D. Golash (Ed.), *Freedom of Expression in a Diverse World. AMINTAPHIL: The Philosophical Foundations of Law and Justice 3* (pp. 13–25). Washington, D.C.: Springer.

Leech, G. (1983). *Principles of pragmatics.* New York, NY: Longman.

Li, X., Dong, X.L., Lyons, K., Meng, W. & Srivastava, D. (2012). Truth finding on the deep web: Is the problem solved? *Proceedings of the VLDB Endowment, 6*(2), 97–108.

MacDougall, C.D. (1968). *Hoaxes.* Lincoln, UK: Dover Publications Inc.

Maréchal, G. (2010). Autoethnography. In A.J. Mills, G. Durepos & E. Wiebe (Eds.), *Encyclopedia of Case Study Research Vol. 2.* (pp. 43–45). Thousand Oaks, CA: Sage Publications.

Martin, P.Y. & Turner, B.A. (1986). Grounded theory and organizational research. *The Journal of Applied Behavioral Science, 22*(2), 141–157.

Nockleby, J.T. (2000). Hate speech. In L.W. Levy & K.L. Karst (Eds.), *Encyclopedia of the American Constitution* (pp. 1277–1279), *3*(2). Detroit, MI: Macmillan Reference.

Ralph, N., Birks, M. & Chapman, Y. (2014). Contextual positioning: Using documents as extant data in grounded theory research. *SAGE, 4*(3): 1–7. doi:10.1177/215824401455 2425.

Renkema, J. (1993). *Discourse studies: An introductory textbook.* Amsterdam, The Netherlands: John Benjamins.

Roleff, T.L. (2001). *Hate crimes.* San Diego, CA: Greenhaven Press.

Rosenfeld, M. (2017). The new challenges of hate speech in constitutional jurisprudence: A comparative analysis. *Peter Allan Memorial Lecture*, 17 April 2017, Academic Conference Centre, The University of Hong Kong Faculty of Law.

Saladana, J. (2012). *The coding manual for qualitative researchers.* London, UK: Sage.

Sifianou, M. (2013). The impact of globalisation on politeness and impoliteness. *Journal of Pragmatics, 55*, 86–102.

Silverman, C. (2017). Lies, damn lies, and viral content: How news websites spread (and debunk) online rumors, unverified claims, and misinformation. *A Tow Center for Digital Journalism, A Tow/Knight Report.* Columbia Journalism School. Retrieved from https://towcenter.org/research/lies-damn-lies-and-viral-content/.

Strauss, A. & Corbin, J. (1990). *Basics of qualitative research: Grounded theory procedures and techniques.* Newbury Park, CA: Sage.

The United States Holocaust Memorial Museum. (2009). *Hate speech and group-targeted violence: The role of speech in violent conflicts.* Washington, DC: The United States Holocaust Memorial Museum.

White, A. (2017). Truth-telling and ethics remain the keys to open democracy. In A. White (Ed.). *Ethics in the News: EJN Report on Challenges for Journalism in the Post-truth Era* (pp. 4–5). London, UK: The Ethical Journalism Network.

Can students develop self-regulated learning through worked examples?

S. Nurhayati
SMPN 5 Depok Sleman Yogyakarta, Yogyakarta, Indonesia

E. Retnowati & Y.A. Alzuhdy
Universitas Negeri Yogyakarta, Yogyakarta, Indonesia

ABSTRACT: This study aimed at comparing the effectiveness of learning through worked example pairs and problem-solving strategies in solving word problems using Linear Equation System with Two Variables (SPLDV) with regard to Self-Regulated Learning (SRL). Fifty junior high school students who had learned word problems about two-variable linear equation systems using graph, substitution, and elimination methods participated in the study. There were 24 students in a worked example pairs group and 26 students in a problem-solving group. To measure the self-regulated learning level, a questionnaire for Measured Strategies for Learning Questionnaire (MSLQ) was adapted. Analysis used independent sample t-test, and the result showed that there was no significant difference of self-regulated learning level between the worked example pairs strategy and the problem-solving strategy. It may be said that both strategies could facilitate the development of self-regulated learning.

1 INTRODUCTION

One character which is important for students to possess and that is closely related to the assessment of learning outcomes is namely *Self-Regulated Learning* (SRL). SRL is very closely related to the way a person regulates himself toward his learning environment so that it will make him an expert (master) in learning (Zimmerman, 1989). According to a study by Alsa (2005), there is a positive correlation between learning based on *self-regulated learning* and the achievement of assessment results in learning mathematics.

SRL is an active and constructive process of a learner in determining the learning goals, monitoring, regulating, controlling all the cognition, motivation, and their behavior guided by learning objectives as well as the context of the environment (Pintrich, 2000). The main key to the success of SRL is in setting the goals to be achieved. When someone learns mathematical symbols, he will organize the knowledge he has so that he can write the correct symbols, because he has the confidence and sense of responsibility in what he is doing. This suggests that the SRL greatly affects the attainment of assessment of the learning being carried out.

Learning occurs because of an experience (Schunk, 2012). This implies that learning through the process is not solely results-oriented. There might be varied activities designated to learning mathematics, but the main activity of learning mathematics is widely agreed as being problem-solving (Retnowati et al., 2010). A problem-solving activity essentially takes place when the students perform a systematic procedural operation step-by-step with actions in sequence to solve a problem (Wena, 2009).

Below is an example of a math problem-solving:

> '*Hilda has two buckets with no indicated scale, each measuring 7 liters and 4 liters. What can Hilda do to get exactly 6 liters of water from a reservoir just by using the two buckets?*'

The example above shows that the problem presented cannot be solved in a straightforward manner or in only one step. It needs some understanding of mathematical concepts underlying the problem. Such problems, like the example above, cannot be solved through a familiar (routine) procedure, and may become a challenge for students. According to Kilpatrick (1985) this kind of problem is called a non-routine problem.

In learning mathematics, the problem is often presented in a realistic way, such as by a story problem (or is also called a *word problem*). According to Killen (2009), learning word problems will be appropriate if carried out through *problem-solving*. Intuitively, a word problem that is contextual is easier to solve because it provides some information that may be used to comprehend the required knowledge to solve the problem. However, Cummins et al. (1988) stated that the word problems will be more difficult to solve than the problems presented in the form of symbols since they do not always provide clear hints of what the problem is. Converting the meaning and interpretation of sentences into symbols (mathematical model) is the main key to solving the word problem (Mahmudi, 2010).

Turning to the self-regulated learning, according to Schraw et al. (2006) the main components of SRL are (1) motivation (beliefs and attitudes that affect the use and development of cognitive and metacognitive abilities); (2) cognition (the ability needed in coding, remembering, and recalling information); and, (3) metacognition (the ability to understand and monitor their cognitive processes).

Learning by problem-solving is assumed to be useful to develop SRL. Through problem-solving strategy teachers can: (1) improve student's learning motivation; (2) improve metacognition ability; (3) take an approach to learning; (4) improve the ability to analyze situation; (5) apply prior knowledge to new situations; (6) organize the differences of facts and opinions to make a goal-oriented decision; and, (7) direct students to be more responsive to learning instruction (Killen, 2009). Problem-solving strategy is a learning strategy that lets students solve problems in their own way. This seems more suitable for students with some prerequisite knowledge, in order to successfully solve and learn the problem solution (Kalyuga, 2007).

Novice students (i.e. students with minimum or no prior knowledge) will likely face learning difficulties, particularly when solving complex problems. They are unable to construct new knowledge effectively because they have to use their working memory load (where cognitive process occurs in our mind) to achieve the solution of the problem-solving without necessarily learning it. It may cause students to have a lack of confidence in their own ability to solve the problems because of the failure they may experience when solving the problem using random knowledge (Killen, 2009).

Worked example pairs, on the other hand, have been empirically proven to be effective in assisting novice students by providing sufficient guidance that directs the working memory load to construct new knowledge based on the problem solution; hence they are able to build relevant knowledge structure in their long-term memory (Paas & Van Gog, 2006). Worked example pairs is a learning strategy that presents a problem statement with a detailed step-by-step example and complete problem-solving procedure from the beginning to the end (Sweller et al., 2011).

Previous studies indicate several designs of worked examples. The design of worked example pairs is shown to be more effective than block design (i.e. a design that presents a set of new samples and then follows with a set of similar problems to solve) (Trafton & Reisser, 1993). Worked example pairs provides learning material that is presented sequentially where a problem with the example of the solution is given first, followed by a similar problem without an example (Sweller & Cooper, 1985). It should be noted here that the pairs of example-problems are identical or isomorphic. There may be several pairs to be learned by students; however, it is suggested that variation of contexts and procedures are in accordance with the student's level of expertise.

In this study, the worked example pairs strategy was tested with regard to the level of SRL after the study period. In the *introduction phase*, the students first activated the *prior knowledge* they already had, by recalling prerequisite material relevant to the learning material being studied and by completing problem-solving. It was then followed by the *acquisition phase,* which is the learning phase itself. In this phase, the formation of acquisition capability is formed through three phases: *declarative knowledge* (phase 1), *compilation knowledge* (phase 2), and *procedural knowledge* (phase 3) (Kanfer & Ackerman, 1989). *Declarative knowledge* is the earliest phase of the formation of skill acquisition, which involves the process in order to obtain an understanding of the task. *Compilation knowledge* is the phase in which students integrate the sequence of cognitive processes and monitor what is required to complete the task. *Procedural knowledge* is the ability of how to perform various cognitive activities as a result of the compilation knowledge by rehearsing the constructed knowledge. It should be noted that when learning worked example in the *acquisition phase,* students are given an identical (isomorphic) problem to be solved, simply by using their newly acquired understanding without looking at the learning resource (Retnowati & Marissa, 2018). It was predicted that following this worked example strategy, students may develop *self-regulated learning* because they have to internalize independently and monitor their own understanding toward the knowledge presented by the worked example, and then apply it into the isomorphic problem-solving.

The material of Linear Equation System with Two Variables (SPLDV) is a material that is taught to eighth graders, and has never been given in the previous grades. SPLDV presented in the form of word problems is a difficult and complex material (Cummins et al., 1988), because it requires some basic concepts that become the prerequisite materials (*prior knowledge*) (i.e. the linear equation with one variable and mathematical model on the system of two linear equations).

2 RESEARCH METHOD

An experimental research with a *post-test non-equivalent control group* design was used (Cohen et al., 2007; Creswell, 2012). The participants of this research were junior high school students who had never learned about word problems on SPLDV material. Fifty students of class VIII participated in this study, divided into two experimental groups, namely *working example pairs* consisting of 24 participants and *problem-solving* consisting of 26 participants. Random grouping was used; nevertheless, all students had the same learning resources, teacher, as well as the same number of learning hours.

The research was conducted from March 1 to April 27, 2017 in Grade VIII of Junior High School in Sleman Yogyakarta.

2.1 *Procedure*

The study was conducted in eight lesson meetings: two initial meetings were held for classical prerequisite learning, three meetings were held for material learning (graph, substitution and elimination), and three meetings for the test phase. The allocated learning time was divided as follows: Meeting-1 (3 × 40 minutes); Meeting-2 (2 × 40 minutes); Meeting-3 (3 × 40 minutes); Meeting-4 (2 × 40 minutes); Meeting-5 (3 × 40 minutes); Meeting-6 (2 × 40 minutes); Meeting-7 (3 × 40 minutes); and Meeting-8 (2 × 40 minutes). At the 8th meeting (the last meeting), the learning ended 15 minutes earlier, as the time was used for a Measured Strategies for Learning Questionnaire (MSLQ).

Procedures undertaken in this study were: (1) selection of experimental groups conducted randomly and equally; (2) learning prerequisite material 1 (solving Linear Equations with Two Variables) classically through direct learning; (3) learning prerequisite material 2 (mathematical model) classically through worked example pairs; (4) implementing learning with worked example pairs strategy for experimental group 1 and learning with problem-solving strategy for experimental group 2 with graph material; (5) giving post-test classically for graph material; (6) conducting learning with worked example pairs strategy for experimental group 1 and learning problem-solving strategy for experimental group 2 with substitution material; (7) giving post-test classically for substitution material; (8) implementing learning with worked example pairs strategy for experimental group 1 and learning with problem-solving strategy for experimental group 2 for elimination material; (9) giving post-test classically for elimination material; and, (10) giving post-test classically to measure the SRL variable and MSLQ.

This study involved two independent variables: learning strategies (worked example pairs and problem-solving) and materials (graphs, substitutions, and elimination), and one dependent variable: SRL. *Problem-solving* strategy is a learning strategy used to solve new problems for students by linking the prior knowledge to new situations through problem identification, problem-solving strategy planning, implementing the planned strategy, and checking results. Worked example pairs strategy is a problem-solving learning strategy, with an example of a step-by-step problem-solving designed in pairs between *example* and *problem,* which is given as soon as the student studies the example. *SRL* is a process of managing and activating the cognitive aspects and students' learning behavior, systematically oriented to the achievement of learning objectives through the management of metacognition, motivation, and behavior conducted by students to achieve the objectives of learning mathematics.

There are two main components of SRL, namely motivation (consisting of three aspects) and learning strategy (consisting of two aspects). These five aspects are spelled out in 13 indicators (sub-aspects), developed according to the theory of SRL (Zimmerman, 1989). The complete aspects of SRL are presented in Table 1. Using the MSLQ (Pintrich & DeGroot, 1990), there were 44 statements divided into two main components, i.e. motivation (22 items) and learning strategy (22 items).

The MSLQ instrument was validated using construct validity, by factor analysis and content validity by expert judgment. The result of factor analysis on each aspect of SRL was obtained using Kaiser-Meyer-Olkin (KMO) value, as shown in Table 2. The coefficients of Cronbach's alpha on graph, substitution and elimination materials, and SRL were 0.769; 0.775; 0.888; and 0.886 with SEM 3, 65; 3.63; 3.18; and 8.25.

Table 1. SRL aspects and indicators.

No	Aspect	Indicator	Item
1	Self-efficacy	a. Self-efficacy for learning b. Self-efficacy for performance c. Control of learning	a. I believe I will receive a good grade in this class. b. I am certain I can understand the ideas taught. c. I am sure that I can do an excellent job on the problems and task assigned for this class.
2	Intrinsic value	a. Intrinsic interest b. Important of course work c. Challenge and mastery goal	a. I think what we are learning in this mathematics class is interesting. b. It is important for me to learn what is being taught in mathematical class. c. I prefer class work that is challenging so I can learn new things.
3	Test anxiety	a. Anxiety in having a mathematics test b. Anxiety related to the result of test	a. I am so nervous during a test that I cannot remember facts I have learned. b. When I take a test I think about how poorly I am doing.
4	Cognitive strategy use	a. Rehearsal b. Elaboration c. Organizing d. Critical thinking	a. I say the word over and over to myself to help me remember. b. I put important ideas in to my own words. c. I outline the chapters in my book to help me study. d. When I read I try to connect the things I am reading about with what I already know.
5	Self-regulation	a. Metacognition b. Management effort	a. I find that when the teacher is talking I think of other things and don't really listen to what is being said. b. When the work is hard I either give up or study only the easy part.

Table 2. KMO of SRL aspects.

No	Aspect	KMO value	Number of indicators
1	Self-efficacy	0.517	3
2	Intrinsic value	0.561	3
3	Test anxiety	0.504	2
4	Cognitive strategy use	0.716	4
5	Self-regulation	5.65	2

Table 3. Qualitative to quantitative data conversion.

Interval score (X)	Category
$242 < X \leq 308$	Very high
$198 < X \leq 242$	High
$154 < X \leq 198$	Medium
$110 < X \leq 154$	Low
$44 < X \leq 110$	Very low

The obtained SRL scores were then categorized, based on the converted standardized intervals. The categorization of SRL score was done by referring to the ideal average classification score and ideal standard deviation according to Azwar (2005 p. 108), as presented in Table 3.

The inferential analysis was carried out using *t-test independent sample*. Prior to hypothesis testing, the assumption test was done first. The assumption test consisted of the normality test (*K-S test*) and homogeneity test of variance (*F test*). The assumption test showed that the obtained SRL data in this study was normally and homogeneously distributed.

3 RESULT AND DISCUSSION

Data descriptions of the worked example pairs and problem-solving group on SRL are presented in Table 4, while the description of percentage data of the students' category in both groups is presented in Table 5.

Table 4 shows that the SRL mean of worked example pairs group is higher than that of the problem-solving group. However, the highest questionnaire score was achieved in the problem-solving group. Standard deviation in problem-solving group is also larger than in the worked example pairs group. From Table 5 it is found that the percentage of SRL in the worked example pairs group is high, while that in the problem-solving group is medium. From the description of the data it is also found that both groups have the same category of the average score (i.e. medium).

The hypothesis to be tested in this study is that there is no significant difference of SRL level after studying by worked example pairs strategy or problem-solving strategy. The analysis shows that t (48) = 0.489; p value = 0.627; Cohen's d = 0.139239; and f = 0.06962.

Because p value > 0.05 and *effect size* = 0.06962 < 0.1, it can be said that there is no significant difference between each strategy group and SRL. In other words, we can say that worked example pairs and problem-solving strategies have the same effectiveness toward SRL.

Retnowati et al. (2010) explain that worked example is problem-solving with additional examples. The worksheet given to students differs only in whether or not there is explicit guidance. So, in the affective aspect it is assumed that both strategies have the same effect on SRL. In addition, there may be many factors affecting the SRL, while the different used

Table 4. Description of *SRL data score*.

Description	WE	PS
Average	194.42	190.75
Max	240.00	261.00
Min	136.00	128.00
SD	25.73	34.75

Table 5. SRL percentage category (%).

Description	WE (%)	PS (%)
Very High	0	3.85
High	37.5	26.92
Medium	50	61.54
Low	12.5	7.69
Very Low	0	0
Average	Medium	Medium

learning strategies is only one factor in SRL development, which both includes metacognition and effort management.

Motivation is one of the main components of SRL (Pintrich et al., 1991). The isomorphic problems with the example given in worked example pairs will motivate students, who feel guided and satisfied to develop SRL because they are able to solve similar problems with the examples. According to Killen (2009), problem-solving makes students more responsive to learning forms and instructions. This element is also strongly suspected as one of the causes of both strategies having the same effectiveness with regard to SRL. Students who are knowledgeable (*expert*) indeed have a higher SRL. They tend to be able to manage the time, environment, and emotion in learning (Sweller & Kalyuga, 2011), so they will always be motivated to overcome obstacles they face (Ruliyanti, 2014). Students with good SRL will always be positive and have the confidence that they can solve the problems through their analysis, equipped with the experience they already have in problem-solving (Foshay & Kirkley, 2003).

From metacognitive viewpoints, the knowledge and experience obtained by students (their prior knowledge) becomes the basis for them to develop appropriate steps and strategies in solving the problems they are facing (Kalyuga, 2007) so that problems will be easily solved. Worked example pairs strategy gives examples in an effort to provide prior knowledge for students. In problem-solving strategy, the adequacy of prior knowledge of students is also a key to the success of learning (Sweller et al., 2011). Students become experts, not only because they have strong confidence, but also a good SRL skill in mastering the learning materials. This is very important to achieve success in learning (Ruliyanti, 2014, Wangid, 2004). Students who have low confidence in learning will tend to choose easy tasks and avoid difficult and challenging tasks. The element of confidence is an indicator of self-efficacy, and one of the benefits of problem-solving learning strategy (Killen, 2009).

Through problem-solving, students who are able to complete tasks and discover new knowledge by themselves will have great satisfaction (Killen, 2009), so that students will have a high self-confidence (self-efficacy). With the many examples that they have learned through the learning process, students will have many strategic viewpoints to solve the problem. They will have the confidence that they are capable of solving the given problem (self-efficacy). Pintrich and DeGroot (1990) say that the skills students achieve in the classroom are strongly influenced by students' cognition and learning behaviors in the learning process. Both of these are important components in forming SRL. In mathematics learning, a student will learn well if he has enough prior knowledge relevant to the material being studied (Sweller & Kalyuga, 2011).

4 CONCLUSION

Basically, a worked example pairs strategy is similar to a problem-solving strategy but with a different level of guidance. Both has focus on assisting students to acquire problem-solving skill. Although they have different characteristics, both strategies have the same advantages in developing *self-regulated learning.*

REFERENCES

Alsa, A. (2005). *Program belajar self-regulated learning, dan prestasi matematika siswa SMU di Yogyakarta [Study program, self-regulated learning, and mathematics achievement of senior high school students in Yogyakarta]* (Dissertation), Faculty of Psychology Universitas Gadjah Mada, Yogyakarta, Indonesia.

Azwar, S. (2005). *Penyusunan skala psikologi [psychology scale up]* (2nd ed.). Yogyakarta, Indonesia: Pustaka Pelajar.

Cohen, L., Manion, L., & Morrison, K. (2007). *Research methods in education.* New York, NY: Routledge.

Creswell, J. (2012). *Educational research: Planning, conducting, and evaluating quantitative and qualitative research* (4th ed.). Boston, MA: Pearson Education.

Cummins, D.D., Kintsch, W., Reusser, K. & Weimer, R. (1988). The role of understanding in solving word problems. *Cognitive Psychology, 20*(4), 405–438.

Kalyuga, S. (2007). Expertise reversal effect and its implications for learner-tailored instruction. *Educ Psychol Rev, 19*(4), 509–539. doi:10.1007/s10648-007-9054-3.

Kanfer, R. & Ackerman, P.L. (1989). Motivation and cognitive abilities: An integrative/aptitude-treatment interaction approach to skill acquisition. *Journal of Applied Psychology Monograph.*

Killen, R. (2009). *Effective teaching strategies: Lesson from research and practice* (5th ed.). Newcastle: Social Science Press.

Kilpatrick, J. (1985). A retrospective account of the past 25 years of research on teaching mathematical problem solving. In E. Silver (Ed.), *Teaching and Learning Mathematical Problem Solving: Multiple Research Respective* (Vol. A, pp. 1–16). Hillsdale, New Jersey: Lawrence Erlbaum Associates.

Foshay, R., & Kirkley, J. (2003). Principles for teaching problem solving. Retrieved from http://vcell. ndsu.nodak.edu/~ganesh/seminar/2003_Foshay_PLATO%20 Learning%20Inc._Tech%20 Paper%20 %234_Principles%20for%20Teaching%20Problem-Solving.pdf.

Mahmudi, A. (2010). *Pengaruh pembelajaran dengan strategi MHM berbasis masalah terhadap kemam-puan berpikir kreatif, kemampuan pemecahan masalah, dan disposisi matematis, serta persepsi terhadap kreatifitas [Influence of learning with problem-based MHM strategy on creative thinking ability, prob-lem solving ability, mathematical disposition, and perception on creativity].* (Dissertation), Universitas Pendidikan Indonesia, Bandung, Indonesia.

Paas, F. & Van Gog, T. (2006). Optimising worked example instruction: Different ways to increase ger-mane cognitive load. *Learning and Instruction, 16,* 87–91.

Pintrich, P.R. (2000). The role of goal orientation in self-regulated learning. In M. Boekaerts & P.R. Pintrich (Eds.), *Handbook of Self-Regulation* (pp. 451–502). San Diego, CA: Academic Press.

Pintrich, P.R. & DeGroot, E.V. (1990). Motivational and self-regulated learning components of class-room academic performance. *Journal of Educational Psychology, 82*(1), 33–40.

Pintrich, P.R., Smith, D.A.F., Garcia, T. & McKeachie, W.J. (1991). *A manual for the use of motivated strategis for learning questionnaire (MSLQ).* [College students; Learning strategies; Questionnaire; Research Methodology; Scientific and Technical Information; Students educational Objective; Stu-dents motivation]. (NCRIPTAL-91-B-004). Ann Arbor, Michigan.

Retnowati, E., Ayres, P. & Sweller, J. (2010). Worked example effect in individual and group worked set-tings. *Educational Psychology, 30*(3), 349–367. doi:10.1080/01443411003659960.

Retnowati, E., & Marissa. (2018). Designing worked examples for learning tangent lines to circles. *Journal of Physics: Conference Series, 983*(1), 012124. doi: doi:10.1088/1742–6596/983/1/012124.

Ruliyanti, B.D. (2014). Hubungan antara self-efficacy dan self-regulated learning dengan prestasi akademik matematika siswa SMAN 2 Bangkalan *[The relationship between self-efficacy and self-regulated learning with mathematics achievement of students in Senior High School 2 in Bangkalan].* *Character, 3*(2): 1–7.

Schraw, G., Crippen, K.J. & Hartley, K. (2006). Promoting self-regulation in science education: Metacognition as part of a broader perspective on learning. *Research in Science Education, 36*(1–2), 111–139.

Schunk, H.D. (2012). *Teori-teori pembelajaran perspektif pendidikan [Learning theories of educational perspective].* (E.H.R. Fajar, Trans). New Jersey, NJ: Pearson Education.

Sweller, & Kalyuga, S. (2011). The worked example and problem completion effect. In *Cognitive Load Theory* (pp. 99–109). New York: Springer.

Sweller, J., Ayres, P. & Kalyuga, S. (2011). *Cognitive load theory, explorations in the learning sciences, instructional systems and performance technologies.* New York, NY: Springer.

Sweller, J., & Cooper, G.A. (1985). The use of worked examples as a substitute for problem solving in learning algebra. *Cognition and Instruction, 2*(1), 59–89.

Trafton, J. & Reisser, B. (1993). *The contribution of studying examples and solving problems to skill acquisition.* Paper presented at the fifteenth annual conference of cognitive science society, Hillsdale, NJ.

Wangid, M.N. (2004). Peningkatan prestasi belajar siswa melalui self regulated learning [Improvement of student's achievement through self regulated learning]. *Cakrawala Pendidikan, XXIII*(1): 2–19. doi: 10.21831/cp.v1i1.4858.

Wena, M. (2009). *Stategi pembelajaran inovatif kontemporer [Contemporary innovative learning strategies].* Jakarta, Indonesia: Bumi Aksara.

Zimmerman, B.J. (1989). A social cognitive view of self-regulated academic learning. *Journal of Educational Psychology, 81*(3), 329–339.

Presenting Indonesian character in English language teaching materials: Is it possible?

S. Sudartini
Universitas Negeri Yogyakarta, Indonesia

ABSTRACT: The fact that foreign language instruction cannot be separated from the foreign cultural domain and that the students come from other cultural domains, needs to be seen as an entry point to promote intercultural communication in the practice of teaching a foreign language in Indonesia, including the English language. This particular study, however, describes the process of integrating one aspect of the Indonesian cultural domain, that is Indonesian character, into English teaching materials. In particular, it tries to answer the questions of types of character to be integrated and how to integrate Indonesian and Western character into English language teaching materials. This was a qualitative study conducted in Yogyakarta Province. This study explores the curriculum and commonly used English textbooks for 10th Grade of senior high schools. The study revealed that the character types being integrated are those containing Indonesian moral values. These moral values are integrated both implicitly and explicitly into the teaching materials by using texts and pictures. In conclusion, presenting Indonesian character in English teaching materials can be one way of introducing foreign language character and Indonesian character, to promote intercultural communication when teaching English as a foreign language in Indonesia.

1 INTRODUCTION

Discussing the practices of teaching a foreign language, including English, in Indonesia as a multicultural country is always interesting, especially when it comes to the notion of international and intercultural communication (Choudhury, 2014; Honna, 2005; Zhang & Zhang, 2015) that commonly accompanies each of the processes. Every individual involved in this kind of educational practice needs to consider this and find ways of conducting the teaching effectively, by presenting not only the linguistic elements in the classroom but also the cultural elements of both the source and target cultures. It is in line with the current trend of teaching English that English is considered as being an international, auxiliary language for intercultural, intercountry and regional communication or *lingua franca* (Woodford & Jackson, 2003; Mauranen & Ranta, 2009; Fiedler, 2011).

Therefore, the practice of English teaching and learning processes can be seen as the process of teaching linguistic elements to learners, but at the same time providing a real context of the language. What we mean by the word 'context' here is the situational contexts of the language use.

In doing so, teachers and material developers need to present real situations of language use. Contexts in English language teaching need to cover teaching the linguistic elements and teaching cultural elements. It is for this reason that teaching English as a foreign language in Indonesia cannot be separated from teaching Western cultures to those Indonesian learners who have their own distinctive cultural beliefs, norms and values. Therefore, presenting Indonesian cultural aspects, not to mention moral values accompanying Western cultural aspects, becomes a great challenge to those involved in the instructional processes.

As one part of educational processes conducted in Indonesia, the teaching of English as a foreign language in Indonesia should be in line with the national goal of education.

It needs to be considered as one of the conscious and also well-planned processes to enable learners to develop not only their capabilities and intelligence in understanding knowledge, but also to develop their personality and morals, and develop character for their own sake, and for the sake of their society and country as well (Salkind & Rasmussen, 2008; Daniels et al., 2012; Daneshmand, 2013). It is in line with the definition of the national education system proposed by the Indonesian government, saying that it refers to every educational process in the country that is based on *Pancasila* and the 1945 Constitution, and is rooted in the religious values, national cultures of Indonesia, and is one that is responsive to the needs of the ever-changing era (RoI, 2003).

In other words, the practice of teaching English as a foreign language in Indonesia as a process of presenting two different cultures must be able to maintain Indonesian character taken from the local Indonesian cultural norms and values. Therefore, the practice of teaching English as a foreign language in Indonesia cannot be value-free or focus mainly on teaching linguistic elements and function. Cultural content, including the moral values, need to be integrated into the instructional processes and particularly into the materials, as one of the important aspects in conducting the teaching (UNESCO Asia and Pacific Regional Bureau for Education, 2002; Hadi, 2015). In doing so, the integration of character education into English teaching materials is necessary.

The currently used curriculum known as Curriculum 2013 has accommodated the integration of character education. Regarding this, questions concerning types of character to be integrated and ways of integrating that character into the instructional materials used in the classrooms need to be discussed. This particular writing focuses on answering two questions concerning the integration of character education into the practice of teaching English as a foreign language in Indonesia, particularly in terms of the teaching materials. The first one is related to the question of types of character to be integrated into English teaching and learning materials, and the second describes ways of integrating that character into English teaching learning materials.

2 METHOD

This piece of writing comes from a qualitative research, which is a naturalistic inquiry (Bowen, 2008; Tesch, 2013), with a hermeneutic approach that focuses on activities on making interpretations (Lindseth & Norberg, 2004; Bulhof, 2012), conducted at Yogyakarta, Indonesia. This study explores the curriculum and commonly used English textbooks for 10th Grade of senior high schools. The data was taken from those two types of sources and became the main source of interpretation and discussion of the proposed questions for this study. The analysis began with the reading of the curriculum to find types of Indonesian character to be integrated into teaching materials. Then it continued with the reading of the English textbooks to find both Western and Indonesian cultural aspects accompanying the presentation of the linguistic elements in textbooks. The result of this analysis leads to the answer of possible ways of integrating Indonesian character into English textbooks.

3 RESULTS AND DISCUSSION

3.1 *Types of character*

There have been some definitions proposed concerning the term 'character' and 'character education'. One of the definitions mentions that the term 'character' can be defined as attitudes, representations of belief as the result of the internalization process of various virtues that people believe and use as their foundation to think and behave, as well as to act (RoI, 2010). Character is defined as behavior and attitudes that typify or exemplify an individual's moral and ethical strength (LeCorn, 2013). By this definition, the word character is closely related to the word 'virtues' (Baumgarten, 2011; Kristjánsson, 2013; van Domselaar, 2015).

On the other hand, the word 'virtues' is defined as a set of values, morality and norms, such as honest, trustful, and respectful to other people. As humans interact with others, they will form the society character and in turn will form the nation's character.

The word 'character' is also similar to the word 'moral' or values (Bass & Steidlmeier, 1999; Sekerka, 2015), determining the qualities of a person that make him/her different from other people (Nucci & Narvaés, 2008). The term 'character' in Islam is known as *akhlaq* (Heck, 2009; Jamaluddin, 2013; Jackson, 2014). Battistich (2005) emphasizes that character includes attitudes such as the desire to do one's best and being concerned about the welfare of others; intellectual capacities such as critical thinking and moral reasoning; behaviors such as being honest and responsible, and standing up for moral principles in the face of injustice; interpersonal and emotional skills that enable us to interact effectively with others in a variety of circumstances; and the commitment to contribute to one's community and society. Moreover, character, in other words, is the realization of one's positive development as a person, intellectually, socially, emotionally, and ethically.

Meanwhile the term 'character education' is commonly used when referring to the process of character formation and is considered as being similar to moral education (Althoh & Berkowitz, 2006; Çubukçu, 2012). Character education can be defined as a movement held by educational institutions and promoted by the government to foster ethics responsibility and care of young generations by modeling and teaching good character, emphasizing mainly on the universal values such as caring, honesty, fairness, kindness, generosity, courage, freedom, responsibility, equality, and respect for self and others, which are believed by the society (Berkowitz & Bier, 2005).

In the Indonesian context, there have been some scientific meetings and discussions on the importance of integrating character education into schools and those discussions deal with the formations of types of character to be integrated into the school practices. One of the ideas proposed by Zuchdi et al. (2010) covered some values that in general can be divided into those related to the human and God relationship, human and human relationship, the human and the government, and humans' relation to their social environment. Those values cover religious obedience, honesty, responsibility, integrity, work ethos, independence, synergy, being critical, being creative and innovative, having a good vision, care and empathy, generosity, fairness, being simple, nationalism and internationalism. In short, character education is an effort conducted to integrate morals or virtues into the educational practices.

In relation to the definition of education proposed by the government and the definition of character education, the term 'character education' in Indonesian contexts needs to be seen as efforts to integrate a set of values and norms taken from universal values that are in line with Indonesian local cultural beliefs and values, to develop Indonesian people's personality and morals as well as character for their own sake, for the sake of their society, as well as for their country.

These types of values and norms to be integrated into the teaching materials are explicitly mentioned in the currently used curriculum known as *Curriculum 2013*, which in general can be divided into two main elements that make up its structure. Those are Main Competences and Basic Competences (RoI, 2013). The main competences cover four types of main competences, namely Main Competence 1 or those related to religious competence, Main Competence 2 or those related to social attitude, Main Competence 3 or those concerning the knowledge and understanding of language elements, and Main Competence 4 that is related to the skills in learning the language. Both main competences and basic competences are two main elements that need to be included in each lesson plan made by teachers.

As mentioned earlier, this curriculum accommodates the integration of character education. It is clearly seen from the content of the Main Competences, particularly the Main Competence 1 and Main Competence 2. It is in these two main parts of the curriculum that the character is explicitly mentioned. The Main Competence 1 accommodates the character types that are closely related to religious values while Main Competence 2 is concerned with universal character types considered as being in line with local Indonesian values, such as being honest, disciplined, responsible, caring, cooperative, tolerant, peaceful, responsive, polite and having good manners (RoI, 2013).

Although these values accommodated in the curriculum only cover some moral values, in some points it has showed the great commitment of the government in integrating character education into instructional practices, including in teaching English as a foreign language in Indonesia. Every individual involved in the practice of teaching English as a foreign language in Indonesia will have the same perspectives on the importance of integrating character into their teaching as being explicitly mentioned in the curriculum. They need to start presenting both the target culture and local culture in their classroom materials equally, as the character types mentioned adapted in the curriculum are taken from both universal and local Indonesian cultural norms and values. By doing so, they will have participated in the efforts of forming the next generation of Indonesia who have a good understanding of their own cultural beliefs and norms, and this understanding is manifested in them as having good character.

3.2 *Ways of integrating character into English language teaching materials*

The previous part has discussed the types of character to be integrated into teaching materials. This part focuses its discussion on the ways of integrating those character types or moral values into teaching materials.

As one of the main components in teaching and learning practices, materials or textbooks play the central role in conducting instructional purposes (Sykes & Wilson, 2015; Kane, 2017; Mills & Wake, 2017). Teaching materials that are commonly in the form of textbooks are often used as the instructional guide covering language skills (Richards & Schmidt, 2002; Kawaguchi, 2005; Burden, 2016). This study reveals that most English teachers in Yogyakarta, Indonesia like to use ready-made textbooks as supplementary materials in the classroom. Regarding this, it is necessary to have a closer look at the commonly used textbooks. A former study on cultural contents in English textbooks conducted in 2013 revealed that cultural content is an integral part of the teaching materials, in particular the textbooks. It was found that the cultural content is presented either explicitly or implicitly in the textbooks. In terms of the representation, the cultural content is presented both in the form of pictures (either the main pictures—those that become the input texts—or those that become the illustrations) and texts (Sugirin & Sudartini, 2013). There were five books that were used as the sources in that study. Those are books used by most senior high schools in this region, namely *English Alive* (published by Yudhistira), *Real English* (published by Esis-Erlangga), *Look Ahead* (published by Erlangga), *English Zone* (published by Erlangga), and *Bahasa Inggris* (published by Yrama Widya).

Although this study did not focus on the integration of character education into English materials, it can be a source of reference on ways of integrating character education into English teaching materials. It shows that similar to the integration of cultural contents into textbooks, the integration of character education into teaching materials can also be conducted. In fact, character can be said to be one domain of culture. Yet, it will be easier to integrate culture into teaching materials.

In doing so, there are two main points to be noted. The first one is related to the pattern of integration. Writers or authors of those books use the explicit or implicit patterns of integration; for example, they present a story portraying moral values as a way of introducing certain character types implicitly to their readers. They may use one of the local Indonesian legends or stories as one way of introducing local Indonesian culture and character.

The second is related to the media used to integrate the character. The exploration conducted in all of the books revealed that writers use texts as well as pictures as the media to integrate Western and Indonesian cultural domains. In the Gramscian perspective, this phenomenon is categorized as a hegemony (Howard & Gaztambide-Fernandez, 2010; Huppatz et al., 2015).

Regarding this, it is possible to conduct the integration of character or moral values into English teaching materials. As a part of cultural domains, moral values can also be implicitly and explicitly embedded into the content materials, not only by using texts but also pictures (the main pictures as well as those used as illustrations in the books). Writers and material developers can start presenting Indonesian character and moral values together with Western cultural domains equally and meet the ideal of intercultural communication in foreign language teaching practices. The government has provided a wider chance and opportunities to do so by mentioning character education in the curriculum.

4 CONCLUSION

Regarding the discussion in the previous part, there are two important points to be noted. First, the study revealed that the character types being integrated are mainly those containing or in line with Indonesian moral values. The curriculum has explicitly mentioned the types of character that need to be included in every lesson plan made by teachers before conducting their teaching.

Second, in relation to the possible ways of integrating those moral values into English teaching materials, there are two patterns of integration that can be adapted: the explicit and the implicit integration. These moral values can be integrated both implicitly and explicitly into the teaching materials by using texts and pictures. In other words, presenting Indonesian character in English teaching material can be done as one way of introducing foreign language character and Indonesian character to promote intercultural communication in teaching English as a foreign language in Indonesia.

REFERENCES

Althoh, W. & Berkowitz, M.W. (2006). Moral education and character education: Their relationship and roles in citizenship education. *Journal of Moral Education, 35*(4), 495–518.

Bass, B.M. & Steidlmeier, P. (1999). Ethics, character, and authentic transformational leadership behavior. *The Leadership Quarterly, 10*(2), 181–217.

Battistich, V. (2005). *Character education, prevention, and positive youth development.* Washington, DC: Character Education Partnership.

Baumgarten, E. (2011). Curiosity as a moral virtue. *International Journal of Applied Philosophy, 15*(2), 169–184.

Berkowitz, M.W. & Bier, M.C. (2005). *What works in character education: A research-driven guide for educators.* St. Louis, Missouri: University of Missouri Press.

Bowen, G.A. (2008). Naturalistic inquiry and the saturation concept: A research note. *Qualitative Research, 8*(1), 137–152.

Bulhof, I.N. (2012). *Wilhelm Dilthey: A hermeneutic approach to the study of history and culture.* Berlin: Springer Science & Business Media.

Burden, P. (2016). *Classroom management: Creating a successful K-12 learning community.* Hoboken, NJ: John Wiley & Sons.

Choudhury, R.U. (2014). The role of culture in teaching and learning of English as a foreign language. *International Journal of Multi Disciplinary Research, 1*(4), 1–20.

Çubukçu, Z. (2012). The effect of hidden curriculum on character education process of primary school students. *Educational Sciences: Theory and Practice, 12*(2), 1526–1534.

Daneshmand, B. (2013). Emotional intelligence and its necessity in teaching training. *International Journal of Economy, Management and Social Sciences, 2*(10), 899–904.

Daniels, H., Lauder, H. & Porter, J. (2012). *Educational theories, cultures and learning: A critical perspective.* London, UK: Routledge.

Fiedler, S. (2011). English as a lingua franca—a native-culture-free code? Language of communication vs. language of identification. *Apples—Journal of Applied Language Studies, 5*(3), 79–97.

Hadi, R. (2015). The integration of character values in the teaching of economics: A case of selected high schools in Banjarmasin. *International Education Studies, 8*(7), 11–20. doi:10.5539/ies.v8n7p11.

Heck, P.L. (2009). *Common ground: Islam, Christianity, and religious pluralism.* Washington, DC: Georgetown University Press.

Honna, N. (2005). English as a multicultural language in Asia and intercultural literacy. *Intercultural Communication Studies, XIV*(2), 9–16.

Howard, A. & Gaztambide-Fernandez, R.A. (2010). *Educating elites: Class privilege and educational advantage.* Lanham, MD: R & L Education.

Huppatz, K., Hawkins, M. & Matthews, A. (2015). *Identity and belonging.* London, UK: Palgrave Macmillan.

Jackson, R. (2014). *What is Islamic philosophy?* London, UK: Routledge.

Jamaluddin, D. (2013). Character education in Islamic perspective. *International Journal of Scientific & Technology Research, 2*(2), 187–189.

Kane, S. (2017). *Literacy and learning in the content areas.* Oxford, UK: Taylor & Francis.

Kawaguchi, Y. (2005). *Linguistic informatics: State of the art and the future.* Amsterdam, The Netherlands: John Benjamins Publishing.

Kristjánsson, K. (2013). Ten myths about character, virtue and virtue education—plus three well-founded misgivings. *British Journal of Educational Studies, 61*(3), 269–287. doi:10.1080/00071005. 2013.778386.

LeCorn, B.W. (2013). *What ifs of life: Second guessing the decisions that we made.* Bloomington, CA: WestBow Press.

Lindseth, A. & Norberg, A. (2004). A phenomenological hermeneutical method for researching lived experience. *Scandinavian Journal of Caring Sciences, 18*(2), 145–153. doi:10.1111/j.1471–6712.2004.00258.x.

Mauranen, A. & Ranta, E. (2009). *English as a lingua franca: Studies and findings.* Newcastle, UK: Cambridge Scholars Publishing.

Mills, M. & Wake, D. (Eds.). (2017). *Empowering learners with mobile open-access learning initiative.* Hershey, PA: IGI Global Publishing.

Nucci, L.P. & Narvaés, D. (2008). *Handbook of moral and character education.* New York, NY: Routledge.

Richards, J.C. & Schmidt, R. (2002). *Longman dictionary of language teaching and applied linguistics* (3rd ed.). Edinburgh, UK: Pearson Education Limited.

RoI. (2003). *Undang-undang Republik Indonesia Nomor 20 Tahun 2003 Tentang Sistem Pendidikan Nasional BAB 1 Pasal 2* [Law Number 20 Year 2003 Regarding National Education System, Chapter 1 Article 2]. Jakarta, Indonesia: Repubic of Indonesia.

RoI. (2010). *Pengembangan Pendidikan Budaya dan Karakter Bangsa* [Development of Cultural Education and Nation Character]. Jakarta, Indonesia: Curriculum Center of the Ministry of National Education of Republic Indonesia, Republic of Indonesia.

RoI. (2013). *Peraturan Menteri Pendidikan dan Kebudayaan Republik Indonesia Tahun 2013 No. 69 Kompetensi Dasar dan Struktur Kurikulum Sekolah Menengah Atas* [Regulation of the Minister of Education and Culture of the Republic of Indonesia of 2013 No. 69 Basic Competencies and Structure of High School Curriculum]. Jakarta, Indonesia: Ministry of Education and Culture, Republic of Indonesia, Republic of Indonesia.

Salkind, J.N. & Rasmussen, K. (2008). *Encyclopedia of educational psychology.* London: Sage Publication.

Sekerka, L.E. (2015). *Ethics is a daily deal: Choosing to build moral strength as a practice.* Berlin, Germany: Springer.

Sugirin, S. & Sudartini, S. (2013). *Developing a model of character education integration in English teaching and learning in senior high school.* Report of National Strategic Research. Institute of Research and Community Service of Universitas Negeri Yogyakarta, Indonesia.

Sykes, G. & Wilson, S. (2015). *How teachers teach: Mapping the terrain of practice.* Princeton, NJ: Educational Testing Service.

Tesch, R. (2013). *Qualitative research: Analysis types and software.* London, UK: Routledge.

UNESCO Asia and Pacific Regional Bureau for Education. (2002). *Learning to be: A holistic and integrated approach to values education for human development: Core values and the valuing process for developing innovative practices for values education toward international understanding and a culture of peace.* Bangkok, Thailand: Author.

Van Domselaar, I. (2015). Moral quality in adjudication: On judicial virtues and civic friendship. *Netherlands Journal of Legal Philosophy, 44*(1), 24–46.

Woodford, K. & Jackson, G. (Eds.). (2003). *Cambridge advanced learner's dictionary.* Cambridge, UK: Cambridge University Press.

Zhang, X. & Zhang, J. (2015). English language teaching and intercultural communication competence. *International Journal for Innovation Education and Research, 3*(8), 55–59.

Zuchdi, D. Zuhdan, K.P., & Muhsinatun S.M. (2010) Pengembangan Model Pendidikan Karakter Terintegrasi dalam Pembelajaran Bidang Studi di Sekolah Dasar [Development of Integrated Character Education Model in Teaching in Primary School]. *Cakrawala, XXIX* (Edisi Khusus Dies Natalis UNY), 1–12.

Questioning western character hegemony in Indonesian aesthetics books

K. Kasiyan
Universitas Negeri Yogyakarta, Indonesia

ABSTRACT: One crucial question in the discussion of aesthetic discourse in Indonesian context is its construction of being under hegemony of Western aesthetic character. This can be easily verified, not to mention, in various aesthetic books. This phenomenon leads to the destruction of a particular authentic entity of Indonesian aesthetics, as one of the strategic capitals of arts and cultural development. For this reason, this particular study focuses on: describing the construction of Western character hegemony found in aesthetic books in Indonesia; and identifying various factors leading to that. The hermeneutic approach was used to explain that phenomenon. The analysis of data revealed the following results. First, the construction of Western character hegemony is mainly in terms of 'substance' and 'person'. In relation to the substance, this is easily seen from the content materials of the books that show Western character bias. Meanwhile, in terms of the 'person', writers tend to present Western philosophers that are considered as those who construct the theories of art philosophy. Second, the main factor underlying the Western bias is the strong hegemony of orientalism construction in the knowledge system of most Indonesians, not to mention in art performance today.

1 INTRODUCTION

Discussing aesthetics, people begin questioning the fundamental part of arts with its big narration. Stecker (2005, p. 2) once said that "the concept of the aesthetic is the key to understanding the nature and value of art." In line with Stecker, Hartoko (1986) claims that the discussion of arts cannot be separated from aesthetics.

The term aesthetics is commonly defined as a branch of philosophy that explores the nature of art, beauty and taste, together with the creation and appreciation of beauty (*Merriam-Webster Dictionary Online* 2017) that can be called "the science of sensory cognition" and that the goal or end of aesthetics, in Baumgartenian perspective, is the perfection of sensory cognition as such (Shusterman 2012). Aesthetics can also be considered as the study of subjective and sensori-emotional values that can sometimes be called judgments of sentiment and taste (Zangwill 2017). Meanwhile many scholars in the field define it as "critical reflection on art, culture and nature" (Milbrath & Elliot 2010, p. 190; White 2009, p. 7; Hulatt 2013, p. 139; Kelly 2003, p. x). It was a German philosopher, Alexander Baumgarten, who introduced the concept of aesthetics for the first time in his dissertation entitled *Meditationes Philosophicae De Nonnullis ad Poema Pertinentibus* (Philosophical Considerations of Some Matters Pertaining the Poem) in 1735 (Guyer 2005).

Regarding this, discussions about aesthetics are considered important owing to the act that those lead to other discussions on reasons why art has always existed, and why human beings need art through the ages to have different and better views of the world (Chand 2017). For Süzena & Mamur (2014), the term aesthetics is seen as art philosophy, that similar to the study of philosophy itself, focuses on the discussion of human, nature and universe with its own perspective. Therefore, a well-developed society will be created if art plays its significant roles in forming the culture of the society as they work together with philosophy (Süzena & Mamur 2014).

Thus, the combination of arts and aesthetics will be a good means of forming a better mankind and society (Isrow 2017). Both arts and aesthetics play significant roles for forming better civilization for every society (Listowel 2016). In this sense, the meaning of the term aesthetics needs to be understood not only by a small number of individuals, but also by the society as a whole through continual dialectical discussion and interaction (Süzena & Mamur 2014).

It is for this reason that the main topic in discussing the existence of arts and aesthetics is the importance of considering the socio-cultural contexts of each society or nation that tend to be different and pluralistic. By doing so, the discussion will lead to the forming of so called societal or cultural aesthetics that mainly deal with each distinct environment (Howard 2005). Meanwhile, the term "culture" is said to have a dynamic meaning in the late twentieth and early twenty-first centuries, culture is said to be intercultural (Leuthold 2010). On the other hand, any relation or bond on the basis of local particularities, not to mention the one in arts and aesthetics, is unavoidable. This local cultural bond is important in arts as it can create a sense of significant identity (Hellman 2003, Tung 2014, O'Neill 2017).

In other words, the existence of arts and aesthetics in the society is always *culturally bound* (Omwake 2012, Carter 2015), that its meaning can be derived from the context where it belongs. This can be in the form of either literal and direct or symbolic and imitative derivation. It is for this reason that in Aristotelian perspectives, arts is said to be *mimesis* (Paddison 2010, Babuts 2017, Dutton 2009, Şenol 2014, Scaramuzzo 2016), and becomes a medium that reflects how humanity lives, containing a sets of essential projects to continuously find solutions for conflicts of values or cultural ideas. Therefore, arts are always based on its location on the matrix of cultural contexts (Sayuti 2014). It is very common then that different cultures and social groups may have distinct aesthetic experiences (Palmer et al. 2013). Then, Fong (2001 p. 256) argues that "knowledge and art are both historical and cultural in character and that as evidence of history, works of art are a palpable, physical presence of the past, a manifestation of culture and history that words alone cannot describe." Similar to Wang, Bao et al. (2016), states that aesthetic processing can only be understood, if it is also seen as being embedded in cultural contexts and being modulated by social conditions. On the other hand, Karl Marx defines the term arts as being a social product (Foster & Blau 2011).

Consequently, in Indonesian context as it is seen from its long history in the past, the existence of arts and aesthetics basically is a part of particularities of Eastern aesthetic contexts that in some sense are different from those of Western. As mentioned by Bao et al. (2016), there are further claims that Western and Eastern artists even seem to believe and use different perspectives. One of the features of Eastern aesthetics that makes it different from Western, is the fact that it is based on "transcendence-sense" (Sutrisno 2005, p. 81; Hamersma 2008, pp. 3–5; Rouner 2012, p. 346; Morgan 2014, p. ix), compared to the Western aesthetics belief that tends to be based on "logic-immanence" (Haddock & Wakefield 2015). The existence of Eastern arts and aesthetics at present is mainly seen from many sites of China, Japan, India and Arabic countries (Ruttkowski 2007, Sasaki 2011), not to mention Indonesia. Then, Western aesthetics refers to European and American cultural settings (Hussain & Wilkinson 2006, Marra 2013).

This particularity of the two cultural settings is mainly the result of each construction of natural consciousness. Western natural consciousness can be said to follow instrumentalism logic (Liu 2000,; Long 2003), and on the contrary, Eastern aesthetics puts emphasis on senses, emotional and even religious feeling (Leaman 2002, Hardiman 2003). In addition, Western philosophy is said to be more personal-individual compared to Eastern philosophy that puts more respects on social togetherness (Chan 2011). The reflection of those different philosophical values can be easily recognized from Eastern and Western arts and aesthetic awareness as mentioned by Schellekens & Goldie (2011 p. 74), that "aesthetic preferences may vary between Eastern and Western culture with Western preferring to depict focal individuals and Easterners preferring more encompassing scenes."

Yet, the main problem lies in the current Eastern arts and aesthetics awareness that seems to be in unpleasant condition. This condition is particularly related to its condition of being under the dominant hegemony of Western aesthetics. This Western hegemony is also reflected in aesthetic books. This study in particular focuses on this reflection of Western hegemony found in aesthetic books used in this country.

Therefore this study is important and strategic to be conducted. It focuses on identifying and investigating: 1) representation of Western philosophy hegemony in aesthetic books used in this country; and 2) various factors leading to the strong hegemony of Western philosophy in the aesthetics books.

2 METHOD

This study is a *library research* (Mann 2015), that specifically discusses some Aesthetics books written by Indonesian writers and became the main resources of aesthetic sciences in Indoensian art schools. The books are written by Djelantik (1999), Sumardjo (2000) and Sutrisno (2005).

The main method used in this study is qualitative-naturalistic method (Lapan et al. 2011), with hermeneutic approach that focuses on activities on making interpretations (Bulhof 2012). The main instrument is the researcher himself (Brown & Baker 2007). Then, the data analysis technique used is descriptive qualitative particularly the interactive model proposed by Miles et al. (2013).

3 RESULTS AND DISCUSSION

3.1 *Hegemony of Western values in aesthetics books in Indonesia*

Based on the analysis conducted from those books, it revealed that there are at least two essential problems concerning Western hegemony in the content of the aesthetic books. The first is related to the "substance" domain and the second is related to the "subject" domain.

First, the substance domain in this context is closely related to the discussion of the substance of the content material or the subject matter of the aesthetics books. It is in line with the term "substance" itself, that in Hegelian perspective is defined as focus of study or subject matter of a course (Bunnin & Yu 2008, Rotenstreich 2012, Hickman & Porfilio 2012).

From the substance, the whole narration of the books talk about the essential study of aesthetics theories, starting from the notion of "what" (ontology), "how" (epistemology), and "why" (axiology) referring to a conception formed within Western aesthetic awareness.

In the part of ontology for instance, it is clear that the existing arts and aesthetics knowledge construction, such as the concept of beauty, scope, elements, etc. is based on the construction of Western art ontology. The clear picture of that problem can be seen from the extract of Aesthetics books written by Djelantik (1999). When it discusses art and beauty theories it merely refers to the origin of Western civilization, particularly Greece. It considers that no place except Western countries has art history.

Then, in the epistemology part, it is clearly seen that the knowledge construction of arts, for example, in terms of the discussion of the why of 'being' either in the form of a signifier system (form or physical appearance) or in the form of signified system (content or substance), is based on Western concept dictum. This kind of picture and story of hegemony in the level of epistemology is not merely found in the form of concepts but also in its entire history that contains the whole narration of art histories that happened in the Western culture. The historiography of arts philosophy in Indonesian aesthetics covers the periods of arts history starting from Western pre-modern and mainly discusses its history starting from the Greece and old Roman era up to the Classic period in the centuries before, up to the current era that is the 20th century and 21st century. One more thing that seems important in the discussion of arts epistemology with Western bias is why the writer puts the art knowledge construction based mainly on those of natural science perspectives based on objectivity.

In addition, in the axiology part, the underlying basis used are also values covering Western bias. One example of this phenomenon can be seen in the discussion of arts and aesthetics in connection with a values system, particularly ethics, that is based on the division proposed by a German philosopher Baumgarten (Djelantik 1999).

This phenomenon implies that Eastern countries do not have any diction related to the concepts or terminology of arts. In fact, Eastern countries have more than that of Western, especially when arts becomes the topic of discussion. Eastern countries are considered being on the opposite side of Western in the way they are dominated by senses compared to Western that are dominated by "logic".

Second, in relation to the content, particularly in terms of the "subject", the discussion focuses on the important figures becoming the central narration of the books (Smith 2010). The discussion of this variable focuses on the figures that are common of philosophers or individuals that have great attention and influence in arts and the development of arts, always becoming the main resources concerning aesthetics and arts. This study reveals that in terms of the "subject", these books also reflect the Western hegemony seen from the use of Western figures, namely: Plotinus, Plato, Aristoteles, Socrates (in the discussion of classic Greece); Shaftesbury, Hutchesson, David Hume, Immanuel Kant (Enlightenment Age); Edward Bullough, Stolniz, Aldrich (The Age of XIX Century); and Susane Langer, Collingwood, Clive Bell, Morris Weitz, George Dickie (The Age of XX).

On the contrary, there is not even a single figure coming from local Indonesia, not to mention those having great narration and influence during the kingdom era in the past who are also commonly called *empu* (expert), such as *Empu* Kanwa, Panuluh, Prapanca, and Ranggawarsita. These important figures have proposed great ideas in relation to the aesthetics concepts through their masterpieces as magnificent as those proposed by Western philosophers.

Consequently, it is no wonder that the current understanding and knowledge of aesthetics can be said to be the result of receiving Western arts philosophical concepts, which cannot be used to encourage critical perspectives to find 'authentic' Indonesian art philosophy with all its typical Indonesian character. This condition shows that Indonesian aesthetics needs to be taken care of and reconsidered (Sumardjo 2000). The problem of Western values hegemony essentially could lead to negative impacts and even be destructive in terms of socio-cultural life of Indonesians.

3.2 *Factors underlying Western philosophy hegemony in Indonesian aesthetics books*

The complex hegemony of Western bias aesthetic books is not natural but consciously constructed. It means that this problem is closely related to cultural complexity that is inextricably intertwined with "power". The term "power" is closely connected to the Western orientalism system. Said (2014) once claimed that the term orientalism needs to be defined as modes or ways of thinking constructed by a certain power and the colonial knowledge. Said (2014) further mentions that orientalism is the result of imperial power that the discourse products tend to have more political interest.

Orientalism is Western historical construction towards Eastern to make them "the other" that Western people will consider "exotic aliens" (Said 2014). On the contrary, conventional colonialism is not similar to rude and brutal racism. Therefore, it is important to understand the term orientalism as a discourse showing the essential differences and discrimination between two opposite sides of "we as Western people" and "they as Eastern." Colonialism is dominative while orientalism is hegemonic emphasizing on the superiority of European culture and identity (Carbin 2016).

Actually, orientalism is a kind of ethnocentrism meaning that the center of awareness is the researchers' civilization, their ethnicity. They are reluctant to use their local knowledge that might be very phenomenological. They prefer to be in the big framework of colonization that the reflection of their study in many occasions seems bias and unstable. Finally, orientalism can be said to be a study that doesn't fulfill the requirement to be a science. This ethnocentric orientalism in many occasions is trapped in Western ego. For these people there is only Western that exists and the existence of Eastern people is the result of Western awareness that came to them. Without this Western awareness-ocean exploration, Eastern will not exist. It is for this reason that orientalism spirit likes truth claims, awareness claims, existentialism claims. The truth claim coming from Western has been accepted in the historical development of the world civilization. In addition, having supported by a certain power, finally what has been claimed by orientalism becomes the final truth, unquestionable and undebatable.

What happen to the construction of aesthetics as a science mentioned earlier is one example of the fact that the role of Western orientalism is very strong in defeating knowledge, not to mention in this context of the knowledge of aesthetics and arts. It can be said that everything comes from Western countries and that Eastern countries have nothing. In fact, it is not true. Similar to Western countries, Eastern countries also have a great amount of knowledge, not to mention aesthetics and arts, and even have more valuable knowledge compared to Western.

Orientalistic attitude that tends to be egocentric and ethnocentric further will destroy their truth claims that they believe until today. Hanafi (2003) mentions that orientalism will get their credibility as modern orientalists refuse European centric and no longer divide the world into the center and peripheral, modern and traditional civilizations. They are honest in explaining cultural characteristics of those nations and also their creativity through the ages, and are fair and proportional in writing the history of humanity.

One important thing to be noted is those who want to prove their credibility, they need to be able to get out from the idea of 'center-peripheral' for it is this idea that leads to a relation of power that place Western as the center of awareness and Eastern as a matter of its reflection. If the discussion of 'the other' conducted by Western is a kind of mutual discourse that enables each side to learn and understand each other, orientalism will lead to ideal humanism that respects cultural pluralities (Hanafi 2003).

Factors influencing the Western orientalism hegemony are not only coming from external factors that like the Western influence, but also coming from the internal factors, that so called "Eastern orientalism" (Bamyeh 2012, Ward 2004). It is for this reason that in its latest development, the truth construction of Western orientalism is not mainly constructed by Western orientalists but also Eastern ones, not to mention the one found in this study. This phenomenon has made the problem of Western hegemony towards Eastern more complicated as the main subjects of colonization and hegemony are not merely coming from Western countries, but also those of Eastern countries. It turns out to be a perfect hegemony that becomes almost absolute in which finally East suffers from not only Western colonization but at the same time suffers from Eastern colonization that Etkind (2013) and Deloria & Salisbury (2004 p. 460) call as "internal colonization."

4 CONCLUSION

In relation to the previous part, there are some important points to be noted.

First, the construction of Western hegemony in Indonesian aesthetic books, was found in at least two categories: the substance and the subjects. Substantively, the construction of hegemony can be easily found in the content, message and subject matter of the books that can be said to cover Western bias.

Then, in relation to the subject category, it is found that most figures whose ideas became the main resources to construct the truth of knowledge explained in the books, are mostly taken from Western art philosophers.

Second, the Western aesthetic hegemony as mentioned is the result of what is so-called the Western post-colonial syndrome coming from the strong hegemony of Western orientalism that remains persistent in ex-Western colonies, not to mention Indonesia. Therefore, it is necessary to have a new awareness to get away from this cultural, aesthetic and arts colonization in the future.

REFERENCES

Babuts, N. 2017. *Mimesis in a cognitive perspective: mallarme, flaubert, and eminescu.* London: Routledge.

Bamyeh, M.A. 2012. *Intellectuals and civil society in the Middle East: liberalism, modernity and political discourse.* London and New York: I.B. Tauris.

Bao, Y. et al. 2016. Aesthetic preferences for Eastern and Western traditional visual art: identity matters. *Frontiers in Psychology.* 7(1596). doi: 10.3389/fpsyg.2016.01596.

Brown, B.J. & Baker, S. 2007. *Philosophies of research into higher education.* London: A & C Black.

Bulhof, I.N. 2012. *Wilhelm Dilthey: A hermeneutic approach to the study of history and culture.* Berlin: Springer Science & Business Media.

Bunnin, N. & Yu, J. 2008. *The blackwell dictionary of Western philosophy.* New York, NY: John Wiley & Sons.

Carbin, M. 2016. Postcolonialism, theoretical and critical perspectives on. *The Wiley Blackwell Encyclopedia of Gender and Sexuality Studies.* 21 April 2016. doi: 10.1002/9781118663219.wbegss202.

Carter, R. 2015. *Language and creativity: the art of common talk.* London: Routledge.

Chan, D.W. 2011. East vs. West. *Encyclopedia of Creativity* (Second Edition). https://doi.org/10.1016/B978-0-12-375038-9.00082-0.

Chand, B. 2017. *Advance philosophy of education.* Chennai, Tamil Nadu: Notion Press.

Deloria, P.J. & Salisbury, N. 2004. *A companion to American Indian history.* New York, NY: John Wiley & Sons.

Djelantik, A.A.M. 1999. *Estetika: Sebuah pengantar.* [Aesthetics: An introduction]. Bandung, Indonesia: MSPI.

Dutton, D. 2009. *The art instinct: beauty, pleasure & human evolution.* Oxford: Oxford University Press.

Etkind, A. 2013. *Internal colonization: Russia's imperial experience.* New York, NY: John Wiley & Sons.

Fong, W. 2001. *Between two cultures: Late-nineteenth and twentieth-century Chinese paintings from the Robert H. Ellsworth collection in the Metropolitan Museum of Art.* New York, NY: Metropolitan Museum of Art.

Foster, A.W. & Blau, J. R. 2011. *Art and society: Readings in the sociology of the arts.* Albany: SUNY Press.

Guyer, P. 2005. *Values of beauty-historical essays in aesthetics.* Cambridge: Cambridge University Press.

Hamersma, H. 2008. *Pintu masuk ke dunia filsafat.* [Entrance to the world of philosophy]. Yogyakarta, Indonesia: Kanisius.

Hanafi, H. 2003. *Oposisi pasca tradisi.* [The post-tradition opposition]. Yogyakarta, Indonesia: Syarikat.

Hardiman, F. B. 2003. *Melampaui positivisme dan modernitas.* [Beyond positivism and modernity]. Yogyakarta, Indonesia: Kanisius.

Hartoko, D. 1986. *Manusia dan seni.* [Human and art]. Yogyakarta, Indonesia: Kanisius.

Hellman, J. 2003. *Performing the nation: cultural politics in new order Indonesia.* Copenhagen: Nordic Institute of Asian Studies Press.

Hickman, H. & Porfilio, B.J. 2012. *The new politics of the textbook: critical analysis in the core content areas.* Berlin: Springer Science & Business Media.

Hulatt, O. 2013. *Aesthetic and artistic autonomy.* London: A & C Black.

Hussain, M & Wilkinson, R. 2006. *The pursuit of comparative aesthetics: an interface between East and West.* Farnham: Ashgate Publishing, Ltd.

Isrow, Z. 2017. Defining art and its future. *Journal of Arts & Humanities* 6(6): 84–94. doi: 10.18533 / journal.v6i6.1207.

Kelly, M. 2003. *Iconoclasm in aesthetics.* Cambridge: Cambridge University Press.

Lapan, S.D., et al. 2011. *Qualitative research: an introduction to methods and designs.* New York, NY: John Wiley & Sons.

Leaman, O. 2002. *Eastern philosophy: key readings.* London: Routledge.

Leuthold, S. 2010. *Cross-cultural issues in art: frames for understanding.* London: Routledge.

Listowel, O.E. 2016. *A critical history of modern aesthetics.* London: Routledge.

Liu, K. 2000 *Aesthetics and marxism: Chinese aesthetic marxists and their Western contemporaries.* Durham, NC: Duke University Press.

Long, E.T. 2003. *Twentieth-century Western philosophy of religion 1900–2000.* Berlin: Springer Science & Business Media.

Mann, T. 2015. *The Oxford guide to library research.* Oxford: Oxford University Press.

Marra, M. 2010. *Essays on Japan: between aesthetics and literature.* Buckinghamshire, England: BRILL.

Merriam-Webster Dictionary. 2017. Definition of aesthetics. *Merriam-Webster Dictionary Online.*

Milbrath, C. & Lightfoot, C. 2010. *Art and human development.* London: Psychology Press.

Miles, M.B.A., et al. 2013. *Qualitative data analysis.* London: Sage Publication.

O'Neill, R. 2017. *Art and visual culture on the French Riviera, 1956-971: The ecole de nice.* London: Routledge.

Paddison, M. 2010. Mimesis and the aesthetics of musical expression. *Music Analysis,* 29(1–3): 126–148. doi:10.1111/j.1468–2249.2011.00333.x.

Rotenstreich, N. 2012. *From substance to subject: studies in Hegel.* Berlin: Springer Science & Business Media.

Ruttkowski, W. 2007. *East and West and the concept of literature.* München: GRIN Verlag.

Said, E.W. 2014. *Orientalism.* New York, NY: Knopf Doubleday Publishing Group.

Sasaki, K. 2011. *Asian aesthetics.* Singapore: NUS Press.

Sayuti, S.A. 2014. Art and art education: construction identity venue, *Paper presented in the 1st International Conference of Arts and Arts Education on Indonesia,* Universitas Negeri Yogyakarta, 5–6 Maret 2014.

Schellekens, E. & Goldie, P. (Ed). 2011. *The aesthetic mind: philosophy and psychology.* Oxford: OUP Oxford.

Scaramuzzo, G. 2016. Aristotle's homo mimeticus as an educational paradigm for human coexistence. *Journal of Philosophy of Education,* 50(2):246–260. doi: 10.1111/1467–9752.12204.

Palmer, S.E., et al. 2013. Visual aesthetics and human preference. *Annu Rev Psychol.* 2013(64): 77–107. doi: 10.1146/annurev-psych-120710-100504.

Şenol, A. 2014. Is art mimesis or creation?. *Procedia-Social and Behavioral Sciences,* 116(2014): 2866–2870. doi: 10.1016/j.sbspro.2014.01.670. Shusterman, R. 2012. Back to the future: Aesthetics today. *The Nordic Journal of Aesthetics,* 43(2012):104–124. doi: 10.7146/nja.v23i43.7500.

Sumardjo, J. 2000. *Filsafat seni.* [Art philosophy]. Bandung, Indonesia: Sunan Ambu Press STSI Bandung.

Sutrisno, M. 2005. *Teks-teks kunci estetika: filsafat seni.* [Aesthetic key texts: arts philosophy]. Yogyakarta, Indonesia: Galang Press.

Süzena, H.N. & Mamur, N. 2014. Reflection of philosophy on art and philosophy of art. *Procedia-Social and Behavioral Sciences,* 122(2014): 261–265. doi: 10.1016/j.sbspro.2014.01.1339.

Tung, W.H. 2014. *Art for social change and cultural awakening: An anthropology of residence in Taiwan.* Lexington Books.

Ward, A. 2004. Eastern others on Western pages: Eighteenth-century literary orientalism. *Literature Compass,* 1(1),. doi: 10.1111/j.1741–4113.2004.00068.x.

White, B. 2009. *Aesthetics primer.* New York, NY: Peter Lang.

Zangwill, N. 2017. Aesthetic judgment. *Stanford Encyclopedia of Philosophy.*

Character Education for 21st Century Global Citizens – Retnowati et al. (Eds)
© 2019 Taylor & Francis Group, London, ISBN 978-1-138-09922-7

Constructing global citizenship: Kindergarten and primary schoolteachers' understanding of globalization and education

H. Yulindrasari & S. Susilowati
Universitas Pendidikan Indonesia, Indonesia

ABSTRACT: Advanced information technology has made global networks of various sectors such as business, government, education, communities, humanitarian acts, and crime easier to organize and strengthen. The world has become increasingly borderless. Thus, it is inevitable that young generations need to be prepared for the challenge of exposure to different cultures, ethics, values, and ways of life. The awareness of global competition has influenced policies in education. The discourse on preparing children for global competition is stated in many government documents related to education. However, there is a lack of research examining how teachers construct knowledge about global citizenship. Using in-depth interviews with six teachers who teach young children aged 4–12 years old, this study explores teachers' understanding of the global citizenship discourse. This study finds complexity and contradictory understanding of global citizenship, especially in terms of ethics and morality.

1 INTRODUCTION

> 'The policy's focus for 2015–2019 is to improve the quality of and access to education to prepare the children for global competitions by understanding differences and diversity, strengthening best practices and increasing innovations.'

The above quote is taken from a government document of the strategic plans of the Ministry of Education and Culture for 2015–2019 (RoI, 2015). Through education, the government of Indonesia aims at creating internationally competitive citizens by 2024. The term *kewarganegaraan global* (global citizenship) occurs 16 times in the document. Interestingly, the term *kewarganegaraan global* is only used in parts where the document discusses society and early childhood education development. The term is absent from the discussion of the development of other levels of education. An assumption of early childhood (from birth to eight years old) as being a period where a sound basis of values and norms is learned and internalized by an individual, underpins the emphasis of planting a seed of global citizenship within the potential future generation of Indonesia. Young children are seen as the future agents of social and economic transformations for prosperity, and education is seen as an investment to realize the imagined future agents (Ailwood, 2008; Adriany & Saefullah, 2015; Formen & Nutall, 2014). However, the document does not explain what it means to be a global citizen, other than to be able to compete at the international level.

An unclear definition and conceptualization of global citizenship potentially creates confusion among teachers in translating the concept into pedagogical practices. This article aims to identify the teachers' understanding of global citizenship, their definition of it, and how they connect education to globalization and vice versa.

2 LITERATURE REVIEW

2.1 *Globalization*

Globalization has commonly been defined as social, economic, cultural, political and ideological mobility, movement and interactions between countries across the globe (Ferguson &

Mansbach, 2012). The driver of globalization is economic development and prosperity. The advances of technology and globalization have brought humans from different parts of the world closer to each other, as if the world is borderless.

For Indonesia, and many other countries, international agencies play important roles in interactions between local and international politics, the economy, society and culture. Although globalization's first and foremost mission is economic development, it has inevitably provoked changes in every sector of human life, including the way in which people see education. Globalization shifts educational goals from fulfilling local needs to global needs. The skills taught to the children should follow international standards and demands. The World Bank and The United Nations Educational, Scientific and Cultural Organization (UNESCO) are the most influential international agencies in the field of education in Indonesia, especially in early education. They set indicators of successful education and impose it on Indonesia and many other countries in the world, aiming to produce global citizens all around the world who will become the agents of the global economy.

2.2 *Global citizenship*

Most scholars who wrote about global citizenship have positioned the term within a neoliberal discourse, which involves global competitions, transnationalism, and cosmopolitanism (see Appiah, 2006; Rizvi, 2009). Global competitions refer to open competition on employment and economic resources. To be competitive, a neoliberal agent must have strong professional skills, be outward looking, and be able to seek opportunities of success and prosperity beyond the border of one's own country, which resonates with transnationalism that refers to the movement of ideas, people, and goods across the globe, which perpetuates easier social, economic, political and ideological interactions across countries (Bauböck & Faist, 2010). Another important aspect of global citizenship is cosmopolitanism, which encourages the individual to think him/herself as being part of a larger social, cultural and political entity or community than just his/her homeland (Riberio, 2015).

The massive interaction between people across the globe through physical and virtual mobility allows individuals to be exposed to something unfamiliar and different from their initial structure of knowledge. Differences can create negotiations, resistances, conflicts, or transformations, or all of them simultaneously. To be able to manage conflict, global citizens should embrace diversity and acknowledge the complexity and interconnectivity of the local and global phenomena (Lilley et al., 2017). Complexity and interconnectivity in global citizenship lead to multiple ways of being a global citizen; there is no one fixed expression and representation of global citizens (Schattle, 2008).

This article utilizes the concept of global citizenship developed by Lilley et al. (2017). Based on the implementation of the idea of global citizenship into learning by Rizvi (2009), Marginson & Sawir (2011), and Lilley et al. (2017), it can be conceptualized four capacities a global citizen should possess: social imaginary, criticality, reflexivity, and relationality. Social imaginary is the ability to empathize with others who come from different cultural, social, political and ideological backgrounds. It also includes openness to ideas and perspectives different from his/her own; and openness to new possibilities in doing things beyond what is commonly done. Criticality refers to the ability to reflect on our perspective and other people's assumptions; learning to understand differences critically, challenging the known and embracing the possibility of changes. Reflexivity is the capability and willingness to challenge pre-existing assumptions and knowledge. Relationality resonates with empathy, the ability to think about ourselves in relation to others and vice versa; the understanding of being part of an entity (Lilley et al., 2017). This article uses the four characteristics of global citizenship as a starting point in analyzing the responses of the respondents.

3 METHOD

This research is a preliminary stage of a bigger research about global citizenship in early education. Aiming at exploring teachers' basic perceptions about globalization, education, and global citizenship, the authors interviewed six teachers (three kindergarten teachers and three primary schoolteachers) in Bandung, the capital city of West Java. All interviewees had more than five years' teaching experience in a private school and had a degree of Bachelor of Education. The author used a convenience-sampling technique in choosing the participants of the study. Before an interview, the authors asked each of the participants for their consent to be interviewed and participate in the study. The interviews were audio-recorded and transcribed by the authors. The transcripts were then thematically coded and analyzed using an interpretative approach. Pseudonyms are used throughout the article to respect the participants' confidentiality.

4 FINDING AND DISCUSSION

Most respondents in this study were unaware of the government's mission as stated in the 2015–2019 education strategic plan document. They were familiar with 'global citizenship'. Most of them claimed that they had heard of the term but not from a government's education policy. However, they did not have a clear idea of what it means to be a global citizen and what characterizes global citizenship. However, as they tried to answer our questions about their understanding of the terms 'global citizenship' and 'global citizen', there are three main themes consistently that came up in their answers. First, global citizenship as being citizens in the world of information technology; we label it as the citizen of the cyber world. Second, freedom is the key to global citizenship. Last, the dangers of western influence.

Most teachers' understanding of the purpose of Indonesian education is '*untuk mencerdaskan kehidupan bangsa*' (translated: to enlighten the nation). None of them mentioned about creating a competitive global citizen. None of them realized what was written in the government's strategic plan for education. Ihsan, a primary schoolteacher, said:

> 'I only know the one that is stated in Indonesian Constitution: education is '*untuk mencerdaskan kehidupan bangsa.*' That is all. I am not following the newest education policies.' (Ihsan, 19 July 2017, Bandung).

The respondents' unawareness of the current policy shows that the information in the policy document is not necessarily read, followed, and understood by the teachers. Although most respondents did not notice that the government has made global citizenship part of its educational goal, they had a limited understanding of global citizenship and agreed that the children should be prepared to be global citizens. The following sections discuss their understanding of global citizenship.

4.1 *The citizen of the cyberworld*

All respondents connected global citizenship with the cyberworld, how easy it is to get information and connect with others. As Ihsan said below,

> 'Global citizenship is where we interact with other people from all over the world without any border. It (global citizenship) occurs in the cyber space interaction which is borderless.' (Ihsan, 19 July 2017, Bandung).

Ihsan's understanding of global citizenship as what only happens in the cyberworld is related to the Indonesian context and the city he lives in. He lives and works in Bandung, a city where expats and foreign workers are still rare. He, like other respondents, is not familiar with direct interactions with people from other countries. Unlike in the case of big cosmopolitan melting-pot cities like Melbourne, where people from all over the world interact in a non-virtual world on a daily basis since more than a third of the population were born outside Australia (Australian Bureau of Statistics [ABS], 2014), Indonesian society, let alone Bandung, is considered homogeneous in terms of citizenship and country of origin. Throughout Indonesia, only 0.03 percent of the people are non-Indonesian (Badan Pusat Statistik, 2011). Therefore, global citizenship is still restricted to the cyberspace where the world has become seemingly borderless.

Ihsan and other teachers acknowledged the positive aspects of Internet-based interaction, especially in terms of the speed and access to information. Most of them were convinced that being a citizen of the cyberworld would widen their network and cross-cultural knowledge, and open up their horizon.

However, they also realized the other side of the coin, as not all information circulated on the Internet would benefit the students. The school where Ihsan worked was aware of that and equipped their students with Internet literacy, including the ethics of Internet interactions and how to use the technology to produce animation so that the children can be the future inventors of technological products, and not only the consumers. Unfortunately, it was not common in other schools in this study. For example, the school where Ria, a primary schoolteacher, worked focused more on teaching the students to use the product of information technology, as she explained,

> 'Yes, students need to understand how the contemporary technology works. In my time young children were taught to post a mail to a post office, now teachers should teach the children to send an email and use WhatsApp.' (Ria, 24 July 2017, Bandung).

The other five respondents in this study did not share forward-thinking like Ihsan's school. The definition and implementation of education for future global citizens varies and seems to be left to each school's interpretation.

4.2 *Freedom*

Another key term that they used to refer to global citizenship is 'freedom'; being a global citizen means being a free citizen. Ria explained,

> 'The world is open and free now. We can go abroad freely; it is borderless. Also free trade, free market. Anyone from abroad can come to Indonesia freely. With the Internet, the information also freely comes from outside Indonesia.' (Ria, 24 July 2017, Bandung).

Ria's comment started to mention the movement of people or migration. Her understanding of globalization and global citizenship also included the political-economy aspect of globalization, although she could not clearly articulate it.

Similarly, Nana, a kindergarten teacher, also agreed that global citizenship involves freedom, especially freedom to choose nationality, freedom to have as many citizenships as he/she likes, and to live and work wherever he/she wants to live although it is not his/her country of citizenship. Nana believed that a global citizen must have foreign language skills, winning capacities, and cross-cultural understanding to be able to compete in the free labor market. Freedom in many aspects of life comes with consequences, risks, and responsibility, which will be further discussed in the following section (4.3).

4.3 *Filtering western values*

Most respondents argued that the free flow of unfiltered information would jeopardize the local values, ethics and morality. They framed cross-cultural interactions more in a negative tone than a positive one, especially when they talked about western influence on the Indonesian generation, as Sari, a primary schoolteacher, put it,

> 'In the era of global citizenship, there are a lot of incompatible western culture penetrates Indonesia through the Internet, television, and the printed media, which then copied by the people. Free sex is an example. What a loss!' (Sari, 27 July 2017, Bandung).

The West has long been propagated to Indonesian, especially Muslims, as being the enemy of the society's standard morality and ethics (for example see Jones, 2003; Brenner, 1999). All respondents felt the need to filter information, especially that which comes from the West, and filtering the information is the parents' and teachers' responsibility to protect the future generation. They believed that the borderless world should be with boundaries.

They suggested that nationalism is the key to counter western influences. They thought it was important to instill the love for the country (*cinta tanah air*) and to strengthen local identity as early as possible since they did not want the 'genuine' Indonesian identity to be eroded by the West.

5 CONCLUSION

Referring back to Lilley et al.'s (2017) conceptualizations of global citizenship, none of the respondents talked about embracing cross-cultural diversity and differences, nor suggested teaching cross-cultural values to the children. Teachers in this study understood global citizenship in terms of access to global information and building children's technical skills rather than building a cross-cultural social competence. Citizenship homogeneity in Indonesia limits people's exposure to multicultural and multinational interactions on a daily basis. Although one would argue that Indonesia is a multicultural society by itself, it is restricted to local cultures only, not international cultures. This research recommends a discussion and workshop for teachers about globalization and global citizenship. Teachers, academics, and the government can sit together to build a concept of global citizenship that is appropriate for Indonesian education.

REFERENCES

Adriany, V. & Saefullah, K. (2015). Deconstructing human capital discourse in early childhood education in Indonesia. In T. Lightfoot-Rueda, R.L. Peach & N. Leask (Eds.), *Global Perspective on Human Capital in Early Childhood Education* (pp. 159–182). New York, NY: Palgrave.

Ailwood, J. (2008). Learning or earning in the 'smart state': Changing tactics for governing early childhood. *Childhood, 15*(4), 531–551.

Appiah, A. (2006). *Cosmopolitanism: Ethics in a world of strangers.* London, England: Penguin.

Australian Bureau of Statistics (ABS). (2014). *Australian Social Trend, 2014.* Retrieved from http://www.abs.gov.au/ausstats/abs@.nsf/Lookup/4102.0 main+features102014#migrant.

Badan Pusat Statistik (BPS). (2011). *Kewarganegaraan, Suku Bangsa, Agama, dan Bahasa Sehari-hari Penduduk Indonesia: Hasil Sensus Penduduk 2010 [Nationality, ethnic origin, religion, language and everyday residents of Indonesia: Results of a population census 2010].* Jakarta: BPS. Retrieved from http://demografi.bps.go.id/phpfiletree/bahan/kumpulan_tugas_mobilitas_pak_chotib/Kelompok_1/Referensi/BPS_kewarganegaraan_sukubangsa_agama_bahasa_2010.pdf.

Bauböck, R. & Faist, T. (Eds). (2010). *Diaspora and transnationalism: Concepts, theories and methods.* Amsterdam, The Netherlands: Amsterdam University Press.

Brenner, S. (1999). On the public intimacy of the new order: Images of women in the popular Indonesia print media. *Indonesia,* (67), 13–37.

Ferguson, Y.H. & Mansbach, R.W. (2012). *Globalization: The return of borders to a borderless world?* Hoboken, NJ: Taylor and Francis.

Formen, A. & Nuttall, J. (2014). Tensions between discourses of development, religion, and human capital in early childhood education policy texts: The case of Indonesia. *International Journal of Early Childhood, 46*(1), 15–31.

Jones, C. (2003). Dress for sukses: Fashioning femininity and nationality. In S. Niessen, A.M. Leshkowich & C. Jones (Eds.), *Re-Orienting Fashion: The Globalization of Asian Dress* (pp. 185–213). New York, NY: Berg.

Lilley, K., Barker, M. & Harris, N. (2017). The global citizen conceptualized: Accommodating ambiguity. *Journal of Studies in International Education, 21*(1), 6–21.

Marginson, S. & Sawir, E. (2011). *Ideas for intercultural education.* New York, NY: Palgrave.

Reysen, S. & Katzarska-Miller, I. (2013). A model of global citizenship: Antecedents and outcomes. *International Journal of Psychology, 48*(5), 858–870. doi:10.1080/00207594.2012.701749.

Riberio, G.L. (2015). What is Cosmopolitanism? *Brazilian Anthropology, 2*(1–2), 19–26.

Rizvi, F. (2009). Toward cosmopolitan learning. *Discourse: Studies in the Cultural Politics of Education, 30*(3), 253–264.

RoI. (2015). *Rencana Strategis Kementrian Pendidikan dan Kebudayaan 2015–2019 [The strategic plan of the Ministry of education and culture 2015–2019].* Jakarta, Indonesia: Indonesian Ministry of Education and Culture, Republic of Indonesia.

Schattle, H. (2008). *The practice of global citizenship.* Plymouth, UK: Rowman & Littlefield.

The seafarers' characters standard for international shipping industry

W. Pratama
Akademi Maritim Yogyakarta, Indonesia

P. Pardjono & H. Sofyan
Universitas Negeri Yogyakarta, Indonesia

ABSTRACT: The study aims to identify seafarers' characters as required by International Shipping Industry (ISI) standard. Characters need to be developed for the cadets in Indonesian Maritime Colleges (IMCs) that have been well or less developed at the IMCs. Descriptive method was employed in this study. Data was collected using the instrument validated by expert review. Subjects consisted of the shipping industry managers, the experienced seafarers of ocean going vessels, IMC managers, and lecturers. The results of the study were: (1) the ISI required 17 characters: discipline, toughness, commitment, responsibility, creativity, integrity, confidence, cooperation, courage, tenacity, work ethics, fortitude, adaptability, independence, problem-solving, sense of humor and vigilance; (2) determination of the character that needed to be developed, respondents chose 6 characters from the provided such as: discipline, responsibility, confidence, work ethic, problem solving, and cooperation; (3) discipline is a well-developed character in IMCs, on the other hand, responsibility, self-confidence, work ethic, problem-solving skills, and cooperative ability are less developed.

1 INTRODUCTION

December 2016 was a historic moment for Indonesia when it officially joined the ASEAN Economic Community. Since then, in terms of trade and labor, Indonesia has entered the world of competition among ASEAN countries. Even so, the quality of Indonesian workers, in terms of competence, is very worrying when competing with colleagues from neighboring countries of ASEAN.

Preparing the workforce to meet the requirements of international standard for competencies needs rigorous planning. Likewise, when the government prepares workers for the maritime field, which is required by the International Maritime Organization (IMO) for the qualification of maritime workers, the government needs a strategy and is generally tricky and requires smart effort. To meet the International Standard Certification Training and Watchkeeping for Seafarer (STCW) and its amendments, a person must have a Certificate of Competence (CoC) and a Certificate of Proficiency (CoP).

The STCW requirements were reaffirmed based on the IMO for its implementation of the Quality Standard System (QSS). However, the implementation of QSS in the Indonesian Maritime Colleges (IMCs) is not optimized as only reaching 88.26% as stated by Pratama (2010) based on his research findings, entitled 'Evaluation of the Implementation of the IMO-based-QSS in Akademi Maritim Yogyakarta (AMY)'. It is the fact that the AMY graduates are still at a hard struggle of entering the work market of International Shipping Industry (ISI). The competency gap between those required by the labor market and achieved by IMC graduates in which their colleges have implemented the ISI standards, indicates some factors affect the ability of graduates in handling work in the real world, and particular characters are the factors estimated to affect graduate competence. This conjecture is supported by

Elenora, the Human Rejuvenation Development Officer at PT. Arpeni Pratama Ocean Lines Tbk. In her public lecturing at AMY in 2009, she stated that a component of competence required by seafarers in addition to knowledges and skills, were characters.

Blanpain & Dimitrova (2010) argued that thetypical characteristics of the seafarer's working environment are risky and uncomfortable and associated with monotonous work, remote living, a socially isolated working environment, and a high stressful working condition. They must also prepare themselves at any time for facing bad weather, marine accidents and other unsafe conditions of the ship. These working characteristics, accordingly, need a particular character of seafarers, among others are toughness, self confidence, discipline, adaptability and independence.

The problem is unavailability of information pertaining characters required by the ISI where the graduates of the IMC will be working in. The other problem is to what extent the IMC graduates' character fulfilling standard of seafarers' character is required by the ISI. To what extent the practice of educating cadets' characters in IMC and what seafarers' characters that have been successfully developed in the IMCs, and what characters have not, it is not clearly identified, therefore, need exploring.

To get an initial overview of the character required by the ISI, interviews were held with AMY students and students of Akademi Maritim Nasional (AMN) Cilacap and also the Vice Director of AMN Cilacap. It was reflected from this preliminary interview that the development of character education in IMCs has not been carried out in an appropriate plan. As a result, the seafarers' character that havs been developed has never been properly and clearly identified. At this time, cadets' characters are developed with hierarchical-based approach through physical health fitness and discipline activities. Arthur, et al. (2015: 4) states that one of the most important ethical developments today is the recovery of ancient wisdom about the importance of character. We need good character for an ethical, productive, satisfying life, and creating a just society which is loving, and productive.

Samani & Hariyanto (2012) interpreted character as the basic values that build a person, formed because of the influence of heridity and of the environment, which distinguishes him/her from others, and embodied in attitudes and behavior in everyday life. Character can also be regarded as human behavior related to God Almighty, self, fellow human, environment, and nationality embodied in thoughts, attitudes, feelings, words and deeds based on religious norms, law, etiquette, culture, customs and aesthetics. In addition, one of the shipping companies engaged in Crew Management, Ship Management, Marine Consultation in Jakarta, the capital of Indonesia, when recruiting seafarers used recruitment instrument containing certain elements of character. Industry considers the critical success factors in the workplace, such as responsibility, integrity, professional skills, attitude, team building, personality, general appearance, cheerful, sense of humor, communication skills, problem solving, decision making, and future potential or creativity as part of aspects being interviewed (Altus Anglo-Eastern Crewing Services 2013). Building particular character of a person is not a simple attempt nor easy program, as Helen Keller said, "Character can not be developed easily and simply, only through trial experience and suffering can the soul be strengthened, the vision clears the ambition inspired, and success is achieved." (Soedarsono 2009) Yet, success in character building will positively impact on the following matters such as: (1) a strong soul, (2) a clear far ahead vision, and (3) having an inspiring drive in all effort to achieve true success. More conceptual, the character is the value embedded in human beings through education, experience, experimentation, and sacrifice. The value of humanity when combined with environmental experience affects the power system underlying thoughts, attitudes, and behaviors.

Education is an effective vehicle used to form certain characters of a person and this has long been proven. Therefore, the development of seafarers' character will be effective when it is conducted through a well-designed education program. Kamaruddin (2012) adds that character education is important for human growth as a whole and should be done early on. Character education is intended to promote the development of the character of learners. This character education has been considered, recommended, and/or applied for a variety of reasons (Berkowitz & Bier 2005). Christina (2005) even reminds us that it is important for any university not only to pay more attention on the academic needs of the students but also on character, so the graduates are well prepared both in academics and in character. In other words, educational institutions should not only serve as a place to build students' academic abilities but also to build students' characters.

Brown (2008) stated that character education does have an impact on students' behavior. Research conducted by Rahdiyanto et al. (2017) also found that implementation of character education by integrating it into learning subjects of complex machinery has a positive impact on the improvement of students' attitude and achievement.

Character education is the key to build seafarers' character, so that all people who are involved in this program should be knowledgeable on the principles of education, especially education for character building. Lickona (1992) argues that character education is education for personality involving knowing, feeling, and acting. Without these three aspects, the education of seafarers' character will not be effective. In addition, character education should be carried out in a systematic and sustainable manner. Shield (2011) argues that education should develop intellectual character, moral character, and character of citizenship performance character, along with the collective character of the school.

Davis (2003) stated that character education can be used more broadly to include institutional efforts other than school to form characters. To integrate education, Williams (2010) suggested that we must share our views during the instructional planning process. An individual develops competence by applying the overall aspects of the person, including the ability to reflect on the situation and the actions themselves. Students and workers become the producer of knowledge, therefore, it is essential to the success of knowledge-based work processes (Brockmann et al. 2008).

Based on the several concepts of seafarers' character and character education described earlier, research to find characters that meets the ISI standards needs to be conducted. Which of the identified characters need to be developed for the IMC cadet in compliance with the ISI requirements? Which characters have been considered well developed by the IMCs and which characters are less developed?

2 RESEARCH METHOD

This research was essentially using quantitative inquiry approach with the pattern of presentation of results as well as the analysis of data in descriptive manner. This research was preceded by preliminary research to get initial data about the seafarer characters in general. Starting from the characters obtained, the characters were submitted to the Focus Group Discussion (FGD) forum to be discussed in order to get feedback from the participants and whether they agree or not with the characters. The characters agreed by the FGD participants then were chosen based on the priority and level of importance of characters to be developed in IMC. FGD participants ranked characters from the most important characters to the least important characters. The results were then used to conduct a survey in some predefined IMCs to investigate which characters have been developed and which have not.

2.1 *Data collection and respondents*

Primary data was collected through survey technique using questionnaire validated by expert review. Respondents were determined puposively and selected from lecturers and leaders of the IMCs. Four IMCs were selected randomly among the IMCs which had been approved by the Directorate General of Sea Transportation as IMO administrator in Indonesia. The selected four IMCs were: Indonesian Maritime Academy (AMI) Medan, Indonesian Maritime Academy Veteran (IMAV) Makasar, STIMART-AMNI Semarang, and Akademi Maritime Yogyakarta (AMY). There were 14 participants involved in the FGD comprising of 12 experienced seafarers who had been involved as maritime officers in various international shipping companies and two managers of international shipping companies.

Table 1. Investigation process in a brief.

Research step	Target	Method
Preliminary	Initial data of seafarers'character values; 16 character values were elicited	Interviewing AMY lecturer who experienced as seafere officers.
Focus Group Discussion (FGD)	Deciding agreed seaferers'character values from the 16 and others	Discussing for international maritime industry.
Survey	Ranking the agreed character values	Analyze and evaluate the agreed character values to bring up priority character values using questionnaire.
Survey	Obtaining seafarers'character values that had been developed and that had not.	Evaluating and responding to obtain seafarers' character values that had been developed and that had not.

2.2 *The procedures of eliciting characters*

The procedure for eliciting seafarers' characters was divided into three stages. The first was to identify the seafarers' characters in general through preliminary research. The second was to identify the seafarers' characters that must be developed in meeting the ISI requirements. The third was to identify the characters that have been effectively developed in the IMCs and those that have not.

A preliminary study has been conducted to explore the seafarers' characters, and sixteen characters have been identified. The second step, the identified characters were then submitted to the FGD forum to gain input and agreement from the participants based on the urgency and the importance of character for the seafarer. There was one character of the 16 that was not approved by the participants, but there were additional two seafarer characters that were considered being important. So the number of approved characters was 17 (seventeen). The FGD participants were then asked to rank these seventeen characters from the most important to the least. The researchers then determined a number of urgent and feasible characters to develop in the IMCs, and 6 (six) characters of seafarer were determined. Table 1 shows a brief description of the entire investigation process.

3 RESEARCH FINDINGS AND DISCUSSION

3.1 *The characters of seafarers*

The initial data provided 16 (sixteen) characters, such as discipline, toughness, commitment, responsibility, integrity, confidence, cooperation, courage, tenacity, work ethics, fortitude, adaptability, independence, problem solving, sense of humor, and corsa spirit. This data of seafarers' characters was then submitted to the FGD forum to have agreement from the participants concerning seafarers' characters that meet ISI requirements. Then the selected sixteen characters were completed with the descriptors of each and used as the materials for FGD.

3.2 *The seafarer characters developed by the IMC*

The FGD discussions came to the conclusion that the characters of the korsa spirit was not chosen by any participant. The reason for not choosing the corsa spirit because it should only be owned by a cadet as long as he/she is a student. When he/she has already working as seafarer a corsa spirit is hypothetically having a negative effect. Instead, there are two additional character values that seafarers must have, namely alertness and creativity. The reason is that when a seafarer is on the ship, vigilance is needed to respond quickly to dangerous and emergency situations. Creativity is also considered being important for seafarers by which they can find ways and tasks within facilities, infrastructure and limited time to properly save the ships, passengers and cargo. After FGD session, the 17 (seventeen) characters were found to become seafarer characters that have been recommended by the participants. Those character values were discipline, toughness, commitment, responsibility, creativity, integrity, confidence, cooperation, courage, tenacity, work ethic, fortitude, adaptability, independent, problem solving, sense of humor, and vigilance.

In the next step, the FGD participants were then asked to rate the 17 character values to specify 6 (six) characters of seafarers that need to be developed in the IMCs. Those six characters were then promoted as the target of seafarers' character development in the IMCs.

These six seafarers' characters must be prioritized to be developed in the IMCs to improve quality of graduates. Table 3, shows six types of seafarers' characters sorted from the ones that get the highest priority scores to the lowest.

3.3 Developing seafarer characters in IMCs

After finding seafarers' characters required by the shipping industry, the researcher investigated the six characters that have been developed as mentioned in Table 3. The data obtained from the questionnaire was then analyzed to find the characters in percentage. The results

Table 2. Scores of each character value and the descriptor.

No	Character values	Descriptors	Scores
1	Discipline	Attitudes and behaviors that arise as a result of training or habit of obeying rules, laws and commands	12
2	Responsibility	Know, understand, and do what must be done as expected by others	10
3	Problem solving	Create or design problem-solving strategies for nonroutine problems as well as the problems encountered in everyday life.	6
4	Courage	Stay steadfast in the truth, no matter negative pressure, not afraid to fail, not afraid to voice the conscience, dare to do right.	4
5	Creativity	Carry out the fulfillment of needs, complete tasks, or realize ideas with new perspectives	4
6	Self confidence	Believe in self in terms of ability and self-efficacy, a mental attitude that believes fully and depends on one's own ability	8
7	Integrity	Always try to do right thing, bring about what has ever been said or promised, life based on ethics, willing to learn from problems and failures	5
8	Teracity	Stay in action in defending the goal or a situation especially in the case of many obstacles, challenges or disappointing things	4
9	Toughness	Personal unyielding in any situation	4
10	Cooperation	Actions and ttitudes willing to work with others to achieve common goals and mutual benefits	6
11	Commitment	Feel boun to the duty and call of a strong soul to exercise it emotionally, intellectually, and physically	3
12	Fortitude	The power of the mind and heart to withstand stress because of the magnitude of trials and hindrances so as to do the best	3
13	Work ethics	Confidence in the existence of moral benefits and the inherent skills derived from work that can strengthen the caracter	9
14	Adaptibility	Easily adjust to his environment and learn something from his past experience	2
15	Independence	Able to meet the needs of oneself with his own efforts and not depend on others	2
16	Sense of humor	Ready to laugh and make others laugh without disturbing and offending others	1
17	Vigilance	Be aware of what is going on aroung and respond appropriately and correctly	1

Table 3. Prioritized seafarer characters.

No	Characters	Score
1	Discipline	12
2	Responsibility	10
3	Work ethics	9
4	Self confidence	8
5	Problem solving	6
6	Cooperation	6

Table 4. Distribution of responses for each character.

No	Character values	Score for the character values that have not been developed		Score for the character values that have been developed		Average
		Learning	Campus culture	Learning	Campus culture	%
1	Descipline	0	0	48	48	100
2	Responsibility	5	1	43	47	93.75
3	Work ethics	12	8	36	40	79.17
4	Confidence	28	32	20	16	37.50
5	Problem solving	31	26	17	22	40.62
6	Cooperation	27	23	21	25	47.92

of the percentage analysis are shown as follows. Discipline has been fully developed (100%); responsibility has been developed through learning 89.58% and through campus culture 97.92%, and average is 93.75%; confidence has been developed through learning 41.67%, through campus culture 33,33%, and the average is 37.50%; work ethics have been developed through learning 75%, through campus culture of 83.33%, and the average is 79.17%; problem-solving through learning has been developed 35.42%, through campus culture 45.83%, and the average is 40.62%; cooperation through learning has been developed 43.75%, through campus culture 52.08%, and the average is 47.92%.

Table 4 shows the response distribution and the percentages of each character and the activity through which each character is acquired. It also shows that development of discipline in the IMCs has been done well as it achieved 100% both through learning process and campus culture. While the characters of responsibility, work ethics, cooperation, problem solving and confidence have not been fully developed.

Self-confidence, for example, has not been effectively developed since only at 28% or 53.3% of learning and through campus culture at 32% or 67%. Similarly, problem solving, through learning with 31% or 65% and campus culture 26% or 54%. Confidence and problem solving skills are the two characters that have not been developed effectively both through learning activities and campus culture.

4 CONCLUSION

The seafarers' character values required by the shipping industry are 17 characters which are: discipline, toughness, commitment, responsibility, creativity, integrity, confidence, cooperation, courage, tenacity, work ethics, fortitude, adaptability, independent, problem solving, sense of humor, and vigilance.

Six seafarers' characters are recommended to be developed by IMCs in accordance with the demands of ISI such as: discipline, responsibility, confidence, work ethics, problem solving, and cooperation.

The characters of seafarers that have been developed effectively for the cadets in IMCs is only discipline, and those that are less developed are responsibility, confidence, work ethics, problem solving, and cooperation.

The recommendations based on the findings then are submitted. It is quite numerous, however there are characters considered being important to be developed in the IMCs, since these 17 characters are also relevant to the needs of the ISI and need to be well developed in IMCs. Responsibility, confidence, work ethic, problem solving, and cooperation are less developed in the IMCs, therefore, it needs to develop learning model that is capable of developing these five characters.

REFERENCES

Altus Anglo-Eastern Crewing Services. 2013. *Proposal for recruitment*. Jakarta: author.

Arthur, J., et al. 2015. *Character education in UK Schools Research Report The Jubilee Centre for Characters and Virtues*. London: University of Birmingham.

Berkowitz, M.W. & Melinda, C.B. 2007. *What works in character education: A research-driven guide for educators*. Washington: University of Missouri—St Louis.

Blanpain, R. & Dimitrova, D.N. 2010. *Seafarers' rights in the globalized maritime industry*. Netherland: Kluwer Law International.

Brockmann, M., et al. 2008. Knowledge, skills, competence: European divergences in vocational education and training (VET): The English, Germany and Dutch Cases. *Oxford Review of Education* 34(5): 547–567.

Brown, C. 2008. *Teachers' perception of character education and its impact on student behavior*. Dissertation. Walden University.

Christina, W. 2005. Upaya penerapan pendidikan karakter bagi taruna: studi di Jurusan Teknik Industri UK [Attempts to apply character education for cadets: study at the UK Department of Industrial Engineering]. *Petra Jurnal Teknik Industri* 7(1): 83–90.

Davis, M. 2003. What's Wrong with Character Education? *American Journal of Education* 110(1): 32–57.

Elenora, E. 2009. *Prespektif kebutuhan SDM pelayaran sekarang dan masa yang akan datang* [Perspective of needs of human resource current and the future]. Paper presented at guest lecture at Akademi Maritim Yogyakarta.

Internatioal Maritime Organization (IMO). 2011. *STCW convention and STCW code*. London: Author.

Kamaruddin, S.A. 2012. Character education and students social behavior. *Journal of Education and Learning* 6(4): 223–230.

Lickona, T. 1992. *Educating for character: How our schools can teach respect and responsibility*. New York, NY: Bantam Books.

Pratama, W. 2010. *Evaluasi implementasi QSS berbasis IMO di Akademi Maritim Yogyakarta* [Evaluation of QSS iImplementation based on IMO at the Yogyakarta Maritime Academy]. Unpublished research report at Universitas Negeri Yogyakarta.

Rahdiyanto, P. et al. 2017. Character-based collaborative learning model: Its impacts on students' attitudes and achievement. *Jurnal Pendidikan Teknologi dan Kejuruan* 23(3): 227–234.

Samani, M. & Hariyanto, M.S. 2012. *Konsep dan model pendidikan karakter* [The concept and model of character education]. Bandung: Remaja Rosdakarya.

Shield, D.L. 2011. Character as the aim of education. *The Phi Delta Kappan* 92(8): 48–53.

Soedarsono, S. 2009. *Karakter pengantar bangsa dari gelap menuju terang* [The character of the nation's introduction from dark to light]. Jakarta: Gramedia.

Williams, H. & Stiff, R. 2010. Widening the lens to teach character education alongside standards curriculum. 83(4): 115–120.

Preparing teachers for integrated values education

The importance of *Halal* education in forming the civilized and exemplary global citizen

B. Kartika
International Islamic University Malaysia, Malaysia

ABSTRACT: Morality and ethics have become the fundamental questions facing humanity in the current climate, as how is more effective moral and ethical education to be accomplished? The global education system needs to be designed for the refinement of the moral values and to prevent the immorality. The question is, what is best for global human rights and the highest human good? Can the current educational system lead to a world where fundamental decisions could be based upon trust, humility and wisdom, and where interactions are focused on what is best for all human beings; and, in which interactions are characterized by compassion, loyalty, honesty, respect and forgiveness. This paper discusses the integration between *Naqli* (revealed knowledge) and *Aqli* (rational knowledge) in achieving the goal of citizens to being civilized, educated and ethical at the same time. One of the disciplines which falls under the integration of *Naqli* and *Aqli* in Islam is the rulings of *Halal* (lawful) and *Haram* (unlawful). *Halal* and *Haram* are universal terms that apply to all aspects of human life, whether related to his *ibadat* (worship), *muamalat* (related to worldly matters such as business/trading/commercial transactions, lending, borrowing contracts and civil acts or dealings under Islamic law) or*muasharah* (the way he treats others). The concept considers all the physical and spiritual advantages of the product to the humankind. For this, the article also proposes that the *Halal* knowledge is to be transferred to the students as early as at the Primary Level of their education, due to its importance in forming the good characteristics of the global citizens.

1 OVERVIEW OF ISLAMIC EDUCATION; INTEGRATION BETWEEN NAQLI AND AQLI

Islamic education is not limited to the study of the Qur'an, as it is the foundation of all knowledge which guides the behavior of the believing Muslim. It integrates modern scientific knowledge into the Islamic worldview as early Muslims did during the first few centuries of Islam. For example, the methods of science in medicine, genetics and biology are integrated with the Qur'an and Sunnah. The Qur'an and Sunnah provide a rich resource of knowledge, solutions and ethics in the fields of science, medicine, economics and the humanities. If we refer to the beginning of the revelation, the very first verse of God revealed to the Prophet was the imperative verb: *Iqra'* (read). It is interesting to note that verb 'to read' is indeed classified as a transitive verb (*al muta'addy*), which means that it needs objects.

The verses 1–5 of Qur'an Surah al 'Alaq (96) says:

اقْرَأْ بِاسْمِ رَبِّكَ الَّذِي خَلَقَ

خَلَقَ الْإِنْسَانَ مِنْ عَلَقٍ

اقْرَأْ وَرَبُّكَ الْأَكْرَمُ

الَّذِي عَلَّمَ بِالْقَلَمِ

عَلَّمَ الْإِنْسَانَ مَا لَمْ يَعْلَمْ

Read in the name of your Lord who created –
Created man from a clinging substance.
Who taught by the pen –
Read, and your Lord is the most Generous –
Taught man that which he knew not.

The first verse says: *Read in the name of your Lord who created.* Word 'read!' – it is an imperative verb, whereby God has commanded the Prophet to read; this goes to his followers as well, but indeed it is a command to the entire humankind, all citizens. The command is for the citizens to become knowledgeable and educated. We note that after the verb 'read' there is no single object mentioned. This shows that we are required and are allowed to read anything, provided the foundation and the goal of this reading are to seek God's pleasure, and in the way guided by God, since the verse says: *Read in the name of your Lord who created.* Indeed the signs of the Greatness of God are in the forms of His sacred Book, Al-Qur'an, and in His creations of the whole entire universe. The verb is then repeated in the third verse: *Read, and your Lord is the most Generous.* This command to read has a different approach, as it is for us to read, to acquire knowledge as the form of our gratitude for the generosity of God who had provided us with everything we need in this life. This goal of reading is in line with the goal of the creation of everything in this universe, especially the creation of humankind. Man was created to worship and serve Almighty God; the Holy Qur'an informs us of man's duty with the verse 56 of Surah al Dzariyat (51):

وَمَا خَلَقْتُ الْجِنَّ وَالْإِنسَ إِلَّا لِيَعْبُدُونِ

I created not jinn and man except that they might worship.

Humans are created for a few main purposes, which are:

1. Recognizing God, using knowledge, as the first stage.
2. Belief in Him, is the second stage after recognizing Him with knowledge. This is analyzed as Belief by investigation, and not by imitation.
3. Worshipping Him, is the last stage after recognizing and believing in Him with knowledge

Thus, here the ultimate aim of knowledge is the worship and servitude of God. The above verse also tells us that the creation of humans is not without purpose: it also applies for the creation of everything. Based on this belief, human beings are charged with certain duties, and are placed on the temporary world, which is the transient earth, for a trial and examination. They are also given faculties to discover God's attributes that are manifested in the universe in different kinds of forms. Thus the humanity's highest aim and purpose is to complete its faith with knowledge and love of God. So worshiping God takes many different forms too. The goal of the creation of the human being is portrayed in him being a vicegerent of God on earth, to be in relation with the rest of creation as a caretaker. The verses 30–33 of Qur'an Surah al Baqarah (2) say:

وَإِذْ قَالَ رَبُّكَ لِلْمَلَائِكَةِ إِنِّي جَاعِلٌ فِي الْأَرْضِ خَلِيفَةً ۖ قَالُوا أَتَجْعَلُ فِيهَا مَن يُفْسِدُ فِيهَا وَيَسْفِكُ الدِّمَاءَ وَنَحْنُ نُسَبِّحُ
بِحَمْدِكَ وَنُقَدِّسُ لَكَ ۖ قَالَ إِنِّي أَعْلَمُ مَا لَا تَعْلَمُونَ

وَعَلَّمَ آدَمَ الْأَسْمَاءَ كُلَّهَا ثُمَّ عَرَضَهُمْ عَلَى الْمَلَائِكَةِ فَقَالَ أَنبِئُونِي بِأَسْمَاءِ هَٰؤُلَاءِ إِن كُنتُمْ صَادِقِينَ

قَالُوا سُبْحَانَكَ لَا عِلْمَ لَنَا إِلَّا مَا عَلَّمْتَنَا ۖ إِنَّكَ أَنتَ الْعَلِيمُ الْحَكِيمُ

قَالَ يَا آدَمُ أَنبِئْهُم بِأَسْمَائِهِمْ ۖ فَلَمَّا أَنبَأَهُم بِأَسْمَائِهِمْ قَالَ أَلَمْ أَقُل لَّكُمْ إِنِّي أَعْلَمُ غَيْبَ السَّمَاوَاتِ وَالْأَرْضِ وَأَعْلَمُ مَا
تُبْدُونَ وَمَا كُنتُمْ تَكْتُمُونَ

Behold, your Lord said to the angels: 'I will create a vicegerent on earth.' They said:

'Will You place therein one who will make mischief therein and shed blood? – whilst We do celebrate Your praises and glorify Your holy (name)?' He said: 'I know what you know not.'

And He taught Adam the nature of all things; then He placed them before the angels,
and said: 'Tell Me the names of these if ye are right.'
They said: 'Glory to You, of knowledge We have none, save what You have taught us:
In truth it is You Who are perfect in knowledge and wisdom.'
He said: 'O Adam! Tell them their names.' When he had told them their names, Allah
said:
'Did I not tell you that I know the secrets of heaven and earth, and I know what you
reveal and what you conceal?'

Adam's comprehensive disposition was one of the reasons that he was taught the names of all things and was made vicegerent of the earth. Adam's superiority lay in his ability to acquire complete knowledge of the names. Knowledge is thus 'the pivot of the vicegerency.' For the execution of God's ordinances and the application of His laws, which constitute the vicegerent's function on the earth, is dependent on full knowledge.

In interpreting what the Qur'an is alluding to with this verse, *Nursi*, one of the thematic exegetes, posits that since men are the descendants of Adam and the inheritors of his abilities, it is a must for them to learn the names, that is, to explore the universe and through their combined efforts develop the sciences that reveal its functioning. As holders of the Supreme Trust, they have also to prove their worthiness before other creatures. It is through human development and scientific technological progress that man gains mastery over the universe; also it is only by such progress that man actually rises to the vicegerency rank. The pursuits of modern science can be a way of developing the knowledge of the names that was given to Adam. Fulfillment of this duty is achieved by acquiring knowledge of the Divine names and attributes (Nursi 2008).

Barguth says that man's ultimate duty in his temporary life is to examine and search for the names presented in the universe under the form of knowledge. Man's intellectual, mental, and spiritual faculties are the means by which the wisdom, consciousness, and good are opened to him and which the horizons of civilization and the vicegerency unfold. He also says that it is these faculties which represent the crown of man's abilities, and enable him to be God's vicegerent on earth. For God almighty deposited in man a vast innate capacity and potential vicegerency which renders him capable of assuming the Trust and struggling with its requirements and conditions by means of consciousness, reason and thought, which affect the universe and life (Barguth 2010).

The primary importance of the above-mentioned verse is in the designation of man as an agent of God on earth. It serves as a manifestation of the greatest of God's blessings to humanity that has been placed on earth under human dominion to be utilized in the manner humans desire.

From the discussion on the main aim of humans' creation and their vicegerency, it can be concluded that what is meant by education here is **learning, acquiring knowledge and its application in one's life**. Through Divine revelation the wisdom behind the creation of the universe and human beings is learned. From here it is understood that one of the reason why man has been sent to this world is to attain to perfection by way of worship backed up by knowledge and belief; he is then made answerable and accountable for the Trust given to him. The whole process of being caretaker on earth can be achieved by applying the integration between *Naqli* and *Aqli*. The integration of *Naqli* and *Aqli* knowledge in the curriculum can be a strategic plan. The term *Naqli* refers to Divine knowledge; this derives from the Qur'an, Sunnah and references of respected books by previous religious scholars, while the term *Aqli* refers to modern knowledge that is gained through research and discoveries in the present. By having this integration in its curriculum on both of these disciplines, it is hoped that the educational system will serve humanity to reach its goal of being civilized, educated and ethical; thus it becomes an exemplary one. This is because the primary objective of the Islamic rulings in *Naqli* is a mercy to the worlds (Q.S. al Anbiya', 21:107). The Qur'an, as the first source of Islamic rulings, has the characteristics of being a healing to the spiritual ailment of hearts, and a guidance and mercy for the believers and mankind (Q.S. Yunus, 10:57). Thus, Islamic Laws/*Shariah* aim to benefit individuals and the community. Its laws are: to protect these benefits and to facilitate improvement and perfection of the conditions of human life on earth (establish justice, eliminate prejudice, alleviate hardship); they are meant to promote cooperation and mutual support within the family and society at large and to manifest the realization of benefits/*maslahah*. Another important objective of *Shariah* is educating the individual/*tahdhib al-fard*, which seeks to make every individual a trustworthy agent and carrier of the values of the *Shariah,* so that the social objectives will be gained. *Tahdhib al-fard* comes before justice and *maslahah*, which are socially oriented and require social relations. The overall purpose of *Shariah*, especially in devotional matters/*ibadat* and moral teaching/ *akhlaq*, is to train the individual who is mindful of the virtues of *taqwa* and becomes an agent of benefit to others (Kamali 2008).

The general definition of civilization is as follows:

- The stage of human social development and organization which is considered most advanced (Oxford, 2017).
- The process by which a society or place reaches an advanced stage of social development and organization (Oxford, 2017).
- An advanced state of intellectual, cultural, and material development in human society, marked by progress in the arts and sciences, the extensive use of record-keeping including writing, and the appearance of complex political and social institutions (Houghton 2016).
- Cultural or intellectual refinement; good taste (Houghton 2016).
- Modern society with its convenience (Houghton 2016).

For this, the term 'civilized' refers to having an advanced or humane culture and society.

It also refers to being polite, well-bred and refined, as well as easy to manage or control, well organized or ordered.

Humanity's highest aim and purpose is to complete its faith with knowledge and love of Allah. So worshiping God takes many different forms too. Indeed, each and every activity of man in his daily life is a form of worship. God has set up all the guidelines for a man to utilize them so he will be always worshiping God in any form he does. But how does one keep this goal in mind while going about one's daily business? When one is involved in the routine of daily life, it may appear difficult to maintain the spiritual level required to win Allah's love. Allah himself provided mankind with the means to do this. He revealed the Holy Qur'an to the Prophet Muhammad (peace and blessings of Allah be on him) and raised him up to be the 'Perfect Leader' and teacher of the Quranic law to all people. Prophet Muhammad (peace and blessings of Allah be on him) was called on to furnish an example through following Allah's guidelines.

2 *HALAL* EDUCATION

Halal is an Arabic word meaning lawful and permitted. This *Halal* concept comes from the Al-Quran and is used to describe objects and actions. Islam is a natural way of life and encompasses the concept of an economic system based on human cooperation and brotherhood, which is based on the consultation and dietary laws for all humanity. *Halal* is a Quranic term that means permitted, allowed, lawful or legal. Its opposite is *haram* (forbidden, unlawful or illegal). According to *Shariah*, all issues concerning *Halal* or *Haram*, and even all disputes, should be referred to Qur'an and Sunnah. *Halal* and *Haram* are universal terms that apply to all aspects of human life, whether they be related to his *ibadat* (worship), *muamalat* (related to worldly matters such as business/trading/commerce transactions, lending, borrowing contracts and civil acts or dealings under Islamic law), or *muasharah* (the way he treats others). *Halal* may be defined as an act, object or conduct over which the individual has freedom of choice. *Halal* may have been identified by explicit evidence in the *Shariah* or by reference to the presumption of permissibility (*ibahah*).

There are two key words in this verse, which are *Halal* (Permitted and Lawful by Islamic Law) and *Toyyib* (Hygiene, Safe, Good, Clean, Wholesome).

In Surah al Baqarah verse 168, Allah states:

O mankind, eat from whatever is on earth [that is] lawful and good and do not follow
the footsteps of Satan. Indeed, he is to you a clear enemy.

Adopting a *Halal* lifestyle is not only important for Muslims, but also for non-Muslims. Even though the verse mentions about consumption, the application covers the application of *Halal* lifestyle in general. It implies that the lifestyle we adopt and choose for ourselves is a considered one. Thus, our chosen *Halal* lifestyle is an ethical one. It is a choice of lifestyle that is without force, discrimination, bias or prejudice. It would therefore be a chosen lifestyle that is pleasing to us, conducive and tolerable for all people and for every circumstance, place or venture. Indeed, we have contentment of heart and an exemplary natural disposition toward others, too. A *Halal* lifestyle guarantees and manages the morality of whatever we do, say or become involved in. Once we have embraced such a *Halal* lifestyle for ourselves, we would find happiness and contentment, as opposed to perplexity and anxiety in our lives. At the same time, it safeguards and preserves our dignity and honor, our self-respect and self-control, and our integrity and individuality. Maintaining a *Halal* lifestyle offers us modesty, sustainability, stability and safety in our lives. In particular, *Halal* is a right of every Muslim, but it is indeed a privilege for every non-Muslim.

For a person to be aware about what is allowed and what is not is very important; it will control one's life conduct to be always on the right path. *Halal* and *Haram* have become two key ethical aspects. Understanding these two will lead the person to possess good characteristics and thus become an exemplary citizen of the universe. For this, it is very important that the *Halal* education is introduced to everyone at their early education level. Some of the proposed elements to include in *Halal* education are: The Awareness of *Halal* Lifestyle in general; *Halal* Consumption in particular; and *Halal* logo.

3 *HALAL* AWARENESS

The word 'awareness' means the knowledge or understanding of a particular subject or situation. The word 'awareness' in the context of *Halal* literally means having special interest in or experience of something, and/or being well informed of what is happening at the present time regarding *Halal* foods, drinks and products. As such, awareness describes human perception and cognitive reaction to a condition of what they eat, drink and use. Awareness provides the raw material to develop subjective ideas about one's experience related to something (Nizam 2006). Awareness about something is therefore a basic part of human existence. On top of everything is the self-awareness. Different people have different levels of awareness about something. So, awareness in the context of *Halal* can be conceptualized as the informing process to increasing the levels of consciousness toward what is permitted for Muslims to eat, drink and use. *Halal* affords one the subconscious self-awareness of hygiene and health, safety and security, independence and self-determination.

This knowledge of *Halal* awareness should be transferred to both Muslims and non-Muslim alike. When Allah refers to the command of *Halal* consumption for example, He addresses three different groups of people: first is all mankind, second is the Believers, and third is the Messengers.

The detail is as follows:

1. To all mankind (Q.S. al Baqarah, 2: 168)

يَا أَيُّهَا النَّاسُ كُلُوا مِمَّا فِي الْأَرْضِ حَلَالًا طَيِّبًا وَلَا تَتَّبِعُوا خُطُوَاتِ الشَّيْطَانِ ۚ إِنَّهُ لَكُمْ عَدُوٌّ مُبِينٌ

O mankind, eat from whatever is on earth [that is] lawful and good and do not follow
the footsteps of Satan. Indeed, he is to you a clear enemy.

2. To the Believers (Q.S. al Baqarah, 2: 172)

يَا أَيُّهَا الَّذِينَ آمَنُوا كُلُوا مِنْ طَيِّبَاتِ مَا رَزَقْنَاكُمْ وَاشْكُرُوا لِلَّهِ إِنْ كُنْتُمْ إِيَّاهُ تَعْبُدُونَ

O ye who believe! Eat of the good things wherewith We have provided you, and render
thanks to Allah if it is (indeed) He whom ye worship.

3. To the Messengers (Q.S. al Mu'minun, 23: 51)

يَا أَيُّهَا الرُّسُلُ كُلُوا مِنَ الطَّيِّبَاتِ وَاعْمَلُوا صَالِحًا ۖ إِنِّي بِمَا تَعْمَلُونَ عَلِيمٌ

O ye apostles! enjoy (all) things good and pure and work righteousness: for I am well-acquainted with (all) that ye do.

When we talk about the *Halal* awareness, it is not limited to consumption. Even in the Qur'an some of the issues are mention in regard to what is permitted (*Halal*) and what are not. Some of these examples are as follows.

3.1 Food

Say: I do not find in that which has been revealed to Me anything forbidden for an eater to eat of except that it be what has died of itself, or blood poured forth, or flesh of swine—for that surely is unclean—or that which is a transgression, other than (the name of) Allah having been invoked on it; but whoever is driven to necessity, not desiring nor exceeding the limit, then surely your Lord is Forgiving, Merciful (Q.S. al An'am, 6: 145).

Forbidden to you is that which dies of itself, and blood, and flesh of swine, and that on which any other name than that of Allah has been invoked, and the strangled (animal) and that beaten to death, and that killed by a fall and that killed by being smitten with the horn, and that which wild beasts have eaten, except what you slaughter, and what is sacrificed on stones set up (for idols) and that you divide by the arrows; that is a transgression. This day have those who disbelieve despaired of your religion, so fear them not, and fear Me. This day have I perfected for you your religion and completed My favor on you and chosen for you Islam as a religion; but whoever is compelled by hunger, not inclining willfully to sin, then surely Allah is Forgiving, Merciful.

They ask you as to what is allowed to them. Say: The good things are allowed to you, and what you have taught the beasts and birds of prey, training them to hunt—you teach them of what Allah has taught you—so eat of that which they catch for you and mention the name of Allah over it; and be careful of (your duty to) Allah; surely Allah is swift in reckoning (Q.S. al Maidah, 5: 3–4).

3.2 Gambling

O you who believe! Intoxicants and games of chance and (sacrificing to) stones set up and (dividing by) arrows are only an uncleanness, the Shaitan's work; shun it therefore that you may be successful (Q.S. al Maidah, 5: 90).

3.3 Trading

It explains the difference made between usury and trading, and how to conduct trade:

Those who swallow down usury cannot arise except as one whom the Shaytan has prostrated by (his) touch does rise. That is because they say, trading is only like usury, while Allah has allowed trading and forbidden usury. To whomsoever then the admonition has come from his Lord, then he desists, he shall have what has already passed, and his affair is in the hands of Allah; and whoever returns (to it) – these are the inmates of the fire; they shall abide in it (Q.S. al Baqarah, 2: 275).

Give a full measure and be not of those who diminish; And weigh (things) with a right balance, And do not wrong men of their things, and do not act corruptly in the earth, making mischief (Q.S. al Syu'ara, 26: 181–183).

4 *HALAL* CONSUMPTION AS A WAY OF WORSHIP

Within the Islamic tradition, our spirituality and purity are tied closely to the food and drink we consume. In an age of increasing commercialization, the food we consume can be destitute of Allah's presence and the reverence for His creation that Islam requires.

Allah says (Q.S. al Baqarah, 2:35):

وَ قُلْنَا يَا آدَمُ اسْكُنْ أَنْتَ وَزَوْجُكَ الْجَنَّةَ وَكُلَا مِنْهَا رَغَدًا حَيْثُ شِئْتُمَا وَلَا تَقْرَبَا هَٰذِهِ الشَّجَرَةَ فَتَكُونَا مِنَ الظَّالِمِينَ

We said, 'O Adam, dwell, you and your wife, in Paradise and eat there from in [ease and] abundance from wherever you will. But do not approach this tree, lest you be among the wrongdoers.'

This verse teaches us very important lessons:

1. Man is created to obey Allah in any form.
2. It shows that the act of eating from the permissible sources is mandatory and obligatory. Eating in Islam is not merely for the sake of overcoming hunger, but more than that, it has the Divine order, thus it lies the Divine reason behind the order.
3. Eating is a form of obeying Allah/worshiping Allah, for this, its guidelines must be followed.
4. Man is at the same time tested by Allah in any form Allah wills.
5. The prohibition of approaching to the certain tree in this verse shows the test given to man, and it has been laid under the Divine reason as well.
6. It shows that to approach to the forbidden things will lead one to do the forbidden acts.
7. Anyone goes beyond the permissible limit, drags his own self to be wrongdoer, uncivilized citizen and disobedient to Allah. It leads to the disaster, backwardness, ignorance, shamefulness and punishment.

As Allah's servant, one will always be tested by Him, in both favorable and unfavorable ways to determine the level and sincerity of faith as stated in the Qur'an (Q.S. al Anbiya', 21:35):

كُلُّ نَفْسٍ ذَائِقَةُ الْمَوْتِ ۗ وَنَبْلُوكُمْ بِالشَّرِّ وَالْخَيْرِ فِتْنَةً ۖ وَإِلَيْنَا تُرْجَعُونَ

'Every soul will taste death. And We test you with evil and with good as trial; and to Us you will be returned.'

The prohibition of approaching the tree is a great symbol of a test. Where a human is provided with an abundance of permissible things to do and to use, permissible food and drink to be consumed, and at the same time he has to abstain from unlawful things to do and to use, and unlawful foods and drink to consume, all of these have a Divine purpose behind them, which is obedience. This verse has clearly stated that Adam and his wife were provided with an abundance of lawful foods to eat from wherever they would, and that they should not approach a certain tree, otherwise they would be wrongdoers which implies the disobedience. That is the main reason of prohibition.

5 IMPLICATION OF *HALAL* OR PERMISSIBLE AND LAWFUL IN ISLAM

Halal stands for values that are embedded in Divine Values, and is explained as follows:

- It is a Divine prescription that has its own inherent and distinct worth. It is part of a belief system and moral code of conduct. What is deemed *Halal* is ultimately governed by Divine Laws. It is directly related to the identity and lifestyle of a Muslim.
- *Halal* is prescribed by Allah. It is intrinsically pure and wholesome. 'He commands them what is right and forbids them what is wrong, he makes lawful the things that are wholesome and makes unlawful the things that are bad and lifts from them their burdens and the yokes that were upon them (Q.S. al A'raf, 7:157).' It is intrinsically pure and wholesome. The word *Tayyibat* has a spiritual and physical meaning which includes the physical, mental and spiritual well-being.

- *Halal* creates the capacity within the individual to act righteously. It creates an appetite to be upright and righteous. Abu Hurairah (May Allah be pleased with him) reported: The Messenger of Allah (*pbuh*) said, 'O people! Allah is Pure and, therefore, accepts only that which is pure. Allah has commanded the believers as He has commanded His Messengers by saying: "O Messengers! Eat of the good things, and do good deeds." (23:51) And He said: "O you who believe (in the Oneness of Allah—Islamic Monotheism)! Eat of the lawful things that We have provided you..."' (Muslim 2000).
- *Halal* purifies the heart and mind and opens the doors for the graceful acceptance of our prayers. Prophet Muhammad (*pbuh*) made a mention of the person who travels for a long period of time: 'His hair is disheveled and covered with dust. He lifts his hand toward the sky and thus makes the supplication: "My Rabb! My Rabb!" But his food is unlawful, his drink is unlawful, his clothes are unlawful and his nourishment is unlawful, how can, then his supplication be accepted' (Muslim 2000).Further, Imam Ghazali mentions in *Ihya' Ulumiddin* that once Sa'ad said to Rasulullah (*pbuh*): 'Pray to Allah that He may accept my invocation.' Rasulullah (*pbuh*) said: 'Eat lawful food, your invocations will be accepted.' He also said:'If a man eats lawful food for forty days, Allah illumines his heart and lets flow wisdom from his heart through his tongue' (al Ghazali 1982).
- *Halal* also signifies: Health—Well-being, Wholesomeness—Quality—Hygiene, not only with regards to the end product but with regards to all related activity from the 'farm to the fork'.
- Consuming *Halal* and avoiding suspicious things saves one's religion and honor. Narrated An-Nu'man bin Bashir RA:'I heard Allah's Apostle saying, both legal and illegal things are evident but in between them there are doubtful (suspicious) things and most of the people have no knowledge about them. So whoever saves himself from these suspicious things saves his religion and his honor. And whoever indulges in these suspicious things is like a shepherd who grazes (his animals) near the sanctuary (private pasture) of someone else and at any moment he is liable to get in it. (O people!) Beware! Every king has a sanctuary and the sanctuary of Allah on the earth is His illegal (forbidden) things. Beware! There is a piece of flesh in the body if it becomes good (reformed) the whole body becomes good but if it gets spoiled the whole body gets spoiled and that is the heart' (al Bukhari2000).
- Consuming *Haram* is one way of following the steps of Satan, it leads one to the disaster and even shirk. Imam Nursi says regarding the verse in Surah al Baqarah verse 36: 'But Satan caused them to slip out of it (by being successful in provoking Adam and Eve to eat from the forbidden and unlawful tree) and removed them from that [condition] in which they had been. And We said, "Go down, [all of you], as enemies to one another, and you will have upon the earth a place of settlement and provision for a time (Q.S. al Baqarah, 2: 36)."

6 *HALAL* LOGO

Halal lifestyle is no longer a mere religious obligation or observance for Muslims. *Halal* has become a powerful market force, becoming increasingly a worldwide market phenomenon for both Muslims and non-Muslims alike. The *Halal* logo is actually the assurance for the consumers, to state that the product is lawful according to Islamic law, and not only that it is permissible but it is also healthy, good, clean and wholesome. It is an assurance that indeed it is safe to use or to consume. To know about the recognized *Halal* logo one can refer to the website of Jakim (*Jabatan Kemajuan Islam Malaysia* or Islamic Progress Position Malaysia) Malaysia, where at least 67 logos worldwide including MajlisUlama Indonesia (MUI) are recognized. Moreover, knowing and understanding the *Halal* logo needs further discussion, as the *Halal* Certification Bodies have conditions, rules, regulations and standards to follow.

7 CONCLUSION

Halal promotes good morality and decency. *Halal* Education relates all aspects of life, and one of its important aspects is consumption. It has a strong relationship with worship and civilization. Allah had forbidden Prophet Adam and his wife from approaching the tree so that they would not become wrong doers. It also assures that keeping oneself within the limit of permissible things shows one's obedience and modesty. The first instance of human lapse resulted in the exposure of nudity, as Allah said in Qur'an Surah al A'raf:

1. فَدَلَّاهُمَا بِغُرُورٍ ۚ فَلَمَّا ذَاقَا الشَّجَرَةَ بَدَتْ لَهُمَا سَوْآتُهُمَا وَطَفِقَا يَخْصِفَانِ عَلَيْهِمَا مِن وَرَقِ الْجَنَّةِ ۖ وَنَادَاهُمَا رَبُّهُمَا أَلَمْ أَنْهَكُمَا عَن تِلْكُمَا الشَّجَرَةِ وَأَقُل لَّكُمَا إِنَّ الشَّيْطَانَ لَكُمَا عَدُوٌّ مُّبِينٌ

2. So he made them fall, through deception. And when they tasted of the tree, their private parts became apparent to them, and they began to fasten together over themselves from the leaves of Paradise. And their Lord called to them, 'Did I not forbid you from that tree and tell you that Satan is to you a clear enemy? (Q.S. al A'raf, 7:22).

3. *This verse establishes the nexus between Halal food and modesty. The consumption of Haram resulted in the exposure of their private parts, this exposure is the symbol of losing the respect, as well as being uncivilized at the same time, which were the result of disobeying Allah by consuming unlawful things.*

The Hadith mentioned before has also explained that the supplication of a person who is nourished with unlawful things is not accepted and his life is not blessed; any effort made in this life to be civilized will be in vain. The basis for the advancement of civilization of a nation is to stick to the Lord's guidance. Once His guidelines are neglected and no longer followed, there will be no civilization on earth. To form civilization which includes the exemplary citizens, the integration between *Naqli* and *Aqli* is needed. There is an Arabic statement that clarifies this conclusion best: كلما امر به الشرع امر به العقل و كلما امر به العقل امر به الشرع which implies that anything that Islam orders to, intellectual power will also prescribe and anything that intellectual power orders to, Islam also prescribes. We know that the intellectual power is based in heart and the statements to be true need the heart to be healthy.

REFERENCES

Al Qur'an. (2005). *The Qur'an: New Modern English Edition*. (Abdullah Yusuf Ali, Trans.), London, UK: The Young Muslims.

Al Bukhari, M.I.A.A. (2000). *Sahih Al Bukhari*. In Salih bin 'Abd al Aziz bin Muhammad bin Ibrahim Ali Al Shaikh (Ed.). Riyadh, Kingdom of Saudi Arabia: Darus Salam.

Al Ghazali, A.H. (1982). *Ihya'ulumiddin* (revives the religious sciences), (Fazlul Karim, Trans.). Karachi, Pakistan: Darul-Ishaat Publisher.

Barguth, A.A. (2010). *The place of the theory of knowledge in the vicegerency and civilizational process in the thought of Bediuzzaman Said Nursi*. Istanbul, Turkey: Sozler Publication.

Kamali, H. (2008). *Maqasid shariah made simple*. Kuala Lumpur, Malaysia: IAIS.

Muslim, A.H. (2000). *Sahih Muslim*. In Salih bin 'Abd al Aziz bin Muhammad bin Ibrahim Ali Al Shaikh (Ed.) (3rd ed.). Riyadh, Kingdom of Saudi Arabia: Darus Salam.

Nizam, A. A. (2006). *Perception and awareness among food manufacturers and marketers on halal food in the Klang Valley.*(Dissertation). Universiti Putra Malaysia, Kuala Lumpur, Malaysia.

Nursi, B.S. (2008). *The signs of miraculousness* (Sukran Vahide, Trans). Istanbul Turkey: Sozler Publication.

Houghton. (2016). *American heritage dictionary of the English language*, Fifth Edition. Boston, MA: Houghton Mifflin Harcourt Publishing Company. http://www.thefreedictionary.com/civilization.

Oxford. (2017). *Oxford dictionaries*. http://www.oxforddictionaries.com/definition/english/civilization.

Online learning as an innovative model of Teachers' Professional Development (TPD) in the digital era: A literature review

W. Wuryaningsih, M. Darwin & D.H. Susilastuti
Universitas Gadjah Mada, Yogyakarta, Indonesia

A.C. Pierewan
Universitas Negeri Yogyakarta, Yogyakarta, Indonesia

ABSTRACT: This paper serves as a literature review of online learning as an innovative model of Teachers' Professional Development (TPD) in the digital era. TPD is an important element for the quality of education because it affects teacher performance and student achievement. Therefore, TPD is designed to prepare teachers to perform their duties and profession well. Teachers in the 21st century encounter new challenges as a consequence of the Information and Communication Technology (ICT) integration in all areas. Online learning is developed using a Web 2.0-based digital technology. This is in order not only to improve the quality of education in the digital age but also to bring positive outcomes. This can be done by making information more easily accessible, helping to increase knowledge and skills easily, and providing an affordable training more available for teachers and overcomes the barrier of time, space, and cost at the same time. Several studies have been conducted to figure out the different results of TPD and its impact on students by comparing the traditional learning versus online learning, although more studies have been conducted on courses at the university level which have more diverse levels of studies and methods. In general, there was no significant difference on the teachers' performance in the traditional learning versus online learning, but there was an increase in knowledge, performance and beliefs as well as interaction in the online and blended communities, which had an impact by increasing the TPD effectiveness.

1   INTRODUCTION

In 2016, the Indonesian government made an innovation by developing the Teachers' Professional Development (TPD) model, in an effort to continuously improve teachers' competence (as mandated by law number 14 of 2005) as well as being a response to the need for the teachers to have knowledge and skills in the digital era. The innovation was conducted by developing online learning as a part of the learning model in the teachers' training; this is in addition to Face-to-Face (F2F) learning, which is divided into three different modes, namely face-to-face, full online and blended (mixed learning). Online learning was conducted using Web 2.0-based digital technology that enables synchronous and asynchronous learning to be done through a learning management system application.

The online learning in TPD becomes an innovation for the improvement of teachers' competence, from being the conventional face-to-face model to being a model in which learning is done through network-connected computers (Larreamendy-Joerns & Leinhardt, 2006). This is in line with technological advances (Walker, 2007) because teachers become agents of change in continuous education, and so their knowledge and competence should be constantly developed to improve innovation in learning (Rieckmann & Gardiner, 2017) that positively improves the teachers' knowledge, classroom practice and student achievement (Singh & Stoloff, 2007; Uzunboylu, 2007; Yoon et al., 2007).

However, there are many teachers' professional development programs that are not qualified (Borko, 2004) and unable to provide ongoing support to implement a new curriculum or teaching technique (Barnett, 2002). Therefore, there needs to be an effective TPD that not only provides short-term, episodic, and disconnected so that does not influence teaching and learning (DeMonte, 2013). An effective TPD focuses on the teachers' pedagogical knowledge, the continuous use of learning methods, continuous learning from time to time, attention to the school environment context, and collective participation as well as collaborative participation among fellow teachers (Desimone, 2009).

2 LITERATURE REVIEW

2.1 Teachers' professional development

'TPD is the professional growth a teacher achieves as a result of gaining increased experience and examining his or her teaching systematically.' (Glatthorn, 1995). Professional development undertaken by teachers can be successful if there was a change of attitude, improvement of knowledge and skills enhancement (Marfu'ah et al., 2017). TPD covers both formal and informal practice (Ganser, 2000) which makes it professional. It becomes a new platform for teachers and educators to promote the concept of 'teaching as a profession' (Villegas-Reimers, 2003). TPD can be in the form of in-service teacher education (Tan, 2014), which refers to the process of improving the teachers' skills and competency to provide outstanding education for students (Hassel, 1999). Teachers' professional development also contributes to changing teaching methods that have a positive impact on learning (Villegas-Reimers, 2003). TPD has no direct impact on students, but its influence is shown through the increase of the teachers' knowledge and its impact on teaching. The teachers' knowledge is a direct and important result of all TPD, and become the fundamental factors affecting the integration of TPD increased student learning (Golob, 2012). The results of several studies (Smith & Gillespie, 2007; Yoon et al., 2007) show that when teachers have more professional knowledge, it impacts on the students' higher achievement. Therefore, teachers should have the motivation to follow the TPD and then apply the innovations they experience to their teaching (Gorozidis & Papaioannou, 2014), improve their knowledge, develop new instructional practices (Gore & Ladwig, 2006), and need to relate teaching and learning to this world (Grosch, 2017) that lead to the improvement of student learning (Alberta, 2010; Blank et al., 2008).

2.2 TPD in the digital era

It is then needed to balance the form of traditional TPD using both face to face long session and full day conference with the process of learning through technology web 2.0 which are more interesting (Davis, 2009). It also creates more meaningful learning structures (Huber, 2010) in which Information and communication technology (ICT) should be seen as an opportunity to introduce new goals, structures and roles that support changes and enhance digital skills that are expected to have a positive impact on education (Twining et al., 2013); the use of ICT in learning will develop the students' thinking skills and learning opportunities, and improve their learning outcomes.

However, the main constraint is on the teachers' readiness as they are going to be the ones who will adopt the technology, in addition to the lack of administrative support and technological infrastructure at school (Kachelhoffer & Khine, 2009). Therefore, ICT integration into TPD is done so that teachers possess good digital literacy skills for classroom learning.

The assessment of the digital competence of teachers in the information era should be one of the considerations in building the TPD model to change perceptions of ICT use (Krumsvik, 2008). TPD should be collaborative, experimental and reflective (Baumfield et al., 2008; Fraser et al., 2007; Murchan et al., 2009).

There needs to be a program for knowledge and skills development using Web 2.0-based technology to create opportunities for using ICT in learning. This was a recommendation drawn from results of a survey conducted on 91 students, showing that the application of the Web 2.0-based technology has made TPD effective in improving learning (Kachelhoffer & Khine, 2009), despite the open challenge to make TPD more effective in a variety of contexts (Tondeur et al., 2016).

The shift of TPD in digital era which is in online form innovates the pedagogic practice of traditional condition. This shift is then called as a contemporary changing (Brooks & Gibson, 2012). Shulman & Shulman (2004) assured that the use of web 2.0 could support the teachers performance through some of the great features such as network and access. It is a motivation and reflection on changes and community learning to overcome the isolated teaching constraint (Hargreaves, 2010). Although this new pattern is designed to prepare teachers to be able to apply pedagogical innovations in the classroom, there are other factors that will influence the teachers' adoption of ICT, such as the teachers' beliefs related to the socio-cultural factors (Twining et al., 2013), resource capacity, infrastructure sustainability, as well as the teachers' skills and attitudes (Sang et al., 2010), and teachers are not ready to adapt with/ learn ICT, especially for teachers at age 50 or older (Santoso, 2014)

2.3 Online learning in TPD

Sun, et al. (2008) explain e learning as the utilization of communication technology to deliver information through web system which does not have any limit or limitless both in time and distance. It originated from the tradition of long-distance education and evolved from asynchronous activity (Kruger-Ross & Waters, 2013) where the development of technology, distance, and online learning provide opportunities for interactive learning (Joksimović et al., 2015). Online learning is divided into three categories, namely: synchronized (real-time) activities, asynchronous (different times for participants) activities and hybrid activities (take place as part of a larger in-person learning opportunity) (Bates et al., 2016). In fact, there is no standard for an e-learning model. E-learning can include the use of technology to obtain better impact on learning, or to create a learning environment which is more efficient in terms of the cost by showing the addition of 'e' value in the learning and pedagogical principles (Mayes & de Freitas, 2004).

A study on the development of the e-learning theory framework from 1960 to 2014 illustrates the concepts related to e-learning, such as Computer-Assisted Instruction (CAI), Computer-Based Education (CBE), Computer-Assisted Learning (CAL), Learning Management Systems (LMS), Computer-Managed Instruction (CMI), Computer-Assisted Education (CAE), Electronic Learning (e-learning), Artificial Learning, Mobile Learning, Self-Regulatory Efficacy, Computer Support for Collaborative Learning (ALE), mobile learning (m-learning), self regulatory efficacy (SRE), computer support for collaborative learning (CSCL), Rich Environments for Active (REAL), Mega university computer facilitated learning (Mega University CFL), Learning content management systems (LCMS), blended learning (B-learning), connective MOOC (c-MOOC), self directed learning (SDL), internet based learning (ILM), Massive open online coures (MOOC), MITx & EDX MOOC (x-MOOC), Little Open Online Course (LOOC), and Small Private Online Course (SPOC) (Aparicio et al., 2016).

Online learning can fill the gap in providing quality materials such as video or interactive software, and also counter the limitations of textbooks because it is more mobile and easily accessible. In addition, it is also able to develop the information content, and also interactive, critical and creative learning approaches, as well as to develop technology skills which are in accordance with the 21st century demands (Olson et al., 2011). Changes in the TPD model from the traditional model to an online model provide solutions to some constraints encountered in the face-to-face learning model, such as time, finance, geography, and content (Appana, 2008), in which money savings can be in line with the goal of improving the impact of education (Hughes et al., 2015).

Online TPD emerges as a response to the need for professional development that can be adjusted to the teachers' busy schedule and their access to unlimited information resources (Dede, 2006), as well as providing other potential benefits of an online community among teachers, reflections obtained from asynchronous interaction, interactions through online media which transform teachers to be more active in voicing their opinions, and the ability to learn virtually (Dede, 2004). However, technology-based TPD that becomes popular is not a replacement for face-to-face experience (Brooks & Gibson, 2012), but is more an alternative learning model in addition to the traditional face-to-face model (Keegan, 1996; Nipper, 1989). Although there is no direct difference in results between traditional face-to-face learning and online learning (Moon et al., 2014), online learning promises more time and space for the students, in which the students' persistence is more challenged besides the non-intimidating participation and the quality of increased interaction in online learning (Ni, 2012).

3 METHOD

Several studies have been conducted to examine the comparison of learning in different modes (either in the traditional face-to-face model, purely online or blended learning in teachers training), and to examine the impact on the students. The studies were conducted either quantitatively, qualitatively, or experimentally with different methods, as seen in the Table below, although the number of studies is not as many as those that have been done at the university level with more diverse methods.

3.1 *Comparing face to face (F2F), online & blended learning (BL) in teachers' training*

Table 1. Comparing F2F, online & blended in teachers' training.

Comparing	Method
F2F vs. online	A cluster randomized experiment of secondary teachers from across the United States; a total of 49 teachers (24 F2F, 25 online) (2013).
	125 students with randomized controlled trial design (2010).
	Data were collected using questionnaires, tests, and interviews of teacher training between June 2002 and May 2003 from F2F training of 80 teachers and 108 teachers in online training (2005).
F2F vs. online vs. BL	Survey responses collected over two years from 4,832 students of 75 K-12 teachers who participated in either online, blended, or face-to-face professional development design (2016).
F2F vs. BL	Quasi-experimental study over one year of K-12 schoolteachers in Auckland, New Zealand (11 BL, 19 F2F) (2013).
	Survey data from 26 online communities for secondary education teachers in the Netherlands (2013).
	Quasi-experimental study with a nonequivalent group design from 117 BL teachers and 60 F2F teachers in teachers' training (2016).

Table 2.   Comparing F2F, online & blended in higher education.

Comparing	Method
F2F vs. online	The dataset (information of courses and its demographics) of students taking high school courses for 2007/08–2010/11 at Florida high school from Florida Department of Education (2015).
	A survey with both closed- and open-ended questions from students at the university (2015).
	Empirical review from various articles (2015).
	Students surveyed on their preference at immunization delivery course (2014).
	Quantitative datasets from students at the university (2014).
	Using data from students in the psychology course on distinct set of variables (pre, course, and post) for the semester (2014).
	Quasi-experimental study from students in English at the Thai International College (2013).
	Questionnaire classified using the Bloom's taxonomy from students in chemistry course (2013).
	Student performance records, as well as student survey responses, both from public administration students in university (2012).
	Survey of 156 students in hospitality and recreation majors from a university in northern California (2011).
	A causal-comparative research design from students at the university (2010).
	A combination of both quantitative and qualitative methods from training programme with total 250 students participated (129 online and 121 face-to-face), plus three teachers who took part in both types of teaching (2009).
	A meta-analysis from the empirical literature (2009).
	Empirical review of literature from students and degree programs in higher education (2008).
F2F vs. BL	Group experiment (traditional & blended) with the same instructor from students in university (business and economic statistics course) (2017).
	Experimental group student in introductory economic courses (2014).
Online vs. BL	Questionnaire (motivated strategies for learning). Undergraduate students from the university of Melbourne in 2014–2016 (2017).
	Instrument over 750 university students (2015).
F2F vs. (online + BL)	A meta-analysis from the empirical literature (2013).

4   RESULTS

Several studies on the learning model on **TPD** have been done to determine the differences in learning outcomes and the impact on the students by comparing the learning in three different modes, which are face-to-face, online, and blended The studies found that in general there is no significant difference (Webb & Bush, 2016), although there is a positive influence on the improvement of teachers' knowledge and performance (Fisher et al., 2010), as well as beliefs (Fishman et al., 2013) that influence the curriculum implementation. Online teachers' training was more cost-effective than face-to-face teachers' training, mainly due to the lower opportunity cost of the participants (Jung, 2005), although Fishman et al. (2013) reveal that this is a context-dependent case, in which online learning will be cost-effective for teachers who are geographically dispersed and vice versa. Although there is no difference in performance, there is still a difference (diversity of preference) between those modes, where online learning increases student perception and social interaction (Smith, 2013) and blended communities have an impact toward an effective TPD.

In general, the studies conducted at the university level show similar results on the course comparison in the three modes as the ones conducted in the teachers' training, which show no significant difference. However, studies conducted at the university level are wider and more diverse. Some of the results that tend to be different are shown by Lundberg et al. (2008), who reveal the tendency of better results shown by students in online course and in linguistic students (Johnson & Palmer, 2015), and Helms (2014) who states otherwise, that online students have a greater chance of failing the course compared to the students in the face-to-face learning. Demographic data also contributes to the performance gap (Xu & Jaggers, 2014).

5 DISCUSSION

Online learning in TPD exists as a consequence of the skills demand in the digital era, in which learning media also develop along with the development from the traditional era to provide learning with a touch of modern technology. However, it does not affect the teacher performance and student achievement, because the traditional models and models which are based on digital technology have their own advantages and disadvantages. Also, in further development blended learning has begun to exist as a new trend (Kaur, 2013) to answer the phenomenon where blended learning is the form of e-learning with a touch of face-to-face learning. McMurray, O'Neill & Thompson (2016) developed an innovative TPD model showing that in addition to online learning, seminars or classes are presented to deepen the knowledge and practice as well as interaction by the instructors, by identifying the constraints in online learning and also by strengthening the interaction between students (Means et al., 2013).

6 CONCLUSION

Online learning on the teachers' professional development program conducted by the Indonesian government in 2016 is an innovation of the Indonesian government policy toward teachers' training pattern to create professional teachers. It is in line with the demands in the digital era through the use of Web 2.0-based technology in the learning, which is expected to be able to overcome the constraints of space (geographical), time, and cost, as well as being able to fill in the digital gap for teachers. Teachers are always required to improve their knowledge, practice, and competence in order to improve student achievement through innovative learning.

The improvement of teacher quality both as individual and profession is expected. Teachers should be able to contribute more in the quality of education with some kinds of creative and innovative way. All factors related to some policies and financial condition are important to consider. Government should consider policies which support the application of TPD in order to reach the final goal called education quality.

REFERENCES

Alberta. (2010). *Background paper: Professional learning for teachers in K-12 Alberta's system*. Retrieved from https://open.alberta.ca/dataset/f28d7e18-891e-4107-b63e-867e7127e4a9/resource/55d48d8a-9196-4d9a-8196-6c3cf15bf5cc/download/5456292-2010-09-background-paper-professional-learning-for-teachers-in-alberta-eduaction-system.pdf.

Aparicio, M., Bacao, F. & Oliveira, T. (2016). An e-learning theoretical framework. *Journal of Educational Technology & Society*, 19(1), 292–307.

Appana, S. (2008). A review of benefits and limitations of online learning in the context of the student, the instructor, and the tenured faculty. *International Journal on E-Learning*, 7(1), 5–22.

Barnett, M. (2002). Issues and trends concerning electronic networking technologies for teacher professional development: A critical review of the literature. *Paper presented at the American Educational Research Association, New Orleans, LA.*

Bates, M.S., Phalen, L. & Moran, C.G. (2016). If you build it, will they reflect? Examining teachers' use of an online video-based learning website. *Teaching and Teacher Education, 58*, 17–27.

Baumfield, V., Hall, E. & Wall, K. (2008). *Action research in the classroom.* London, UK: Sage.

Blank, R.K., de las Alas, N. & Smith, C. (2008). *Does teacher professional development have effects on teaching and learning?* Evaluation findings from programs in 14 states. Washington, DC: Council of Chief State School Officers.

Borko, H. (2004). Professional development and teacher learning: Mapping the terrain. *Educational Researcher, 33*(8), 3–15.

Brooks, C. & Gibson, S. (2012). Professional learning in a digital age. *Canadian Journal of Learning and Technology, 38*(2), 17.

Dede, C. (2004). Enabling distributed learning communities via emerging technologies. *Technological Horizons in Education Journal, 32*(3), 16–26.

Dede, C. (2006). A research agenda for online teacher professional development. *Journal of Technology and Teacher Education, 14*(4), 657–661.

DeMonte, J. (2013). *High-quality professional development for teachers: Supporting teacher training to improve student learning.* Washington, DC: Center for American Progress.

Desimone, L.M. (2009). Improving impact studies of teachers' professional development: Toward better conceptualizations and measures. *Educational Researcher, 38*(3) 181–199.

Fisher, J.B., Schumaker, J.B., Culbertson, J. & Deshler, D. (2010). Effects of a computerized professional development program on teacher and student outcomes. *Journal of Teacher Education, 61*(4), 302–312.

Fishman, B., Konstantopoulos, S., Kubitskey, B.W., Vath, R., Park, G., Johnson, H. & Edelson, D.C. (2013). Comparing the impact of online and face-to-face professional development in the context of curriculum implementation. *Journal of Teacher Education, 64*(5), 426–438.

Fraser, C., Kennedy, A., Reid, L. & Mckinney, S. (2007). Teachers' continuing professional development: Contested concepts, understandings and models. *Journal of In-Service Education, 33*(2), 153–169.

Ganser, T. (2000). An ambitious vision of professional development for teachers. *NASSP Bulletin, 84*(618), 6–12.

Glatthorn, A. (1995). Teacher development. In L.W. Anderson (Ed.), *International Encyclopedia of Teaching and Teacher Education* (2nd ed.) (pp. 41–46). Oxford, UK: Pergamon.

Golob, H.M. (2012). The impact of teacher's professional development on the results of pupils at national assessment of knowledge. *Procedia-Social and Behavioral Sciences, 47*, 1648–1654. Retrieved from https://doi.org/10.1016/j.sbspro.2012.06.878.

Gorozidis, G. & Papaioannou, A.G. (2014). Teachers' motivation to participate in training and to implement innovations. *Teaching and Teacher Education, 39*, 1–11.

Grosch, M. (2017). Developing a competency standard for tvet teacher education in asean countries. *Jurnal Pendidikan Teknologi dan Kejuruan, 23*(3), 279–287.

Hargreaves, A. (2010). Presentism, individualism, and conservatism: The legacy of Dan Lortie's schoolteacher: A sociological study. *Curriculum Inquiry, 40*(1), 143–154.

Hassel, B.C. (1999). *The charter school challenge.* Washington, DC: Brookings Institution.

Helms, J.L. (2014). Comparing student performance in online and face-to-face delivery modalities. *Journal of Asynchronous Learning Network, 18*(1), 147–160.

Huber, C. (2010). Professional learning 2.0. *Educational Leadership. 67*(8), 41–46.

Hughes, J., Zhou, C. & Petscher, Y. (2015). *Comparing success rates for the general and credit recovery courses online and face to face: Results for Florida high school courses (REL 2015-095).* Washington, DC: U.S. Department of Education, Institute of Education Sciences, National Center for Education Evaluation and Regional Assistance, Regional Educational Laboratory Southeast. Retrieved from http://ies.ed.gov/ncee/edlabs.

Johnson, D. & Palmer, C.C. (2015). Comparing student assessments and perceptions of online and face-to-face versions of an introductory linguistics course. *Online Learning, 19*(2), 1–18.

Joksimović, S., Gašević, D., Loughin, T.M., Kovanović, V. & Hatala, M. (2015). Learning at distance: Effects of interaction traces on academic achievement. *Computers and Education, 87*, 204–217.

Jung, I. (2005). ICT-pedagogy integration in teacher training application cases. *Educational Technology & Society, 8*(2), 94–101.

Kachelhoffer, A. & Khine, M.S. (2009). Bridging the digital divide, aiming to become lifelong learners. In A. Tatnall & A. Jones (Eds.). Education and technology for a better world. *IFIP Advances in Information and Communication Technology*, (Vol. *302*, pp. 229–237). Berlin, Germany: Springer.

Kaur, M. (2013). Blended learning-challenges and its future. *Procedia-Social and Behavioral Sciences,* *93*(21), 612–617.

Keegan, D. (1996). Definition of distance education. In L. Foster, B. Bower & L. Watson (Eds.), *Distance Education: Teaching and Learning in Higher Education*. An ASHE Reader Series (Association for the Study of Higher Education) Series. Needham Heights, MA: Simon & Schuster.

Kruger-Ross, M.J. & Waters, R.D. (2013). Predicting online learning success: Applying the situational theory of number to the virtual classroom. *Computers and Education, 61*(1), 176–184.

Krumsvik, R.J. (2008). Situated learning and teachers' digital competence. *Education and Information Technologies, 13*(4), 279–290.

Larreamendy-Joerns, J. & Leinhardt, G. (2006). Going the distance with online education. *Review of Educational Research, 76*(4), 567–605.

Lundberg, J., Castillo-Merino, D. & Dahmani, M. (2008). Do online students perform better than face-to-face students? Reflections and a short review of some empirical findings. *Revista de Universidad y Sociedad del Conocimiento, 5*(1), 35–44.

Marfu'ah, S., Djatmiko, I.W. & Khairuddin, M. (2017). Learning goals achievement of a teacher in professional development. *Jurnal Pendidikan Teknologi dan Kejuruan, 23*(3), 295–303.

Mayes, T. & de Freitas, S. (2004). Review of e-learning theories, frameworks and models. London, UK: Joint Information Systems Committee.

McMurray, S., O'Neill, S., & Thompson, R. (2016). An innovative model for professional development. *Journal of Research in Special Educational Needs, 16,* 145–149. https://doi.org/10.1111/1471-3802.12139.

Means, B., Toyama, Y., Murphy, R. & Baki, M. (2013). The effectiveness of online and blended learning: A meta-analysis of the empirical literature. *Teachers College, 115*(3), 1–47.

Means, B., Toyama, Y., Murphy, R., Baki, M. & Jones, K. (2009). *Evaluation of evidence-based practices in online learning: A meta-analysis and review of online learning studies.* Washington, D.C.: US Department of Education.

Moon, J., Passmore, C., Reiser, B.J. & Michaels, S. (2014). Beyond comparisons of online versus face-to-face PD: Commentary in response to Fishman et al., 'Comparing the impact of online and face-to-face professional development in the context of curriculum implementation'. *Journal of Teacher Education, 65*(2), 172–176.

Murchan, D., Loxley, A. & Johnston, K. (2009). Teacher learning and policy intention: Selected findings from an evaluation of a large-scale programme of professional development in the Republic of Ireland. *European Journal of Teacher Education, 32*(4), 455–471.

Ni, A.Y. (2012). Comparing the effectiveness of classroom and online learning: Teaching research methods. *Journal of Public Affairs Education, 19*(2), 199–215.

Nipper, S. (1989). Third generation distance learning and computer conferencing. In R. Mason & A. Kaye (Eds.), *Mindweave, Communication, Computers and Distance Education* (pp. 63–73). Oxford, UK: Pergamon Press.

Olson, J., Codde, J., deMaagd, K., Tarkelson, E., Sinclair, J., Yook, S., & Egidio, R. (2011). *An Analysis of e-Learning Impacts & Best Practices in Developing Countries: With Reference to Secondary School Education in Tanzania* (Information & Communication Technology for Development). Michigan, USA: Michigan State University.

Rieckmann, M.L. & Gardiner, S. (2017). *Education for sustainable development goals learning objectives.* Paris, France: United Nations Educational, Scientific and Cultural Organization.

RoI. (2006). *Undang-undang Nomor 14 tahun 2005 tentang Guru dan Dosen* [Law Number 14 Year 2005 regarding on Teachers and Lecturers]. Jakarta, Indonesia: Sinar Grafika.

Sang, G., Valcke, M., Van Braak, J. & Tondeur, J. (2010). Student teachers' thinking processes and ICT integration: Predictors of prospective teaching behaviors with educational technology. *Computers & Education, 54*(1), 103–112.

Santoso, D. (2014). Need assessment pengembangan keprofesionalan berkelanjutan guru smk teknik audio video [Need assessment of continuous professional development for audio video vocational high school teachers]. *Jurnal Pendidikan Teknologi dan Kejuruan, 22*(2), 148–154.

Shulman, L.S., & Shulman, J.H. (2004). How and What Teachers Learn: A Shifting Perspective. Journal of Curriculum Studies, *36*(2), 257–271.

Singh, D.K. & Stoloff, D.L. (2007). Effectiveness of online instruction: Perceptions of pre-service teachers. *International Journal of Technology, Knowledge & Society, 2*(6), 121–124.

Smart, K.L & Cappel, J.J. (2006). Students' perceptions of online learning: A comparative study. *Journal of Information Technology Education, 5,* 201–219.

Smith, N.V. (2013). Face-to-face vs. blended learning: Effects on secondary students' perceptions and performance. *Procedia-Social and Behavioral Sciences, 89,* 79–83.

Smith, C. & Gillespie, D. (2007). Research on professional development and teacher change: Implications for adult basic education. In J. Comings, B. Garner & C. Smith (Eds.), *Review of Adult Learning and literacy* (pp. 205–244). New Jersey, NJ: Lawrence Erlbaum Associates.

Sun, P.C., Tsai, R.J., Finger, G., Chen, Y.Y. & Yeh, D. (2008). What drives a successful e-Learning? An empirical investigation of the critical factors influencing learner satisfaction. *Computers & Education, 50*(4), 1183–1202.

Tan, A.-L. (2015). In-service teacher education. In R. Gunstone (Ed.), *Encyclopedia of Science Education, 516–518.*

Tondeur, J., Forkosh-Baruch, A., Prestridge, S., Albion, P. & Edinsinghe, S. (2016). Responding to challenges in teacher professional development for ICT integration in education. *Journal of Educational Technology and Society, 19*(3), 110–120.

Twining, P., Raffaghelli, J., Albion, P. & Knezek, D. (2013). Moving education into the digital age: The contribution of teachers' professional development. *Journal of Computer Assisted Learning, 29*(5), 426–437.

Uzunboylu, H. (2007). Teacher attitudes toward online education following an online inservice program. *International Journal on E-Learning, 6*(2), 267–277.

Villegas-Reimers, E. (2003). *Teacher professional development: An international review of the literature.* Paris, France: International Institute for Educational Planning.

Walker, B.K. (2007). *Bridging the distance: How social interaction, presence, social presence, and sense of community influence student learning experiences in an online virtual environment* (Doctoral dissertation). Retrieved from. https://libres.uncg.edu/ir/uncg/listing.aspx?id=136.

Webb, D.C., Nickerson, H., & Bush, J.B. (2017, March). A comparative analysis of online and face-to-face professional development models for CS education. *Proceedings of the 48th ACM SIGCSE Technical Symposium on Computer Science Education* (pp. 621–626). Seattle, WA.

Xu, D. & Jaggers, S.S. (2014). Performance gaps between online and face-to-face courses: Differences across types of students and academic subject areas. *Journal of Higher Education, 85*(5), 633–659.

Yoon, K.S., Duncan, T., Lee, S.W.Y., Scarloss, B. & Shapley, K.L. (2007). *Reviewing the evidence on how teacher professional development affects student achievement: Issues & answers report REL 2007–No. 033.* Washington, DC: U.S. Department of Education.

The practice of early childhood musicality education in Germany

L. Kurniawati
Universitas Pendidikan Indonesia, Indonesia

ABSTRACT: This article describes the practice of early childhood musicality education based on the concepts of Carl Orff, one of Germany's leading composers, who focused on developing the concept of early childhood musicality education. The focus of this research is on how the musicality concept through the unity of music, language, and movement can stimulate the potential of the children. The results indicated that musicality education in early childhood is directed to build the competence of a child's musicality in order to train the musical sensitivity, and implicitly to positively develop the children's basic potential. The result of the research through its *praxis analytics* shows that *Orff-Schulwerk* can be implemented in musicality education for early childhood in Indonesia, which does not purposively apply the European music system, but is based on the introduction of local musical aspects that can be found in Indonesia.

1 INTRODUCTION

Carl Orff (1895–1982) is a pioneer in how children are supposed to start their musical activity. From the overall activity in music education for children through seminars, courses, and various experiences of teaching music, rhythms, and dance in *Günther-Schule*, München, he documented his experiences to be a '*Schulwerk*'. Between 1950 and 1954, Carl Orff along with Gunild Keetman published the documents to become *Orff-Schulwerk für Früherziehung 'Musik für Kinder'* [Orff-Schulwork for early education "musik for children"] (Orff & Keetman, 1950).

The most common thing is that children are positioned as they should be, as a child whose musical competence is on a developing stage, or that they are taught a music instrument as adults. Therefore, *Das Orff-Schulwerk* offers enlightenment for these mistakes, strengthens the positive things that have been established pedagogically, and also highlights the limitations in how to design music education for a children's program.

Das Orff-Schulwerk, which was written in the 50s, contains five chapters. It is plausible that at that time it was only Carl Orff who had made a revolution in how children learn elementary aspects of music. For example, how children can play music using parts of their body, or including the expression or motion which is also an elementary aspect.

Why is it called elementary? Because it is a basic skill that children have to know and master, for the sake of developing their self-potential in the future. At this time, it is not considered important to start mastering a musical instrument conventionally, except for those who can be introduced to music's basic parameter through every medium that can be sounded around them. This is why Carl Orff considered that *Elementare Musik* for children is something important (Jungmair, 1992).

2 DISCUSSION

Carl Orff further formulated *Elementare Musik*, based on a deep study about children's music education practice, that elementary music is natural, physically, to learn and to get an experience for and form everyone (Böhm, 1975).

In that context, music cannot be defined as knowledge and a skill that is hard to be learned by anyone, for people who do not have music playing skill, or even for people with zero experience in playing music. Elementary music does not require someone to have skill in a specific field, especially in music, but instead develops something that people have, or what their environment has. The most important thing in this context is that people have an interest or will to respond in any form to musical stimulation that the teacher or the facilitator gives.

Through that process, musical skill in children will gradually be touched on and built, and will even automatically grow based on their self-potential. This can be a foundation to grow an interest in developing music playing skill. This is the nature of something elementary in music. Related to this context, in order to teach children something elementary in music, music itself cannot stand alone. Music is integrated with other fields, especially motion, dance, and language. These three elements are regarded as being close to music. Therefore any elementary expressions in music through these three mediums can ease children into expressing or learning the elementary skills.

A colleague of Carl Orff, Keetman (who became Orff's partner), explains how he packaged elementary material and any basic musical learning possibility for children. Orf is always looking for new sounds and filled with some new ideas, he constantly brings instruments to try and puts us in for extraordinary enthusiasm (...) we are all so impressed with the new discovery that the lessons became more focus (Keetman, 1998).

The statement shows how Carl Orff applied *Elementare Musik* learning as an ongoing, progressing, and developing process in line with the idea that comes out in a music activity process with children. The music learning process can be an activity of exploration with various musical instruments, motion exploration, and vocal exercise so that the music learning parameter can be accomplished by students happily and passionately.

As explained above, Carl Orff's *Schulwerk* becomes one of the musical approaches that is established as a learning model orientation by Early Musicality Education teachers. This becomes a proof that the Early Musicality Education program is a main concern and important part of early childhood education. Music lessons for children (four to six years old) serve to prepare training instrumental and vocal in music school (Ribke, 1995).

Early Musicality Education is practiced by various educational institutions in early childhood education in Germany. In line with it, various learning models are developed by practitioners and pedagogues so that the quality of Early Musicality Education learning is well-guaranteed. In fact, many material contents are tested before becoming a book in order to develop a model that is suitable for the situation and condition.

This proves that the concern in the field is increasing. Most important is early musicality education has moved to the first year of life since the birth of a child, and can be done in groups of children and parents (Gruhn, 2010). It may be confusing for adults who do not understand the concept of basic musicality education in how can we make music for baby? They do not have a skill as children in Early Musicality Education. They are open and natural, unbiased and free from prejudiced judgments, they are always curious and accepting of all things, happy to experiment and adventurous (Gruhn, 2010).

Parents and music can be a bridge to develop children's basic skill. This statement becomes a consideration and an idea to make musicality education for early childhood one of the pedagogic stimuli for children. Children do not need a lesson, but need a children friendly guide (Gruhn, 2010).

Therefore, children's skill can be built and developed through an arranged learning model. Learning models that are specially made for Early Musicality Education have also been prepared by some experts and educational practitioners, in order to solve the phenomenon that happened in music education for early childhood.

Noll (1992) has also made a critical reassessment of the model in Early Musicality Education that is related to the five areas. The fact shows that the music learning process for children always catches the attention because of its always altering and developing situation and condition.

2.1 *Musicality learning process*

Singing is considered as one of the important musicality activities in early childhood, because it can arouse feeling (*weckt Gefühl*) and make children happy. Singing is also easy to do, both individually or in a group. The song is an elementary musical form: simple and basically (Gerg, 1994).

Those assumptions may cause a wrong perception that music playing activity for children is enough with only singing. Various interesting activities through a song can give a lot of benefits for children, because the learning approach can be a stimulus to develop the children's self-potential. For example, a song can be a game to know each other's names. This introduction song model is very good to be conducted at the beginning of an activity when children first meet in the room. So the activity process can happen well and effectively because they do not feel awkward with each other. It can create a safe feeling for the children because it gradually grows trust and a feeling of safety in their new environment.

Therefore, this model can be repeated at the next meeting with little variations. For example, with the same model but children can replace their name with animals, things, or anything else they like. It can give a different impression, so it will grab their attention and motivate them to do the next activities. A game like this, as well as planting and developing their musicality potential, it also can be learn about social awareness (Ribke, 1995).

2.2 *Language aspect in musicality education*

Every child who can talk naturally has musical potential and they indirectly master it through speaking practice. If this skill is not well-stimulated, the potential will not be developed maximally although indirectly it is taught as part of their development. Children's language skill can also become basic material of teacher analysis for their basic musicality skill, because language is considered as being a singing activity. Language and talk are a skill to express musicality things.

This can be a reason why singing and talking become a main material in music education for early childhood. Singing is considered to be an easy and fun music activity because singing is the most private music instrument; when children learn about sound in music, the main point is how to connect the expression of the music itself to the sound language.

From the example above, we know that singing has huge benefits to build a will to play, develop fantasy, learn to deepen self-expression skill, and also stimulate body development. The children also learn to play with hard and soft words, and when to use specific language in singing. The most important thing is that children can develop their basic musicality skill through music, and also how they develop their spectrum of self-expression skill.

Various games using singing and talking are good for the development of children's vocabulary. Word games through musical parameters such as duration (long-short), dynamic (hard-soft), and talking tempo (fast-slow) helps children's articulation; even the combination of movement and simple music instrument playing can help children to develop their hard and soft motoric skills.

2.3 *Movement aspect in musicality education*

Movement and dance are a 'game' for children. So that whenever music is playing, movement will be an important part, not only to understand the material but also to help their brain development. As we know that the concentration level of children is limited, so we also know that they always want to move. They need to move to release their energy.

Moving becomes a way in which they train their agility. Various stimuli of movement games are the beginning stage for children to make contact using body language. These activities build the confidence in their body and makes them confident to dance as a way to soften their movement. Therefore, their hard and soft motoric skills will be well-stimulated and well-developed.

2.4 *Playing elementary instruments*

Activity that is related to the introduction of Elementary Instrument Playing has been conducted in the previous activities, which are singing and talking. Elementary Instrument Playing in that context is more functioned as an addition to singing and talking activity, and also as an addition to moving and dancing activity.

Elementary Instrument Playing should be done in a group because it will be more fun for them. It is possible if there are more than three people in the class. Everyone gets a chance to choose an instrument, hold and play it. They do not have enough patience to wait their turn to play the instrument, so there needs to be enough music instruments to be played by the children.

The choice of a simple and lightweight music instrument should be a consideration when they are about to play a music instrument in a group. The music instrument should be one which can be played together by children while they actively move; big music instruments, such as conga, percussion, can only be played in one place.

3 CONCLUSION

Early Musicality Education in early childhood can be well-developed if there is good stimulation. Hearing skill as a practice of sound sensitivity is important for children in the future in their social life. Respect for the social environment will be needed so that they can be people who are useful for themselves and for the community.

REFERENCES

Böhm, S. (1975). *Spiele Mit Dem Orff-Schulwerk, Photographiert Von Peter Keetman* [Play with Orff-Schulwork, Fotograf from Peter Keetman]. Stuttgart, Germany: Metzler.

Gerg, K. (1994). *Singen und Musizieren mit Kindern. Elementare Begleitformen auf Orff-Instrumenten* [Elementry of accompanying form in Orff-instrument]. Donauwörth, Germany: Auer.

Gruhn, W. (2010). *Anfänge des Musiklernens. Eine lerntheoretische und entwicklungspsychologische Einführung* [Starting music learning: Introduction of learning theory and development of psychology]. Hildesheim u.a., Germany: Olms.

Jungmair, U.E. (1992). *Das Elementare. Zur Musik- und Bewegungserziehung im Sinne Carl Orffs. Theorie und Praxis*, Mainz u.a., Germany: Schott.

Keetman, G. (1998). *Elementaria, Erster Umgang mit dem Orff-Schulwerk* [The elementry Music education and movement education related to Carl Orff. Theory and practice]. Stuttgart, Germany: Klett.

Noll, G. (1992). *Musikalische Früherziehung. Erprobung eines Modells, unter Mitarbeit von Adam Kormann (Materialien und Dokumente aus der Musikpädagogik; 20)* [Early chilhood musical education. Testing a model, with cooperation from Adam Kormann (materials and documents from music education, 20]. Regensburg, Germany: Bosse.

Orff, C. & Keetman, G. (1950). *Orff-Schulwerk: Musik für Kinder* Band I [Orff-Schulwork: Music for children]. Mainz u.a., Germany: Schott.

Ribke, J. (1995). *Elementare Musikpädagogik. Persönlichkeitsbildung als musikerzieherisches Konzept* [Elementry Music education. Character building as a musical educational concept]. Regensburg, Germany: ConBrio.

Constraints on the physics practicum for visually impaired students in inclusive junior high schools

J. Arlinwibowo & H. Retnawati
Universitas Negeri Yogyakarta, Indonesia

R.G. Pradani
SMA Negeri 1 Gebog, Indonesia

ABSTRACT: This study aims to describe the constraints in the Physics practicum on visually impaired students in inclusive junior high school, along with the alternative solutions. This was a descriptive explorative study. The data in this study was qualitative and was obtained by observation and in-depth interviews with six visually impaired students, and five Physics teachers who taught the visually impaired students in Yogyakarta, Indonesia. The data was analyzed by using the following steps: reducing the data, grouping the data into specific themes, linking the themes and drawing conclusions. The results of this study show that there have been obstacles in the Physics laboratory for the visually impaired students in the inclusive junior high schools and that these obstacles are due to the lack of special Physics practicums for visually impaired students, the non-availability of the practicum instruments for visually impaired owned by the schools, the limited procurement process of the practicum instruments, and the teachers' incapability to independently manufacture the practicum instruments in order to minimize the impact caused by the limited facilities. The alternative solutions for dealing with these obstacles can be 1) pursuing a synergy between the schools and the government in procuring the practicum instruments; 2) holding the training programs intended to improve the teachers' competencies; and 3) establishing cooperation with researchers/academicians.

1 INTRODUCTION

Verse 31 Article 1 from the Constitution of 1945 (RoI, 1945) states that every Indonesian citizen has the equal opportunity to attain education; therefore, the government is obliged to provide educational facilities to citizens, including the ones with disabilities. The results of the National Social Economic Survey that is displayed by RoI (2014) show that the number of disabled people in Indonesia in 2012 was 6,008,661. From this figure, 1,780,200 people are visually impaired, 472,855 people are hearing-impaired, 402,817 are learning-disabled, 616,387 people are physically disabled, 170,120 people are self-maintenance-impaired and 2,401,592 people suffer from double impairment.

According to RoI (2016), there are only 2,070 special schools that have been providing learning facilities for people with disabilities over the past several decades. The educational development results in a system of inclusion that has been considered as a solution to the limited number of special schools. The concept of inclusive education has been proposed as an integrated school system. In inclusive schools, all students learn in the same environment. Through the concept of integrated education, the students with disabilities can develop their social skills so that they can have good socialization (Bowen, 2010).

In response to the inclusive education, the government (RoI, 2009) regulates the continuation of inclusive schools in order to facilitate the students with special needs so that these students can pursue their education in general schools. In this regulation, it is mentioned clearly in Article 4 Verse 1 that one district should have at least one inclusive school that can facilitate the learning process for the students with special needs. Thereby, it is expected that all students with disabilities can pursue their schooling activities easily.

Many schools changed into an inclusive school. Astono et al. (2010) said that although from the philosophical, judicial, pedagogic, and empirical aspects this inclusive education has been strong, the technical setting of this education in the schools has still been very weak. Facilities have been a very important issue. Arlinwibowo and Retnawati (2015) said that ideally a school should have sufficient facilities in order for students to learn well.

One of the subjects that heavily depend on the availability of facilities is Physics. The reason is that Physics involves many practicum activities. Practicum activities or environments cannot be separated from the Physics learning in schools. Both the Educational Unit Level Curriculum (KTSP, *Kurikulum Tingkat Satuan Pendidikan*) and the 2013 Curriculum in Indonesia demand that the students have scientific work competencies and that these competencies are taught through the scientific experiments. In addition, El-Rabadi (2013) elaborated that Physics learning through laboratory practices has enormous influence on the students' learning achievement. Practicums can show the implementation of theory in the real problem. Therefore, practicums can help students in learning Physics. Retnawati et al. (2017) said that learning should connect many concepts and be contextual with real matter around them that makes the concepts easy to remember or understand.

The limitation of vision makes a significant difference between the visually impaired and the sighted students in accessing information (Roe & Webster, 2003). Nowadays, most of the Physics practicum instruments are accessed visually. According to Delthawati et al. (2011), the visually impaired students have limited vision that presents an obstacle in performing the Physics practicums because the majority of the Physics practicums involve observing and reading scales.

The different students' conditions in an inclusive classroom makes the learning process become more complicated, especially in the practicum sessions that demand the students to actively explore many things. Ideally, a teacher should master the characters of all students so that they can devise a conducive practicum for them. This condition certainly becomes a peculiar obstacle. Therefore, the researchers are interested to review this situation in order to describe the obstacles in performing Physics practicums for visually impaired students in inclusive junior high schools, along with the alternative solutions.

2 METHOD

2.1 *Design and participants*

The study was a descriptive explorative research using qualitative approach that was aimed at understanding the obstacles in performing Physics practicum by the visually impaired students in inclusive junior high schools. This study involved six visually impaired students (S1, S2, S3, S4, S5, and S6) and five teachers who taught the visually impaired students (T1, T2, T3, T4, and T5).

2.2 *Data collection techniques*

The data in this study was obtained from the field observation and the in-depth interview. The researchers started the data gathering activities by performing the field observation along with the in-depth interview in order to clarify the findings and to find more in-depth information.

2.3 *Data analysis techniques*

The results of the observation and the in-depth interview were gathered and then were reduced. The data was analyzed through the following steps: reducing data, grouping the data into specific themes, linking the themes, and drawing conclusions. The researchers defined the inter-theme relationship in order to attain an understanding and this definition made use of the Bogdan and Biklen Model (1982).

3 RESULTS

The Physics learning process in inclusive schools had not been ideally conducted, especially in relation to the students with disabilities. In conducting this study, the researchers found the data of obstacles that had occurred during the implementation of the Physics practicum for the visually impaired students in inclusive schools. These are displayed in Table 1.

When the three themes were linked, the researcher concluded that there had been problems in the implementation of the Physics practicum that had been caused by the instruments' availability. Inclusive schools had not constructed a standardized Physics laboratory for the visually impaired students.

The practicum had been one of the strategies for providing an in-depth understanding toward a Physics concept because the students viewed natural phenomena that had been related to the materials that were taught. However, the data showed that there had been less ideal situations in performing the Physics practicum within the inclusive schools, especially for the visually impaired students. In general, the practicum was only performed for the sighted students. When others were practicing, visually impaired students were given material by teachers based on remembering, such as explaining about substances and practices. The situation that usually occurred in the inclusive schools was that the teacher described the practicum process using lectures in order that the visually impaired students could imagine the activities that the sighted students performed. Many statements that implied such situations were given as follows.

'... As far as we remember we have not ever performed any Physics practicum. ...' (S1 and S4)

'... Physics learning for the visually impaired students is only theoretical. ...' (T3)

'... The learning obstacles often cause us to consider "Literary Science" because the Physics learning is only based on memorization. ...' (T2).

Unfortunately, when the regular school had been changed into an inclusive one, the facilities had not been adapted to provide access for students with disabilities. One of the facilities that had not been adapted to accommodate the students with disabilities was the laboratory equipment, for example the Physics laboratory. The laboratory in the inclusive school was still considered old.

Table 1. The data reduction on the main situations regarding Physics practicum in inclusive schools.

Data reduction	Theme
A practicum in Physics learning has rarely been implemented for the visually impaired students. The visually impaired students have rarely attended the practicum. The Physics practicum activities have been changed to description.	The implementation of the Physics practicum for the visually impaired students in inclusive schools has been very low.
Inclusive schools have not had standardized laboratories for the visually impaired students. The Physics practicum instruments have been vision-based. Most of the Physics practicum instruments have not been accessible for the visually impaired students.	Most of the Physics practicum instruments have been based on vision and, therefore, they cannot be accessed by the visually impaired students.
The Physics practicum instruments for the visually impaired students have been very few. There have been rare grant programs for the procurement.	The availability of special Physics practicum instruments for the visually impaired students have been few.

Most of the Physics practicum instruments that had been possessed by the inclusive schools were based on vision. The visually impaired students were very disadvantaged because they could not access the visual-based devices. This situation was the main obstacle in the implementation of the Physics practicum.

According to T1, basically in terms of accessibility for visually impaired students, the practicum instruments in the laboratory could be classified into three categories, namely directly operated, modified, and totally modified instruments. The first category was the practicum instruments that could be accessed without involving vision, such as magnets and electricity; however, the practice should be under the supervision of the teacher. Then, the second and third categories demanded a new innovation for the visually impaired students. Unfortunately, the special practicum instruments for visually impaired students had been rarely found in the school.

SMP Negeri 2 Sewon and MTs Yaketuniswas used to receive the grant from the government or the partnering institutions. The grant from the government included an audio player device, typewriter, printer, and multiple products that had been sold in the market. On the other hand, the partnering institutions were dominated by the academicians and research-granted innovative instruments (specifically instruments for the visually impaired) that supported the learning process. The grant was admitted to be very helpful for the learning process; however, in terms of quantity, the grant had been rare and as a result, the problems related to the Physics practicum still remained.

Several teachers clarified the problems that had been mentioned above through the following statements:

'... The teachers cannot operate the laboratory instruments for assisting the visually impaired students in their learning process. ...' (T2)
'... Our laboratory facilities have not supported the learning process for the visually impaired. ...' (T4)
'... In our school, the visually impaired students can only perform the practicum of magnets, electricity, and resonance. ...' (T3).

The solution for the absence of Physics practicum instruments was procurement. However, the field data showed that the procurement of Physics practicum instruments for the visually impaired students had been relatively more difficult than for other equipment. The reduced data for the obstacles in the procurement is displayed in Table 2.

The link between the two themes showed that the procurement efforts for the Physics laboratory equipment had been disturbed by the problems of funding, instrument availability in the market, the low number of instrument developers, and the relatively low level of the teachers' innovative capacity.

The first factor that became an obstacle in the procurement of Physics practicum instruments was funding. T2 stated that his school had limited funding for procuring the instru-

Table 2. The obstacles in the procurement of Physics practicum instruments for the visually impaired students.

Data reduction	Theme
School has limited budgeting. The Physics practicum instruments for disabled students are rarely available in the market. Developers of the Physics practicum instruments for disabled students are rare.	Schools have difficulties in procuring the special Physics practicum instruments for the visually impaired students.
The teachers' pedagogic competencies are limited. The teachers have difficulties in developing instrument innovation in order to facilitate the visually impaired students. The teachers have difficulties in designing the practicum.	Most of the teachers have not been able to innovate the practicum instruments and to perform the practicum activities that can facilitate the visually impaired students.

ments since most of the Physics practicum instruments were expensive. The second factor was instrument availability; the special Physics practicum instruments for visually impaired students had been difficult to find in the market. As a result, if the school had received a grant in the form of a fund for procuring the practicum instruments, then the school would have had difficulties in allocating the fund. T3 stated as well that she had never found a store that provided special practicum instruments for visually impaired students. This situation was caused by the low number of learning media developers that paid attention to the visually impaired students as their segment. In addition to the fact that the segment of visually impaired students had been limited, not many people had the competencies in developing instruments or media for the visually impaired.

One of the alternatives for solving the problems in difficult procurement was forging the teachers' creativity in manufacturing the practicum supporting instruments independently. However, such an ideal situation was still difficult to bring into reality because the subject teachers in the inclusive schools did not have the scientific background of teaching students with disabilities. All of the teachers were the graduates of teaching and education departments who had been dealing with students without disabilities; at the same time, all of the teachers who had graduated from inclusive education departments were assigned in the special schools. This condition influenced the sufficiency of the teachers' pedagogic competencies while they were teaching in an inclusive classroom. T3 stated that she did not have sufficient experience to solve such problems because during her undergraduate education she had not been equipped with the skills for teaching students with special needs.

The problems of low pedagogic competencies did not only influence the teachers' skill to create the innovative instruments but also influenced the maximizing of the existing instruments. Teachers often had difficulties in designing the practicum sessions in such a way that both the sighted students and the visually impaired students in their class could attend the practicum. This condition was apparent from the following statements that had been provided by the respondents.

'... The idea of performing a practicum used to be proposed but it has never been brought into reality. ...' (S2 and S3)
'... So far the Physics practicum has been difficult to design and to implement. ...' (T3).

4   DISCUSSION

The implementation of Physics practicum learning in inclusive junior high schools has been the main issue in this study. The data shows that the implementation of the practicum learning has been very minimal because the availability of accessible practicum instruments for the visually impaired has been very low. This data has been confirmed by the results of a research by Astono et al. (2010) which states that the learning service for visually impaired students, especially the ones related to the practicum activities, has been less.

This case is very fundamental because up to now, the laboratory facilities of inclusive schools have been oriented toward the students without disabilities; whereas, there has been a very significant difference between the students with disabilities and the students without disabilities in terms of information access. Referring to the data that 80% of information that is retrieved by human beings comes from their vision (Department of Education, 2011), the visually impaired students have limitations in grasping their environment (Hadi, 2005); as a result, the instruments that will be used should be comfortable and easily accessed. The learning instruments for students with disabilities should have proportional design (not be too big or too small), should be strong, and should be rigid so that the instruments can be explored in a maximum manner. Such instruments will be comfortable for the visually impaired students when they identify their environment through touch; at the same time, the visually impaired students are also protected from disturbing objects (sharp objects). If the instruments should be folded, then an introduction to the folding sequences should be performed (Mani et al., 2005). In addition, the instruments should be well-observed; in other words, the instruments could be explored through touch (Hersh & Johnson, 2008) or hearing (Lennie & Van Hemel, 2002). Based on the special criteria, it is likely that most of the practicum instruments for Physics learning activities in the inclusive junior high schools cannot be accessed by the visually impaired students.

In response to this case that has occurred in the implementation of the Physics practicum learning process for the visually impaired students in the inclusive schools, there should be a mapping toward the problems along with the solutions. The mapping is displayed in Table 3.

The obstacles in the practicum activities can be classified into two categories, namely the teacher competency and the less optimum synergy with the partners. The first factor can be solved by strengthening the teachers' pedagogic competencies. This solution is a response to the condition of the inclusive schoolteachers who do not have a scientific background in teaching the classes in which there are students with disabilities. The expectation is that the subject teachers in the inclusive school have a better understanding toward the teaching activities in an inclusive class so that they will be more sensible in looking at the problems. Thus, they will be able to respond appropriately when they deal with the problems related to the students with disabilities, as has been stated by Haseena and Mohammed (2015) that the teacher quality determines the learning process quality. In a straightforward manner, Hakim (2015) stated that pedagogy determines teaching materials mastery, classroom management, and teacher commitment.

Strengthening the pedagogic competencies is the main strategy for maximizing the teachers' role. The results of a study by Mahmood et al. (2013) showed that teachers' pedagogic competencies heavily influence the improvement of the learning process, and in fact they are considered to be at the heart of the students' learning activities. Having been equipped with good pedagogic competencies, a teacher should be able to serve as a student facilitator in the learning process. The concept of a facilitator in an inclusive classroom is an individual who should be able to provide facilities to the students' needs, including the ones with disabilities. This situation demands teachers to be innovators because the cases in an inclusive classroom are very complex. The importance of innovation is creating possibility out of impossibility.

An example of innovation that can be undertaken for providing facilities to the visually impaired students is a research by Delthawati et al. (2011) in this study. She could encourage the visually impaired students to perform the practicum of measuring length, mass, weight, volume, energy, and density. A similar study was also performed by Astono et al. (2010). In this study they could encourage the visually impaired students to perform the practicum of measuring temperature. Advanced pedagogic understanding and good innovative skills can be an ideal combination for overcoming the obstacles in the implementation of Physics practicums.

The second factor is the synergy with multiple parties, as having been proposed by Haseena and Mohammed (2015). The government has a central role in eliminating the obstacles and this is one of the results of a study by Jacob and Ludwig (2009). In this study, they found that the government at all levels has the capacity to improve the educational equality. The first role is as initiator in the provision of multiple programs such as training sessions. In another study by Lunenburg (2011), it was found that training has been effective to improve a teacher's skills in solving multiple educational problems. The government, as the given authority, has a crucial role in supporting the implementation of training programs at the school level, the

Table 3. The mapping of the problems and the alternative solutions.

Category of problems	Alternative solutions
Teachers' competencies	Strengthening the subject teachers' pedagogic competencies. Equipping the subject teachers with innovation capacities.
Synergy with multiple parties	Strengthening the communication system with the government. Establishing cooperation with universities.

regency level, the provincial level, and to the national level. Then, the second role is as grant provider; through the grant, the government can help provide practicum instruments in order to complete the learning facilities because some inclusive schools have a limited budget. It is urgent for the government to provide the grant because an inclusive school has limited funds and such limited funds inhibit the school to free themselves from many problems, especially the ones related to procurement, as has been stated by Jacob and Ludwig (2009). In order for the government to execute both roles well, there should be good communication between the government and the school, especially a discussion regarding multiple learning process-related problems in the inclusive school. According to Sutapa (2006), communication is a step to developing a harmonious relationship with stakeholders in order to understand one another. By doing so, each problem can be detected and the solution can be found as soon as possible.

Another party who can provide great contribution in the inclusive school development is universities. A university is a warehouse of knowledge and academic reviews that can assist in solving the problems and providing the scientific foundation in the school's policymaking process. Jacob and Ludwig (2009) in their research proposed that research and development both can overcome problems and improve the quality of a school. In addition, still according to them, research-based school policies can assist a school to leave their problems behind.

The facts in the field show that several schools have established cooperation with Universitas Negeri Yogyakarta and Universitas Islam Negeri Sunan Kalijaga. However, the quantity of such cooperation is still low. This partnership should be improved and be expanded. Thereby, more alternative solutions can be found in order to solve multiple problems related to the learning process in inclusive schools.

5 CONCLUSION

There have been obstacles in the Physics practicum for the visually impaired students in inclusive schools, namely 1) the number of special Physics practicum instruments for the visually impaired students is limited; 2) most of the practicum instruments demand vision and therefore they cannot be accessed by the visually impaired students; 3) the procurement process for the practicum instruments is limited; and 4) teachers have not been able to manufacture the practicum instruments independently. Therefore, the alternative solutions for overcoming these problems are 1) pursuing a synergy between the school and the government within the practicum instruments procurement; 2) holding training programs for improving the subject teachers' competencies in inclusive schools; and 3) establishing cooperation with academicians/researchers.

REFERENCES

Arlinwibowo, J. & Retnawati, H. (2015). Developing audio tactile for visually impaired students. *International Journal on New Trends in Education and Their Implications*, 6(4), 78–90.

Astono, J., Rosana, D., Sumarna, & Maryanto, A. (2010). Pengembangan model praktikum sains untuk siswa tunanetra melalui pendekatan konstruktivis serta aplikasinya pada pendidikan inklusif [Developing science practicum model for blind students through constructivist approach and its application to inclusive education]. *Cakrawala Pendidikan*, *XXIX*(1), 43–54.

Bogdan, R.C. & Biklen, S.K. (1982). *Qualitative research for education: An introduction to theory and methods*. Boston, MA: Allyn and Bacon, Inc.

Bowen, J. (2010). Visual impairment and self-esteem. *The British Journal of Visual Impairment*, 28(3), 235–243.

Delthawati, I.R., Supriyani, R., Pertiwi, U.I., Badru, T., & Arlinwibowo, J. (2011). *Inovasi alat ukur besaran fisika berhuruf braille untuk meningkatkan kemampuan psikomotorik siswa tunanetra melalui praktikum IPA* [Innovation of measuring instruments of physics braille to improve the psychomotor ability of students with visual impairment through science lab]. Prosiding Seminar Nasional Penelitian, Pendidikan dan Penerapan MIPA, Fakultas MIPA, Universitas Negeri Yogyakarta.

Department of Education. (2011). *Teaching children who are blind or visually impaired inside, Februari 2011*. Newfoundland, CA: Government of Newfoundland and Labrador.

El-Rabadi, E.G.S. (2013). The effect of laboratory experiments on the upper basic stage students' achievement in physics. *Journal of Education and Practice*, 4(8), 62–70.

Hadi, P. (2005). *Kemandirian tunanetra: orientasi akademik dan orientasi sosial* [Blind autonomy New-foundland, CA: Academic orientation and social orientation]. Jakarta, Indonesia: Departemen Pendidikan Nasional.

Hakim, A. (2015). Contribution of competence teacher (pedagogical, personality, professional competence and social) on the performance of learning. *The International Journal of Engineering and Science*, 4(2), 1–12.

Haseena, V.A. & Mohammed, A.P. (2015). Aspects of quality in education for the improvement of education scenario. *Journal of Education and Practice*, 6(4), 100–106.

Hersh, M.A. & Johnson, M.A. (Eds.). (2008). *Assistive technology for visually impaired and blind people*. London, UK: Springer.

Jacob, B.A. & Ludwig, J. (2009). Improving educational outcomes for poor children. *Focus*, 26(2), 56–61.

Lennie, P. & Van Hemel, S.B. (Eds.). (2002). *Visual impairments: Determining eligibility for social security benefits*. Washington, DC: National Academy Press.

Lunenburg, F.C. (2011). The comer school development program: Improving education for low-income students. *National Forum of Multicultural Issues Journal*, 8(1), 1–14.

Mahmood. T., Ahmed, M. & Iqbal, M.T. (2013). Assessing the pedagogical competences of teacher educators in the teacher education institution of Pakistan. *Academic Journal of Interdisciplinary Studies*, 2(1), 403–416.

Mani, M.N.G., Plernchaivanich, A., Ramesh, G.R., & Campbell, L. (2005). *Mathematics made easy: For children with visual impairment*. Philadelphia: International Council for Education of People with Visual Impairment (ICEVI).

RoI. (1945). *Undang-undang dasar negara Republik Indonesia Tahun 1945* [1945 Constitution of the Republic of Indonesia]. Jakarta, Indonesia: Republik of Indonesia.

RoI. (2009). *Peraturan Menteri Pendidikan Nasional nomor 70, Tahun 2009 tentang Pendidikan Inklusi Bagi Peserta Didik yang Memiliki Kelainan dan Memiliki Potensi Kecerdasan dan/atau Bakat Istimewa* [Regulation of the Minister of National Education number 70, Year 2009 on Inclusive Education for Students Who Have Disorder and Has Special Intelligence and/or Talent Density]. Jakarta, Indonesia: The Ministry of National Education, Republic of Indonesia.

RoI. (2014). *Situasi Penyandang Disabilitas* [Situation of people with disabilities]. Jakarta, Indonesia: The Ministry of Health, Republic of Indonesia.

RoI. (2016). *Statistik Persekolahan PLB 2016/2017* [Inclusive educational school statistics 2016/2017]. Jakarta: The Ministry of Education and Culture, Republic of Indonesia.

Retnawati, H., Kartowagiran, B., Arlinwibowo, J. & Sulistyaningsih, E. (2017). Why are the mathematics national examination items difficult and what is teachers' strategy to overcome it? *International Journal of Instruction*, 10(3), 257–276.

Roe, J. & Webster, A. (2003). *Children with visual impairments, social interaction, language and learning*. New York, NY: Routledge.

Sutapa, M. (2006). Membangun komunikasi efektif sekolah [Construct effective school communication]. *Jurnal Manajemen Pendidikan*, 2(2), 69–76.

Teachers' intention to implement instructional innovation: Do attitudes matter?

B. Basikin
Universitas Negeri Yogyakarta, Indonesia

ABSTRACT: Attitudes towards behavior have long been suggested to be significant predictors of behavioral intention. Individuals with better attitudes are believed to have a higher intention to do a behavior. Using the Theory of Planned Behavior (Ajzen 1991), this paper presents evidence where specific characteristics of a sample group, and the existence of a psychological constraint lead to different findings from a majority of research and theory. From data collected among 202 Indonesian junior school English teachers, analyzed using the Structural Equation Modeling (SEM) using AMOS, findings of a three-factor model of teachers' intention suggest that teachers' attitudes did not significantly predict intention. Such findings implied the effects of a volitional restriction among the teachers on teachers' psychology. It also confirms the low sense of autonomy among the Indonesian teachers indicated by Bjork (2004), resulting in low levels of decision making among teachers.

1 INTRODUCTION

1.1 *Teacher professional development*

Teacher professional development (PD) has become an important focus in much educational research for the last decades. The areas researchers examine are, for example, factors and criteria of successful PD programs (Gordon 2004, Little 1993, Lobman & Ryan 2008), teacher PD programs and teacher change (Guskey 2002, Stein & Wang 1988), effects of PD on teachers' beliefs, attitudes and practices (Buumen 2009, Guskey 2002, Karabenick & Noda 2004, Ross & Bruce 2007), and models of PD programs (Gordon 2004, Harland & Kinder 1997, Mulholland & Wallace 2001, Stein & Wang 1988, Vulliamy 1998).

In much of the literature, teacher PD is considered an effective tool to promote teachers' innovation, where on the one hand of teachers' participation in professional development is a change in teachers' instructional practices. Particularly when a teacher PD program addresses a huge number of teachers like in Indonesia, failing to stimulate changes is a waste of time and energy. However, it is only a high quality professional development program that may result in changes and is the central component in most proposals of teacher PD aiming at teachers' change (Guskey 2002).

Teachers are normally interested in professional development because they believe it will expand their knowledge and skills, contribute to their growth, and enhance their effectiveness with students, so they tend to be quite pragmatic about their attendance. Through professional development, they expect to get specific, concrete, and practical ideas that directly relate to their daily classroom teaching (Fullan & Miles 1992). Therefore, when a particular professional development program fails to address these needs, it is unlikely to be successful in changing teacher behavior in terms of teaching practices and activities.

In addition, although a number of researchers have suggested that there are effects of teacher professional development programs on teachers' individual and collective efficacy beliefs (Borko 2004, Buumen 2009, Gordon 2004, Guskey 2002), there is little information on how teachers' adopt and implement the results of teacher PD programs they have attended.

A number of researchers have provided insights related to the effects of teacher development programs on a number of teachers' life aspects. For example, it has been reported that changes in teachers' attitudes and beliefs resulted from a teacher development project (Guskey 1989). Yan (2014), using the theory of planned behavior, has reported on changes in teachers' intention to implement school-based assessments.

1.2 Teacher change

High quality teacher professional development is considered a central component in nearly every modern proposal for improving education, with most teacher PD programs designed to initiate changes in teachers' attitudes, beliefs, and perceptions (Guskey 2002). These changes are presumed to be preceded by changes in classroom behavior and practices, which in turn result in improved student learning—see Figure 1.

In terms of the nature of teacher change, referring to findings in the Turkish context, Buumen (2009) suggests that changes in instructional classroom practices after attending a teacher PD program is limited individually and not a part of a systemic school-wide change. This effect most likely happens not only in Turkish context, but also in other countries, such as in Indonesia. In the Indonesian context, it is believed that after returning from a PD program, many teachers do not implement the new methods of teaching they experienced during the teacher PD programs in their classroom practices. Instead, they go back to their old practices that provide them with security (Bjork 2004).

Changes in teacher practices can be a long-term process. To provide an explanation of how changes occur in teacher practices, it is perhaps useful to start by considering Guskey's model of teacher change (Guskey 2002). The model in Figure 1 assumes that there are three major outcomes that might occur among teachers after attending a professional development program. These include changes in teachers' classroom practices, students' achievement, and teachers' beliefs and attitudes (Guskey 2002). This model suggests that these changes happen in a particular sequence, with changes in teachers' attitudes and beliefs occurring only after they have evidence of improvements in student learning. These improvements are the result of changes in teacher classroom practices, particularly in terms of using new instructional approaches, materials or curricula, teaching procedures classroom format (Guskey 2002).

Guskey's model seems to make sense in terms of the linear order of how the changes happened. However, such a model is rather simplistic in that it focuses on a PD program as specific activities, processes or program (Opfer & Pedder 2011) without taking into account the complex nature of teaching and learning in the context in which the teacher lives and works. There is a need to provide a more comprehensive explanation of how such changes might occur. A better explanation is proposed by Opfer and Pedder (2011). Following Davis and Sumara (2006), they argue that "teacher learning tends to constituted simultaneously in the activity of autonomous entity (teachers), collective (such as grade level and subject groups) and subsystems within grander unities (schools within school systems within socio-political educational context" (Opfer & Pedder 2011 p. 379).

1.3 Teacher intention regarding their professional development

Teacher intention in this paper is investigated under the framework of the theory of planned behavior or TPB (Azjen 1991, 2002). In this framework, intention to do a behavior or action

Figure 1.   Model of teacher change (*Guskey 2002b*).

is predicted by three predictors as attitudes, subjective norms, and perceived behavioral control.

TPB (Ajzen 1991, 2002) is developed from the theory of reasoned action or TRA (Fisbein & Ajzen 1975, Fisbein 1979), by incorporating the measure of perceived behavior control (PBC) in the model as another antecedent of intention. Such an addition was a response to the lack of empirical evidence of the predictive power of the theory of reasoned action, particularly in situations where complete volitional control was absent (Ajzen 1991, Armitage & Corner 2001).

According to TRA, intention that is an aggregate of attitude towards the behaviors and subjective norms provide sufficient prediction for behavior (Ajzen & Fishbein 1972). However, in situations where there are perceived constraints on an action, intention alone will fail to predict behaviors. In such situations, it is PBC that helps explain the prediction of behavior (Ajzen 1991). The introduction of PBC addresses the fact that even when one really intends to perform a certain action and the pressure of significant others are favorable, actual behavior might still fail due to causes related to factors beyond one's control. An example of such a situation is related to institutionally mandated behavior. Measuring behavior control, therefore, is useful.

According to TPB, control over behavior influences whether or not a person performs the behavior. However, as suggested by TRA, the actual attainment of the behavior depends on the levels of actual control within the individual. A maximum level of control will create a complete volitional state where failure is not a probability. Where there is restriction regarding the degree of volitional status, where control is not at its maximum level, intending to perform the behavior does not always result in performance of the actual behavior. Addiction is a good example of where intention does not turn into actual behavior. The failure to perform the behavior is due to insufficient control over the behavior. However, that subjective perception of control affects the attempts to perform—it predicts the individual's behavioral intentions. As Ajzen has observed:

> "Subjective perceptions of control may, of course, influence attempts to perform behaviour regardless of their accuracy ... perceived control will usually correlate with behavioural performance. Again, however, this correlation will tend to be strong only when perceived control corresponds reasonably well to actual control" (Ajzen 1985).

Adding the PBC to TRA, the theory of planned behavior claims that intention to perform a behavior is influenced by three factors: 1) attitudes toward the behavior, 2) subjective norms, and 3) perceived behavioral control (Ajzen 1991). Attitudes toward a particular behavior derive from the aggregate of one's behavioral beliefs, which consist of the beliefs about the possible outcomes of the behavior and the evaluation of these outcomes. Subjective norms are made up of the normative expectations of others with respect to the behavior, and one's motivation to comply with those expectations. Perceived behavioral control is an

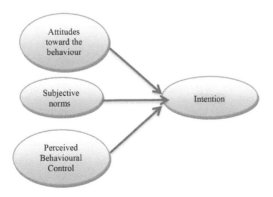

Figure 2. Theory of planned behaviour (Ajzen 1991).

aggregate of one's beliefs about the presence of factors that might either facilitate or impede the performance of behavior, and the perceived power of these factors. Figure 2 provides a schematic explanation of behavioral intention and its antecedents proposed by theory of planned behavior. In general, the more favorable one's attitudes and the subjective norms toward performing the behavior are, and the greater control one perceives when performing the behavior, the stronger the intention to perform the behavior (Ajzen 1991).

TPB is expected to better predict teachers' intention to implement results of a PD program in classroom practices. In the context of teachers in Indonesia, theory of planned behavior is preferable over theory of reasoned action because of the absence of the volitional condition. There is no choice available to the participants except the requirement to respond to the invitation or assignment to attend the PD program.

Constraints are also present due to problems among Indonesian teachers, in which this study was contextualized, with regard to their levels of autonomy (Bjork 2004). It is therefore considered that the complete volitional condition is absent. Following the theory of planned behavior, the predictors of intention, that is attitudes toward behavior, subjective norms, and perceived behavioral control, are measured either directly or indirectly.

Research using the theory of planned behavior in education confirms the predictive roles of attitudes, subjective norms, and perceived-behavioral control on intentions. For example, it is suggested that attitudes toward behavior, subjective norms, and perceived behavioral control predict teacher's intention to use educational technology (Lee, Cerreto & Lee 2010). They also argue that attitudes toward behavior have twice the influence of subjective norms and three times that of PBC. They therefore recommend that "teachers must have positive attitudes about using computers to create and deliver lessons. They are less concerned about what others think of this practice, and far less bothered by internal or external constraints" (Lee et al. 2010). In the area of inclusive education, Masud (2013) finds that teacher attitudes, teacher efficacy and perceived school support were significant predictors of teachers' intention to include students with disability into their classes, with 40% of the variance in teachers' intentions explained.

2 METHODS

Data about teacher's intention to implement the results of their PD program was collected using the Intention to Implement the Genre-based English Teaching Scale (IIGETS), consisting of four sub-scales of 1) teacher intention to implement the Genre-based English teaching, 2) teacher attitudes toward the implementation to the Genre-based English teaching, 3) subjective norms, and 4) perceived behavioral control, with three items in every scale.

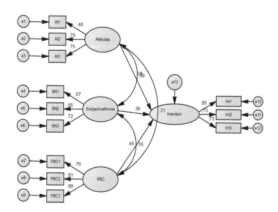

Figure 3.   Three-factor model of teachers' intention.
Note: $\chi2/df = 1.301$, RMR = 0.140, GFI = 0.954, AGFI = 0.925, CFI = 0.983, RMSEA = 0.039.

The survey was given to secondary school English teachers from four districts and one municipality in Yogyakarta Province. Among 260 teachers who regularly attended the English teacher forum (MGMP), 210 teachers returned the completed survey and their consents. Among the 210 teachers, 156 teachers were female and 54 were male. One hundred and eleven teachers have been in the teaching profession for 5–15 years, while the other 99 teachers have been teaching for more than 15 years. One hundred and seventy three teachers were certified and 37 teachers were non-certified teachers.

Preliminary screening of the data found that 8 responses were highly patterned and were dropped from the analyses. Data was then analyzed using the structural equation modeling (SEM) using a three factor model of teacher intention, as in Figure 3.

3 FINDINGS AND DISCUSSION

Teachers' intention to implement the GbET is generally high (M = 6.74; SD = 1.06) on a zero to nine Likert-like scale. This means that most participants state that they do intend to implement the GbET in their classroom teaching practices. Findings from the present study, however, did not confirm the findings of other research using TPB. Unlike what was suggested by TPB that attitudes, subjective norms, and PBC significantly predict intention, findings from this study revealed that only subjective norms ($R^2 = 0.360$; $p < 0.001$) and PBC ($R^2 = 0.454$; $p < 0.001$) significantly predicted participants' intentions. Attitudes were not significant predictors of intention ($R^2 = 0.049$; $p < 0.508$).

The majority of researchers investigating intention have suggested attitudes as significant predictors of intention (Ajzen 1991, 2002). However, in the present study with its particular group of teacher participants, attitudes towards the implementation of GbET do not significantly predict teachers' intention to implement GbET. No matter how favorable or unfavorable the attitudes towards the implementation of GbET are, they do not affect teachers' intention to implement GbET. More interesting is that the findings differ from findings of numerous research studies about intention, suggesting that attitudes predict intention more convincingly than subjective norms. They claim that personal considerations, including attitudes and PBC, overshadow the influence of perceived social pressures (Ajzen 1991).

Although not in line with the findings from the earlier versions of TPB (Ajzen 1991), the findings of the present study support the more recent findings using TPB (Brown et al. 2002, Yan 2014). That attitudes do not significantly predict teachers' intention indicates no matter how favorable or unfavorable the participants' attitudes towards the implementation of GbET are, they do not lead to differences in participants' intention to implement GbET.

Similar findings were reported by Brown et.al. (2002) when researching the employees' acceptance of the implementation of a standardized computer banking system. They suggested that attitudes failed to significantly predict intention because it was in an institutional mandated context. The volitional condition was absent so that participants were left with no option but to comply.

Yan (2014) has reported similar findings when researching teachers' intention to implement school-based assessments. He reported that although instrumental attitudes, which are operationalized as participants' appraisal of how advantageous the consequences of performing the behavior would be, predicted teacher intention, affective attitudes, which are feelings or emotions arising from the prospect of performing the behavior, did not. Yan's study was also situated in a non-volitional context. Implementation of school-based assessments was institutionally mandated.

The fact that teachers' attitudes towards the implementation of GbET in their classroom were not significant was expected in the context of the present study. It makes sense in the specific context of Indonesian teachers and the teaching profession where volitional condition is restricted. There are at least three explanations supporting the insignificant predictive power of attitudes towards intention in the context of the present study.

The first explanation is related to the nature of the implementation of GbET as a newly introduced instructional model in Indonesia. Although the context of the present study was different from that of Yan's and Brown's, where implementation of GbET was not institutionally mandated, it was widely believed that teachers perceived a new instructional model introduced by the government as 'compulsory'. This was due to a long tradition of centralization causing a strong dependency among teachers on central government policy. From this point of view, it could be expected that attitudes would not matter. Whether or not teachers liked GbET, they perceived no choice available to them in intending to implement the curriculum.

Secondly, Bjork (2003, 2004, 2006) has suggested that teachers in Indonesia are low in their sense of autonomy. As already discussed, this is caused by a number of factors—for example the dominant roles of school principals, supervisors, others superiors. The low sense of autonomy and the excessive external pressures put on teachers have left them with no option but to implement the newly introduced genre-based teaching approach. In this case, they did not consider their own attitudes toward the approach.

The lack of autonomy among teachers has led to significant predictive power of subjective norms on teachers' intention to implement GbET. Subjective norms which were related to what significant others said or expected about implementing GbET functioned as social pressures for teachers. When teachers are able to be highly autonomous, social pressures would not overwhelmingly affect them. However, because teachers were low in their sense of autonomy, whatever their superiors said was what they thought they had to do.

Thirdly, earlier TPB itself has actually anticipated circumstances where subjective norms and attitudes are not adequate enough to predict behavioral intention and actual behavior (Ajzen 1991), particularly when volition is restricted. That is why in such a context introducing PBC as predictor of intention is necessary.

4   CONCLUSION

In a situation where volitional condition was perceived to be absent, attitudes do not significantly contribute to teachers' intention to implement the results of a PD program in their classroom teaching. The fact that teachers fail to perceive options is due to the low levels of autonomy, which is also signaled by the strong predictive power of subjective norms.

Findings imply that, 1) there is problem with teacher's professional life in Indonesia where volition is absent which leads to the low levels of autonomy among teachers in the region; and 2) there should be changes in terms of the conduct of teacher PD programs for Indonesian teachers in terms of promoting teacher's involvement, particularly in the decision making.

Findings also open a new direction of research related to teacher's management and empowerment, particularly in the case of promoting teacher professional development that also promotes teacher autonomy and leadership.

REFERENCES

Ajzen, I. (1985). From intentions to action: A theory of planned behavior. In J. Kuhl & J. Beckman (Eds.), *Action control: From cognitions to behaviors* (pp. 11–39). New York, NY: Springer.
Ajzen, I. (1991). The Theory of Planned Behavior. *Organizational Behavior and Human Decision Processes, 50,* 179–211.
Ajzen, I. (2002). Constructing a TpB Questionnaire: Conceptual and Methodological Consideration.
Ajzen, I., & Fishbein, M. (1972). Attitudes and normative beliefs as factors influencing behavioral intentions. *Journal of Personality & Social Psychology, 21*(1), 1–9.
Armitage, C.J., & Corner, M. (2001). Efficacy of the Theory of Planned Behaviour: A meta-analytic review. *British Journal of Social Psychology, 40,* 471–499.
Bjork, C. (2003). Local responses to decentralization in Indonesia. *Comparative Education Review, 47*(2).
Bjork, C. (2004). Decentralisation in Education, Institutional Culture and Teacher Autonomy in Indonesia. *International Review of Education, 50,* 245–262.

Bjork, C. (2006). Transferring Authority to Local School Communities in Indonesia: Ambitious plans, mixed results. In C. Bjork (Ed.), *Educational Decentralization: Asian Experiences and Conceptual Contributions* (pp. 129–148). Dordrecht, The Netherlands: Springer.

Borko, H. (2004). Professional Development and Teacher Learning: Mapping the Terrain. *Educational Research, 33*(8), 3–15.

Brown, S.A., Massey, A.P., Montoya-Weiss, M.M., & Burkman, J.R. (2002). Do I really have to? User acceptance of mandated technology. *European Journal of Information Systems, 11*, 283–295.

Buumen, N.T. (2009). Possible effects of professional development on Turkish teachers' self-efficacy and classroom practice *Professional Develeopment in Education, 35*(2), 261v – 278.

Davis, B., & Sumara, D. (2006). *Complexity and education: Inquiries into learning, teaching and research.* London, UK: Lawrence Erlbaum.

Fullan, M.G., & Miles, M.B. (1992). Getting reform right: what works and what doesn't. *Phi Delta Kappa, 73*(10), 745–752.

Gordon, S.P. (2004). *Professional development for school improvement: Empowering learning communities.* Boston: Allyn and Bacon.

Guskey, T.R. (1989). Attitude and perceptual change in teachers. *International Journal of Educational Research, 13*(4), 439–453.

Guskey, T.R. (2002). Professional development and teacher change. *Teachers and Teaching, 8*(3), 381–391.

Harland, J., & Kinder, K. (1997). Teachers' Continuing Professional Development: framing a model of outcomes. *British Journal of In-service Education, 23*(1), 71–84.

Karabenick, S.A., & Noda, P.A.C. (2004). Professional development implications of teachers' beliefs and attitudes toward English language learners. *Bilingual Research Journal, 28*(1), 55–76.

Lee, J., Cerreto, F.A., & Lee, J. (2010). Theory of Planned Behavior and Teachers' Decisions Regarding Use of Educational Technology. *Educational Technology & Society, 13*(1), 152–164.

Little, J.W. (1993). Teachers' professional development in a climate of educational reform *Educational Evaluation and Policy Analyses, 15*(2), 129–151.

Lobman, C., & Ryan, S. (2008). Creating an effective system of teacher preparation and professional development: Conversation with stakeholders. *Educational Policy, 22*(4), 215–244. doi:10.1177/0895904807307064.

Masud, A. (2013). *Exploring predictors of teachers' intentions towards the inclusion of students with disabilities in regular primary schools in Bangladesh.* (Ph. D Doctorate thesis), Monash University, Melbourne.

Mulholland, J., & Wallace, J. (2001). Teacher induction and elementary science teaching: Enhancing self-efficacy. *Teaching and Teacher Education, 17*(2), 243–261.

Opfer, V.D., & Pedder, D. (2011). Conceptualizing teacher professional development. *Review of Educational Research, 81*(3). doi:10.3102/0034654311413609.

Ross, J.A., & Bruce, C.D. (2007). *Effects of professional development on teacher efficacy: Results of randomized field trial.* Paper presented at the annual meeting of the American Educational Research Association 2007, Chicago.

Stein, M.K., & Wang, M.C. (1988). Teacher development and school improvement: The process of teacher change *Teaching and Teacher Education, 4*(2), 171–187.

Vulliamy, G. (1998). Teacher development in primary schools: Some other lesson from Taiwan. *Teacher Development, 2*(1), 5–16.

Yan, Z. (2014). Predicting teachers' intentions to implement school-based assessment using the theory of planned behaviour. *Educational Research and Evaluation, 20*(2), 83–97. doi:10.1080/13803611.2013.877394.

The determinants of supervision success for the improvement of educators' professionalism in the era of the Association of Southeast Asian Nations Economic Community (AEC)

S. Slameto
Universitas Kristen Satya Wacana Salatiga, Indonesia

ABSTRACT: This study aimed to: 1) describe steps that were taken by a seminar to apply a Classical Model to ensure success; 2) to extend the rate of success of the seminar for the development of the teachers' professionalism; and 3) to determine among the ten independent variables, which ones are determinants. The seminar development uses the Classical Model, which consists of five steps that are assessed for their efficiency and effectiveness. The success of the seminar and the determinant factor were measured on the basis of the data from 71 persons (out of 183 seminar teacher participants) by way of a self-rating scale. This model is proved to be efficient and effective. The analysis revealed that the success of the seminar is at a high level category. It was found there were two determinant models that affected the variables relating to the success of the seminar: provisioning received as a means to develop teachers' professionalism, and quality of presentation and materials.

1 INTRODUCTION

The management for the upgrading of the teachers' competence leads to human growth and teachers' professionalism (Mantja, 2002). The improvement management, or teachers' competence development, is an effort to advance and to improve the teachers' quality, expertise, ability, and skills for the perfection of their work (Santyasa & Ganesha, 2008) toward teachers' professionalism.

To maintain the consistency of teachers' professionalism in keeping with the development of science, technology, and the arts, a sustainable effort to improve professionalism is needed. Prescriptively speaking, the support of management competence, empowering strategy, and development supervision serve as alternative theoretical dimensions in order to upscale teachers' professionalism.

The teachers' certification program is actually the government's effort to identify qualified teachers. Teachers who are proven to be qualified from the result of certification become the basis for receiving professional allowances. Teachers who receive a professional allowance are categorized as being professional teachers. Supervision of teachers' competence development in the certification program is based on the following considerations: 1) science and technology development, especially the waves of globalization and information; 2) covering invisible weaknesses at the time of selection; 3) developing professional attitude; 4) developing professional competence; and 5) growing emotional ties between teachers and principals. In order to build professional teachers who are characterized by their being certified as professionals, we need to have Sustainable Professional Development (SPD) (Slameto, 2011). However, in its implementation of points 3 and 4, some unsatisfactory facts are found.

Some studies reported that teachers' certification in Indonesia has not brought about significant changes in the quality of teachers. These researches concerned among others, the World Bank's research in 2009, 2011, and 2012, which found facts that teachers' certification does not affect the way in which teachers teach, nor their behavior. It was concluded that the certification program has not been successful in improving the teachers' quality or improving the students' achievement (Hayadin, 2015). However, it is different from the findings by Azwar & Murniati (2015), who found that there is a significant relationship between the certification and the teachers' performance in increasing the students' study results, namely: 1) there is a positive effect between the teachers' certification and performance; and 2) there is a positive effect between the teachers' certification and teachers' performance on the students' study results (Azwar & Murniati, 2015).

It is for this purpose that the Association of Indonesian Educators (ISPI), Central Java Province, carried out a seminar under the theme 'The Role of ISPI in the Improvement of Educators' Professionalism in the AEC Era', considering that one of the 12 activities for teachers' professional building (instructional supervision) which is technically feasible to improve the teachers' competence is the seminar (Santyasa & Ganesha, 2008). The academic seminar requires interaction among participants, aided by professors or other experts (Kurniawan et al., 2016). The implementation of the scientific seminar is parallel to the finding of Kartowagiran (2011), who concluded that the efforts and activities of most certified teachers who have received professional allowances are not yet satisfactory, one of which especially is in participating in a scientific forum or seminar. Seminars are one of the activities which must be attended and fulfilled for the SPD. It is one form of teacher educational and training model.

The function of seminars being conducted is to convey ideas or new thoughts for the solution of a problem faced by participants or members in the future (Surya, 2016). Seminars which were carried out by ISPI Central Java are hoped to equip or inform the participants in the AEC era about teachers' professionalism, in order that the seminars may have a positive effect on the development of teachers' professionalism. In addition, such seminars give an added value for new ideas or inspirations.

In the perspective of management or supervision as a system (Miarso, 2008), the success of a seminar, which is the added values to inspire new ideas for the development of teachers' professionalism (Y), is directly affected by its own process (punctuality for the agenda as planned X_1, quality of performance by resource person (X_2), quality of materials (X_3), relevance of materials to participants' needs (X_4), provisioning received for the development of teachers' professionalism (X_5), input from participants (such as, years of service (X_6), rank/category (X_7), title (X_8), status of certification (X_9), and gender (X_{10}).

An educational and training model is considered effective when it is based on the curriculum, approach and strategy, which are appropriate to the participants' needs and problems occurring among them. For this purpose, a special requirement is needed in building an efficient and effective educational and training model. The requirement is among others, the participants' need for learning.

A Classical Model is used in the management of teachers' competence development, by way of supervision in the seminar for the added values in the form of inspirations and new ideas for the development of teachers' professionalism. This model is meant to make the designated seminar materials appropriate to the learning needs of the training participants as the target. In this model, the resource persons have a manual in the form of curriculum (scope and seminar materials), seminar design, material, handouts/PowerPoint slides, and other material. The identification of seminar participants' needs is carried out openly and directly with the existing teacher participants as the target. The resource persons identify a gap between abilities already possessed by the teachers (target) and learning materials to be presented (Kamil, 2010).

The aim of the Classical Model is to make the abilities already possessed closer to the abilities to be developed or learned, in such a way that the teachers (target) will not experience a gap and difficulty in learning new material. The advantages of this model is to make it easy for the teachers (target) to learn the material, and that their present abilities will become a starting point to understand new learning material.

The five teaching-learning principles in this education and training are: 1) the learning objective is perceived, meaning that learners should understand why they have to learn something; 2) gradual sequence, meaning learners study step by step, starting from easy to difficult material; 3) respect to individual differences, meaning each learner must be given opportunity to learn in his or her comfortable way; 4) appropriate practices, meaning that all learners must practice or exercise based on the description of the instructional objectives; and 5) results should be known immediately, by telling the participants at any time, whether they do something correctly or not.

The problems in this study are: 1) What steps are taken in the seminar, which was held by ISPI Central Java in cooperation with the Faculty of Teacher Training and Education, Satya Wacana Christian University, Salatiga, Indonesia that applies the Classical Model, that make the seminar successful? 2) How high is the rate of success of the seminar in its effort to develop added values to inspire new ideas in the development of teachers' professionalism? (Y), and 3) Among the ten variables which are assumed to affect the success of the seminar, which one or ones become determinants?

Although the role of developmental supervision is no more than a facility and a starting point for the teachers to increase their professional commitment, this research will produce a supervision model through a seminar on the basis of success determinants. In this way, the management of teachers' competence improvement which leads to their professionalism becomes more effective and efficient, because they have firm theoretical and empirical grounds.

2 METHOD

The steps in carrying out the scientific seminar using Classical Model (Otto dan Glaser in Kamil, 2003) are as follows: 1) analyzing education and training problems; 2) formulating and developing education and training objectives; 3) selecting materials for education and training, learning media, methods and techniques (i.e. the scientific seminar); 4) devising a curriculum, units, education, and training subjects/topics; and 5) assessing the results of the education and training.

The scope of the scientific seminar is 'the Role of ISPI in Improving Educators' Professionalism in the ASEAN Economic Community (EAC) Era'. The materials were conveyed by Professor H. Sunaryo Kartadinata and Professor H. Ravik Karsidi Both are from public universities in Indonesia). This scientific seminar was attended by 183 Primary School teachers (5%), Junior High School teachers (65%), High School teachers (16%), and vocational school teachers (14%) from Demak Regency and the surrounding area. It was held on January 5–6, 2017.

For the second problem statement, this research is an evaluation, and for the third problem, it is a causality quantitative research, which is designed to explain the cause and result relationship among the variables, that is, testing the effects of the process variables (punctuality on the agenda as planned) X_1, quality of presentations by the resource persons X_2 together with the quality of the materials presented X_3, relevance of the materials to participants' needs X_4 provisioning received for the development of professional teachers X_5, and the input variable from the seminar participants (years of service X_6, rank/category X_7, title X_8, the status of certification X_9, and gender (X_{10}) upon the seminar success, i.e. added values to inspire new ideas in developing teachers' professionalism (Y).

Data was screened by self-rating scale which consists of 15 items, and attribute data which consists of six items. The self-rating scale has been tested valid (0.235–0.513) and reliable (0,705) after one item was dropped. The data source is 71 seminar participants who returned the rating scale concerned. Data was then analyzed descriptively and inferentially with the help of the SPSS program for Windows version 20.

3 MEASUREMENT AND RESULTS

3.1 *The steps of the seminar with the classical model which are proven successful*

The seminar, which consists of five activities as explained above, is divided into three groups:

1. Preliminary study, which consists of the analysis of the training problems;
2. Planning and model development, which covers the following steps: formulating and developing training objectives; selecting training materials, learning media, training methods and techniques; devising curriculum and units; training subjects; training topics; and implementing;
3. Model validation, which includes assessing the results of training, which then becomes the basis in the education and training model, and the strategy.

After the first and the second steps are finished, the model validation is complete. The step for the Classical Model validation is done by measuring the process and the result of the seminar, which consists of measuring the rate of efficiency and the effectiveness or success of the seminar. This activity is carried out by self-assessment by participants on the process of the seminar that they participated in and the result which they experienced. The result of the self-assessment in Table 1 below is the proof of the efficiency and effectiveness of the training.

Based on the data presentation in Table 1 above, it is clear that the completion of the seminar in the three stages has achieved a high rate, which is seen from its completion, facilities, and relevance to the two sessions. Therefore, the supervision on the development of teachers' professionalism through this Classical Model seminar which was done in three stages has fulfilled the five principles (perceived purpose, gradual sequence, individual differentiation, appropriate practice, and knowledge of results), as supported by the data.

3.2 The success of supervision on the development of teachers' professionalism through the Classical Model seminar

The success of the supervision on the development of teachers' professionalism through the Classical Model seminar is measured from two indicators, that is, efficiency and effectiveness on the basis of seminar participants' assessment. Based on the descriptive analysis, the results are presented in the following Table 2.

The data presented in Table 2 above shows the success of the Classical Model seminar, and the participants' inspiration is seen from the rate of efficiency that is of a high category, and the rate of effectiveness which is also of a high rate. Therefore, the supervision on the development of teachers' professionalism through the Classical Model seminar is efficient and effective, as supported by data.

3.3 Determinant of the success of the Classical Model seminar

After the data was screened by a self-rating scale, which consists of 14 items with five items for the teachers' attribute, they were then analyzed descriptively. The results are as follows.

Table 1. Validation of the Classical Model seminar.

Var.	Mean	Median	Sd.	Min.	Max.
1. Seminar is conducted as the planned agenda	2.8310	3.0000	0.41355	1.00	3.00
2. Seminar facility is appropriate as planned	2.4930	3.0000	0.67344	0.00	3.00
3. Seminar material in session 1 is relevant	2.9577	3.0000	0.20260	2.00	3.00
4. Seminar material in session 2 is relevant	2.9155	3.0000	0.40520	0.00	3.00

Table 2. Description of the Classical Model seminar which inspires participants.

Var.	Mean	Me.	Sd.	Min.	Max.
1. Efficiency	2.8451	3.0000	0.46745	0.00	3.00
2. Effectiveness	2.9155	3.0000	0.32717	1.00	3.00

The results of the descriptive analysis presented in Table 3 above show that the seminar process as perceived by the participants as provisioning for teachers' professionalism is at the high rate. The teachers' years of service moves from the shortest two years to the longest 35 years; the average is 17 years, with most (65%) having under 20 years of service. The participants' rank/category varied from III/A to IV/D; most of them (40%) are at III/A, and the rest (30%) are at IV/A. Most of them (76%) are S1 graduates, and the rest are Magister (S2). Most of them (62%) have enjoyed certification allowances, so to speak, and they are already professional teachers. Most of them (83%) are female teachers.

Next, to find determinants or determining factors for the success of the seminar, that is, the added values to inspire new ideas in developing teachers' professionalism (Y), a multiple regression statistical test using a Step Wise model was administered. The results are shown in Tables 4 and 5 below.

Based on the result of the regression test as shown in Table 3 above, two model determinants were found; Model 1 provisioning of professional teachers with $R = 0.741$ and Adjusted R Square $= 0.542$ or 54.20%, and Model 2 quality of resources persons with $R = 0.764$ and Adjusted R Square $= 0.571$ or 57.10%. To ensure whether the size of R^2 of this model is significant, the following Table 4 Analysis of variance (ANOVA) can be checked.

The Table 5 ANOVA above shows that Model 1 has $F = 78.018$ at the significance 0.00, and Model 2 has $F = 44.221$ at the significance 0.00. The rate of significance 0.00 is smaller than 0.5. Therefore, the success of the seminar is 54.20%, determined only by the process of the seminar as provisioning professional teachers (Model 1). However, if it is followed by

Table 3. Description of process and input variables for the success of the Classical Model seminar.

Var.	Mean	Me.	Sd.	Min.	Max.
X_5 Provisioning of prof. teachers	2.9718	3.0000	0.16663	2.00	3.00
X_6 Years of service	17.1493	14.0000	9.01052	2.00	35.00
X_7 Rank	2.9143	2.0000	1.93175	1.00	6.00
X_8 Title	B.Ed = 76%		M.Ed. = 24%		
X_9 Certification	No = 38%		Yes = 62%		
X_{10} Gender	Male = 17%		Female = 83%		

Table 4. Model summary of success determinants for Classical Model seminar.

Model	R	R square	Adjusted R square	Std. error
1	0.741[a]	0.549	0.542	0.22911
2	0.764[b]	0.584	0.571	0.22186

[a]Predictors: (Constant), Provisioning of professional teachers.
[b]Predictors: (Constant), Provisioning of professional teachers, quality of resource persons.

Table 5. ANOVA[c] determinants for the success of the Classical Model seminar model.

Model	Sum of squares	df	Mean square	F	Sig.
1	Regression	4.095	1	4.095	78.018	0.000[a]
	Residual	3.359	64	0.052		
	Total	7.455	65			
2	Regression	4.353	2	2.177	44.221	0.000[b]
	Residual	3.101	63	0.049		
	Total	7.455	65			

[a]Predictors: (Constant), Provisioning professional teachers.
[b]Predictors: (Constant), Provisioning professional teachers, quality of resource persons.
[c]Dependent Variable: Added Values for inspiration.

the quality resource persons (Model 2) it becomes = 57.10%, supported by data. It means that punctuality on the agenda as planned (X_1), the quality of the materials presented by the resource persons (X_3), relevance of the materials to participants' needs (X_4), years of service (X_6), rank/category (X_7), title (X_8), status of teachers' certification (X_9), and gender (X_{10}), do not become the determinants for the success of the seminar, which functions as a form of supervision on the development of teachers' professionalism.

4 DISCUSSION

The seminar being developed here consists of five activity steps, which are divided into three (preliminary study, planning and model development, and model validation), has been proved to have a high rate of efficiency and effectiveness. It also has a high rate of success as well as a high rate of completion, facilities, and relevance. The two session materials also achieve a high rate. Why? One of the things which can be used to answer this is the fact that the seminar fulfilled the principles of: 1) perceived purpose; 2) gradual sequence; 3) individual differentiation; 4) appropriate practice; and 5) knowledge of results. The developed model is parallel to the findings of Slameto et al. (2016). Through this Classical Model seminar, the participants (who consist of teachers, school superintendents, and principals who belong to the teachers' workgroups (namely KKG) in Demak Indonesia) need improvement in their professionalism in the form of article-writing skills so that they can publish their writings in the academic journals. This need was met by Primary School Teacher Education Study Program and Education, Satya Wacana Christian University team, who has facilitated them with the Scientific Article Writing Training program up to the delivery of the articles to the scientific journals. The training participants felt the advantages of the program concerning both the knowledge and the writing techniques, so that they were able to improve their compositions to become ready-to-publish articles in journals. Therefore, the supervision on the development of teachers' professionalism through the Classical Model seminar, being developed into three or five stages here, is a success with data support.

Viewed from the high rate of efficiency and effectiveness of the seminar which indicates its success, the seminar has also inspired the teacher participants to further develop their professionalism in the AEC era. Based on the result of his research on the Effect of Goal Setting on Individual Personal Development of the Alumni from Indonesia in the International Seminar at Haggai Institute, Tarigan (2011) suggested the importance of creating individual inspiration, since it gives motivation to us as leaders. The Faculty of Teacher Training and Education, Satya Wacana Christian University Salatiga, Indonesia, has implemented its services with integrity and has high accountability, so that the Association of Indonesian Educators in the Central Java of Indonesia put their trust in this purpose.

This causality quantitative research, which is designed to explain the cause and result relationship among the variables, tests the effect of process variables (X_1 punctuality of agenda as planned, X_2 quality of presentation by resource persons, X_3 quality of materials presented, X_4 relevance of materials to participants' needs, and X_5 provisioning received for the development of professional teachers) and input variables of seminar participants (X_6 years of service, X_7 rank/category, X_8 title, X_9 status of certification, and X_{10} gender) on the success of the seminar, that is, the added value of inspirations for new ideas in the development of teachers' professionalism (Y). The result of this research evidently found only two determinant models that affect the independent variable (X) on the success of the seminar (Y), that is: 1) X_5 provisioning received for developing professional teachers; and 2) X_2 quality of presentation by resource persons. The effect X_5 is 54.20% and if followed, X_2 becomes 57.10%.

The provisioning for the development of professional teachers as a determinant for the success of the seminar, which gives added values to inspire new ideas in the development of teachers' professionalism, is enhanced by the analysis of Prastowo (2016). He affirmed that in the development of professional teachers' capacity in the Islamic Elementary School, teachers need to be given provisions which enable them to create contexts, or an atmosphere which supports the development of Islamic way of life, to be realized in the skills in everyday life.

Further, the holistic-integral development of teachers in the area needs to be implemented by the principals, school superintendents, and other development agents. Some ideas which can be explored to develop teachers' professionalism are building a discussion forum for the sharing of ideas, opinions, or research findings among teachers of various circles. Another idea is to practice, get used to and enjoy writing and sharing ideas or research findings through scientific writings, such as publishing in journals, books, or newspapers, including of course seminars, let alone being supported by quality resources.

Although the role of supervision for the development of teachers' professionalism is no more than facilities and a starting point for teachers to improve their professional commitment, this research produces two model determinants for the success of the seminar, which take the form of added values to inspire with new ideas in the development of teachers' professionalism. As a form of supervision on the development of teachers' professionalism, the management for increasing teachers' competence which leads to their professionalism becomes more effective and efficient because they possess a strong basis both theoretically and empirically.

5 CONCLUSION

Based on the results and discussion of the research outcome above, it can be concluded that the seminar steps which applied the Classical Model are evidently successful in terms of efficiency and effectiveness. The success of this seminar is to develop added values to inspire new ideas in developing teachers' professionalism at a high level, and only two model determinants or determiners for the effect of variables: 1) X_5 provisioning received as a professional teacher development (54.20%); and 2) X_2 quality of presentations by resource persons (becomes 57.10%). It means that all other variables: punctuality of agenda as planned, quality of materials presented by resource persons, relevance of materials to participants' needs, years of service, rank/category, title, gender, and status of teachers' certification as seminar participants are not determinants for the success of the seminar as a form of supervision on teachers' professionalism. As a form of supervision on the development of teachers' professionalism, the management of teachers' competence improvement which leads to their professionalism becomes more effective and efficient, because they possess a strong basis, both theoretically and empirically.

REFERENCES

Azwar, K. & Murniati, A.R. (2015). Pengaruh sertifikasi dan kinerja guru terhadap peningkatan hasil belajar siswa di smp negeri 2 Banda Aceh [The influence of certification and teacher performance on improving student learning outcomes in public junior high school 2 Banda Aceh]. *Jurnal Administrasi Pendidikan, 3*(2), 138–147.

Hayadin. (2015). *Kerja besar untuk mutu guru* [Great work for teacher quality]. Retrieved from https://balitbangdiklat.kemenag.go.id/posting/ read/925-postingreadkerja-besar-untuk-mutu-guru.

Kamil, M. (2003). *Model-Model Pelatihan* [Training Models]. Bandung: UPI.

Kamil, M. (2010). Model Pendidikan dan Pelatihan [Education and Training Model]. Bandung: Alfabeta.

Kartowagiran, B. (2011). Kinerja guru profesional (guru pasca sertifikasi) [Performance of professional teachers (post-certification teachers)]. *Cakrawala Pendidikan, XXX*(3), 463–473.

Kurniawan, A., Lestari, S. & Martha, R. (2016). Efektifitas teknik paper seminar untuk mengajar menulis ditinjau dari aktualisasi diri mahasiswa [The effectiveness of paper seminar techniques for teaching writing in terms of self-actualization of students]. *Jurnal Penelitian Lembaga Penelitian dan Pengabdian kepada Masyarakat IKIP PGRI Madiun, 3*(1), 23–30.

Mantja, W. (2002). *Manajemen pendidikan dan supervisi pengajaran* [Education management and teaching supervision]. Malang: Wineka Media.

Miarso, Y. (2008). Peningkatan kualifikasi guru dalam perspektif teknologi pendidikan [Increased teacher qualification in educational technology perspective]. *Jurnal Pendidikan Penabur, 7*(10), 66–76.

Prastowo, A. (2016). Kapasitas guru profesional di pendidikan dasar islam [Professional teacher capacity in Islamic basic education]. *Jurnal Ekonomi Syariah Indonesia, 4*(2), 233–254.

Santyasa, I.W. & Ganesha, J.P.F.U.P. (2008). Dimensi-dimensi teoretis peningkatan profesionalisme guru [Theoretical dimensions of teacher professionalism increase]. *Jurnal pendidikan dan Pengajaran UNDHIKSHA, XXXXI*(Mei), 473–494.

Slameto. (2011). Pengembangan profesi guru berkelanjutan berdasar permennegpan No. 16 Tahun 2009 [Development of continuing teacher profession based on the ministerial No. 16 Year 2009]. Retrieved from https://herrywidayat.files.wordpress.com/.

Slameto, Wardani, N.S. & Kristin, F. (2016). Pengembangan model pembelajaran berbasis riset untuk meningkatkan keterampilan berpikir aras tinggi [Development of research-based learning model to improve high-level thinking skills]. *Prosiding Konser Karya Ilmiah Nasional 2016 (Vol. 2,* pp. 213–228). Salatiga, Indonesia: Fakultas Pertanian dan Bisnis Universitas Kristen Satya Wacana.

Surya, R. (2016). *Pengertian seminar beserta fungsi seminar lengkap* [Understanding seminar and complete seminar function]. Retrieved from http://www.seputar pengetahuan.com/2016/09/pengertian-seminar-beserta-fungsi-seminar-lengkap.html.

Tarigan, N.P. (2011). Dampak goal setting terhadap individual personal development alumni seminar internasional Haggai Institute yang berasal dari Indonesia pada periode tahun 2000–2002 [Impact goal setting on individual personal development alumni of international seminar Haggai Institute originating from Indonesia in the period 2000–2002]. *Humaniora, 2*(2), 1069–1083.

Teacher competency to design and evaluate the craft and entrepreneurship learning based on the Indonesian national curriculum

E. Mulyani, A. Widiastuti & T. Nurseto
Universitas Negeri Yogyakarta, Indonesia

ABSTRACT: This research aims at investigating: 1) teachers' competence in designing learning; 2) teacher's' competence in implementing learning designs; 3) teachers' competence in designing evaluation; and 4) teachers' competence in implementing the evaluation designs of craft and entrepreneurship learning. This research is an explorative descriptive research. The research population includes the teachers of craft and entrepreneurship subjects consisting of 17 teachers and the research sample comprises of seven teachers from seven participated high schools. These schools implement the Indonesian national curriculum and located in Sleman Regency, Yogyakarta, Indonesia. The data was collected using questionnaires, observations and documentations. Data analysis technique used in this research is a quantitative descriptive analysis technique. The findings show that: 1) teachers' competence in designing learning is included in the medium category; 2) teachers' competence in implementing the learning designs is classified in the medium category; 3) teachers' competence in designing evaluation belongs to the medium category; 4) teachers' competence in implementing the evaluation designs is classified in the medium category.

1 INTRODUCTION

Entrepreneurship lately becomes a popular topic to talk about as stated by Fayolle & Gailly (2008): "entrepreneurship has become an important economic and social phenomenon as well as a popular research subject."

Entrepreneurship can be learned as well as widely internalized through the education process. For example, Finland has promoted an entrepreneurship education. "Finland in particular has extensively promoted entrepreneurship education in curricula reforms undertaken at all education levels. For example, entrepreneurship education has been one of the so-called cross-curricular themes for basic education since 1994 and for upper secondary education since 2003" (Leino et al. 2010).

Entrepreneurship education should not only be taught in higher education but it is supposed to be introduced from basic education. "Enterprise education should not be equated solely to any specific institutions but throughout all phases of the education system. The educational institutions will require much restructuring to enhance skill development for entrepreneurship." (Ahmad 2013).

Entrepreneurship, which is not fully accommodated in the 2006 curriculum, is realized by the introduction of craft and entrepreneurship learning subjects as compulsory subjects at high schools in the 2013 curriculum design.

Craft and entrepreneurship are new subjects in the curriculum structure of 2013. Therefore, this requires teachers to have new qualifications for teaching craft and entrepreneurship subjects. Based on the letter issued by the Head of Human Resources Development Agency (locally abbreviated as PSDMPK-PMP) on teacher's certificate and authority referring to 2013 curriculum, the teachers of craft and entrepreneur subjects include skill, physics, chemistry, biology, economics and vocational teachers (handicraft, engineering, cultivation and processing) (Kemdikbud 2014).

Many teachers today do not have an educational background in the field of craft and entrepreneurship. This for sure affects the implementation of craft and entrepreneurship learning.

The learning of craft and entrepreneurship is the activity of changing the students' behavior in order to have understanding, initiative, innovation, ability to form capital material, be social, have intellect, a religious mentality, ability to familiarize themselves and the skill in producing a worth selling product. Thus, students' entrepreneurship skills will increase.

Craft and entrepreneurship subjects basically exist in order to improve the students' skills in producing products in the form of goods or services that have economic value (worth selling). In order for goods or services to be sold, the craft and entrepreneurship subjects are integrated. Based on the information obtained from the teachers of craft and entrepreneurship subjects, they are still confused how the subjects are taught. Therefore, it is necessary to conduct research to know the conditions in the field related to the implementation of craft and entrepreneurship learning. Through this research it is expected to know the teachers' competence in designing, implementing, and conducting evaluation of learning, so that a recommendation for improvement can be formulated.

2 RESEARCH METHODS

This research is a descriptive explorative research employing a quantitative approach carried out at high schools that implement the Indonesian national curriculum called Curriculum 2013, located in a district in Yogyakarta, Indonesia. The research population includes 17 teachers of craft and entrepreneurship subjects. The sample consists of seven teachers using purposive sampling techniques by selecting one teacher from each school as a representative. The data was collected by using questionnaire, observation and documentation techniques utilizing a questionnaire instrument, observation sheet and documentation checklist. They were analyzed by employing a descriptive statistical analysis technique through the steps of editing, coding, tabulating, and analyzing the data.

3 RESEARCH FINDINGS AND DISCUSSIONS

3.1 Teacher's competence in designing a lesson plan

"Lesson planning is a process that came prior to lesson delivery in which the teacher delineated the instructional procedures that would be implemented" (Amador 2010, p. 10).

Table 1 shows that based on total score, teachers who have less than 10 years of teaching have competencies in designing medium and high category of lesson plans, whereas teachers with 10 to 20 years of teaching have competencies in designing medium category of lesson plans, and teachers with more than 20 years of teaching have competence to design a lesson plan in high category.

Table 1. A cross-table of teacher competencies in designing a lesson plan according to total score based on teaching duration.

| | | <10 years | | 10 to 20 years | | >20 years | |
| | | Frequency | | Frequency | | Frequency | |
Category	Score range	Absolut	%	Absolut	%	Absolut	%
Low	17–39,67	0	0	0	0	0	0
Medium	39,68–62,34	2	50	2	100	0	0
High	62,35–85	2	50	0	0	1	100
Total		4	100	2	100	1	100

The research findings show that teachers have a good understanding of the lesson plan format, the components of a lesson plan, and are able to prepare the lesson plan based on the curriculum of 2013. This is indicated by the teacher-prepared lesson plans that meet the standard of lesson plan format and components as required by Curriculum 2013 (which includes identity, core competencies, basic competencies, competency achievement indicators, learning materials, evaluation, remedial learning, enrichment, media, tools and learning resources), therefore it can be concluded that teachers are able to develop lesson plans based on the curriculum of 2013. However, the competence achievement indicators are inappropriately formulated. In this case, the competence achievement indicators only address the knowledge competence, while the competence of attitudes and psychomotor have not been formulated properly. This shows that the learning implemented by the teachers only develops the cognitive aspects of the students and gives less attention to the affective and psychomotor aspects of the students. This is not exactly right, as Peklaj (2015, p. 186) said, "In order to achieve the best outcomes in students, teachers have to be focused on all their levels of processes in their teaching, taking each of them into account". The teacher should design the lesson plan which can develop three competencies; knowledge competency, attitude competency and psychomotor competency. Specifically, teachers should plan effective lessons by including learning goals in plans; including classroom discussions, creating an appropriate classroom environment, and making instructional decisions based on knowledge of students (Amador 2010).

3.2 Teacher's competence in implementing a lesson plan

Lesson plans that have been developed by teachers will be implemented in classroom learning activities. The following presents the score of teachers' competence in implementing the learning of craft and entrepreneurship subjects.

Table 2 shows that based on total scores, teachers who have less than 10 years of teaching have competencies in implementing medium and high category of lesson plans, teachers with 10 to 20 years of teaching have competencies in implementing lesson plan designs in high categories, and teachers who have more than 20 years of teaching have competence to design a lesson plan in a medium category.

Many teachers still focus on teaching how to produce products but they do not have adequate capacity to teach how to manage these products into high-value products (worth selling). In addition, teachers do not master the scientific approach. Teachers have been able to encourage students to carry out activities such as observing, collecting information, associating and communicating but they still have difficulty in the questioning stage, that is, how to encourage students in formulating questions. Teachers have been able to apply various learning approaches recommended in the 2013 curriculum which include inquiry/discovery learning, problem-based learning, and project-based learning.

Lesson planning is an important thing to achieve successful learning. The result of Amador's research (2010) indicates that standardized testing influenced the lesson planning process by mediating curricula use, teacher collaboration, and other lesson planning elements.

Table 2. A cross-table of teacher's competence in implementing a lesson plan design according to total score based on teaching duration.

Category	Score range	<10 years Frequency Absolut	%	10 to 20 years Frequency Absolut	%	>20 years Frequency Absolut	%
Low	17–39,67	0	0	0	0	0	0
Medium	39,68–62,34	2	50	0	0	1	100
High	62,35–85	2	50	2	100	0	0
Total		4	100	2	100	1	100

3.3 Teacher's competence in designing evaluation

There is widespread concern that assessments which have no direct consequences for students, teachers or schools underestimate student ability, and that the extent of this underestimation increases as the students become ever more familiar with such tests (Baumert & Demmrich 2001). Teachers should design and implement the evaluation so that the successfulness of learning can be achieved.

The data on teachers' competence in designing an evaluation is obtained from questionnaires.

Table 3 shows that teachers' competence in designing an evaluation is as follows. Teachers who have a teaching duration of less than 10 years are included in the medium category, teachers who have a teaching duration of 10 to 20 years belong to a medium and high category, and teachers who have teaching duration of more than 20 years are included in the high category.

Based on the research findings, some teachers state that they do not understand the procedure of learning evaluation according to Curriculum 2013. Teachers also lack understanding of the evaluation components that must be arranged. Although they have been able to design evaluations based on the 2013 curriculum, they still miss the scoring guidelines for each evaluation.

Most teachers are able to design an evaluation for cognitive aspects (evaluation of knowledge), affective (attitude evaluation), and psychomotor (skill evaluation). However, based on the Bloom's taxonomy regarding the knowledge evaluation, teachers only perform C1, C2, and C3, while C4, C5, and C6 are not implemented by the teachers. Many teachers only focus on how to memorize, understand and apply, but how to analyze, evaluate, and create are not realized by the teachers. Also, there are some teachers who are not able to develop an affective evaluation (attitude evaluation), especially the evaluation for spiritual attitudes. For assessing student's skill (psychomotor aspect), teachers have been able to arrange them quite well, although the evaluation criteria remain unclear. Evaluation of skills is performed by teachers to assess students' product and skills of presenting the results of group discussions.

3.4 Teacher's competence in implementing evaluations plan

The teachers' competence in implementing the evaluation design can be viewed from the implementation of classroom learning.

Table 4 shows that based on total scores, teachers who have less than 10 years of teaching have competencies in implementing evaluation design in a medium category, teachers with 10- to 20-years of teaching have competencies in implementing medium and high category of lesson plan design, and teachers who have more than 20 years of teaching have competencies to design a lesson plan in a medium category.

Based on the research findings, all evaluation designs developed by the teachers as stated in the lesson plan are not comprehensively implemented. Evaluations that are mostly carried out by teachers include the evaluations of knowledge, while attitude and skill evaluations are

Table 3. Teacher's competence in designing an evaluation based on teaching duration.

| | | <10 years | | 10 to 20 years | | >20 years | |
| | | Frequency | | Frequency | | Frequency | |
Category	Score range	Absolut	%	Absolut	%	Absolut	%
Low	5–11,67	0	0	0	0	0	0
Medium	11,68–18,33	3	75	1	50	0	0
High	18,34–25	1	0	1	50	1	100
Total		4	100	2	100	1	100

Table 4. A cross-table of teacher's competence in implementing a lesson plan design according to total score based on teaching duration.

Category	Score range	<10 years Frequency		10 to 20 years Frequency		>20 years Frequency	
		Absolut	%	Absolut	%	Absolut	%
Low	12–28	0	0	0	0	0	0
Medium	29–44	4	100	1	50	1	100
High	45–60	0	0	1	50	0	0
Total		4	100	2	100	1	100

not optimally performed. The implemented evaluation is an outcome-evaluation and does not reflect the evaluation process. Teachers are not able to apply attitude evaluations. The implementation of skills evaluation is not optimal as well. Teachers do not carry evaluation sheets during the learning process, therefore they only rely on their memory. This is certainly inappropriate because authentic evaluation requires concrete evidence in the form of an observation sheet of a learning process containing an evaluation rubric that has been written in the lesson plans. If teachers only rely on memory, the evaluation can be influenced by their subjectivity. Also, teachers do not always carry out the evaluation of knowledge as stated in the lesson plan.

4 CONCLUSIONS

The conclusions of this research are as follows:

a. The competence of high school teachers who teach craft and entrepreneurship subjects in Sleman Regency in designing a lesson plan is included in a medium category. Teachers have a good understanding of the lesson plan format, components, and are able to prepare the lesson plan based on the 2013 curriculum but the formulation of indicators for attitude and skills are still incorrect.
b. The competence of high school teachers in Sleman regency who teach craft and entrepreneurship subjects in implementing a lesson plan is included in a medium category. Teachers do not have a good mastery on scientific approach but have been able to apply some of the recommended teaching methods in the 2013 curriculum.
c. The competence of high school teachers in Sleman regency who teach craft and entrepreneurship subjects in designing evaluation is included in a medium category. Teachers have been able to design cognitive evaluations, but they have not formulated affective and psychomotor evaluations properly.
d. The competence of high school teachers who teach craft and entrepreneurship subjects in Sleman regency in implementing the evaluation plan is included in the medium category. Teachers have been able to carry out a knowledge evaluation but are not yet optimal in performing an attitude and skill evaluation.

REFERENCES

Ahmad, S., Z. 2013. The need of Inclusion of Entrepreneurship Education in Malaysia Lower and Higher Learning Institutions. *Education and Training.* 55 (2): 191–203.
Amadore, J., M. 2010. *Affordances, Constraints, and Mediating Aspects of Elementary Mathematics Lesson Planning Practices and Lesson Plan Actualization.* Dissertation of University of Nevada.
Baumert, J. & Demmrich, A. 2001. Test Motivation in the Evaluation of Student Skills: The Effects of Incentives on Motivation and Performance. *European Journal of Psychology of Education.* 16 (3): 441–462.

Fayolle, A. & Gailly, B. 2008. From Craft to Science: models and learning processes to entrepreneurship education. *Journal of European Industrial Training* 32 (7): 569–593.

Kemdikbud. 2014. *Surat Kepala BPSDM dan PMP tentang Sertifikat Pendidik dan Kewenangan Mengajar Guru Berdasarkan Kurikulum 2013* (Document of Chairman BPSD and PMP for certified teachers and teaching authority based on curriculum 2013). Accessed from http://www.plpg.unimed.ac.id. Tuesday 3rd November 2015 at 10.10 pm.

Leino, et al. 2010. Promoting Entrepreneurship Education: the role of the teacher. *Education and Training.* 52 (2): 117–127.

Chances for the Indonesian Qualification Framework (IQF) based recruitment of human resources in the industries and professions as productive teachers of Vocational High Schools (VHS)

B. Kartowagiran, A. Jaedun & H. Retnawati
Universitas Negeri Yogyakarta, Indonesia

F. Musyadad
IKIP PGRI Wates, Indonesia

ABSTRACT: The objective of the research was to describe the chances for human resources in the industries and professions with rare expertise the fields become productive teachers of Vocational High Schools (VHS). A survey approach was used with a cross-sectional survey design. The data was collected utilizing documentation, field observation, interviews, instrument filling and answering, FGDs (focus group discussions) and seminars. The research results indicate that: (1) VHS with studies of rare expertise need productive teachers with high-level practice experience, (2) most human resources in the industries and professions with rare expertise needed by VHS are willing to be productive teachers at vocational high school as long as they do not have to quit their primary job, and (3) their chances for recruitment as productive teachers at VHS are sufficiently great. One recruiting strategy is called *Bakti Negeri* ('National Dedication'), a mechanism conforming to the principles of *gotong royong* ('cooperation and collaboration') by emphasizing education as the responsibility of not only the government but also all society members. A mutually beneficial collaboration among the Government passing by Directorate of VHS Ministry of Education and Culture, LPTK ('Institutions of Teacher Education'), and industries or professional associations.

1   INTRODUCTION

A nation's success greatly depends on the quality of the education in it. Therefore, efforts should always be made to improve the quality of education in Indonesia. The efforts to improve the quality of education could be made in various ways. One way to make any educational quality improvement effort is to improve the quality of the learning conducted. The main factor in any effort to improve the learning quality is the teacher. The higher the level of a teacher's quality, the higher the quality level of the learning that the teacher conducts. It is in line with the opinion expressed by Marzano (2011), that the greater the number of the teacher's positive activities in class, the higher the level of students' learning achievement. In line with Marzano et al. (2006) said that the teacher's role becomes highly significant for each success of the learning process (Jones et al. 2006).

Meanwhile, Barber & Mourshed (2012) say that students' learning achievement starts from the effective teacher and principal. In fact, in another part, Barber & Mourshed even explain that students placed with high-performing teachers would progress three times as fast as those placed with low-performing teachers. It means that if the graduates of the Vocational High School (VHS) that are as desired as those with high-level skills, then the quality of the productive teacher managing the learning of practices should also be high in level. The term *productive* here is used in the sense related to, among others, how VHS differs from regular High School.

Chapter 2 of Governmental Regulation No. 74/2008 concerning teachers explains that it is obligatory that teachers have the academic qualification, competence, and certificate as an educator, be physically and mentally/spiritually fit, and possess the ability to actualize the objective of national education. Further, governmental regulation is explained that the competence as meant in the afore-mentioned is a set of knowledge, skills, and behaviors that teachers should possess, deeply understand, master, and actualize in performing their professional tasks.

When related to Indonesian National Qualification Framework (IQF), teachers that are good should: (1) be able to apply their field of expertise and utilize science, technology, and art in their field in problem solving and adapt to situations that they are faced with, (2) master the theoretical concepts of certain fields of knowledge in general ways, the theoretical concepts of particular parts in the said fields of knowledge in profound ways, and be able to formulate solutions to procedural problems, (3) be able to accurately make decisions based on analyses of information and data and be able to give guidance in choosing from various solution alternatives as individuals and as group members, and (4) be responsible for their own work and be able to be given the responsibility for the achievement of organizational work outcomes.

The good teachers in demand have not yet fully become a reality because the data at BPSDMP (*Badan Pengembangan Sumber Daya Manusia Pendidikan*, 'A Governmental agency dealing with the development of educational human resources') & LPMP (*Lembaga Penjaminan Mutu Pendidikan*, 'a governmental institution dealing with educational quality assurance') indicate that teacher quality still requires improvement. Teachers in Indonesia are 2,925,676 in number and around 49% or 1,434,513 of them do not meet the requirements for academic qualification yet (Gultom 2011).

A problem that is arising is the way to make teachers in general and productive teachers in particular able to attain such competence as government regulations. For that to be the question at this point is understandable because to have good productive teachers is highly expensive. One reason is that they have to be enrolled as participants in various training. To lessen the expenses for participation in various skill training for teachers, it appears necessary to explore the possibility of recruiting productive teachers from industrial and professional fields in accordance with IQF.

Another matter to be concerned is there are fields of expertise highly needed at VHS but teachers who are qualified to teach those fields are not supplied yet by any LPTK. Those fields of expertise are: (1) aircraft technology and engineering, (2) agribusiness and agrotechnology in general and particularly animal welfare, (3) social work, (4) *pedalangan* ('traditional Javanese puppetry with *dalang* as the name of the single puppeteer of a puppet play meant here'), and (5) fish-catching engineering. Therefore, the five particular expertise programs simplified above are the ones to be given priority in efforts of recruiting productive teachers from industrial and professional fields.

The problems arising in the recruitment of human resources employed in industrial and professional fields as productive teachers at VHS are: (1) what the conditions of the VHSs requiring productive teachers of rare expertise are like, (2) whether there are employees or personnel in the industrial and professional fields who are willing to become teachers, (3) what chances the industrial and professional practitioners have to be recruited as productive teachers at VHS.

The requirements for becoming a teacher in Indonesia are not mild ones. The law of the Republic of Indonesia referred to as Law No. 14/2005 about teachers and lecturers explain that the teachers' and lecturers' professions belong to a special field of work that requires certain professional principles. The teachers and lecturers should: (1) possess suitable aptitude/talent, interest, calling, and idealism, (2) possess the educational qualification and educational background that suits their professional field, (3) possess the necessary competence that suits their professional field, (4) obey the code of ethics of the profession, (5) have rights and obligations in the performance of their professional work, (6) receive an income determined in accordance with their work performance, (7) have opportunities to develop themselves professionally and sustainably, (8) be given legal protection in the performance of their professional work, and (9) have professional organizations that have the status of a legal body.

Meanwhile, as previously mentioned, in *Chapter 2* of Government Regulation No. 74/2008, it is stated that it is obligatory for teachers to have the academic qualification, competence, and certificate as an educator, be physically and mentally/spiritually fit, and possess the ability to actualize the objective of national education.

To determine whether there are any chances or not for practitioners in industrial or professional fields who possess rare expertise to be recruited as productive teachers at VHS, an in-depth analysis should be made on various teacher-related rules. In addition, research on several VHSs with rare expertise taught needs to be made to reveal how greatly those schools require productive teachers and research on several industrial and professional fields also needs to be made to reveal the willingness of practitioners in the industrial and professional fields with rare expertise to become productive teachers at VHS.

2 RESEARCH METHOD

The research was conducted in post-graduate educational at *Universitas Negeri Yogyakarta,* Indonesia and six vocational high schools in need of teachers with rare expertise. The research was also made in nine companies, institutions, or organizations in the industrial or professional fields with rare expertise at VHS. The research was from September-December 2015. The research data was collected by field observation, documentation, interview, questionnaire, FGD (focus group discussion), and seminars. The documentation was done to collect data of the legal basis and regulation concerning teachers and workforce matters and the existence, qualification, and certification of teachers of five fields of rare expertise. The observation was on teachers' performance, including their works, and the condition and learning environment of the visited with rare expertise taught.

The interviewing was done to collect data from the school principals and teachers concerning the need for productive teachers of especially packages of rare expertise and concerning teacher performance at VHS. The interviewing was also done to collect data from the direct superiors and colleagues of the practitioners in the industrial or professional fields who were the teacher candidates that would be recruited as productive teachers. Meanwhile, the instrument filling and answering was done to collect information related to the willingness of the practitioners in the industrial and professional fields to be recruited as productive teachers at VHS. The FGD and seminars were used to collect input concerning the research report draft.

The data analysis was mix sequential explanatory quantitative and qualitative descriptive. The quantitatively descriptive analysis was used to answer the question regarding how the lack of need teachers at the VHS with rare expertise taught and how many practitioners in industrial or professional fields were willing to become productive teachers at VHS. The qualitative descriptive analysis was used to describe the legal basis that could support the recruitment of practitioners from the industrial and professional fields as productive teachers at VHS. The qualitative descriptive analysis was also used to describe the chances for practitioners in the industrial and professional to be recruited as productive teachers at VHS.

3 RESEARCH RESULTS

3.1 *Conditions of productive teachers at VHS rare expertise*

The conditions of several VHS that require teachers with rare expertise still make one feel greatly concerned. The principal of VHS1 explains that his school of aviation is greatly in need of productive teachers. Further, the VHS1 principal explains that a school as his could have an accredited status when it has six productive teachers while there is not yet any LPTK generating teachers with aircraft expertise.

The principal of VHS2 explains that for the fish-catching there are not yet any teachers who are civil servants, most of the teachers who originally graduated from VHS Depok have expertise in ship machinery. The teaching done by the teacher bachelor of Fish Catching is limited to being theoretical and has insufficient practical experience, so that is a paradox as the students are more experienced because their parents are fisherman who often go to sailing with them.

The principal of VHS 3 explains that concerned has three expertise program packages, namely, those respectively of *seni tari* ('dance arts'), *karawitan* ('traditional Javanese music arts'), and *pedalangan* ('mastermind of javanese puppet show').

Further, the principal of VHS3 explains that at the minimum, four teachers are needed for each expertise program but often a rule from the Directorate of Vocational High School Development restricts the number to only two per expertise program package. Therefore, VHS 3 greatly needs productive teachers and especially for the packages of expertise in *karawitan* and *pedalangan*.

Meanwhile, the principal of VHS 4 stated that he doesn't receive any attention from either the government and the professional world. Further, that principal VHS 4 stated greatly that he needs productive teachers with social health as their field of expertise. In the meantime, the principal of VHS 5 stated the students do not go as far as doing productive practice because the laboratories are incomplete, and the school does not have experienced productive teachers. VHS 5 greatly needs productive teachers in the fields of agriculture and animal welfare.

The exposition above illustrates that almost all VHSs that have study program of rare expertise need experienced productive teachers but unfortunately LPTKs do not supply them yet. Therefore, it is not unusual for such schools to conduct some recruitment of practitioners in the industrial or professional fields as productive teachers.

3.2 *Willingness of practitioners in the industrial and professional fields to become productive teachers at VHS*

Results of interviews with practitioners in the industrial or professional fields indicate that in general they are willing and interesting to work as productive teachers at VHS in the hours between their primary work hours. Some employees at Animal Welfare Center, Animal Disease Investigation Center, and Veterinary Association say that they do not mind giving assistance by teaching at VHS. In that hall, there are veterinarians and employees with expertise in animal welfare who are ready to become productive teachers at VHS. Something in line with it is also shown in the aircraft expertise study program package.

The occupation of productive teacher positions available at the VHS running the aircraft engineering study program at the moment is dominated by PNS *Pegawai Negeri Sipil* ('civil servant') retirees from various institutions like Ministry of Transportation, the Air Force Army, and practitioners in the field of aviation. Most teachers of aircraft engineering expertise do not come from the field of education because the study program for this expertise is not yet available at LPTKs.

Vocational high schools with the expertise package of fish catching collaborate with partners from the Navy Army. For the learning of the theory in that collaboration, the involved personnel from those collaborator institutions come to the school, while for the practice, the students go to the headquarters of the Navy because the facilities there are more complete than at the school. It means that practitioners in industrial or professional fields in relation with the expertise package of fish catching are already willing to become vocational high school teachers.

Meanwhile, VHS 3 requests expert *dalangs* and people from art galleries to teach. As previously mentioned, a *dalang* is the single puppeteer performing one traditional Javanese puppet play. The *dalang*, *karawitan*, and traditional Javanese dance arts there should conform to those of the Yogyakarta style. At the moment, VHS3 also already conducts the recruitment of professional *dalangs/empu*. These *dalangs'* abilities are quite extraordinary, and they are willing to receive a very low salary.

The above exposition indicates that generally employees in industrial fields or practitioners in professional fields are willing to become productive teachers at VHS but in the hours between primary work hours because they do not want lose their occupation as a professional. A part of them are also willing to become fulltime productive teachers at VHS. There are even employees in the industrial fields or practitioners in the professional fields who are willing to become productive teachers at VHS with low pay.

3.3   *Chances for recruitment as productive teachers at VHS*

Even now, the Indonesian government still tries hard to make sure that all the teachers are professional, namely, teachers that fulfill the academic qualification of the bachelor level and have professional certificates. The productive teachers at vocational high schools, who are recruited from industrial or professional fields, should also fulfill the academic qualification of the bachelor level and have the educator's certificate. Therefore, relevant regulations need to be studied in order that the efforts to procure productive VHS teachers of quality at low cost through recruitment as teachers from industrial and professional fields achieve success.

Several relevant legal rulings studied in the research concerned here are, Indonesian Act No. 43/1999 concerning a modification on Indonesian Act No. 8/1974 concerning the main points about employee matters, Indonesian Act No. 20/2003 concerning the national education system, Indonesian Act No. 14/2005 concerning teachers and lecturers, Government Regulation of Republic Indonesia No. 74/2008 concerning teachers, Government Regulation of Republic Indonesia No. 32/2013 concerning a modification on Government Regulation of Republic Indonesia No. 19/2005 concerning the national education standard, Presidential Regulation No. 8/2012 concerning Indonesian Qualification Framework, Minister of National Education's Regulation No. 16/2007 concerning Standard of Qualification and Competence of Teacher. Minister of National Education's Regulation No. 10/2009 concerning certification for professional teachers, and No. 9/2010 concerning teacher professional education for teachers, and Minister of Education and Culture's Regulation No. 73/2013 concerning the IQF application on the field of higher education and No. 62/2013 concerning teacher certification.

In the results of the study on the regulations and others above, it could be concluded that there are chances for employees in industrial fields and practitioners in professional fields to be recruited as productive teachers at VHS. There are three recruitment types, namely, Types I, II, and III. The teacher candidates of Type I Academic Educational are members of the workforce at industries or members of professional circles having the qualification as graduates of the bachelor level in education but who have not yet possess the educator's certificate. They have very great chances to be recruited as productive teachers at VHS. It is understandable because, besides having the academic qualification required, they also possess the four competence types of teachers, namely, professional, pedagogical, personal, and social competence. As for those among them who do not have the educator's certificate, they could register as participants in the teacher certification at the nearest LPTK.

The teacher candidates of Type II Academic Non-Educational, who are members of the workforce in industries or members of professional circles who have the qualification as graduates of the bachelor level but not from LPTKs, have sufficiently great chances to be recruited as productive teachers at VHS. They already have the academic qualification as required and the professional, personal, and social competence though not yet having the pedagogical competence as teacher candidates. They need to become participants in pedagogical training for six months at LPTK before participating in the teacher certification program.

The teacher candidates of Type III Competence Non-Academic, who are members of the workforce in industries and members of professional circles who do not have yet the academic qualification as required, also have chances to be recruited as productive teachers at VHS, though the chances are low. They could fulfill the academic qualification through: (a) regular education; and/or (b) recognition of self-study achievement measured through equivalence testing (the term used is RPL, recognition of past experience done by means of a comprehensive examination done by accredited universities or other equally accredited higher educational institutions in Government Regulation No. 74/2008.

Furthermore, the academic qualification fulfilment for these teacher candidates of this Type III could also be done through verification of expertise certificates that they have and equivalence testing on expertise competence by accredited higher educational institutions. The educator's certificate could be obtained through feasibility testing by LPTKs that are accredited and appointed by the government through the teacher certification program in Government Regulation No. 74/2008 after the individuals concerned fulfill their academic qualification.

Candidates of productive teachers at VHS coming from the business/industrial and the professional of this Type III could be recruited through three paths of recruitment, namely, Paths A, B, and C. **Path A** refers to the ordinary category as explained above. Candidates of productive teachers at VHS who are less than 50 years of age could fulfill their academic qualification through regular education while those who are more than 50 years of age could do it through RPL. In the meantime, the educator's certificate could be obtained through the teacher certification.

Path B refers to extraordinary category, the way to use for candidates of productive teachers coming from the industrial or the professional who are not interested in becoming teachers by Path A. These teacher candidates recruited through Path B are given several conveniences; (1) they remain working as members of the workforce (or employees, that is) in industries and as practitioners in professional fields while becoming teachers (a) every Saturday, or (b) a week at the end of every semester, or (c) two weeks at the end of each academic year; (2) they could fulfill their academic qualification through a recognition of their self-study achievement measured through equivalence testing by accredited higher educational institutions. Meanwhile, the educator's certificate could be obtained through the teacher certification program after the persons concerned fulfill the academic qualification.

Path C is the path of *Bakti Negeri* ('National Dedication'), it is a mechanism with the principles of *gotong royong* ('cooperation/collaboration'), which emphasizes that education is not only the responsibility of the government but also all society members. This path is also in line with the vision 2015–2019 of the Ministry of Education and Culture, it is the period of the formation of the human individual and ecosystem of education and culture with character by being based on *gotong royong*. This way of recruitment requires a note of agreement between the Ministry of Education and Culture, who could be represented by the Directorate of PSMK) and relevant industrial and professional associations.

The existence of the aforementioned note of agreement demands that industrial and/or professional associations have a share in taking the responsibility for the accomplishment of education that is of quality but within the reach of the lower classes in society. In this case, it is hoped that the industrial and professional associations could design programs assigning their employees or members that are needed by the schools concerned here to teaching jobs lasting for 1 until 2 weeks at those schools. The industrial companies could allocate a part of their product promotion funds through the schools or through assigning their employees to teaching at the schools, which indirectly promotes the companies where these teachers still work. Furthermore, the government also needs to give positive points to industrial or professional fields that have done what is called *Bakti Negeri*. In addition, the government also gives these Path-C or *Bakti-Negeri*-path candidate teachers conveniences in the fulfillment of academic qualification and in getting the educator's certificate.

The fulfillment of academic qualification could be done through recognition of self-study achievement measured by means of equivalence testing by an accredited higher educational institution. The educator's certificate could be obtained through feasibility testing by an accredited LPTK appointed by the government.

4 CONCLUSION AND RECOMMENDATION

4.1 *Conclusion*

All VHS with study programs of rare expertise need productive teachers with high-level practice experience. Most members of the workforce in industries and members of professional association with rare expertise and needed by vocational high schools are willing to become productive teachers at VHS as long as they do not have to leave their main line of work.

The chances for the recruitment as productive teachers at VHS of the workforce in industries and professional are sufficiently great. Teacher candidates of Type I, namely, members of the workforce in industries or professional fields who have the qualification as graduates of the D-IV or S1 level in education, have very great chances to be recruited as productive teachers at VHS.

Teacher candidates of Type II, who have the qualification as graduates of the D-IV or S1 level from non-LPTKs have sufficiently great chances to be recruited as productive teachers at VHS.

Teacher candidates of Type III, the workforce in industries and professional fields who do not have yet the academic qualification as required but are highly competent, also have chances to be recruited as productive teachers at VHS. For them, there is more than one way of recruitment. They could be recruited through one of three possible recruitment paths, namely, Paths A, B, and C. Each path is adjusted to the respective characteristics of the teacher candidates.

4.2 *Recommendations*

The Directorate of vocational high schools should collaborate with accredited LPTKs appointed by the government to recruit productive teachers at VHS from industries and professional fields by using the identification into three types as follows.

Type I, Teacher candidates should possess the educator's certificate. They could obtain the educator's certificate through feasibility testing by accredited LPTKs which are appointed by the government and do it by enrolling these teacher candidates in a Teacher Certification Program.

Type II, Teacher candidates have to obtain recognition of having pedagogic competence and the educator's certificate. The attainment of pedagogic competence recognition and the possession of the educator's certificate are done by the teacher candidates being enrolled in Teacher Certification Program.

Type III, Teacher candidates should be get (1) pedagogic competence, (2) academic qualification, and (3) the educator's certificate. There are three recruitment paths for this group of teacher candidates. Especially for **Path C** *Bakti Negeri* category, the industries and professional fields could allocate part of their product promotion funds through schools or through the assignment of their employees to teaching jobs at schools, which indirectly promote the companies where the teachers' work as employees. The requirements for their academic qualification are fulfilled through the recognition of the self-study achievement measured through equivalence testing by accredited higher educational institutions and their attainment of the required educator's certificate is done by the teacher candidates being enrolled as participants in Teacher Certification Program.

The Directorate of PSMK needs to make a note of agreement with LPTKs and industries or professional associations in order that they make it compulsory for their staff and experts to be participants in the recruitment as productive teachers in the *Bakti-Negeri* way.

The government should give appreciation to the industrial/professional companies whose employees become participants in the recruitment as productive teachers in the *Bakti-Negeri* category. The Directorate of PSMK should give more attention to SMKs with rare expertise competence programs and especially to the number and quality of the teachers concerned.

It would be better for the government to build an ecosystem of education and culture with character by having *gotong royong* as the basis of development.

REFERENCES

Barber, M & Mourshed, M. 2012. Professional development international. New York: Pearson.

Governmental Regulation No. 74/2008 about Teachers.

Government Regulation of Republic Indonesia No. 32/2013 about Sertification for Teachers' Managing and Equity.

Gultom, S. 2011. *Strategi pembinaan profesionalisme guru (Teacher development strategy of teache)*. Delivered in the Workshop for the Initial Competence Test Item Development at Hotel Sentul, Bogor, 2–4 February2–4, 2012.

Indonesian Act No. 20/2003 about National Education System.

Indonesian Act No. 43/1999 as revision of No. 8/ 1974, about Employment.

Indonesian Act No. 14/2005 Tentang Guru dan Dosen.

Jones, J. et al. 2006. *Developing effective teacher performance*. London: Paul Chapman Publishing.

Marzano, R.J. et al. 2011. *Effective supervision*. Alexandria: ASCD.

Minister of Education and Culture's Regulation No. 73/2013 about Implementation IQF of Higher.

Minister of National Education's Regulation No. *9/2010 about In-Service Teacher Education.*

Minister of National Education's Regulation No. 16/2007 about Standard of Teachers' Qualification and Competence.

Teacher professional development for enhancing character education

Character-based reflective picture storybook: Improving students' social self-concept in elementary school

A. Mustadi, S. Suhardi, E.S. Susilaningrum, R. Ummah, P.E. Wijayanti &
M. Purwatiningtyas
Universitas Negeri Yogyakarta, Indonesia

ABSTRACT: This research aims: (1) to produce a character-based reflective picture storybook to improve social self-concept and (2) to determine the effectiveness of the character-based reflective picture storybook in improving the social self-concept. This research was R & D developed by Borg & Gall (1983). The subjects were the fifth grade students of elementary schools in Yogyakarta, Indonesia. The data collection techniques were interview, observation, scale, and questionnaire. Data was analyzed by a paired t-test and independent t-test at 0.05 significance level. The result of the research shows that: (1) the character-based reflective picture storybook has passed the criteria of feasibility to improve student's social self concept based on expert validation result, teacher response scale, and student response scale with very good criteria; (2) the character-based reflective picture storybook is effective to improve the social self-concept based on the t-test result with significance level < 0.05, that is 0.000. Through pictorial stories and reflection activities developed in the character-based reflective picture storybook, elementary students can internalize the values of characters in a more fun way, acquire knowledge in accordance with the curriculum, and add insight into how to interact with others in the surrounding environment.

1 INTRODUCTION

Character education is an idea that has long been applied in the education system in Indonesia. Through the implementation of character education, noble characters that match the identity and characteristics of the Indonesian nation can be instilled early on in Indonesian children. By having these noble characters, it is expected that the children can have a good personality so that they can interact with the wider society in a global context.

Early character cultivating in children also plays an important role in the development of social self-concept. Therefore, by having a good character, every child can establish good social interaction with others in the surrounding environment. An important character in the globalization era is a person who not only becomes a life-long learner, but must also be able to play a role in relation to society (Zuchdi et al. 2010).

Children's social self-concept plays an important role in the development of a positive personality. However, reality shows that some students do not have a positive social self-concept. This is indicated by the involvement of students in violence or bullying, drug abuse, student gangs, and brawls. In addition, data from the Indonesian Children Protection Commission (*Komisi Perlindungan Anak Indonesia* or *KPAI*) through a rapid survey of 1,026 elementary, junior and senior high school students from nine provinces scattered throughout Indonesia, showed that the percentage of children who became perpetrators of violence reached 78.3%. Therefore, the formation of positive social self-concept in children through the implementation of character education requires more attention.

Piaget said that one of the obstacles faced in the formation of positive social self-concept through the implementation of character education in Primary Schools, is the lack of learning media that can facilitate the characteristics of elementary students who are still at the concrete operational stage (Santrock 2011). During this time, the available learning media generally emphasizes the aspects of knowledge and skill. Learning media that can help the cultivation of character values are still very limited.

The development of the character-based reflective picture storybook, was based on the results of need analysis. The results show that in the implementation of Curriculum 2013, the availability of textbooks and learning media which support integrative-thematic learning is still limited. Teachers and students need character-based learning media that can make learning more meaningful and have a positive impact on the formation of students' social self-concept. Learning media featuring reflective illustrations and reflection activities such as the character based reflective picture storybook are not yet available in schools. Most of the learning media is still textbooks that have not reflected daily life and have not integrated character values.

The character-based reflective picture storybook was developed based on an implementation of reflective thinking (Gillespie 2005, Harisson & Dymoke 2009, Shambaugh & Magliaro 2006, Dervent 2015) in picture storybooks. The picture storybooks can be utilized in the formation of positive social self-concept. Picture storybooks can help children to learn to understand others, to understand the relationships between people and the environment, and to develop feelings. Therefore, picture storybooks can help students to understand how to interact with others (Mitchell 2003).

This research was conducted to produce an appropriate character-based reflective picture storybook to improve social self-concept and to determine the effectiveness of the media in improving the social self-concept of the fifth grade students of elementary schools in Bantul, Yogyakarta.

2 LITERATURE REVIEW

2.1 Social self-concept

Self-concept becomes a reference for individuals to clarify and understand themselves. Self-concept is a frame of reference for every individual to interact with his environment (Fitts 1971). Therefore, life experiences and the environment play a very important role in the development of individual self-concept (Laryea et al. 2014).

Social self-concept is one aspect of self—concept which relates to the interaction of individuals with their social environment. Social self-concept refers to the individual's view of the abilities to relate to others that cause behavioral change. Therefore, social self-concept will result in a change of behavior, when it is in a positive or negative direction (Bakhurst & Sypnowich 1995).

Social self-concept consists of three aspects. These three aspects include social acceptance, social competence, and social responsibility (Stump et al. 2009, Fernández-Zabala et al. 2016). Leary said that the first aspect is social acceptance. Social acceptance is an individual's perception of how he or she is accepted by the group and in social interaction (DeWall & Bushman 2011). The second aspect is social competence. Semrud-Clikeman (2007) points out that "Social competence is an ability to take a perspective concerning a situation and to learn from past experiences and apply that learning to the ever-changing social landscape." The third aspect is social responsibility. Social responsibility is the individual's perception of the demands and role of the individual in the social context whether it is with friends, family or society (Berman 1990).

2.2 The character-based reflective picture storybook

The picture storybook is composed of narrative text and illustrative images that are interconnected to convey the message of the story to the reader. Picture storybooks are books that convey messages through two ways, namely illustrations and writings. The illustrations and writings used to convey the message do not stand alone, but are a unity and are mutually supportive to express a message to the reader (Huck et al. 1987).

Picture storybooks are one of the media that can be utilized as a learning media in accordance with the characteristics of elementary students who are still at the concrete operational stage. Piaget (Santrock 2011) suggests that elementary students are still at a concrete operational stage, so that students can understand abstract concepts through concrete and specific examples contained in their daily lives.

In the context of the implementation of educational characters, picture storybooks can be utilized as the learning media that helps the formation of a positive social self-concept in students. Picture books help children to learn to understand others, to understand the relationships that occur between people and the environment, and to develop feelings (Mitchell 2003).

The character-based reflective picture storybook is a form of picture storybook development that implements reflective thinking process. Through the application of reflective thinking processes, students can reflect the pictorial story presented in the media with the students' daily lives. Reflective thinking is a meaningful process in which one will have new experiences, which will then relate to other experiences. This makes a person learn continuously so that the person will have more extensive experience and take better action moving forward. Furthermore, students developmental levels and teaching approaches through their experiences and their reflections on these experiences were determined (Dervent 2015).

The character-based reflective picture storybook gives stories that contain learning materials according to the curriculum reflected in the daily life of students, and children's stories that reflect the characters in the daily life of students. Through illustrations and stories based on character values, students can obtain role models from the characters, and can reflect stories and materials into daily life (Richter & Calio 2014, Turan & Ulutas 2016, Zaky 2016). Through the utilization of this media in the learning process, students' social self-concept is expected to develop.

3 METHOD

3.1 Approach

This research used the development research approach from Borg & Gall (1983). The character-based reflective picture storybook media was developed based on ten stages of product development, including: 1) research and information collecting, 2) planning, 3) developing a preliminary form of product, 4) preliminary field testing, 5) main product revision, 6) main field testing, 7) operational product revision, 8) operational field testing, 9) final product revision, and 10) dissemination and implementation.

3.2 Subjects

The subjects of this study were teachers and fifth grade students of elementary schools in Bantul District, Yogyakarta, Indonesia. The subjects of the preliminary field testing were fifth grade students and the teacher in three public primary schools.

3.3 Data collection techniques and instruments

Data collection techniques in this research included interviewing, observation, questionnaire, and scale. Instruments of data collection in this research included interview and observation guidelines for need analysis, self-assessment and peer assessment, product validation sheet of material expert and media expert, teacher response scale, student response scale, and social self-concept scale.

3.4 Data analysis techniques

Data on the results of the need analysis at the research and information collecting stage were analyzed by using the descriptive analysis technique. The feasibility of the character-based reflective picture storybook on the product validation sheet of material expert and media expert, and the results of teacher response scales and student response scales were analyzed by conversion of scores into four quantitative categories Mansyur et al. (2015).

The effectiveness of the character-based reflective picture storybook in improving students' social self-concept was analyzed by t-independent and t-paired tests. The t-test was carried out after fulfilling the prerequisite tests, namely a normality test and homogeneity test.

4 RESULTS

4.1 Description of the developed product

The character-based reflective picture storybook developed in this study consists of several main components, namely: 1) a story matter that contains learning materials and reflection on everyday life; 2) reflective story-based values of the characters in the student's daily life; 3) reflective activity in the form of reflective questions, reflective stories, reflective journals for daily conditioning, and students' self-stories in the form of student ratings related to their social relationships. Story material in the form of material that is packaged in the form of stories and reflection of material in the daily life of students. A reflective story is a storybased on values of the characters that reflect in the daily life of students. Reflective questions are questions of preferred characters, the values found in the story, and how the student becomes a good person according to the person in the story presented. A reflective journal is structured to condition students. Self-story is a space given to students so that they can tell their experiences, and they can asses their self when dealing with others at school, home, and in the environment.

The character-based reflective picture storybook was designed and developed with Microsoft Word, Corel DRAW X7, Adobe in Design, and Adobe Photoshop programs. The developed medium was printed with Ivory paper for the cover and HVS paper for the content in A4 size (21 cm × 29.7 cm). The type of images in this media are cartoon images.

The learning materials that are packaged in the story are tailored to the theme and subtheme on the fifth grade subject matter of the Indonesian Curriculum 2013. The character values reflected are adapted to the material and core competence of social attitudes.

4.2 Trial result

4.2.1 Results of material and media expert validation

Prior to trials in preliminary field testing, main field testing, and operational field testing, the character-based reflective picture storybook was validated by a material expert and a media expert. Aspects assessed by the material expert include aspects of substance, learning criteria, and story elements. Table 1 shows the results of the material expert assessment.

Based on the assessment of the material expert, the character-based reflective picture storybook was declared feasible as a learning media that can improve the social self-concept of fifth grade students of elementary schools. The total score of all aspects is 126, with an A, and is included in the very good criteria.

Aspects assessed by the media expert include aspects of clarity of usage instructions, media readability, material systematics, image display quality, color composition, narrative quality, format and layout. The table below shows the results of the media expert's assessment of the character-based reflective picture storybook.

Table 1. Results of the material expert assessment.

No.	Aspect	Total score	Value	Criteria
1	Substance material	11	A	Very Good
2	Lesson criteria	57	A	Very Good
3	The elements of the story	58	A	Very Good
The Whole Aspect		126	A	Very Good

Based on the assessment of the media expert, the character-based reflective picture story-book wase declared feasible as a learning media that can improve the social self-concept of fifth grade students of elementary schools. Total score of the whole aspect is 120, with an A, and is included in the very good criteria.

4.2.2 *Trial result of the product*

After it had been declared feasible by the material expert and the media expert, the character based-reflective picture storybook was tested in the preliminary field testing, main field test-ing, and operational field testing to determine teacher and student responses to the devel-oped media. Media aspects that were well responded by teachers and students include aspects of material substance, learning criteria, story elements, and media appeal.

The result of the teacher response scale shows that the overall score of the media aspect is 160, with the value of A, and is included in the very good criteria. The result of the student response scale shows that the overall score of the media aspect is 72.2, with an A value, and is included in the very good criteria. The results of the teacher and student response scale indicate that the character-based reflective picture storybook was considered feasible to be tested in the operational field testing stage.

Table 2. Results of the media expert assessment.

No.	Indicator	Score total	Value	Criteria
1	Usage instructions clarity	4	A	Very Good
2	Media readability	10	B	Good
3	Material systematics	12	A	Very Good
4	Image display quality	31	A	Very Good
5	Color composition	12	A	Very Good
6	The quality of narration	16	A	Very Good
7	Formats and layouts	35	A	Very Good
Overall Aspect		120	A	Very Good

Table 3. Social self-concept of the control class and experiment class.

	Experiment class		Control class
	Pre-test	Post-test	
Approximately	174.04	190.08	186.43
Lowest Value	163	181	181
Highest Value	186	196	191

Table 4. Independent t-test results.

Group	Condition	df	Significance value	Explanation
Control	Post-test	52	0.000	Significant
Experiment	Post-test	49364		

Table 5. Paired t-test results.

Group	Condition	Significance value	Explanation
Experiment	Pre-test Post-test	0.000	Significant

The operational field testing stage was carried out to determine the effectiveness of the product in improving the social self-concept of the fifth grade students. The table below shows the results of the social self-concept scale of the experiment class and the control class.

The effectiveness of the character-basedreflective picture storybook in improving students' social self-concept was analyzed by t-independent and t-paired tests. The table below shows the independent t-test result.

Based on these calculations, the significance value is less than 0.05. Thus, it is concluded that there is a significant difference of social self-concept between the control class and the experiment class.

The table below shows the paired t-test results of the experimental self-concept social data on pre-test and post-test.

Based on these calculations, the significance value is less than 0.05. Thus, it can be concluded that there is a significant difference of social self-concept between pre-test (before using media) and pos-ttest (after using media).

The independent t-test results and paired t-test results show that the significance value is < 0.05. Therefore, the developed media storybook is effective in improving students' social self-concept.

5 DISCUSSION

The results of the validation of material experts, media experts, the teacher response scale, and the student response scale indicate that the character-based reflective picture storybook was declared feasible as a learning media that can improve the social self-concept of the fifth grade students of elementary schools. Media is declared feasible when it meets the minimum category of "good". After the analysis of the validation of the media expert and the material expert, the criteria of the character-based reflective picture storybook is "very good".

Based on validation results of the material expert, the character-based reflective picture storybook is included in the very good criteria. Besides the substance material and the lesson criteria, the material feasibility in the character-based reflective picture storybook is also reviewed from the elements of the story builder. The elements of the story builder include storylines, characterizations, themes, backgrounds, styles, and illustrations (Nurgiyantoro 2013, Tompkins & Hoskisson 1995, Lukens 1999). These elements determine the story in the character-based reflective picture storybook is interesting and meaningful for children.

Furthermore, the expert media validation obtained that the character-based reflective picture storybook is included in the very good criteria. Media feasibility in the aspect of formatting and layout is supported by some expert opinions which suggest that the format and layout is one of the components to be considered in the development of print-based media (Kemp & Dayton 1985, Nurgiyantoro 2013).

Once declared feasible as a media of learning by the material expert and the media expert, the character-based reflective picture storybook was tested at the stage of preliminary field testing and main field testing. The classroom teachers and the fifth grade students responded to the character-based reflective picture storybook media through a teacher response scale and a student response scale. The results of the teacher response scales and student responses indicated that the character-based reflective picture storybook was considered feasible to be tested at a later stage.

Feasibility of the character-based reflective picture storybook can be viewed from the media component consisting of component images and narrative text (Lukens 2003, Mitchell 2003, Huck et al. 1987). The existence of narrative text and images in developed media can overcome the limitations of elementary student in understanding abstract concepts as they are still at a concrete operational stage.

The story with the drawings will appeal to the children and is appropriate for elementary school age (Colwell 2013, Mourao 2016). Elementary students are in a concrete operational stage that requires something concrete like images to help understand the story's content. With the interesting images in a story, it will be easier to digest its meaning.

The result of operational field testing shows that the character-based reflective picture storybook is effective to improve students' social self-concept. Independent t-test results and paired t-tests show that the significance value is smaller or lower than 0.05 (0.000).

From the results above, it can be seen that the character-based reflective picture storybook can improve students' social self-concept. This is in accordance with the results of research from Richter, & Calio (2014) that by reading the story, it can improve students' self-concept. Students will be attracted by the media because the contents of the story can be applied in everyday life at home, in school, and in the community. In other words, students will understand how to deal with others at home, in school, and in society, so students have a positive social self-concept.

A character-based reflective picture storybook that can improve student's social self-concept is supported by Mitchell (2003), as picture storybooks help children to learn to understand others, to understand the relationships that occur between people and the environment, and to develop feelings. Thus, picture storybooks can help create a positive social self-concept by helping students understand relationships and how to establish good social relationships with others in their social environment.

Through the utilization of the character-based reflective picture storybook in the learning process, students' social self-concept can be developed. The character-based reflective picture storybook gives stories that contain learning materials according to the curriculum reflective of the students' daily life, and children's stories that reflect the characters in the daily life of students. Through illustrations and stories based on character values, students can obtain role models from the characters, and can reflect stories and materials into daily life (Richter & Calio 2014, Turan & Ulutas 2016, Zaky 2016).

6 CONCLUSION

The character-based reflective picture storybook developed in this research has passed the criteria of feasibility to improve students' social self concept based on the expert validation result, teacher response scale, and student response scale in the very good criteria. The character-based reflective picture storybook is effective to improve students' social self-concept. This is based on operational field testing that has significance value < 0.05, which concluded that there is a significant difference in the social self-concept of students following learning using the character-based reflective picture storybook and students who do not use the character-based reflective picture storybook. Through pictorial stories and reflection activities in the character-reflected picture storybook, students can internalize the values of characters in a more fun way, acquire knowledge in accordance with the curriculum, and add insight into how to interact with others in the surrounding environment.

REFERENCES

Bakhurst, D. & Sypnowich, C. 1995. *The social self.* London: Sage.
Berman, S. 1990. Educating for social responsibility. *Service Learning General* 48(3): 75–80.
Borg, W.R. & Gall, M.D. 1983. *Educational research.* New York: Longman.
Colwell, C. 2013. Children's storybooks in the elementary music classroom: a description of their use by orff-schulwerk teachers. *Journal Music Therapy & Special Music Education* 5(2): 174–187.
Dervent, F. 2015. The effect of reflective thinking on the teaching practices of preservice physical education teachers. *Issues in Educational Research* 25(3): 260–275.
DeWall, C.N. & Bushman, B.J. 2011. Social acceptance and rejection: the sweet and the bitter. *Current Directions in Psychological Science* 20(4): 256–260.
Fernández-Zabala, A., et al. 2016. The structure of the social self-concept (ssc) questionnaire. *Anales de Psicología* 32(1): 199–205.
Fitts, W.H. 1971. *The self-concept and self actualization.* Los Angeles, California: Western Psychological Services.

Gillespie, A. 2005. *Becoming other: from social interaction to self-reflection*. London: Information Age Publishing.

Harisson, J. & Dymoke S. 2009. *Reflective teaching and learning; a guide to professional issues for beginning secondary teachers*. London: Sage.

Huck, C.S., et al. 1987. *Children's literature in the elementary school*. New York: Holt, Rinehart, and Winston.

Kemp, J.E. & Dayton, D.K. 1985. *Planning and producing* instructional *media* (Fifth ed.). New York: Harper & Row.

Laryea, J., et al.. 2014. Influence of students self-concept on their academic performance in the elmina township. *European Journal of Research and Reflection in Educational Sciences* 2(4): 1–10.

Lukens, R.J. 1999. *A critical handbook of children's literature* (Sixth ed.). New York: Addison-Wesley.

Mansyur, et al. 2015. *Asesmen pembelajaran di sekolah: panduan* bagi *guru dan calon guru* [Assessment of Instruction in School: a guideline for Teachers and Pre-service Teachers]. Yogyakarta: Pustaka Pelajar.

Mitchell, D. 2003. *Children's literature, an invitation to the world.* Boston: Ablongman.

Mourao, S. 2016. Picture books in the primary efl classroom: authentic literature for an authentic response. *CLELE Journal* 4(1): 25–43.

Nurgiyantoro, B. 2013. *Sastra anak: pengantar pemahaman dunia anak* [Children literature: introduction to children world understanding]. Yogyakarta: Gadjah Mada University Press.

Richter, T. & Calio F. 2014. Stories can influence the self-concept. *Social Influence* 9(3): 172–188.

Santrock, J.W. 2011. *Educational psychology* (Fifth ed.). New York: The McGraw-Hill.

Semrud-Clikeman, M. 2007. *Social competence in children*. New York, PA: Springer Science & Business Media.

Shambaugh, N.N. & Magliaro, S.G. 2006. *Instructional design; a systematic approach for reflective practice*. Boston: Pearson Education.

Stump, K., et al. 2009. *Theories of social competence from the top-down to the bottom-up: a case for considering foundational human needs*. Lawrence: University of Kansas.

Tompkins, G.E. & Hoskisson, K. 1995. *Language arts: content and teaching strategies (3rd ed.)*. Englewood Cliffs, New Jersey: *Prentice*-Hall.

Turan, F. & Ulutas, I. 2016. Using storybooks as a character education tools. *Journal of Education and Practice* 7(15): 169–176.

Zaky, E.A. 2016. Once upon a time, we were all little kids too! Influence of cartoons on children's behavior; is it just a *world* of fantasy or a nightmare?. *International Journal of Science and Research* 5(5): 1296–1298.

Zuchdi, D., et al. 2010. Pengembangan model pendidikan karakter terintegrasi dalam pembelajaran bidang studi di sekolah dasar [Development of Character Education Model Integrated in Primary School Lesson]. Cakrawala Pendidikan 1(3): 1–12.

Perception of pre-service SM3T teachers of access, participation and control: A gender analysis

W. Setyaningrum, L. Nurhayati & K. Komariah
Universitas Negeri Yogyakarta, Indonesia

ABSTRACT: The aim of the study was to describe the perception of pre-service teachers in the *Sarjana Mendidik di Daerah Terdepan, Terluar, dan Tertinggal* (SM3T) program toward gender differences. This is a descriptive qualitative study involving 111 participants. The data was collected using survey and interview. Purposive random sampling was employed to select the participants. The quantitative data was collected and analyzed by using the Harvard Analytical Framework. The data from interview was analyzed by using interpretive analysis, comprising of reduction, display, conclusion drawing, and verification. The results show that based on female perception about access, there has been a tendency that access for males is different from that of females. The males have more access to school funds, sport facilities and electronic devices, while females have more access to libraries, laboratories and books. In terms of participation, males are still dominant in self-development and have a better chance to be leader. In terms of control, males are dominant in the decision-making process.

1 INTRODUCTION

Indonesia is a huge country that has a complex education problem. One crucial problem is the disparity in education quality among regions. Recently the term 3T, which stands for *Terdepan Terluar dan Tertinggal or*, has been created to represent the frontier, outermost, and disadvantaged areas in the country, spreading from the Aceh Special Region to the Papua provinces. The education quality in a number of regions categorized as 3T zones in Indonesia is relatively less advanced; therefore, the government has launched the *Sarjana Mendidik di Daerah Terdepan, Terluar, dan Tertinggal* (SM3T) program in order to fulfill the needs of teachers in those areas (Nurhayati & Septia, 2015). This program is offered to those who are willing to serve for a year in the 3T zones, and as the reward they would be given the right to join the professional training program (PPG-SM3T) for free in the following year. However, this program is very competitive.

Teacher training universities/institutes have a significant role in shaping national education. Teacher professional program is a strategy of the government to encourage teachers to develop professionalism (Marfu'ah et al, 2017). In relation to SM3T, some of these are assigned officially and legally to select and prepare the pre-service teachers who are willing to participate in the program. Having been selected, the participants should stay for a year to teach in a school in a 3T area. This program, initially managed by the Directorate General of Ministry of Education and Higher Education with Teacher Training Universities/Institutes, started in 2011 to help people in 3T zones to receive a better education. It is expected that in the future all citizens in all areas could access education easily so that they could empower themselves and improve their life quality.

SM3T is open to both male and female pre-service teachers. In fact, the number of female participants is very significant. SM3T for female pre-service teachers is an interesting but challenging program. In general, an Indonesian family will not easily let female family members go far from home, particularly to 'unknown' or unsafe zones. Therefore, those who participate must be physically and mentally ready to successfully complete the program. The challenges faced by male and female pre-service teachers in 3T zones are not the same. In real life, especially in certain cultures, males and females are treated differently. SM3T is based on egalitarian principles and is open to both genders; however, this opportunity does not result in 'egalitarian' output/outcome.

Each government program or policy must have an impact on the participating parties. Gender issues appearing due to role differences and social relationships must be identified. This is because those differences are a potential to cause discrimination between females and males in terms of experience, needs, knowledge, and attention. Therefore, a gender analysis study to identify problems faced by female pre-service teachers during the SM3T program is needed.

Gender analysis is a tool to examine differences in women's and men's lives (Kumar, 2016:1). This analysis is an important step in identifying the existing gender balance and to examine the impact of this program on women and men. Many frameworks have been developed to analyze gender differences, such as the Harvard Analytical Framework (Rao et al., 1991), the Moser Gender Planning Framework (Moser, 1993), the Gender Analysis Matrix (Parker, 1993), the Women's Empowerment Framework (Longwe, 1995), and a social relations approach (Kabeer, 2005). The data discussed in this paper was collected using a Harvard Analytical Framework. This framework has been used to reveal formal and informal discrimination and stereotyping in schools (UNGEI, 2012:7).

Using the Harvard Analytical Framework, this paper provides a perception of women participants (pre-service teachers' perception) in SM3T with regard to equality in practice; therefore, it offers a critical review for improvement of this program. There has been no previous study in relation to female perceptions toward gender differences in the SM3T program. Therefore, this study focuses on how female pre-service teachers perceive access to and control over resources, and how they perceive participation during the SM3T program. In this study, control refers to power over decision-making and resources (Hunt, 2004) in school. Access involves having equal access (Hunt, 2004) to educational resources. Meanwhile, participation level defines equality participation in decision-making processes related to policy making, planning and administration (Hunt, 2004) at school.

2 METHOD

This is a descriptive qualitative study. The subjects were 111 female pre-service teachers joining the SM3T program in 2015–2016, and who were doing their *Program Profesi Guru* (PPG) (Teacher Professional Program) in Yogyakarta State University. In general, the data was collected by using two techniques that complement one another, namely survey and interview. A questionnaire was distributed to the respondents. The respondents were randomly interviewed in order to reduce bias of the use of *purposive sampling* in this study.

The survey aimed to get data related to perception and opinion about gender issues in 3T zones; the in-depth interview was done to gain more and deeper information, especially about data that was not covered in the survey, and also to confirm the answer in the questionnaire. Semi-structured interview was employed so the researchers could develop questions based on respondents' responses.

Table 1. Example of a Harvard checklist.

Resources	Access		Control	
	Women	Men	Women	Men
Funding		√	√	
Rooms			√	√
Books	√			√
Electronic devices		√		
Sport facilities		√		√
Laboratory		√		√
Public facilities		√		√

Qualitative data gained through the questionnaire were analyzed by using descriptive statistics, based on the Harvard Analytical Framework suggested by Overholt et al. (1985), as seen in Table 1. Meanwhile, qualitative data resulting from the interviews was analyzed by using the Miles and Huberman's analysis technique (1994), comprising of data reduction, data display, conclusion drawing, and verification.

3 RESULTS AND DISCUSSION

3.1 *Results*

The data of this study focused on three main aspects of gender analysis, namely access, control and participation. Therefore, the data presentation followed these three main aspects.

3.1.1 *Access*

Access is defined as the opportunities or chances for both females and males to utilize available resources. In this research, the resources are limited to natural resources, social resources, and school facilities. Data from the survey shows that there was a tendency for access difference between women and men, as seen in Figure 1. Females tended to access or use school facilities such as libraries, laboratories and books. They accessed these to support the teaching and learning in the classroom. They mentioned that male teachers rarely used those facilities, especially libraries and books. However, some participants said that they had no opportunity to access laboratories or libraries because neither were available.

On the other side, some of them said that male teachers used school facilities and sport equipment more often than female teachers. Female teachers rarely used them, nor electronic devices such as computers, loudspeakers and tape recorders. It can be seen that there is a trend for men to engage more in sport and electronic things, rather than in more serious things such as books and laboratory equipment. From the interview, it was known that males use the resources for both school and non-school activities. Some female participants also reported that people in the area did not trust them to deal with electronic device because they are women. The society tends to believe that men would do better than women in handling electronic equipment. Nevertheless, sometimes they also could access the devices. So, although very limited, women have been given an opportunity to use the devices.

Figure 1 also shows that there was a tendency for males to have more opportunities than females to access school funds. It is clear from the chart that 70 percent of the female respondents reported that they often were not involved in the discussions of the use of school funds. Usually, it would become the authority of the headmaster or influential teacher in the school, and usually men. Some participants said they were often asked to give an opinion or view about how to use the school funds, but it was only for the sake of formality and not really for making decisions.

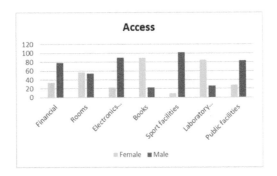

Figure 1.   Perspectives of female participants toward access to resources.

In relation to the use of rooms in the schools, it can be seen that there was no significant difference between the genders. They could use all rooms for school activities. Some participants used the rooms for housing as ordered by the head teachers. However, the interview reveals that male pre-service teachers are favored in the staffroom and they have more access to all rooms at school on an almost 24-hour basis, whereas female access is restricted during the day.

In short, there are differences in terms of access in 3T zones. Males are perceived as being a superior group. People give more opportunity and put more trust in males rather than females in relation to accessing financial and sport matters as well as electronics devices. Meanwhile, females are trusted in managing academic facilities such as libraries and laboratories. The participation aspect has to do with the equal involvement or representation in program, activity and organization. Figure 2 below shows data about female participants' perception of participation in the 3T zone. It is clear from the Figure that males are more dominant in self-development activities and school development, and they have more opportunity to be selected as teachers and headmasters.

The Figure also reveals that the most significant difference was in the opportunity to be a society leader. Males are more dominant and have more chance to be a leader. More than 80 percent of respondents said that leaders in the society were dominated by males. It was revealed that people do not put full trust in women to lead. They think that males are stronger and more persevering than women. Some people also assume that women cannot carry big burden and responsibility.

In terms of participation in public activities, the data shows that there was no significant difference. Both male and female teachers could participate equally. However, there were some traditions that could be done by males only. So, although having equal opportunity to join public activities in some zones, females only took part in 'kitchen-related' activities such as preparing meals and doing the cleaning/washing. From the interview, it was confirmed that most of the female activities were in the kitchen. All of these show that in the 3T zones some societies have distinguished roles based on gender that the SM3T participants need to adjust.

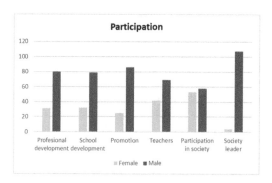

Figure 2.   Female participants' perception of participation aspect.

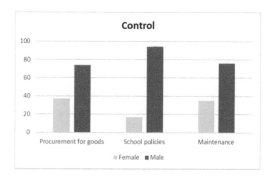

Figure 3.   Female participants' perception on control aspect.

3.2 Control

The aspect of control covers the authority to decide how to use resources and who has access to the use of the resources. Figure 3 shows that males were more dominant in the decision-making process, such as in deciding procurement for goods, the maintenance of school facilities, and school policies. There is a significant difference between males and females in deciding the school policies.

From the interview it was found that decisions about providing materials and tools depended a lot on males. The decision-making process involved many males as most of school stakeholders are male. The same things happened when the leader is a female. When a school is led by females and most of the stakeholders are female, then they would control the decision-making process. However, as the number of females that become headmaster in 3T zones is still limited, males still take the most control.

4 DISCUSSION

The data shows that there is gender issue with regard to access, participation and control. In general, males have higher status than females. Males have more access to finance, more opportunity to be leaders and to have other professional jobs, as well as more opportunity to make crucial decisions. Data also reveals that males are more trusted to be teachers. Sunnari (1997) and Lahelma (2011) pointed out similar stereotypes in the Finnish pre-service teachers.

Meanwhile, females are trusted to take care of school facilities such as libraries and laboratories. The interview revealed that people give more opportunities for females to do domestic jobs such as preparing food or decorating places.

From the evidence above, males are perceived as being a superior group. They can have access to some places without time limitation. They can perform some traditional ceremonies while females cannot. This presumption might be influenced by a patrilineal system that still practices in many parts of Indonesia (Loeb, 2009). In the patriarchy system, males tend to have the power to protect, to provide, to dominate and to control (Elam, 2010).

The result of the analysis also shows that SM3T participants felt they received a lot of benefits. Besides getting general benefits, they also enjoyed pedagogical benefits to improve their teaching professionalisms. General benefits, such as getting new experience of living in a new place and culture, interacting with new people, friends and 'families', have made them feel happy and proud. They knew Indonesia better and wanted to do more than they can to improve the nation's education. The limitations they faced did not make them surrender. They came to understand that managing the education sector in a huge and diverse country like Indonesia is never easy and needs lots of effort and strong will. In short, the SM3T program has changed some of participants' views about the relationship between culture, education and the country's advancement. They also felt that SM3T has helped in nurturing some good characteristics, such as perseverance, discipline, responsibility and caring. The love for the new place and its people has started to grow and become stronger as days pass. That is why they felt uneasy when the time came to leave. In terms of pedagogic benefits, they admitted that SM3T helped them to know the students' characters better and shape their teaching skills. They improved their communication and socializing skills, class management skills, school management knowledge and skills, and administrative skills. Therefore, the SM3T program seems to be beneficial for women; this is evidence of women power in educational achievement (Ringrose, 2007).

During the program, they faced some challenges such as regarding geographical situation, communication, water availability, tasks as teachers and socializing with society. Natural conditions in most 3T zones are quite challenging, especially for women. Potholed roads are a common sight across rural areas in Indonesia. This environment could possibly hamper the female teachers in doing their jobs. They could not drive motorbikes themselves on the poor roads. Limited road access often required them to walk through forests when going to schools or going to the shop to buy basic needs such as food or personal care products. In other areas some of them had to cross the sea or rivers by using small boats or ships, to get into cities and buy basic needs. As the ships sometimes only come once a week, they had to leave earlier and start walking to the port from 2 a.m. Those who could not swim did not feel easy while traveling on the water. Some of them also felt frightened every time they went out in the night for safety reasons. Other challenges were landslides and floods. Some of the areas are generally prone to landslides and flooding. So, when storms came they had to stay alert and awake, be sensitive to strange voices, and watch for possible land movement after the rain or storm. Before coming to the areas, they took survival workshops and training that involved the Indonesian army force. However, these are not enough. In the field, they had to adjust themselves to the weather and environments, including patterns of rain and storms, and be familiar with the land around them.

Language differences were also a problem for some of the female participants. Communication problems existed because they speak different languages. All teachers speak the national language, Bahasa Indonesia, but many people in the 3T zones do not speak it. Some of the students at schools also do not speak it. As a result, the communication process between teachers and the people around them did not run smoothly. For example, due to language constraints some Muslim teachers could not explain well about things that they cannot eat. They were afraid if the way they explained were considered impolite or rude. Due to language problems, the teaching and learning process was not really effective. It needed more time to make teachers and students understand each other's explanation. Nurhayati and Septia (2015) stated that the teachers' inability to speak the students' local languages creates problems as both the teachers and the students cannot find a good way to understand each other.

Limited access to water is another big challenge for female teachers. They need it for washing, cleaning and taking a bath. They need more water, especially when they have their period. They felt really uncomfortable when the water supply was limited or when water was not available. Some of them had to take the water in the night. To reach the water source they had to go through a forest alone. Doing such activity while having a period is perceived to be the hardest thing to do. Some other areas have no water source so the people make special places to collect the rainwater. Sometimes water is very limited, or even there is no water at all. In this situation, they must stand in a line to buy water. At worst, they must wait for four weeks for the water to come.

Their challenges are not only from the cultural and geographical situation; another one is the difference of acceptance. The majority of participants said that the society accepted them very well. Only a few said the opposite. However, there was an indication that in some parts male teachers are more respected than female teachers by the society and students. Thirty percent of participants admitted that male teachers are accepted better than female. Twenty percent of them said that appreciation for the male teachers was higher. In schools, students tend to listen to the male teachers rather than the female teachers. These results are in line with the result of the study of female teachers in Tanzania conducted by Tao (2017), which pointed out the existence of the perception that male and female teachers have different capabilities. This phenomenon is also pointed out in the previous research that males are favored when applying for teacher positions (Sunnari, 1997, 2003; Lahelma, 2006).

Almost all participants feel challenged to be a creative teacher although they teach in a place with limited facilities, infrastructures and learning resources. The hardest part is creating or finding a learning approach that best suits the students' needs and characteristics.

5 CONCLUSION

During the SM3T program, the young female teachers face different situations and meet many challenges. Difficulties and complexity of problems faced by pre-service teachers, especially female, often trigger stress, low self-esteem and burnout. However, it is also possible that these issues would cause certain positive feelings, such as confidence and obsessions toward professional development, in order to grow. From the data, it can be concluded that based on the female's point of view, gender issues related to access, participation, and control still exist in the SM3T sites. In terms of access, there is a tendency for males and females to get different levels and types of access. In terms of participation, males are still dominant in activities projecting self-development and school development. They get higher opportunities to be accepted as teachers, headmasters and society leaders. However, both males and females participate in the public activities equally. In terms of control aspects, males take more control and power in the decision-making process.

The results suggest that the SM3T program should consider gender issues, as this exists in the site. Future study in this field could consider male perspectives to provide a balanced picture on gender differences in the SM3T program. Others gender studies focusing on empowering female participants would also be beneficial as there has been limited study on this issue in Indonesia.

REFERENCES

Elam, P. (2010). *Patriarchy for dummies. The spearhead.* Retrieved from http://www.the-spearhead.com/2010/07/25/patri archy-for-dummies/.

Hunt, J. (2004). Introduction to gender analysis concepts and steps. *Development Bulletin, 64,* 100–106.

Kabeer, N. (2005). Gender equality and women's empowerment: A critical analysis of the third millennium development goal. *Gender & Development, 13*(1), 13–24. doi:10.1080/13552070512331332273.

Kumar, A. (2016). Complementing gender analysis methods. *Journal of Evidence-Informed Social Work, 13*(1), 99–110. doi:10.1080/15433714.2014.997097

Lahelma, E. (2006). Gender perspective: A challenge for schools and teacher education. In R.J. Teoksessa & H. Niemi (Eds.), *Research-Based Teacher Education in Finland – Reflections by Finnish Teacher Educators* (pp. 153–162). Turku, Finland: Finnish Educational Research Association.

Lahelma, E. (2011). Gender awareness in Finnish teacher education: An impossible mission? *Education Inquiry, 2*(2), 263–276. doi:10.3402/edui.v2i2.21979.

Loeb, E.M. (2009). Patrilineal and matrilineal organization in Sumatra: The Batak and the Minang-kabau. *American Anthropologist, 35*(1), 16–50.

Longwe, S. (1995). Gender awareness: The missing element in the third world development program. In C. March & T. Wallace (Eds.), *Changing Perception: New Writings on Gender and Development* (pp. 149–157). Oxford, England: Oxfam.

March, C., Smyth, I. & Mukhopadhyay, M. (1999). *A guide to gender-analysis frameworks.* London, UK: Oxfam Publishing.

Marfu'ah, S., Djatmiko, I.W., & Khairudin, M. (2017). Learning Goals Achievement of A Teacher in Professional Development. *Jurnal Pendidikan Teknologi dan Kejuruan. 23*(3). 295–303.

Miles, M.B. & Huberman, A.M. (1994). *Qualitative data analysis* (2nd ed.). Thousand Oaks, CA: Sage Publications.

Miske, S., Meagher M. & DeJaeghere, J. (2010). Gender mainstreaming in education at the level of field operations: The case of CARE USA's indicator framework. *Compare: A Journal of Comparative and International Education, 40*(4), 441–458.

Moser, C. (1993). *Gender planning and development: Theory, practice, and training.* London, England: Routledge. doi:10.4324/9780203411940.

Nurhayati, L & Septia, S.D. (2015). *When theories do not work well: Voices from former young teachers in remote areas.* Proceedings of ITELL Conferences.

Overholt, C., Anderson, M.B., Cloud, K.& Austin, J.E. (1985). Women in development: A Framework for project analysis. In C. Overholt, M.B. Anderson, K. Cloud & J.E. Austin (Eds.), *Gender Roles in Development Projects: A Case Book* (pp. 3–15). West Hartford, CT: Kumarian Press.

Parker, R. (1993). *Another point of view: A manual on gender analysis training for grassroots workers.* New York, NY: UNIFEM.

Rao, A., Anderson, M.B. & Overholt, C. (1991). *Gender analysis in development planning: A case book.* West Hartford, CT: Kumarian Press.

Ringrose, J. (2007). Successful girls? Complicating post-feminist, neoliberal discourses of educational achievement and gender equality. *Gender and Education, 19*(4) 471–489.

Sunnari, V. (1997). Gendered structures and processes in primary teacher education – Challenge for gender sensitive pedagogy. *Northern Gender Studies.* University of Oulu. Oulu, Finland: Oulun Yliopistopaino.

Sunnari, V. (2003). Training women for the role of the responsible other in primary teacher education in Finland. *European Journal of Teacher Education, 26*(2), 217–228.

Tao, S. (2017). Female teachers in Tanzania: An analysis of gender, poverty, and constrained capabilities. *Gender and Education, 29,* 1–17.

United Nations Girls' Education Initiative (UNGEI). (2012). *Gender analysis in education: A conceptual overview.* A working paper.

Character Education for 21st Century Global Citizens – Retnowati et al. (Eds)
© 2019 Taylor & Francis Group, London, ISBN 978-1-138-09922-7

Teacher's understanding of gaining children's consent in the research process: A phenomenology approach

L. Lutfatulatifah, V. Adriany & E. Kurniati
Universitas Pendidikan Indonesia, Indonesia

ABSTRACT: The purpose of this paper is to explore children's rights in the research process. Children's rights in the research process are very often overlooked, particularly in the Indonesian context. When conducting research, gaining consent from young children is always perceived to be highly problematic. On the one hand, children are seen as fragile individuals who need constant protection from adults. On the other hand, children are also perceived as individuals with their own voices. This paper attempts to explore the complexity of adults undertaking research with very young children. This paper adopts a phenomenological method, with five teachers selected as participants for this study. The findings suggest that all of the participants have attempted to observe the ethical issues when conducting research with young children. Although there are still problems related to an imbalance of power relations, this is because the participants play two roles, as both researcher and teacher.

1 INTRODUCTION

Conducting research with young children is always considered to be highly problematic and might lead to the exploitation of the children if the research does not take into consideration the children's voices (Alderson, 2008; Thomas & O'Kane, 1998; Warin, 2011; Woodhead & Faulkner, 2008). As suggested by the Indonesia Children's Protection Law no. 23, 2002, and the United Nations's Convention of Children's Rights, it is highly important to include children's voices in all circumstances, so that the children's best interests can be fulfilled.

One of the most important ethical aspects of doing research is how to seek children's consent. This aspect has been regulated by the various research institutions in Northern countries, such as the British Educational Research Association and the American Educational Research Association, as well as a research committee in higher education institutions (BERA, 2011; AERA, 2011). It is unfortunate that, despite the fact that laws for the protection of children exist in Indonesia, bodies that are specifically concerned with children's rights during the research process do not yet exist (Lutfatulatifah, 2017).

For the past seven years there has been pressure on Early Childhood Education (ECE) teachers to conduct research as part of their professional assessment (RoI, 2010). As a result, the amount of research on ECE has been increasing. We, as the authors, are interested to know the extent to which the teachers have understood the notion of children's consent in their research and, most importantly, whether or not they have taken the concept into consideration.

There has been much research that has attempted to explore how children's consent is gained in the research process, yet most of the research is based on the authors' ethnographic approach. Previous papers show a reflective approach that illuminates the researchers' experiences (Englander, 2012). Papers that attempt to explore other researchers' understanding of children's consent remain limited. Thus, this paper attempts to fill the gap in the previous literature on children's consent. It is expected that the findings of this study will deepen our understanding on children's rights. It is hoped that by observing children's rights, the children's well-being will be better appreciated in the context of research.

2 UNDERSTANDING CHILDREN'S RIGHTS

There are at least two major perspectives with regards to understanding children's rights. The first perspective is the protectionist approach. This approach goes back to the 15th century, where children were constructed as evil that carried original sin (Adriany, 2013a). The protectionist approach was also influenced by Locke and Rousseau, who viewed children as a blank piece of paper (Gianoutsos, 2006). From this perspective, children are seen as individuals who are in need of constant guidance from adults (Adriany, 2013a; Lundahl et al., 2006; Skolnick, 1975).

This paradigm is likely to be challenged by the liberalist approach, which sees children as rational beings who can decide for themselves (O'Reilly & Tushman, 2013). This perspective believes that children are capable of voicing their opinion.

Both the protectionist and liberalist approach influence the way that children are viewed in the research process. In one sense, children are seen as helpless individuals who need to be protected, but in another sense, they are perceived as individuals who can make decisions (Morrow & Richards, 1996; Unicef, 2012). The tension between these two approaches makes undertaking research with young children more problematic.

3 RESEARCH METHOD

The method used in this paper is the phenomenology approach, which aims to explore the participants' deep understanding and experiences in conducting research with young children (Giorgi, 1994; Aspers, 2004). Hence, the purpose of this research is to illuminate the understanding of early childhood education teachers regarding children's consent. The participants for this research are five teachers who have done research that involves young children. Since this is a qualitative research, this paper does not aim to seek for generalization (Silverman & Marvasti, 2008). Rather, this paper is interested in exploring the complexity of the narrative. Information was collected using non-formal interviews. The interviews were conducted by the first author. Each participant was interviewed twice or three times and each interview lasted for approximately one to two hours. The first, second and third authors also analyzed the diaries written by the participants. This was aimed at a triangulation process. We wanted to ensure that the information gathered from the participants was trustworthy and reflected their practices. All of the information was then analyzed by Interpretative Phenomenological Analysis (IPA). Three themes emerged during the analysis. They are the construction of childhood, parental consent and power relations between researchers and child participants.

4 FINDINGS

4.1 *Construction of childhood*

Gaining consent from a child participant is indeed very important in any research process. It is interesting that almost all of the participants in this study perceive that it is not important. They often ignore the process and, thus, the research is progressed without consent from the children.

> *I do not ask for consent from the children. I often told them that I am going to play with them, that I will be here with them. It might appear that I don't want to explain, but...this is my custom. I am used to it (saying I am playing with the children)*

(Interview with Aica)

The fact that Aica replaces the word "research" with "playing" is interesting. On the one hand, this could be seen as the researcher's attempt to comprehend the children's level of understanding. On the other hand, this could also be seen as a form of dishonesty. Aica is, in fact, not the only researcher who does not use the word research when explaining what they are doing to the children. The other participants of this study also use various words, such as "teaching".

From the perspective of children's rights, it is very problematic when a researcher does not disclose that the main reason she is in the school is for research purposes (Dockett et al.. 2009). It appears that the researchers view the children as individuals who are not yet able to understand complex issues such as the research process. This perspective is likely to be very much influenced by the protectionist paradigm, in which they see young children as naive and innocent individuals. This perspective is likely to be challenged. A sociological approach to childhood, for instance, has demonstrated how children can not only give or refuse their consent during the research process, but they can also understand the whole process (Bell, 2008). Research conducted by Warin (2011) and Adriany (2013b), for instance, illuminates children's ability to be involved in the research process. Their studies also show how children can negotiate their consent during the research process.

4.2 *Parental consent*

The second finding is related to how the teachers gain consent from the children's parents. The analysis yields that they also skip getting parental consent. They assume that when they are doing research with young children, it is sufficient to get consent only from the school. However, in fact, consent from the school does not automatically equal parental consent, and most definitely is not the same as getting the children's consent. This could be due to the fact that, in most schools, parents are often seen as having inadequate and deficient parenting skills (Kim, 2009). Parents are seen as less knowledgeable than teachers when it comes to understanding children's behavior.

The second reason why the teachers do not seek parental consent might be because the researchers feel that the parents do not mind. One researcher, Yani, for example, claims that the parents in fact feel proud knowing that their children have been selected as a research participant. She argues:

> *I never ask for consent from parents. I am sure parents are excited knowing that their children's names would appear in a thesis. They will be in fact happier if their children's photographs will be there too...It means their children exist.*

> (Interview with Yani)

Yani's belief is not without any reason. She once received a complaint from parents because she selected one child as her research participant and left out another child. As she says:

> *It's actually quite strange. Parents never complain when their children are selected as research participants. They often ask question. "why do you choose that child and not my child?". So instead of asking why do I involve their children, they are asking me why do I not include their children in a research process.*

> (Interview with Yani)

This finding seems to suggest cultural differences in the understanding of the notion of consent. In her research, Adriany (2013b) also shows how parents are very excited to have their children chosen as a participant for her research. Instead of feeling anxious, they see this as a form of privilege. However, despite the cultural differences in acknowledging the notion of consent, a researcher nevertheless still needs to obtain consent from the parents. Without doubt, a researcher needs to understand that seeking consent is part of their responsibilities, while at the same time it is also a right of the participants. (Wiles et al., 2005; Thompson, 1990).

4.3 *Power relations*

The third finding of this paper illuminates the power relations between the teachers, as researchers, and the children. It has been acknowledged that there exists an imbalance in the power relations between the adult researchers and the child participants (Meloni et al., 2015). The fact that consent is not formally gained from both the parents and the children perpetuates the imbalance of power relations in the research process.

The imbalance of power relations is also strengthened by the double roles played by the teachers. They are both a researcher and a teacher. In almost every society, teachers are viewed as having more power than their pupils. Yani, one of the teachers, recognized this, as she admits in her diary that sometimes she would force the children to participate in her research, something that later on she regrets. Nevertheless, Yani's diary seems to suggest that there is a lack of understanding from the teachers that, as a researcher, they should not impose their power as adults and teachers on their child participants. A researcher should, in fact, be able to negotiate their power and ensure that the whole research process is in the best interests of the children.

5   CONCLUSION

The findings of this paper suggest that apparently there is a lack of understanding from the teachers about children's rights in the research process. The teacher does not realize that, by overlooking the issue of consent from the children and the parents, it might jeopardize the trustworthiness of the research (Heath et al., 2007). More importantly, by taking children's and parental consent for granted, they might violate the principle of children's rights promoted by the United Nations (UNGA, 1989).

The teachers' lack of understanding highlights potential problems. Nowadays in Indonesia, it has become compulsory for teachers to conduct research and to write reports as part of their professional assessment. It is unfortunate that teachers are not equipped with sufficient skills on how to conduct research with young children.

This finding, therefore, is a call for teacher training institutes to include discussions on children's rights in the research process in their curriculum. The research institutions also need to develop a code of ethics that can guide researchers when doing research with young children.

REFERENCES

Adriany, V. (2013a). Hak anak dalam konteks penelitian [Child rights in research context]. In T. Hartati, M. Agustin, & M. Somantri (Eds.), *Menyongsong generasi emas: Conference Pendidikan Anak Usia Dini Dan Pendidikan Dasar Sps Upi*. Bandung: Program Studi Pendidikan Dasar.

Adriany, V. (2013b). *Gendered power relations within child-centered discourse: An ethnographic study in a kindergarten in Bandung, Indonesia* (PhD Thesis). Department of Educational Research, Lancaster University, UK.

Alderson, P. (2008). *Young children's rights* (2nd ed.). London, UK: Jessica Kingsley Publishers.

American Educational Research Association. (2011). Code of ethics American Eductional Research Association. *Educational Researcher, 40*(3), 145–156.

Aspers, P. (2004). Empirical phenomenology: An approach for qualitative research, Paper in *Social Research Methods, Qualitative series 9*. London, UK: LSE.

Bell, N. (2008). Ethics in child research: rights, reason and responsibilities. *Children's Geographies*, 6(1), 7–20.

British Educational Research Association. (2011). *Ethical guidelines for educational research*. UK.

Denzin, N.K. & Lincoln, Y.S. (2005). *Qualitative research* (3rd ed.). London, UK: Sage.

Dockett, S., Einarsdottir, J., & Perry, B. (2009). Researching with children: Ethical tensions. *Journal of Early Childhood Research, 7*(3), 283–298.

Englander, M. (2012). The interview: Data collection in descriptive phenomenological human scientific research. *Journal of Phenomenological Psychology, 43*, 13–35.

Franklin, B. (2002). *The new handbook of children's rights: Comparative policy and practice*. New York, NY: Routledge.

Gianoutsos, J. (2006). Locke and Rousseau: early childhood education. The *Undergraduate Journal of Baylor University*, 4(1).

Giorgi, A. (1994). The theory, practice, and evaluation of the phenomenological method as a qualitative research procedure. *Journal of Phenomenological Psychology, 28*(2), 235–260.

Heath, S., Charles, V., Crow, G., & Wiles, R. (2007). Informed consent, gatekeepers and go-betweens: Negotiating consent in child and youth-orientated institutions. *British Educational Research Journal, 33*(3), 403–417.

Kim, Y. (2009). Minority parental involvement and school barriers: Moving the focus away from deficiencies of parents. *Educational Research Review*, 4(2), 80–102.

Lundahl, B.W., Nimer, J., & Parsons, B. (2006). Preventing child abuse: A meta-analysis of parent training programs. *Research on Social Work Practice, 16*(3), 251–262.

Lutfatulatifah, L. (2017). The concept of protection and the rights of the children involved in research. *Advance in Social Science, Education and Humanities Research (ASSEHR), 3rd International Conference on Early Childhood Education (ICECE) 58* (pp. 288–291).

Meloni, F., Vanthuyne, K., & Rousseau, C. (2015). Towards a relational ethics : Rethinking ethics, agency and dependency in research with children and youth. *Anthropological Theory, 15*(1), 106–123.

Morrow, V. & Richards, M. (1996). The ethics of social research with children: An overview. *Children & Society, 10*(2), 90–105.

O'Reilly, C.A., & Tushman, M.L. (2013). Organizational ambidexterity: Past, present, and future. *Academy of Management Perspectives*, 27, 324–338.

RoI. (2010). Peraturan Manteri Pendidikan Nasional Nomor 35 Tahun 2010 tentang Petunjuk Teknis Pelaksanaan Jabatan Fungsional Guru dan Angka Kreditrnya [Regulation of The Minister of National Education Number 35 Year 2010 on Teacher Gudeline]. Jakarta, Indonesia: Ministry of Education, Republic of Indonesia.

Silverman, D. & Marvasti, A. (2008). *Doing qualitative research*. London, UK: Sage Publications.

Skolnick, A. (1975). The limits of childhood: Conceptions of child development and social context. *Law and Contemporary Problems, 39*(3), 38–77.

Thomas, N. & O'Kane, C. (1998). The ethics of participatory research with children. *Children & Society, 12*(5), 336–348.

Thompson, R.A. (1990). Review vulnerability in research: A developmental perspective on research risk. *Child Development, 61*(1), 1–16.

Unicef. (2012). *What are child rights ? An introduction to the convention on the rights of the child.* UNICEF.

United Nations General Assembly. (1989). *Convention on the rights of the child.* Adopted by General Assembly of The United Nations, 20 November 1989 (pp. 1–15).

Warin, J. (2011). Ethical mindfulness and reflexivity: Managing a research relationship with children and young people in a 14-year qualitative longitudinal research (QLR) study. *Qualitative Inquiry, 17*(9), 805–814.

Wiles, R., Heath, S., Crow, G., & Charles, V. (2005). *Informed consent in social research : A literature review*. Southampton, UK: ESRC National Centre for Research Methods.

Woodhead, M. & Faulkner, D. (2008). Subjects, objects or participants? Dilemmas of psychological research with children. In: Christiansen, Pia and James, Allison (*Ed.*). *Research With Children: Perspectives and Practices*. (pp. 10–87) London, UK: Falmer Press.

The improvement of independent learning through the use of a virtual laboratory in chemistry hybrid learning

J. Ikhsan, K.H. Sugiyarto & D.T. Kurnia
Universitas Negeri Yogyakarta, Indonesia

ABSTRACT: Information and Communication Technology (ICT) can support the implementation of a virtual laboratory in chemistry hybrid learning (ViCH-Lab). The use of a ViCH-Lab on the topic of the hydrolysis of salt was studied, by which the improvement of the students' independent learning was measured and compared to that of students who did not use it. The virtual laboratory was developed in HTML format with animations. The ViCH-Lab mainly contained menus of learning materials, laboratory work topics, and evaluation. The method of research was a mixed method with embedded experimental design. The total number of samples were 68 students from class XI, consisting of 34 students in an experiment group and 34 students in a control group. The students in the experiment group learned chemistry through both regularly scheduled face-to-face sessions and teacher supervised online practices through the ViCH-Lab on a website, while those in the control group were only taught through regularly scheduled face-to-face sessions without the online component. The data regarding the improvement in the students' independence was collected at the beginning and at the end of the learning activities using a questionnaire, and during the process of learning using an observation checklist. The data was analyzed statistically by an independent sample t-test and descriptive analysis. The results showed that the improvement in students' independence in the experiment group was significantly higher than that of the control group.

1 INTRODUCTION

A teacher plays an important role in education. Beside teaching, they are also educators (Yuswono et al, 2014). They are required to create creative and innovative learning processes, facilitating students to be as active as the curriculum requires. Many learning processes were dominantly controlled by teachers in one-directional learning or teacher-centered learning. The observation of learning processes in some schools in Yogyakarta, Indonesia showed that some learning activities were still done through teacher-centered learning. Teacher-centered learning was closely related to the culture of Indonesia. Indonesian teachers, whose youngsters usually tend to follow older people's directions, should ensure that students are challenged and active in the learning process. According to Nugroho (2012), students should be active in learning activities, while teachers should be active in preparing learning packs, motivating students, and making the learning process more effective.

Chemistry learning materials are wide-ranging and comprehensive, consisting of macroscopic and microscopic concepts, which challenge students to find an appropriate learning strategy. Since chemistry is learned as a separate subject for the first time at senior high school level, many students find chemistry to be a difficult lesson as it contains intangible concepts. In fact, the concept must be built and constructed by students. For that reason, learning chemistry requires visualization. This visualization can be achieved through working in the chemistry laboratory. In the laboratory, students will be shown the phenomena of a chemical concept through the experiments that are conducted by the students themselves, from which they are expected to be able to build the concept on their own. One of the topics in chemistry that needs practice in the laboratory is salt hydrolysis. The topic is scheduled in the 2nd semester of Year XI in senior high schools. Therefore, to learn about salt hydrolysis, practices in the lab or integration in the theory should be facilitated.

Based on the observations in a senior high school in Yogyakarta, Indonesia, students had a poor understanding of chemistry concepts. The students' learning difficulties might occur because there are too many concepts of chemistry, and they are intangible and systemically interconnected. It might also be due to having a limited time for learning and the poor preparation of the students. Other problems come from the fact that most laboratory activities might not completely support the theory studied in the classroom due to a lack of chemicals. As reported by Yennita et al. (2012), some problems could be faced when conducting practical chemistry lessons in Indonesian schools, such as the time available to provide laboratory work activities for students, the lack of equipment, and the ability of the teachers to design and develop inquiry learning for the needs of the students.

The development of digital technology is very rapid nowadays, which significantly affects most aspects of life, including education. Education in the digital age requires the integration of ICT into all subjects. To update the educational challenges in this digital age, teachers and students in the 21st century must be able to communicate and keep up with the times, including adjusting to the development of technology. The development of ICT is also supported by the progress of the internet network, which has been widely spread to most areas in Indonesia, even the remote areas. From the internet and ICT, a lot of information can be gained easily at any time and anywhere. The development of the internet network can result in a dramatic improvement in the sources of learning, including for chemistry. Therefore, students need to learn how to use technology in order to get the maximum impact from the learning process.

Based on the results of the study of technological advances and observations from the participating senior high school, SMA Negeri 4 Yogyakarta, it was necessary to develop a model of learning and the types of media that support internet-based learning in the classroom and in the laboratory, as well as to foster the learning interests that motivate students to improve their independence both inside and outside the classroom. Kurniawan and Zulkaida (2013) stated that interests had an effect on independent learning, so that students who have a high interest in learning further explore lessons independently. Based on the observation, SMA Negeri 4 Yoyakarta is a school that provides Wi-Fi connection facilities that are supposed to be used freely by the students. In addition, most students from the science group of Year XI in the SMA Negeri 4 Yogyakarta have their own laptops and smartphones. Wi-Fi, laptops, and smartphone facilities were the tools that would be useful to support another model of learning called hybrid learning.

The hybrid learning model is one of the learning models supported by technological development, by integrating innovation and technological advancement through online learning systems, providing interaction and requiring the participation of traditional learning models (Thorne, 2003). Husamah (2014) stated that the hybrid learning model can reduce face-to-face activities but it does not eliminate them, thus enabling students to learn by online methods. Through a hybrid learning based model, students are expected to be able to learn independently and sustainably. As a consequence, learning is going to be effective, efficient, and interesting. Computer technology for education can be utilized for learning media, such as Microsoft PowerPoint, Adobe Flash, Digital Comics, Construct 2, and other programs that can be used both in online and offline mode.

Kemp and Dayton stated that the use of instructional media could make learning interesting, interactive, effective, efficient, improved, and flexible (Susilana & Riyana, 2008). Moreover, the use of digital media was flexible, being used whenever and wherever necessary. One of the digital learning media that can be developed is a virtual chemistry laboratory (ViCH-Lab). Ikhsan and Hadi (2015) stated that the development of virtual lab media could be achieved using HTML5 programming language, which has many advantages. This study utilizes HTML5-based learning media, which was packaged in the form of a ViCH-Lab media and whose implementation was integrated in chemistry hybrid learning. The use of the media was integrated into an interesting learning model, which was one of the efforts used to improve the students' interests independently. Rahmawati (2015) stated that computer-based learning media supported the independent learning of students.

This research investigated the effect of the use of a ViCH-Lab on the students' independent learning of the topic of hydrolysis of salt, from which the improvement in independent learning of those students who used the ViCH-Lab in chemistry hybrid learning was measured and compared to that of the students who did not use the ViCH-Lab.

2 LITERATURE REVIEW

2.1 Virtual laboratory

A laboratory, as defined by Hornby (2010), is a room or building used for scientific research, experimentation, and testing. Woolnough and Allsop (1985) state that the activities of the laboratory are an integral part of learning activities, serving as a vehicle to raise the motivation to learn, develop basic skills to experiment, learn scientific approaches, and to support the subject matter. Dkeindek et al. (2012) state that the activity of the laboratory is a learning environment that establishes the concept of science to learners. Laboratory activities can be summarized as practical activities by learners that can be done either directly in the classroom or field, or indirectly by using a virtual laboratory simulation using a computer device.

The concept of the virtual laboratory, according to Harms (2000), can be divided into two main concepts: (1) trials were replaced with computer models, in the form of a simulation that represented the real laboratory experiments as closely as possible, in the form of a so-called virtual laboratory; (2) it can be called a virtual laboratory experiment when the experiment is controlled through a computer, which is connected to the actual laboratory equipment through a network called a remote lab. A virtual laboratory is a supporting motivational factor for reproducing the experiment and developing the skills of the learners in the trials (Dobrzanski & Honysz, 2011; Tatli & Ayas, 2012). A virtual laboratory can be summarized as a series of computer programs that can be used to visualize microscopic and macroscopic phenomena in a certain scale so that learners can observe the salt hydrolysis phenomena on the material clearly and easily.

2.2 Hypertext Markup Language version 5 (HTML5)

Hypertext Markup Language Version (HTML) is the most common language used in web technology today. Microsoft and Google are using HTML on websites (Yibin, 2006). The latest version of HTML is HTML5, and this will become the new standard of design applications, though not all browsers should use it.

HTML5 is a programming language that structures the contents of The World Wide Web, and is a major technology on the internet (Zamroni et al., 2013). HTML5 will be the trend of the future for internet technologies because it is so enriched with features that it will surely be seen as standard web-based information media development.

2.3 Hybrid learning

Hybrid learning is learning that integrates innovation and technological advancement through online learning systems with the interaction and participation of traditional learning (Thorne, 2003). Hendrayati and Pamungkas (2007) state that the program hybrids produced today are the incorporation of one or more of the following dimensions: (1) Learning Face-to-Face (in-person learning) is used for learning activities in the classroom, in the laboratory, or for mentoring. (2) Synchronous Virtual Collaboration is a teaching format that is collaborative, involving interactions between educators and learners delivered at the same time. (3) Asynchronous Virtual Collaboration is a teaching format that is collaborative, involving interactions between educators and learners delivered at different times. (4) Self-Paced Asynchronous is a model of independent learning where people learn in their own time. This research uses a learning system that combines the dimensions of learning face-to-face with self-paced asynchronous learning.

The stages of hybrid learning, as shown by Lalima and Dangwal (2017), include: (1) Face-to-face teaching; (2) Student interaction with the course; (3) Peer group interaction; (4) Group discussion and exchange of ideas; (5) Accessing e-library; (6) Virtual classrooms; (7) Online assessment; (8) e-tuitions; (9) Accessing and maintaining tasks; (10) Webinars (application of hybrid learning); (11) Viewing expert lectures on YouTube; (12) Online learning through videos and audios; (13) Virtual laboratories.

2.4 *Independent learners*

Independent studying, according to the explanation of Weinstein et al. (2011), is the ability of learners to control their cognitive processes through planning, setting goals, and monitoring and evaluating their understanding of the subject matter. This is consistent with the explanation of Schunk (2012), that independent learning involves the potential for learners to possess self-control, self-observation and self-evaluation, so as to create individuals who understand their capabilities. White and Harbaugh (2010) also explain that independent learning is a form of consciousness and is the ability of learners to receive information, combine information, and be able to link information together.

Studies that have been done by Bernacki et al. (2011) found that applying a technology-based learning atmosphere can train a learner's ability to regulate their own learning. The application of a computer-based learning log is closely linked with the increase of independent learning. This is also expressed by Winters et al. (2008), who found that, through computer-based learning, learners are given the opportunity to organize, plan, and control their learning activities effectively with a high level of flexibility. This is supported by Greene and Azevedo (2009), who stated that independent learning in the monitoring aspects can be trained through the use of technology-based learning. Based on the above explanation, it can be concluded that independent learning is an attitude that leads to the activity and initiative of learners, enabling them to search for information to study, analyze and evaluate without being dependent on others. They are able to take overall responsibility in the learning process, mainly according to the competence and knowledge they possess in chemical materials salt hydrolysis. Active learners can be helped by the use of technology-based learning as a medium of learning. The aspects of independent learning used in this study are aspects that are dependent on others: responsibility, initiative, discipline, and confidence.

3 RESEARCH METHOD

This research was a mixed method research including both qualitative and quantitative research with embedded design (Creswell, 2012). The purpose of embedded design was to collect the quantitative and qualitative data simultaneously and sequentially. The reason for collecting these two types of data was to supply and support the primary data. The supporting data could be either quantitative or qualitative (Creswell, 2012). The qualitative data was obtained from observations and the quantitative data was collected from the questionnaire on independent learning. The questionnaire on independent learning used in this study was adapted from Hidayati and Listyani (2010), whose outline is shown in Table 1.

This research was conducted in the SMA Negeri 4 Yogyakarta, Indonesia. The population of this research were students in the science group in Year XI of SMA Negeri 4 Yogyakarta during the academic year 2016/2017. The samples were students from class XI-Sci-4 as the control group and class XI-Sci-5 as the experimental group, chosen randomly. Learning in both the control and experimental groups was carried out on the same topic: laboratory works and evaluations. Learning in the control group was face-to-face using the Direct Instruction (DI) method, where the delivery of materials was assisted by PowerPoint media and through laboratory work. While the learning in the experimental group was the same as that in the control group, the delivery system used was the hybrid learning method, in which face-to-face learning using Direct Instruction (DI) was combined with e-learning in a website-based learning system assisted by HTML5-based ViCH-Lab (Virtual Chemistry Laboratory) media.

Table 1. Content outline of questionnaire on independent learning.

No	Aspect	Number of statements
1	Confidence	4
2	Initiative	5
3	Motivation	4
4	Responsibility	5
Total		18

Table 2. Research design.

		Treatment of-				
Class	Before treatment	1	2	3	4	After treatment
Control	O11	P11	P12	P13	P14	O12
Experiment	O21	P21	P22	P23	P24	O22

where
O11: Independence of control class learning before learning process
O12: Independence of control class learning after learning process
O21: Independence of experimental class learning before learning process
O22: Independence of experimental class learning after learning process
P1i: Independent learning by using the face-to-face learning method, using Direct Instruction (DI), PowerPoint media, and laboratory work
P2i: Independent learning by using the hybrid learning method, which is the combination of face-to-face learning with Direct Instruction (DI) and website-based learning using ViCH-Lab media.

Independent learning was measured using a questionnaire with Likert 1–5 ratings and an observation sheet.

The measurement of independent learning using the questionnaire was carried out at both the beginning and the end of the learning process in order to collect quantitative data. The observations on the independent learning of the students were completed during each face-to-face learning activity by using the observation sheet to collect the qualitative data. The design of the data collection is presented in Table 2.

The data taken from each group before and after treatment, in both the control and experimental groups, was analyzed using normalized gain to measure the improvement of the students' independent learning. The normality test of the gain was undertaken by using the following equation (Hake, 1998).

$$g = \frac{S_f - S_i}{Sm - S_i}$$

where
g = gain
Sf = post-test score
Si = pre-test score
Sm = maximum score

The difference in the improvement of the students' independent learning between the control group and the experimental group was analyzed using an independent sample t-test with the requirement of data normally distributed and homogeneous. The statistical test was performed by using the SPSS V24.0 computer program. The hypotheses were as follows:

H_o: There is no significant difference in the improvement in independent learning between the students from the control group and those from the experimental group.

H_a: There is a significant difference in the improvement in independent learning between the students from the control group and those from the experimental group.

The improvement in independent learning during the learning process from 4 times face-to-face teaching method was collected by using an observation sheet, and the data was then analyzed by Least Significance Difference (LSD) in order to determine whether the improvement in independent learning in each session of regular face-to-face learning was significant or not.

4 RESULTS AND DISCUSSION

ViCH-Lab media, based on HTML5 media integrated in hybrid learning, included: (1) Competence; (2) Materials; (3) Laboratory work; and (4) Evaluation. The learning materials presented in the learning media were on the topic of hydrolysis for a chemistry lesson in the 2nd semester of Year XI. The ViCH-Lab interface can be seen in Figure 1.

In this research, HTML5-based ViCH-Lab media was used as a chemistry learning resource, which was used in chemistry hybrid learning. The purpose of the implementation of hybrid learning was to get a better learning model by combining the face-to-face learning model and the online learning model. The hybrid learning in this research was packed in a LMS (Learning Management System). Online learning was done outside the classroom by providing online materials that could be accessed through LMS, while face-to-face learning could be used as a question and answer process by teachers and students for certain topics

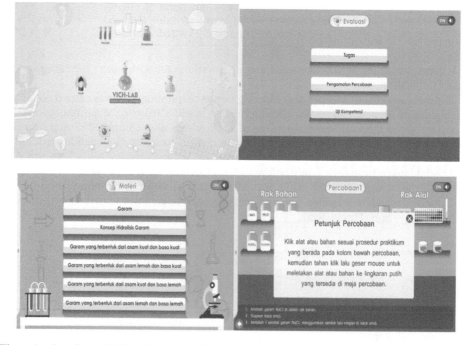

Figure 1.   Interface of HTML5-based ViCH-Lab media on hydrolysis.

that were not clearly understood during the online learning. One kind of LMS that was used in this study was Moodle. The advantages of LMS in hybrid learning were the convenience for students in accessing online learning and the support for students when conducting the online model of hybrid learning. The learning processes were carried out in five sets of regular two hour face-to-face learning sessions in the control group, and its equivalent for the hybrid learning model in the experimental group.

Improvements in independent learning were noted from the data collected by the questionnaire at the beginning and at the end of the learning process. The improvement was expressed by the score of normalized gain. The analysis was conducted with an SPSS V24.0 program at 95% of confidence level. The average gain in independent learning in both classes can be seen in Table 3.

From Table 3 it can be seen that the average gain in the experimental group was higher than that in the control group. Therefore, it can be concluded that the improvement in independent learning of the students in the experimental group was better than that of those in the control group. The improvement in independent learning from time to time during the learning process was also observed during learning activities. The results of the observation can be seen in Table 4.

Table 4 shows that there was an increase in independent learning for every aspect measured in both the control and the experimental groups. The data analysis to determine the significance of improvement diversity through the observation result in every meetings for controled and experimented group using LSD test, while the independent sample t-test was done to study the significance of difference in the gain score in independent learning in the control group and the experimental groups. The LSD test and independent sample t-test were done with data that was normally distributed and homogenous. Normality test results can be seen in Table 5.

Table 3. Data of students' independent learning.

No	Group	Number of students	Score average Before learning	Score average After learning	Gain average	Gain category
1	Control	34	67.353	74.118	0.299	Low
2	Experiment	34	68.301	79.477	0.515	Medium

Table 4. The result of independent learning observation.

No	Aspect	Control				Experiment			
		I	II	III	IV	I	II	III	IV
1	Confidence	230	238	252	259	232	245	266	275
2	Initiative	333	348	363	375	342	373	386	398
3	Motivation	228	240	240	245	229	239	257	280
4	Responsibility	246	248	253	257	248	257	263	278

Table 5. Normality test results.

Variable	Group	Kolmogorov-Smirnov Statistic	df	Sig.
Learning independence	Control	0.104	34	0.200
	Experiment	0.110	34	0.200

Table 6. Test results of independent samples t-test.

		t	df	Sig. (2-tailed)	Mean difference	95% Confidence interval of the difference	
						Lower	Upper
Independent learning	Equal variances assumed	5.651	66	0.000	0.224	0.357	0.453
	Equal variances not assumed	5.651	63.301	0.000	0.224	0.145	0.304

Table 7. Test results of LSD (Least Significance Difference).

	(I) Session	(J) Session	Mean difference (I − J)	Std. error	Sig.
Control	1	2	−4.056	2.418	0.103
	2	3	−3.944	2.418	0.113
	3	4	−2.963	2.418	0.229
Experiment	1	2	−6.963	2.873	0.021
	2	3	−6.426	2.873	0.032
	3	4	−6.611	2.873	0.028

The gain data for the control and the experimental groups were normally distributed; hence a t-test can be done to the gain data in the control and the experimental groups. The results of the t-test for both groups can be seen in Table 6.

The analysis shows that $t_{(66)} = 5.651$, and Sig. p was $0.000 < 0.05$, therefore H_0 is rejected. This result suggested that there was a significant difference in the increase in independent learning between the students in the control class and those in the experimental class. The significant differences in increasing independence can also be seen from the results of the LSD test, taken from the observations made during the learning activities. The LSD test results can be seen in Table 7.

Based on the results of the LSD analysis of the observation of independent learning in the control class, the probability (Sig.) for sessions 1–2, 2–3, and 3–4 are 0.103, 0.113, and 0.229, respectively. The result of the probability (Sig.) > 0.05, hence H_0 is accepted, which means that there is no increase in independent learning during sessions 1–2, 2–3, and 3–4. Meanwhile, the probability (Sig.) results of the LSD test of the observation of independent learning in the experimental class for sessions 1–2, 2–3, and 3–4 respectively are 0.021, 0.032, and 0.028. The probability result (Sig.) < 0.05, therefore H_0 is rejected, which means there is an increase in the students' independent learning during sessions 1–2, 2–3, and 3–4.

Based on the analysis, it can be concluded that ViCH-Lab media based on HTML5, when integrated in hybrid learning, affected the independent learning on the topic of salt hydrolysis of the students who were in Year XI during the 2nd semester. In line with the findings of this research, Bernacki et al. (2011) stated that a technology-based learning atmosphere led to learners being well trained in the ability to manage their own learning. The application of computer-based learning, which is closely related to the improvement in the students' independence, was also expressed by Winters et al. (2008), who reported that through computer-based learning, students were given the opportunity to organize, to plan and to control their learning activities effectively with high levels of flexibility. This was supported by Greene and Azevedo (2009), who stated that monitoring aspects in independent learning can be trained through technology-based learning. Tatli and Ayas (2012) stated that a virtual laboratory was able to make learning more meaningful through virtual experience, thereby enhancing the understanding of concepts, principles, and processes. Through a virtual laboratory, students had the opportunity to repeat the experiments in order to understand the topic independently. Tuysuz (2010) also reported that virtual laboratory applications brought positive effects on the attitudes and achievements of students when compared to traditional teaching methods. The positive effects of a virtual laboratory are: (1) the existence of new learning materials in the digital format of ViCH-Lab media, therefore students feel more excited and interested in attending the learning process; (2) the students' learning becomes fun and relaxing through animation or simulation and they find it easier to understand the concept of learning. The media would be helpful and enable students to decrease the difficulties found due to the monotone modes of learning, which were reading or listening to the explanation of teachers; (3) it lengthened learning duration by being able to utilize cyberspace technology at any time; and (4) through ViCH-Lab, students were expected to be independent learners. In addition, HTML5 can be used individually and provides students with more opportunities to be independent learners outside of the classroom.

5 CONCLUSION AND RECOMMENDATIONS

The HTML5 based ViCH-Lab media that was integrated in the hybrid learning process has been successfully applied in chemistry lessons on the topic of salt hydrolysis in the science class of Year XI at the SMA Negeri 4 Yogyakarta during the academic year 2016/2017. There was a significant difference in the improvement in independent learning in the students from the experimental class, who learned the topic of salt hydrolysis in chemistry lessons through the learning model of hybrid learning, which is a combination of regularly scheduled face-to-face sessions and teacher supervised online practices through ViCH-Lab online, and those in the control group, who only learned the topic through regularly scheduled face-to-face sessions without the online components.

The best practices of the hybrid learning model, which can improve students' performance, would be an interesting investigation for future researches in other topics of chemistry.

REFERENCES

Bernacki, M.L., Aguilar, A., & Byrnes, J. (2011). *Self-regulated learning and technology-enhanced learning environment: An opportunity-propensity analysis*. New York, NY: IGI Global.

Creswell, J.C. (2012). *Education research, planning, conducting and evaluating quantitative and qualitative research* (4th ed.). Boston, MA: Pearson.

Dkeidek, I., Mamlok-Naaman, R., & Hofstein, A. (2012). Assessment of the laboratory learning environment in an inquiry-oriented chemistry laboratory in Arab and Jewish high schools in Israel. *Learning Environment Research, 15*, 141–269.

Dobrzański, L. A. & Honysz, R. (2011). Virtual examinations of alloying elements influence on alloy structural steels mechanical properties. *Journal of Achievements in Mechanical and Materials Engineering, 49*(2), 251–258.

Greene, J.A. & Azevedo, R. (2009). A macro-level analysis of SLR processes and their relations to the acquisition of a sophisticated mental model of a complex system. *Contemporery Educational Psychology, 34*(1), 18–29.

Hake, R.R. (1998). Interactive-engagement versus traditional methods: A six-thousand-student survey of mechanics test data for introductory physics courses. *American Journal of Physics, 66*(1), 64–74.

Harms, U. (2000). Virtual and remote labs in physics education. *German Institute for Research on Distance Education at the University of Tuebingen.*

Hendrayati, H. & Pamungkas, B. (2007). Implementasi model hybrid learning pada proses pembelajaran mata kuliah statistika II di prodi manajemen FPEB UPI [Implementation of a hybrid model of learning in the learning process statistics II courses in management department Universitas Pendidikan Indonesia]. *Jurnal Penelitian Pendidikan, 13*(2), 181–184.

Hidayati, K. & Listyani, E. (2010). Improving instruments of students' self-regulated learning. *Jurnal Penelitian dan Evaluasi Pendidikan, 4*(1), 85–99.

Hornby, A. S. (2010). *Oxford advanced learner's dictionary*. London: Oxford University Press.

Husamah, H. (2014). *Pembelajaran bauran [Blended learning]*. Jakarta, Indonesia: Prestasi Pustaka.

Ikhsan, J. & Hadi, Y.S. (2015). Delivering science-engineering virtual labs using the new web technologies (HTML5), *Konferensi Internasional (ICERI2015)* (pp. 507–513).

Kurniawan, B. & Zulkaida, A. (2013). Contribution of emotional intelligence to autonomy of students of official college x. *Proceeding PESAT (Psikologi, Ekonomi, Sastra, Arsitektur & Teknik Sipil)* (pp. 53–60).

Lalima, L. & Dangwal, K. L. (2017). Blended learning: An innovative approach. *Universal Journal of Educational Research, 5*(1), 129–136.

Nugroho, A. (2012). Model development of web-based distance learning. *Jurnal Transformatika, 9*(2), 72–78.

Rahmawati, L. (2015). Pengembangan media pembelajaran berbasis komputer untuk mendukung kemandirian belajar siswa SMP [Developing computer-based learning media to support the independence of junior high school students]. *Jurnal Fisika dan Pendidikan Fisika, 1*(2), 29–36.

Schunk, D.H. (2012). *Learning theories: An educational perspective* (6th ed.). Boston, MA: Pearson Educational.

Susilana, R. & Riyana, C. (2008). *Media pembelajaran [Instructional media]*. Bandung, Indonesia: FIP UPI.

Tatli, Z. & Ayas, A. (2012). Virtual chemistry laboratory: Effect of constructivist learning environment. *Turkish Online Journal of Distance Education (TOJDE), 13*(1), 183–199.

Thorne, K. (2003). *Blended learning: How to integrate online and traditional learning*. London, UK: Kogan Page.

Tuysuz, C. (2010). The effect of the virtual laboratory on student achievment and attitude in chemistry. *International Online Journal of Educational Sciences (Iojes), 2*(1), 37–53.

Weinstein, C.E., Acee, T.W., & Jung, J. (2011). Self-regulation and learning strategies. *New Directions for Teaching and Learning Journal, 126*, 45–53.

White, J.H.D. & Harbaugh, A.P. (2010). *Learner-centered instruction*. California, CA: Sage Publication.

Wiersma, W. & Jurs, S. (2009). *Research methods in education* (9th ed.). Boston, MA: Pearson Educational Inc.

Winters, F.I., Greene, J.A., & Costich, C.M. (2008). Self-regulation of learning within computer-based learning environments: A critical analysis. *Educational Psychological Review Journal, 20*, 429–444.

Woolnough, B.E. & Allsop, T. (1985). *Practical Work in Science*. Cambridge, UK: Cambridge University Press.

Yennita, Y., Sukmawati, M., & Zulirfan, Z. (2012). Hambatan pelaksanaan praktikum IPA fisika yang dihadapi guru SMP Negeri di Kota Pekanbaru [Obstacles in the implementation of science physics practice faced by junior high school teachers]. *Jurnal Pendidikan, 3*(1), 1–11.

Yibin, S. (2006). *HTML-QS: A query system for hypertext markup language documents* (Unpublished Master's Thesis). University of Regina, Saskatchewan, Canada.

Yuswono, L.C., Martubi, & Sukaswanto. (2014). Profil Kompetensi Guru Sekolah Menengah Kejuruan Teknik Otomotif di Kabupaten Sleman [The Competencies profile of Automotive Engineering Vocational High Scholl Teacher in Sleman Regency]. *Jurnal Pendidikan Teknologi dan Kejuruan, 22*(2), 173–183.

Zamroni, M.R., Suryawan, N., & Jalaluddin, A. (2013). Rancang Bangun Aplikasi Permainan Untuk Pembelajaran Anak Menggunakan HTML 5 [Designed application game for learning children use HTML5]. *Jurnal Teknika, 5*(2), 489–490.

Being professional or humanistic? Teachers' character dilemmas and the challenge of whether or not to help their students

A. Suryani, S. Soedarso, Z. Muhibbin & U. Arief
Institut Teknologi Sepuluh Nopember, Indonesia

ABSTRACT: Teachers frequently encounter various dilemmas in certain situations. One of those experiences is an ethical decision of whether to increase a student's grade or to let them fail. Teachers may feel that they should not increase their student's grade because it is not fair on the other students and teachers; they should be professional by telling the truth and respecting their professional code. However, they may also feel that helping their students is good conduct, since it reflects care, kindness and empathy. This study aims to explore teachers' ethical decisions and behaviors regarding when they should or should not increase their students' grade. This is a qualitative case study grounded on phenomenology. This study is developed by our own experiences as teachers (educators). The data shows that an ethical dilemma forces teachers to weigh up some of their values. Teachers are frequently encouraged to engage in humanistic conduct by prioritizing the values of respect, harmony, care, and not hurting other people. However, as teachers, they are also expected to follow the values of being fair, principled, respecting ethical codes, integrity, and commitment. Teachers' decisions are affected by some social constraints, including culture, social environment, and organizational cultures and systems. The study also shows that character training also substantively involves situated learning.

1   INTRODUCTION

Teaching is a complex endeavor. Teachers are expected to not only be capable of mastering teaching materials, but also to understand how to solve social and cultural issues that emerge during their teaching practices. It may be easier to answer *yes* or *no*, *right* or *wrong* for content or material questions, but deciding which action to take for handling social and cultural teaching problems can be difficult. Solving social and cultural issues around teaching can be very problematic. Selecting which values should be prioritized over other conflicting values can be confusing. This situation can become even more of a dilemma when it is related to teachers' professional and social competencies. The course of action that should be taken by teachers tends to reflect the multifaceted aspects of teaching: moral development, emotional capacity, professional identity, integrity, and virtuous character. Thus, undertaking a moral action can cause a dilemma for teachers when they come to the intersection between professionalism (following the rules of their profession) and humanism (following social rules and the teacher's own heart).

One of many problematic situations that emerges during teachers' interactions with students is when teachers need to decide whether or not to increase their students' grades. This situation may be experienced by many teachers. Teachers may find that a diligent student, who regularly performs excellent academic work every day in class, is suddenly failing down and breaking. His/her failure may have severe effects and long-term consequences for the student's future academic life if he/she is not helped. This situation may lead to teachers facing a dilemma between helping the student, but violating the teacher's professional code, or obeying their professional commitments but letting the student fail. This situation requires teachers to think and feel. They are expected to weigh up and balance multifaceted aspects:

what will be the consequences for both the student and themselves, what about their principles, should the teacher ignore what their heart is telling them, and what about their commitment to the teachers' code of ethics.

Teachers' decisions to either help or not to help may have an impact on their attributed characters. It can signify how other people, for example other teachers, other students, parents, headmasters, and the student that they helped, perceive the qualities of that particular teacher's character. It can also affect the quality of the teacher-student relationship. It is not only the relationship with the assisted student, but also the interaction with the other students who were not helped. Moreover, a teacher's behavior and character can provide implicit character training for the students. As role models and educational figures, teachers should educate their students, not only through what they teach, but also by how they behave. Students are not only learning from *what their teachers give*, but also *who their teachers are*. This indicates that teachers' behavior can affect students' character training.

2 ISSUES OF THE STUDY

There are several issues that we aim to explore:

1. How do teachers react/perform when they face a dilemma; how do they choose between either helping or not helping their students; how do they decide between principles, consequences, agreements, and virtues?
2. What factors affect teachers' decisions and their behavior?

3 THEORETICAL FRAMEWORK

3.1 *Teacher-student relationship*

Students' success not only depends on how well teachers deliver their teaching materials, but also on how well teachers establish and maintain a harmonious relationship with their students. Relationships are the foundation of learning, even more basic than reading and writing (Riley, 2011). Moreover, Riley (2011) states that teachers' anxieties are not only related to the worry of failing to master their teaching material, but also to the fear of not inspiring their students, and not being loved and respected by them. This shows that teachers may face problems beyond cognition.

Harmonious relationships may determine students' learning outcomes. These outcomes are not only in the form of academic success, but also gaining positive academic experiences. Students need their teachers to reassure them and motivate them to explore and face a challenging environment in order to broaden their learning experiences (Riley, 2009). Harmonious teacher-student relationships can save students from a failure to learn (Pianta, Steinberg, & Rollins, 1995, cited in Riley, 2009). Positive relationships and bonding with teachers and schools can strengthen the students' social and emotional capacity (Fowler et al., 2008). This bonding relates to certain aspects: engagement, commitment, belief, and connection (Fowler et al., 2008). Teacher-student relationships are also a foundation for shaping students' behavior, especially for students in elementary grades (Fowler et al., 2008). From the perspective of the attachment theory, teachers are academic attachment figures, who should be motivated, assistive, encouraging, provide security and trust, open connections, and be responsible for students' socio-emotional development and academic demands (Fowler et al., 2008).

Teacher-student relationships are a complex dyad. Teacher-student relationships involve different responsibilities, legal statutes, and authorities (Riley, 2009). Teacher-student relationships tend to be bidirectional. They need each other and one party's behavior can influence the other party's success. How teachers treat their students may also influence the teachers themselves. Students confirm their teachers' professional identity, as there is no teacher without students, but there can be students without teachers (Riley, 2009). Therefore, the roles of caregiver and careseeker can be applied to both teachers and students (Riley, 2009).

Teacher-student relationships can be challenged by teachers' decisions both during and after processing students' grades. The function of students' grades is not only to measure the students' academic outcomes, but also to give them motivation, rewards, and to educate students to follow the systems of learning (Fowler et al., 2008). Fowler et al. (2008) mention that teachers feel that it is difficult to arrive at a decision on students' grades. This can be because, in grading, teachers may confuse the move from subjective to objective assessments (Fowler et al., 2008). Teachers should also be judges who evaluate and decide on students' grades based on hard work, the impact of the grades, and the results of the work (Fowler et al., 2008).

3.2   Teachers' ethical dilemmas, morals and integrity

Teachers' professional responsibility and identity demands that they develop a moral/ethical character. Teachers are expected to behave wisely, based on the moral/virtuous core values. Lickona et al. (2007) define character in education as a character that embodies understanding, feeling concern and interest, and contains central ethical principles, which can be grown through a holistic approach by integrating cognition, emotion, and behavior. Character is also perceived as "doing the right thing when no one is looking" (Lickona et al., 2007). This means that an individual with a strong character tends to hold firmly to values that they believe are right and important.

When an individual is facing a dilemma, he/she is at the intersection of choosing the right thing from other (right) things. The individual should give a great deal of thought to their choice and have reasons for their actions. Teachers' moral reasoning can be influenced by several factors. Jewell et al. (2007) identify three influential factors in an individual's moral reasoning and behavior: cognitive intelligence, affective intelligence, and dispositions toward morality and context (including social, personal, and situational environment). This moral reasoning becomes the grounding for one's moral/ethical behavior (Jewell et al., 2007). Cognitive and affective intelligence allow an individual to be aware of the broad choices that he/she has to make, but it is the value of morality that guides him/her to select the right choices (Jewell et al., 2007). These moral choices tend to be influenced by social, personal, and situational surroundings (Jewell et al., 2007). Jewell et al. (2007) suggest four ways to think and select grounds for moral/ethical character: principles (considering duties and rules), consequences (by weighing up the impacts for many parties), agreements (based on the approval of all parties), and virtues (following one's heart). Bommer et al. (1987), cited in Ehrich et al. (2011: 10), mention several factors that intervene in one's ethical decisions: work surroundings, procedures/systems and governmental environment, collective environment, professional context, home and peer environment, and individual character.

The teachers may have an ethical dilemma when their professional commitment and demands are conflicting with what they personally believe is right. As mentioned by Campbell (1997), cited in Ehrich et al. (2011: 7), teachers and administrators are frequently forced to confront their professional ethics, which may generate uncomfortable feelings. Tirri (1999) categorizes the ethical dilemmas that teachers face: 1) teachers are working, including treating students and making relationship with their collagues; 2) students' attitudes relating to school and home, dishonest behavior; 3) the same opportunities and treatment for minority students, especially relating to religion; and 4) teachers' inconsistency in implementing discipline. Preston and Samford (2002), cited in Ehrich et al. (2011: 11), suggest several procedures that can be used to make an ethical decision: evaluating the context, identifying the alternative decisions, assessing possible choices, determining, and providing reasonable proof/evidence.

The situational/contextual concept and the many possible ways of assessing decisions means that different people may have different justifications and decisions for their ethical/moral actions. The stage theory of moral judgment suggests that different situations may need different judgments, decisions and behaviors (Bandura, 1991). How moral obligation is implemented is contingent on the issues of the dilemma, and the place and forces of social norms/influences (Bandura, 1991). Different people develop their moral standards from different forces, including how they perceive their own behavior, based on other influential people around them (Bandura, 1991). Kitchener (1984), cited in Miller and Davis (1996), suggests that several principles should be taken into account when considering decisions regarding an ethical dilemma: autonomy, non-maleficence, beneficence, justice, and fidelity.

A teacher is expected to be a person with integrity. Integrity is defined as an individual's character, the ability to focus on his/her work and to become whole by integrating emotion, self-interest, and self-awareness (Jacobs, 2004). Integrity involves multidimensional values (Jacobs, 2004). Solomon (1992), cited in Jacobs (2004), argues that integrity embodies professional obedience, moral autonomy, and correlates with moral humility.

4 METHOD OF THE STUDY

To explore this phenomena, we adopt a qualitative method. Since ethical dilemmas, moral justifications and character development are situational (contextual), we use a case study as the research method. We used our experience as teachers when we were facing a dilemma and had to decide whether or not to increase our students' grades. The experiences are framed by following some procedures: describing the experience, balancing and justifying the moral action by considering the principles, agreement, virtue and consequences (Jewell et al., 2007), and deciding and supplying the moral justification.

5 FINDINGS AND DISCUSSION

The data shows that teaching is not only related to curriculum content and teaching practices, but also concerns social dimensions in relation to students, parents and other social parties around teaching. The diversity of social contexts and the teachers' characters and dispositions regarding moral/ethical reasoning lead to them taking different decisions and actions toward their ethical dilemmas. Below are several data excerpts indicating teachers' moral/ethical reasoning and factors influencing teachers' ethical/moral character and behavior.

5.1 *Findings*

Teacher 1:
Below is the ethical issue faced by teacher 1:
  Ethical dilemma and context:

> "When I worked in a certain university, I was ever phoned by one of leaders in institution where I worked. He asked me to increase a student's grade, who is his nephew."

How teacher 1 thinks and feels when considering the ethical dilemma, by weighing up principles, consequences, agreement and virtues, is presented in Table 1.

Based on teacher 1's weighing up process, as presented in Table 1, teacher 1 makes a decision as follows:

Final decision: I decide to follow virtues, what my heart says. I am just a guest lecture. As a guest lecture, I should respect him who already helps me. However, I show my disagreement by resigning.

Teacher 2:
Below is the ethical issue faced by teacher 2:
  Ethical dilemma and context:

> "I ever had a student in his/her fifth semester in a certain university. I am in a dilemmatic situation since if I did not increase his/her grade, he/she will not be allowed to continue his/her study, but if I increased his/her grade, then I am not being consistent with grading rules/norms."

Table 1.   The weighing up of principles, consequences, agreement, and virtues of teacher-participant 1.

Principles:
Yes – I feel that increasing student's grade is not right, but I was forced to do that because I can't select what I can do independently. If I don't follow his order, maybe other people in that institution see me as disrespecting leader in host institution.
No – If I follow his order, he misused his power and this is not right. If I follow his wrong behavior, I am also wrong. Even though he is responsible for my wrong behavior since he orders me.

Agreement:
There is no agreement between me and the leader in that unit. What happens is just dilemmatic situation between organizational ethic, in which I have to conform to his order and academic ethic, in which I have to be fair and objective

Consequences:
The leader will see me as doing the right thing by helping him.
As the consequence, I resign and leave that institution to avoid the similar thing happens again.

Virtues:
I am in clash between following/conforming the leader and being objective and fair.
I am in moral dilemmatic situation in which there are two choices which are conflicting each other, but still I have to select one of them, and for this dilemmatic situation, I decide to conform to organizational ethic in which I conform the leader's order in order to minimize and avoid more/bitter conflict, but then I decide to resign.

Table 2.   The weighing up of principles, consequences, agreement, and virtues of teacher-participant 2.

Principles:
Basically, it is not right, it violates the principle of following the system.

Agreement:
Increasing student's grade means breaking agreement on grading rules/norms.

Consequences:
If I do not help the student, the student will be inflicted from academic sanction. He/she will be not allowed to continue his/her study in that institution.

Virtues:
Increasing student's grade can be right since it is just for emergency situation. It is based on teacher's wisdom when teacher faces certain academic cases.

How teacher 2 thinks and feels when considering the ethical dilemma, by weighing up principles, consequences, agreement and virtues, is presented in Table 2.

Based on teacher 2's weighing up process, as presented in Table 2, teacher 2 makes a decision as follows:

Final decision: I keep on following grading rules and trying to be objective and avoiding subjectivity. Increasing student's grade is breaking that rules, but teacher can help student in particular emergency situation. It should also be grounded on acceptable reasons and teacher should give more/extra assignment/tasks and re-test the student to increase his/her score to the minimum level, grade C.

Teacher 3:
Below is the ethical issue faced by teacher 3:
    Ethical dilemma and context:

> "I had a student, who failed to pass the topic several times in university x, because he is very lazy, indiscipline, unmotivated and did not show bright intelligence. Last semester, he repeated his failed topic and at the end of the semester, he failed again because of again the same reason as several previous semesters. To my surprise, I got a call and message from one of my friend, explaining that this student is his son and asking a help to help his son by increasing his son's grade. If I did not help it is possible that his son cannot his study in this particular university since based on its system, student who cannot pass required accumulated grade for certain semesters, cannot continue his/her study in that university. This situation put me in a certain dilemma."

How teacher 3 thinks and feels when considering the ethical dilemma, by weighing up principles, consequences, agreement and virtues, is presented in Table 3.

Based on teacher 3's weighing up process, as presented in Table 3, teacher 3 makes a decision as follows:

Final decision: Increasing the student's grade because of some reasons: being empathic after consulting to some senior lecturer and social push from a friend, but I feel uncomfortable since it violates value of being committed to professional ethical code, integrity and honesty. To reduce feeling of uncomfortable I make a new agreement to give him/her some additional assignment.

Teacher 4:
Below is the ethical issue faced by teacher 4:
   Ethical dilemma and context:

   "If I do not increase the student's grade, it is possible that the target of certain institution can't be reached. However, I feel uncomfortable with that."

How teacher 4 thinks and feels when considering the ethical dilemma, by weighing up principles, consequences, agreement and virtues, is presented in Table 4.

Based on teacher 4's weighing up process, as presented in Table 4, teacher 4 makes a decision as follows:

Final decision: Increasing student's grade, trying to be humanistic but still being professional. I expect that I can be good person during and after my retirement, safe from any threatening issues.

Table 3. The weighing up of principles, consequences, agreement, and virtues of teacher-participant 3.

Principles:	*Consequences:*
Yes – I should help because helping a friend and help our student is a good deed.	Yes – I should help because if I don't help, the student will not be able to continue his/her study and it will affect his/her future life.
No – I should not help since by helping him/her, I treat all of my students un objectively and it means I am cheating.	No – I should not help him/her since the student will not get any lesson/bitter experience and he/she will continue his/her indiscipline behavior.
Agreement:	
No – I should not help because before the topic is started, we (I and all of the students) have agreement to follow the academic system.	*Virtues:*
	Yes – I should help because I took a pity on this student if he/she cannot continue his/her study.
	No – I should not help because it violates values of honesty and objectivity.

Table 4. The weighing up of principles, consequences, agreement, and virtues of teacher-participant 4.

Principles:	*Consequences:*
I increase the grade of student, but I still uphold the universal value, in which I don't change the position of other students in that class. I only change the final grade of that particular student (for instance B).	I believe lying by increasing student's grade can be good, since if I don't increase it, the consequences are worse than if I don't increase it.
	Virtues:
Agreement:	I feel it is all right to increase student's grade, but it should be reasonable and not over grading. I remember a slogan: *"Ikut rasa binasa, ikut hati mati."*
I think increasing student's grade is still acceptable, as long as the teacher has reasons and is grounded on theoretical foundation and certain rules/norms.	

The data shows that teachers' dispositions, or how they think, feel and behave to solve their ethical dilemmas, are not only influenced by their personal/individual values, but also by social forces. These pressures include social values (friendship, reciprocity, respect), and systems and social identity (professional reputation, image, social expectations). These ethical dilemmas encourage the teachers to weigh/balance their own values and the social consequences of their decision. There is a blurred area in moral decisions. Justifying moral characteristics and decisions is quite different from answering geometry questions (Noddings, 2013). Teachers' individual choices may conflict with social expectations and their professionalism may clash with ideas of humanity and care. Educators are expected by society to be trustworthy and accept responsibilities, use ethical behavior, and be dedicated to their students, profession, community and the parents (Team, 2012). Teaching involves moral behavior (Lishchinsky, 2010).

The data shows that when the teachers face an ethical dilemma regarding whether or not to increase their student's grade, the teachers tend to decide to *help their student*. After consideration, the teachers tend to choose consequences and virtues rather than principles and agreement. Their decisions tend to be grounded on social factors, even though this conflicts with their inner/individual preference. The teachers are aware that increasing a student's grade is violating the principle values of honesty, fairness and objectivity, but the teachers are also aware of the direct consequences of their decision on their student, their relationship with other people and their own individual and social selves. The dominant value embraced by the teachers is care for their students. Frequently, the teacher-student relationship relating to punishment and grading decisions is managed with care (Clark, 1995, cited in Ehrich, et al., 2011: 5). Teachers may face options between either following rules/systems or following their heart. The teaching profession contains elements of an emotional roller coaster, in which teachers should not only manage students' emotions, but also their own feelings (Palmer, 1997).

Virtues, or what teachers feel in their hearts, also tend to be the teachers' guide when undertaking moral behavior rather than agreement (professional code of conduct). The code of conduct frequently does not provide a detailed solution for different contexts/situations (Ehrich et al., 2011). Thus, there is no exact way to determine ethical issues (Kakabadse et al., 2003). Furthermore, the professional code may not include a clear relationship between teachers, students, and other social parties. As moral agents, teachers may face a moral strain in relation to their relationship with their students, their students' parents and their teaching colleagues (Ehrich et al., 2011). The specific and different context of emerging ethical issues leads to no fixed solution, which can only be managed (Lyons, 1990) and is in 'a gray zone' (Kakabadse et al., 2003).

The teachers' dilemmas in certain situations indicate that teachers may put aside their principled values. They tend to be more aware of the direct consequences on their students. This means that they are still considering or taking into account the social aspects (how their decision will affect other people/their students in the future). This indicates the dominant influence of social power over the individual to choose between what is right and what is wrong. Thus, they tend to be forced to do things that they actually do not like (do not want to do). When an individual faces a dilemma, Hohlberg and his colleagues argue that social determinants and surroundings tend to affect that individual's moral justification, rather than his/her moral competence (Higgnis et al., 1984, cited in Bandura, 1991: 48). The teachers also increase a student's grade because of social evaluation and empathy. Individuals decide their moral standards from their own preferences and how the people around them behave toward those preferences (Bandura, 1991). The determinants of their moral standards are frequently based on social influences (Bandura, 1991). The teachers also prefer to help the students by increasing their grade. Helping is part of the culture and social norm, and it involves certain expected roles (Triandis, 1977). The teaching profession demands that teachers motivate/support their students. By not helping the students, it is possible that the students cannot continue their studies and this endangers not only the student's academic path, but also their future life. As mentioned by Triandis (1977: 66) *"when people are told that it is their job to help, and they accept that it is, the self-attribution becomes a cause for action. The person experiences definite pressures to help."* Even though the teachers finally

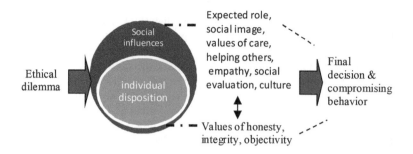

Figure 1. Factors affecting teachers' decisions.

increase their students' grade, they compensate for it with treatment that they believe will apply justice/fairness.

The study shows that the teachers choose to be humanistic by conforming to the norm of helping, which is part of the culture. However, they still make professionalism as a part of humanistic. In spite of the emotions that we feel, for example fear, anger, or hate, we are conditioned to conform to the norm of caring (Noddings, 2013). The way that students behave will affect their teachers' decisions and teaching method To minimize moral misconduct, teachers should insert values and moral education within their teaching. Teachers should develop an affective relationship regarding classroom structure in order to improve students' internal motivation for being autonomous, and to allow them to feel a sense of belonging and competence (Nucci et al., 2014). Some of the data shows that some students do not have the self-awareness to fight for their own future by getting the desired marks. The students should be mindful and use their motivation to gain the desired marks. The students should possess the emotional awareness and passions needed to reach their expectations and develop social relationships (Nucci et al., 2014).

Teachers' ethical dilemmas that emerge from students' issues call for the need for character training, not only for teachers, but also for students. The globalization era may present different values that challenge the firmness of students' characters. The infiltration of outside values through the development of technology and the internet may expose social issues (Widyahening & Wardhani, 2016). Thus, it is the teachers' responsibility to educate their students through their own behavior. Teachers should participate in forming their students' character through the activities of the teaching profession (Chowdhury, 2016; Fatoni, 2017). When teachers have to rescue their failing students, the teachers should also be sensible and be aware that the student's failure could be caused by the failure of the teachers in teaching and forming the positive character of the student. Teachers should try to create an environment that supports character training and models the correct moral conduct, even though teachers cannot force the students' to develop a certain attitude (Nillsen, 2004). There are six main characteristics that should be learned by teachers and students: mindfulness, curiosity, courage, resilience, ethics, and leadership (Bialik et al., 2015). The influence of the social environment on teachers' decisions and students' behavior indicates the need to involve all of the components of society in character training. The macro strategy of character training includes all elements of education (Fatoni, 2017). How teachers decide and behave tends to be affected by both social influences and their individual dispositions, which are presented in Figure 1.

6 CONCLUSION

In the teaching profession, teachers are not only dealing with subject content, but also with social-cultural issues around teaching. Whether or not to increase a student's grade is a frequent teaching dilemma, and there is no clear cut way out since the decision is contextual and highly influenced by social surroundings. When deciding whether or not to increase the grade, the teachers are confused between the social impacts and their personal/individual voice. All of the teachers tended to increase their student's grade because of social factors: expected role, social image, values of care and helping others, empathy, social evaluation, and culture. These factors contradict their individual values of being honest, commitment to maintaining integrity, and being fair and objective to other students. They tended to conform to social values, but compensated for it by other types of behavior.

REFERENCES

Bandura, A. (1991). Social cognitive theory of moral thought and action. In W.M. Kutines & J.L. Gewirtz (Eds.), *Handbook of moral behavior and development: vol. 1: Theory*. New Jersey, NJ: Lawrence Erlbaum Associates Publishers.

Bialik, M., Bogan, M., Fadel, C., & Horvathoa, M. (2015). *Character education for the 21st century: What should students learn?* Massachusetts, MA: CCR.

Chowdhury, M. (2016). Emphasizing morals, values, ethics and character education in science education and science teaching. *The Malaysian Online Journal of Educational Science, 4*(2), 1–16.

Ehrich, L.C., Kimber, M., Millwater, J., & Cranston, N. (2011). Ethical dilemmas: A model to understand teacher practice. *Teachers and Teaching: Theory and Practice, 17*(2), 173–185.

Fatoni, A. (2017). The strategy of character education in globalization era. *International Journal of Scientific & Technology Research, 6*(4), 112–114.

Fowler, L.T.S., Banks, T.I., Anhalt, K., Der, H.H., & Kalis, T. (2008). The association between externalizing behavior problems, teacher-student relationship quality and academic performance in young urban learners. *Behavioural Disorders, 33*(3), 167–183.

Jacobs, D.C. (2004). A pragmatist approach to integrity in business ethics. *Journal of Management Inquiry, 13*(3), 215–223.

Jewell, P.D., Webster, P., Henderson, L.C., Dodd, J., Patterson, S., & McLaughlin, J. (2007). *Care, think and choose. A curriculum based approach to ethical thinking*. Victoria, Australia: Flinders University.

Kakabadse, A., Kakabadse, N.K., & Kouzmin, A. (2003). Ethics, values and behaviours: Comparison of three case studies examining the paucity of leadership in government. *Public Administration, 81*(3), 477–508.

Lickona, T., Schaps, E., & Lewis, C. (2007). *CEP's eleven principles of effective character education*. Washington, DC: Character Education Partnership.

Lishchinsky, O.S. (2010). Teachers' critical incidents: Ethical dilemmas in teaching practice. *Teaching and Teacher Education, xxx*, 1–9.

Lyons, N. (1990). Dilemmas of knowing: Ethical and epistemological dimensions of teachers' work and development. *Harvard Educational Review, 60*(2), 159–180.

Miller, H.F. & Davis, T. (1996). *A practitioner's guide to ethical decision making*. Virginia, VA: American Counseling Association.

Nillsen, R. (2004). *The concept of integrity in teaching and learning*. A paper presented at the Symposium Promoting Academic Integrity, Newcastle, NSW.

Noddings, N. (2013). *Caring. A relational approach to ethics and moral education*. Los Angeles, CA: University of California Press.

Nucci, L., Krettenauer, T., & Krettenauer, T. (2014). *Handbook of moral and character education*. London, UK: Routledge.

Palmer, P.J. (1997). The heart of teacher. Identity and integrity in teaching. An essay adapted from P.J. Palmer, *The courage to teach: Exploring the inner landscape of a teacher's life*. San Fransisco, CA: Jossey-Bass.

Riley, P. (2009). An adult attachment perspective on the student-teacher relationship and classroom management difficulties. *Teaching and Teacher Education, 25*, 626–635.

Riley, P. (2011). *Attachment theory and the teacher-student relationship. A practical guide for teachers, teacher educators and school leaders*. New York, NY: Routledge.

CSDE (2012). *Facilitator's guide. Ethical and professional dilemmas for educators*. Connecticut, CT: Connecticut's Teacher Education and Mentoring Program.

Tirri, K. (1999). Teachers' perceptions of moral dilemmas at school. *Journal of Moral Education, 28*(1), 31–47.

Triandis, H.C. (1977). *Interpersonal behaviour*. California, CA: Wadsworth Publishing Company, Inc.

Widyahening, E.T. & Wardhani, N.E.W. (2016). Literary works and character education. *International Journal of Language and Literature, 4*(1), 176–180.

Developing picture storybook based on scientific approach through inquiry method

N. Hidayatunnisa & Z.K. Prasetyo
Universitas Negeri Yogyakarta, Indonesia

ABSTRACT: This paper presents results of research and development of picture storybook based on scientific approach through inquiry method. This study aimed to produce a valid picture storybook based on scientific approach through inquiry method to be used by elementary school students as their learning material. Research and development procedures in this study were adapted by Borg & Gall procedures with steps including: (1) research and information collecting; (2) planning; (3) developing preliminary form of product; (4) preliminary field testing; (5) main product revision; (6) main field testing; and (7) operational product revision. This development research was motivated by the unavailability of learning materials based on scientific approach through inquiry method for students. Developed picture storybooks are filled with stories based on a scientific approach that builds students' knowledge through the process of observing, questioning, experimenting, socializing and communicating. Implementation of the scientific approach in the story was presented by applying the steps of inquiry method consisting of engaging, exploring, explaining, elaborating, and evaluating. The results showed that picture storybook based on scientific approach through inquiry method developed is valid to use as a learning material based on expert assessment and teacher/student responses.

1 INTRODUCTION

Education is a very important factor in human life. One effort to improve the quality of education conducted by the government in creating a learning process that can accommodate the potential of the learners is through the implementation of the Curriculum 2013. The underlying issue of the previous curriculum compared with the 2013 curriculum is the use of learning materials. In the curriculum of 2013, the learning materials have been prepared by the government. The learning materials in the form of textbooks provided by the government, are in accordance with Peraturan Menteri Pendidikan dan Kebudayaan RI (2013) Number 71 which suggests that the learning materials used in the 2013 curriculum are buku siswa (students' book) and buku pegangan guru (teachers' book), and that learning materials are used as the main learning materials in the learning process. However, the use of these learning materials still has various problems. The results of preliminary studies based on interviews with teachers, observation, student needs analysis and document analysis found a number of problems that show the development of learning materials should be done as a perfect teaching material that already exists.

The definition of learning materials is all forms of materials used to assist teachers in carrying out teaching and learning activities (Majid 2009). Learning materials include print and non-print/ electronic media that contain information and can help students in achieving learning objectives (Sitepu 2014). Associated with Curriculum 2013, used learning materials used should contain the characteristics of thematic learning to optimize the learning process. Thematic learning materials are all materials (whether information, tools, or texts) that are arranged systematically and display the whole figure of the competencies to be mastered by students through a learning process that encourages active and fun student involvement.

Learning materials do not merely encourage students to learn, but learn to do, learn to be, and learn to live together, as well as holistic and authentic for the purpose of planning and reviewing the implementation of learning (Prastowo 2014). Based on this, opinion it can be understood that learning materials are everything that contains information such as facts, concepts, principles, procedures, or generalizations in the form of print and non-print that has a purpose to help students in learning.

Learning materials should be in accordance with the characteristics of students. Learning materials especially for elementary students should pay attention to various aspects. One such aspect is the interest of students to read books with interesting stories, diverse illustrations, and views that motivate students to read. Based on that, the researcher assessed the picture storybook as an appropriate form of learning material according to the characteristics of elementary students.

2 LITERATURE REVIEW

Picture storybooks are part of children's literature. Picture storybooks are a work of fiction in which text and pictures are at least of equal importance, the pictures being an integral part of each page opening or 'double spread', or a book in which pictures alone tell a story (Saxby & Winch 1991). Picture storybooks are storybooks that display verbal narrative texts and are accompanied by illustrative drawings so that the narrative text and illustrations work together to make the story more impressive for the reader (Nurgiyantoro 2013). Based on the opinion, picture storybooks are an integral part of understanding the content of the book so as to clarify the opinion that the two components of text and images are interconnected with each other.

In terms of the characteristics of picture storybooks, they have always required readers to be interactive as they move back and forth between the visual and verbal text (Pantaleo 2004). Picture storybooks have a unique way of telling stories, where children gain knowledge in the game of images and words and construct the meaning of the presented story texts, so that they not only read but gain useful knowledge as well (Goouch & Lambirth 2011). In addition to the interaction between textual narrative and illustrations, picture storybooks also contain messages in the story presented. The picture storybook conveys its messages through two media, in the art of illustrating and the art of writing (Hucket et al. 1987). Based on the explanation of picture storybooks that has been presented, it can be synthesized that picture storybooks are books that present pictures and text, in which the images or illustrations in the book support the text of the story to be more concrete and vice versa. The message to be conveyed can be understood by the reader, in this case the children, but it also can develop the imagination and creativity of children in thinking.

In particular, the characteristics of an appropriate picture storybook for elementary students are seen and adjusted based on five aspects: 1) intellectual development; (2) moral development; (3) emotional and personal development; (4) language development; and (5) the growth of the concept of the story. (Huck et al. 1987; Nurgiyantoro 2013). Picture storybooks are part of children's literature that have various essential components. Many of the experts who expressed their opinion related to the components of picture books. Broadly speaking, the components of picture storybooks can be summarized into several components. The components are storyline, character and characterizations, themes, background, point of view, style, story content, messages and illustrations (Huck & Young 1961; Tompkins & Hoskisson 1995; Lukens 1999; Mitchell 2003; Strickland et al. 2004; Donoghue 2009).

Picture storybooks are used because the format is child friendly and invites a beneficial joint attentional learning context (Kelemen et al. 2014). The development of picture storybooks as learning materials serves as a transmitter of information that is considered appropriate to facilitate logical thinking and students' needs for learning materials as fun and motivating. Based on the aspect of learning materials preparation put forward by the above experts, the development of a picture storybook requires an evaluation that determines whether the picture storybook developed is feasible for use as a learning material. Evaluation components that include the criteria set forth in the instructional development materials are: (1) the feasibility of the content; (2) linguistics; (3) presentation; and (4) graphical display (Depdiknas 2008). So based on this, the developed picture storybook must follow the reference.

The 2013 curriculum emphasizes the importance of learning by using a scientific approach. This is described in Permendikbud Number 65 on the Standard Process of Primary and Secondary Education, which implies the need for a learning process guided by scientific approach principles. A scientific approach is a learning process designed in such a way that students actively construct concepts, laws or principles through observing stages (to identify or find problems), formulating problems, proposing or formulating hypotheses, collecting data with various techniques, analyzing data, drawing conclusions and communicating concepts, laws or principles that are "discovered" (Hosnan 2014). A scientific approach is a learning process that guides students to solve problems through careful planning, careful data collection, and careful data analysis to generate a conclusion (Abidin 2012). The scientific approach is intended to provide an understanding to learners in knowing, understanding the various materials using a scientific approach, that information can come from anywhere, anytime, regardless of teacher's in-line information (Daryanto 2014).

Based on the above explanations, the activities in the scientific approach are carried out in various learning activities because the current 2013 curriculum encourages the use of a scientific approach in the implementation of learning. Then the picture storybook with a scientific approach base is the right material to be developed.

The methods considered in line with the scientific approach are problem based learning, project based learning, inquiry, discovery, and group investigation (Kemendikbud 2013). The target of this research is 5th graders, so the more appropriate method to use is the method of inquiry. This is because the application of inquiry method is more suitable for high grade student characteristics. Children are curious and eager to grow and inquiry training capitalizes on their natural energetic explorations, giving them specific direction so that they explore new areas more forcefully (Joyce & Weil 1972).

Inquiry is the activity of students in which they develop knowledge and understandings of scientific ideas, as well as an understanding of how scientists study the natural world (Abell & Lederman 2007). The inquiry methods contained in the lesson provide many benefits. The purpose of these inquiries is to introduce students to problem situations, suitable for their levels of competence and of inquiry (Kahle 1979). Inquiry-based teaching is more broadly conceived to help students develop understanding about the ways the physical and social world works and the processes used to investigate it (Arends & Kilcher 2010). Therefore, it can be understood that inquiry can enable students to understand and apply important concepts, skills or knowledge in learning.

In relation to the developed picture storybook, Magee & Wingate (2014) found that the inquiry experience can be generated through the use of visual literacy sources and language activities. With the use of visual literacy sources, it can find additional ways to engage students with science. Although some of the best scientific ideas come from activities that involve hands-on, many other ideas require a text source for further explanation. Therefore, in this research, the inquiry method is presented in a picture storybook with the delivery of material from a picture storybook using the stages of inquiry method. This is in line with the opinion of Cox (2012) which states that one of the activities that students can do if they want to do inquiry is to read books and other sources of information to find out what they want to know. In addition, a study by Gray & Marilyn (2006) suggests that students who do inquiry combined with the use of language and images can learn independently and collaboratively. The use of images combined with spoken and written language and mathematical expressions extends the opportunity to develop and represent their understanding and create a double meaning for higher levels of understanding.

Inquiry method has several stages in its application. Stages of inquiry method consist of Engage, Explore, Explain, Elaborate and Evaluate, also known as 5E (Moore 2015). Summaries of each stage of the inquiry method and the form of implementation of the inquiry method in the picture storybook are illustrated in Table 1.

Based on the explanation put forward in the various theoretical studies above, synthesis of the picture storybook based on the scientific approach through inquiry method is one

Table 1. Stages of inquiry method.

Stages	Activity in learning	Implementation in picture storybook
Engage	Students find and identify problems.	Read the story and discover the problems the character encounters through the conversation in the story and the invitation to find out through the activity of "Ayo Amati!"
Explore	Students are directly involved with phenomena and learning resources.	The invitation of a character to find answers based on the information presented by the figure through the activity of "Tahukah Kalian?" and "Ayo Cari Tahu!"
Explain	Students analyze their exploration results by placing the abstract experience into a communicable form.	The invitation to cultivate understanding becomes a form that can be communicated through questions about conceptual understanding through activity of "Aku Tahu!" dan "Ayo Lakukan!".
Elaborate	Students present the findings or what they understand.	A call to express what is understood through activities of "Ayo Simpulkan!"
Evaluate	The teacher evaluates and determines whether the student has reached an understanding of concepts and knowledge.	Presentation of concept comprehension evaluation through activity of "Latihan".

form of literary works that combines elements of images and text as a meaningful unity to make the presented story become more real. This picture storybook is in practice developed as a learning material, so that the material presented in the picture story book is adapted to the 2013 curriculum that has lessons on the topic "Human and Animals Organs" for the 5th graders.

3 METHOD

3.1 *Type of research*

The type of research used in this research is research development, also known as Research and Development by Borg & Gall (2003) which consists of 10 steps. However, in this study, researchers conducted seven steps to produce a valid product that is a picture storybook based on scientific approach through inquiry method which is used as a learning material.

3.2 *Participants*

Participants in this research and development were as follows: preliminary field test involving six students and one teacher from elementary school A in Yogyakarta. The next main field testing involved 12 students and one teacher from school B in Yogyakarta.

3.3 *Development procedures*

This study designed and modified the development model into seven out of the 10 stages proposed by Borg & Gall. The next stage has not been done by researchers because it will be part of the next research. The development procedures carried out are as follows: (1) preliminary studies, this stage collected various preliminary information with library studies, interviews, observations, student needs questionnaires and document studies. Such information is required as a guide in product development planning; (2) planning, this stage was based on the development plan of the preliminary study. The product plan developed in the form of picture storybook based on scientific approach through inquiry method for the 5th graders

Table 2. Categorization of average score.

Score interval	Category
>3,25–4,00	Excellent
>2,50–3,25	Good
>1,75–2,50	Fair
1,00–1,75	Poor

was adapted to the 2013 curriculum; (3) developed preliminary form of product, this stage was to arrange the initial product in the form of a picture storybook based on scientific approach through inquiry method based on the predetermined theme, "Organ Tubuh Manusia dan Hewan". The development of the initial product draft refers to the study of theory regarding the structure and contents of Curriculum 2013 on elementary school learning materials, which is a picture storybook based on scientific approach that contains the steps of inquiry methods. After the product was developed, there was the assessment by experts; (4) preliminary field testing, this stage was done to see some weaknesses that exist in the developed product by analyzing student and teacher response to the book developed; (5) main product revision, this stage was done after obtaining the findings and input from teacher and students in preliminary field testing then revision based on the findings; (6) main field testing, at this stage the revised products according to inputs and critiques in the preliminary field testing were tested on different and more subjects (7) operational product revision, this stage was done after obtaining the findings and input from teacher and students in main field testing, then revision based on the findings to make the picture storybook based on scientific approach through inquiry method declared as valid as a learning material that students can use.

3.4 *Data collection*

Instruments in this research were divided into two parts. The instruments to measure the validity of the product included: (1) validation sheet of material subject expert and language expert, and (2) student and teacher responses.

3.5 *Data analysis*

Qualitative data in this research was criticism and suggestion by expert. This data was collected and summarized to improve the product being developed while the quantitative data was in the form of scores on each instrument item that has been filled by the expert. The validation result was calculated to get the average score. The scores are grouped in categorization. Categorization can be seen in Table 2.

4 RESULT

4.1 *Data analysis from experts*

In the data analysis of product validation results of this subject material expert and language expert.

The assessment of the material expert and language expert was conducted in two stages, that is the initial assessment before the revision and the final assessment after the revision. The results of the material experts and language expert is as follows:

Based on Table 3, it can be said that the result validation by subject experts' final assessment indicate that all three aspects fit into the "excellent" category.

The criteria indicates that the picture storybook meets the "ideal" picture storybook used for learning as reviewed by the subject expert, but there were suggestions for punctuation. The picture storybook is worthy of being used for the initial field test after a revision.

Table 3. The results validation by subject experts.

Assessment aspect	Initial assessment		Final assessment	
	Score	Criteria	Score	Criteria
Feasibility of content	3,47	Excellent	3,71	Excellent
Presentation	3,22	Good	3,78	Excellent
Graphical display	3,43	Excellent	3,86	Excellent

Table 4. The results validation by language experts.

Assessment aspect	Initial assessment		Final assessment	
	Score	Criteria	Score	Criteria
Language	3,25	Good	3,42	Excellent
Characteristics	3,55	Excellent	3,73	Excellent
Component	3,18	Good	3,71	Excellent

Based on Table 4, it can be said that the result validation by the language expert indicates that all three aspects fit into the "excellent" category.

The criteria indicates that the picture storybook meets the "ideal" storybook used for learning by language expert, but there are suggestions for punctuation. The picture story book is worthy of being used for the initial field test after a revision.

4.2 Data analysis from field testing

Based on expert validation results, the picture storybook entered the field testing stage. Field testing in this research was done twice through preliminary field testing and main field testing.

4.2.1 Result of preliminary field testing

The result of preliminary field testing in the form of teacher and student responses related to the effectiveness of the picture storybook as a learning material. The results of student and teacher responses were used as material for the revision of the picture storybook to produce a valid picture storybook. Explanation of teacher and student response results is presented as follows.

Based on the result of teacher response in Table 5, it can be seen that teacher response to picture storybook quality is in "excellent" category.

Based on the result of students' response in Table 6, it can be seen that students' response to picture storybook quality is in "excellent" category. So this picture storybook can be used for further steps.

4.2.2 Result of main field testing

The result of main field testing are student and teacher responses related to the effectiveness of the picture storybook as a learning material. Teacher and teacher response outcomes are used for revision considerations before the picture storybook is used for operational testing.

Based on the result of teacher response in Table 7, it can be seen that teacher response to the quality of picture storybook is in the "excellent" category.

Based on the result of students' response in Table 8, it can be seen that students' response to the quality of the picture storybook is in the "excellent" category. Therefore, the picture storybook is declared valid for use as a learning material.

Table 5. Teacher response in preliminary field testing.

Indicator	Score	Category
Feasibility of material aspect	3,60	Excellent
Feasibility of linguistic aspect	3,43	Excellent
Feasibility of presentation aspect	3,60	Excellent
The effects of picture storybook on learning	3,00	Excellent
The practicality of picture storybook	3,75	Excellent

Table 6. Students response in preliminary field testing.

Respondent	Score	Category
Respondent 1	4,00	Excellent
Respondent 2	4,00	Excellent
Respondent 3	4,00	Excellent
Respondent 4	3,73	Excellent
Respondent 5	4,00	Excellent
Respondent 6	4,00	Excellent

Table 7. Teacher response in main field testing.

Indicator	Score	Category
Feasibility of material aspect	3,80	Excellent
Feasibility of linguistic aspect	3,86	Excellent
Feasibility of presentation aspect	3,80	Excellent
The effects of picture storybook on learning	3,75	Excellent
The practicality of picture storybook	4,00	Excellent

Table 8. Students' response in main field testing.

Respondent	Score	Category
Respondent 1	3,73	Excellent
Respondent 2	3,73	Excellent
Respondent 3	4,00	Excellent
Respondent 4	3,47	Excellent
Respondent 5	3,47	Excellent
Respondent 6	4,00	Excellent
Respondent 7	4,00	Excellent
Respondent 8	4,00	Excellent
Respondent 9	3,73	Excellent
Respondent 10	4,00	Excellent
Respondent 11	4,00	Excellent
Respondent 12	4,00	Excellent

5 DISCUSSION

The current 2013 curriculum encourages the use of scientific approaches in the implementation of learning. Therefore, picture storybooks based on scientific approach are the right instructional materials to develop. This is supported by the findings of Endah &

Mukminan (2016) that the various stages in the scientific approach facilitate students to be able to think hierarchical that begins from observing and asking questions so that they simply think based on what is seen, then proceed with reasoning found facts that encourage students to think more complex. Furthermore, students can communicate well if the concept has been obtained.

The scientific approach that is implemented in literary-shaped picture book has a virtue. DeBoer (2000) notes that scientific literature implies an understanding of a scientific approach to a more general educational purpose so that scientific literacy is not only applied to science learning alone but other learning in accordance with the objectives to be achieved.

In this research, the steps of the scientific approach are implemented in picture storybooks. By reading a picture storybook, the students will be able to carry out a scientific process that can develop various aspects of cognitive, affective and psychomotor. In the application of a scientific approach through this method of inquiry, students can optimize their cognitive abilities and process skills from engage, explore, explain, elaborate and evaluate.

6 CONCLUSION

Picture storybook based on scientific approach through inquiry method has been assessed by subject matter expert and language expert. Both experts declared that it is valid to be used in classroom learning. This can be seen from the result of expert judgement which shows that the developed product gets category of "excellent".

Based on the field test, the result of the preliminary field test shows that the product developed is valid. It is indicated by the result of teacher and student response on the picture storybook which got the "excellent" category. Futhermore, the results of main field testing also show that the product developed is valid to use in the learning based on teacher and student's response is in the "excellent" category.

Based on these results, it is concluded that the picture storybook based on scientific approach through inquiry method has been worthy to use in learning.

REFERENCES

Abell, S.K. & Lederman, N.G. 2007. *Handbook of research on science education*. New York, NY: Routledge.

Abidin, Y. 2012. *Pembelajaran membaca berbasis pendidikan karakter* [Reading Instruction Based on Character Education]. Bandung: Refika Aditama.

Arends, R.I. & Kilcher, A. 2010. *Teaching for student learning: becoming an accomplished teacher*. New York, NY: Routledge.

Cox, C. 2012. Literature-Based teaching in the content areas. California: SAGE Publications.

Daryanto. 2014. *Pendekatan pembelajaran saintifik kurikulum 2013* [The scientific approach in curriculum 2013]. Yogyakarta: Gava Media.

DeBoer, G.E. 2000. Scientific literacy: Another look at its historical and contemporary meanings and it's relationship to science education reform. *Journal of Research in Science Teaching* 37 (6): 582–601.

Depdiknas. 2008. *Panduan pengembangan bahan ajar* [The guideline in developing teaching aid]. Jakarta.

Donoghue, M.R. 2009. *Language arts: Integrating skills for classroom teaching*. Los Angeles, LA: Sage.

Endah, A. & Mukminan. 2016. Implementasi pendekatan saintifik dalam pembelajaran IPS di *Middle Grade* SD Tumbuh 3 Kota Yogyakarta [Implementation of scientific approach in social science in Middle Grade Elementary School Tumbuh 3 Yogyakarta]. *Jurnal Prima Edukasia* 4 (1): 20–31.

Goouch, K. & Lambirth, A. 2011. *Teaching early reading and phonics: Creative approaches to early literacy*. London: Sage.

Gray, E.C. & Marilyn, C. 2006. Children's use of language and picture in classroom inquiry. *Language Art* 15(4): 227–237.

Hosnan, M. 2014. *Pendekatan saintifik dan kontekstual dalam pembelajaran abad 21: Kunci sukses implementasi kurikulum 2013* [The Scientific and Contextual Approach in 21st Century Instruction: A Success Key of Curriculum 2013 Implementation]. Jakarta: Ghalia Indonesia.

Huck, C.S. & Young, D.A. 1961. *Children's literature in the elementary school.* New York, NY: Holt, Rinehart and Winston.

Huck, C.S., et al. 1987. *Children's literature in the elementary school* (4th ed.). Boston, MA: McGraw-Hill.

Joyce, B. & Weil, M. 1972. *Models of teaching* (2nd Ed.) Boston: Allyn and Bacon.

Kahle, J.B. 1979. *Teaching science in the secondary school.* New York, NY: D. Van Nostrand.

Kelemen, D., et al. 2014. Young children can be taught basic natural selection using a picture storybook intervention. *Psychological Science* 25(4): 893–902. doi: 10.1177/0956797613516009.

Kemendikbud. 2013. Peraturan Menteri Pendidikan dan Kebudayaan RI Nomor 81a, Tahun 2014, tentang Implementasi Kurikulum [Indonesian Education and Culture Ministrial Rule on Implementation of Curriculum].

Lukens, R.J. 1999. *A critical handbook of children's literature* (6th Ed.). New York: Longman.

Magee, P.A. & Wingate, E. 2014. Using inquiry to learn about soil: A fourth grade experience. *Science Activities* 51, 89–100.

Majid, A. 2009. *Perencanaan pembelajaran mengembangkan standar kompetensi guru* [Instruction planning on developing teachers' competency standard]. Bandung: Remaja Rosdakarya.

Mitchell, D. 2003. *Children literature, an invitation to the World.* Boston, MA: Ablongman.

Moore, K.D. 2015. *Effective instructional strategies: From theory to practice (4th ed).* Los Angeles, LA: SAGE.

Nurgiyantoro, B. 2013. *Sastra anak: Pengantar pemahaman dunia anak* [Children Literature: An Introduction to Chldren World]. Yogyakarta: Gadjah Mada University Press.

Pantaleo, S. 2004. Young children and radical change characteristics in picture books. *The Reading Teacher* 58 (2): 178–187.

Prastowo, A. 2014. *Pengembangan bahan ajar tematik* [Developing Thematical Teaching Aid]. Jakarta: Kencana.

Saxby, M. & Winch, G. 1991. *Give them wings: The experience of children's literature* (2nd ed.). Melbourne: The MacMillan Company of Australia.

Sitepu, B.P. 2014. *Pengembangan Sumber Belajar* [Developing Learning Resources]. Jakarta: Raja Grafindo Persada.

Tompkins, G.E. & Hoskisson 1995. *Language arts: Content and teaching strategies (3rd ed).* New Jersey, NJ: Prentice Hall.

Character Education for 21st Century Global Citizens – Retnowati et al. (Eds)
© 2019 Taylor & Francis Group, London, ISBN 978-1-138-09922-7

Higher order of thinking skill in chemistry: Is it a character education?

S. Handayani & E. Retnowati
Universitas Negeri Yogyakarta, Indonesia

ABSTRACT: Science is a well-structured subject. Despite its advantage to technology development, however, there is a rumor that scientists do not nurture values. Moreover, there has been an issue of character education and all teachers are expected to implement this character education at schools, including in chemistry. It seems that chemistry in high school is regarded as an impossible subject to use for character education because it is unlikely to be contributing to the development of the students' character. However, this paper argues that a good character can be acquired while studying chemistry. Chemistry learning involves a heavy cognitive process and its problem-solving requires a higher order of thinking ability. This cognitive ability guides students to think critically and creatively when solving chemistry problems. Since it demands high cognitive load, students learn how to be motivated during learning in order to generate answers. Having these characters may be important not only as a scientist, but also as a community member. This paper presents a topic in high school chemistry directing students to analyze chemical reactions. Then, how studying chemistry can facilitate students to develop a higher order of thinking ability and be scientific (respectful to procedure), and the motivation to learn are discussed.

1 INTRODUCTION

The main objective of education is to assist our generation in becoming good citizens by acquiring and implementing values in their community (Lickona, 1991). School, that is, a formal institution of education, has responsibility for this education value which is also known as character education. There might be many ways to describe character attributed to a person. It is widely agreed that good character education is not only to help students to become smart but also, among other things, to be responsible, respectful, caring, loving, and always happy to do good (Halstead & Taylor, 2010). According to Lickona (2006), characters should be taught although there is no single method of teaching characters. Hence, teachers should use various methods for character education.

Taking into account one expected character is to be responsible. This value is important, particularly when our community deals with problems. Responsible means devoting some efforts to solve problems. Problem-solving refers to an activity to solve problems which cannot be solved directly without a sufficient knowledge base (Miri et al., 2007). On the other hand, being irresponsible means giving up easily. Indeed, being responsible in solving problems also needs motivation and persistence since it may involve a long procedure or, as said by Miri et al. (2007), it may need a higher order of thinking skills.

Science is a compulsory subject in schools. This means that there should be character education in science learning. However, science is often regarded as being a well-structured knowledge which limits the possibility for dialog about values. For example, chemistry in high schools offers learning about the reaction of compounds with certain procedures. Nevertheless, it is arguable that teachers can employ an inquiry method of teaching which emphasizes on problem-solving activity or other teaching methods that could possibly facilitate an open-ended problem-solving activity (Chin & Brown, 2002). Accordingly, chemistry teachers may contribute to character education.

Moreover, chemistry learning may involve a heavy cognitive process and its problem-solving requires a higher order of thinking skill. Problem-solving is when students have to apply knowledge about facts and procedures under certain conditions (Bruning et al., 2011). This higher order of thinking skill guides students to think critically and creatively when solving problems in a particular domain (Newman, 1990). Since it demands high cognitive load (Retnowati et al,. 2016), students should implicitly learn how to be scientific and motivated during learning. Having these characters may be important, not only as a scientist, but also as a community member. Accordingly, this paper argues that there is character education when students learn to solve chemistry problems. The following discussion is elaborated to support this argument.

2 HIGHER ORDER OF THINKING SKILLS

Thinking is a conscious cognitive process to assimilate new knowledge and automate the use of the knowledge in various situations (Bruning et al., 2011). It may be categorized as being a skill since thinking efficiently and effectively takes practice. Subsequently, learning at school may be described as a practice of thinking. Hence, a learning strategy should be designed by teachers to assist students in developing their thinking skills.

Recently, many educators have promoted a higher order of thinking skills as being an important achievement of learning (Jailani et al., 2017). Anderson and Krathwohl (2001) affirmed that the cognitive process dimension can be identified according to its complexity; from the lowest to the highest is: remember, understand, apply, analyze, evaluate, and create. The first three (remember, understand, apply) are categorized as being a lower order of thinking, while the last three (analyze, evaluate, create) are the higher order of thinking (Aktamis & Yenice, 2010; Zohar & Schwartzer, 2005). Nevertheless, this categorization might not be rigid. In other words, complexity of cognitive process is also relative to the level of prior knowledge possessed by the learners and the dynamic interaction of elements of the learning material.

It is asserted that a problem-solving activity could facilitate students in applying knowledge into a more complex cognitive dimension (Bruning et al., 2011; Lati et al., 2012). Some chemistry problem-solving indeed gives specific instruction to analyze and interpret the problem solution, and not just simply to find a solution. Some others ask students to inquire into everyday life problems relevant to chemistry, formulate a hypothesis for the solution, collect empirical data, and make a judgment to accept or reject the hypothesis (Lati et al., 2012).

In fact, problem-solving is a challenging task which requires teachers to be innovative when designing the instruction. For learning by problem-solving, teachers may use individualized or collaborative grouping of students, with explicit guidance (such as by worked example) or by less guided instruction. In deciding which instruction to use, teachers should identify the complexity of the problem-solving and whether students possess the knowledge base to complete and study the problem (Retnowati et al., 2016). Without such consideration, the high demand of problem-solving activity may hinder learning. This situation can create a lack of learning motivation which in turn causes learning failure.

3 PROBLEM-SOLVING IN CHEMISTRY

In high schools, students learn how to analyze and interpret chemical reactions. This task is often categorized as being a higher order of thinking activity since it requires application of their understanding of characteristics of every compound, the relationship among elements, and the procedure of their reaction. However, students could practice a higher order of thinking skills while applying this knowledge base to reaction problem-solving.

Below is an example of problem-solving in chemistry for high school. The task is to analyze possible reactions between cyclopentanone (a) with p-hydroxybenzaldehyde (b).

First of all, students have to evaluate possible reactions of the two compounds, as can be seen in Figure 1, which can be predicted using these four theories:

1. Self-aldol condensation reaction
2. Crossed-aldol condensation reaction
3. Hemiacetal formation
4. Tandem of aldol and hemiacetal formation.

In Figure 1, the compound **a** has Hα, that is acidic because of the induction effect from the oxygen of the carbonyl. Therefore compound **a** can turn into nucleophile by releasing Hα, which may react with other compounds by self-aldol condensation (1), or crossed-aldol condensation (2). The compound **b** in Figure 1 has hydroxil groups, which can act as a nucleophile by reacting with oxygen carbonyl by hemiacetal formation (3). Alternatively, it can be a tandem reaction of those three reactions (Bruice, 2007).

Then, students have to elaborate possible results of the reactions, by interpreting and justifying possible occurrences. Based on the theorems, there are six steps that could be performed, which can be described as follows.

1. Self-aldol condensation, the reaction (1), within compound **a** forming compound **c**, as shown in Figure 2.
2. Hemiacetal reaction inter **b** compounds, which results in hemiacetal (Reaction (3)).
The hemiacetal forming as seen in Figure 3 is possible, as found by Handayani (2014). The compound **d1** may dehydrate to form a more stable compound, called ketone (alkenone) (Figure 4). Dehydration of compound **d1** into **d2** is part of the reaction between cyclohexanone and p-hydroxybenzaldehyde (Handayani et al., 2017).

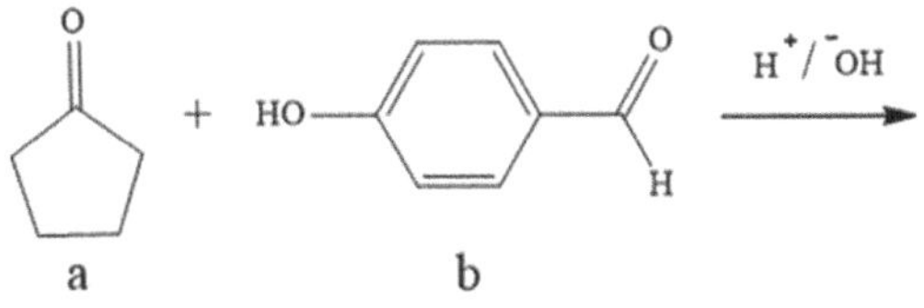

Figure 1. Reaction between **a** and **b** with acid/base catalyst.

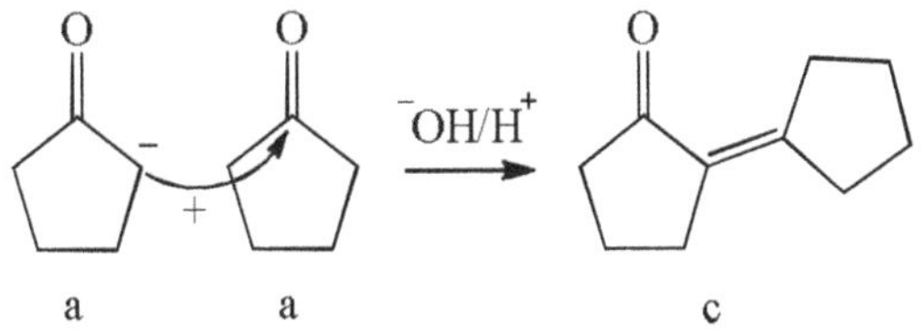

Figure 2. Cyclopentane self-aldol condensation to form [1,1'-bi(cyclopentylidene)]-2-one (**c**).

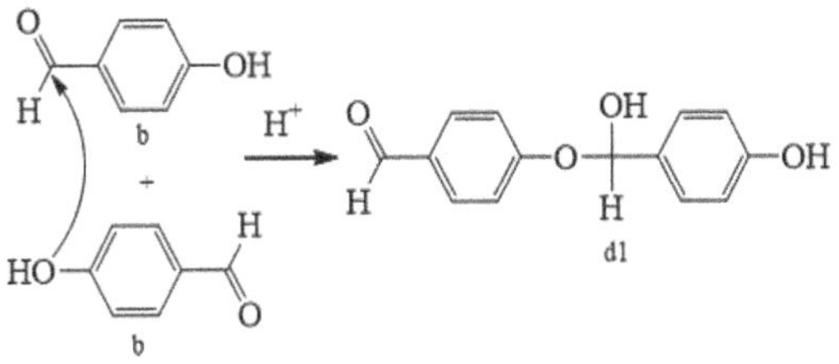

Figure 3. Hemiacetal formation of p-hydroxybenzaldehyde (**b**) to form 4-(hydroxyl (4-ydroxyphenyl) methoxy) benzaldehyde (**d1**).

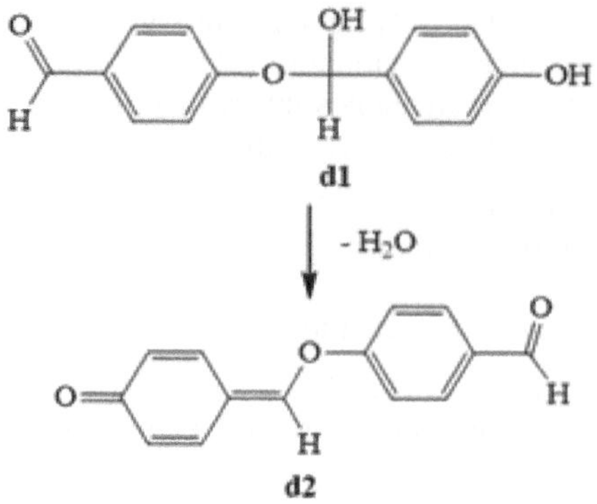

Figure 4. Dehydration of compound **d1** to form 4-((4-hydroxyphenoxy) methylene)cyclohexa-2,5-dienone (**d2**).

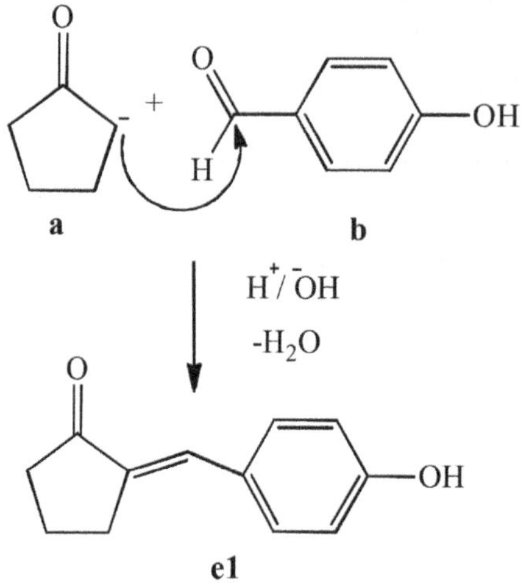

Figure 5. Crossed-aldol condensation between **a** and **b** to form (E)-2-(4-hydroxybenzylidene)cyclopentanone (**e1**).

3. Crossed-aldol reaction between **a** and **b** (Reaction (2)).

The main target of reaction **a** and **b** is usually an **e1**, which is synthesized by crossed-aldol condensation, as indicated in Figure 5. If there is a double crossed-aldol condensation between **a** and **b**, then compound **e2** is formed, as described in Figure 6.

Compound **d2** in Figure 4 remains containing carbonyl, which is very susceptible of nucleophile action. A reaction of **d2** and **e1** in Figure 6 presumably could perform a reaction that is similar to novel synthesis of substituted benzylidene cyclohexanone (Handayani et al., 2017). The occurred reaction is the crossed-aldol condensation where compound **e1** as nucleophile forms the compound **f** (Figure 7).

4. Hemiacetal formation between **a** and **b**.

Another possibility of the targeted reaction between **a** and **b** is hemiacetal formation (**g**), as depicted in Figure 8. This reaction occurs because the hydroxyl on the para position is very reactive.

5. Tandem reaction between self-aldol condensation followed by hemiacetal formation of **b** and **c**.

It is known that p-Hydroxybenaldehyde (**b**) has a hydroxyl group. Meanwhile **c** has a ketone group, so **b** and **c** could perform a hemiacetal formation (Figure 9).

178

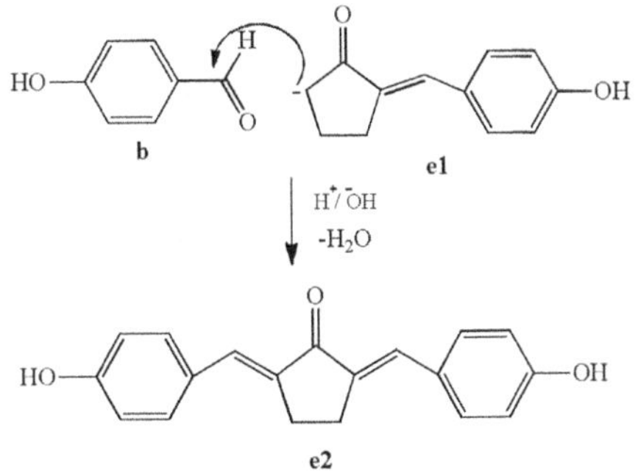

Figure 6.   Double crossed-aldol condensation between **a** and **b** to form (2E,5E)-2,5-bis(4-hydroxybenzylidene)cyclopentanone (**e2**).

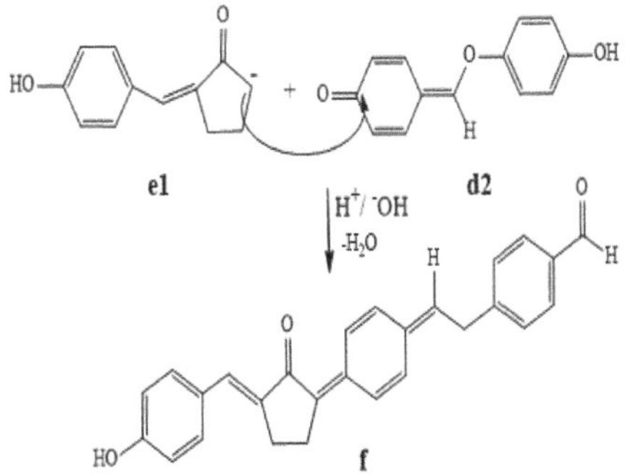

Figure 7.   Tandem reaction between hemiacetal formation, dehydration and crossed-aldol condensation to form 4-((E)-2-((E)-4-((E)-3-(4-hydroxybenzylidene)-2-oxocyclopentylidene) cyclohexa-2,5-dien-1-ylidene)ethyl)benzaldehyde (**f**).

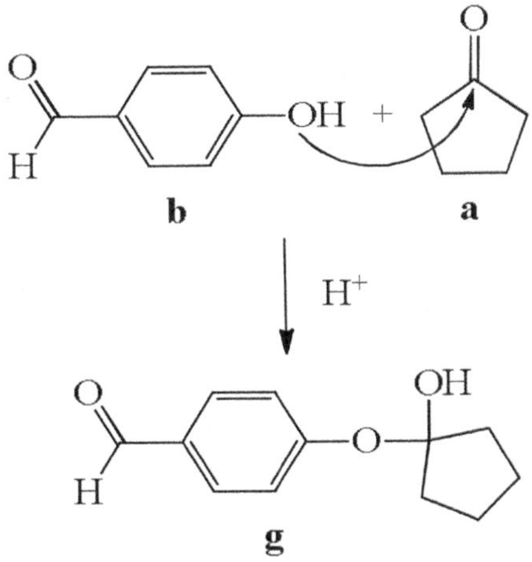

Figure 8.   Synthesis of 4-((1-hydroxycyclopentyl)oxy) benzaldehyde (**g**) by hemiacetal formation.

6. Double crossed-aldol condensation reaction between **a** and **e1** compounds.

Compounds **a** and **e1** have Hα, hence they both can form nucleophiles. If compound **a** becomes nucleophile, then it will react with the **e1**, forming compound **i** (Figure 10). However, if compound **e1** turns into nucleophile and reacts with compound **a**, then it forms compound **j** (Figure 11).

179

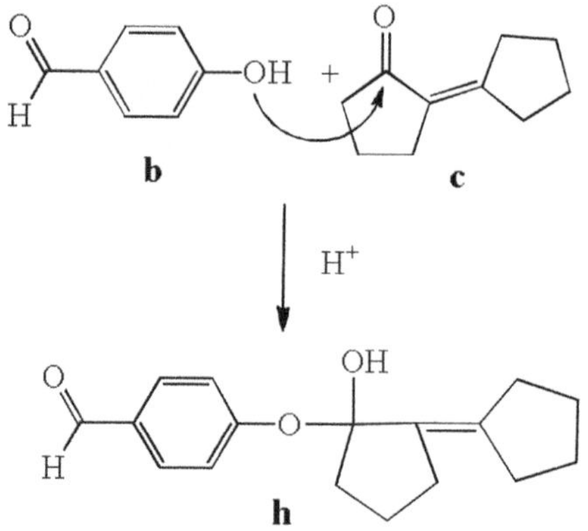

Figure 9. Tandem reaction to form 4-((2-hydroxy-[1,1'-bi(cyclopentylidene)]-2-yl)oxy)benzaldehyde (**h**).

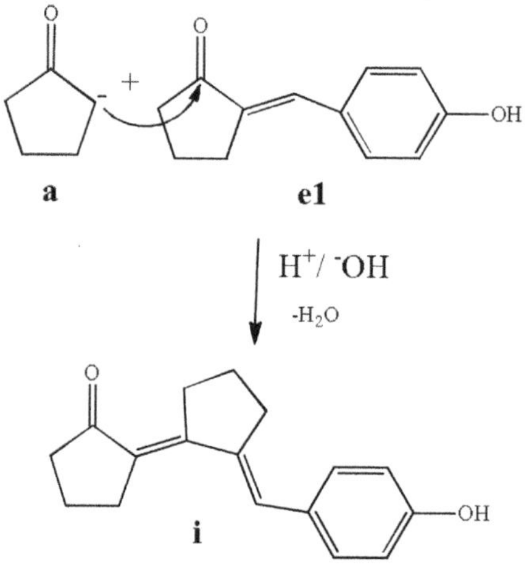

Figure 10. Cyclopentanone as nucleophile to form (1E,2'E)-2'-(4-hydroxybenzylidene)-[1,1'-bi(cyclopen-tylidene)]-2-one (**i**).

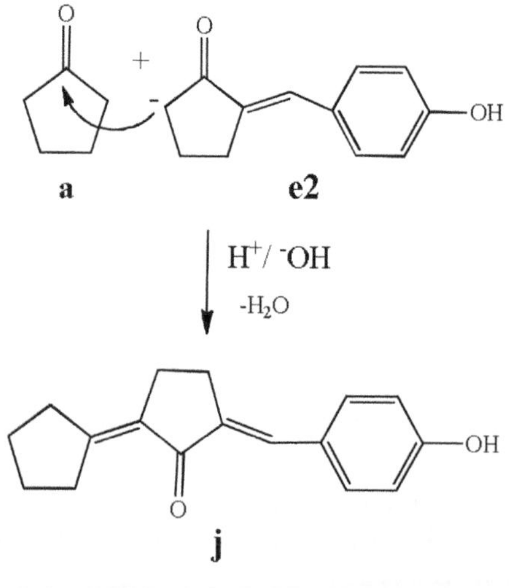

Figure 11. Compound **e** as nucleophile to form (E)-3-(4-hydroxybenzylidene)-[1,1'-bi(cyclopenty lidene)]-2-one (**j**).

180

Generally, there might be a following reaction from each reaction results as shown above, forming new compounds either by crossed-aldol condensation or hemiacetal formation. Accordingly, such problem-solving is an example of an open-ended problem which often appears in chemistry (Lati et al., 2012). Nevertheless, students must be able to identify relevant theories for the given reaction tasks, and apply the correct procedure to yield the reaction results. Being respectful to knowledge is a scientific character widely recognized in chemistry. When students always use a scientific approach in solving reaction problems, they may also nurture the value of being respectful.

4 DISCUSSION

The above discussion presents a topic in high school chemistry that may be directing students to use their cognitive resource to analyze chemical reactions. The main question raised in this paper is how studying chemistry can facilitate students to develop a higher order of thinking skill, and how this can provide alternative ways of nurturing characters expected by our society (Halstead & Taylor, 2010; Jailani et al., 2017; Lickona, 1991, 2006).

Learning chemistry by problem-solving of compound reactions directs students to analyze concepts of compounds and apply procedures of reactions by considering the relevant theories. The relevant theories work as the knowledge base, which enable students to comprehend the given problem-solving context (Miri et al., 2007; Shwartz et al., 2006). This leads students to be critical when applying certain theories and interpreting the reaction results. Students also learn that procedures must be used appropriately in order to yield scientific results. From the above description of solving reaction problems, it can be seen that such chemistry problem-solving involves many possible solutions. There are at least six steps which constitutes 11 reaction Figures. To perform these, students must be able to decide what theories are to be applied as well as the properties of the compounds.

The high degree of cognitive process during problem-solving may cause heavy cognitive load (Retnowati et al., 2016). However, having a sufficient knowledge base and collaborative instruction may stimulate the motivation to learn. The reaction problem-solving itself might be motivating since it opens up the possibility of different answers. Where students do not have an appropriate knowledge base, Retnowati et al. (2016) indicated that students could be able to develop their knowledge transfer after studying pairs of worked examples and problem-solving. The worked examples provide a knowledge base, and the paired problem-solving motivates students to apply the knowledge.

Stated by Aktamis and Yenice (2010), developing complex cognitive ability in science is also indicated by how the instruction is evaluated. Consequently, after learning chemical reactions, the students' achievement may be measured using a test that also requires students to use their higher order of thinking skills. Achievement recognition may be useful for both learners and teachers as it provides information about whether or not the instruction is comprehensive.

5 CONCLUSION

The community can expect chemistry students to develop expected characters while accomplishing chemistry problem-solving. Problem-solving in chemistry requires a higher order of thinking skill, by being critical and creative in thinking of problem solutions. The complex material in chemistry may demand high cognitive load; hence students learn to be respectful to procedures and to be motivated in applying knowledge. Therefore, this paper recommends that chemistry learning should be designed in such a way that problem-solving is the main activity.

REFERENCES

Aktamiş, H. & Yenice, N. (2010). Determination of the science process skills and critical thinking skill levels. *Procedia—Social and Behavioral Sciences, 2*(2), 3282–3288. doi:10.1016/j.sbspro.2010.03.502.

Anderson, L.W. & Krathwohl, D.R. (2001). *A taxonomy for learning, teaching, and assessing: A revision of Bloom's taxonomy of educational objectives.* New York, NY: Addison Wesley Longman.

Bruice, P.Y. (2007). *Organic chemistry* (5th ed.). New York, NY: Pearson Prentice Hall.

Bruning, R.H., Scraw, G.J. & Norby, M.N. (2011). *Cognitive psychology and instruction* (5th ed.). Boston, MA: Pearson.

Chin, C. & Brown, D.E. (2002). Student-generated questions: A meaningful aspect of learning in science. *International Journal of Science Education, 24*(5), 521–549.

Halstead, J.M. & Taylor, M.J. (2010). Learning and teaching about values: A review of recent research. *Cambridge Journal of Education, 30*(2), 169–202. doi:10.1080/713657146

Handayani, S. (2014). Study of acid catalysis for condensation of 4-hydroxybenzaldehyde with acetone. In *Proceedings of ICRIEMS Universitas Negeri Yogyakarta*, 18–20.

Handayani, S., Budimarwanti, C. & Haryadi, W. (2017). Novel synthesis of substituted benzylidene-cyclohexanone by microwave assisted organic synthesis. In *AIP Conference Proceedings, 1823*(1), 020116. doi:10.1063/1.4978189

Jailani, J., Sugiman, S., & Apino, E. (2017). Implementing the problem-based learning in order to improve the students' HOTS and characters. *Jurnal Riset Pendidikan Matematika, 4*(2): 247–259.

Lati, W., Supasorn, S. & Promarak, V. (2012). Enhancement of learning achievement and integrated science process skills using science inquiry learning activities of chemical reaction rates. In G.A. Baskan, F. Ozdamli, S. Kanbul & D. Ozcan (Eds.), *4th World Conference on Educational Sciences, 46*, 4471–4475.

Lickona, T. (1991). Educating for character: How our school can teach respect and responsibility. New York, NY: Bantam Books.

Lickona, T. (2006). Eleven principles of effective character education. *Journal of Moral Education, 25*(1), 93–100. doi:10.1080/0305724960250110

Miri, B., David, B.C. & Uri, Z. (2007). Purposely teaching for the promotion of higher-order thinking skills: A case of critical thinking. *Research in Science Education, 37*(4), 353–369.

Newman, F.M. (1990). Higher order thinking in teaching social studies: A rationale for the assessment of classroom thoughtfulness. *Journal of Curriculum Studies, 22*(1), 41–56.

Retnowati, E., Ayres, P. & Sweller, J. (2016). Can collaborative learning improve the effectiveness of worked examples in learning mathematics? *Journal of Educational Psychology, 109*(5), 666. doi:10.1037/edu0000167

Shwartz, Y., Ben-Zvi, R. & Hofstein, A. (2006). Chemical literacy: What does this mean to scientists and school teachers? *Journal of Chemical Education, 83*(10), 1557–1561.

Zohar, A. & Schwartzer, N. (2005). Assessing teachers' pedagogical knowledge in the context of teaching higher-order thinking. *International Journal of Science Education, 27*(13), 1595–1620.

Need analysis for developing a training model for Office Administration Vocational School teachers

M. Muhyadi, S. Sutirman & Rr.C.S.D. Kusuma
Universitas Negeri Yogyakarta, Indonesia

ABSTRACT: The role of a teacher in the teaching and learning process is very important, but there are many teachers lacking competence in mastering pedagogic skills or subject matter. This study aims to identify the need for the development of a training model for Office Administration Vocational Schools (OAVS) teachers in order to increase their competence. This research study was conducted by a survey method using quantitative descriptive analysis. Research subjects consisted of OAVS teachers in Yogyakarta Special Region, Indonesia. Data were collected through a questionnaire and document analysis. Data analysis was done descriptively. The results showed that: a) the OAVS teachers realized that their teaching and learning competence was still low and that it needed to be improved through training, b) the training desired by OAVS teachers included training in the fields of pedagogy and learning material mastery, c) learning aspects that needed to have training included: development and use of learning media, the practice of learning strategy, development of learning materials, classroom action research, evaluation of learning outcomes, and learning theory, d) the form of training desired by OAVS teachers is a combination between theory and practice that is done simultaneously.

1 INTRODUCTION

Vocational education, including Office Administration Vocational Schools (OAVS), is intended to produce graduates who have the competency either to work or to undertake further study. Vocational education is a way of mastering the basic skills, which is essential for fair competition in the job market (Thompson, 1973). According to Prosser and Quigley (1968), vocational education will be effective if the teachers have adequate experience and consistently apply their ability and skill in teaching. Therefore, the OAVS teachers need to have good knowledge, attitudes and skills in order to produce good graduates.

Preparing teachers for the teaching profession is conceived as being a high priority in any country, since this profession is considered to be challenging and is critical to the nation's development and progress in the different domains. As a huge enterprise, education has great importance in building strong and developed societies, and the teacher is one of the primary agents for achieving that. For these reasons, it is always an urgent educational need that teachers should receive adequate educational and professional training in order to possess adequate knowledge and teaching skills and to be able to dedicate themselves to the teaching profession (Boudersa, 2016).

The professional quality of teachers is an absolute requirement for creating a good educational system. Without good teachers there cannot be good education. This means that teachers must continuously improve their knowledge and skills. Various programs have been undertaken either by the government or by non-government institutions in order to improve the competence of teachers in Indonesia. According to Nursyam (2013), a research study on the effectivity of teacher certification held by UNESCO showed no significant differences in teaching quality between those teachers who have received certification and those who have not. The efforts to improve the educational qualifications of teachers have not yet had any significant impact on the performance of teachers, because most of the further education carried out was not in accordance with their previous educational background, so that the training had very little influence on the improvement of their mastery of the subject matter (Dekawati, 2011; Marfu'ah et al., 2017). In addition, based on information from some OAVS teachers in Yogyakarta, it is known that they rarely have the opportunity for training or professional development. The government training program is often limited to a technical counseling model for the head of the study program, not for the teacher.

Based on the above description, it can be noted that many of the efforts that were made to improve the competence of teachers have not run effectively. The failure is assumed to have resulted from irrelevant training. That is, there has been no specific training model that is operational and practical for increasing the competence of teachers, especially OAVS teachers. Thus, it is important to analyze the need for the development of a training model for OAVS teachers. The results of this study will be used as the basis for developing a training model to increase the competence of OAVS teachers.

Training is closely related to the concept of education. Training is a model of educational process-oriented development of the competence of human resources. Noe (2005) defines training as an activity that is planned by the institution in order to facilitate the employee in learning the competencies of a particular task. Based on this definition, training leads to an improvement in work competency, including knowledge, skills, and attitudes that can support the success of the work.

Along with the above opinion, Dessler (2009) stated that training is an effort to provide the skills that are needed by the employee to do their job. Wexley and Latham (1991) state that training and development refers to a planned effort by an organization to facilitate the learning of job-related behavior on the part of its employees. This training and development is an effort planned by the organization in order to facilitate the learning that is related to the behavior needed by an employee in a work unit.

Training is a planned activity and its purpose is to make a change in the participants' behavior. The purpose of the training is to improve the capabilities of the human resources and the participants' understanding of the work that will become their responsibility. Wexley and Latham (1991) argued that the purpose of training is to increase an employee's self-awareness, skills, and motivation. Thus, it can be concluded that the purpose of conducting training is to improve the knowledge, skills, and attitude of the employees in accordance with the needs and the changes that occur in the workplace. According to Kirkpatrick and Kirkpatrick (2006), there are four levels of criteria for evaluating the effectiveness of training: reaction, learning, behavior, and result. Reaction is the degree to which participants find the training favorable, engaging, and relevant to their jobs. Learning is the degree to which participants acquire the intended knowledge, skills, attitude, confidence, and commitment, based on their participation in the training. Behavior is the degree to which participants apply what they have learned during training when they are back on the job, and result is the degree to which targeted outcomes occur as a result of the training and the support and accountability package.

2 METHODS

This study is a survey research that uses a quantitative approach. The population consisted of OAVS teachers in Yogyakarta Special Region, Indonesia. The research subjects were selected based on area; each district of Yogyakarta Special Region was represented by a sample. Based on this technique, 24 teachers were selected to form the sample. Data were collected through a questionnaire and documentation analysis. Data were analyzed using descriptive technique, presented narratively and followed by presentation in data grouping table based on percentage.

3 RESULT

The teacher's competence to be covered in the training was classified into two kinds: professional competence and pedagogical competence. Professional competence is a teacher's mastery of the subject matter to be delivered to the student, as displayed in Tables 1 and 2. Pedagogical competence relates to the strategy of teaching and the learning processes that should be mastered by teachers when delivering subject matter, as displayed in Tables 3 and 4.

3. 1 The need for training

The results of the data analysis of the need for training materials are summarized in Table 1. This table shows that subjects that are grouped in the basic vocational field (Introduction to

Table 1. The need for training for each of the subjects (N = 24).

Subject matter	Doesn't need (%)	Need (%)	Very need (%)	Not responding (%)
Introduct to business and economic	20.83	54.17	20.83	4.17
Introduct to office administration	25.00	41.67	29.17	4.17
Introduct to accountancy	12.50	62.50	25.00	0.00
Office automation	0.00	25.00	75.00	0.00
Correspondence	0.00	16.67	58.33	0.00
Record and document	0.00	20.83	75.00	4.17
Digital simulation	0.00	16.67	79.17	4.17
Personnel administration	0.00	66.67	33.33	0.00
Finances administration	0.00	54.17	41.67	4.17
Office supplies and equipment	0.00	50.00	50.00	0.00
Public relation and protocol	0.00	37.50	58.33	4.17

Table 2. The training model that is most suitable for each subject (N = 24).

Subject matter	Theory (%)	Practice (%)	Theory & practice (%)	Not responding (%)
Introduct to business and economic	20.85	20.85	37.50	20.85
Introduct to office administration	25.00	20.75	37.50	16.67
Introduct to accountancy	0.00	8.33	75.00	16.67
Office automation	0.00	16.67	83.33	0.00
Correspondence	0.00	12.50	87.50	0.00
Record and document	0.00	12.50	87.50	0.00
Digital simulation	4.17	12.50	83.33	0.00
Personnel administration	8.33	33.33	50.00	8.33
Finances administration	0.00	20.85	79.17	0.00
Office supplies and equipment	4.17	20.85	75.00	0.00
Public relation and protocol	8.33	12.50	79.17	0.00

Table 3. Respondents' opinions about the need for training in various learning aspects (N = 24).

Learning aspect	Does not need (%)	Need (%)	Very need (%)	Not responding (%)
Learning media	0.00	41.67	58.33	0.00
Learning strategy	0.00	54.17	45.83	0.00
Learning material	0.00	58.33	41.67	0.00
Classroom action research	0.00	75.50	62.50	0.00

Table 4. The appropriate learning model for every learning aspect (N = 24).

Subject material	Theory (%)	Practice (%)	Theory and practice (%)
Learning media	4.17	16.67	79.17
Learning strategy	4.17	8.33	87.50
Learning material	4.17	8.33	87.50
Classroom A.R.	0.00	8.33	91.67
Student assessment	0.00	4.17	95.83
Learning theory	0.00	4.17	54.17

Economics and Business, Introduction to Office Administration, and Introduction to Accountancy) have a necessity for training. The mean percentage of the respondents' opinions for those three subjects are 52.78% for the need to be trained and 25.00% for very necessary to be trained. This means that either most teachers lack the competence to master basic vocational subject matters, or they want to improve aspects of their competence through training by third parties.

For the subject matters that are included in the group of basic vocational competencies, namely: Office Automation, Correspondence, Records and Document, and Digital Simulation, most of the respondents expressed that it was either necessary or very necessary to be trained. The mean percentage of the respondents' opinions for those four subjects are: 26.04% for need to be trained, 71.88% for very necessary to be trained, and 2.01% did not respond. These data indicated that the teachers' mastery of the material for basic vocational competencies is still proving to be inadequate, so therefore training is required.

The subject matters that are included in the vocational competency group, namely: Personnel Administration, Finances Administration, Office Supplies and Equipment, and Public Relations and Protocol, also required training. Most respondents expressed that they needed, or even greatly needed, the training. The mean percentage of the respondents' opinions for those four subjects are: 52.08% for need to be trained and 45.83% for very necessary to be trained, while 2.22% did not respond. The data also shows that the teacher does not completely master the teaching materials. Teachers feel that they need training.

3.2 Type of training

The results of the analysis of the data for the type of training are summarized in Table 2. This table shows that, according to the teachers, a model exercise for subjects is different from one subject to another. Most respondents stated that the most appropriate training model for the subject of Office Administration was theory and practice simultaneously. Only a few respondents stated that the most suitable training model was theory. The data shows that the subject matters that should be trained using a model of both theory and practice simultaneously are: Introduction to Accountancy (75.00%), Office Automation (83.33%), Correspondence (87.50%), Records and Document (87.50%), Digital Simulation (83.33%), Personnel Administration (79.17%), Office Supplies and Equipment (75.00%), and Public Relations and Protocol (79.17%).

3.3 The learning aspects that need training

Table 3 shows the results of the data analysis regarding those aspects of learning that required training. This table shows that, according to the teachers, all aspects of learning (learning media, learning strategy, learning materials, classroom action research, student assessment, and learning theory) need to be trained. The detailed data regarding the degree of necessity of various aspects are shown in Table 3.

The data in Table 3 shows that, according to the teachers, all aspects of learning need to be trained. The mean percentage of the respondents' opinions about the need for training for those six aspects of learning are: 49.31% for need to be trained, 50% for very necessary to be trained, and 0.69% did not respond.

3.4 The training model of learning aspects

The data regarding the type of training that should be applied toward various learning aspects is presented in Table 4.

The data in Table 4 shows that most of the respondents stated that the most appropriate model of training for those six learning aspects was the use of theory and practice simultaneously. The mean percentage of the respondents' opinions on the training for those six aspects of learning was that it should be delivered in the following models: 2.01% theory, 8.33% practice, and 82.64% theory and practice simultaneously.

4 DISCUSSION

4.1 The importance of training

Teacher training usually involves providing training in curriculum subjects for teachers by organizing workshops over a period of time. The main objective of such workshops is to keep teachers up to date in their subject area (Boudersa, 2016).

The eleven areas of training content have different weights, as seen from the indicator of the different number of hours of training needed for each of the subjects. For the basic vocational fields group, each subject is given a weight of 6 hours, basic vocational competency groups between 6–12 hours, and vocational competency groups between 18–24 hours.

The training content in the vocational fields is perceived to cover: Introduction to Economics and Business, Introduction to Office Administration, and Introduction to Accountancy; some of the respondents (16.67%) stated that they do not need training, but most of them (83.33%) expressed that it was either necessary or very necessary to be trained. This is an indicator that either their mastery of the material is still weak, or that the teachers require the development of materials that are expected to be gained through the training program.

For the subject matter included in the basic vocational competency group, namely: Office Automation, Correspondence, Records and Document, and Digital Simulation, all of the respondents (100.00%) expressed that it was either necessary or very necessary to be trained. The proportion of respondents who expressed the need for considerable training was 71.88%, while the rest (28.12%) expressed that it was necessary.

This was almost the same for the basic vocational competency group. The subject material included in this group, Personnel Administration, Financial Administration, Office Supplies and Equipment, and Public Relations and Protocol, also required training. All of the respondents (100.00%) expressed the need for training, and even that it was very necessary. The proportion of respondents who expressed such opinions amounted to 52.08% (need the training) and 47.92% (very necessary to have the training). The data also showed that the mastery of the learning material by the teachers was inadequate or that they were eager to develop material through training. This data is supported by Wexley and Latham (1991), who stated that training is needed to increase an employee's self-awareness, skills, and motivation. Martinet et al. (2000) also stated that industry demands increased knowledge, versatility and autonomy from its personnel. These new expectations are held not only by employers, but also by unions and individuals who want to prepare for the new realities of the workplace, meaning that the vocational education sector must also make the switch to a knowledge-based economy. Vocational education teachers must be aware of the challenges and adapt their educational and pedagogical strategies accordingly. Some aspects of learning, namely: learning media, learning strategy, learning materials, classroom action research, student assessment, and learning theory, are also shown to be important in the training of teachers.

4.2 The training models

The incoming data showed that for all subject matters, including learning aspects, the majority of respondents wanted the training model to include theory and practice simultaneously. There was only one aspect that most of respondents required to be organized as a theory

187

model, and that was learning theory. In a detailed form, the appropriate model of training according to the respondents can be described as follows.

A few respondents (6.43%) stated that for a number of subjects, namely: Introduction to Economics and Business, Introduction to Office Administration, Introduction to Accountancy, Digital Simulation, Personnel Administration, Office Supplies and Equipment, and Public Relations and Protocol, the training should be conducted using the theory model. Of the remainder, 17.42% of respondents stated that the training should preferably be conducted using the practice model, and 70.46% stated that it should be conducted using the theory and practice models simultaneously. This data indicated that the respondents already understood the concept or theory of nearly all of the subjects they are taught. The thing that they needed was the skills related to their respective subjects.

In addition to the material aspects of learning, namely: learning media, learning strategies, learning materials, classroom action research, student assessment, and learning theory, the training models that were desired by the respondents were not much different from the training of material learning. Of the learning aspects, there was one point (learning theory) where almost half of the respondents (41.66%) desired to do training using the theory model. This can be understood because the content of the subject is theory.

Overall, the number of respondents who wanted the training regarding the subject matters and learning aspects to be organized using only the practice model was 8.33%, and those who wanted it to be organized using the theory and practice models simultaneously was 83.33%. In line with this finding, Grollmann (2008) stated that work experience is often required in Technical and Vocational Education and Training (TVET) as a precondition to employment as a vocational teacher. Even in Germany, which maintains the highest formal level in terms of academic requirements for entering the vocational teaching field, there is usually an amount of real work experience prescribed through the university curricula. These findings suggest that training for OAVS teachers should be given in the form of theory and practice simultaneously.

4.3 Need for training in learning aspects

For a number of learning aspects, more than half of the respondents stated that it was very necessary to give training. It was considered very necessary to receive training in learning media by 58.33% of respondents, in classroom action research by 62.50%, and in student assessment by 62.50%. Meanwhile, it was considered very necessary to receive training in learning strategy by 45.83% of respondents, in learning materials by 41.67%, and in learning theory by 29.17%. In addition to those who expressed the opinion that training was very necessary, most respondents expressed the need to be trained. The proportion of respondents who expressed that training was needed for learning aspects are as follows: 41.66% for learning media, 54.17% for learning strategies, 58.33% for learning materials, 37.50% for classroom action research, 37.50% for learning student assessment, and 66.67% for learning theory. These findings suggest that training for all aspects of learning, including: learning media, learning strategies, learning materials, classroom action research, assessment of the result of learning, and learning theory, needs to be given to OAVS teachers.

4.4 The training model of learning aspects

Most of the respondents stated that training regarding learning aspects should be delivered using theory and practice models simultaneously. For learning media, there were only 4.17% of respondents who chose theory as a model of training, 16.67% chose the practice model, whereas using the practice and theory models simultaneously was chosen by 79.17%. Next, for the aspect of learning strategies, 4.17% chose theory, 8.33% chose the practice model, and 87.50% chose the theory and practice models simultaneously. Similarly to learning strategies, for learning materials there were only 4.17% who chose theory as the training model, 8.33% chose the practice model, and 87.50% chose the theory and practice models simultaneously. For classroom action research, none of the respondents chose theory as the training model,

8.33% chose the practice model, and 91.67% chose the theory and practice models simultaneously. For student assessment, none of the respondents chose theory as the training model, 4.17% chose the practice model, and 95.83% chose the theory and practice models simultaneously. This finding is in line with Engestron's statement (Oviawe et al., 2017) that the ingredients for effective learning include: (i) ensuring that individuals have access to theoretical and experimental knowledge; (ii) the opportunity to engage in authentic tasks and interaction with others; (iii) the chance to develop critical and intellectual capacities through the application of concept and theory in practice; and (iv) the opportunity to have their thinking and understanding enhanced through the guidance and teaching of others. These findings show that, for a number of aspects of learning, the training model that was considered to be most appropriate by the OAVS teachers was the model that combines theory and practice simultaneously.

5 CONCLUSION

The findings above lead to the following conclusions: (1) the OAVS teachers realize that their teaching and learning competence is still low and that it needs to be improved through training, (2) the training desired by the OAVS teachers includes training in the fields of pedagogy and learning materials mastery, (3) learning aspects that need to be trained include: development and use of learning media, the practice of learning strategy, development of learning materials, classroom action research, evaluation of learning outcomes, and learning theory; (4) the form of training desired by OAVS teachers is a combination between theory and practice that is done simultaneously.

REFERENCES

Boudersa, N. (2016). *The importance of teachers' training programs and professional development in the Algerian educational context: Toward informed and effective teaching practices* (Web log message). Retrieved from https://www.researchgate.net/publication/309430087.

Dekawati, I. (2011). Manajemen pengembangan guru [The Management of teacher development]. *Jurnal Cakrawala Pendidikan, [Journal of Educational Firmament] XXX*(2), 203–215. doi: 10.21831/cp.v0i2.4228

Dessler, G. (2009). *Human resources management* (12th ed.). Upper Saddle River, NJ: Pearson Education, Inc.

Grollmann, P. (2008). The quality of vocational teachers: Teacher education, institutional roles and professional reality. *European Educational Research Journal, 7*(2), 535–547.

Kirkpatrick, D.L. & Kirkpatrick, J.D. (2006). *Evaluating training program* (3rd ed.). San Francisco, CA: Berrett Koehler Publisher, Inc.

Marfu'ah, S., Djatmiko, I.W., & Khairudin, M. (2017) Learning Goals Achievement of a Teacher in Professional Development. *Jurnal Pendidikan Teknologi dan Kejuruan, 23*(3). 295–303

Martinet, M., et al. (2000). *Teacher training in vocational education orientations professional competencies.* Quebec, CA: Ministere de l'Education.

Noe, R.A. (2005). *Employee training and development* (3rd ed.). New York, NY: McGraw-Hill.

Nursyam, N. (2013). *Litbang Perlu Teliti Efektifitas Program Sertifikasi Guru* [Government should examine the effectiveness of teacher sertification program] (Web log message). Retrieved from http://diktis.kemenag. go.id/NEW/index.php? berita=detil&jenis=news&jd=78#.WYBlUum.TPIU

Oviawe, J.I., Uwameiye, R., & Uddin, P. S. O. (2017). Bridging skill gap to meet technical, vocational education and training school-workplace collaboration in the 21st century. *International Journal of Vocational Education and Training Research, 3*(1), 7–14.

Prosser, C.A. & Quigley, T.H. (1968). *Vocational education in a democracy* (2nd ed.). Chicago, IL: American Technical Society.

Thompson, J.F. (1973). *Foundations of vocational education: Social and philosophical concepts.* Englewood Clifts, NJ: Prentice-Hall, Inc.

Wexley, K.N. & Latham, G.P. (1991). *Developing and training human resources in organization* (2th ed.). New York, NY: Harper Collins Publishers Inc.

Developing a picture storybook based on the scientific approach through project-based learning

L.Y. Nawangsih & Z.K. Prasetyo
Universitas Negeri Yogyakarta, Yogyakarta, Indonesia

ABSTRACT: This study aimed to produce a picture storybook, based on the scientific approach of using the project-based learning method, as learning material for grade students. This research and development referred to the steps proposed by Borg and Gall (1983), that is: information collection, planning, developing a preliminary form of product, preliminary field testing, main product revision, main field testing, and operational product revision. The data was collected by means of an assessment sheet for the picture storybook product, a teacher response questionnaire, and a students' response questionnaire. The results of this research showed that the picture storybook, based on the scientific approach of using the project-based learning method, was valid as a learning material for grade students.

1 INTRODUCTION

Education is one of the factors that determines the future of the next generation, and education therefore becomes one of the benchmarks of human resources. Changes have been made by the government in order to improve education in Indonesia. One of these was making changes to the curriculum, and the current curriculum used in Indonesia is Curriculum 2013.

Curriculum 2013 is synonymous with learning characteristics that are implemented using a scientific approach and undertaken in an integrated/thematic way. Success in implementing a Curriculum 2013 lesson does not escape the use of supporting objects, such as the learning materials used in the learning process. Prastowo (2014) states that learning materials are all materials (information, tools, or texts) that are organized systematically, and they display the overall competence to be achieved.

Based on observations, teaching materials are still minimal and only the teaching materials provided by the government, the textbook for Curriculum 2013, are used. In the Curriculum 2013 textbook there is still a lack of fit between the image and the text. Also, the number of images is limited and the size of them is very small, which means that they are often not clearly visible.

Based on the results of interviews, students feel less interested in the teaching materials that are used because of the amount of writing. Students love books that have lots of colors and great pictures.

Based on these needs, this study developed picture storybook material. This study developed a picture storybook because of the needs of the students, as discovered through the questionnaires, and the results showed that students like picture books. The picture storybook in this study presents many pictures, interesting stories, and materials presented through the story, with the goal of attracting the attention of the students reading the picture books.

The picture storybook that was developed has a purpose as a supplement book for Curriculum 2013, so it can be used in the learning process. Curriculum 2013 is synonymous with a scientific approach. It emphasizes learning-focused learning and uses five scientific activities: observing, questioning, gathering information, reasoning, and communicating. This picture storybook was developed by using a scientific approach using the Project Based Learning (PjBL) method. The PjBL method was chosen because it is flexible and workable in accordance with the scientific approach. The PjBL method has advantages because it contains scientific activities, which are project planning, project launch, guided inquiry and product creation, and project conclusion (Mergendoller et al., 2006). It can be said that the PjBL method has the same characteristic of focusing on students. So the picture storybook in this research uses a scientific approach using the PjBL method.

2 LITERATURE REVIEW

The picture storybook is a teaching material that contains stories with pictures and exercises that help students learn. Picture books are books that convey information in two ways, through writing and illustrations (Huck et al., 1987).

The picture storybook in this research is intended as a teaching material; that is, a book that can be used in the learning process. Picture books contain messages about learning materials presented through stories. This is supported by Rothlein and Meinbach (1991) who state, "a picture storybook containing its message through illustrations and written text; both elements are equally important to the story". Based on this it can be understood that the contents of a picture book contain a message that is conveyed through picture illustrations and writing.

Picture storybooks not only display pictures and stories as entertainment for readers, but they can be used as teaching materials. Instructional pictorial storybooks not only contain stories that aim to entertain, but the story content can also be used for learning materials. The discovery of the content of the story should be made with a clear and logical narrative, as shown in Rossiter et al. (2008).

Picture storybooks can be used as teaching materials if they qualified materials. RoI (2008) stated that teaching materials must meet four components, namely: the content, language, presentation, and graphs. In addition, the picture book should contain two components, namely the quality content component and the visual display component (Nurgiyantoro, 2013). The content component consists of themes, characters, background, plot, point of view, language style, message, and content. The visual display component consists of text, image, page, color selection, shape, size, and texture.

The picture storybook in this study uses a theme that relates to the themes in Curriculum 2013, particularly theme 6: "Water of Earth and Sun". The characters used in this study were the four main characters, "Hanung, Cika, Bagas, and Anisa". The setting in this study uses backgrounds, times, and situations, and the plot uses an advanced groove. The point of view used in this illustrated storybook gives an all-round perspective, and the style of the language uses a simple sentence structure that begins with a capital letter and ends with a full stop. The message in this story uses an imitative technique that is conveyed by the author through the character; the story is presented through the daily life of the child by linking activities to the scientific approach using the PjBL method.

The instructional material contained in the picture storybook in this study uses a scientific approach. There are five activities in the scientific approach, namely: observing, questioning, gathering information, reasoning, and communicating (Kurniasih & Sani, 2014). The scientific activities are in accordance with the scientific activities used in the PjBL method. The PjBL method is a learning method in which there are scientific activities, such as project planning, project launch, guided inquiry and product creation, and project conclusion (Mergendoller et al., 2006; Rina, 2017). This scientific activity is in accordance with the activities of the scientific approach as given in Curriculum 2013.

The picture storybook was created using the PjBL method because this method is in accordance with Curriculum 2013, which emphasizes student-focused activities. Also, the PjBL method can be used in various curricula because it is flexible. This is in accordance with the opinion of Westwood (2008), who states that project methods can be embedded in various curricula. Based on these statements, it is evident that the PjBL method is flexible and applicable to learning by the scientific approach used in Curriculum 2013.

Learning by the PjBL method is student-centered, focusing on analyzing activities to solve a problem by applying the concepts learned. PjBL activities require students to conduct investigations to find facts, and to be realistic (Gulbahar & Hasan, 2006). The characteristics of learning with the PjBL method are described by Larmer and Johnas (2010), who states that learning by using the PjBL method is student-centered. These characteristics correspond to a scientific approach.

There are various picture storybooks that can be found. However, they cannot be used as the material because those books are only for fun. Picture-illustrated books have the advantage of helping children to learn independently by developing their skills. This is supported by Donoghue (2009), who states that, in a good picture book, it is necessary to write down the action/activity of a credible and sensible character, and to develop the natural behavior of the character to be highlighted.

The picture storybooks in this study are used as teaching materials that can help students learn. The material from picture storybooks was developed by using scientific approach with PjBL method. The activity is done by a character in the story, who invites the reader to follow up or perform the same activities as the character by inviting them to develop their thinking skills, so it is not done directly through the children's picture book. Therefore it can be interpreted that picture books can help students learn. Research found that illustrated storybooks can improve thinking skills, as shown in Kelemen et al. (2014).

This study develops storybook material that can be used to help in the students' learning process. This is supported by findings by Anastasia and Mukminan (2016, pp. 20–31) on the use of a scientific approach to improve students' deeper thinking skills. This could mean that learning by using direct activities can help students to learn in more depth.

Picture storybooks not only focus on the picture, but can be incorporated in scientific activities, such as scientific approaches and PjBL methods, through the activities performed by the character. This is because, during the course of the plot, the story develops naturally and makes sense.

The picture storybook in this research was developed based on two components: quality of contents and visual quality. In addition, the pictorial storybook has four aspects that show that this picture book is a teaching material: the content feasibility components, components of presentation of material, language, and graphs. The four components of instructional materials are included in the components of a picture storybook. In those picture storybook, content and materials are categorized in the quality of content. The teaching materials components of language and graphs that are included in the picture books relate to the visual display quality component. In addition, picture books should have the benefit of developing personal value and educational value. This pictorial book study is based on a scientific approach using the PjBL method, because in picture books there are five scientific activities using the PjBL method. The PjBL method can be included in a scientific approach, namely: (1) the preparatory stage is included in observing and questioning activities; (2) the planning stage is included in information gathering activities; (3) the implementation stage is included in the activities of associating; and (4) the communicating stage is included in communicating activities. Clearly, pictorial storybook indicators are based on a scientific approach using the PjBL method.

3 RESEARCH METHOD

The development of this resource used the Borg and Gall models. The stages used were: (1) research and information, (2) planning, (3) develop preliminary form of product, (4) preliminary field testing, (5) main product revision, (6) main field testing, and (7) operational product revision.

The research instrument in this development is divided into two parts. The first instrument is used to measure the feasibility of the teaching materials, including: (1) the validation questionnaire of the linguist, (2) the expert material validation questionnaire, (3) the teacher response questionnaire for the developed product, and (4) the student response questionnaire for the developed product. The data in this research is 1) validation data by a media expert and a material expert about product feasibility, developed in the form of an instructional picture book based on a scientific approach using the PjBL method, 2) teacher questionnaire response data toward the teaching materials developed, and 3) student response questionnaire regarding the teaching materials developed.

4 RESULTS

Data analysis was conducted on the questionnaire regarding the picture storybook assessment by a material expert, and on the questionnaire regarding the picture storybook assessment by a language expert.

Based on Table 1, it is seen that the analysis of the results of the validation by a material expert is in the category "Excellent". Indicators of presentation standard aspects are, 1) conformity with theme 6, and the integration of the subjects; 2) aspects of material standards of the indicators of conformity regarding the order of presentation and the completeness of information; 3) steps of PjBL are corresponding to scientific approach; 4) suitability of figures in observing activities, questioning, gathering information, reasoning, and communicating, and 5) usefulness aspects consisting of indicators of conformity with children's cognitive development, involving students actively showing concern for the environment. The results of the validation by a language expert are described in Table 2.

Based on Table 2, it is seen that the analysis of the results of the validation by a language expert is in the category "Excellent". Language aspects consist of indicators of readability, timeliness of time and place, completeness of sentence structure, punctuation appropriateness, consistency of capital letters, and suitability of font type. The graphics aspects consist of indicators of the suitability of the book size, the appropriateness of the cover design, the suitability of the number of pages, the suitability of color selection, the suitability of the illustration with the text, the suitability of the illustration with the meaning, texture conformity, and the suitability of the book form. The advantage aspect consists of an indicator of conformity with the child's emotional development.

The results of the student responses to the product in preliminary testing are described in Table 3.

Based on Table 3, it is seen that the analysis of the results of the student responses is in the category "Excellent." The results of the teacher and student responses to the products in main field testing are described in Tables 4 and 5.

Based on Table 4 it is seen that the analysis of the results of the teacher responses is in the category "Excellent".

Based on Table 5, it is seen that the analysis of the results of the student responses is in the category "Excellent". Reviewing the overall results of the research that has been done, the next step is to revise the product in accordance with the operational product revision step.

Table 1. The results of validation by a material expert.

Aspects	Average	Category
Presentation standards	3.8	Excellent
Material standards	3.7	Excellent
PjBL and scientific approach	3.6	Excellent
Advantage	3.9	Excellent

Table 2. The results of the validation by a language expert.

Aspects	Average	Category
Language standards	3.8	Excellent
Graphics	3.8	Excellent
Advantage	4	Excellent

Table 3. The result of student responses.

Student respondent	Average	Category
Respondent 1	3.8	Excellent
Respondent 2	3.7	Excellent
Respondent 3	3.8	Excellent
Respondent 4	3.9	Excellent
Respondent 5	4	Excellent
Respondent 6	3.7	Excellent

Table 4. The results of teacher responses.

Aspects	Average	Category
Material standards	3.7	Excellent
Presentation standards	3.6	Excellent
Language standards	3.8	Excellent

Table 5. The results of student responses.

Student respondent	Average	Category
Respondent 1	3.5	Excellent
Respondent 2	3.6	Excellent
Respondent 3	3.8	Excellent
Respondent 4	4	Excellent
Respondent 5	3.6	Excellent
Respondent 6	3.6	Excellent
Respondent 7	4	Excellent
Respondent 8	4	Excellent
Respondent 9	3.9	Excellent
Respondent 10	3.8	Excellent
Respondent 11	3.9	Excellent
Respondent 12	3.7	Excellent

5 DISCUSSION

This study developed teaching materials in the form of picture-based illustrated books that were based on a scientific approach using the PjBL method, which can be used for grade II elementary school students. The development of this resource goes through seven stages, in accordance with the design of Borg and Gall. This teaching material is suitable for use in learning activities, as it has been validated by material experts and linguists, teacher responses, student responses, and validation by Focus Group Discussion (FGD).

Based on the linguists and expert judgment, the developed materials were eligible to be applied. Based on the expert assessment of the pictorial storybooks, based on a scientific approach using the PjBL method, the eligibility criteria for its use as a teaching material has been met. In the picture book, there are elements that can be categorized as teaching materials, namely: (1) content conformity aspects, (2) linguistic aspects, (3) aspects of material presentation, and (4) aspects of graphs (RoI, 2008).

The content feasibility component includes: conformity with Kompetensi Inti/main competence (KI) and Kompetensi Dasar/basic competence (KD), conformity with child development, conformity with material requirements, the truth of learning materials substance, benefits for additional insights, conformity with moral values, and social values. The linguistic component includes: readability, clarity of information, conformity with good and correct Indonesian language rules, and effective and efficient use of language (clear and concise).

The presentation component includes: clarity of objectives (indicator) to be achieved, arrangement of information presentation, giving motivation and attractiveness, interaction (provision of stimulus and response), and completeness of information. The keypad component includes: font usage and type and size, layout, illustrations and drawings, and display design. The instructional material of the picture storybook already contains all of the elements, stating that the scientific storybook is based on the scientific approach using the PjBL method and is a teaching material.

Based on both teacher and student responses, the textbook materials, based on a scientific approach using the PjBL method, are in very reasonable categories. The teaching materials in the picture books have advantages that distinguish them from other picture books. Picture storybook contains the activities done by the story characters as the scientific and PjBL method activities. Through the activities of the character, the book indirectly invites the reader to come along to solve the problems. As Gulbahar and Hasan (2006) stated, project-based learning has a unique focus on students by focusing on an event or problem and requiring the students to complete it by applying the concepts and principles they have learned. These project methods involve the students engaging in constructive activities, and therefore learning is realistic.

The picture storybook material contains a scientific approach using the PjBL method. In the scientific approach there are five scientific activities: observing, questioning, experimenting, associating, and communicating. In the PjBL method there are four scientific activities: preparation, design, execution, and communicating. Scientific activities of scientific ascent are combined with scientific activities in the PjBL method, and the incorporation is contained in the picture books. This study developed a picture book that was very suitable for use as a teaching material. The present invention is in accordance with the discovery of instructional material by Purnomo and Wilujeng (2016), who examined the teaching materials.

Instructional materials for storybooks, in addition to assessments by linguists and materials experts, are also assessed by FGD. The results of the FGD assessment were in the form of suggestions for improvements to the teaching materials. The suggestions for improvement were: (1) changing the font type, (2) giving a twinkle to the character's eyes, (3) adding illustrations, and (4) the addition of color variation. The suggestions for improvement made by the assessment of linguists, materials experts, and FGDs were implemented and, following the completion of the improvements, the teaching material of a science-based storybook based on a scientific approach using the PjBL method could be tested by teachers and students to get their response. This response aims to find out the opinions of teachers and students relating to teaching using pictorial storybooks based on a scientific approach using the PjBL method.

Curriculum 2013 is synonymous with a scientific approach in learning activities. Activities using scientific approaches and PjBL methods, such as defining themes, developing questions for design projects, undertaking project development activities, or applying and presenting project outcomes, have an impact on child cognitive patterns. In the process of learning, using scientific approaches and PjBL methods, students gain the ability to interpret, analyze, make conclusions, evaluate, reflect, and follow up. This is in accordance with Fascione (2011), who suggests that in gaining more in-depth knowledge, the student must be able to undertake interpretation, analysis, inference, evaluation, explanation, and self-regulation.

The incorporation of scientific activities on scientific approaches and communication skills works like this; for observing and questioning, activities may be included in the preparatory activities; for information gathering, activities may be included in the preparatory activities; for association, activities may be included in the execution activity; and for the communication activities, both the scientific approach and the PjBL method have communication activities. These activities are featured in picture books through stories presented by characters. As Brown and Tomlinson (1999) state, "Many trade books contain information that is relevant to the topics studied in school. Moreover, this information is presented through captivating, sometimes beautifully illustrated, narratives." This statement can be interpreted to mean that the contents of stories that contain a variety of relevant and logical information can be used to help with learning. This shows that the material can be presented through the story content in a picture book and can be packed with interesting and beautiful illustrations, because children like pictures that show beautiful things. The picture storybook in this research was developed and qualified as a teaching material.

The result of this research is a science-based picture book based on a scientific approach using a valid PjBL method. The PjBL method itself deals with the activity of producing project-based products/activities in groups. To get a deep understanding, the activities are not merely performed individually or independently, but can also be done in groups. This is supported by Schwab (2008), who states that obtaining a deep understanding should be integrated with other competencies, such as activities carried out by group work. This can be interpreted to mean that using a group activity to complete the project can indirectly generate deep thinking patterns in students. This activity is shaped by picture books.

6 CONCLUSION

The result of the research is a picture book, based on the scientific approach using the PjBL method, which is suitable for use as a learning material for 2nd grade students. The results of the tests, which were obtained based on validation by a linguist, a materials expert, and the response of teachers and students, showed the criterion to be "very feasible".

Based on this result, it is concluded that the picture storybook, based on the scientific approach using the PjBL method, was valid as a learning material for 2nd grade students.

REFERENCES

Anastasia, E. & Mukminan, M. (2016). Implementasi pendekatan saintifik dalam pembelajaran IPS di *middle grade* SD Tumbuh 3 Kota Yogyakarta [Implementation of scientific approach in social science instruction in middle grade of SD Tumbuh 3 Yogyakarta]. *Jurnal Prima Edukasia, 4*(1), 20–31. doi:http:journal.uny.ac.id/inex.php/jpe/article/view/7691/pdf.

Borg, W.R. & Gall, M.D. (1983). *Educational Research An Introduction.* New York, NY: Longman.

Brown, C.L. & Tomlinson, C.M. (1999). *Essentials of children's literature.* Boston: Allyn and Bacon.

Donoghue, M.R. (2009). *Language arts: Integrating skills for classroom teaching.* California, CA: SAGE Publication.

Fascione, P.A. (2011). *Critical thinking: What it is and why it counts.* Millbrae, CA: Measured Reasons and The California Academic Press.

Gulbahar, Y. & Hasan, T. (2006). Implementing project-based learning and e-portfolio assessment in an undergraduate course. *Journal of Research on Technology in Education, 38*(3), 309–327.

Huck, C.S., Hepler, S., Hickman, J., & Kiefer, B.Z. (1987). *Children's literature in the elementary school.* New York, NY: Holt, Rinehart and Winston.

Kelemen, D., Emmons, N.A., Schillaci, R.S., & Ganea, P.A. (2014). Young children can be taught basic natural selection using a picture storybook intervention. *Journal of Psychological Science, 6*(2), 1–10.

Kurniasih, I. & Sani, B. (2014). *Sukses mengimplementasikan kurikulum 2013: memahami berbagai aspek dalam kurikulum 2013 (Successfully implementing curriculum 2013: Understanding the various aspects of curriculum 2013).* Surabaya: Kata Pena.

Larmer, J. & John, R.M. (2010). Seven essentials for project-based learning. *Journal of Educational Leadership, 68*(1), 34–37.

Mergendoller, J.R., Maxwell, N.L., & Bellisimo, Y. (2006). *Scaffolding project based learning: Tools, tactics, and technology to facilitate instruction and management.* Novato, CA: Buck Institute for Education.

Nurgiyantoro, B. (2013). Penilaian pembelajaran bahasa berbasis kompetensi [Assessment of bahasa learning based on competence]. Yogyakarta, Indonesia: Badan Percetakan Fakultas Ekonomi Universitas Negeri Yogyakarta.

Prastowo, A. (2014). *Pengembangan bahan ajar tematik: tinjauan teoretis dan praktik (Development of thematic teaching aid: Theoretical and practical point of view).* Jakarta: Kencana Prenadamedia Group.

Purnomo, H. & Wilujeng, I. (2016). Pengembangan bahan ajar dan instrumen penilaian IPA tema indahnya negeriku penyempurnaan buku guru dan siswa kurikulum 2013 [Development of teaching aid and assessment intrument of science subject with the theme wonderful my country. Improvement of teachers' and students' book in curriculum 2013]. *Jurnal Prima Edukasia, 4*(1), 67–78. doi: http://dx. doi.org/10.21831/jpe.v4i1.77697

Rina F. (2017). The Effectiveness of project based learning on students' social attitude and learning outcomes. *Jurnal Pendidikan Teknologi Kejuruan, 23*(4), 374–382.

RoI. (2008). *Panduan pengembangan bahan ajar [Guideline of development of teaching aid].* Jakarta, Indonesia: National Education Departmen, Repiblic of Indonesia.

Rossiter, M.J., Derwing, T.M., & Jones, V.M.L.O. (2008). Is a picture worth a thousand words? *Research Issues, 42*(2), 325–329.

Rothlein, L. & Meinbach, A.M. (1991). *The literature connection: Using children's books in the classroom.* Glenview, Illinois: Scott, Foreman and Company.

Schwab, C.V. (2008). Perception of students' learning critical thinking through debate in a technology classroom: A case study. *Journal of Technology, 34*(5), 1–34.

Westwood, P. (2008). *What teachers need to know about teaching methods.* London, UK: Acer Press.

Character Education for 21st Century Global Citizens – Retnowati et al. (Eds)
© 2019 Taylor & Francis Group, London, ISBN 978-1-138-09922-7

The social competence of in-service teacher at inside and outside school communities

S.I. Indriani
Pelita Harapan University, Indonesia

ABSTRACT: In order to prepare the future teachers with a good quality both in skills and character, they were trained in many activities inside and outside the class where they can practice on their social skills. This study investigates deeper into the social competence of in-service teachers in their communication and collaboration skills as well as identifying factors supporting or impeding their involvement at inside and outside the school communities. This study uses a qualitative research method. The research subjects are five in-service teachers, three principals, and 60 students. The data were collected through interviews, FGD, questionnaires, and field notes. The results show that the in-service teachers are opened to suggestions, willing to collaborate and communicate with the school community, but their contributions outside the school communities are still limited. Teaching load and time management skills are the main issues impeding their contributions outside the school communities.

1 INTRODUCTION

One of determiners of a great nation is the quality of the education system and human resources. The education quality in Indonesia is still considered low because of many reasons. One of the problems faced by Indonesian education system is related to the teacher welfare and teacher quality as the tangible human resource in educational contexts (Sulisworo 2017). Therefore, teacher education programs are expected to improve education system and prepare the future teachers with good competencies in all aspects, including in teaching, professionalism, personality, and social skills. The government really takes it seriously to improve the quality of teachers' competencies in order to provide a good education for Indonesia (Undang–undang Republik Indonesia Tahun 2005 No. 14). Teachers College (TC) – Universitas Pelita Harapan (UPH), a private university in Indonesia, also carries out the same mission, preparing the future teachers who are not only professional in teaching skills, but also have a good quality in social skills. The in-service teachers are expected to be able to become significant members of the community, able to relate well and adapt within the society with multi-cultural, social and spiritual background.

Will the in-service teachers be able to answer the challenges in this 21st century education system? how to build up communication, collaboration, and critical thinking to engage the learners to solve problems creatively and participate actively in the community that will lead to transformative thinking skills and attitudes towards the learning process?

The in-service teachers are expected to be ready to bring impacts towards the school communities started from the smallest scope, the students in class to the biggest scope within the school contexts, the school administrators, the parents, and the society outside the school community.

It is also hoped that there will be more opportunities for the in-service teachers to apply their knowledge and skills to respond to the needs of the society.

1.1 *Purposes of the study*

This study investigates deeper into the social competencies had by the in-service teachers from Biology Education study program who have been at least 1 year teaching in one of the schools in Medan, Indonesia. This study also explores both factors that support or impede them to improve their social competencies outside the school communities.

2 LITERATURE REVIEW

A teacher must possess a pedagogic, professionalism, personality, and social competence in order to achieve a successful learning process. Unfortunately, a social competence is often neglected (Bihim & Rustiyarso 2013). Therefore, a teacher professional development program is supposed to enhance the whole aspects of a teacher's competence, including pedagogic, professional, personality and social competencies (Undang–undang Republik Indonesia Tahun 2005 No. 14). A teacher is someone who builds, leads, and someone who is a role model in establishing orderliness, sustaining the society welfare (Mulyasa 2007).

A teacher's social competence is closely related to the society sustainability. Therefore, a teacher's social competence is predetermined by social and cultural values imbedded in the demand of the society where we live. It is represented in the willingness to collaborate, interact, communicate, make coordination with and appreciate other people's ideas (Nur & Ahmad 2009). An ability to build up a communication and collaboration with the environment inside and outside the school community is called as a social competence (Wibisono 2012).

Both internal and external factors influence the development of a teacher's social competence. The factors that come from the teachers' themselves that might hinder them are the lack of teacher's creativity and openness to the new information and technology innovations, a low interest in learning collaboratively with other teachers or researchers. While some factors influencing a teacher's social competence development is the supportive environment, such as the school administrators, teachers' collaboration, learning media and technology supported system, and other facilities (Suprihatiningrum 2013). In order to create a condusive environment where there is a growth in the learning community, the school should be the first evidence. Hoaglund et al. (2010) stated that professional learning communities should provide the structure that must exist within a school in order to be more effective. DuFour (2004) also added that a learning community is a powerful way of working together that profoundly affects the teachers and the students.

Teachers and the school community have the same commitment which is to bring Shalom community for students, parents, and the community outside the school. It is a supportive learning community in which all students are able to use their talents and contribute in their community in different ways. The shallom community also develops students' sense of responsibility with high expectations through love, encouragement, and support (van Brummelen 1998). Schools should be the place where the community support each other and teachers as well as students should be a part of it. It is important to notice that a teacher should be able to respond to the needs of the society. It is a part of their respond to their role as a community member in the broken society both inside and outside the school community.

3 METHODOLOGY

This study uses a qualitative research method. The research subjects are five in-service teachers from Biology Education Study Program who have been teaching in a high school in Medan for at least 1 year. The data were collected from interviews with the Principals, Focus Group Discussion (FGD) with the five in-service teachers, and questionnaires filled out by the students whom the in-service teachers taught. The research process is as follow.

After all the data have been collected, each data will be analyzed and reduced based on the focus of the research. Coding will make the data easier to be categorized according to the research purposes. The study focuses in investigating the in-service teachers' social competences

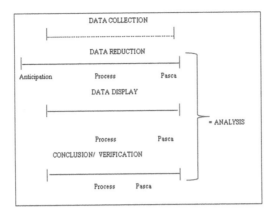

Figure 1.   Procedures of a qualitative research method.

that include their abilities in collaborating and communicating with the school community. This study will also find out factors supporting and impeding their involvement in the communities, both inside and outside the school. Then, all data will be verified to make conclusion.

4   RESULTS AND DISCUSSIONS

The data gathered from various instruments will be displayed, analyzed, and discussed here.

4.1   *A teacher's social competence: Collaboration and communication skills*

Based on the semi-structured interview three principals, it was found that the five teachers who were graduated from Biology education study program, they are open to suggestions, willing to help and serve each other. However, there are several aspects that must be considered about the in-service teachers which are as follows.

"The in-service teachers are opened to suggestions, but they need to be more initiative to share ideas in problem solving."

"The in-service teachers are willing to develop professionalism in following all activities held by school. However, they need to improve their communication with other teachers who were not graduated from TC."

"The in-service teachers have heart to serve, but they need to be more flexible in giving re-ponse to parent's reactions, so that the communication with the parents will flow more smoothly."

Based on the statements above, within 1 year working period, they have been gradually improved their communication skills with the parents. They are also more initiative in dealing with the problems in class. This happens because they also found that the school also supports them by providing a condusive learning community. Based on the result from FGD, the five teachers appreciated that the school principals have given such a big support for teachers to improve their professionalism and social competence as well through many activities, such as Professional Development meetings, group discussions, and other meetings. There are also other activities that support the relational and social engagement among teachers and school administrators, for example team meetings, reguler meetings held to support every teacher with the time to share their burdens and problems in class with the Principal or the Curriculum Coordinator/Teacher Trainer (CCTT). Hoaglund et al. (2010) stated that professional learning communities in schools should provide the structure that must exist within a school in order to be more effective.

201

The school commits to cultivate 'shallom' community or a community that brings peace for everyone which functions as a supportive learning community in which all students are able to use their talents and contribute in their community in different ways. The shallom community also develops students' sense of responsibility with high expectations through love, encouragement, and support (van Brummelen 1998). It is also hoped that teachers will also pass it on to the class, so that the students will also experience the condusive learning atmosphere in the class. According to DuFour (2004), a learning community is a powerful way of working together that profoundly affects the teachers and the students. The activities below show the school's commitments to grow the learning community.

"Team leaders organize with his team members."

"There are meetings held regularly with the Principals to discuss about problems occurred in the class. The principals will give suggestions for alternative solutions."

"There are activities that support spiritual growth, such as chapel, group discussion where teachers and staffs can share ideas to support each other spiritually."

Based on the data gathered above, it has been confirmed that the in-service teachers are supported by the school system that do not only support their social competence, but also their spiritual growth. It is a condusive environment where learning community is supported. One of the factors influencing a teacher's social competence development is the supportive environment, such as the school administrators, teachers' collaboration, learning media and technology supported system, and other facilities (Suprihatiningrum 2013). Through the activities done in the school that support both the social and spiritual growth of the school community, it is hoped that teachers can also bring joy and excitement of learning to the students both in their classroom community and outside the classroom, such as to the parents and other community groups outside the school communities.

In regard to the in-service teachers' the teacher graduates' social competence practiced in classes, 89.15% students from two selected classes in Primary level reported that they were happy being taught by their teachers. They found their teachers as people who give attention while they are studying, people whom they are comfortable to interact with, people who are good to everyone. However, there was one specific statement that reach the highest precentage from both classes. From total 60 students from both classes, there were 14 or 31,67% students who disagreed that their teachers and parents are friends. This was confirmed by one teacher who stated that this could happen because the parent-teacher meetings were only held in certain time period, so the students rarely see their parents and teachers interact as if they were friends.

The complete result of the questionnaires given to Primary students in both classes is displayed in the diagram below.

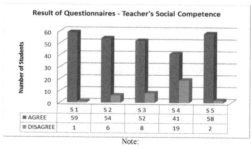

Note:
S1 (Statement 1): I am happy to be taught ny my teacher
S2 (Statement 2): My teacher pays attention on me while I'm studying
S3 (Statemen 3): I feel comfortable talking to my teacher
S4 (Statement 4): My teacher and my parents are friends
S5 (Statement 5): My teacher is good to everyone

Figure 2. Result of questionnaires on primary teacher's social competence (Result from two selected classes).

Table 1. Summary of the in-service teachers' teacher graduates' social competence that need improvement.

No.	Statements	Class	Number of students/ Percentage	Subject taught
1.	You often talk to your teacher, even outside the class	EP	28 (84%)	Biology
		CR	22 (71%)	Biology
		ED	21 (70%)	ICT
2.	Your teachers builds up a good communication with your parents	ED	25 (83%)	ICT
3.	Your teachers speaks in an appropriate manner with your parents.	ED	18 (60%)	ICT

In addition to the questionnaires result from Primary students, there were also results from questionnaires filled out by the secondary students. Based on the result of questionnaires from 3 selected classes, 80.56% from total 94 students agreed that their teachers who are the in-service teachers are able to interact and build a good relationship with them. Meanwhile, there were still 19.44% students disagreed. It means that there were still some aspects of the in-service teachers' social competence that were not optimally implemented.

The table below will show the summary of aspects of social competence that the in-service teachers need to improve on.

From the results above, it was found that 50% of the total numbers of students in each class stated that the social competence of the in-service teachers was not optimal. Based on the interview with the Principals and FGD with the five in-service teachers, it was found that the in-service the teachers have a tight teaching load. Another reason is that there is one in-service teacher (ED) who is not teaching Biology according to his expertise. Instead, he teaches Information & Communication Technology (ICT) who needs to teach parallel classes that makes it impossible for him to spend time to have a talk with the students outside the class. This condition also makes it more impossible for him to have a chat with parents about the students; progress as what homeroom teachers usually do.

4.2 A teacher's involvement in the community inside and outside the school

As what was expected from a Teacher Education Program by the Indonesian Government that a preparation program for future teachers must be oriented to equip the future teachers to be able to apply theories into practices and be skillful in building up a good interaction and communication with students, colleagues, and the surrounding community (Undang–undang Republik Indonesia Tahun 2005 No. 14), one of the teacher's profiles aims for obtaining a competent teacher as a significant member of the community, able to relate well with the community, able to adapt with the environment where multi culture, faith, and social backgrounds, and able to build up a community based on grace, peace, hope, and love. If a teacher's role can be actually practiced and people outside the school community can really experience it, a teacher can maximize their potentials to build, lead, and be a role model who put things in order, sustain the stability of the society (Mulyasa 2007). Knowing that they are a part of the society, every individual in the community should be aware of their limited capacities to reach the goal. Therefore, they should support each other. It involves much more than intellrectual comprehension and analysis towards problems happened. It must result in a committed response and action to take care of the world (van Brummelen 1998).

Teachers and the school community have the same commitment which is to bring sense of responsibility with high expectations through love, encouragement, and support for the entire community, students, parents, and other community groups outside the school.

Therefore, there must be a way to assure that the teachers' competence in social contexts continues to develop along with their teaching practices in schools. One of the ways to find out is by involving teachers in activities where teachers, students and other school community members participate simultaneously within the school environment or outside the school community.

Based on the interview with the Principals, there are school activities that support teachers' contribution in the community both inside and outside the school environment. Some of those activities are summarized below.

"Community service involves students, teachers and the school community to go outside and serve other people, such as serving kids in the orphanage, visiting people with disabilities and others."

"Teachers and students are involved collaboratively to clean up the environment with the people outside the school community, such as *gotong royong*, GPS (*Gerakan Pungut Sampah*) organized with the Head of the environment, cleaning up the ditches along the road around the school."

From the FGD with the in-service teachers, they stated that actually there are numbers of activities where they can participate in the community.

"Participating in the church activities could help them open their views and they will be able to find more friends."

"By knowing more people around the school community through coming to their gatherings, it opens way to be the Light and share the Gospel."

"Being involved in the community outside the school can bridge up more opportunities to participate in other activities. The head of the environment responded very positively to the students and teachers' participation in *gotong royong* and expects participation in the next activities."

Based on the results from FGD and interview, it can be concluded that actually the in-service teachers are willing to contribute more in the community outside the school and there are many activities they can join. Those are factors that support their contribution in the community outside the school community. While there are also factors that hinder the in-service teachers to contribute more in the community, such as the time management, priorities standard which could be different from one to another. The result from FGD showed that the in-service teachers often found themselves trapped in a limited time to prepare materials to deliver in the next morning, to choose which activities need to be prioritized to attend. The summary of those struggles are as follow.

"Actually, I want to join in more activities in the church, but I just don't have time. Usually the choir team practices at the time when I still need stay at school and finish my job."

"Sometimes I participate in a gathering held by the community near the school, but I cannot do it all the time because I have things to prepare for the lessons tomorrow."

"We cannot force someone to do what we think good for them to do, such as joining in fellowship with friends or neighbors, and others. Every person has his own right to make decision,"

In addition to what the in-service teachers have shared through the FGD, one of the Principals also confirmed that one of the factors that hinder teachers to get more involved in the community outside the school is a security matter. There have been numbers of criminals happened in the neighborhood, such as robbery and mugging. Therefore, the Principals need to warn everyone as the school community to stay alert with the situations outside the school environment. However, the school still supports teachers to build a good relationship with the community both inside and outside the school.

5 CONCLUSION

According to all the data results, it can be concluded that the five in-service teachers who have been teaching in the participated private high school in Medan, Indonesia have been able to show their social competences in their daily practices, such as in building a good interaction, communication, and relationship with the students and colleagues through many activities that involve the school community all together within the school community and outside the school environment.

The activities that support the in-service teachers' contribution within the school community are the coordination meeting between the Principal and every teacher to discuss problems occurred in class, team meeting where teachers can support each other, Chapel and a small group to share and support each other spiritually.

There are also activities where the in-service teachers involve that also support their contribution to the community outside the school environment, such as Community Service where students and teachers serve other people who need love and care, children with disabilities, *Gotong Royong* and GPS (*Gerakan Pungut Sampah*) where teachers and students participate with people outside the school community to clean up the environment around the school. These activities have been given a positive response from the community outside the school. The leader of the community hopes that there will be more activities that involve both the school community members and people outside the school community.

Though there are many activities that support the in-service teachers to contribute more in the community outside the school, there are some obstacles that impede their contribution both within and outside the school community. Based on the data gathered from interviews with the Principals, FGD with the in-service teachers, and questionnaires given to the students, it was found that teaching load and time management are the two main problems. The in-service teachers are willing to join in activities that might involve them more in the community outside the school environment, such as gathering, meetings with the head of environment to discuss about the community issues, joining in the church activities, and others. These activities open up ways to contribute more in the social community as well as an opportunity to bring transformation in the society.

ACKNOWLEDGEMENT

A high appreciation is given to the Faculty of Education, Teachers College and Research Institute of Pelita Harapan University for the funding support of the study No P-004-FIP/X/2016.

REFERENCES

Bihim, Y.B., & Rustiyarso. 2013. Pelaksanaan kompetensi sosial guru dalam pembelajaran sosiologi kelas XI IPS, SMAN 3 Teluk Keramat [Implementation teachers'social competencies in sociology grade XI, Public Senior High School 3 Teluk Keramat] *Jurnal Pendidikan dan pembelajaran* 2(10): 239–245.

Dick, W. et al. 2009. *The systemic design of instruction*. New Jersey: Pearson.

DuFour, R. 2004. Schools as learning communities. *Educational Leadership* 61(8): 6–11.

Hoaglund, A.E. et al. 2010. Professional learning communities. *Education* 134(4): 156–162.

Miles, M.B. & Huberman, A.M. 1992. *Analisis data kualitatif [Qualitative data analysis]*. Jakarta: Penerbit Universitas Indonesia.

Mulyasa, E. 2007. *Standar kompetensi dan sertifikasi guru [Standard competencies and ceritification teachers]* Bandung: Rosdakarya.

Nur, N. & Ahmad, G. 2009. *Integrated Human Resources Development*. Graasindo, Jakarta.

Sanjaya, W. 2016. *Strategi pembelajaran: berorientasi standar proses pendidikan [Learning strategy: Standard-process-education-oriented]*. Jakarta: Kencana.

Sulisworo, D. et al. 2017. Identification of teachers' problems in Indonesia on facing global community. *International Journal of Research Studies in Education* 6(2): 81–90.

Sumardi. 2007. *Password menuju sukses: Rahasia membangun sukses individu, lembaga, dan perusahaan* [*Password to success: The secrets to build individual success, institution, and company*]. Jakarta: Erlangga.

Suprihatiningrum, J. 2013. *Strategi Pembelajaran Teori dan Aplikasi* [*Strategy of learning theory and aplication*]. Yogyakarta: AR Ruzz Media.

Undang–undang Republik Indonesia Nomor 14 Tahun 2005 tentang Guru dan Dosen, pasal 8 [*Regulation Constitution of Indonesia number 14 Year 2005 about Teachers and Lecturer, verse 8*].

Van Brummelen, H. 1998. *Walking with God in the classroom: approachers to teaching and learning.* Seattle: Alta Vista College Press.

Wibisono, H. 2012. *Menjadi guru berkarakter: Strategi membangun kompetensi dan karakter guru* [*Being characteristic teacher: strategy to construct teachers' competencies and teachers' character*]. Jakarta: Pustaka Belajar.

Building the nation through the evaluation of the curriculum for teacher education

R. Rosidah
Universitas Negeri Yogyakarta, Indonesia

ABSTRACT: Education is one of the most vital aspects of nation building. Teachers, as the main actors in education, have a strategic role in creating human resources that support the nation's development. In the 21st century, teachers are challenged to have several qualifications, including leadership, digital literacy, communication, emotional intelligence, entrepreneurship, global citizenship, problem solving, and teamwork. Teachers are also required to be more responsive in the face of competitive global education. In the process of teaching and learning, in addition to having social competence, pedagogy, personality, and professionalism, the teacher must also be able to: teach the learning process, illustrate, demonstrate, and inspire students for their future life. To ensure the various qualifications required of teachers in the 21st century, the evaluation of the curriculum for teacher education is required. This paper will discuss the demands required of qualified professional teachers, which contribute to the development of the nation, and their relation to the evaluation of the curriculum for teacher education in the 21st century.

1 INTRODUCTION

One of the responsibilities of every citizen in building a nation is to contribute through their occupation. When a person works and produces an income it will become a development resource and indirectly reduce the burden of the state. Conversely, when citizens do not have jobs (are unemployed) it will be a burden on the state, and consequently hamper the process of building a nation.

Most ASEAN member countries are currently facing a lack of skilled labor which is jeopardizing their further economic development (Grosch, 2017). To prevent the occupation problem, which may impede development, the nation should attempt to empower human resources. One of these efforts is through human development that purposely accelerates the achievement of international, national and regional government programs. Human development plays an important role in producing good quality human resources and this can be reached through education.

Educational institutions are needed as instruments that can play a role in developing human resources. Indonesian Law (*Undang-Undang*) number 20 in 2003 on the National Education System, Article 3, states that the purpose of national education is to develop the potential of learners as follows: to be human beings who believe and are cautious to God Almighty, have a noble character, are healthy, knowledgeable, capable, creative, independent, and will become citizens of a democratic and responsible state. To achieve this, the quality of education is a priority that needs to be pursued.

Education is expected to affect the future of society, respond to social change, and have the power to change conditions, even forming a cultural identity. As stated in the Report to UNESCO of The International Commission of Education (1996): To create tomorrow's society, imagination will have to keep ahead of technological progress in order to avoid further increases in unemployment and social exclusion or inequalities in development. It seems that the concept of an education pursued throughout life, with all its advantages in terms of flexibility, diversity and availability at different times and in different places, should command wide support.

One important element in improving the quality of national education is teachers. Ball and Forzani (2009 stated that improvements in student learning depend on substantial, large-scale changes in how we prepare and support teachers. Agreement is widespread that teachers are the key to student learning, and efforts to improve the quality of teachers have prolifer-ated. Most initiatives, however, have focused on teacher recruitment and retention and on developing new pathways to teaching.

Teachers play an important role in building the quality of human resources. Professional teachers are able to give learners potential that can be empowered and synergized with other resources, so as to be productive in building the nation. Teachers have the power to bring meaningful values to their students' lives. Teachers also play a role in the process of bringing change to learners and giving people dignity. Teachers should have the skills to adapt to the circumstances and be capable of integrating the learning process so that it is understood by students. Therefore, a teacher's education needs to be designed in order to enrich teachers with certain skills.

The success of learning will depend on the ability of teachers to teach, and teaching exper-tise covers various aspects, including: material mastery, suitability of methods and media, the ability to manage classes, the ability to explain the material systematically, the ability to main-tain intonation and volume, and the ability to perform the role of motivator and inspirer. These can be useful when considering the content of the curriculum for teacher education.

Facing globalization, it is indisputable that various elements of education, which include educational facility, supportive environment, educational policy, teacher training, and the curriculum for teacher education, have been evolved along with technological advancement, industrial development, and the improvement of human civilization. However, in fact, the curriculum for teacher education has rarely shown any changes. This has been the reason of this study as the curriculum plays a significant role in producing highly qualified teachers that can support national development.

Curriculum adjustment is needed along with the dynamics of a change in education policy and the development of the industrial world as a consequence of globalization. The curricu-lum is an important element for giving instructional guidance in education and for directing learners in accordance with the expertise they desire. A good curriculum should be able to accommodate state development missions, educational missions, college missions, and study program missions, which are constantly evolving and changing over time. Therefore, the eval-uation and development of the curriculum for teacher education needs to be reviewed and performed continuously.

The quality of the curriculum will be largely determined by whether or not it is adequately capable of solving national problems through education. The curriculum needs to be devel-oped in order to meet the specific qualifications needed for an occupation in a specific nation. Its development includes the learning outcome orientation and educational objectives, and it should cover both local and national needs. This aims to correspond to the needs and changes occurring on a national and international scale.

Along with the dynamics of development, the output of education is not only expected to meet the demands of the job market (demand driven), but also to meet the developing market demand (market driven). Job markets require well qualified human labor to cope with the industrial world. Therefore, it is obviously important for the development of the curriculum to be adjusted to the demands of industrial development.

2 TEACHER QUALIFICATION REQUIREMENTS IN THE ERA OF GLOBALIZATION

Globalization is an indispensable condition in national development. The flow of globaliza-tion has an impact on teacher qualifications. The globalization era has forced teachers to gain better qualities that cover widely complex aspects, including: communication, emotional intelligence, entrepreneurship, global citizenship, problem solving, and teamwork. A teacher needs to be capable of adjusting the skills of human resources to meet the needs of industrial and technological development. Marfu'ah et al (2017) stated that teachers have to implement the knowledge and skills acquired in continuous learning. Trilling and Fadel (2009) stated that, in the 21st century, a teacher is required to be able to master certain aspects: learning and innovation skills (learning to learn and be innovative, critical thinking and problem solving), digital literacy skills (information literacy, ICT literacy), and career and life skills. In Indonesia Government Regulation no. 74 of 2008 about Teachers, it is mentioned that in the management of learning, teachers must have the following competencies: pedagogy competency, personality competency, social competency, and professional competency. These are holistic and are obtained through professional education. Pedagogy competence involves: curriculum development, learning design, utilization of learning technology, and evaluation of learning outcomes. Personality competence covers having a noble character, being wise, democratic, faithful, and 'taqwa'. Social competence comprises communicating, writing, using communication and information technology appropriately, and also social manners. Professional competence covers the ability of scientific substance. These competencies, which must be mastered by teachers, require comprehensive learning materials.

In the global era, the demands for quality teachers are increasingly complex, including: communication skills, emotional intelligence, entrepreneurship, global citizenship, problem solving, and teamwork. Teachers are required to adapt to technological and industrial developments. To meet the demands of the 21st century it is necessary to develop a curriculum for teacher education, especially with regards to content. Berry (2003) suggests that the conditions for today's learners are different from how they will be in the future. This is the subject of discussion among others: "Unfortunately the focus of today's education is mostly about a sophisticated but still 20th-century blueprint for learning." Instead, the key conversation needs to be about creating a system that supports the learning environment and preparing teachers that are built to learn profoundly. It is a reflection that what is given in the learning process is what the preparation of future learners will obtain. So the curriculum should be evaluated and designed to meet future needs.

In order to face the global era and the development of information technology, the adjustments concerning how to teach students are being developed. Lau (2001) explained that curriculum should be continuously developed. Curriculum development is not an entity that stops before going into classrooms, and the curriculum is not a package that stops developing in the classroom. It is a continuous process of constructing and modifying. Some of the shifts that have occurred in the global era are expressed by Trilling and Fadel (2009, p. 136): "A curriculum based on a blend of these learning methods with more direct forms of instruction is what is now needed to build knowledge, understanding creativity, and other 21st century skills". This implies the mastery of teaching skills in the learning approach. The learning process needs to be modified in an effort to give students the understanding, knowledge, skills, and thinking ability to face life in the future. In the future, teachers will be expected: 1) to adjust their role in the learning process, in the development of information technology, and in the development of the global environment; 2) to align and organize the learning process. The Institute embodies a conducive environment for the development of an academic atmosphere. School is a community of learning for teachers as well as for students, through collaboration and sharing; 3) to manage learning using various development facilities and to take into account the cultural differences of learners; and 4) to conduct assessments and evaluations of appropriate learning in order to really measure the success of students in accordance with the objectives of learning. The evaluation of the curriculum for teacher education needs to be adjusted to take into account the shift in the teacher's role.

3 DEVELOPMENT OF THE CURRICULUM FOR TEACHER EDUCATION

The evaluation of the curriculum for teacher education is an effort to adjust to the needs of qualified teacher candidates and to improve the quality of teachers in order to fulfill the needs of future learners and the demands of work in the global era. Globalization has an impact on free trade, free flow capital, and the strict competition of the foreign labor market. This also increases the need for innovation in the adjustment of the learning process, which ends with the development of the curriculum for teacher education. Focusing on learning innovation, it is necessary to improve teacher competence with relation to how to teach materials that can be used as a provision of students in entering the world of work. Berry (2003), in his book *Teaching 2030*, explains "some of the challenges ahead, including: differences in career field, 'teacher-preneurism', digital era and cyberspace". The UNESCO report of the International Commission on Education for the Twenty-first Century explains the four pillars of education: "1) Learning to know; 2) Learning to do (from skill to competence, the 'dematerialization' of work and the rise of the service sector, work in the informal economy); 3) Learning to live together, learning to live with others, discovering others, working towards common objectives; 4) Learning to be". Responding to the dynamics of the development of teacher roles and the needs of learners in the digital era, the types of skills taught to prospective teachers needs to be developed. In response to this, the development of the curriculum for teacher education must be adaptive and anticipatory. This means that the components of the curriculum that are related to the substance of the field materials, both scientific and educational, are able to adapt to any developments that occur.

Teacher education aims to produce prospective teachers who are able to educate and prepare the future generations to meet the demands of the workplace. The curriculum is a set of plans and arrangements concerning graduate learning achievements, study materials, processes, and assessments, which is used as a guideline for the implementation of study programs. The curriculum for teacher education is comprehensively designed to meet the educational goals in general, and specifically to produce teachers in each area of expertise. Brady (1992, p. 20) explained that "evaluation of curriculum covers student learning outcomes and programs. Evaluation program involves judgements about the achievement of goals, the satisfaction of needs and the implementation of policies". As stated by Campbell et al. (1993, p. 3): "frequently, curriculum guides are prepared by a curriculum committee composed of educators, curriculum specialists, and business and industry representatives either at the state, regional, or local school system levels". Evaluation of the teacher curriculum development process is done in an effort to adjust to the development of science, technology and information, and the influence of globalization. The purpose of curriculum development impacts the fulfillment of the material needs of both the field of science and the field of education in the curriculum, which include: 1) scope of material to be taught during teacher education, 2) how to teach the material, 3) how the stages of the material are taught, 4) the impact of learning on the learners, and 5) how to teach learners in order to make them understand, become competent, survive in society, and decide what they want to be. Trilling and Fadel (2009) described that innovations in learning require teachers to master: 1) learning and innovation skills: critical thinking and problem solving, communication and collaboration, and creativity and innovation; 2) digital literacy skills: information literacy, media literacy, and ICT; and 3) career and life skills: flexibility and adaptability, initiative and self-directing ability, social and cross-cultural interaction, productivity and accountability, and leadership and responsibility. Therefore, the evaluation and development of the curriculum is directed at the scope of the material that is related to students' skills that are necessary when encountering various job tasks and problems, the cross-cultural working situation, and in response to the education policy. On the other hand, the evaluation and development of the curriculum leads to material related to the above skills, which are needed in order to fit the needs of learners.

Curriculum development is a complex process. Preparation of the curriculum ideally requires synergies from various institutions. There are at least three related institutions, namely the industrial world, educational institutions (school and university), and government institutions (Ministry of Labor). There is a synergy between the educational institutions, who produce the graduates, industry as the user, and the Ministry of Labor, which undertakes school liaison with industry and the career development of learners. Each institution contributes both directly and indirectly to the curriculum content. The improvement in the capacity of industry to engage in education cannot be denied. Education that is based on pragmatism

profoundly tends to empower their curriculum about the world of work, like industrial institutions relevant to the graduates. The preparation of the curriculum is not limited to the
content that relates to areas of expertise, but also to the content relating to the mission of
the development of the nation, the requirements of the workplace, and the development of
the community environment. The proportion of curriculum content is largely determined
by the differing interests and by the awareness of each contributor. Referring to Finch and
Crunkilton (1999), with regards to thinking that is based on the Curriculum Pedagogy
Assessment Model, the development approach includes: 1) the knowledge, skills, and affects
required to exit a course are communicated in advance; and 2) the course content drives the
model. To evaluate the curriculum, four aspects were questioned: 1) content of the lesson
must match with the expected learning outcome, 2) whether the teacher education curriculum accommodates sustainable development program, 3) efficacy of the curriculum in terms
of developing student's pedagogy knowledge, and 4) whether the curriculum contribute to
the development of capable citizen in the education field.

The process of national development is affected by issues spreading in society. This could
not be omitted, as it would hinder the achievement of curriculum goals, or even the national
development itself. Teachers need to eliminate these issues through some integrated response
in the learning process. The developing issues and how to overcome them seems to be the
hidden curriculum. The hidden curriculum issues that seep into curriculum implementation
are highlighted in curriculum development. If the values that exist in the hidden curriculum
strongly influence the quality of the learning outcome, then these will be further considered
in the preparation of the curriculum, as stated by Jerald (Merfat, 2015). He noted, "When
they are aware of the importance and influence of the hidden curriculum, they will always
review their personal attitudes with their students in the classroom. Moreover, teachers may
use the hidden curriculum in their teaching as a strategy or method to send a specific message to the students through these approaches, such as cooperative learning". The hidden
curriculum that has been running in the learning process, which has been relatively successful in improving the quality of learning, will be study material in the completion of the next
curriculum.

Factors affecting the evaluation of the curriculum can also be seen from the situational
analysis that is considered in the evaluation. Skilback (Brady, 1992) describes which factors
constitute the situational analysis: 1) External factors: cultural and social change and expectation, educational system requirements and challenges, the potential contribution of teacher
support systems, and the flow of resources into the school, and 2) Internal factors: pupils,
teachers, and perceived and felt problems and shortcomings in the existing curriculum. Ball
and Forzani (2009) explained that building a practice-focused curriculum for teacher education requires specifying the content—what teachers need to learn to do and unpacking it for
learning. It requires the development of instructional approaches to help teachers to learn to
do these things for a particular purpose in a certain context. Particularly challenging is how
to design ways to teach education practice that do not reduce it to propositional knowledge
and beliefs. This implies that various things need to be implemented in the development of
the curriculum for teacher education, which include: the scope of the material, the orientation
of the outcome, meeting the development missions of both central and regional government,
contextual learning methods, evaluation, and the involvement of industrial institutions.

The implementation of a qualified curriculum will contribute to nation building. Basically
the curriculum contains both national and regional development missions, which will linearly
affect the acceleration of development. As the results of research by Hussain et al. (2011),
among others, regarding the Evaluation of Curriculum Development Process show: "1) the
contents of the curriculum issue mentioned in the national education policy, 2) scientists
contribute in a proper way for the science subjects". In curriculum planning, the finishing
elements, based on Finch and Crunkilton (1999 p. 46), are: "1) involvement of stakeholders,
2) conduction of environmental scans, 3) identification of factors related to a program's success and failure, 4) development of vision and mission statement, 5) identification of career
and future resources/restraints, 6) development of realistic goals and objectives, 7) formulation of plans of action, 8) monitoring and follow up activities". The evaluation of the cur-

riculum for teacher education, therefore, looks at how curriculum planning has an impact on the success of the expected curriculum objectives.

4 CONCLUSION

The evaluation of the curriculum for teacher education is important for achieving national development. It needs to adjust the learning outcomes to the requirements of the job market, which are related to the acceleration of the national vision. Development programs became a guide to developing the curriculum so that it matches the needs of the job market and can meet these needs with the human resources produced. In the twenty-first century, teachers are required to be able to skillfuly and capably teach students to obtain further benefits, either for the job market or for their career. Education policy should accommodate the necessity of curriculum evaluation, especially the parts related to a teacher's scope, which can produce teachers that are able to educate the next generation to meet the needs of national development. Curriculum development needs to adjust to the developments and changes in the global era. The analysis covers the changing nature of subjects, cultural and social changes, and also the related expectations. Through a periodical evaluation of the curriculum for teacher education, it is expected that intellectual and skillful teachers will be produced, hence the acceleration of national development can be conducted.

REFERENCES

Ball, L. & Forzani, F.M. (2009). The work of teaching and the challenge for teacher education. *Journal of Teacher Education, 60*(5), 497–511.

Becker, G.S. (1975). *Human capital, a theoretical and empirical analysis, with special reference to education.* New York, NY: National Bureau of Economic Research.

Berry, B. (2003). *Teaching 2030.* London, UK: Teacher College Press.

Brady, L. (1992). *Curriculum development.* New York, NY: Prentice Hall.

Campbell, C.P., et al. (1993). *Improving vocational curriculum.* South Holland, IL: The Good-heart-Willcok. Finch, C.R. & Crunkilton, J.R. (1999). *Curriculum development in vocational and technical Education: Planning, content, and implementation* (5th ed.). Needham Heights, MA: Allyn and Bacon.

Hussain, A., Dogar, A.H., Azeem, M., & Shakoor, A. (2011). Evaluation of Curriculum Development Process. *International Journal of Humanities and Social Science, 1*(4), 263–271.

Grosch, M. (2017). Developing a competency standard for tvet teacher education in Asean countries. *Jurnal Pendidikan Teknologi dan Kejuruan, 23*(3), 279–287.

Lau, D.C.-M. (2001). Analysing the curriculum development process: Three models. *Pedagogy, Culture & Society, 9*(1), 29–44. doi: 10.1080/14681360100200107.

Merfat, A.A. (2015). Hidden curriculum as one of current issues of curriculum. *Journal of Education and Practice, 6*(33), 125–128.

Psacharopoulos, G. (1985). *Education for development: An analysis of investment choices.* Oxford, UK: Oxford University Press.

Salis, E. (2010). *Total quality management in education.* Yogyakarta: IRCiSoS.

Syarifudin, S. et al. (2004). *Pendidikan Indonesia Masa Depan [Future education of Indonesia].* Jakarta, Indonesia: UNJ Press.

Trilling, B. & Fadel, C. (2009). 21st century skills: Learning for life in our times. San Francisco, CA: John Wiley & Sons.

UNESCO. (1996). *The international commission on education for twenty-first century.* Paris, France: UNESCO Publishing.

Marfu'ah, S., Djatmiko, I.W. & Khairuddin, M. (2017). Learning goals achievement of a teacher in professional development. *Jurnal Pendidikan Teknologi dan Kejuruan, 23*(3), 295–303.

Yu, A. (2015). A study of university teachers' enactment of curriculum reform in china. *Journal of International Education Studies, 8*(11), 113–122. doi: 10.5539/ies.v8n11p113.

Character Education for 21st Century Global Citizens – Retnowati et al. (Eds)
© 2019 Taylor & Francis Group, London, ISBN 978-1-138-09922-7

Integrating values in EFL teaching: The voice of student teachers

A. Ashadi
Universitas Negeri Yogyakarta, Indonesia

ABSTRACT: This paper examines how English Foreign Language (EFL) teacher candidates perceive value integration in their experiences and how it is realized through their practice. An online survey followed by in-depth interviews with 15 student teachers in a teacher professional program was held in a state-owned teacher training college. This study is expected to provide early insights to what extent 18 values in character education enacted by the minister of education can be and have been instilled in a language learning setting. The initial result is projected to benefit teachers with respect to the curricular burden in terms of value integration in their instructional practice. Further implications on English language teaching at large is discussed in the light of 21st century skills.

1 INTRODUCTION

The global society is experiencing a wave of crises such as hate crimes, genocide, injustices to minorities and financial chaos due to unfair practices. Teachers and educational practitioners have responded to the situations by confirming the importance of promoting positive learners' character (Gilead 2011). Berkowitz & Grych (2000) believe that the development of positive character among children and teenagers is a lifelong effort, and should begin at an early age. The effort to pass on desirable character to young children entails shaping their thinking and behaviors so that these become beneficial to their future, their families and friends, their local communities, and the global society. Lickona (2000) addresses the endeavor as "one of our most basic responsibilities as adults is to sustain our civilization by passing on the values that are foundation of our society."

Schools play a very important role in the attempt to instill values as the foundation of modern societies. However, Althof & Berkowitz (2006) argue that there have been intersections among moral, character and citizenship education in the role of schools. While moral education tends to be theory-based, character education is more a-theoretical and a very eclectic discipline incorporating the virtues of moral education. An important facet of democratic citizenship is the capability to surpass personal interest and to be devoted to the welfare of some larger society in which she belongs to (Sherrod 2002). It, thus, entails the motivation and ability to engage in debates on current community issues and to participate in local and national policymaking. With these junctures, moral education provides the guidelines and a theoretical base for character education for school teachers and educational practitioners to explore in more attractive and effective instructional practices. These are, therefore, aimed to help the young generations to participate actively in the nondiscriminatory development of the society.

This article reflects how teachers and educational practitioners make sense of character education through stages of instructional practices they conduct in schools. It begins with exploration of their understanding in the initiative, then it investigates their perspective on the implementation of teacher education at schools. Eventually, this mini research seeks to understand what student teachers have done in their classrooms and how they incorporate teacher education in their instructional practices along with the rationale.

1.1 *Character education and teacher education*

The prominence of character education is gaining popularity among educators and politicians. Regardless of all the given attention in implementing character education at school level, it is still questionable on how teacher education programs are purposefully preparing preservice teachers for the mission. The virtual abandonment of moral character education in the formal preservice teacher curriculum has at least two related reasons (Narvaez & Lapsley 2008). The first is the overwhelming load of training purposes that have packed the subject matter's academic curriculum, and the second is the confusing stance in which teachers, parents and other stakeholders expect schools to address the character of students, without being caught teaching values directly. The later may not be the case of the current study's context as the government have stipulated specifically 18 positive characters to be instilled along with classroom instructions. More problematic than those two is the dilemma on how character education should be introduced through classroom instruction.

To address the problem, Narvaez & Lapsley (2008) suggested two approaches of minimalist and maximalist stances. The first approach requires teacher educators to make clear the hidden moral education curriculum and to reveal the complex relation between best practice instruction and moral character results. This entails the incorporation of character education through the mastery of a set of instructional instruments. The maximalist strategy requires preservice teachers to master a tool kit of pedagogical strategies that target moral character directly as a curricular goal. This strategy suggests the introduction of character through curricular activities stated clearly in the policy document. This strategy is what has been initiated in the context of the current study.

The perspectives of student teachers have been unnoticed in research on schooling and education in general (Lanier 1986) although their assertiveness towards moral education, and more particularly character education, is significant because of the association between their understandings as students and their future instructional practices as professionals in the classroom (Richardson 1991). There has been an ongoing discussion about the relationship between student teachers' attitudes and their behavior as teachers, and the impact of education courses and the training that teachers receive as students. This study is therefore important to detect these student teachers' opinions, attitudes and related behaviors in relation to the introduction of moral education in EFL classrooms, although, research (Hollingsworth 1989, Zeichner & Gore 1990) has shown mixed results on whether teacher education courses have impacts on teachers' beliefs and attitude about teaching.

Popkewitz (1987) argues that teacher professional character is attached not only with their subject knowledge, but also with their perceived character in the public. Such a role is tangled with concepts of moral authority and social responsibility. This notion of professionalism is boosted through the word education itself as a moral endeavor (Pring 2001). Student teachers who had no relevant training and no expectation that they will perform as expected and improve their own practice in line with the curricular demand are barely reliable. Revell and Arthur (2007) showed that while students were dedicated to various forms of character and moral education less visibly, in the context of classroom practice they became less confident in their instructional practices. With the absence of knowledge regarding the incorporation of character education in teacher education, this paper seeks to understand from the lens of student teachers who underwent teacher training in a state teachers college.

2 METHOD

Employing a case study approach (Yin 2013), the research investigated 15 student teachers through a set of online questionnaires, document analysis of the participants' lesson plans, and individual in-depth interviews. Further observations of their teaching performance were also conducted to examine and corroborate what they stated in the interviews and wrote in the lesson plans.

It took around six months (January to June 2017) to conduct the aforementioned stages. Initially, the participants were asked to fill in an online form containing a questionnaire.

Table 1. Data collection & analysis.

Tech	Modes	Analyses
Questionnaires	Online	Downloaded, Coded
Observations	Direct	Noted, Coded
Documents	Copy	Highlighted, Coded
Interviews	Direct	Recorded, Transcribed, Coded

As initial patterns were shaped from the thematic analysis of their answers in the questionnaire, observation, and teaching documents, the significant themes were then combined to construct 11 interview questions related to character education in their teacher training and instructional activities.

Constantly comparing the answers from different data gathering techniques, the study found several patterns that reappeared and became stronger. These were then selected (Strauss & Corbin 1990) to determine the core category.

3 FINDINGS AND DISCUSSION

3.1 *The need for more introduction*

With 78% of the participants admitting to have sufficiently known the incorporation of character education in teaching and the rest showing their limited knowledge in the matter, character education is likely to seem novel among the participants. Many of them commented '*the absence of socialization in schools*', others mentioned that they '*don't really understand the initiative*', and the rest declared '*just read at a glance but never attended any trainings/workshops on it*'.

In relation to this, Jones, Ryan, and Bohlin (1999) proposed a very strong argument: 'In order for teachers to fully embrace the moral dimension of their profession, institutions which prepare teachers must become more deliberate and clear about their own overarching goals and mission. Character education needs to become a more explicit, intentional focus across the curriculum of teacher education, not merely left to the informal efforts of individual professors'. Hence, visibility of the goals and clarity of what are expected from the teacher participants is the main concern to make character education effective on them.

The limited access to knowing the initiatives caused some of them to '*guess*' what the government meant and wanted. The speculation is reflected in the way the participants constructed their lesson plans.

3.2 *The prominence of collaboration*

The data also showed that most participants regard the eminence of integrating cooperation as values in their teaching. Almost three quarters of the participants found it easy and applicable in any instructional settings.

Ann, one of the participants, believed that learning foreign language would be more effective through collaborative activities in the instructions. She stated, "Collaboration allows everyone in the classroom to be more engaged and learn faster." Her statement was confirmed by most of her colleagues who repeatedly expressed the associated terms such as '*team-work*', '*cooperation*', and '*group-work*' in the interviews. These expressions also appeared repetitively in most of their lesson plans.

Slater & Ravid (2010) believed that collaboration is 'the human dimension that it can most effectively change. People working with people, garnering trust, and sharing expertise and experiences are important indicators of change'. The importance of collaboration has also been well documented in research and is clearly stated in the 21st century skills as a part of learning and innovation skills. If education means change, this signifies the importance collaboration in education and in character education particularly.

Figure 1.   The stage of teaching in which character values are best integrated.

The other values such as independence, nationalism, and religiosity follow the prominence of collaboration. However, it is not clear whether the order is due to limited knowledge of the participants or other inevitable causes.

3.3   *Activities matter most*

With regard to the practice of integration in the teaching process, the majority of the participants argued that character education is best incorporated through the teaching activities. Most of them also believed that it should appear in the lesson plan design in which they could focus on strengthening the character values.

The figure shows reflection and evaluation are second to the instructional design. It is surprising that basic competence analysis was perceived to enable character integration while authentic learning assessment came later. This indicates that teachers believed that character values are more practical than conceptual.

As character involves cognitive, affective and behavioral traits (Lickona et al. 2002), character education should embrace the three domains in order to be successful. Emphases need to be put accordingly in the three spheres to allow students to think, feel, and behave in desirable characters.

4   CONCLUSION

In general, the results of this mini research show the dilemma between what the student teachers learned minimally in their teacher training and the expectation to perform as character educators. When character education is not of the main concern of teacher education, empty talks in character education will persist. The study has also indicated the need to reinforce and prioritize as a more practical issue rather than an abstract concept of character education in schools.

Such requires clarity in terms of what education character education goals are and what teachers are expected to do in the curriculum documents. Therefore, what Walker et al. (2015) suggest concerning winning character education in the official policy does make sense and needs to be supported. This would give teachers clear guidance, provide them with alternative ideas and actions, and prevent them from speculating.

In short, the wholeness of character education needs to be well introduced to potential teachers. When it exists in the teacher education, student teachers would also think, feel, and behave accordingly. They would understand, make sense, and execute character education with their students.

REFERENCES

Althof, W. & Berkowitz, M.W. 2006. Moral education and character education: their relationship and roles in citizenship education. *Journal of moral education* 35 (4): 495–518. doi: 10.1080/ 03057240601012204.

Berkowitz, et al. 2000. Early character development and education. *Early Education and Development* 11(1): 55–72.

Gilead, T. 2011. Countering the vices: on the neglected side of character education. *Studies in Philosophy & Education* 30(3): 271–284.

Hollinsworth, S. 1989. Prior beliefs and cognitive change in learning to teach. *American Educational Research Journal* 6(2): 160–189.

Jones, et al. 1999. Character Education & Teacher Education: How are Prospective Teachers Being Prepared to Foster Good Character in Students? *Action in Teacher Education* 20(4): 11–28.

Lanier, J. 1986 Research on teacher education, in: M. Wittrock (Ed.) *Handbook of research on teaching*. London: MacMillan.

Lickona, T. 2000. Talks about character education. *Scholastic Early Childhood Today* 14(7): 48–49.

Lickona, T., et al. 2002. *Eleven principles of effective character education*. Washington: Character Education Partnership.

Popkewitz, T.S. 1987 The formation of school subjects and the *political* context of schooling, in: T.S. Popkewitz (Ed.) *The formation of the school subjects: the struggle for creating an American institution*. New York, NY: Falmer Press.

Revell, L. & Arthur, J. 2007. Character education in schools and the education of teachers. *Journal of Moral Education* 36(1): 79–92.

Richardson, V. 1991. The role of attitudes and beliefs. In: J. Sikula (Ed.) *Learning to Teach, Handbook of Research on Teacher Education*. London: Macmillan.

Sherrod, L., et al. 2002 Dimensions of citizenship and opportunities for youth development: the what, why, when, where, and who of citizenship development. *Applied Developmental Science* 6(4): 264–272.

Slater, J.J. & Ravid, R. (Ed.). 2010. Collaboration in education. New York, NY: Routledge.

Strauss, A. & Corbin, J. 1990. *Basics of qualitative research: grounded theory procedures and techniques*. Newbury Park, CA: Sage.

Walker, D.I., et al. 2015. Towards a new era of character education in theory and in practice. *Educational Review* 67(1): 79–96.

Yin, R.K. 2013. *Case study research: design and methods*. Thousand Oaks: Sage.

Zeichner, K.M. & Gore, J.N. 1990. Teacher socialisation, in W. R. Houston (Ed.) *Handbook of Research on teacher Education*. New York, NY: Macmillan.

Pancasila education system and religious skills

S. Rochmat
Universitas Negeri Yogyakarta, Indonesia

ABSTRACT: Religious skills have been eliminated from the revised Curriculum 2013, except for the subject of religion. This indicates that the government has not formulated clearly what it is the core values of Pancasila education system. This education system is oriented towards the establishment of nation and character building. This paper aims to formulate the religious skills that have been mandated by Pancasila. The national education system tends to adopt corrupted modern Western education, as education does not really train students' critical thinking abilities. This kind of education system has a negative contribution to the establishment of the modern state of Indonesia. The success of the West in building modern nation states is due to education systems that rely on the aspect of ratio, supported by strong law enforcement. Actually it is possible to produce religious skills from modern knowledge, as long as the latter is synergized with religion, which functions as a source of values. In this regard I agree with Albert Einstein's statement that 'religion without science is lame, science without religion is blind'.

1 INTRODUCTION

The Pancasila education system should develop religious skills as the first pillar of Pancasila. The belief in God, the only one God, should guide in the understanding and implementing of the other pillars, which advocate the principles of humanity, nationalism, democracy, and justice. The Pancasila education system has been mentioned since the establishment of the Republic of Indonesia in 1945, but it has not been elaborated clearly.

Indonesian education tends to follow the steps of the modern education system, emphasizing the role of ratio, and accordingly Indonesia is also suffering from the crisis of the modern era. This crisis is more acute as Indonesian education has not succeeded in developing critical thinking or in executing law enforcement. This lack of law enforcement will not become a big problem if the people are practicing religious skills, namely the ability to relate the religious system of knowledge to the secular one (Rochmat and Hadi, 2004: 148).

Roi (2003) mentions what is meant by the Pancasila education system, as well as the goals of the national education system, but does not elaborate on how the Pancasila education system works. I can understand why Indonesian people have difficulties in formulating the Pancasila education system, as Indonesian elites are divided into two opposing systems of knowledge, which are the religious system of knowledge and the secular one. The former is organized under the Ministry of Religious Affairs and the latter is organized under the Ministry of Education and Culture.

The Ministries of Religious Affairs and of Education and Culture have developed different systems of knowledge. The former has developed knowledge whose original source comes from the heart, meanwhile the latter has developed secular knowledge, whose original source comes from ratio. In fact, at school, religious subjects tend to follow the same methods as secular subjects, which teaches students religious knowledge, namely its rational aspects, rather than training students how to implement religious values and how to relate religious values to the secular subjects (Tirtosudiro, 1995). Accordingly, a religious environment has not been manifested in the life of students. Students suffer from moral decadency, as they do not know the goals of their life.

Schools tend to focus on providing the students with subject materials that develop rational thinking but do not relate to the knowledge of the heart. Furthermore, this rational thinking has not been fully developed in order to train the students in critical and analytical thinking. Schools tend to fill the students with subject materials that must be memorized if they want to pass the test. Accordingly, many students get stressed, as they do not understand the logic of the things that they memorize. Quite high levels of stress occur among students, as they must study a lot of subjects. This level of stress increases with the introduction of the concept of the mastery of learning, which students must reach in order to gain a high pass mark. Last, but not least, the ranking system in schools is also not sensitive to the individual characteristics of their students, such as in terms of talent and motivation. This kind of education just appreciates the academic skills and ignores the humanity aspects of education, which should take care of all the students. The school should provide the students with practical skills that can be used in the workplace if they do not want to pursue higher education. Actually, the most essential element of education is to produce a good student, namely, a student who does good deeds and has proper attitudes. This article aims to formulate the Pancasila education system.

2   NATIONAL EDUCATION SYSTEM: SYNERGIZING TRADITION AND MODERNIZATION

Tradition and modernization constitute two different systems of knowledge. The former represents the religious system of knowledge and the latter represents the modern system of knowledge. Tradition and modernization constitute the fundamental elements for building the national culture. For the Japanese, the formation of a national culture has been quite successful, as the Japanese have never lived under colonial rule and, accordingly, the traditional elites were able to transform into modern elites. As a result, they were able to modernize their country in a short space of time. They also had a strong commitment to developing education as the key element for the advancement of the nation.

Japan did not experience acute tensions between religion and nationalism, as the religious leaders were able to synergize the religious system of knowledge with the modern one. In line with this, these religious leaders continued their role of educating the people. In this regard, the Japanese government did not develop its own education system, but supported the process of transformation from the traditional system of knowledge into the modern one. It is not surprising that teachers regularly bring students to visit the Shinto shrine, as there exists a continuation between tradition and modernization.

Indonesia is different from Japan. Indonesia has difficulties in formulating its national education system, as its modern elites are different from its traditional ones. In other words, Indonesia has difficulties in formulating its national culture. Different social and political groups compete with each other to dominate the national political system and to secure their systems of knowledge. To simplify, I can categorize two different systems of knowledge that have manifested into two ministries of education. The Ministry of Religious Affairs organizes religious systems of knowledge, while the Ministry of Education and Culture organizes modern systems of knowledge.

The dichotomy of the education system happened because the traditional elites failed to synergize the Islamic system of knowledge with the modern system of knowledge (Nasution, 1995). What happened was that the traditional ruling elites supported the *modern system of knowledge*. They also became the modern ruling elites. In this regard, Geertz (1960), in his book *The Religion of Java*, successfully categorized three different systems of knowledge, which are "*Santri, Abangan, dan Priyayi*" (religious people, lay people, and the ruling elites). In this regard, some Islamic religious leaders are often confronted with the modern ruling elites, as they were the first to create their own modern education system to secure their political interests. This implies that the government tried to monopolize national political life and, accordingly, Indonesia failed to build a democratic state under the rule of the Soekarno and Soeharto regimes. A democratic government is a kind of social capital that constitutes the superstructures of a modern state.

Social capital is very important in sustaining a modern state. Another type of social capital that contributes to the establishment of a modern state is a national education system. Indonesia has not been able to formulate a Pancasila education system, which would unify both the religious system of knowledge and the modern system of knowledge. In this regard, Tony Barnet's evaluation of the Third World, including Indonesia, finds its relevance:

> "The main problems in the Third World are not, by and large, the absence of technical specialists—countries such as India and Pakistan have these aplenty; … The main problems are sociological and political problems, the contexts within which apparently "technical" decision are taken" (Barnet, 1995, p. vii).

Social capital should be related to religious skills, which can be socialized by students through several different subjects, such as character education, civic education, and religion. Dwiningrum (2013) has tried to formulate the nation's character education based on social capital, which includes participation and social networks, trust, reciprocity, social norms, social values, and proactive action.

It is difficult to achieve the Japanese phenomenon of the integration of religious teachings and nationalism, as we have not been able to formulate the Pancasila education system. President Soekarno and President Soeharto were not able to bridge the gap between the Islamic system of knowledge and the modern system of knowledge. They tended to adopt a purely modern system of knowledge, as they were occupied by the concept of an integral state. Based on the concept of an integral state, presidents tended to rule autocratically so that the people were afraid of criticizing the government. Accordingly, this concept of a governmental system did not produce a system of education that was conducive for developing democracy.

Actually, the management dichotomy of religious and modern education systems can be maintained as long as both are concerned with the Pancasila education system, which synergizes knowledge of both the ratio and heart resources. These two departments can function to provide checks and balances on each other, as monopolies tend to corrupt.

3   NATIONAL EDUCATION AND RELIGIOUS SKILLS

In line with the nature of Pancasila as an open ideology, Indonesians have developed an eclectic, incorporative, harmonious dynamic to their life, such as that suggested by Notonagoro. Eclectic refers to the act of selecting the best approach from various sources, meanwhile incorporative means to include as a part or to combine into a unified whole. Harmonious means forming a pleasing or consistent incorporation with the whole, and dynamic means the strength to produce a fast and fully enthusiastic movement (Siswoyo, 2013). This approach is fundamental in underlying the discovery of truth or reality. Consistent with it, Indonesians try to synergize the modern mode of knowledge into the traditional mode of knowledge in order to produce religious skills.

The national education system should produce religious skills among the students, as has been mandated by the first pillar of Pancasila, which should guide the understanding of the other four pillars. This means that religion should guide secular knowledge, namely, the knowledge of the rational product. By doing so, secular knowledge will not be used to degrade humanity. Our founding fathers had learned from the history of the West, so that they did not imitate the purely secular system of knowledge. They understood that Western civilization had suffered from modern crises with the outbreak of World War I and World War II. The origins of these crises can be traced from the modern system of knowledge, which relies on the role of ratio. One of the characters of ratio is that it always tries to gain benefit. For that purpose, ratio often does not function properly if it is not guided by the heart. For example, people often make use of ratio to trick others for their own interests.

The government has formulated the definition of the national education system since its independence in 1945, but what is meant by the national education system has not been formulated. Even recently, RoI (2003) has not elaborated on the Pancasila system of knowledge.

The Reformation era, which forced President Soeharto to step down, has given the country the opportunity to formulate an education system that develops critical thinking as well as liberated education. However, this has not been formulated successfully. The same tune also happens with the education system which tries to develop religious skills, it also seems hard to be formulated. Indeed, religious skill is a new concept with the issuance of National Curriculum of 2013. It was not easy to implement religious skills in education and, accordingly, it was dissolved with the issuance of the Revised National Curriculum 2013, except for the subject of religion. Ironically this aspect of religious skills should be maintained in the *Rencana Pelaksanaan Pembelajaran* (Preparation of Teaching Practice) of any subjects. Teachers may teach religious skills either during intra-curriculum activities or extra-curriculum activities. This scheme, however, is out of context with the true religious skills that train the students to relate their secular knowledge to their religious system of knowledge. It is also possible for teachers to teach religious skills during the end part of their teaching activities, that is, during the section of reflection. However, this has not fulfilled the ideal model of teaching religious skills, which should prevail throughout the teaching–learning process.

Religious skills education is at a cross section, as the policy of regional autonomy has not been followed by the true policy of autonomy. In the education sector, the issuance of the National Curriculum (*Kurikulum Nasional*) seems to be continuing the centralistic legacy of the New Order Regime. This National Curriculum has dictated either the subject matters or their contents, which tend to fill the students' minds with memorizing the subject materials. This National Curriculum should provide the minimal standards of education, while at the same time guaranteeing the implementation of religious skills.

On other hand, teachers of the subject of religion have also not been able to train their students in religious skills, as the subject materials tend to be dogmatic, and do not include training critical thinking through contextual teaching–learning. Religion should relate to the actual problems of society as well as to the secular subjects. The Curriculum for religious education should incorporate an empirical approach into its normative approach so as to train the students' logical capacity, imagination, and creativity. By doing so, the students would be able to give meaning to their real life (Rochmat, 2007).

Autonomous policy gives opportunities for the local government to reformulate history education, which is based on religious local culture, and then transform it into nationalism. Until now, history education has had centralistic characteristics that just emphasize nationalism, so that it was felt to be too abstract for students at elementary school (SD), as well as for students at junior and senior high schools. Moreover, nationalism rooted in Western civilization never pretends to assume to be a source of values. Consequently, we should integrate it with religion, as well as local culture, which indeed take on roles as sources of values.

As there are many systems of knowledge, I questioned the role of an ideological approach to life. However, I agree with the character of knowledge as the product of the heart which evaluates the principle of harmony highly (Wahid, 1997). By so doing, all ideologies should develop the principle of mutual give and take in order to be relevant in the contemporary era, and to solve the problems of human beings. There is no absolute truth in this worldly life, so people are challenged to find the truth in the form of the common good, regarding the truth in the context of relationships. The existence of truth is not independent within the individual's rationality, but it is correlated with social and religious orders. This implies that an individual's conception of truth should be negotiated with the society (Wahid, 2001).

The Pancasila system of knowledge should acknowledge both the empirical method and the consciousness method as a way to acquire science. With this synergy with all of the sciences, it focused upon integrating the sciences into the spirit of God. This Pancasila paradigm is similar to Bassam Tibi's idea of Islamic humanism, which would function as 'cross-cultural international morality', in which he appreciates Ibn Rushd's idea of double truth as follows:

> "Such morality would bring to the fore the work of the first and greatest Islamic political philosopher, al-Farabi.... It would also remind scholars of the value that Europeans conferred on the work of the humanist philosophers ... in particular of Ibn Rushd's teachings about the *Haqiqa al-muzdawaja* (double truth), which differentiated

between philosophical or rational knowledge and religious beliefs or divine revelation and paved the way for modern European rationalism" (Tibi, 2009, p. 65).

I agree with Wahid that modern sciences are crucial to supporting the existence of people in this worldly life and, as a Muslim, I believe that Islamic cosmology directs me about the secular sciences. Indeed, my rationality does not always find the logical arguments from modern sciences, but finds its foundation in the integration of religious sciences (Wahid, 1999). Consequently, I try to create a dialog within my thoughts between religious sciences and the secular ones for the purposes of implementing the truth with the realistic actions in the worldly life.

4 CONCLUSION

The dissolving of religious skill in the revised Curriculum of 2013 means that it does not have the philosophical foundation of the first pillar of the national ideology of Pancasila, namely, the belief in God, which should guide the understanding and implementation of the other four pillars. This implies that the Pancasila education system develops a religious paradigm to this worldly life. In other words, secular knowledge should be viewed through the religious paradigm rooted in the heart. This does not mean that religious sciences are more important than secular sciences, because both are interrelated. In line with this, in life we should find a balance between the normative aspects of religion and the freedom of thinking in matters of secular affairs. Accordingly, we should understand life as the ability to relate the spiritual vision of God with the actions of human beings in worldly life by making use of secular sciences.

REFERENCES

Barnet, T. (1995). *Sociology and development*. London, UK: Routledge.
Dwiningrum, S.I.A. (2013). Nation's character education based on social capital theory. *Asian Social Science, 9*(12), 144–155.
Geertz, C. (1960). *Religion of Java*. Chicago, IL: Chicago University Press.
Nasution, H. (1995). Perlunya menghidupkan kembali pendidikan moral [Revitalizing the moral education]. In S. Mujani & A. Subhan (Eds), *Pendidikan Agama dalam Perspektif Agama-Agama* [Religious education in the view of religions]. Jakarta, Indonesia: Logos Publishing House.
Rochmat, S. & Hadi, B. S. (2004). Tradisi dalam Pembentukan Identitas Bangsa Indonesia di Era Modern [The role of tradition in the formation of National Character Building at Modern Era], *Cakrawala Pendidikan, 1*(XXIII), 147–164.
Rochmat, S. (2007). *Paradigma Historis Pendidikan Agama agar Doktrin Agama Fungsional di Era Modern* [Historical approach to Religious Education]. Yogyakarta, Indonesia: FIS UNY.
RoI. (2003). *Undang Undang Nomor 20 Tahun 2003, Sistem Pendidikan Nasional* [Law Number 20 Year 2003, National Education System]. Jakarta, Indonesia: Republic of Indonesia.
Siswoyo, D. (2013). Philosophy of education in Indonesia: Theory and thoughts of institutionalized state (Pancasila). *Asian Social Science, 9*(9), 136–143.
Tibi, B.T. (2009). Bridging the heterogeneity of civilizations: Reviving the grammar of Islamic humanism. *Theoria: A Journal of Social & Political Theory, 56*(120), 65–80.
Tirtosudiro, A. (1995). Menciptakan suasana pendidikan yang agamis [Conditioning religious context for education]. In S. Mujani & A. Subhan (Eds), *Pendidikan Agama dalam Perspektif Agama-Agama [Religious education in the view of religions]*. Jakarta, Indonesia: Logos Publishing House.
Wahid, A. (1997). Dimensi kehalusan budi dan rasa [The dimention of feeling]. In I.M. Anshori (Ed.), *Islam, Negara, Dan Demokrasi* [Islam, politics, and democracy] (pp. 79–82). Jakarta, Indonesia: Erlangga.
Wahid, A. (1999). *Kyai nyentrik membela pemerintah* [The controvertial clerics supports to the governments]. Yogyakarta, Indonesia: LKiS.
Wahid, A. (2001). *Menggerakkan Tradisi: Esai-Esai Pesantren* [Revitalizing the traditions of Islam]. Yogyakarta, Indonesia: LKiS.

Character Education for 21st Century Global Citizens – Retnowati et al. (Eds)
© 2019 Taylor & Francis Group, London, ISBN 978-1-138-09922-7

The prevention strategies needed to avoid the negative impacts of using social media and a humanistic approach to character education

S. Nurhayati
Universitas Pekalongan, Indonesia

ABSTRACT: The rapid development of information technology is one indicator of the rapid advancement of education in Indonesia. Many people use the internet for everyday purposes, including education. For the students, advances in technology have had both positive and negative impacts. The negative impacts may influence the moral development of the students. Many strategies regarding the prevention of the negative impacts of the use of social media have been discussed, but most of these have not had a specific humanistic approach. Therefore, in this study the author proposes a strategy to prevent the negative impacts of using social media, using a humanistic approach. The humanistic approach will distinguish between the strategies applied for the character education of both school-age children and college students. There are differences in the methods of character formation between school-age children and college students.

1 INTRODUCTION

It cannot be denied that the development of information technology today is growing rapidly. This has an impact on all aspects of life, both positive and negative. Kaplan and Haenlein (2010) define "social media as an internet-based application group that builds on the foundation of Web 2.0 ideology and technology, and which enables the creation and exchange of user generated content". Various forms of social media are now an integral part of community life, from toddlers, children, and adolescents, up to adults. The use of social media by the community covers many aspects, from the worlds of education, health, and government, to the business world. As users of social media we can freely edit, add, and modify text, images, video, graphics, and various content models. Internet-based social media has several features, including those mentioned below:

a. Messages are not only conveyed to one person but can be seen by many people, for example messages via SMS or other internet sites
b. Messages are conveyed freely, without having to go through a Gatekeeper
c. Messages tend to be conveyed faster than by other media
d. The recipient of the message determines the time of the interaction.

Therefore social media users have the freedom to send messages or images or upload an article, and sometimes this is used by irresponsible people who want to spread "something negative". These negative messages will be easily accessible to everybody, so it often leads to a media war.

Moral education or character education in the present context is very relevant to help overcome the current moral crisis in our country. The crisis is in the form of increased promiscuity, violence of children and adolescents, crimes against friends, teenage theft, cheating (plagiarism), drug abuse, pornography, destruction of the property of others, and so on.

2 THEORETICAL REVIEW

2.1 *The development of social media*

The development of social media began with the birth of the Geo Cities site in 1995. This site provided a rental service data storage website, so that website pages could be accessed from anywhere. The emergence of Geo Cities was a milestone for the establishment of other websites.

The first social networking site that emerged was six degree.com in 1997, then the Blogger site appeared, which provided personal blogging services by offering users the ability to create their own website pages. In 2002 Friendster began, which was a social networking site that boomed at the time. In 2004 Facebook started up, which is still a famous social networking site today and has the most users among other social networking sites. On the other hand, there is also Twitter, which began in 2006. This is a social networking site that is different from the others, because Twitter users are limited to only using 140 characters to update their status, also called a tweet. The latest social networking development is Google+, which began in 2011.

Almost everyone uses social media as a means of information and entertainment, such as Twitter, Facebook, Instagram, YouTube and WhatsApp (WA).

2.2 *Basic human personality*

Every human being is born with a distinct personality. Many theories regarding personality have been developed by experts, but, in general, human personality is grouped into four types: the Choleric, Sanguine, Melancholic, and Phlegmatic personalities (Littauer, 2000).

1. Choleric personality
   Choleric personalities have characteristics that mean they will complete their job in their own way (my way). They are really creative; even if there is a manual, they do not like to obey it. However, the choleric personality will try to finish the job thoroughly.
2. Sanguine personality
   Sanguine is an easy-going personality. They will finish their job in the way that they think is the most fun (fun way). For them, the job is fun if they forget the time. They work without a plan and tend to underestimate whatever they do. Their attitude tends to be casual.
3. Melancholy personality
   Melancholy personalities are regular, well organized and systematic types of workers. In completing their work, a melancholy character will choose the best way. If there is a guide, then they will follow the guide 100 percent.
4. Phlegmatic personality
   Phlegmatic is the most pleasant personality for everyone. The phlegmatic personality is hardly ever angry. They have a sincere smile. It is just like someone who has no ambition. These people are peaceful, and do not like to quarrel.

2.3 *The importance of character education*

In article 3 of Law No. 20 Year 2003 on Indonesian National Education System (RoI, 2003), it is mentioned that the national education system has the function of developing the ability and forming the character and civilization of a dignified nation in order to educate the nation. National education aims to develop the potential of learners to become human beings who believe and are cautious to God Almighty, have a noble character, are healthy, knowledgeable, capable, creative, and independent, and who become citizens that are democratic and responsible.

Character education is a type of education that forms one's personality through character, whose results are seen in a person's actual actions, such as: good behavior, honesty, responsibility, respect for the rights of others, hard work, and so on (Samani & Harianto, 2011). This movement is expected to create a superior Indonesian human in the field of science and technology. The five basic things are:

a. Indonesians must be moralistic, and have a good character and good behavior. Therefore people have to live in a religious society and be opposed to violence.
b. The nation of Indonesia must become an intelligent and rational nation, be knowledgeable, and have a high sense of reason.
c. The Indonesian people must become an innovative nation, and pursue progress and work hard to change the situation.
d. They must be able to strengthen their spirit. No matter how difficult the problem is that needs to be faced, the answer is always there.
e. Indonesian people must be true patriots who love their nation, country, and homeland.

So the essence of character education in Indonesia aims to form a nation that is tough, competitive, noble, moral, tolerant, cooperative, patriotic, dynamic, science-oriented, knowledgeable, and technological, inspirited by faith and a belief in Almighty God based on Pancasila. Character education should be instilled in children from an early age to enable them to fortify themselves against possible negative influences originating from various sources, including those from social media (internet), remembering that the messages, images, or information presented in social media are not necessarily all good or educational.

2.4 *Humanistic theory*

Learning based on humanistic theory is suitable to be applied to learning materials that involve the formation of personality, conscience, attitude changes, and the analysis of social phenomena. Indicators of the success of this application are that the students feel excited, take the initiative in learning, and have a change of mindset, behavior and attitude that comes from their own will.

The humanistic approach regards learners as a whole person. In other words, learning does not only teach the targeted materials, but also helps learners to develop as human beings. This belief has led to the emergence of a number of learning techniques and methodologies that emphasize the humanistic aspects of learning. The humanistic approach prioritizes the role of learners and is needs oriented. Like teachers, learners are people who have emotional, spiritual, or intellectual needs. Learners should be able to help themselves in the learning process and not just be the recipients of passive science (Purwo, 1989).

Some of the principles of the humanistic learning theory are as follows:

1. Humans have natural learning ability
2. Significant learning occurs when the subject matter is perceived by the students to have relevance for a particular purpose
3. Students' learning should be as self-directed as possible
4. Students learn best in a non-threatening environment
5. When the threat is low, the learners gain experience from the learning process
6. Meaningful learning occurs when the students learning by doing
7. Learning goes well if the learners are involved in the learning process
8. Learning that involves the whole learner can give profound results
9. Self-confidence in learners is grown by self-introspection
10. Social learning is learning about the learning process.

3 THE IMPACTS OF USING SOCIAL MEDIA

Already the use of the internet as a medium of fast connections cannot easily be separated from our lives. All sorts of communication problems can be solved with the internet. The use of social media (internet) has both a negative and a positive impact, which can be felt by all people, ranging from children and adolescents to parents. Students are the group who have the widest use of social media.

The survey conducted by The Royal Society for Public Health (RSPH) in the UK (2017) indicates that there are positive impacts, because social media helps to maintain social relationships. Also, the research report reveals that YouTube is the platform with the most positive impact. This means that YouTube has the ability to have a better impact on the mental health of young people.

Meanwhile, the negative impact is that, in many cases, social media is rated by the respondent as a cause of depression or anxiety. Sleep quality is also affected because of going to sleep late and waking up to check messages on their smartphone.

4 PREVENTION STRATEGY THROUGH THE HUMANISTIC APPROACH

4.1 *Prevention strategy of the use of social media for school-age children (primary school)*

1. Give a detailed explanation of the benefits and dangers of using social media. Tell them that getting into a negative (pornographic/violent) site is taboo
2. Create a schedule of things to do together online
3. Assistance by parents in the use of social media. Do not let children access the internet alone
4. Install negative content software to block negative sites
5. Encourage the children to socialize with other family members and people in the community
6. Do not let the child be alone; keep an eye on their movements
7. Develop games that teach character education for children
8. Strengthen the politeness and ethics education for students.

4.2 *Prevention strategy of the use of social media for high school and college students*

1. Use internet protection software
2. Recommend them to the Law on Information and Electronic Transactions
3. Give sufficient explanation of the problems of plagiarism and sanctions
4. Teachers / lecturers should not set tasks where the material is only easily obtained through the internet
5. Optimize the role of teachers /lecturers to make a familial approach with students
6. Optimize character education processes in school / college
7. Involve students in various social and religious activities in the school/college environment.

5 CONCLUSION

The negative impacts from the use of social media for students have been found, but the strategy based on the humanistic approach is considered to be more effective. The humanistic approach is intended to provide a more focused study on the formation of students' characters, so it is expected that the behavior, habits, thoughts, and the willingness of students to learn will be changed by their own will. With the humanistic approach, learners are expected to have the freedom to undertake activities in a responsible manner without breaking the rules, norms, discipline, or ethics.

The implementation of a negative impact prevention strategy for social media usage should be differentiated between elementary school and junior high school students. This needs to be done because these two groups of students have different powers of thinking, emotions, and needs. Various strategies that have been pointed out in advance by authors can be used as a reference for educators (teachers and professors) in implementing the learning process in schools/colleges.

REFERENCES

Kaplan, A.M. & Haenlein, M. (2010). Users of the world, unite! The challenges and opportunities of social media. *Business Horizons, 53*(1).

Littauer, F. (2000). *Personality plus*. Grand Rapids, MI: Revell.

Purwo, B. (1989). *Pertemuan linguistik lembaga bahasa atma jaya* [*Linguistic institution atma jaya's meeting*]. Jakarta, Indonesia: Lembaga Bahasa Unika Atma Jaya.

RoI. (2003). Undang-Undang Republik Indonesia Nomor 20 Tahun 2003 Tentang Sistem Pendidikan Nasional [Regulation of Indonesia number 20 year 2003 about System of National Education]. Jakarta, Indonesia: Ministry of Education, Republic of Indonesia.

RoI. (2010). *Bahan pelatihan penguatan metodologi pembelajaran berdasarkan nilai-nilai budaya untuk membentuk daya saing dan karakter bangsa* [*Module of training learning methods based on culture for competetitiveness and nation character*]. Jakarta, Indonesia: National Education Department, Republic of Indonesia.

Royal Society for Public Health. (2017). *The use of social media can disrupt mental health*. London, UK: RSPH.

Samani, M & Harianto, H. (2011). *Pendidikan karakter, konsep dan model* [*Character education, its concept and model*].Bandung, Indonesia: Remaja Rosdakarya.

Character Education for 21st Century Global Citizens – Retnowati et al. (Eds)
© 2019 Taylor & Francis Group, London, ISBN 978-1-138-09922-7

Financial well-being among college students: The role of financial literacy and financial coping behavior

R.C. Sari, D. Priantinah & M.N. Aisyah
Universitas Negeri Yogyakarta, Indonesia

ABSTRACT: College students are in a transition phase from financial dependence to financial independence. Current phenomena, however, show that college students have low levels of financial literacy and tend to be impulsive buyers. The first objective of this present study was to examine all possible financial socialization agents that have the potential to affect financial literacy. The link between financial literacy and financial well-being has obtained very limited attention. Lower financial well-being/high financial distress affects health and is related to absenteeism and loss of productivity which means that students miss classes and have poor academic performance. However, the mechanism to improve financial well-being in adolescents is not yet clear, thus providing an opportunity for researchers to identify appropriate strategies to improve financial well-being in adolescents. The second objective of this present study was to understand the effects of financial literacy and well-being through coping behavior. This study included the responses of 206 students from public and private universities who joined a financial seminar held by the Financial Services Authority. Hypothesis testing used a structural equation modeling with Partial Least Squares (PLS). Based on the results of the factor analysis test, financial socialization agents affecting financial literacy were parents and family, teachers, financial professionals and formal education. The results support the hypothesis that financial literacy is positively associated with financial well-being and that financial literacy significantly influences financial well-being directly as well as through coping behaviors. Implications and suggestions for future research are proposed.

1   INTRODUCTION

College students are in a transition phase from financial dependence to financial independence. In this phase, parental supervision is reduced and students begin to achieve financial autonomy and financial responsibility. They are required to be responsible for paying bills or making monthly spending plans. Their ability to meet these challenges depends on the financial knowledge and behavior they have before entering college (Lyons et al., 2006). Current phenomena, however, show that college students have low levels of financial literacy and tend to be impulsive buyers (Hira & Brinkman, 1992). Lack of financial literacy will lead to increased conspicuous consumption behavior such as lavish spending on goods and services for the purpose of impressing others, poor financial management and premature affluence. Financial habits will be carried over into adulthood and may cause life-long difficulties.

Several previous studies indicated that satisfaction with financial security or financial well-being is an important component of overall life satisfaction (Shim et al., 2009; Hira, 1997). Financial well-being is an overall satisfaction with one's financial situation (Joo, 2008). Behaviors can improve financial well-being, i.e. managing resources effectively, planning ahead and making informed decisions (Drever et al., 2015). To make effective financial decisions, one must have good financial literacy (Lusardi & Mitchell, 2007) as well as critical thinking skills (Consumer Financial Protection Bureau, 2015). Financial literacy is the ability to use knowledge and skills to manage financial resources effectively to achieve financial well-being. Studies show that college students have low levels of financial literacy (Chen & Volpe, 1998).

As suggested by previous studies, financial socialization has effects on financial literacy (Lucey & Giannangelo, 2006). Financial socialization is a process by which young people acquire the standards, values, norms, skills and knowledge, and attitudes that foster their financial independence and subsequently facilitate their successful transition into young adults (Kim & Chatterjee, 2013). Financial socialization agents, including parents, teachers, family members, peers, the media, schools and religion, are typical sources of financial information (Koonce et al., 2008). However, most previous studies on predicting financial literacy have been focused on the influences of parents rather than those of other agents. The first objective of this present study was to examine all possible financial socialization agents that have the potential to affect financial literacy. Obtaining specific predictors of college students' financial literacy is necessary to design effective financial socialization programs, especially in countries with no specific mandates on financial education.

Lower financial well-being/high financial distress affects health (Mills et al., 1992) and is related to absenteeism and loss of productivity (Garman & Sorhaindo, 2005), which means that students miss classes and have poor academic performance. However, the mechanism to improve financial well-being in adolescents is not yet clear and thus provides an opportunity for researchers to identify appropriate strategies to improve financial well-being in adolescents. Therefore, the second objective of this study was to understand the effects of financial literacy and well-being through coping behavior.

2 LITERATURE REVIEW AND HYPOTHESIS DEVELOPMENT

2.1 Agents of financial socialization

College students become a lucrative market segment as their lifestyle changes; in addition, students have greater financial independence from their parents to make their own consumption decisions. Poor financial knowledge leads to poor financial management (Chen & Volpe, 1998), furthermore, it can affect academic performance, mental and physical well-being as well as the ability to get a job (Bodvarsson & Walker, 2004; Lyons, 2003). Consumer behaviors among adults is learned during pre-adult years through financial socialization agents (Moschis et al., 1984).

Financial socialization covers the development of attitudes, values and standards that ultimately will support or inhibit financial capabilities and well-being (Drever, 2015). Family life cycle theorists posit that the success of the transition to the next stage of the life cycle depends on the achievements and skills gained from the previous stage (McGoldrick & Carter, 1999).

Based on consumer socialization theory, individuals, especially children and adolescents, build consumer skills through interaction with various socialization agents to acquire knowledge, attitudes and values which eventually affect their consumption behavior (Moschis, 1987). Financial socialization agents include parents and family, peers, teachers, formal education and the mass media.

According to family life cycle and individual theories (McGoldrick & Carter, 1999), one of the main tasks in the transition to adolescence is to become financially independent. The development of financial independence among adolescents is related to financial socialization from agents. Financial values, norms, knowledge and skills are obtained from financial socialization that facilitates the financial literacy and competence of the young adult (Kim & Chatterjee, 2013).

The present study takes on board the overall conclusion from the above discussion and develops the following hypothesis:

H1: Financial socialization from parents, people other than parents, formal education and the mass media has an influence on financial literacy.

2.2 Financial literacy on well-being

Financial well-being is a continuum of an individual's perception and response to his/her financial state (Prawitz et al., 2006). Financial distress is on one side and financial well-being is on the other side of the continuum. It is crucial to note that financial distress is not directly related to a person's financial situation, such as individual incomes or high debts, but rather to the individual's perception of financial conditions that produce feelings of distress/well-being. A person may have high debts but feels a low level of financial distress (Prawitz et al., 2006). Financial distress is not only experienced by those on low incomes, but it is also related to poor money management (Garman & Sorhaiondo, 2005).

The effect of financial literacy on financial well-being still receives limited attention. Shim et al. (2009) examined the effect of financial literacy on the financial well-being of students but did not find empirical evidence for such an influence. Financial well-being in their study was defined as satisfaction with the current financial status. In another study, however, Joo and Grable (2004) indicated that financial literacy had an effect on financial well-being.

2.3 *Coping behavior on well-being*

Schwarzer and Knoll (2003) defined coping behavior as a process of ongoing adaptation to life including both reactive processes (dealing with problems) and future-oriented processes (preventive coping such as following budgets, tracking monthly expenses), as well as attaining future goals (proactive coping such as saving and investment).

Greenglass and Fiksenbaum (2009) and Gan et al. (2007) found that coping behavior is related to distress. A positive financial behavior will minimize financial stress (Kim et al., 2003). Xiao et al. (2006) found that consumers who engaged in positive financial behavior such as reducing debts, following budgets and practicing positive financial behaviors had a low level of financial distress.

In the present study, we predict that financial literacy can affect financial well-being indirectly by developing responsible financial behaviors; namely, preventive and proactive financial coping behaviors. This study uses the overall conclusion from the above discussion and develops the hypothesis below:

H2: Financial literacy has an effect on financial well-being through coping behaviors.

3 RESEARCH METHOD

3.1 *Sample*

This study included the responses of 206 students from public and private universities who joined a financial seminar held by the Financial Services Authority.

3.2 *Measurement variable*

a. Financial literacy
Financial literacy is the ability to use one's knowledge and skills to manage financial resources effectively to achieve financial well-being. Its measurement is adopted from OECD INFE (2011) which includes a test about investment, stock, bonds and interest rates.
b. Financial coping
Financial coping behavior is an attitude to respond to financial problems as well as an adaptation process having a future orientation to attain future goals. Financial coping behavior includes preventive financial behaviors and proactive financial behaviors.
c. Financial well-being
Financial well-being is students' perception of their financial state. It is the continuum on which financial distress is on one side and financial well-being is on the other. Financial well-being is measured by three constructs; namely financial stress, psychological distress and subjective well-being.

3.3 *Data analysis*

Hypothesis testing used a structural equation modeling with Partial Least Squares (PLS).

4 RESULTS

4.1 Descriptive statistics

Table 1. Descriptive statistics of the research variables.

Variable	Mean	SD
Financial literacy	2.385	0.438
Financial coping behavior	2.803	0.076
Financial well-being	3.182	0.125

4.2 Model measurement analysis

A model measurement is used to evaluate the relationship between indicators and constructs by assessing the reliability and validity of the constructs relating to specific constructs. The overall analysis shows that the measurement model is reliable and valid.

Before testing the inner model, it is necessary to evaluate the outer model. The indicator weight shows that all indicators are significant and the Variance Inflation Factor (VIF) value does not exceed 2.5, thus indicating that the financial literacy dimension, coping behavior dimension and financial well-being dimension have fulfilled the validity requirement for normative constructs.

4.3 Factor analysis

Factor analysis was used to determine which factors affect the level of financial literacy; namely parent and family, people other than family, formal education and mass media. Based on the results of the factor analysis test, three significant factors explain 33% of financial literacy, they were: parent and family, people other than family (teacher, financial professional) and formal education.

4.4 Structural model analysis

The structural model was used to test the hypothesis, particularly to examine whether the effect of financial literacy on well-being is direct or indirect (i.e. mediated by coping behaviors). We used a two-step approach (Baron & Kenny, 1986). Firstly, we tested whether financial literacy affects financial well-being directly by testing hypothesis one (H1). Secondly, we ran the PLS by introducing financial coping behaviors as the mediating variable in hypothesis two (H2).

The results (Table 3, Panel A) show that financial literacy is positively associated with financial well-being (coefficient: 0.17; $p < 0.05$, $R^2 = 0.03$). Therefore, hypothesis one (H1), which states that financial literacy is positively associated with financial well-being, is supported. Further analysis by introducing financial coping behavior as a mediating variable

Table 2. Factor analysis for item comprising source of influence on financial literacy.

Item description	Factor loading		
	Factor 1	Factor 2	Factor 3
Parent and family			
–Parental socialization	0.768		
People other than family			
–Teacher		0.769	
– Financial service (bank, financial professional)		0.769	
Financial education			
–School			0.864
–Financial training			0.864

Table 3. PLS results (path coefficient, t-statistics, R^2).

Variable	Path to	
	Coping behavior	Financial well-being
Panel A: Direct effect		
Financial literacy		0.17**
R^2		0.03
Panel B: Indirect effect		
Financial literacy	0.59***	
Coping behavior		0.32***
Indirect effect		0.19 ***
R^2	0.35	0.08

***p 0.01; **p 0.05; *p 0.10.

Table 4. Indirect, direct and total effect on financial well-being.

Path (financial literacy–coping)		0.59
Path (coping–well-being)		0.32
Indirect effect	0.59×0.32	0.19
Direct effect: path (financial literacy–well-being)		0.17
Total effect		0.36

reveals that financial literacy is positively associated with financial coping behavior (coefficient: 0.59; $p < 0.01$) and financial coping behavior is also positively associated with financial well-being (coefficient: 0.32; $p < 0.01$). However, the association between financial literacy and financial well-being remains significant (coefficient: 0.19; $p < 0.10$) (Table 3, Panel B). This means that financial coping behavior only *partially* mediates the relationship between financial literacy and financial well-being. In other words, while there is an indirect effect of financial literacy on financial well-being via financial coping behavior, financial literacy itself still has a direct effect on financial well-being.

In this study, the total indirect effect was 0.36 which was calculated based on the path coefficients among variables as shown in Table 4.

5 CONCLUSION AND DISCUSSION

This study aimed to examine the financial socialization agents that affect financial literacy and financial well-being in adolescents. This study is important for education in Indonesia since currently, financial literacy education is not included in the school curriculum. It is hoped that the findings of this study can be used to design effective programs to improve youth's financial literacy and provide empirical evidence on the development of intervention for financial well-being in adolescents.

Hypothesis 1 examined which financial socialization agents have significant effects on financial literacy. This is consistent with previous studies that parents become agents of financial socialization (Lawrence et al., 2003; Allen et al., 2007). In addition to parents, some financial socialization agents playing a role in improving financial literacy in adolescents include teachers, financial professionals, financial education in school and financial training. This finding is consistent with Sari et al. (2017) that voluntary financial education in Indonesia—where financial literacy education is not mandatory—has the effect of increasing financial knowledge. This implies the importance of financial literacy education for adolescents either formally in the school curriculum or informally through training and short courses outside of school.

The second hypothesis examined whether financial literacy influences financial well-being through coping behaviors. Coping behaviors partially mediate the effect of financial literacy on financial well-being. Financial literacy significantly influences financial well-being directly as well as through coping behaviors. These predictors explain only 8% of the variance. One plausible reason for this is the fact that the study was conducted in Yogyakarta, one of the provinces in Indonesia that has an index of happiness of 70.77 on a 0–100 index scale. The aspects used to measure the happiness index are health, education, employment, household income, family harmony, leisure time availability, social relations, housing conditions and assets, environmental conditions and security conditions. On average, people in Yogyakarta do not experience financial stress. Moreover, financial well-being is only one part of life qualities (Sirgy et al., 2007). Based on the survey, the level of people's satisfaction reaches the highest (81.03%) in family harmony and the lowest in education (62.63%), whereas the financial condition is not a major consideration of people in Yogyakarta.

6   LIMITATIONS AND FUTURE RESEARCH

There are limitations to be considered in interpreting the results of this study. First, the sample was a non-probability, convenience sample. As such, the results should be interpreted with caution. Second, there are other factors that have not been included in this study, and hence it is strongly recommended for future researchers to use other variable sets to explore further relationships among the variables. Based on the finding that the predictors explain only 8% of the variance, we suggest that future studies would include cultural variables in determining financial well-being and examine whether cultural differences are able to exploit financial well-being.

REFERENCES

Allen, M.W., Edwards, R., Hayhoe, C.R., & Leach, L. (2007). Imagined interactions, family money management patterns and coalitions, and attitudes toward money and credit. *Journal of Family and Economic Issues, 28*(1), 3–22.

Baron, R.M. & Kenny, D.A. (1986). The moderator-mediator variable distinction in social psychological research: Conceptual, strategic and statistical considerations. *Journal of Personality and Social Psychology, 51*(6), 1173–1182.

Bodvarsson, O.B. & Walker, R.L. (2004). Do parental cash transfers weaken performance in college? *Economics of Education Review, 23*(5), 483–495.

McGoldrick, M. & Carter, B. (1999). *The expanded family life cycle: Individual, family, and social perspectives* (3rd ed.). Neeedham Heights, MA: Allyn & Bacon.

Chen, H. & Volpe, R.P. (1998). An analysis of personal financial knowledge among college students. *Financial Services Review, 7*(2), 107–128.

CFPB (Consumer Financial Protection Bureau). (2015). *Financial well-being: The goal of financial education.* Retrieved from http://www.consumerfinance.gov /dataresearch/research-reports/financial-well-being/.

Drever, A.I., Odders-White, E., Kalish, C.W., Else-Quest, N.M., Hoagland, E.M., & Nelms, E.N. (2015). Foundations of financial well-being: Insights into the role of executive function, financial socialization, and experience-based learning in childhood and youth. *The Journal of Consumer Affairs, 49*(1), 13–38.

Gan, Y., Yang, M., Zhou, Y., & Zhang, Y. (2007). The two-factor structure of future-oriented coping and its mediating role in student engagement. *Personality and Individual Differences, 43*(4), 851–863. doi: 10.1016/j.paid.2007. 02.009.

Garman, E.T. & Sorhaindo, B. (2005). Delphi study of experts' rankings of personal finance concepts important in the development of the in charge financial distress/financial well being scale. *Consumer Interests Annual, 51,* 184–194.

Greenglass, E.R. & Fiksenbaum, L. (2009). Proactive coping, positive affect, and well-being: Testing for mediation using path analysis considerations. In J.P. Ziegelmann & S. Lippke (Eds.), *Invited paper in European Psychologist, Special section on theory-based approaches of stress and coping, 14,* 29–39.

Hira, T.K. (1997). Financial attitudes, beliefs, and behaviours: Differences by age. *International Journal of Consumer Studies, 21*(3), 271–290.

Hira, T.K. & Brinkman, C. (1992). Factors influencing the size of student debt. *Journal of Student Financial Aid, 22*(2), 33–50.

Joo, S. (2008). Personal financial wellness. In J.J. Xiao (Ed.), *Handbook of consumer finance research* (pp. 21–33). New York: Springer.

Joo, S. & Grable, J.E. (2004). An exploratory framework of the determinants of financial satisfaction. *Journal of Family and Economic Issues, 25*(1), 25–50.

Kim, J. & Chatterjee, S. (2013). Childhood financial socialization and young adults' financial management. *Journal of Financial Counseling and Planning, 24*(1), 61–79.

Kim, J., Garman, E.T., & Sorhaindo, B. (2003). Relationships among credit counseling clients' financial well-being, financial behaviors, financial stressor events and health. *Financial Counseling and Planning, 14*(2), 75–87.

Koonce, J.C., Mimura, Y., Mauldin, T.A., Rupures, A.M., & Jordan, J. (2008). Financial information: Is it related to savings and investing knowledge and financial behavior of teenagers. *Journal of Financial Counseling and Planning, 19*(2), 19–28.

Lawrence, F.C., Christofferson, R.C., Nester, S.E., Moser, E.B., Tucker, J.A., & Lyons, A.C. (2003). *Credit card usage of college students: Evidence from Louisiana State University*. Baton Rouge, LA: Louiana Agricultural Center.

Lucey, T.A. & Giannangelo, D.M. (2006). Short changed: The importance of facilitating equitable financial education in urban society. *Education and Urban Society, 38*(3), 268–287.

Lusardi, A. & Mitchell, O.S. (2007). Financial literacy and retirement preparedness: Evidence and implications for financial education. *Business Economics*, 35–44.

Lyons, A.C. (2003). *Credit practices and financial education need of Midwest college students, Champaign, IL*. Department of Agricultural and Consumers Economics. University of Illinois at Urbana-Champaign.

Lyons, A.C., Palmer, L., Jayaratne, K.S.U., & Scherpf, E.F. (2006). Financial education and communication between parents and children. *The Journal of Consumer Education, 23*, 64–76.

Mills, R.J., Grasmick, H.G., Morgan, C.S., & Wenk, D. (1992). The effects of gender, family satisfaction, and economic strain on psychological well-being. *Family Relations, 41*(4), 440–445.

Moschis, G.P. (1987). *Consumer socialization: A life cycle perspective*. Lexington, MA: Lexington Books.

Moschis, G.P., Moore, R.L., & Smith, R.B. (1984). The impact of family communication on adolescent consumer socialization. *Advances in Consumer Research, 11*, 314–319.

OECD INFE. (2011). *Measuring financial literacy: Questionnaire and guidance notes for conductiing an internationally comparable survey of financial literacy*. Paris, Franch: OECD.

Prawitz, A.D., Garman, E.T., Sorhaindo, B., O'Neill, B., Kim, J., & Drentea, P. (2006). In charge financial distress/financial well-being scale: Development, administration, and score interpretation. *Financial Counseling and Planning, 17*(1), 34–50.

Sari, R.C., Fatimah, P.L.R., & Suyanto, S. (2017). Bringing voluntary financial education in emerging economy: Role of financial socialization during elementary years. *The Asia Pacific Education Researcher, 26*(3–4), 183–192.

Shim, S., Xiao, J.J., Barber, B.L., & Lyons, A.C. (2009). Pathways to life success: A conceptual model of financial well-being for young adults. *Journal of Applied Developmental Psychology, 30*(6), 708–723.

Sirgy, M.J., Grzeskowiak, S., & Rahtz, D.I (2007). Quality of College Life (QCL) of students: Developing and validating a measure of well-being. *Social Indicators Research, 80*, 343–360.

Schwarzer, R. & Knoll, N. (2003). Positive coping: Mastering demands and searching for meaning. In S.J. Lopez & C.R. Snyder (Eds), *Positive psychological assessment,* pp. 393–409. Washington, WA: American Psychological Association.

Xiao, J.J., Serido, J., & Shim, S. (2006). Financial behavior of consumers in credit counseling. *International Journal of Consumer Studies, 30*(2), 108–121.

Building professional characters based on a teacher's understanding of professionalism

R.E. Kusmaryani
Universitas Negeri Yogyakarta, Indonesia

J.R. Siregar, H. Widjaja & R. Jatnika
Universitas Padjajaran, Indonesia

ABSTRACT: This research aims at exploring an understanding of professionalism and the professional character frequently needed in the teaching profession. The research methods were quantitative and qualitative approaches involving descriptive research. The data was collected using open questionnaires and interview protocol. The subjects of this research consisted of 74 teachers chosen from 3 (three) areas, namely Yogyakarta Municipality, Sleman and Bantul Regency. This research found that, according to the perception of the teachers, professionalism comprises a teacher's expertise, social ability, self-integrity, and productive behavior. These four criteria are the benchmarks of a teacher's competence in building their professional characters through professional development.

1 INTRODUCTION

Human beings are essentially bound by working activity, which is considered to be the most important thing in human life. They spend most of their time at work. This activity is very important, to the extent that some experts have tried to define the meaning of 'work'. Steers and Porter (1983) defines 'work' as a kind of activity designed to produce something and to achieve intrinsic goals. The more interesting and fun the job is, the more self-directed a person is in doing the work. Shershow (in Yeoman, 2014) describes that there are two different types of reasons for 'work', namely 'working to live' and 'living to work.' On the other hand, 'work' is a source of expressive human actions, which, in this case, is carried out to meet the hopes of civilization. People tend to work in order to obtain social status, affection, and decent jobs.

These demands are inevitable for most people. Some people work for survival and others work to gain meaningful experiences in their lives. Human beings initially work in order to fulfill their needs and live their life. Nevertheless, the meaning of 'work' changes along with the development of human civilization. This change occurs as a result of environmental change. Therefore, these changes lead to changes in determining the quality of work and improvements in work performance.

Being professional is necessary in order to meet the demands of getting an occupation, job or profession. Nowadays, being a professional is obligatory for those who are in certain professions. These professionals are expected to be able to provide the best quality of work in accordance with the standards of professional practice. The question that therefore arises is "What does a professional mean?". Being a professional means having special knowledge and skills (Stern, 2000). A professional also derives certain rights, duties and responsibilities based on people's views and trust. However, in practice, some job holders are often considered as professionals, for example, a painter whose work is mostly satisfying is called a professional painter. Meanwhile, some teachers or doctors who have a professional membership are not considered professional, since their service is not satisfying or good for others. Therefore, to answer the question above, it is necessary to clarify the meaning of the term 'professional' itself.

The term 'professional' is defined in two ways: 'being a professional' or 'behaving professionally' (Englund, in Frelin, 2013; Day, 1999). 'Being a professional' refers to those who are members of a particular profession and are able to meet professional standards, while 'behaving professionally' refers to a person's success in doing a job that requires skill, even though they are not members of a particular profession. In this case, certain professions do not guarantee that a person can accomplish a job professionally. However, it is possible that those people who are not a member of any profession are able to accomplish their work professionally. Based on the previous description, it could be described that the term 'professional' refers to both those who are able to acquire the requirements of professionalism in their work and those who show the characteristics of working in a particular profession.

Teachers are a crucial resource in the educational sector and they are expected to carry out their duties professionally. In relation to these duties, teachers have personal, social, intellectual, moral, and spiritual responsibilities (Good, 2008). A teacher's personal responsibility is indicated by their ability to understand, accept, manage, and develop themselves. Social responsibility is indicated by their ability to understand that they are part of the social environment and being able to interact effectively within it. Intellectual responsibility is manifested in their ability to master their knowledge and use it objectively. Moral responsibility includes the teacher's ability to understand and obey the moral rules in their environment. Spiritual responsibility could be manifested in a teacher's behavior as a divine being.

The professional character is closely related to the requirements of the teaching profession (Suwandi, 2016). This character needs to be internalized by an individual so that it becomes part of him/her. A person would never possess any professional character unless she/he went through the process of building a character. The term 'building a character' consists of two main words, namely 'build' and 'character.' The term 'build' means the process of constructing something, while 'character' means a mental or moral quality that distinguishes one person from another. According to *Kamus Besar Bahasa Indonesia* (Indonesian Dictionary), character means behavior, the psychological traits, morals, or manners distinguishing one person from another. Pala (2011) defines character as the number of qualities possessed by a person, including values, thoughts, words, and behavior. Moreover, Singla (Pala, 2011) states that a person is often judged based on their character, whether they like it or not. This character describes a person's either good or bad behavior. In other words, a person would be judged to be either good or bad based on their character. Ryan and Bohlin (1999) define a good person as someone who understands, loves, and is virtuous. Thus, a person's character involves the aspects of thoughts, feelings, and behavior that they possess. Based on the previous meaning, 'building a character' in this case is a continuous process or effort to construct mental qualities in the form of a human's thoughts, feelings, and behavior, so as to show their good behavior and attitudes based on noble values.

The process of building a character also requires the involvement of all parties, their surroundings and oneself. It should not only be taught but also habituated (Jaedun et al., 2012). According to Pala (2011), a good character is not automatically created, but it is developed over time through an ongoing teaching process, such as learning and practicing. In this case, the role of the environment is manifested in the presence of education. Character is developed through education both formally and informally (Pala 2011). Education is a process of internalizing culture so as to create a civilized individual and society. Education is not only a means to transfer knowledge, but, in a broader sense, it is also a means of culture and value distribution (enculturation and socialization). Character education should reach the basic dimensions of humanity. The Law of the Republic of Indonesia No. 20 of 2003, on National Education System Act, Article 3, states that "The National Education functions to develop the capability, character, and civilization of the nation for enhancing its intellectual capacity, and is aimed at developing a learner's potential so that they become persons imbued with human values who are faithful and pious to one and only God; who possess morals and noble character; who are healthy, knowledgeable, competent, creative, independent; and as citizens, are democratic and responsible." The aim of national education is the formulation of Indonesian qualities, which should be established by every single educational unit. This character education becomes a process of emphasizing ethics, responsibility, and concern toward others by modeling or teaching a good character through an emphasis on universal values. As a key to realizing character education, teachers have to gain the students' trust and demonstrate their commitment to character education. This is the main reason for the importance of building the teacher's character, in order to realize the Javanese philosophy of the word 'teacher', namely *digugu* (believed) and *ditiru* (exemplified).

With regards to the importance of the involvement of teachers and the environment, building a professional character could be done through both character education and professional development for teachers. In order to be professional, teachers need to complete a process of self-development. Self-development is a process of personal development, with the person taking primary responsibility for their own learning and choosing the means by which to achieve it (Pedler et al., 2007). This is based on the idea that human beings possess a unique psychological ability for self-reflection (Hartung & Subich, 2010). This means that in order to possess a professional character, teachers potentially have the responsibility to develop themselves through self-reflection. Self-reflection would be the teachers' principal means of developing themselves in order to be professional and possess a professional character.

Hartung and Subich (2010) emphasize that, today, workers are required to be self-directed by adjusting and managing their own work lives, which can change greatly and be unpredictable, rather than solely relying on the organizations to direct and support them. Through this self-direction, a person is going to consciously involve themselves in continuous learning and reflection in order to acquire the required ability, skills, and knowledge (Hartung & Subich, 2010). Ultimately, self-development is about increasing their capacity and willingness to take control over and be responsible for events (Pedler et al., 2007).

Regarding the teachers' personal, social, intellectual, moral, and spiritual responsibility, as presented by Good (2008), they certainly need to build their professional character and ability through self-development. Professional character should be closely related to professionalism, which corresponds to the ideas of Kanes (2011), who states that professionalism refers to the quality of professional work. In other words, by fulfilling the standards of professionalism, a teacher would be seen as having a professional quality of work. A teachers' professionalism would be seen from their behavior in performing their duties, which would distinguish them from the non-professionals.

Professionalism generally comprises five qualifications: a) providing public services; b) having a theoretically as well as a practically grounded expertise; c) having an ethical dimension; d) requiring organization and regulation for purposes of recruitment and discipline; and e) requiring a high degree of individual autonomy, especially in the independence of judgment for effective practice (Carr, 2000). Meanwhile, Robson (2006) states that professionalism includes autonomy, professional knowledge, and responsibility. Both experts emphasize different aspects when viewing professionalism, but these still have similarities and are even related to each other.

Professional character refers to the signs of professionalism. Each profession has different qualifications for assessing the quality of work, according to the requirements of the profession. A further question arises along with the problems in this research. That is "How do teachers understand and perceive professionalism?" This understanding and perception would become the teacher's benchmark for developing themselves into a professional. Given the importance of teacher resources as the key to determining the quality of education, research needs to be conducted in order to explore the teacher's understanding of what is required. The aim of this research is to obtain the qualification of professionalism as a professional practice standard. The findings of this study are expected to be able to contribute toward building a professional character for teachers.

2 RESEARCH METHODS

This research uses both a quantitative and qualitative approach. The type of research includes survey research, which aims at obtaining the exploratory data of the teachers' perceptions on professionalism.

Table 1. Examples of the research subjects' answers on the qualifications of professionalism.

No.	Sex	Age (y/o)	Working experience	Answer
1	F	34	12 years	a. Having a pedagogic ability b. Having a professional ability c. Having a social ability d. Having a technological ability e. Having critical thinking f. Having effective time management g. Able to be a role model for the students, colleagues, and society
2	F	44	16 years	a. Being able to be responsible for his/her duty b. Able to teach the students well (ethics/knowledge) c. Teach something beneficial for the students d. Disciplining the students e. Controlling the class and delivering the material appropriately f. Guiding students to be successful
3	M	40	24 years	a. Being noble and virtuous b. Being a qualified educator c. Having an inner self-motivation, aware of the meaning behind the word "teacher" d. Having a good ability to teach e. Constantly developing themselves f. Constantly accepting constructive changes g. Being active, creative, and innovative h. Mastering the learning materials i. Always on time during teaching and learning activities j. Observing the students development

This research involved 74 teachers chosen incidentally from the Senior High Schools in Yogyakarta Special Region, especially Bantul Regency and Yogyakarta Municipality. The data was collected using an open questionnaire and interview. The instrument validity involved a professional judgment. The questionnaires consisted of seven questions representing the understanding of professional development, professionalism qualifications, professional development activities, and frequency aspects. The subjects of the research were asked to answer each question in accordance with their perceptions. Examples of the subjects' answers can be seen in Table 1. The collected data was further analyzed through a coding process. The data was tabulated through identification, and grouped into specific categories. Furthermore, the data was analyzed using descriptive quantitative analysis by percentage. After that, interviews were conducted with three teachers; two with teacher certification managers in the Special Region of Yogyakarta, and one with the Vice Chairman of the Board of PGRI. In order to control the validity of the data, triangulation of sources and methods were used so that the research data could genuinely represent the subject's answers.

3 RESULTS

Based on the subjects' answers, Table 2 presents the findings of this exploratory research.

In this study, there are four criteria of teacher professionalisms, namely teachers' expertise, social ability, self-integrity, and productive behavior. Overall, resulting from these explorations, there are three criteria of appropriate professionalisms that have been found to have a relationship with the concepts of Carr (2000), namely teacher expertise, social ability, and personal integrity.

Table 2.   Results of teachers' perceptions of professionalism.

No.	Criteria of teacher professionalism	Percentage
1	Teachers' expertise	43%
2	Social ability	18%
3	Self-integrity	17%
4	Productive behavior	22%
		100%

4   DISCUSSION

Building a teacher's professional character is a continuous process of constructing the quality of the teacher's character by acquiring the important professional requirements needed to carry out professional practices. This process could be performed with character education supported by both the environment and the teacher, one of which is by professional development. In continuous character building, encouragement is certainly needed from oneself, as well as a deep understanding of the meaning of professionalism for teachers.

Table 2 presents data showing the teachers' understanding of professionalism, comprising the criteria of teachers' expertise, social ability, self-integrity, and productive behavior. Teachers' expertise could be clearly described as having the most dominant answers. Expertise is the teachers' behavior, which represents the mastery of their professional duties, in the form of mastering the learning strategies and the subject material being taught. Teachers' capacity is a form of expertise, which Carr states is one of the professionalism criteria that should be possessed by a professional. This criterion includes both the theoretical and practical. The expertise that should be possessed by teachers is related to the mastery of the materials and the process of implementing the learning and subjects being taught by the teacher. This is the principal criterion for teachers in conducting professional practices.

For teachers, expertise is considered as a crucial criterion of professionalism. Teachers cannot perform their professional practices without mastering the instructional strategies, even less so if they do not master their subject material. In fact, in order to be able to master these two things, teachers need to take a course of study relevant to their field of expertise. Thus, it is not everyone who is able to conduct the professional practice of teaching.

The second important criterion in the professionalism of teachers is productive behavior. Out of the three criteria of professionalism, productive behavior is the only behavior that ensures a change for the better. Productive behavior refers to the constructive and imaginative behavior of a teacher who provides a tangible and significant contribution to self-potential and the environment. These productive behaviors include self-development and adaptation to the environment. By this behavior, teachers are able to improve both themselves and their environment. By these qualifications, teachers are able to develop themselves and their environment by creating innovations that support their profession as a teacher. Therefore, teachers actively participate in professional development activities, and are even encouraged to make some changes in order to improve their work and their professional qualities.

The less dominant criteria are social ability and self-integrity. Social ability represents the teachers' behavior in interacting with students, parents, and other teachers, especially in socializing and building effective communication. This capability involves socialization and effective communication with students, parents, and other teachers. In Carr's definition, this criterion has much to do with the public service criterion. In accordance with professional elements, public services are performed by teachers in order to meet other's needs, which are satisfaction oriented. According to Gastelaars (2009), a service is focused more on client interaction. In practice, the services performed by teachers are bound with social interaction through educating, teaching, guiding, briefing, training, assessing, and evaluating the students. In addition to the services toward students, teachers also perform services for parents, as persons who are responsible for the students, along with services among other teachers. Therefore, social ability is important for teachers in conducting professional services.

Self-integrity represents teachers' behavior based on noble and moral values, including maintaining consistency, being responsible, and upholding the value of rules. In Carr's concept (2000), the teachers' self-integrity is closely related to the code of ethics and teachers' moral behavior criterion. Based on this, ethics relates to the ethical standards and how ethics itself affects one's behavior (Dutelle, 2011). Furthermore, Dutelle also states that ethics comprises of right and wrong, good and bad, and benefit and loss. As a professional, a teacher is able to be right, good and beneficial, such as being responsible, honest, disciplined, obedient to the rules, wise, and trustworthy, which psychologically portrays the integrity of a teacher.

The small percentage shown in the results with regards to social ability and self-integrity in this study indicates that these two criteria are considered by the teachers not to be necessary for professionalism. In fact, both of them are no less important than the others. In other words, those having poor social ability and a lack of self-integrity are still able to practice the profession, even though they possess a poor quality of mental character.

The four criteria of professional character, according to this understanding, could be a reference for teachers in developing themselves as professionals. Other than by formal and informal character education, building a professional character could be undertaken by professional development. Professional development is the personal responsibility of the teachers to develop themselves in order to improve the sustainable quality of their work, rather than being an environmental responsibility.

5 CONCLUSION

From this study, it could be concluded that professionalism, according to the perception of teachers, includes the teachers' skills, social ability, self-integrity, and productive behavior. These four criteria could be the benchmarks for teachers' competency in building a professional character through professional development. Based on the conclusion, it could suggest that: 1) teachers' perception of professionalism could be used in further research by examining the teachers' quality of work, and as a reference for building a teacher's professional character based on their understanding; 2) researchers who are interested in teachers' professionalism could further develop the findings regarding professionalism discussed in this study into a larger area. In addition, it could be applied to other topics of research in relation to teachers' professionalism.

REFERENCES

Cai-feng, W. (2010). An empirical study of university teachers based on organizational commitment, job stress, mental health and achievement motivation. *Canadian Social Science*, 6(4), 127–140.
Carr, D. (2000). *Professionalism and ethics in teaching*. London, UK: Routledge Taylor & Francis.
Day, C. (1999). *Developing teachers: The challenges of lifelong learning*. London: Falmer Press.
Dutelle, A.W. (2011). *Ethics for the public service professional*. New York, NY: Taylor & Francis Group, LLC.
Frelin, A. (2013). *Exploring relational professionalism in schools*. Rotterdam, Netherlands: Sense Publishers.
Gastelaars, M. (2009). *The public services under reconstruction: Client experiences, professional practice, managerial control*. New York, NY: Routledge.
Good, T.L. (2008). *21st century education: A reference handbook*. Los Angeles, CA: A Sage Reference Publication.
Hartung, P.J. & Subich, L.M. (2010). *Developing self in work and career: Concepts, cases, and contexts*. Washington, WA: American Psychological Association.
Jaedun, A., Purwaningsih, E., Novita, F., & Wiranata, M. (2012). Implementasi pendidikan karakter terintegrasi dalam kegiatan pembelajaran pada SMK jurusan bangunan di Daerah Istimewa

Yogyakarta [Implementation of Character Education Integrated in Learning Activities at Vocational High Schools of Building Engineering in Yogyakarta Special Region]. *Jurnal Pendidikan Teknologi dan Kejuruan, 21*(1), 74–82.

Kanes, C. (2011). *Elaborating professionalism. studies in practices and theory*. New York, NY: Springer Science and Business Media.

Pala, A. (2011).The need for character education. *International Journal of Social Science and Humanity Studies, 3*(2), 23–32.

Pedler, M., Burgoyne, J. & Boydell, T. (2007). *A manager's guide to self development* (5th ed.). London, UK: McGraw-Hill Companies.

Robson, J. (2006). *Teacher professionalism in further and higher education: Challenges to culture and practice*. London, UK: Routledge Taylor & Francis Group.

Ryan, K. & Bohlin, K. (1999). *Building character in school: Bringing moral instruction to life*. San Francisco, CA: Jossey-Bass.

Steers, R.M. & Porter, L.W. (1983). *Motivation and Work Behavior*. New York, NY: Acadaemic Press.

Stern, D.T. (2000). The development of professional character in medical students. Special Supplement. *Hastings Center Report, 30*(4), S26–S29.

Suhardjono, S. (2006, November). *Development of teacher profession and scientific writing*. Paper presented at the Consultation Meeting in the Framework of Coordination and Development of Personnel Educators and Education Personnel, Ministry of National Education, Bureau of Personnel.

Suwandi, S. (2016). Analisis studi kebijakan pengelolaan guru SMK dalam rangka peningkatan mutu pendidikan [An analysis of vocational high school teachers management policy for enhancing the quality of education]. *Jurnal Pendidikan Teknologi dan Kejuruan, 23*(1), 90–100.

Townsend, T. & Bates, R. (2007). *Handbook of teacher education: Globalization, standards and professionalism in times of change*. Dordrecht, Netherlands: Springer.

VandenBos, G.R. (2007). *APA dictionary of psychology*. Washington, DC: American Psychological Association.

Yeoman, R. (2014). *Meaningful work and workplace democracy: A philosophy of work and a politics of meaningfulness*. London, UK: Palgrave Macmillan.

Curriculum syllabus lesson plan learning materials development for integrated values education

Character Education for 21st Century Global Citizens – Retnowati et al. (Eds)
© *2019 Taylor & Francis Group, London, ISBN 978-1-138-09922-7*

Creative problem solving for improving students' Higher Order Thinking Skills (HOTS) and characters

E. Apino & H. Retnawati
Universitas Negeri Yogyakarta, Indonesia

ABSTRACT: In order to achieve success, an individual should have good thinking skills and character. Both matters can be achieved through a well-qualified educational system. In the mathematics' learning process, a Creative Problem Solving (CPS) model might be selected as one of the tools for improving students' thinking skills and character. Therefore, this study aimed at describing students' Higher Order Thinking Skills (HOTS) and character improvement in the mathematics' learning process through a CPS model. This study itself was a design research. The subjects in this study were tenth grade students of the Natural Science Program from an Islamic senior high school (namely *Madrasah Aliyah Negeri*/MAN 3) in Yogyakarta, Indonesia. The data were gathered through test and observation, while the data analysis was conducted in a descriptive manner both quantitatively and qualitatively. The results of this study showed that: (1) the students' HOTS score improved after the CPS model had been implemented in the mathematics' learning process; (2) the students' character score improved after the CPS model had been implemented in the mathematics' learning process; and (3) the characteristics that might be trained through the implementation of the CPS model in the mathematics' learning process included hard work, curiosity, responsibility, teamwork, tolerance, care and self-confidence.

1 INTRODUCTION

The rapid advancement of knowledge and technology demands each individual to improve his or her self-competencies as their equipment to deal with more complex global competition. The results of various studies show that in order to achieve success within global competition, an individual should have strong thinking skills and character. Specifically, there are four skills that should be possessed as the equipment to deal with twenty-first century competition and these skills are also known as twenty-first century skills. The four skills include communication, collaboration, critical thinking and creativity (Partnership for 21st Century Skills, 2002). The four skills might be integrated and trained through an educational process that starts from elementary until higher education level.

Indonesian education started implementing a curriculum that leads to the development of these skills, namely Curriculum 2013. Referring to the content standards (Regulation of the minister of education and culture number 20 on 2016) (RoI, 2016), Curriculum 2013 sets higher priority on the development of character aspects (social and spiritual attitudes), Higher Order Thinking Skills (HOTS) and various skills both abstract and concrete. In Curriculum 2013, character development aspects might be regarded as the means for developing communication and collaboration skills, while a learning process that leads to the development of HOTS might be turned to as the means for training critical and creative thinking skills. Therefore, the content standards of the governing curriculum in Indonesia is currently relevant to the four skills contained in the twenty-first century skills.

One of the aspects that has become the focus in the current Indonesian educational process is the development of HOTS. HOTS are skills that involve critical thinking and creative thinking skills (Conklin, 2012; King et al., 2010; Krulik & Rudnick, 1999; Presseisen, 1988;

Yen & Halili, 2015), problem solving skills (Brookhart, 2010; Presseisen, 1988; Yen & Halili, 2015), logical and reflective thinking skills (King et al., 2010), metacognitive thinking skills (King et al., 2010; Yen & Halili, 2015) and decision-making skills (Presseisen, 1988; Yen & Halili, 2015`). If these skills are associated with the revised Bloom taxonomy (Anderson & Krathwohl, 2001), then HOTS will involve cognitive process namely analyze, evaluate and create (Liu, 2010). In addition, the Indonesian educational process not only focuses on developing thinking skills solely but also on shaping character values that are integrated into the learning process. Therefore, developing various thinking skills and characters at the same time might be regarded as the main objective of the current Indonesian educational process.

The aspects that will be achieved in the educational process are still contradictory to the facts in the today's field. The results of several studies reported that students' HOTS in Indonesia still currently fall into the low category (Nurina & Retnawati, 2015; Riadi & Retnawati, 2014; Susanti et al., 2014). In addition, the indication that students' HOTS in Indonesia is still low might also be found in Indonesian achievement in various international studies such as PISA (Program for International Student Assessment) and TIMSS (Trends in International Mathematics and Science Study). The latest release of both international studies puts Indonesia in the lower rank (Mullis et al., 2012; Mullis et al., 2016; OECD, 2014, 2016). Other indications might be seen from the difficulties of students in working on the equivalent problems of PISA which began to be accommodated in the Mathematics National Examination in 2014 (Retnawati et al., 2017).

In terms of character, there are still various problems related to students' moral degradation such as student brawls, adolescent misbehavior and negative phenomena among students such as cheating during exams, not doing homework and being less confident in learning mathematics (Musfiqi & Jailani, 2014). These matters should be handled seriously and solutions to these matters determined; one of which might be improving learning quality within classrooms. In addition, the obstacles faced in the implementation of Curriculum 2013, such as the learning process using a scientific approach that has not been maximized, as well as difficulties in the implementation of the assessment, especially for attitude assessment (Retnawati, 2015a; Retnawati et al., 2016), impacts the implementation of character education which has not been maximized.

One of the efforts that could be conducted in order to improve learning quality is implementing a learning process that might facilitate students in developing their HOTS and also shaping their character. A learning process that might manifest this effort is one that involves problem solving activities (Apino & Retnawati, 2016, 2017; Jailani & Retnawati, 2016; Musfiqi & Jailani, 2014; Saido et al., 2015; Susanto & Retnawati, 2016; Haryanto, 2015) and cooperative learning (Matchett, 2009; Zakaria et al., 2010). One learning model that involves problem solving activities and might be used to improve students' HOTS is Creative Problem Solving (CPS) (Apino & Retnawati, 2016, 2017; Bohan & Bohan, 1993). CPS is a learning model that encourages students to solve problems through good and systematic thinking skills, and includes creative and critical thinking skills (Isaksen et al., 2011; Treffinger, 1995; Tseng et al., 2013). Operationally, Osborn and Parnes (Giangreco et al., 1994) explained that CPS might be implemented through the following stages: (1) objective-finding, namely proposing problems and determining learning topics; (2) fact-finding, namely analyzing important information that has been relevant to the problems; (3) problem-finding, namely identifying the key questions of the problems; (4) idea-finding, namely searching as many ideas as possible that might be used to solve the problems; (5) solution-finding, namely implementing each idea that has been found in order to solve the problems; and (6) acceptance-finding, namely selecting the best solution in order to solve the problems based on certain criteria. By paying attention to these stages, CPS implementation in mathematics learning might also be used as a tool for training characteristics such as hard work, curiosity, responsibility, care, teamwork, tolerance and self-confidence. Some reasons why implementation of a CPS model in learning might improve students' character are: (1) CPS encourages and strengthens many academic and attitude skills (Giangreco et al., 1994); (2) the learning process conducted in groups (Bohan & Bohan, 1993; Giangreco et al., 1994) trains students' collaborative skills useful for improving cooperation, tolerance, care and responsibility.

Implementation of a CPS model is not only able to improve students' HOTS but also to shape students' positive characters. In order to shape students' characters, several actions that might be taken include role modeling (Lumpkin, 2008; Thornberg & Oğuz, 2013; Zuchdi et al., 2011; Partawibawa et al, 2014); consistent habituation (Choudhury, 2016; Zuchdi et al. 2011); intervention and strengthening (Zuchdi et al., 2011). These four actions might certainly be facilitated through the implementation of CPS in mathematics learning with a hope that students' HOTS and characters will improve. Therefore, this study aimed at describing students' HOTS and character improvement in mathematics learning through the use of a CPS model.

2 METHOD

The study was a design research. The stages in the study were adapted from the Plomp model (Plomp, 2013) which consisted of: (1) needs analysis; (2) intervention development in the form of learning design for improving students' HOTS and characters; and (3) evaluation. The subjects in the study were tenth grade students from the Natural Science Program in MAN 3 Yogyakarta, the Province of Yogyakarta Special Region, Indonesia. The data were gathered through test and observation. A test was administered in order to gather data regarding students' HOTS before and after the CPS model was implemented in mathematics learning. Then, observation was conducted in order to gather data regarding students' characters before and after the CPS model was implemented in mathematics learning. The HOTS' data were analyzed in a descriptive, quantitative manner by comparing the pretest and the post-test scores. On the other hand, the character data were analyzed in a descriptive, qualitative manner by identifying and describing the characters that appeared before and after the CPS model was implemented in mathematics learning.

3 RESULTS AND DISCUSSION

3.1 *Needs analysis*

Needs analysis was conducted by interview with mathematics teachers and observation of the learning process in classrooms. Interviews were conducted with three mathematics teachers from MAN 3 Yogyakarta. From the results of the interviews, the researchers found that: (1) teachers still had difficulties in implementing learning processes that trained HOTS within mathematics learning; (2) there were limited examples of HOTS development oriented learning sets such as lesson plans and student worksheets; (3) teachers still had difficulties in creating HOTS test items; (4) in group activities, there were still some students who had not been able to cooperate with one another as group members; (5) the students had a lack of responsibility in completing the assignments that had been provided; and (6) the students had not been confident when they were asked to do a presentation in front of the class. In order to support these findings, the researchers observed the mathematics learning process directly in the classroom. From this observation, the researchers might conclude that information that had been attained in the interview was in accordance with the condition that had occurred within the classroom, in which the learning process had not been oriented to the development of students' HOTS and still lacked character values that should be trained during the learning process. The reason was that the teachers still implemented a conventional learning model in which the teacher's role was very dominant during the learning process.

3.2 *Learning design for improving students' HOTS and characters*

The information attained from the needs analysis was followed up by creating a learning design in order to train HOTS and characters. Based on the theoretical review, a CPS learning model was selected in order to train HOTS and characters during the mathematical learning process. The learning process was formulated into a lesson plan that contained the CPS model syntax and that had been equipped with student worksheets. The characteristics of the lesson plan designed were: (1) the learning objectives led to the development of critical thinking and creative thinking skills; (2) the design contained stimulus-providing activities from the teachers by asking "why" and "how" questions; (3) the design contained activities that led to the discovery of multiple answers and or multiple ways to solve given problems as a response to the given stimulus; (4) the design maximized the active role in students' interaction; and (5) the design facilitated group activities. The design of the student worksheet implemented had the following characteristics: (1) the content of the student worksheet had an association between materials and concepts; (2) the student worksheet contained reflection activities; (3) the student worksheet contained discovery and investigative activities that had been adjusted to the characteristics of the learning materials; (4) the student worksheet applied creative problems that demanded multiple answers or solutions; and (5) the problems implemented were associated with the actual context and had ill-structure (unstructured).

3.3 Evaluation

After the learning design and the learning support kits had been designed, the researchers gathered the data through the CPS model implemented into the mathematical learning process in the tenth grade of MAN 3 Yogyakarta. Before the learning process took place, the students were provided with a pretest in order to measure their initial ability in completing HOTS test items. The CPS model was implemented in six meetings and the materials applied were trigonometric comparison in right triangle. During the learning process, the researchers also observed the character values that appeared. Eventually, the students were provided with a post-test in order to measure their HOTS after the CPS model had been implemented in the mathematics' learning process. The evaluation in this study targeted viewing the impacts of the CPS model implementation in the improvement of students' HOTS and character.

3.3.1 HOTS data

The following are the data from the students' pretest and post-test in completing the HOTS test items.

From the data in Table 1, the researcher attained information that there was an improvement in students' HOTS scores after the CPS learning model was implemented in the mathematical learning process. The mean score of students' HOTS improved by approximately 45.69% after the implementation. The percentage of settlement also experienced a drastic increase from 0.00% to 74.07%. These findings show that CPS implementation in the mathematics' learning process might improve students' HOTS. This achievement is in accordance with the opinion proposed by Apino and Retnawati (2017), Bohan and Bohan (1993) and Susanto and Retnawati (2016) that learning by means of a CPS model might be performed in order to improve students' HOTS. In addition, the use of creative problems through student worksheets caused the students to be more challenged in solving these problems. Loewen (1995) proposed that the use of creative problems might increase students' awareness that not all problems had one appropriate solution. This is a matter that might trigger and train students' creativity in mathematics learning and this creativity is part of HOTS development.

Another factor that caused the improvement of students' HOTS was the meaningful learning activities where the students were actively engaged in a discussion process in order to construct their knowledge and use multiple relevant sources for gathering the desired insight. This is in accordance with the opinion of Bohan and Bohan (1993) that the learning process by

Table 1. Data analysis results of students' HOTS.

Description	Pretest	Post-test
Highest score	56.52	93.33
Lowest score	17.39	46.47
Mean score	31.72	77.41
Number of settled students	0.00	20.00
Number of unsettled students	27.00	7.00
Percentage of settlement (%)	0.00%	74.07%

means of a CPS model might present meaningful learning activities for students. Furthermore, students might be enthusiastic because they are challenged by the problems provided. This matter is certainly able to improve students' motivation to learn. These findings are in accordance with the opinion of Loewen (1995), which stated that the presentation of challenging creative problems in a CPS model might improve students' interest and motivation to learn. Thereby, there will be an improvement in students' interest and motivation to learn and this improvement becomes one of the decisive factors within the improvement of students' HOTS.

3.3.2 *Character data*

Data regarding the development of students' characters were gathered through observation. Observation was conducted before and during the CPS model implementation in the mathematics' learning process. Results of the observation are presented in Table 2 and Table 3.

From Table 2 it is apparent that the implementation of a CPS model in a mathematical learning process impacts students' attitudes and behaviors during the learning process. These results provide evidence that learning intervention by means of CPS model implementation might train students' character values. Based on the observation that was conducted during the learning process, the character values that might be trained through the implementation of a CPS model are presented in Table 3.

Based on the qualitative data in Table 2 and Table 3, the researchers concluded that the implementation of a CPS model in the mathematics' learning process might train and improve students' characters altogether in the same time. The character values that might be trained and be improved namely curiosity, hard work, teamwork, care, responsibility, tolerance and self-confidence. The implementation of a CPS model in the mathematical learning process is one of the interventions that teachers might perform in order to train and to improve students' characters. These results are certainly in accordance to the opinion of Zuchdi et al. (2011) which stated that one of the ways to improve character is by providing intervention. Furthermore, the CPS model is one of the variations in problem-based learning models and the results of previous studies (Arofah, 2015; Jailani & Retnawati, 2016; Musfiqi & Jailani, 2014; Wardani, 2014) showed that problem-based learning models might improve students' character. The collaborative aspects in the implementation of the CPS model also played an important role in improving the students' characters (Bohan & Bohan, 1993; Giangreco et al., 1994). The most prominent characteristic that might be trained through the implementation

Table 2. Description on the observation results before and during CPS model implementation.

Before CPS model implementation	During CPS model implementation
Students have not been able to perform task division within the group discussion.	Students have been able to perform task division within the group discussion under the teacher's direction.
Students select group members based on gender and familiarity.	Students belong to heterogeneous groups according to the group division that the teacher has determined.
Students with high academic performance tend to work alone, while students with low academic performance tend to wait and be passive.	There is an interaction among students with high, moderate and low academic performance.
Students have not been confident in performing their presentation; they will perform their presentation after the teacher appoints them.	The representatives of each group become confident in performing their presentation.
Students have not been encouraged to share their opinions when the teacher asks them to respond their peers' presentation.	There are several students who respond to the results of the presentation by the speaking groups.
Only few students were encouraged to raise questions to the teacher if they have not understood some parts in the learning materials.	Some students are encouraged to raise questions, both to the teacher and to their peers, if they have not understood some parts in the learning materials.

Table 3. Character values that might be trained through the implementation of a CPS model.

CPS syntax	Learning activities	Character values that might be trained
Objective-finding	Proposing problems (by teachers) and determining learning topics based on the problems that had been proposed.	Curiosity
Fact-finding	Analyzing important information that has been relevant to the problems and analyzing other supporting information (these activities might be performed through group discussion and question and answer sessions among group members).	Teamwork, care, tolerance
Problem-finding	Finding the key question of the problems that have been proposed (this activity might be performed through group discussion, question and answer sessions among group members and decision-making process).	Teamwork, care, tolerance
Idea-finding	Finding ideas, as many as possible, that might be used to solve the problems (this activity might be conducted through group discussion, question and answer sessions among group members and decision-making process).	Hard work, responsibility, teamwork
Solution-finding	Implementing each idea that has been found in order to solve the problems (this activity might be performed through group discussion and task division).	Teamwork and responsibility
Acceptance-finding	Selecting the best solution in order to solve the problems based on certain criteria (this activity might be performed through an analysis of the strength and the weakness of each solution that has been found, through the presentation of each group member's point of view and final decision-making process).	Tolerance, responsibility and self-confidence

of a CPS model was hard work and this finding is in accordance with the opinion of Bohan and Bohan (1993) and Loewen (1995) which stated that CPS has been able to trigger students' struggle in accomplishing the given challenges.

Implementation of a CPS model in learning, especially mathematics learning, is one of the efforts to support the character education program that has been designed by the government. Implementation of a CPS model certainly does not touch all the expected character values, so mathematics learning through a CPS model might only train some of the character values mentioned earlier. Thus, the CPS model might not be fully used as one of the main tools to instill character values in students. In other words, the CPS model is just one of the variations in mathematics learning and other learning subjects to support the character education process.

In addition, to enhance the attractiveness of the implementation of the CPS model in learning, it is necessary to innovate in the application of the learning model. Such innovations may be related to a combination of using Information and Communication Technology (ICT), both in the process and in the assessment of learning (Retnawati, 2015b). Thus, in the future, for the CPS model to be more interesting for students and teachers, the implementation of the CPS model could be combined with the use of ICT media such as computers and other supporting devices. Likewise the assessment techniques also need to involve the use of ICT such as Computer-Based Testing (CBT). Retnawati (2015b) argues that the use of CBT is more accurate than Paper and Pencil Testing (PPT). Through computer-based assessment, it is expected to motivate and enhance students' self-confidence, both in mathematics and other subjects.

4 CONCLUSIONS AND RECOMMENDATIONS

Based on the results of the study and the discussion, the researchers concluded that: (1) the students' HOTS score improved after the CPS model was implemented into the mathematics' learning process; (2) the students' character score improved after the CPS model was implemented into the mathematics' learning process; and (3) the characteristics that might be trained through the implementation of a CPS model in the mathematics' learning process include hard work, curiosity, responsibility, teamwork, tolerance, care and self-confidence. Based on this conclusion, the researchers would like to provide the following suggestions: (1) a CPS model should be implemented by teachers within their classroom teaching as one of the alternative learning models for the implementation of Curriculum 2013; (2) CPS models might be used by teachers to train students' HOTS, both in mathematics and other subjects; (3) a CPS model might be selected as one of the interventions in character education; and (4) in relation to future studies, there should be a similar study involving a larger population so that generalization might have wider coverage.

REFERENCES

Anderson, L.W. & Krathwohl, D.R. (2001). *A taxonomy for learning teaching and assessing: A revision of Bloom's taxonomy of educational objectives.* New York, NY: Addison Wesley Longman.

Apino, E. & Retnawati, H. (2016). Creative Problem Solving to improve students' Higher Order Thinking Skills in mathematics instructions. In *Proceedings of 3rd International Conference on Research, Implementation and Education of Mathematics and Science* (pp. 339–346). Yogyakarta, Indonesia: Yogyakarta State University.

Apino, E. & Retnawati, H. (2017). Developing instructional design to improve mathematical higher order thinking skills of students. *Journal of Physics: Conference Series, 8*(12), 1–7.

Arofah, L. (2015). Implementing character education through problem based learning in sociology subjects for the development of social capital. *Vidya Karya Jurnal Kependidikan, 30*(1), 1–12.

Bohan, H. & Bohan, S. (1993). Extending the regular curriculum through creative problem solving. *The Arithmetics Teacher, 41*(2), 83–87.

Brookhart, S.M. (2010). *How to assess higher order thinking skills in your classroom.* Alexandria, VA: ASCD.

Choudhury, M. (2016). Emphasizing morals, values, ethics, and character education in science education and science teaching. *The Malaysian Online Journal of Educational Science, 4*(2), 1–16.

Conklin, W. (2012). *Higher-order thinking skills to develop 21st century learners.* Huntington Beach, CA: Shell Educational Publishing.

Giangreco, M.F., Cloninger, C.J., Dennis, R.E., & Edelman, S.W. (1994). Problem-solving methods to facilitate inclusive education. In J.S. Thousand, R.A. Villa, & A.I. Nevin (Eds.), *Creativity and collaborative learning: A practical guide to empowering students and teachers* (pp. 321–346). Baltimore, MD: Paul H. Brookes Publishing.

Haryanto, U. (2015). Peningkatan kemampuan memecahkan masalah melalui media komputer dalam pembelajaran matematika pada siswa SMKN 1 Ngawen [Improved ability to solve problems through computer media in learning mathematics on students SMKN 1 Ngawen]. *Jurnal Pendidikan Teknologi dan Kejuruan, 22*(4), 432–442.

Isaksen, S.G., Dorval, K.B., & Treffinger, D.J. (2011). *Creative approach to problem solving: A framework for innovation and change* (3rd ed.). Thousand Oaks, CA: Sage.

Jailani, J. & Retnawati, H. (2016). Efektifitas pemanfaatan perangkat pembelajaran berbasis masalah untuk meningkatkan HOTS dan karakter siswea [Effectiveness of using learning media based on problem solving to improve HOTS and student's character]. *Jurnal Pendidikan dan Pembelajaran, 23*(2), 111–123.

King, F., Godson, L., & Rohani, F. (2010). *Higher order thinking skills: Definition, teaching strategies, assessment.* Retrieved from www.cala.fsu.edu.

Krulik, S. & Rudnick, J.A. (1999). Innovative task to improve critical and creative thinking skill. In L.V. Stiff & F.R. Curcio (Eds.), *Developing mathematical reasoning in grades K-12* (p. 138). Reston, VA.: NCTM.

Liu, X. (2010). *Essentials of sciences classroom assessment.* Thousand Oaks, CA: Sage.

Loewen, A.C. (1995). Creative problem solving. *Teaching Children Mathematics, 2*(2), 96–99.

Lumpkin, A. (2008). Teacher as role models: Teaching character and moral virtues. *Journal of Physical Education, Recreation and Dance, 79*(2), 45–50.

Matchett, N.J. (2009). Cooperative learning, critical thinking, and character. *Public Integrity, 12*(1), 25–38.

Mullis, I.V.S., Martin, M.O., Foy, P., & Arora, A. (2012). *TIMSS 2011 international results in mathematics.* Chestnut Hill, MA: IEA.

Mullis, I.V.S., Martin, M.O., Foy, P., & Hooper, M. (2016). *TIMSS 2015 international results in mathematics.* Chestnut Hill, MA: IEA.

Musfiqi, S. & Jailani, J. (2014). Pengembangan bahan ajar matematika yang berorientasi pada karakter dan higher order thinking skill (HOTS) [The developing of mathematics learning materials based on character and higher order thinking skill (HOTS)]. *Pythagoras: Jurnal Pendidikan Matematika, 9*(1), 45–59.

Nurina, D.L. & Retnawati, H. (2015). Keefektifan pembelajaran menggunakan pendekatan problem posing dan pendekatan open ended ditinjau dari HOTS [Effectiveness of problem posing and open ended approach regarded HOTS]. *Pythagoras: Jurnal Pendidikan Matematika, 10*(2), 129–136.

Organisation for Economic Cooperation and Development (OECD). (2014). *PISA 2012 results: What students know and can do-student performance in mathematics, reading and science (Volume 1)* (Revised). Paris, French: OECD Publishing.

Organisation for Economic Cooperation and Development (OECD). (2016). *PISA 2015 results: Excellence and equity in education (Volume 1)*. Paris, French: OECD Publishing.

Partawibawa, A., Fathudin, S., & Widodo, A. (2014). Peran pembimbing akademik terhadap pembentukan karakter mahasiswa [The role of academic counselor to the formation of student character]. *Jurnal Pendidikan Teknologi dan Kejuruan, 22*(1), 1–8.

Partnership for 21st Century Skills. (2002). *Learning for the 21st century: A report and mile guide for 21st century skills*. Tucson, AZ: Author.

RoI. (2016). Peraturan menteri pendidikan dan kebudayaan Republik Indonesia nomor 20 tahun 2016 tentang standar kompetensi lulusan pendidikan dasar dan menengah [Regulation of the minister of education and culture number 20 on 2016 about the standard competence of primary school and middle school's passing grade]. Jakarta, Indonesia: Ministry of Education and Culture, Republik of Indonesia.

Plomp, T. (2013). Educational design research: An introduction. In T. Plomp & N. Nieveen (Eds.), *Educational design research* (pp. 10–51). Enschede, Netherland: National Institute for Curriculum Development.

Presseisen, B.Z. (1988). Thinking skills: Meanings and models. In A.L. Costa (Ed.), *Developing minds: A resource book for teaching thinking* (pp. 43–48). Alexandria, VA: ASCD.

Retnawati, H. (2015a). Hambatan guru matematika sekolah menengah pertama dalam menerapkan kurikulum baru [The mathematics teacher's obstruction of junior high school to apply new curriculum]. *Cakrawala Pendidikan, 14*(3), 390–403.

Retnawati, H. (2015b). The comparison of accuracy scores on the paper and pencil testing vs. computer-based testing. *Turkish Online Journal of Educational Technology, 14*(4), 135–142.

Retnawati, H., Hadi, S., & Nugraha, A.C. (2016). Vocational high school teachers' difficulties in implementing the assessment in Curriculum 2013 in Yogyakarta Province of Indonesia. *International Journal of Instruction, 9*(1), 33–48.

Retnawati, H., Kartowagiran, B., Arlinwibowo, J., & Sulistyaningsih, E. (2017). Why are the mathematics national examination items difficult and what is teachers' strategy to overcome it? *International Journal of Instruction, 10*(3), 257–276.

Riadi, A. & Retnawati, H. (2014). Pengembangan Perangkat Pembelajaran untuk Meningkatkan HOTS pada Kompetensi Bangun Ruang Sisi Datar [Developing Learning Kit to Improve HOTS for Flat Side of Space Competence]. *Pythagoras: Jurnal Pendidikan Matematika, 9*(2), 126–135.

Saido, G.A.M., Siraj, S., Nordin, A.B., & Al-Amedy, O.S. (2015). Teaching strategies for promoting higher order thinking skills: A case of secondary science teachers. *Malaysian Online Journal of Educational Management, 3*(4), 16–30.

Susanti, E., Kusumah, Y.S., & Sabandar, J. (2014). Computer-assisted realistic mathematics education for enhancing students' higher-order thinking skills (experimental study in junior high school in palembang, Indonesia). *Journal of Education and Practice, 5*(18), 51–59.

Susanto, E. & Retnawati, H. (2016). Perangkat pembelajaran matematika bercirikan PBL untuk mengembangkan HOTS siswa SMA [Mathematics learning kits charactized PBL to develop HOTS of junior high school students]. *Jurnal Riset Pendidikan Matematika, 3*(2), 189–187.

Thornberg, R. & Oğuz, U. (2013). Teachers' views on values education: A qualitative study in Sweden and Turkey. *International Journal of Educational Research, 59*(1), 49–56.

Treffinger, D.J. (1995). Creative problem solving: Overview and educational implications. *Educational Psychology Review, 7*(3), 301–312.

Tseng, K., Chang, C., Lou, S., & Hsu, P. (2013). Using creative problem solving to promote students' performance of concept mapping. *International Journal of Technology and Design Education, 23*(4), 1093–1109.

Wardani, N.S. (2014). Development of student' character through project based learning for social studies subject. *Jurnal Ilmiah Pendidikan Sejarah Dan Sosial Budaya, 16*(2), 103–115.

Yen, T.S. & Halili, S.H. (2015). Effective teaching of Higher-Order Thinking (HOT) in education. *The Online Journal of Distance Education and E-Learning, 3*(2), 41–47.

Zakaria, E., Chin, L.C., & Daud, M.Y. (2010). The effects of cooperative learning on students' mathematics achievement and attitude toward mathematics. *Journal of Social Science, 6*(2), 272–275.

Zuchdi, D., Prasetyo, Z.K., Masruri, M.S. (2011). *Model pendidikan karakter terintegrasi dalam pembelajaran dan pengembangan kultur sekolah*. Yogyakarta: UNY Press.

As linguistics is separated from the literary domain: Questioning the blurred portrait of English language instruction in Indonesian senior high schools

S. Sugirin, K. Kasiyan & S. Sudartini
Universitas Negeri Yogyakarta, Indonesia

ABSTRACT: Language instruction performs its strategic substance when it is used as a medium for teaching students the communicative domain of the language as well as moral values. In teaching the communicative domain, language instruction is closely related to the rational domain of language. Meanwhile, its emotional domain is related to the moral values represented in literary works. The use of both rational and emotional domains of language in language instructional practices is expected to give a valuable contribution in creating a balance in students' mental development. Hence, this study discusses the practice of English instruction conducted in senior high schools focusing on the question of whether it serves the integration of the linguistic and literary domains of the language and factors leading to that. The method used was a qualitative-naturalistic approach by analyzing the available textbooks. The study revealed that: 1) the practice of English language instruction conducted in senior high schools, seen from its textbooks, has put its emotional domain aside and is mainly concerned with the rational domain of the language; 2) this phenomenon was caused by the misconception of the communicative function of language as a result of the *zeitgeist* of this modern era, which tends to propose a rationalism hegemony.

1 INTRODUCTION

One of the important goals of national education mentioned in the Laws of the Republic of Indonesia No. 20, Year 2003, is to improve the quality of Indonesian people who have equal understanding of both intellectual and emotional potentials. Intellectual potentials are related to the critical logic development that exists in the left part of the brain (Marks-Tarlow, 2013). Meanwhile, the emotional potentials are closely related to the development of the sensitivity dimension of the senses and emotion that belong to the right part of the brain (Schore, 2000; Kolb & Whishaw, 2016).

These two domains are of the same importance although some studies show that the domain of the senses and emotion that are closely related to values and morality, provides a greater contribution to humans' success in their lives (Goleman, 2011, 2012). Consequently, it is widely believed that moral education is the center of humanity education (McKernan, 2007).

There are various subjects concerning learners' emotion potential developments under the field of moral education (Salovey et al., 2004), and one of these is language education, not to mention English language education.

Although language education is generally under the field area of moral education, its discussion can be divided into two main parts. The first part is the discussion closely related to logic substances primarily concerning the linguistic domain of the language under the influence of the left hemisphere of the brain (Keiper & Utz, 1997). In addition, Feinstein (2006, p. 512) mentions that "language skills are primarily associated with the left brain". The second part concerns the discussion of topics related to the substance of emotion-sense sensitivity, well known as the literary domain (Henderson, 2014; Mills, 2016). This domain is closely related to the right brain (Williams, 1986). It is in these two domain sides—linguistics and literary—that an ideal practice of language education needs to be conducted. Each of the domains needs to be presented equally in the instructional practices.

However, there is one crucial problem in the practice of teaching English at all levels of education in Indonesia including the practice conducted in senior high schools. The current practices of teaching English in Indonesia, either in elementary school, middle school level or higher education level, tend to merely teach its linguistic substance and neglect or even omit its literary substance. It can be clearly identified, not only from the content of the curriculum but also from the content of commonly used textbooks that focus mainly on the mastery of linguistic aspects of the language represented by the learners' mastery of the four language skills: listening, speaking, reading and writing.

This language education model in turn will merely support one side of learners' brain development—the left side of the brain dealing with logical intelligence. On the contrary, this model does not give enough space for the development of the right side of the brain which deals with emotional intelligence. As mentioned earlier, it is the emotional intelligence that provides better capital of learners' success in their lives than that of logical or intellectual intelligence as it is closely related to the abilities to adapt to other people and also other environments (Mourlas & Germanakos, 2008).

In fact, there have been some scholars stating that the literary dimension has a positive contribution to the linguistic domain. One of the ideas says that the literary domain is an efficient vehicle for foreign language acquisition (O'Sullivan et al., 2015). Another idea is that proposed by Bruns (2011) that the literary domain improves the students' motivation in learning, considering the fact that most literary works are presented using interesting linguistic expressions.

In a wider context, literature can be used to teach values and to provide examples of finding a good solution in life to learners. In addition, Frevert and Olsen. (2014) mention that literature plays a significant role in developing children's emotional intelligence. Another idea was proposed by Bruns (2011, p. 13) saying that "literature serves as a potential source of values, perspectives, or ways of living that may be better than one's own or those available in present society". Meanwhile, Jerome et al. (2016, p. 36) mention that "literature has the potential to enhance readers' understanding of themselves in relations to others in this world through their engagement with literary texts. All forms of literature not only provide a source of enjoyment and satisfaction, but also serve a multiplicity of functions and purposes for different kinds of readers across space and time".

There are various definitions related to the term "literature" on the basis of in which cultural context this term is used. Historically, in Western Europe around the eighteenth century, the term "literature" was used in reference to all books and writing (Leitch, 2010). Generally, the term literature refers to any text having an interesting quality of representation either in the form of fiction or non-fiction, such as poetry, novels, short stories, plays and the like (Eagleton, 2008).

Considering the benefits that learners get from teaching the literary domain, it will be much more beneficial to present the literary domain in the practice of teaching English for all levels of education owing to the fact that schools are the right place to build intellectuality and emotionality of learners (Matthews et al., 2004).

In line with this idea, the founding father of national education in Indonesia, Ki Hadjar Dewantara, long ago proposed "senses" as the main component of the education trilogy with his famous terms *cipta* (create), *rasa* (senses) and *karsa* (intention) (Dewantara, 2013). The synergy of these three elements could enhance the sensitivity of the soul for the sake of forming noble human beings (Kasiyan, 2002).

Regarding this, there have been movements in many parts of the world for a long time, not to mention in England, to consider literature as a means of forming humans' morality or characters—an effective medium to educate people to become sensitive. Many scholars and poets have considered "literature" as a *torch*, a torch that gives light to human life and is considered as having a similar position as *holy books* (Eagleton, 2008). In line with this, Kristeva (McAfee, 2004, p. 50) claims that literature can be the catharsis of human souls. Following

Aristotle, the term catharsis means the purity process or soul purification (Ghezelsofla et al., 2015).

Schmidt and Pailliotet (2008, p. 211) mention that much of the literature read at school is assumed to instill character, morals and citizenship education. Therefore, the discussion of the term literary education in this context is oriented not to enable learners to become poets but needs to be seen under the framework of "education through literature" (Schmidt & Pailliotet, 2008; Cummings et al., 2014). Similar to that idea, Byram (2008, p. 150) states that as language teachers start focusing primarily on skills that are "value-free", language learning has become separated from the teaching of literature because the teaching of literature has the potential to make learners learn the use of language aesthetically and also to make them learn moral values accompanying the use of language.

In discussing the instructional process and its relation to the internalization of various values either those related to the linguistic or literary dimensions, one of the important media that can be used is textbooks that can be defined as a manual of instruction in any branch of study.

In accordance with the background of the problem, a study related to the idea of integrating linguistic and literary dimensions in the practice of English teaching and learning is important and strategic. In relation to that, this study focuses on describing two things: (1) the existence of English textbooks containing the integration of linguistic and literary domains; and (2) factors leading to their separation.

2 METHOD

This study used a qualitative method with naturalistic inquiry (Agostinho, 2005) by analyzing available documents, in this case, English textbooks for senior high schools used in Indonesia. It can be categorized as a library research (McNabb, 2015). In this study, the samples are English textbooks published by the Center for Curriculum and Books, Ministry of Education and Culture, Jakarta and those by other private publishers. The books are *Bahasa Inggris untuk SMA/MA/SMK/MAK* (English for Senior High Schools) written by Widiati, Rohmah and Furaidah (2017); Pathway to English for Senior High Schools: General Program, and Pathway to English for Senior High Schools: Special Program (Sudarwati & Grace, 2014a, 2014b).

Meanwhile, the approach used in this study was hermeneutic focusing on interpretation activities (Kinsella, 2006; Roberge, 2011). The main instrument is the researcher as a *human instrument* (Peredaryenko & Krauss, 2013). The technique for analyzing the data used in this study was the descriptive-interpretative technique (Miles et al., 2013; Sloan & Bowe, 2014).

3 RESULTS AND DISCUSSION

3.1 *The existence of English textbooks for SMA in Indonesia containing the integration of linguistic and literary dimensions*

Before presenting the results and discussion of this study, it is necessary to mention that English is a subject taught in Indonesian schools starting from elementary school (as the local content), junior and senior high schools to university level. For junior and senior high school levels, the English subject is given from the first grade to the third grade. The currently used curriculum is known as Curriculum 2013.

In relation to the textbooks used in schools, the government has provided electronic versions of textbooks for almost all subjects, not to mention English, but most teachers tend to use them together with other books provided by private publishers. The books chosen by most teachers are those published by Erlangga Publishers including Pathway to English for the general and language programs.

The study of these three textbooks for senior high school grades X to XII has revealed some findings as follows. From the whole subject matter contained in the English textbooks, as represented in the content of the books, it can be said that there are no efforts of integrating the two dimensions. This can be seen, for instance, from one of the English textbooks for grade X provided by the government that consists of 15 chapters focusing on lessons and activities related only to the linguistic dimension.

Take for example, the formulation of the learning objectives in Chapter 11 of the same book. It is clearly stated that after studying Chapter 11, learners are expected to be able to: 1) explain the objectives of communication, structures of the texts, and linguistic elements of simple spoken and written narrative texts of folktales based on the context; 2) explain the contents of the spoken and written folktales by considering the objectives of communication, the structures of the texts and linguistic elements of narrative texts based on their contexts; and 3) retell the story by using spoken language and written language by considering the objectives of the communication and so on. The similar formulation of learning objectives as mentioned has also been found in the other chapters of the same book and also in chapters of the other books for different grades.

Regarding this, it can be said that the whole practice of teaching English at the senior high school level, at least on the basis of its content materials in the textbooks, is aimed at teaching the language as a means of communication.

If the entire practice of language education has put the communication competence, both receptive as well as productive as its main orientation, it would eliminate the function of language as a means of developing critical thinking that can be defined as self-guided, self-disciplined thinking which attempts to reason at the highest level of quality in a fair-minded way. It is widely believed that those who think critically consistently attempt to live rationally, reasonably and empathically. They are keenly aware of the inherently flawed nature of human thinking when left unchecked (Elder, 2007).

One of the benefits of having critical thinking skills that are indispensable is that the person may have more empathy and tolerance and tend to be ready for communication in multicultural contexts (Fortanet-Gomez, 2013).

Being critical is the social capital of each individual to become a perfect social citizen. To conclude, we would like to quote Sumner (Paul & Elder, 2014) when he mentions the paramount influence of critical thinking within societies and among human beings: "education in the critical faculty is the only education of which it can be truly said to make a good citizen".

On the basis of critical thinking terminology, the ability to be critical is closely related to the logical thinking developed by linguistic domains and at the same time involving the complexity of emotional senses, meaning that it needs to involve the literary domain (Fitzgerald & Branch, 2002). In line with this, Clark (2010) mentions that the affective domain includes the manner in which we deal with things emotionally, such as feelings, values, appreciation, enthusiasm, motivations and attitudes. Therefore, for the sake of critical thinking development, it is crucial to consider the types of activities from the point of view of how they contribute both to the intellectual and the affective development (Vdovina & Gaibisso, 2013).

This logical construction, once again, clarifies the significance of integrating linguistic and literary domains in every language instructional practice even though this ideal construction has not yet been found in English textbooks provided for senior high schools in Indonesia.

3.2 Factors leading to the separation of linguistic and literary domains in English textbooks for senior high schools in Indonesia

The underlying factors leading to the separation of these two domains in English textbooks for senior high schools in Indonesia can be said to be very complex. However, basically, there are two main factors that are closely related to that phenomenon. First, it is closely related to the need of the global communication function in the twenty-first century, in which the English language plays a significant role and even in its latest development, English has become a kind of "lingua franca" (Spolsky & Moon, 2012; Jenkins, 2013; Mackenzie, 2014).

This strong argumentation related to the function of language as an important medium of communication is also mentioned in the foreword of the textbook for grade X. It is said that the rapid development of information technology and communication in the twenty-first century has placed English as one of the main languages in inter-national and global communication. Curriculum 2013, which was designed to deal with the twenty-first century model of learning, realizes the importance of enabling senior high school graduates to master English and enable them to express their ideas and to get ideas from other people coming from other countries for the sake of their nation (Widiati et al., 2017; Sudarwati & Grace, 2014).

Second, in relation to the hegemony factor of modernism that can be easily identified mainly from the emergence of rationalism philosophy (Knox, 2010; Linehan, 2012; Sparke, 2013), it puts everything outside logic including arts and literature that are considered as useless things (Agassi & Jarvie, 2012).

4 CONCLUSION

The main points discussed in the previous part can be concluded as follows. Firstly, related to the main problem of integrating linguistic and literary domains in English textbooks for senior high schools in Indonesia, it can be concluded that there are no efforts to integrate those two dimensions in the entire content or subject matter of those textbooks. From the entire content of the textbooks, it is clearly seen that the meaning and functions of language education is merely as a communication medium. In this context, language education is considered as having its narrow function and has no capabilities to have more comprehensive roles over its communicative function as it is connected to various humanity domains.

Secondly, related to the factors leading to the separation of the two domains of language education in the English textbooks for senior high school in Indonesia, there are at least two crucial and strategic factors identified. First, it is related to the misconception of the communicative function of language that has put aside its critical thinking. It is only the logical aspects that become the focus of discussion. In fact, the critical concept involving the emotional and sense intelligence coming from literary domain of the language in this construction of communication patterns is also important to be noted. Second, the English language hegemony that focuses mainly on the linguistic domain and does not put any attention on the literary domain may be the result of the *zeitgeist* of this modern era, that tends to put forward rationalism. This rationalism perspective has influenced entire aspects of human life that all cultural expression and science, not to mention in this context language education, tend to follow the same linearity that is based on logic. In this sense, literature is considered as something nonsense and useless. It should not be like this as we go back to the long socio-historical facts showing that literature has been the best part of various civilization and humanity processes. This piece of writing is intended to provide a small contribution on the reorientation and revitalization of English teaching and the learning process in Indonesia in the future by considering the integration of linguistic and literary domains in the content materials.

REFERENCES

Agassi, J. & Jarvie, I.C. (2012). *Rationality: The critical view*. Berlin: Springer Science & Business Media.
Agostinho, S. (2005). Naturalistic inquiry in e-learning research. *International Journal of Qualitative Methods*, 4(1), 13–26. doi: 10.1177/160940690500400102.
Bashir, M. (2017). *Bahasa Inggris untuk SMA/MA/SMK/MAK kelas XI (English for Senior High School)*. Edisi revisi. Jakarta, Indonesia: Kementerian Pendidikan dan Kebudayaan.
Bruns, C.V. (2011). *Why literature? The value of literary reading and what it means for teaching*. New York, NY: The Continuum International Publishing Group.
Byram, M. (2008). *From foreign language education to education for intercultural citizenship: Essays and reflections*. Clevedon, United Kingdom: Multilingual Matters, Ltd.
Clark, D.R. (2010). *Bloom's taxonomy of learning domains*. Retrieved July 10, 2012 from http://www.nwlink.com/~donclark/hrd/bloom.html/.
Cummings, W. K., Gopinathan, S., & Tomoda, Y. (2014). *The revival of values education in Asia & the West*. Amsterdam, Netherlands: Elsevier.

Depertemen Pendidikan Nasional. (2003). *Undang-undang Republik Indonesia nomor 20 tahun 2003, tentang sistem pendidikan nasional Indonesia (Law of The Republic of Indonesia number 20 of 2003)*. Jakarta, Indonesia: Depertemen Pendidikan Nasional.

Dewantara, K.H. (2013). *Ki Hadjar Dewantara: Pemikiran, konsepsi, keteladanan, sikap merdeka: II kebudayaan (Ki Hadjar Dewantara: Thought, conception, exemplary, free attitude: II culture)*. Yogyakarta, Indonesia: Majelis Luhur Persatuan Taman Siswa.

Eagleton, T. (2008). *Literary theory: An introduction*. Malden, MA: Blackwell Publishing.

Elder, L. (2007). A brief conceptualization of critical thinking. Retrieved August 10, 2017, from http://www.criticalthinking.org/pages/defining-critical-thinking/410/.

Feinstein, S. (2006). *The Praeger handbook of learning and the brain*. 2. Wetport, CT: Greenwood Publishing Group.

Fitzgerald, M. A. & Branch, R. M. (2002). *Educational media and technology yearbook* 2002. Englewood, CO: Libraries Unlimited.

Fortanet-Gomez, I. (2013). *CLIL in higher education: Towards a multilingual language policy*. Clevedon, United Kingdom: Multilingual Matters.

Frevert, U.P.E., & Olsen, S. (2014). *Learning how to feel: Children's literature and emotional socialization, 1870–1970*. Oxford: Oxford University Press.

Ghezelsofla, M., Karimpoor, S., & Khosrejerdy, H. (2015). A study of position of Aristotle's catharsis in the Persians Ta'ziyeh. *Indian Journal of Fundamental and Applied Life Sciences, 5*(S1), 622–626.

Goleman, D. (2011). *Working with emotional intelligence*. New York, NY: Random House, Inc.

Goleman, D. (2012). *Emotional intelligence*. 10th anniversary edition. New York, NY: Random House Publishing Group.

Henderson, J. (ed.). (2014). *Reconceptualizing curriculum development: Inspiring and informing action*. London, UK: Routledge.

Jenkins, J. (2013). *English as a lingua franca in the international university: The politics of academic English language policy*. London, UK: Routledge.

Jerome, C., Hashim, R. S., & Ting, S. (2016). Multiple literary identities in contemporary Malaysian literature: An analysis of readers' views on heroes by Karim Raslan. *3L: The Southeast Asian Journal of English Language Studies, 22*(3), 35–47.

Kasiyan, K. (2002). Pendidikan Kesenian dalam pembangunan karakter bangsa. (Art and character development of the nation). *Jurnal Cakrawala Pendidikan*, Lembaga Pengabdian kepada Masyarakat Universitas Negeri Yogyakarta, Indonesia (Terakreditasi Nasional), *21*(1), Februari 2002, Tahun XXI.

Keiper, H.C.B. & Utz, R.J. (1997). *Nominalism and literary discourse: New perspectives*. Amsterdam, Netherlands: Rodopi.

Kinsella, E.A. (2006). Hermeneutics and critical hermeneutics: Exploring possibilities within the art of interpretation. *Forum: Qualitative Social Research, 7*(3), 1–7. Retrieved from http://nbn-resolving.de/urn:nbn:de:0114-fqs0603190.

Knox, P.L. (2010). *Cities and design*. London, UK: Routledge.

Kolb, B. & Whishaw, I. (2016). *Brain and behaviour: Revisiting the classic studies*. London: Sage Publication.

Leitch, V.B. (2010). *The Norton anthology of theory and criticism*. Second Edition. New York, NY: W. W. Norton & Company.

Linehan, T. (2012). *Modernism and British socialism*. Berlin, Germany: Springer.

Mackenzie, I. (2014). *English as a lingua franca: Theorizing and teaching English*. London, UK: Routledge.

Marks-Tarlow, T. (2013). *Psyche's veil: Psychotherapy, fractals and complexity*. London: Routledge.

Matthews, G., Zeidner, M., & Roberts, R. D. (2004). *Emotional intelligence: Science and myth*. Boston, MA: MIT (Massachusetts Institute of Technology) Press.

McAfee, N. (2004). *Julia Kristeva*. Hove, United Kingdom: Psychology Press.

McKernan, J. (2007). *Curriculum and imagination: Process theory, pedagogy and action research*. London, UK: Routledge.

McNabb, D.E. (2015). *Research methods in public administration and nonprofit management*. London, UK: Routledge.

Miles, M.B., Huberman, A.M., & Saldana, J. (2013). *Qualitative data analysis*. London, UK: Sage Publication.

Mills, C. (2016). *Ethics and children's literature*. London, UK: Routledge.

Mourlas, C. & Germanakos, P. (2008). *Intelligent user interfaces: Adaptation and personalization systems and technologies*. New York, NY: IGI Global Publishing.

O'Sullivan, M., Huddart, D., & Lee, C. (2015). *The future of English in Asia: Perspectives on language and literature*. London, UK: Routledge.

Paul, R. & Elder, L. (2014) *The miniature guide to critical thinking. Concepts and tools.* Dillon Beach, CA: The Foundation for Critical Thinking.

Peredaryenko, M.S. & Krauss, S.E. (2013). Calibrating the human instrument: Understanding the interviewing experience of novice qualitative researchers. *The Qualitative Report, 18*(43), 1–17.

Roberge, J. (2011). What is critical hermeneutics? *Thesis Eleven, 106*(1). doi: abs/10.1177/07255136 11411682.

Mayer, J. D., Brackett, M. A., & Salovey, P. (2004). *Emotional intelligence: Key readings on the Mayer and Salovey model.* Chicago, IL: Dude Publishing.

Schmidt, P.R. & Pailliotet, A.W. (2008). *Exploring values through literature, multimedia, and literacy events: Making connections.* Charlotte, NC: IAP (Information Age Publishing, Inc.).

Schore, A.N. (2000). Attachment and the regulation of the right brain. *Attachment & Human Development, 2*(1), 23–47. http://dx.doi.org/10.1080/1461673003 61309.

Simon, R. & Ryan, D. (2017). What is literature? *Foundation: Fundamentals of literature and drama.* Australian Catholic University. Retrieved August 2017.

Sloan, A. & Bowe, B. (2014). Phenomenology and hermeneutic phenomenology: The philosophy, the methodologies and using hermeneutic phenomenology to investigate lecturers' experiences of curriculum design. *Quality & Quantity, 48*(3), 1291–1303. Doi:10.1007/s11135–013–9835–3.

Sparke, P. (2013). *An introduction to design and culture: 1900 to the present.* London: Routledge.

Spolsky, B. & Moon, Y. (2012). *Primary school English-language education in Asia: From policy to practice.* London, UK: Routledge.

Sudarwati, Th. M. & Grace, E. (2014a). *Pathway to English for senior high school: General program.* Surabaya, Indonesia: Erlangga.

Sudarwati, Th. M. & Grace, E. (2014b). *Pathway to English for senior high school: Special program.* Surabaya, Indonesia: Erlangga.

Vdovina, E. & Gaibisso, L. C. (2013). Developing critical thinking in the English language classroom: A lesson plan. *English Language Teachers' Association Journal, 1(1), 54–68.*

Widiati, U., Rohmah, Z., & Furaidah. (2017). *Bahasa Inggris untuk SMA/MA/SMK/MAK* [English for Senior High Schools]. Surabaya, Indonesia: Erlangga.

Williams, L.V. (1986). *Teaching for the two sided mind: A guide to right brain/left brain education.* New York, NY: Simon & Schuster, Inc.

Developing a picture storybook based on a scientific approach through a problem-based learning method

C. Setyaningrum & H. Rasyid
Universitas Negeri Yogyakarta, Indonesia

ABSTRACT: This research aimed to develop a picture storybook based on a scientific approach through a problem-based learning method as a teaching material for fourth grade students.. The steps to develop the picture storybook consisted of: 1) research and information collecting; 2) planning; 3) developing a preliminary form of the product; 4) preliminary field testing; 5) main product revision; 6) main field testing; and 7) operational product revision. The product implementation consisted of the first tryout involving six students of the fourth grade of Primary School A. The second tryout involved 12 students of the fourth grade of Primary School B. Data collection instruments included a rating scale for picture storybook assessment for a material expert, a rating scale for picture storybook assessment for a language expert, a teacher response questionnaire and a student response questionnaire. Data were analyzed using validity test analysis with four scales. The result of this research was a picture storybook based on a scientific approach through a problem-based learning method, which was valid as a teaching material for fourth grade students.

1 INTRODUCTION

The education curriculum in Indonesia has changed from the School-based Curriculum (KTSP) 2006 to Curriculum 2013. Curriculum 2013 aimed to prepare Indonesian citizens to have the ability to live life, both as individuals and citizens who have a belief, are productive, creative, innovative and effective, and are able to contribute to the community, nation, state and civilization of the world (Jaedun et al., 2014). However, the implementation of Curriculum 2013 is in fact not compatible with the aims, thus resulting in an educational gap. The educational gap is reflected in the results of a needs analysis through interviews, observations and questionnaires of students' needs. As an outline, the results of interviews and observations show that the textbooks of Curriculum 2013 are less able to facilitate students to construct their knowledge in thinking for their problem solving and their in-depth curiosity. This fact is clearly contrary to the aims of Curriculum 2013, so that the development of a product in the form of a picture storybook is perceived as a solution that can help the achievement of the aims of Curriculum 2013.

In relation to the above explanation, a needs analysis was also done by distributing questionnaires of students' needs to obtain opinions and conceptions about the picture storybook that fits students' needs. The results of the questionnaires of students' needs indicate that most students are happy if the learning is done by using a picture storybook. Students also need a picture storybook with many pictures, a picture storybook that can help in understanding the subject matter, a picture storybook with a cover that describes the togetherness of children and a picture storybook that talks about the surrounding environment. Based on the overall results of the previous needs analysis, the picture storybook in this research was developed in accordance with the needs of students. The picture storybook was also developed in accordance with the students' needs and Curriculum 2013. Therefore, the picture storybook was developed using a scientific approach, because a scientific approach is required in Curriculum 2013. In addition, the selection of learning method which is relevant to the scientific approach is the Problem-Based Learning (PBL) method. Thus, the product developed in this research is a picture storybook based on a scientific approach through the PBL method as a teaching material for fourth grade students.

2 LITERATURE VIEW

Picture storybooks, according to Mitchell (2003) are described as books in which the pictures and the text are tightly intertwined. A similar explanation was proposed by Huck et al. (1987), where picture books are seen as conveying their messages through two media, the art of illustration and the art of writing. Based on the understanding of the picture storybook, the presentation of the text images allows students to be more interested in understanding the message contained in the story content of the picture storybooks, especially for child readers, where the presence of images can be a great magnet for them.

Continuing with the above exposition, the preparation of picture storybooks requires attention to their constituent elements. First, the theme element. Norton reveals that the theme is "a central idea that connects the plot, characters, and settings" (Barone, 2011, p. 113). The theme as a central idea, according to Nurgiyantoro (2013), can be a theme of family life, between the child, mother, father, brother, sister and neighbor, the child's relationships with peers at school and outside school, sports and cultural arts themes, and others. The reason for choosing these themes was that they are close to the daily life of the students. This statement was reinforced by the opinion of Brown and Tomlinson (1999), who said that the theme in children's books should be worthy of the children's attention and should convey the truth to them. Furthermore, the themes should be used based on high moral and ethical standards.

Second, the character element. Tompkins and Hoskisson (1995) explained that characters are the people or personified animals who are involved in the story. Characters are often the most important element of the story structure, because the story is centered on a character or group of characters. As according to Foster, when viewed from the character dimensions of the characters, children's stories are more likely to figure flat characters than rounded characters (Nurgiyantoro, 2013). The selection of the flat character is assessed by Nurgiyantoro (2013) as being a character who has a character like children in general, that is still plain and not yet has behavior that reflects the opposite character. Brown and Tomlinson (1999) stated that every character should be reasonable. That characterization refers to how the author helps the reader to know a character. The most obvious way is by the character's physical appearance and personality. Third, the setting element. Donoghue (2009) revealed that the setting is the time and place of the action. Thus, the events narrated in a story certainly have an attachment to the time and place where the events in the story take place.

The fourth picture storybook element is plot. Barone (2011) stated that plots involve conflict and resolution. Therefore, a path can be presented through the issues raised into the story, and how the problem can also be overcome. The fifth element is point of view. Related to picture books for children, Lukens (1999) described the omniscient viewpoint (perspective viewpoint) as a point of view that provides any relevant information about each character. Thus, all-round perspective is considered suitable for children's storybooks. The sixth element is style. The quality of writing can be seen from the quality of a book (Donoghue, 2009). While a quality book is a book that has a style. Such styles should reflect the various characteristics of students as a smart individuals. Based on that opinion, it can be interpreted that the quality of writing of children's storybooks should be adjusted to the ability and development of the child, so that the meaning of the story in the book is really conveyed to the child. The meaning of the story can be messages from the author (Kraayenoord & Paris, 1996). Therefore, stories in the book should contain topics and address issues related to daily life, such as the relationship between child, mother, father, brother, sister and others (Nurgiyantoro, 2013).

The seventh element is the images. Huck et al. (1987) stated that many children's books are illustrated in a cartoon style which depends on a lively drawing to create movement and humor. The selection of such image styles can be more interesting for students to learn using a developed picture storybook. The eighth element is related to the letters. An opinion from Sitepu (2015) stated that the letters commonly used for fourth grade are in font size 12–14. Not only that, Wilkins et al. (2009) revealed that most typefaces for children are sans serif. The next element, the ninth, is the page setting. Picture book pages used as teaching materials are divided into four sections: cover, preliminaries, text matter and postliminaries. The tenth element deals with color selection (Prastowo, 2014). Huck et al. (1987) revealed that the choice of color or colors depends on the theme of the picture storybook. Then, the eleventh element is the book size. A suitable size for children's books especially for elementary school students is 21 × 28 centimeters (Prastowo, 2014). The entire constituent elements of the picture storybook described above are the benchmarks that can be used in the measurement of the validity of a developed picture book.

Not only that, the storybook that is used as teaching material in this research must also fulfill the aspects of the preparation of teaching materials; that is material aspect, presentation aspect, and language or legibility aspect (Prastowo, 2014). In Indonesia the preparation of the storybook is also adapted to Curriculum 2013, where a scientific approach is required to be implemented. Barringer et al. (2010) revealed that learning using a scientific process is a learning that requires students to think systematically and critically to solve problems whose solutions are not easily seen. The solution is derived from efforts that must be reflected in a series of scientific stages: observing, asking, collecting data, associating and communicating (Hosnan, 2014). In the scientific steps, the relevant method is highlighted, that is PBL. PBL is a curriculum model designed around real-life problems that are ill-structured, open-ended or ambiguous. PBL itself comes from several stages, such as student organizing for problems, student organizing for learning, assisting independent and group investigations, and the development and presentation of works.

3 RESEARCH METHOD

This research used a research and development approach by Borg and Gall (1983). The development steps of a picture storybook based on a scientific approach through a PBL method consisted of: 1) research and information collecting; 2) planning; 3) developing a preliminary form of the product; 4) preliminary field testing; 5) main product revision; 6) main field testing; and 7) operational product revision.

Instruments in this research involved a rating scale for picture storybook assessment by a material expert, a rating scale for picture storybook assessment by a language expert, a teacher response questionnaire and a student response questionnaire. The data analysis technique used was an analytical technique for validity testing. The product validity was analyzed using the results of the validation by the material expert and language expert through a rating scale for picture storybook assessment, as well as a teacher response questionnaire and a student response questionnaire. Furthermore, the results of the analysis were categorized by using validity test analysis with four scales as described by Widoyoko (2009), shown in Table 1.

The product validity in this research is determined at least on the "Valid" category. That is, if the results of the rating scale for a picture storybook assessment by a material expert,

Table 1.   Product category based on the average of total score.

Number	Score interval	Category
1.	>3.25–4	Very valid
2.	>2.50–3.25	Valid
3.	>1.75–2.50	Valid enough
4.	1.00–1.75	Not valid

the rating scale for picture storybook assessment by a language expert, the teacher response questionnaire and the student response questionnaire are "Valid", then the teaching material products in the form of picture storybooks developed are said to be suitable for use in the learning process.

4 RESULTS

Data analysis conducted on the rating scale for the picture storybook assessment by a material expert and the rating scale for picture storybook assessment by a language expert are described in Table 2.

Based on Table 2, it is seen that the analysis of the results of the rating scale for picture storybook assessment by a material expert and language expert are in the category of "Very valid". The assessment of the developed picture storybook was also validated through a Focus Group Discussion (FGD) conducted by the experts.

The results of teacher and student responses to the product in preliminary testing are described in Table 3 and Table 4.

Based on the Table 4, it can be seen that the results of teacher responses to the product in preliminary testing shows an average of 3.4, so it can be categorized as "Very valid". The results of student responses to the product in preliminary testing shows an average of 3.8, so it can also be categorized as "Very valid".

The results of teacher and student responses to the product in main field testing are described in Table 5 and Table 6.

Based on Table 5 and Table 6, it can be seen that the results of teacher responses to the product in main field testing shows an average of 3.6, so it can be categorized as "Very valid". The results of student responses to the product in preliminary testing shows an average of 3.5, so it can also be categorized as "Very valid".

Table 2. Results of expert data analysis.

Material expert	Language expert	Category
3.9	3.6	Very valid

Table 3. Results of teacher responses to the product in preliminary testing.

Number	Aspects	Average	Category
1.	Material standards	3.1	Valid
2.	Presentation standards	3.4	Very valid
3.	Language standards	3.6	Very valid

Table 4. Results of student responses to the product in preliminary testing.

Number	Respondents (Student)	Average	Category
1.	Respondent 1	3.7	Very valid
2.	Respondent 2	3.7	Very valid
3.	Respondent 3	3.7	Very valid
4.	Respondent 4	3.8	Very valid
5.	Respondent 5	3.8	Very valid
6.	Respondent 6	3.9	Very valid

Table 5. Results of teacher responses to the product in main field testing.

Number	Aspects	Average	Category
1.	Material standards	3.7	Very valid
2.	Presentation standards	3.6	Very valid
3.	Language standards	3.4	Very valid

Table 6. Results of student responses to the product in main field testing.

Number	Respondents (Student)	Average	Category
1.	Respondent 1	3.8	Very valid
2.	Respondent 2	3.3	Very valid
3.	Respondent 3	3.8	Very valid
4.	Respondent 4	3.7	Very valid
5.	Respondent 5	3.3	Very valid
6.	Respondent 6	3.3	Very valid
7.	Respondent 7	3.8	Very valid
8.	Respondent 8	3.4	Very valid
9.	Respondent 9	3.3	Very valid
10.	Respondent 10	3.3	Very valid
11.	Respondent 11	3.3	Very valid
12.	Respondent 12	3.5	Very valid

5 DISCUSSION

The success of learning is not only based on the results achieved, but also on the learning process that is implemented. The learning process itself must also be supported by the determinants of success, one of which is the availability of teaching materials that can help the implementation of the learning process.

Teaching materials, according to the RoI (2008), consist as all forms of materials used to assist teachers/instructors in carrying out teaching and learning activities, in both written and unwritten material. Therefore, the selection of picture storybooks as teaching materials that can be used to help the learning process is reinforced by the opinion of Kelemen et al. (2014) who suggested the use of picture storybooks because the format is child friendly and invites a beneficial joint-attentional learning context. The designed picture storybook is also tailored to the current curriculum, Curriculum 2013, so the picture storybook is designed with a scientific approach. The design of the picture storybook is also done through a method relevant to the scientific approach, the method is PBL. The relevance of the PBL method to the scientific approach is expressed by Sani (2014) who stated that PBL is one suitable method for use in a scientific approach. Thus, the picture storybook is designed to be based on a scientific approach through the PBL method in the hope of helping achieve the objectives of Curriculum 2013.

In its development, the picture storybook went through a validity test phase before the test phase of effectiveness. The validity test undertaken resulted in assessments obtained from material expert validation, language expert validation, FGD, teacher and student response questionnaires in preliminary testing, and teacher and student response questionnaires in main field testing. Continuing the elaboration above, based on the material expert validation results, it is known that the average score for all aspects assessed was 3.9, so it can be said that the quality of the picture storybooks is judged in terms of the material included in the category of "Very valid". The language expert validation results show that the average score for the two elements assessed was 3.6, so it can be said that the quality of the picture books according to the language expert is included in the "Very valid" category.

Not only that, the results of the FGD conducted also obtained input, both criticism and suggestions used as a material revision. The revision from the FGD input is expected to increase the validity of the developed picture storybooks. Reviewing the position of picture storybooks as teaching materials, the assessment of storybooks is based on the validity of teaching materials in terms of preparation of teaching materials by Prastowo (2014) that consisted of: 1) material aspects; 2) presentation standard aspects; and 3) language/legibility aspects. Given the nature of picture storybooks, the assessment of picture storybooks is based on validity in terms of the elements of the picture book itself.

Given the previous explanation, further eligibility is shown in the teacher's response questionnaire in preliminary testing showing that the quality of the picture storybook is included in the "Very valid" category with a final average score of 3.4, and the results of the student response questionnaire also included it in the "Very valid" category with a final average score from six respondents of 3.8. Similarly, the results of questionnaires of teacher and student responses in main field testing indicated that the quality of the picture storybook is included in the "Very valid" category with a final average score of 3.6, as well as the student response questionnaire which also included it in the "Very valid" category with a final mean score from 12 respondents of 3.5.

6 CONCLUSION

Picture storybooks based on a scientific approach through a PBL method are suitable for use as a teaching material for fourth grade students. This is evident from the analysis of the results that show the validity of the developed picture storybook. The results shown that the analysis of the picture storybook by material expert is 3.9, this score on the "Very valid" category. The assessment of picture books by the language expert is 3.6, this score also on the "Very valid" category. Furthermore, the results of the analysis in preliminary test on the teacher responses is 3.4, this score on the "Very valid" category. While the score of student responses is 3.8, this score on the "Very valid" category. Finally, the results of analysis in the main field test on the teacher responses is 3.6, this score on the "Very valid" category, also on the student responses is 3.5, this score on the "Very valid" category.

REFERENCES

Barringer, M.D., Pohlam, C., & Robinson, M. (2010). *Schools for all kinds of minds: Boosting student success by embracing learning variation*. Alexandria, VA: ASCD.

Barone, D.M. (2011). *Children's literature in the classroom: Engaging lifelong readers*. New York, NY: The Guilford Press.

Borg, W.R. & Gall, M.D. (1983). *Educational research an introduction*. New York, NY: Longman.

Brown, C.L. & Tomlinson, C.M. (1999). *Essentials of children's literature*. Boston, MA: Allyn and Bacon.

Donoghue, M.R. (2009). *Language arts: Integrating skills for classroom teaching*. Thousand Oaks, CA: SAGE Publications.

Hosnan, H. (2014). *Pendekatan saintifik dan kontekstual dalam pembelajaran abad 21: Kunci sukses implementasi Kurikulum 2013* [Scientific and contextual approach in 21st century instruction: A key to successful implementation of Curriculum 2013]. Bogor, Indonesia: Ghalia Indonesia.

Huck, C.S., Hepler, S., & Hickman, J. (1987). *Children's literature in the elementary school*. New York, NY: Holt, Rinehart and Winston.

Jaedun, A., Haryanto, L., & Rahardjo, N. E. (2014). An evaluation of the implementation of curriculum 2013 at the building construction Department of Vocational High Schools in Yogyakarta. *Journal of Education, 7*(1), 14–22.

Kelemen, D., Emmons, N. A., Schillaci, R. S., & Genea, P. A. (2014). Young children can be taught *basic* natural selection using a picture-storybook intervention. *Psychological Science, 25(4)*, 851–860.

Kraayenoord, C.E.V. & Paris, S.C. (1996). Story construction from a picture book: An assessment activity for young learners. *Early Childhood Research Quarterly, 11*, 41–61.

Lukens, R.J. (1999). *A critical handbook of children's literature*. New York, NY: Longman.

Mitchell, D. (2003). *Children's literature: An invitation to the world.* Boston, MA: Ablongman.

Nurgiyantoro, B. (2013). *Sastra anak: pengantar pemahaman dunia anak* [Children literature: Introduction to children world understanding]. Yogyakarta, Indonesia: Gadjah Mada University Press.

Prastowo, A. (2014). *Pengembangan bahan ajar tematik: tinjauan teoretis dan praktik* [Development of thematics teaching aid: Theoretical and practical point of view]. Jakarta: Kencana Prenadamedia Group.

RoI. (2008). *Panduan Pengembangan Bahan Ajar* [*Guidelines for materials development*]. Jakarta, Indonesia: Ministry of Education and Culture, Republik of Indonesia.

Sani, R.A. (2014). *Pembelajaran saintifik untuk implementasi kurikulum 2013* [Scientific learning for implementation of Curriculum 2013]. Jakarta, Indonesia: Bumi Aksara.

Sitepu, B.P. (2015). *Penulisan buku teks pelajaran* [Writing text book]. Bandung, Indonesia: Remaja Rosdakarya.

Tompkins, G.E. & Hoskisson, K. (1995). *Language arts: Content and strategies* (3rd ed.). New York, NY: McMillan College Publishing Company.

Wilkins, A., Cleave, R., Grayson, N., & Wilson, L. (2009). Typography for children may be inappropriately designed. *Journal of Research in Reading, 32*(4), 402–412.

Widoyoko, E.P. (2009). *Evaluasi program pembelajaran: Panduan praktis bagi pendidikan dan calon pendidik* [Evaluation of learning programs: A practical guide for education and educator candidates]. Yogyakarta, Indonesia: Pustaka pelajar.

Character Education for 21st Century Global Citizens – Retnowati et al. (Eds)
© *2019 Taylor & Francis Group, London, ISBN 978-1-138-09922-7*

Development of Indonesian national qualification framework based instruments for a custom-made fashion competence test in universities

E. Budiastuti & W. Widihastuti
Universitas Negeri Yogyakarta, Indonesia

ABSTRACT: This research aimed to: (1) identify aspects needed to develop a competence test in the field of custom made fashion, based on a needs assessment of fashion expertise; (2) develop instruments of a custom-made competence test for fashion expertise in universities; (3) develop assessment criteria (rubrics) of a custom-made competence test for fashion expertise in universities; (4) develop assessment guidance for a custom-made competence test of fashion expertise in universities; and (5) examine the characteristics of a custom-made competence test for fashion expertise in universities. This research employed a research and development design. The development model was based on Thiagarajan's 4D model. In the define step, the researcher identified a needs assessment of a KKNI (Indonesian National Qualification Framework)-based competence test for universities. Meanwhile, in the design step, the researcher composed a prototype of a needs assessment-based test instrument. To validate the data, this research employed expert judgment and involved Lembaga Sertifikasi Profesi (LSP)/Professional Certification Institute from Universitas Negeri Yogyakarta and garment industries. Meanwhile, to gain reliability, this research employed the Kappa index. The results show that: (1) a competence test for universities required level 6 on KKNI; (2) the KKNI-based assessments of a competence test for universities were composed of: (a) planning and organizing assessments; (b) developing assessment tools; and (c) assessing competence; 3) assessment rubrics were composed; 4) assessment guidance for a competence test was composed; and 5) the validity measurement involving the Professional Certification Institutes from Universitas Negeri Yogyakarta and garment industries and also the expert judgment suggests that the instruments are valid.

1 INTRODUCTION

The establishment of Asean Economic Community in 2016 has influenced the field of education, particularly vocational education. Therefore, it is necessary to have human resources that possess the ability and skills to compete at national and international levels. Students, as an educational product, have the opportunity to compete and work at the international level.

Vocational education (SMK) has a strongly strategic role in preparing human resources who possess reliable competence, skills and performance characteristics. Besides, vocational education has the responsibility to develop competitive human resources who readily work in their field, either in the workplace or industry. To create competent graduates, higher education (university) has a mission to meet market demands.

The Educational Law of Higher Education number 12, year 2012, states that to meet the national interest and national development of competence, it is necessary to have graduates who master knowledge and/or technology. Furthermore, to find out students' knowledge and technology expertise in accordance with the needs of market demand, a test is necessary. A test conducted to examine students' competence is called a competence test. Related to educational quality, enhancing students' and teachers' competence is accepted as an important action (Grosch, 2017).

Competency test which is particularly for the *custom-made fashion*, is not extensively used For several years, SMK has implemented a competency test for students of custom made fashion design. However, the test has not been implemented in university. Higher education (university) never conducts competence tests to assess students' particular skills. The development of a performance assessment system of making a cloth consists of three aspects: cognitive, affective and psychomotor. However, in developing an assessment system of a competence test of making a cloth, the psychomotor aspect (skill) is the most dominant (Andono et al., 2003).

Vocational education is closely related to industry as the user of its graduates. However vocational education has not worked in synergic collaboration with the industry to determine qualifications of the graduates. Collaboration is necessarily conducted because graduates of vocational education need qualifications and competence recognition to realize the dreams of society and stakeholders.

Based on the President Regulation number 8, year 2012, Indonesian National Qualification Framework of educational qualification, including vocational education, is necessarily conducted to: 1) face global competition and challenges; 2) solve the quality gap and ability of graduates; 3) eliminate the existence of huge unemployment due to the absence of relevance between the number of graduates and the user of the graduates; and 4) face the existence of multiple qualification rules. Since this is necessarily anticipated, vocational higher education should implement the quality assurance management of human resources.

Based on the concept of KKNI, the implications of KKNI in higher education are: 1) managing educational types and levels; 2) standardizing the quality of graduates; 3) developing a system of quality assurance and curriculum development; and 4) facilitating life-long education.

This is crucially conducted because the graduates of vocational education need qualifications and competence recognition to realize the expectations of society or stakeholders. Recognition of human resources is based on comprehensive competence, skills and performance characteristics in accordance with national and international demands. The competences expected by the industry are field skill (hard skill) and attitude competence, cooperation and motivation (soft skill). Indonesian schools have not equipped their graduates with these skills. Consequently, their graduates will have difficultly competing in the workplace (Wibowo, 2016).

Based on the description of KKNI, standardizing the quality of university graduates (undergraduate level) involves: 1) being able to apply their expertise and utilize knowledge and technology in solving problems, and adjust to the situation faced; 2) mastering theoretical concepts of knowledge comprehensively, and being able to formulate procedural problem solving; and 3) being able to make precise decisions based on information and data analysis, and be able to give guidance to select various alternative solutions individually or in a team. The working world expects that the educational world prepares graduates to work, teaches them knowledge as well as skill, and trains them in strong working performance (Slamet, 2011).

2 METHOD

This research used a research and development approach. The development model employed in this research was the 4D model of Thiagarajan involving define, design, development and dissemination. The define and design models aim to: plan and organize the assessment of competence and be implemented as a guidance by assessors in conducting competence-based assessments. The development step aims to test the instrument in a limited scale of Java Island to ensure the validity and reliability of the instruments. The stages of instrument development consist of: (a) planning and organizing assessments, which include: 1) determining the assessment approach; 2) preparing assessment plans; 3) organizing assessments; and 4) contextualizing and reviewing assessment plans; (b) developing assessment instruments which consist of: 1) determining the focus of the assessment instruments; 2) determining the needs of the assessment instruments; 3) planning and developing assessment instruments employed as a guideline by assessors during the competence assessment process; and 4) examining and testing

Figure 1.   Instruments development process.

assessment instruments to create valid and reliable assessment instruments (Badan Nasional Sertifikasi Profesi, 2015b, p. 6). The validity of assessment instruments will be implemented in a certification process of custom-made competence in fashion for Indonesian students in order to receive qualification from the National Professional Certification Agency The phases of developing the assessment instruments are illustrated in Figure 1.

To validate the instrument, this research employed content validity based on expert judgment, consisting of fashion experts (fashion association, fashion academia), the head of Professional Certification Institute of Universitas Negeri Yogyakarta, experts, an from the garment industry and valid research instruments. Meanwhile, to analyze the reliability, this research employed a Kappa analysis with a reliability index of $r \geq 0.84$. The score of the reliability index meets the requirement of $r \geq 0.70$.

To analyze the data, this research employed descriptive analysis to describe necessary aspects to produce the instruments prototype and the characteristics of assessment instruments for custom-made competence of fashion in universities through the validation process and reliability measurement.

3   RESEARCH RESULTS

This research resulted in an instrument in the form of competence test material as the manifestation of competence recognition from the working world to assess a competence test of fashion. The description of assessment instrument development of competence is explained below.

1. Planning and organizing assessments.
    a. Determining the assessment approach, which includes determining the certification supplicant, the aims, context, approach and strategy of assessments as well as comparing instruments.
    b. Preparing assessment plans by determining working criteria, types of proof, assessment method and instruments by considering the competence dimension as follows:

Task Skill (TS)	-Doing individual task
Task Management Skill (TMS)	- Managing a number of different task of one work
Contingency Management Skill (CMS)	-The ability to response and manage irregular event and problems.
Job/Role invironment Skill (JRES)	-The ability to adapt with responsibility and expectation with the working environment.

275

c. Contextualizing and reviewing assessment plans which consist of determining the characteristics of participants and accommodating requirements of industry, examining assessment methods and instruments.

d. Organizing assessments, which include managing human resource assessment, organizing
assessors, participants, committees, as well as supervisors and determining the communication strategy.

2. Developing assessment instruments. In this stage, the required competencies were analysed to develop the assessment instruments.

a. Determining focus of assessment instruments:

In this stage, assessment instruments are determined by considering the certifications
of the participants, the aims of the assessments, and do literature review to determine
competency standards which consist of several units and thus, the determined competency scheme is successfully illustrated.

b. Determining the needs of assessment instruments:

This stage is aimed at determining assessment methods. A competency test for custom
made fashion employs two kinds of tests: written test to reveal cognitive aspects, and
an observation to observe affective and practical aspects which reveal students' skill
reflected from the competence dimensions. The selection of assessment method by
considering assessment context includes the place of test and real working activity in
the workplace to gain assessment quality, while considering valid, reliable, flexible and
fair assessment principles are very crucial.

Table 1. Practical test of competence test.

Unit code and unit title	Working instruction	Check point (√)
1. Implementing health safety and security procedure (K3) for working	Implement and do K3 procedure at workplace	
2. Maintaining sewing machine	(a) identify machine	
	(b) prepare machine	
	(c) operate sewing machine	
3. Draw fashion model	(d) draw a blouse model	
4. Measure customer body in accordance with design	(e) measure customer's body	
5. Make cloth patterns with construction technique	(a) draw basic pattern of a blouse	
	(b) change the pattern in accordance with design	
	(c) check the pattern	
	(d) cut the pattern	
	(e) keep the pattern	
6. Make clothing patterns with construction technique on the fabric	(a) prepare the production of pattern	
	(b) make the patterns on fabric	
	(c) check the pattern	
7. Cut fabric	(a) put the pattern on the fabric	
	(b) cut the fabric	
	(c) move the pattern on the fabric	
8. Sew with machine	(a) prepare place and tools	
	(b) prepare sewing machine	
	(c) operate the sewing machine	
	(d) sew parts of fashion	
9. Finish cloth with hand stitching	(e) sew the blouse with hand stitching	
10. Do pressing process machine used	(a) prepare main and support pressing	
	(b) do pressing process	
11. Finishing	(a) iron the blouse	
	(b) package blouse	

c. Planning and developing the assessment instruments:
   1. Composing assessment instruments of competence;
   2. Composing the guideline of assessment. This phase develops practical assessment instruments in the form of the MIPA form. The assignment aspects of the practical competence test are outlined in Table 1 as follows.
d. Examining and testing assessment instruments:
   1. Conducting trials to validate content and level of user compatibility. Assessment test instruments are composed based on the demands of the working world and industry. Therefore, the developed instruments were validated by an expert from garment industries, the head of LSP UNY, measurement experts, fashion practitioners and *Asosiasi Pengusaha dan Perancang Mode Indonesia* (The Association of Indonesian Entrepreneurs and Designers) through Focus Group Discussions activity 1.
   2. Revising instruments based on feedback from the validators. In the validation phase, some items were revised. After being revised, assessment instruments were in the form of FGD 2 to meet the agreement for either content or assessment structures.
   3. Administering and documenting the assessment instruments. The results of FGD 2 show that, based on expert judgment, the instruments are valid and reliable. The experts state that the depth of the competence test materials reflect the competence units with measurable working performance, and thus, the assessment instruments of the competence test are readily used.

Assessment instruments which are considered valid are ready to be used to test students' competency. The competency certificates are earned by the students who are considered as competent.

The competence criterion on the competence test for the test takers or university students is measured from the students' ability to answer all competence units.

If one of the fifteen assessment units is not answered, the student is considered incompetent. The competent students will receive a competence certificate from the LSP UNY licensed by the BNSP.

4 CONCLUSION AND SUGGESTIONS

To provide skilled workers in the globalization era, and to prepare competent human resources in their fields, competency tests for custom-made fashion are very important. A worker is considered as competent if s/he meets the competency standards required in the competence test. The competence test can be conducted if assessment instruments are readily used. Therefore, the instruments for a competence test were developed, particularly for fashion students in Indonesia. The instrument development involved four stages: 1) determining the focus of the assessment instruments; 2) determining the needs of the assessment instruments; 3) planning and developing the assessment instruments; and 4) examining and testing the assessment instruments. The development of the instruments for Competence Test Materials (MUK) meets validity and reliability thus, it provides practical aspects. To meet the global demands of competence, students must gain competence recognition in the form of a competence certificate from the LSP institution licensed by the BNSP.

REFERENCES

Andono, et al. (2003). *Standar kompetensi bidang keahlian busana "Custom-made"* [Standard competency of fashionexpertise field "Custom-mode"]. Jakarta: PPPG Kejuruan.

Badan Nasional Sertifikasi Profesi. (2015a). *Merencanakan dan mengorganisasikan asesmen* [Planning and organizing assessment]. Jakarta: Author.

Badan Nasional Sertifikasi Profesi. (2015b). Mengembangkan perangkat asesmen [Development of assessment tools]. Jakarta: Author.

Ernawati, Izwerni, & Weni, N. (2008). *Tata busana untuk SMK jilid 3* [Fashion for vocational school vol. 3]. Jakarta: Direktorat Pembinaan Sekolah Menengah Kejuruan.

Grosch, M. (2017). Developing a competency standard for TVET teacher education in ASEAN countries. *Jurnal Pendidikan Teknologi dan Kejuruan, 23*(3), 279–287.

Kerangka Kualifikasi Nasional Indonesia [Indonesia National Qualification Framework]. Retrieved from *http://www.scribd.com/doc/256215177/KKNI-dan-Pendidikan-tinggi-untuk-Tim-pdf.*

Slamet, P.H. (2011). *Pendidikan karakter dalam perspektif teori dan praktek: Implementasi pendidikan karakter kerja dalam pendidikan kejuruan* [Character education in the perspective of theory and practice: Implementation of work character education in vocational school]. Yogyakarta: UNY Press.

Wibowo, N. (2016). Upaya memperkecil kesenjangan lulusan Sekolah Menengah Kejuruan dengan tuntutan dunia industri [Efforts to minimize the gap of Vocational School graduates with the demand of industrial field]. *Jurnal Pendidikan Teknologi dan Kejuruan, 23*(1), 45–50.

Developing learning activities tasks strategies for character education

Study of Indonesian national culture from the civil engineering students' perspective

P.F. Kaming
Universitas Atma Jaya Yogyakarta, Indonesia

ABSTRACT: Culture is very important, especially when meeting with people from different culture countries. among people from various nations. Culture also implies a national identity that is differentiable between countries. This study aims at identifying a national identity, through a survey of university students who have various backgrounds, including ethnic and regional. A sample of 76 civil engineering students from the University of Atma Jaya Yogyakarta (UAJY) in Indonesia were obtained. Five instruments of cultural dimension were adopted from Hofstede and Hall. Results of the study identified that the national culture of Indonesia, based on five dimensions from Hofstede and Hall, has the scale of 35, 58, 41, 49, and 67, for power distance, individualism, uncertainty avoidance, masculinity, and long-term orientation respectively. The study found that there was no significant difference in the five national cultural dimensions between the ethnic groups of Java, China, and other nations. The study also found that there were no significant difference between each cultural dimension against the ethnic groups of Java, China, and other nations. This implies that the national identity has been solid among the ethnic groups. With its ethnic and subcultural diversity in the region, Indonesia is currently considered to have the lowest power distance among the Asian countries.

1 INTRODUCTION

Cultural differences manifest themselves in several ways—symbols, heroes, rituals, and values. Manifestations of culture are at different levels of depth. Symbols are words, gestures, pictures or objects that carry a particular meaning, which are only recognized by those who share the culture. The words in a language or jargon belong to this category, as do dress, hairstyles, soft drinks, and flags. New symbols are easily developed and old ones disappear. Examples of symbol at the University of Atma Jaya Yogyakarta (UAJY) are: logos, and uniforms for students and staffs. Heroes are persons, alive or dead, real or imaginary, who possess characteristics which are highly prized in a culture, and who thus serve as models for behavior. For example at UAJY: Saint Albertus Magnus, Saint Bonaventure, Saint Thomas Aquinas, Saint Alfonsus, Saint Don Bosco, and Mother Theresa. Rituals are collective activities, technically superfluous in reaching the desired ends, but which, within a culture, are considered to be socially essential: they are therefore carried out for their own sake. Ways of greeting and paying respects to others, and social and religious ceremonies are examples. UAJY has rituals such as a first Friday Mass, a yearly Christmas celebration, and anniversary events.

Symbols, heroes, and rituals can be subsumed under the term 'practices'. The core of a culture is formed by its values. Values are broad tendencies to prefer certain states of affairs over others. Values are feelings that leads them: they have a plus and a minus side. They deal with: evil vs. good, dirty vs. clean, ugly vs. beautiful, unnatural vs. natural, abnormal vs. normal, paradoxical vs. logical, and irrational vs. rational.

Values are among the first things children learn—not consciously, but implicitly. Development psychologists believe that by the age of ten, most children have their basic value system firmly in place, and after that age, changes are difficult to make. Because values were acquired so early in our lives, many remain unconscious to those who hold them. Therefore they cannot be discussed, nor can they be directly observed by outsiders. They can only be inferred from the way people act under various circumstances.

As almost everyone belongs to a number of different groups and categories of people at the same time, people unavoidably carry several layers of mental programming within themselves, corresponding to different levels of culture. For example: 1) a national level according to one's country (or countries, for people who migrated during their lifetime); 2) a regional and/or ethnic and/or religious and/or linguistic affiliation level, as most nations are composed of culturally different regions and/or ethnic and/or religious and/or language groups; 3) a gender level, according to whether a person was born as a girl or as a boy; 4) a generation level, which separates grandparents from parents from children; 5) a social class level, associated with educational opportunities and with a person's occupation or profession; and 6) for those who are employed, an organizational or corporate level, according to the way employees have been socialized by their work organization.

Culture has been defined in many ways. Hofstede's shorthand definition is: 'Culture is the collective programming of the mind that distinguishes the members of one group or category of people from others'. It is always a collective phenomenon, but it can be connected to different collectives. Within each collective there is a variety of individuals. If characteristics of individuals are imagined as varying according to some bell curve then the variation between cultures is the shift of the bell curve when one moves from one society to another. Most commonly the term 'culture' is used for tribes or ethnic groups, or for nations such as Indonesia, and for organizations such as UAJY. A relatively unexplored field is the culture of occupations—for instance, of engineers versus accountants, or of academics versus practitioners. The term can also be applied to the genders, to generations, to traits in Indonesia or to social classes. However, changing the level of aggregation studied changes the nature of the concept of 'culture'. Societal, national and gender cultures, which children acquire from their earliest youth onwards, are much deeper rooted in the human mind than occupational cultures acquired at school, or than organizational cultures acquired at work.

The latter is exchangeable when people take a new job. Societal cultures reside in values, in the sense of broad tendencies to prefer certain states of affairs over others (Hofstede, 2001). Organizational cultures reside rather in practices: the way people perceive what goes on in their organizational environment. In various perspectives, students of UAJY normally come from many backgrounds. They consist of almost as many characteristics, such as tribe, birth place, region, religion, and family background. In other words, cross micro-culture among the students is very important for management to transfer the value of the university. Also, internal communication can be more effective among students, and between students and management.

Based on the problems stated in the background, the research questions are as follows: 1) what are the national cultures of civil engineering students based on five dimensions from Hofstede? 2) is there any difference in national culture between ethnic groups in Indonesia? and 3) is there any difference in national culture between Asian countries?

Three objectives were set in this study: 1) to identify the overall national culture of civil engineering students based on five dimensions adopted from Hofstede; 2) to analyze the difference in national culture between ethnic groups; and 3) to discuss the difference in national culture between Asian countries.

Due to a limitation of resources, samples of the study are taken from civil engineering students of the Faculty of Engineering UAJY. National culture was limited to five dimensions based on Hofstede: power distance, individualism, uncertainty avoidance, masculinity, and long-term orientation. Ethnic groups are limited to being from Java, China, across Sumatra, Kalimantan, Sulawesi, Bali, Papua, and Nusa Tenggara.

2   LITERATURE REVIEW

2.1   *Cross-cultural study: Indonesia versus Australia*

Jones (2007) learned about the workers in Indonesia and Australia. Jones stated that Hofstede's work on culture was the most widely cited in existence (Bond, 2002; Hofstede, 1997). Hofstede's observations and analysis provided scholars and practitioners with a highly valuable insight into the dynamics of cross-cultural relationships. However, such a groundbreaking body of work did not escape criticism. Hofstede had been dogged by academics discrediting his work in part or whole. On the other side of this contentious argument were academics who support his work. Far more scholars belong to the pro-Hofstede team than those who do not; most quote Hofstede's work with unabashed confidence, many including his findings as absolute assumptions. Jones's paper took an in-depth look at Hofstede's work, discussed both sides of these arguments, and then recommended areas for further discussion and research. Finally, Jones provided the findings of the study that applied to a practical environment regarding two countries, Australia and Indonesia.

2.2   *Cross-cultural study: Taiwan versus USA*

Wu, from Western Illinois University (2006), studied the cross-culture between workers from Taiwan and the United States, adopting Hofstede's Culture dimensions. Wu mentioned that Hofstede's (1984, 2001) works on work-related culture dimensions had been regarded as a paradigm in the field of cross-cultural studies. Specifically, his country classification on five work-related cultural values (power distance, uncertainty avoidance, masculinity-femininity, individualism-collectivism, and Confucian work dynamics) had been frequently cited by researchers in the past few decades. While his work had been used effectively, his data was collected 30 years ago and has become outdated. By collecting data from one Eastern culture, Taiwan, and one Western culture, the United States, Wu did the study to update and re-examine Hofstede's (1984, 2001) culture dimensions in these two cultures. In addition, the study had extended Hofstede's work by investigating occupational culture in the higher education setting. The results of this study suggested that work-related cultural values in a specific culture were not static and could be changed over time. Wu also argued that 'When the political, societal, and economic environments change, people's cultural values also change. Thus, many cultural theories should be updated and re-evaluated periodically'.

2.3   *Cross-cultural study: China versus America*

Tan and Chow (2009) did a study regarding values and Chinese ethnic workers at PRC (People's Republic of China), Chinese workers working in the US, and American workers in the US. They did cross-cultural comparative studies of national values and ethics. Both of them argued that shared practices, values and ethical standards depend on shared beliefs had been studied. Many findings of such studies had been unable to reach a consensus on the impact of culture on ethic-related attitudes and behavior. Empirically, many 'cross-cultural' differences reported by previous studies might actually stem from cross-national differences. In order to partially fill this gap, Tan and Chow's study advocates an analytical framework that isolates the role of cultural and national differences in order to test their relationship to individual-level variables. Within this framework, we test competing hypotheses based on both cultural and national contexts by comparing groups of Chinese and American respondents, together with a 'bridging group' of Chinese-Americans. Theoretically, this contextual approach helps to resolve the debate on the role of culture, by showing that culture plays a far more important role in shaping value orientations than does the national background. Specifically, the two ethnic Chinese groups had many cultural values in common, and differed significantly from the Caucasian group.

2.4 Cross-cultural study: Five Asian countries

Choy et al. (2007) reported a cross-cultural study in Singapore Management Reviews. The study respondents were workers from Singapore, Malaysia, India, PRC, and Indonesia. They insisted that as a consequence of globalization, business corporations were engaging in increasingly complex and interdependent operations across national borders. There was greater inclusion of multinational staff in the workplace. A major consideration for managers was the wide range of staff behaviors, attitudes, and values, which might affect organizational processes. A quantitative study was conducted to investigate the variations in value priorities of staff from five national groups in a Singaporean MNC (Multinational Corporation). The questionnaire survey involved 240 participants. The findings affirm that the structural components of the respective countries (that is, the ideological superstructure), social structure, and material infrastructure, could have a significant influence on the value orientations of the staff from diverse national backgrounds.

3 METHOD OF STUDY

Respondents would be students in civil engineering from the Faculty of Engineering, University of Atma Jaya Yogyakarta, in Indonesia. Students from UAJY normally come from various regions of Indonesia: Sumatera, Kalimantan, Sulawesi, Maluku, Papua, Nusa Tenggara, Bali, and of course Java Island, which is the main region where the majority of the UAJY students come from.

The first instrument is regarding general information from respondents; this includes gender, trait, and original-region students staying during their childhood. The second instrument is the main questionnaire adopted from Hofstede regarding five culture dimensions.

Hofstede (2009) summarized the content of each dimension opposing cultures with low and high scores. These oppositions are based on correlations with studies by others, and because the relationship is statistical, not every line applies equally strongly to every country, including Indonesia. The following Tables are the instrument of culture dimension adopted from Hofstede and Hall (Hofstede, 2011).

Table 1 shows the elements of the power distance dimension. A respondent who selects the left-hand side, belongs to a small power distance society; otherwise she/he would belong to a large power distance society.

Table 1. Ten differences between small- and large-power distance societies (Hofstede, 2011).

Small power distance	Large power distance
Use of power should be legitimate and is subject to criteria of good and evil	Power is a basic fact of society antedating good or evil: its legitimacy is irrelevant
Parents treat children as equals	Parents teach children obedience
Older people are neither respected nor feared	Older people are both respected and feared
Student-centered education	Teacher-centered education
Hierarchy means inequality of roles, established for convenience	Hierarchy means existential inequality
Subordinates expect to be consulted	Subordinates expect to be told what to do
Pluralist governments based on majority vote and changed peacefully	Autocratic governments based on co-optation and changed by revolution
Corruption rare; scandals end political careers	Corruption frequent; scandals are covered up
Income distribution in society rather even	Income distribution in society very uneven
Religions stressing equality of believers	Religions with a hierarchy of priests

Table 2 shows the elements of the Uncertainty Avoidance. If the preference perception from the respondent tends to the left-hand side, it means that she/he has weak uncertainty avoidance.

On the other hand, the right-hand side shows the respondent as a member of a strong uncertainty avoidance society.

Table 3 shows the elements of the individualism dimension. The left-hand side means that the respondent belongs to an individualist society, while the right-hand side means that she/he belongs to a collectivist society.

Table 4 shows the elements of the quality of life dimension. The left-hand side means that the respondent belongs to a care society (femininity), while the right-hand side means that she/he belongs to a strong role society (masculinity).

Table 5 shows the elements of the orientation dimension. The left-hand side means that the respondent belongs to a short-term orientation society, while the right-hand side means that she/he belongs to a long-term orientation society.

Most collected data would be analyzed using descriptive analysis (such as the mean, standard deviation, and frequency) to describe the data obtained from the first part of the questionnaire. It aims to analyze the different of culture dimension between ethnic. One-way ANOVA (Analysis of Variance) would be used to test means of the value national culture that

Table 2.   Ten differences between weak- and strong-uncertainty avoidance societies (Hofstede, 2011).

Weak uncertainty avoidance	Strong uncertainty avoidance
The uncertainty inherent in life is accepted and each day is taken as it comes	The uncertainty inherent in life is felt as a continuous threat that must be fought
Ease, lower stress, self-control, low anxiety	Higher stress, emotionality, anxiety, neuroticism
Higher scores on subjective health and well-being	Lower scores on subjective health and well-being
Tolerance of deviant persons and ideas: what is different is curious	Intolerance of deviant persons and ideas: what is different is dangerous
Comfortable with ambiguity and chaos	Need for clarity and structure
Teachers may say 'I don't know'	Teachers supposed to have all the answers
Changing jobs no problem	Staying in jobs even if disliked
Dislike of rules – written or unwritten	Emotional need for rules – even if not obeyed
In politics, citizens feel and are seen as competent toward authorities	In politics, citizens feel and are seen as incompetent toward authorities
In religion, philosophy and science: relativism and empiricism	In religion, philosophy and science: belief in ultimate truths and grand theories

Table 3.   Ten differences between collectivist and individualist societies (Hofstede, 2011).

Individualism	Collectivism
Everyone is supposed to take care of him – or herself and his or her immediate family only	People are born into extended families or clans which protect each other in exchange for loyalty
'I' – consciousness	'We' – consciousness
Right of privacy	Stress on belonging
Speaking one's mind is healthy	Harmony should always be maintained
Others classified as individuals	Others classified as in-group or out-group
Personal opinion expected: one person one vote	Opinions and votes predetermined by in-group
Transgression of norms leads to guilt feelings	Transgression of norms leads to shame feelings
Languages in which the word 'I' is indispensable	Languages in which the word 'I' is avoided
Purpose of education is learning how to learn	Purpose of education is learning how to do
Task prevails over relationship	Relationship prevails over task

Table 4. Ten differences between feminine and masculine societies (Hofstede, 2011).

Femininity	Masculinity
Minimum emotional and social role differentiation between the genders	Maximum emotional and social role differentiation between the genders
Men and women should be modest and caring	Men should be and women may be assertive and ambitious
Balance between family and work	Work prevails over family
Sympathy for the weak	Admiration for the strong
Both fathers and mothers deal with facts and feelings	Fathers deal with facts, mothers with feelings
Both boys and girls may cry but neither should fight	Girls cry, boys do not; boys should fight back, girls should not fight
Mothers decide on number of children	Fathers decide on family size
Many women in elected political positions	Few women in elected political positions
Religion focuses on fellow human beings	Religion focuses on God or gods
Matter-of-fact attitudes about sexuality; sex is a way of relating	Moralistic attitudes about sexuality; sex is a way of performing

Table 5. Ten differences between short- and long-term-oriented societies (Hofstede, 2011).

Short-term orientation	Long-term orientation
Most important events in life occurred in the past or take place now	Most important events in life will occur in the future
Personal steadiness and stability: a good person is always the same	A good person adapts to the circumstances
There are universal guidelines about what is good and evil	What is good and evil depends upon the circumstances
Traditions are sacrosanct	Traditions are adaptable to changed circumstances
Family life guided by imperatives	Family life guided by shared tasks
Supposed to be proud of one's country	Trying to learn from other countries
Service to others is an important goal	Thrift and perseverance are important goals
Social spending and consumption	Large savings quote, funds available for investment
Students attribute success and failure to luck	Students attribute success to effort and failure to lack of effort
Slow or no economic growth of poor countries	Fast economic growth of countries up to a level of prosperity

more than or equal three groups. A multiple comparison technique would also be applied to find the statistical significance among the groups under investigation.

The following are the ten elements of masculinity versus femininity dimension of culture as shown in Table 4, and ten elements of long-term orientation dimension of culture in Table 5.

4 RESULTS AND DISCUSSION

The study obtained 76 samples of the civil engineering students from the university. The respondents consisted of 27 Javanese, 22 Chinese, and 27 of other ethnic origin. Figure 1 shows the scale of culture obtained from the study. The results showed that overall respondents had tended to have long-term orientation. So it is similar to Singapore, and other Asian countries like China, Hong Kong, and India.

The following are Hofstede and Hall's terminology for the five culture dimensions. The Power Distance Index (PDI) is defined as 'the extent to which the less powerful members of institutions and organizations within a country expect and accept that power is distributed unequally' (See Tamas, 2007). A High Power Distance ranking indicates that inequalities of

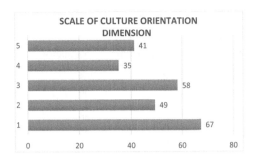

SCALE OF CULTURE ORIENTATION DIMENSION

Figure 1.   Five culture dimensions of the civil engineering students.
Notes: 1 = Long-Term Orientation; 2 = Masculinity; 3 = Individualism; 4 = Power Distance; and 5 = Uncertainty Avoidance.

Table 6.   Ethnic against dimension of cultures.

No.	Culture dimension	Javanese	Chinese	Others
1	Long-term oriented	70	66	68
2	Masculinity	48	51	47
3	Individualism	55	54	56
4	Power distance	36	37	36
5	Uncertainty avoidance	41	40	41

power and wealth exist within the society and that the less powerful members of the society accept this situation (Hofstede, 1994 p. 28). A Low Power Distance ranking indicates that the society de-emphasizes the differences between citizen's power and wealth. In these societies equality and opportunity for everyone is stressed.

Hofstede defines the dimension Individualism (IDV) as follows: 'Individualism pertains to societies in which the ties between individuals are loose: everyone is expected to look after himself or herself and his or her immediate family.' A High Individualism ranking indicates that individuality and individual rights are paramount within the society. A Low Individualism ranking typifies societies of a more collectivist nature with close ties among its members. In these societies '…people from birth onwards are integrated into strong, cohesive in-groups, which throughout people's lifetime continue to protect them in exchange for unquestioning loyalty.' (Hofstede, 1994 p. 51).

Masculinity (MAS) focuses on the degree to which 'masculine' values, like competitiveness and the acquisition of wealth, are valued over 'feminine' values, like relationship building and quality of life. A high masculinity ranking indicates that the society values are assertive and aggressive 'masculine' traits. Low masculinity ranking typifies societies in which nurturing and caring 'feminine' characteristics predominate.

Uncertainty Avoidance Index (UAI) focuses on the level of tolerance for uncertainty and ambiguity within the society. A High Uncertainty Avoidance ranking indicates that the country has a low tolerance for uncertainty and ambiguity. This creates a rule-oriented society that institutes laws, rules, regulations, and controls in order to reduce the amount of uncertainty. A Low Uncertainty Avoidance ranking indicates that the country has less concern about ambiguity and uncertainty and has more tolerance for a diversity of opinions. This is reflected in a society that is less rule-oriented, more readily accepts change, and takes more and greater risks.

Further analysis was then carried out of the multiple comparison using the Scheffé method. The result found was that there were no significant differences of the five national cultures between the Javanese, Chinese, and other ethnic groups. The Scheffé post-hoc analysis also

Table 7. A culture dimension comparison of 2007 survey from Asian countries.

Country	Power distance	Individualism	Uncertainty avoidance	Masculinity	Long-term orientation
Arab	80	38	68	53	
China	80	20	40	66	118
Hong Kong	68	25	29	57	96
India	77	48	40	56	61
Indonesia	78	14	48	46	
Iran	58	41	59	43	
Japan	54	46	92	95	80
Malaysia	104	26	36	50	
Pakistan	55	14	70	50	
Philippines	94	32	44	64	19
Singapore	74	20	8	48	48
South Korea	60	18	85	39	75
Taiwan	58	17	69	45	87
Thailand	64	20	64	34	56
Indonesia 2017 (this study)	35	58	41	49	67

Sources: Part of data adopted from Tamas (2007).

found that there were not significant differences between each culture dimension against the Javanese, Chinese and other ethnic groups.

Table 7 shows the comparison between the results of the current survey and that from the report by Tamas (2007). The current culture dimension obtained in the 2017study could also be compared with that which was reported by Tamas ten years earlier. The sample showed that: Indonesia moves from 1) large power distance (scale of 78) to small power distance (scale of 35); and 2) collectivity (scale of 14) to more individual (scale of 58) society.

Indonesia is considered mostly as low power distance. The scale of 35 implied that Indonesia now exercises as a democratic country. Powers were distributed to executive, legislative as well as judicative. On the other hand, Indonesia was no more a collectivist society. The scale 58 implied that the society tends to be more individual-oriented than before (the scale of 14 in 2007 survey).

Regarding uncertainty avoidance, the scale reduces a little bit from 46 to 41. This implied the Indonesia's preference for low uncertainty avoidance. With the scale of 49 of masculinity of culture dimension from this survey, Indonesia looked like neither a masculine nor feminine society. From the perspective of long-term orientation, Indonesia as an Asian country with a long term oriented society.

5 CONCLUSION AND RECOMMENDATION

This study has identified the national culture of Indonesia based on five dimensions from Hofstede and Hall. The scales of culture dimensions obtained from this study were 35, 58, 41, 49, and 67 for power distance, individualism, uncertainty avoidance, masculinity, and long-term orientation respectively.

The study found that there was no significant difference of the five national culture dimensions between the Javanese, Chinese, and other ethnic groups. The study also found that there were no significant differences between each culture dimension against the Javanese, Chinese and other ethnic groups. This implied that the national identity has been solid among the ethnic groups.

With the diversity of ethnics and subculture in the region, Indonesia is currently considered to have the lowest power distance among the Asian countries.

The study recommends to widely extend the coverage of the study, not only to students, but also to other communities from the society and business entity in Indonesia.

ACKNOWLEDGMENT

The author is grateful to University of Atma Jaya Yogyakarta for funding the study. Many thanks to the students who participated to the study. Thanks to the reviewers for their helpful comments.

REFERENCES

Choy, W.K.W., Adeline, B.E.L., and Ramburuth, P. (2007). Multinationalism in the workplace: A myriad of values in a Singaporean firm. *Singapore Management Review*, 31(1), 1.

Hofstede, G. & Bond, M.H. (1984). Hofstede's culture dimensions: An independent validation using Rokeach's value survey. *Journal of Cross-Cultural Psychology*, *15*(4), 417–433.

Hofstede, G. & Bond, M.H. (1988). The Confucius connection: From cultural roots to economic growth. *Organizational Dynamics*, *16*(4), 5–21.

Hofstede, G. (1980). *Culture's consequences: International differences in work-related values*. Beverly Hill, CA: Sage.

Hofstede, G. (1991a). *Culture and organizations*. New York, NY: McGraw-Hill.

Hofstede, G. (1991b). *Cultures and organizations: Software of the mind*. New York, NY: McGraw-Hill.

Hofstede, G. (1997). The Archimedes effect. In M.H. Bond (Ed.), *Working at the Interface of Cultures: 18 Lives in Social Science* (pp. 47–61). London: Routledge.

Hofstede, G. (1998). Attitudes, values and organizational culture: Disentangling the concepts. *Organization Studies*, *19*(3), 477.

Hofstede, G. (2001). *Culture's consequences: International differences in work-related values*. Thousand Oaks, CA: Sage.

Hofstede, G. (2011). Dimensionalizing cultures: The Hofstede model in context. *Online Readings in Psychology and Culture*, *2*(1), 8. doi:10.9707/2307–0919.1014.

Jones, M. (2007). Hofstede—Culturally questionable? *Oxford Business & Economics Conference. Oxford, UK, 24–26 June.*

Tamas, A. (2007). *Geert Hofstede's dimensions of culture and Edward T. Hall's time orientations.* Retrieved from www.tamas.com.

Tan, J. & Chow, I.H. (2009). Isolating cultural and national influence on value and ethics: A test of competing hypotheses. *Journal of Business Ethics*, *88*(1), 197–210. doi:10.1007/s10551-008-9822-0.

Wu, M. (2006). Hofstede's culture dimensions 30 years later: A study of Taiwan and the United States Western Illinois University. *Intercultural Communication Studies* XV:1.

The impact of cooperative blended learning on higher order thinking skills using a mobile application

D. Suliswaro & M. Mutammimah
Universitas Ahmad Dahlan, Indonesia

P. Parwiti
Madrasah Aliyah Negeri Wonokromo, Indonesia

ABSTRACT: Currently, many schools in Indonesia and also in other countries in the world are encouraging the use of mobile technology in learning for a number of reasons. However, there is limited research examining the appropriate learning strategies for utilizing mobile applications to improve learning performance for today's learning competencies, especially character building. This study aims to determine the impact of cooperative blended learning using a mobile application to increase higher order thinking skills as part of strengthening the value of education in school. The feature of the mobile app supported the cooperative learning activity. The research design was pre-test and post-test one sample design. Data were analyzed quantitatively by using the paired t-test. The results show that there is a significant improvement in Higher Order Thinking Skills (HOTS) after learning.

1 INTRODUCTION

Currently, many schools in Indonesia and also in other countries in the world are encouraging the use of mobile technology in learning for a number of reasons (Hsu et al., 2014; Suliswaro & Toifur, 2016). However, there is limited research examining the relevant learning strategies for utilizing mobile applications to improve learning performance for today's learning competencies (Suliswaro & Toifur, 2016). Educators realize that learning media is important for learning. The most important thing is how to build a productive interaction among students, students and teachers, and students and the broader environment to achieve the expected competencies (Bird & Stubbs, 2015; Suliswaro et al., 2016a; Khaddage et al., 2015). This productive interaction will support the value internalization inserted by the teacher during learning activities.

Various researchers have shown that there is an opportunity for utilizing mobile technology in learning (Ally et al., 2014; Ally & Prieto-Blázquez, 2014). It is not only easy to access, but various learning strategies using mobile technology can also impact on higher order thinking skills in multiple subjects across multiple education levels (Lai & Hwang, 2014). Several factors become a study in learning success also have been studied from the teacher perspective (Mouza & Barrett-Greenly, 2015; Khaddage et al., 2015; Furió et al., 2015) and even from the student side (Looi et al., 2015; Lai & Hwang, 2014; Melero et al., 2015).

From previous research conducted in Indonesia examining mobile technology utilization in high schools, some problems were found such as the low support from school policy to implement mobile learning (Suliswaro & Toifur, 2016). With the implementation of the national curriculum of 2013 referring to twenty-first century learning, education in Indonesia needs to improve the learning system to achieve the current competencies immediately. The higher order thinking skills is very important in the Curriculum 2013. In this curriculum, the learning result is measured by the Higher Order Thinking Skills (HOTS) improvement. It means that HOTS is the important learning objective at high schools in Indonesia.

From the study of mobile technology utilization in schools in Indonesia, it was shown that teachers have not yet used learning media and application effectively to improve student motivation and learning results (Sulisworo et al., 2016a). It has been shown that students with their Information and Communications Technology (ICT) literacy tend to be able to optimally use various mobile apps in learning (Lee et al., 2016; Ciampa, 2014; Mouza & Barrett-Greenly, 2015). Although there has been a shift in learning management that utilizes mobile technology, there are still many teachers who apply learning in the classroom (Keengwe & Bhargava, 2014). This situation becomes the fundamental reason to develop blended learning by using mobile application with specific learning strategy (Furió et al., 2015; Sulisworo et al., 2016a).

This research is a follow-up of a previous study that showed how the impact of cooperative blended learning using mobile application to the student high thinking skill considering previous knowledge and motivation. The result of this research is the justification for the possibility of using mobile applications in blended learning to improve HOTS (Sulisworo, 2017). The results of this study will provide a positive contribution to the possibility of using mobile learning to support character education embedded in learning interaction.

2 THEORETICAL FRAMEWORK

2.1 *Cooperative blended learning*

The use of mobile devices and wireless networks has widely expanded the chances for the existence of a broader learning community (Bernard et al., 2014; Güzer & Caner, 2014). This technology utilization to improve learning quality creates various new learning strategies (Lee et al., 2016; Sulisworo et al., 2016a, 2016b). Cooperative learning is expected to be able to develop multiple social skills which were previously less emphasized in learning at school. When learning started to adopt mobile technology, researchers examined how to integrate cooperative learning by utilizing technology (Zhou & Chua, 2016). Various aspects of learning success with this strategy were also carried out to ensure the vast benefits in cooperative learning using mobile technology (Lee et al., 2016; Sulisworo et al., 2016b).

From another aspect, it has not been possible to conduct several learning activities fully through mobile learning. Some interactions between teacher and students or among students still need to be undertaken in class. However, this teaching strategy has positive opportunities from mobile learning. These benefits are why blended learning is a strategy that can accommodate some learning interests (Bernard et al., 2014; Zhou & Chua, 2016). With those rationalities, studies on cooperative blended learning have become interesting. The expectation of cooperative blended learning implementation is how to get various benefits of cooperative learning utilizing mobile technology while giving the opportunity for interactions in the class and the cyber world.

2.2 *Integrating learning strategy on mobile applications*

Mobile applications or mobile apps have been utilized in various fields including education (Brown & Mbati, 2015; Foldnes, 2016; Bidarra & Rusman, 2017; Francese et al., 2015). Mobile apps in school can be utilized both for individual or group learning where students can interact virtually (Henry, 2014; Su & Cheng, 2015; Wang, 2016; Choi et al., 2014). In education, mobile apps function as a medium to transfer learning information to students. The effectiveness of media in the interaction process between knowledge information and students will be determined by the strategy of the delivery process (Gan et al., 2015; Harris & Greer, 2017; Bidarra & Rusman, 2017). Mobile apps are used as a learning medium to create more exciting learning so that student motivation will increase. The broader and easier access to the information source makes it possible to carry out the activity anywhere according to the learning needs.

Researchers have studied the effectiveness of mobile app utilization in learning on various aspects. Mobile apps can be designed for a one-way directive learning strategy which means students only interact with the application and the approach to understanding knowledge depends on the students while offline (Choi et al., 2014; Kong, 2016). Furthermore, teachers can also design mobile apps by installing a learning strategy in the application. With this approach, students can directly understand the knowledge that provided in the app In blended learning, mobile apps installed with a learning strategy will build teacher–student and student–student interactions (Choi et al., 2014; Foldnes, 2016; Francese et al., 2015). In previous research, researchers have developed mobile apps with a cooperative learning strategy that aimed to improve HOTS.

2.3 Strategy to improve HOTS in learning

Learning that supports HOTS needs clear communication to decrease ambiguity and confusion, and to improve the student's attitude related to the thinking duty (Jensen et al., 2014; Dwyer et al., 2014). This kind of lesson plan needs to include thinking skill modeling, applied thinking samples and adaptation for various student needs (Hwang et al., 2014; Tiruneh et al., 2014; Ashford-Rowe et al., 2014). Scaffolding, by giving students support at the beginning of learning and step by step motivation for students to work independently, will help students to develop their higher learning skills (Ashford-Rowe et al., 2014; Abrami et al., 2015; Hesse et al, 2015).

Several central concepts which are relevant to the process of high order thinking refer to assumptions about thinking and learning. The thinking level has a close correlation to the learning level. The thinking level and learning level are mutually dependent on each other. Studying the thinking order without studying specific material is impossible and only theoretical (Tiruneh et al., 2014; Ashford-Rowe et al., 2014; Abrami et al., 2015; Hesse et al., 2015). In real life, students will learn the excellent content in the community and school experience, no matter what the academic experts conclude. The concept and vocabulary they previously learn will help them to learn the HOTS and new content in the future. Higher order thinking involves various thinking processes implemented in complicated situations and has many variables (Garrison & Akyol, 2015; Goldhammer et al., 2014).

Some learning strategies used in improving HOTS include practice, elaboration, organization and metacognition. Learning needs to be specially designed to teach specific learning strategies. Small group activities such as student discussion, peer-teaching and cooperative learning can be useful in improving thinking skills. The actions need to provide a challenging task, motivation from teachers to keep working and sustainable feedback on group improvement (Goldhammer et al., 2014; Ashford-Rowe et al., 2014; Tiruneh et al., 2014). Communication and instruction mediated by computers can give access to data sources from various places and make collaboration with students in different locations possible. This environment can be efficient in developing skills in fields such as verbal analogy, logical thinking and inductive/deductive reasoning. Therefore, learning is not only the acquisition of knowledge but also the improvement of character as part of being a global citizen.

3 METHODS

3.1 Research design

This is a quantitative research approach with a pretest and post-test of one sample group design, where students experience cooperative blended learning with mobile apps containing physics learning media on heat and temperature materials. The independent variable of the research is the learning strategy, while the dependent one is higher order thinking skill. The sampling technique used was cluster random sampling of grade 11 students with samples of 35 students from an Indonesian school.

The data analysis used some quantitative approaches. Analysis with the paired t-test was used to determine if there was a significant difference between HOTS before and after learning. Additionally, descriptive data analysis was also conducted to identify if there was an average HOTS difference and standard deviation difference from the result of before and after learning.

3.2 *Learning activities*

The learning activity implemented was cooperative learning with the Preview, Question, Read, Reflection, Recite and Review (PR4R) model. The PQ4R learning model is part of innovative learning models oriented to constructivism theory. Constructivism learning theory allows students the opportunity to find and transform complex information, to discover new details with old rules and to revise these regulations when they are no longer relevant. The PQ4R learning steps are as follows:

- Preview: giving students the opportunity to scan the text to find the main idea of learning.
- Question: asking students to create their questions as the result of quick reading.
- Read: giving students the opportunity to read actively and in depth, while providing feedback.
- Reflect: giving students the problems and solving the problems with the information provided by the teacher and the knowledge from reading.
- Recite: asking students to look back in the records or extracts made previously from reading.
- Review: reading back the learning sources and strengthening their understanding of the passage.

Considering the degree of suitability of this application in mobile learning, the researcher did not conduct the preview and question steps, because the students cannot be restricted doing the activity on a mobile device, and they tend to be more flexible in using the learning resources according to their level of needs. The action in the first application is to provide students with reading materials to study. Therefore, this strategy becomes 4R, i.e. Read, Reflect, Recite and Review. Consequently, there is a difference between this learning model compared to the PQ4R. This learning strategy has been embedded on previously developed mobile apps.

3.3 *HOTS evaluation instruments*

The materials used as media to develop HOTS are temperature and heat. These materials can be used for cognitive domain levels from Knowledge to Evaluation. With consideration that research subjects are only expected to master the content at the C4 level, the instruments used are made by adapting the decided competency standard. Therefore, in this learning, HOTS is procedural knowledge mastery measured by a test instrument that emphasizes knowing the student's ability in the C4 cognitive domain level (analysis). The review is the fourth category or level on Bloom's taxonomy about the cognitive domain. The review is the ability to explain material in its constituent parts. Analytical ability includes element analysis (identifying portions of the content), relational analysis (identifying relations) and organizational analysis of principle, concept and information. Questions for a lower level were also given to support the test completeness of the instrument. The instrument provided was a multiple-choice test with four alternative options. The composition of each item based on the cognitive domain is as follows: C1 or knowledge (four items), C2 or comprehension (three items), C3 or application (six items) and C4 or analysis (seven items).

4 RESULT AND DISCUSSION

The result of the research was obtained by using statistical analysis on the data of HOTS level before and after learning. The analytical technique used is the descriptive statistic (Table 1) and paired *t*-test (Table 2). From the statistical analysis, some data become the basis for the explanation in the discussion.

Table 1. Descriptive statistics.

	N	Min.	Max.	Mean	Std. deviation
HOTs leve/l before	35	10.00	65.00	51.1429	12.7236
HOTS level after	35	55.00	85.00	69.2857	9.55963
Valid N (list-wise)	35				

Table 2. Paired samples test: HOTS level before and after.

Paired differences

Mean	Std. deviation	Std. error mean	95% Confidence interval of the difference		t	f	Sig. (2-tailed)
			Lower	Upper			
−18.14286	14.65715	2.47751	−23.17776	−13.10795	−7.323	4	0.000

From the Table 1, there is an increase in HOTS score from an average of 51.14 before learning to 69.29 after learning. Additionally, the standard deviation reduces. This shows that the HOTS level variation is low or student HOTS level tends to spread more evenly.

From Table 2, the HOTS levels before and after learning are significantly different. This means that the independent variable (learning strategy) has a substantial influence on HOTS level increase.

A lot of research shows that there is an improvement in learning achievement when utilizing mobile learning media. Students in the group of the twenty-first century generation have excellent ICT literacy. ICT has become a standard thing in daily student life. However, schools have not efficiently used ICT in learning in Indonesia (Sulisworo & Toifur, 2016). The utilization of ICT, which is still low in school learning, makes students feel isolated from the digital life they have outside of school. This research, which tries to utilize mobile media in learning, is an exciting thing for students. Learning activities make students more comfortable. Many analyses have shown the opportunities and benefits of ICT in learning, especially mobile technology (Mouza & Barrett-Greenly, 2015; Melero et al., 2015; Lai & Hwang, 2014). This is similar to previous research that showed a high level of learning acceptance and comfort of students when their learning utilized mobile devices (Zhou & Chua, 2016; Wang, 2016; Sulisworo et al., 2016b).

There are several reasons when students using the mobile apps, so that their learning outcomes better than usual. First, students can accessing learning activities everytime and everywhere easily. Second, there are a lot of learning sources that students use to acquisition process of knowledge and skills. Third, students virtual interaction makes them more comfortable and confident. Therefore, digital native characteristic of students suitable with learning strategy which implemented by utilizing mobile apps. So that the media features can be use when teacher perform cooperative learning.

Some statistical indicators which have been processed (kurtosis difference, the standard deviation that becomes smaller, paired correlation and paired t-test) can be references to trace knowledge sharing processes in this learning. In learning in this research, the terminology of cooperative learning and collaborative learning tend to have the same meaning. In learning observation, there are characteristics of collaborative and cooperative which appear. The collaborative or cooperative activities will have an effect on learning performance (Hwang et al., 2014; Hesse et al., 2015; Gan et al., 2015; Foldnes, 2016). In terminology, collaborative needs feedback participation of all participants and coordinated work to solve problems, while cooperative obliges individuals to be responsible for a specific part and then coordinate each section with others (Sulisworo et al., 2016a; Lee et al., 2016). In this research, group learning occurs when students need a skills or knowledge in the specific material through the mobile application.

Intensive participation also occurs in learning through intellectual interaction among students or between teachers and students. This participation increases the meaningful learning for students compared to when students learn individually (Dwyer et al., 2014; Choi et al.,

2014; Hesse et al., 2015). With cooperative learning activities, students can learn more skills and materials by mutual interaction with each other. This action develops a learning environment that creates knowledge sharing among them. Students can better obtain information through discussion and have a positive attitude toward learning with cooperation. Therefore, cooperative learning with mobile learning applications supports students to work in small learning groups to have better performance compared to previous experience when students worked individually in direct lesson learning.

Another explanation for the result of HOTS tests is the impact of prior ability of grouping. It seems likely that students with low ability like to work in the mixed group, average students might be preferred homogeneous group, while high ability students might be able to work in any group. Additionally, a knowledge sharing process occurred among students through learning interaction (Lai & Hwang, 2014). In general, this mobile learning strategy also gives the possibility to insert and embed other values to support an improvement in student characteristics. Learning is not only knowledge acquisition but also internalizing the positive values as part of character building.

REFERENCES

Abrami, P.C., Bernard, R.M., Borokhovski, E., Waddington, D.I., Wade, C.A., & Persson, T. (2015). Strategies for teaching students to think critically: A meta-analysis. *Review of Educational Research, 85*(2), 275–314.

Ally, M., Grimus, M., & Ebner, M. (2014). Preparing teachers for a mobile world, to improve access to education. *Prospects, 44*(1), 43–59.

Ally, M. & Prieto-Blázquez, J. (2014). What is the future of mobile learning in education? *International Journal of Educational Technology in Higher Education, 11*(1), 142–151.

Ashford-Rowe, K., Herrington, J., & Brown, C. (2014). Establishing the critical elements that determine authentic assessment. *Assessment & Evaluation in Higher Education, 39*(2), 205–222.

Bernard, R.M., Borokhovski, E., Schmid, R.F., Tamim, R.M., & Abrami, P.C. (2014). A meta-analysis of blended learning and technology use in higher education: From the general to the applied. *Journal of Computing in Higher Education, 26*(1), 87–122.

Bidarra, J. & Rusman, E. (2017). Towards a pedagogical model for science education: Bridging educational contexts through a blended learning approach. *Open Learning: The Journal of Open, Distance and E-Learning, 32*(1), 6–20.

Bird, P. & Stubbs, M. (2015). It's not just the pedagogy: Challenges in scaling mobile learning applications into institution-wide learning technologies. Paper presented at the International Conference on Mobile Learning 2015 (11th, Madeira, Portugal, 14–16 March, 2015).

Brown, T.H. & Mbati, L.S. (2015). Mobile learning: Moving past the myths and embracing the opportunities. *The International Review of Research in Open and Distributed Learning, 16*(2), 115–135.

Choi, S.H., So, H.S., Choi, J.Y., Yoo, S.H., Yun, S.Y., Kim, M.H., & Song, M.O. (2014). Comparison of blended practicum combined e-learning between cooperative and individual learning on learning outcomes. *Journal of Korean Academic Society of Nursing Education, 20*(2), 341–349.

Ciampa, K. (2014). Learning in a mobile age: An investigation of student motivation. *Journal of Computer Assisted Learning, 30*(1), 82–96.

Dwyer, C., Hogan, M., & Stewart, I. (2014). An integrated critical thinking framework for the 21st century. *Thinking Skills and Creativity, 12*(June), 43–52.

Foldnes, N. (2016). The flipped classroom and cooperative learning: Evidence from a randomised experiment. *Active Learning in Higher Education, 17*(1), 39–49.

Francese, R., Gravino, C., Risi, M., Scanniello, G., & Tortora, G. (2015). Using Project-Based-Learning in a mobile application development course—An experience report. *Journal of Visual Languages & Computing, 31*(December), 196–205.

Furió, D., Juan, M.C., Segui, I., & Vivo, R. (2015). Mobile learning vs traditional classroom lessons: A comparative study. *Journal of Computer Assisted Learning, 31*(3), 189–201.

Gan, B., Menkhoff, T., & Smith, R.R. (2015). Enhancing students' learning process through interactive digital media: New opportunities for collaborative learning. *Computers in Human Behavior, 51* (Part B), 652–663.

Garrison, D.R. & Akyol, Z. (2015). Toward the development of a metacognition construct for communities of inquiry. *The Internet and Higher Education, 24*(April), 66–71.

Goldhammer, F., Naumann, J., Stelter, A., Toth, K., Rolke, H., & Klieme, E. (2014). The time on task effect in reading and problem solving is moderated by task difficulty and skill: Insights from a computer-based large-scale assessment. *Journal of Educational Psychology, 106*(3), 608.

Güzer, B. & Caner, H. (2014). The past, present and future of blended learning: An in depth analysis of literature. *Procedia-Social and Behavioral Sciences, 116*(February), 4596–4603.

Harris, H.S. & Greer, M. (2017). Over, under, or through: Design strategies to supplement the LMS and enhance interaction in online writing courses. *Communication Design Quarterly Review, 4*(4), 46–54.

Henry, P.D. (2014). *Using Mobile Apps and Social Media for Online Learner-Generated Content.* Paper presented at the International Conference on Mobile Learning 2014 (10th, Madrid, Spain, Feb 28-Mar 2, 2014).

Hesse, F., Care, E., Buder, J., Sassenberg, K., & Griffin, P. (2015). A framework for teachable collaborative problem solving skills. In *Assessment and teaching of 21st century skills,* pp. 37–56. Dordrecht, Netherlands: Springer.

Hsu, Y.C., Ching, Y.H., & Snelson, C. (2014). Research priorities in mobile learning: An international Delphi study. *Canadian Journal of Learning and Technology, 40*(2), 1–22.

Hwang, G.J., Hung, C.M., & Chen, N.S. (2014). Improving learning achievements, motivations and problem-solving skills through a peer assessment-based game development approach. *Educational Technology Research and Development, 62*(2), 129–145.

Jensen, J.L., McDaniel, M.A., Woodard, S.M., & Kummer, T.A. (2014). Teaching to the test or testing to teach: Exams requiring higher order thinking skills encourage greater conceptual understanding. *Educational Psychology Review, 26*(2), 307–329.

Keengwe, J. & Bhargava, M. (2014). Mobile learning and integration of mobile technologies in education. *Education and Information Technologies, 19*(4), 737–746.

Khaddage, F., Cristensen, R., Lai, W., Knezek, G., Norris, C., & Soloway, E. (2015). A model-driven framework to address challenges in a mobile learning environment. *Education and Information Technologies, 20*(4), 625–640.

Kong, S.C. (2016). A framework of curriculum design for computational thinking development in K-12 education. *Journal of Computers in Education, 3*(4), 377–394.

Lai, C.L. & Hwang, G.J. (2014). Effects of mobile learning time on students' conception of collaboration, communication, complex problem–solving, meta–cognitive awareness and creativity. *International Journal of Mobile Learning and Organisation, 8*(3–4), 276–291.

Lee, H., Persons, D., & Ryu, H. (2016). Cooperation begins: Encouraging critical thinking skills through cooperative reciprocity using a mobile learning game. *Computers & Education, 97*(June), 97–115.

Looi, C.K., Sun, D., & Xie, W. (2015). Exploring students' progression in an inquiry science curriculum enabled by mobile learning. *IEEE Transactions on Learning Technologies, 8*(1), 43–54.

Melero, J., Hernandez-Leo, D., & Manatunga, K. (2015). Group-based mobile learning: Do group size and sharing mobile devices matter? *Computers in Human Behavior, 44*(Part C), 377–385.

Mouza, C. & Barrett-Greenly, T. (2015). Bridging the app gap: An examination of a professional development initiative on mobile learning in urban schools. *Computers & Education, 88*(October), 1–14.

Su, C.H. & Cheng, C.H. (2015). A mobile gamification learning system for improving the learning motivation and achievements. *Journal of Computer Assisted Learning, 31*(3), 268–286.

Suliswaro, D. (2017). Mobile learning application development fostering high order thinking skills on physics learning. *Paper at International Conference on Education, Business and Management (ICEBM-2017),* 102–107. Bali (Indonesia).

Suliswaro, D., Agustin, S.P., & Sudarmiyati, E. (2016a). Cooperative-blended learning using Moodle as an open source learning platform. *International Journal of Technology Enhanced Learning, 8*(2), 187–198.

Suliswaro, D., Ishafit, I., & Firdausy, K. (2016b). The development of mobile learning application using jigsaw technique. *International Journal of Interactive Mobile Technologies, 10*(3), 11–16.

Suliswaro D. & Toifur, M. (2016). The role of mobile learning on the learning environment shifting at high school in Indonesia. *International Journal of Mobile Learning and Organisation, 10*(3), 159–170.

Tiruneh, D.T., Verburgh, A., & Elen, J. (2014). Effectiveness of critical thinking instruction in higher education: A systematic review of intervention studies. *Higher Education Studies, 4*(1), 1–17.

Wang, Y.H. (2016). Could a mobile-assisted learning system support flipped classrooms for classical Chinese learning? *Journal of Computer Assisted Learning, 32*(5), 391–415.

Zhou, M. & Chua, B.L. (2016). Using blended learning design to enhance learning experience in teacher Education. *International Journal on E-Learning, 15*(1), 121–140.

Tembang macapat lyrics-based character education learning materials for secondary school students

D.B.P. Setiyadi & P. Haryono
UniversitasWidya Dharma Klaten, Indonesia

ABSTRACT: The study aims at developing learning materials containing character education integrated in teaching and learning *Bahasa Jawa* (Javanese language) at secondary schools in Central Java. The model is integrated in the teaching-learning process, curriculum, syllabus, lesson plan, and learning materials. The researchers apply a descriptive qualitative method. The objects cover *tembang macapat* (Javanese song) lyrics containing five-character values. Data are collected using library and recording techniques. Data validity is checked using triangulation. Data are analyzed by texts analysis. The result shows that *tembang macapat* texts contain character education and can be used as learning materials integrated in teaching and learning of Javanese Language given to secondary school students in Central Java. This can be implemented by singing *tembang macapat* at the beginning or the end of every lesson. Then, teacher explains the character education beyond the song. Besides, it can be inserted in parts of curriculum, syllabus, lesson plan, and learning materials.

1   INTRODUCTION

Character education has been launched as a national movement since 2010. However, the program has not been strongly enough echoed. It is this condition that makes the government enforce the character education through the National Program of Character Education Reinforcement (CER) instigated by the Ministry of Education and Culture of the Republic of Indonesia starting in 2016. The launch of the CER is the follow-up of the instruction of the President of Republic of Indonesia, Joko Widodo, to the Minister of Education and Culture. The CER is the realization of one of the *Nawacita* items (the current Indonesian president's vision), through the Mental Revolution National Movement. At the beginning of the acknowledgement of this book the Minister of Education and Culture said that a great nation is the one possessing strong characters complemented by high competences growing and developing from pleasing education and environment that apply good and noble values in all aspects of life of the nation and the state (*Kemendikbud*, 2017a & 2017b).

National character building is a fundamental need in the nationhood and statehood process. Since the Indonesian people have declared themselves as a free nation, at least there are three main obstacles encountered, namely the establishment of a unified and sovereignty state, the building of the state and of the character. The inauguration as a unified and sovereignty state raises a lot of obstacles, since Indonesia is a multicultural nation-state. Recently, there have been some signs from a group of people who want to break out the unity of the Unitary State of the Republic of Indonesia. It is caused by the weak foundation of character education underlying Indonesian young generations. Therefore, it is necessary to implant and reinforce character education to the young generation through the field of education. Character education is a planned effort to make the students know, concern and internalize values so that they behave as perfect human beings (*Kemendiknas*, 2011a & 2011b).

Character education should be made as early as possible so that students are used to behave positively and to show noble character. The character should be implanted through various ways namely language and art, history texts, song texts and so on (Setiyadi, 2012,

2012a, 2013, 2014, 2015; Setiyadi & Haryono, 2016; Setiyadi & Wiyono, 2017). If this character building is done in the field of planned and programmed education, Indonesian people will become a nation that is in line with the aim of the programmed national education (Cornelius & Greg 2013; Anwar 2016).

To implement the basic framework of the national character development above, it is important to integrate the character education through learning Javanese language, especially in the areas of Central Java, Special Region of Yogyakarta, and East Java, Indonesia, which conduct local content of Javanese language subject. The learning implementation of character education is made through the integration in the learning process, curriculum, syllabus, lesson plans and learning materials. Learning materials proposed in this discussion are those focused on the *tembang macapat* lyrics containing noble character education values. It is a pity that in the 2013 curriculum, once the Javanese Language subject was omitted, but because of persistent struggles made by the experts and observers of Javanese language, the Javanese language subject still exists.

Therefore, it is necessary to think about the learning model, curriculum, syllabus, and lesson plan related to the reinforcement of the character education so that they will be able to integrate and personalize the values of character education in the students' daily behaviors. From such integration it is expected that students' competence in Javanese language will be developing optimally besides their character and behaviors (Agboola & Tsai 2012; Banicki 2017). There are five main interrelated characters that form value network that should be developed as the priority in the CER movement as follows (1) religious, (2) nationalist, (3) autonomous, (4) *gotong royong* (mutual voluntary cooperation), and (5) integrity. In the movement, the five main values are not the ones which are independent, but interacted one another, which are dynamically developing and forming a personal unity.

From preliminary studies, it is known that the existence of the learning materials of Javanese language containing character education is still lacking. Based on the interviews with some teachers, at present, junior high school students show the following characters: weak motivation, less autonomous in life, easily hopeless in facing problems, less respect to teachers, and fading manners and ethics at school. Some interviewed teachers agreed if the materials of character education should be integrated in the Javanese language learning with materials of *tembang macapat* texts. Therefore, this present research is intended to contribute to solution of problems encountered by Indonesian People, national character education. This research was aimed at answering the following questions: (1) Which *tembang macapat* texts contains character education values that may be used as learning materials of character education integrated in learning Javanese language in secondary school? (2) What is the model of integration in learning, curriculum, syllabus, lesson plan, and learning materials like?

Character education developed in the field of education serves as one of the ways in forming one's character early. Experts in education in general agree with the importance of improving character education through formal education. However, differences in opinions happen among them especially those dealing the approach and mode of education. In terms of approach, some experts suggest the use of moral education approach that has been developed in western countries such as moral cognitive, value analysis, and value classification approaches. Some others suggest the use of traditional approach namely through implanting certain social values in students (Althof & Berkowitz 2006; Aqib 2012; Aqboola & Tsai, 2013; Ashraf et al 2013; Anwar 2016; Banicki 2017).

Based on the grand design developed by *Kemendikbud* (2013), psychologically and socio-culturally, character formation in individuals is the function of all individual potentials of human beings in the contexts of socio-cultural interaction that happens all the time. In 2016, in Indonesia, a national movement of Character Education Reinforcement was launched to strengthen the application *of* previous character education which is necessary to be improved. This movement is then followed up by integrating character education into learning activities. There are some types of texts, one of which is literary texts that covers prose, poetry and drama. *Tembang macapat* is a literary text included into the poetry type (Setiyadi 2012; *Kemendikbud*, 2013; Wodak 2011; Collin 2012; Saj & Sarraf 2013; Pujianto et al 2014; Pino 2016; Merino & Fina 2017).

Tembang macapat is an original poetry of Javanese people. The classification of *tembang macapat* texts as a Javanese poetry text is related to the existence of characteristics of the number of lines in stanza, interlude, or juncture in the musical measure, the number of syllables in a line and the final rhyme for each line. *Tembang macapat* covers *Pangkur, Maskumambang, Sinom, Asmaradana, Dandanggula, Durma, Mijil, Kinanti, Gambuh, Megatruh,* and *Pocung* (Setiyadi 2012).

Macapat was developed by the Walis (the revered saints of Islam in Indonesia, especially on the island of Java). This genre possesses a social function as a proper medium for the activities during their time. It is still very relevant and proper to be used as a medium for implanting character education, especially into the students. It is because the materials are still taught in elementary and secondary schools, especially in Central Java, Special Region of Yogyakarta, and East Java. In the Indonesian 2013 curriculum, the materials of *tembang macapat* are still listed as teaching materials. There is an expert saying that literary works, either classic or modern, possess great potentials in character education, even they serve as the core of character education (Widyahening & Wardhani 2016; Mardikarini & Suwarjo 2016).

2 METHOD

In this present research, a descriptive qualitative method was adopted. The data source was *tembang macapat* texts from *Serat Wulangreh* written by *PakuBuwana IV, Tripama* and *Wedhatama* written by *MangkunagaraIV*, and *tembang macapat* taken from internet. The techniques adopted in collecting data were observations, interviews, documents, audio-visual materials (Creswell 2014). Documents and audio-visual materials technique was used to collect data in the forms of *tembang macapat* texts that have been composed into audiovisual form such as MP3, You Tube, or website such as *ki demang*.com. Interviews technique was employed to obtain information related to *tembang macapat* and anything dealing with the implementation of learning Javanese language. Interviews were made with informants knowing much about *tembang macapat*, and teachers. The data were validated using triangulation, namely data, researchers and methodology triangulation techniques (Flick 2009; Patton 2015). Data from library and audiovisual instruments and website were collected using the library, recording, notes, and interviews techniques. The researchers compared some research results on texts with the theme of character education. Data reduction was made to choose relevant data. Then the data which are in the form of texts were analyzed using the texts analysis technique.

3 RESULTS AND DISCUSSION

3.1 *Tembang texts used as learning materials*

On the basis of the collected data, *tembang macapat* texts used as the teaching materials are those containing values of character education. The values may be integrated in the learning of Javanese language in secondary schools since in the 2013 curriculum it is possible to make such an integration. In the 2013 curriculum for Junior High School, there is a basic competence of "studying texts *Serat Wulangrehpupuh Kinanti*", while for Senior High School, there is a basic competence of "describing the content of *Serat Wedhatamapupuh Pangkur*". Therefore, it is possible to integrate the character education contained in the *pupuh*s to the students. The descriptions of examples of *pupuh*s that may be used as teaching materials are given below.

1. Texts containing the main values of religious character
Both the texts of *Wulangrehpupuh Dhandhanggula* (*Jroning Kurannggoningrāsājati,* 'In the Holy Quran a true feel rests", and *Wedhatama* contains religious education values. Religious value is a value showing attitudes of faith to the One Supreme God by implementing the teachings of the religion/faith believed in, respecting differences in religions, showing tolerance, living harmoniously and peacefully with other believers. The religious value is also

reflected in the relation between human beings and their Creator, other human beings and the universe. Attitudes of loving peace, anti bullying and violence, friendship, sincerity, no coercion of will, cooperation among believers, strong determination, self-confidence, loving the environment, and protecting the weak are noble characters that should be grown. These attitudes are very important for maintaining the peaceful life of Indonesian citizens since the Unitary State of the Republic of Indonesia is a multicultural state. Knowledge of anything dealing religion should be improved so that all members of this country possess the attitude of tolerance since the citizens consist of various tribes, races, groups with various religions

2. Texts containing the main character of nationalist attitude.

Texts of *Tripamapupuh Dhandhanggula* (*Wontên malih tuladhan prayogi,* 'a more good exemplary exists, and *satriyā gung nigari Ngalêngkā,*'a great knight from *Ngalengka* country') contain the main value of nationalist. The value of nationalist covers how one or citizen thinks, behaves and does to his/her country. This attitude is the one showing loyalty, concern, and high respect to language. Indonesian national language is called *Bahasa Indonesia*. Then, physical, social, cultural, and political environments of the nation should also be treated equally. Citizens with nationalist attitude are the one placing the nation and state interests above his/her interests and his/her group's interests. This character also covers some appreciations to the culture of the nation and to the diversity of cultures, tribes, and religions existing in Indonesia by keeping and maintaining the richness of the cultures, instead of destroying them. Attitudes of willingness to sacrifice, loving the homeland, obeying the law for the betterment of the nation and state are also demanded. Any attitude of discipline in all our actions will make this nation to have high achievement and to show superiority.

3. Texts containing the main value of autonomy

In the texts of *Wulangrehpupuh Mijil* (*Pomā kaki pādhādipuneling,* 'my children, remember', *ingpituturingong'* in my teaching') the value of autonomy is contained. Attitude of autonomy is an attitude and behavior that is not dependent upon others. In this *pupuh,* it is portrayed as an attitude shown by a knight possessing a polite attitude in all his actions. A knight is also a symbol of someone that may be dependent upon and that may solve any problems in life by working hard using all his bravery, ability, struggle, professionalism, and creativity. In all his activities he always improves his knowledge so that he will always update his competences to solve any new problems appearing in the era.

4. Texts containing the main character *gotong royong* (mutual cooperation)

The texts of *tembang "BawaPucung"* (*Gugur gunung gotong royong sambaing srawung* 'Implementing mutual cooperation and calling one another'in the form of video recording in You Tube uploaded by *Paguyuban Seni Pustaka Rakyat* contain the value of *gotong royong*. This value reflects attitudes of respect, spirits of cooperation, and help one another, and volunteerism in solving problems, making communication and friendship, and helping others. This attitude shows that there is a commitment to help one another and to show an empathy among the citizens in developing the nation in all aspects of life. Moreover, there is also a value to give priority on deliberation in solving any problems in the life of the nation. This value also contains an attitude of not making any discriminations and of anti violence.

5. Texts containing the main value of integrity

The texts of *tembang macapat Serat Wedhatamapupuh Sinom* (*Nuladhalakuutama,* 'imitate positive behaviors'), contains the value of integrity, namely any effort to create someone with special characters as one whose words, deeds, and works are reliable. He must be assigned to implement state duties because when he becomes a leader he will be able to make his subordinates peaceful. The value of integrity covers attitudes of responsibility as a citizen with rights and duties that should be implemented in a balanced way. Sub values of integrity are attitudes of citizens who are actively involved in social activities, sincere, righteous, loyal, dependable, just, and also someone who respect individual dignities. If all of these attitudes are implanted into students, the sustainability of the Unitary State of the Republic of Indonesia may be fully depended on young generations. These characters may also be imitated by all citizens.

3.2 *Model of integration into learning*

The model of integration into learning may be seen in the local content of the Indonesian 2013 curriculum of Javanese Language for *high schools*. In the curriculum, the *Standar Kelulusan* (Standard of Graduation) was integrated into attitudes that are then described in details in the competences of content and in the basic competences of which on part studies the texts of *piwulang Serat Wulangreh, Tripama,* and *Wedhatama.*

A syllabus that has been described in detail according to what exists in the curriculum was made. The syllabus has been made and published at the same time as curriculum. The syllabus and curriculum contains learning outcomes related to students' character as written in core competence and basic competence.

Dealing with the lesson plan, teachers made the lesson plan together in activities of Deliberations of Subject Teachers. Lesson plan must be based on the existing curriculum and syllabus, and be used as learning material in the classroom.

Teaching materials had been provided by the publishers appointed by the Education Service in each regency or municipality. The textbook that had been made may be used as the learning guidance materials. Therefore, the character education learning model may be made based on the pattern.

The teaching method chosen depended on the topic. Other method is singing the *Tembang Macapat* containing values of character education in each teaching and learning process. The learning medium used is internet in the website of *ki demang.com* or *YouTube* containing examples of *tembang macapat* in the hope that students may sing the *tembang* and memorize the lyrics.

Thus, teaching model of character education can be integrated in Javanese language learning which is based on the existing or current curriculum and syllabus, lesson plan, and learning materials containing character education. This can be done by building it as habit and using relevant media.

Character education has become the national movement in education since 2010. However, the result of the movement has not been pleasing, so that it is necessary to reinforce it through the CER movement launched by the Minister of Education and Culture. The minister's idea is then stated in the CER movement aimed at implanting the main values of character education consisting of the five main values. Based on Kemdikbud (2017a & 2017b), the implanting of the values may be integrated in the learning activities of all subjects.

In the Javanese language learning, the five main values may be integrated into learning, especially through teaching materials in the form of *tembang macapat* texts. The integration may be done since in the 2013 curriculum. The local content of Javanese language for *Indonesian high schools* contains the Basic Competence that facilitates the integration of character education. The syllabus has also been arranged based on the curriculum, so that teachers merely make the Lesson Plan according to the needs of each school. The method and media may also be chosen based on the appropriateness of teaching materials presented (*Dinas Pendidikan Jawa Tengah* 2014a & 2014b; Agboola & Tsai 2013; Okeke & Drake 2014; Silay 2014; Setiyadi 2015a).

Strategy in the implementation of character education in this Javanese language learning is done through materials related to the basic competence. The basic competence in the Indonesian 2013 curriculum facilitates the integration of the character education. The discussion is focused on anything dealing with the basic competence related to the values of education in the *tembang macapat texts.* The contents of character education values are presented to the students through the existing learning materials in the textbooks in each school published by different publishers. In the *tembang macapat* texts, many rare classical expressions may be found. Teachers as facilitators should explain the meanings of the expressions to the students who have difficulties in understanding the expressions. Then the teachers deliver the meanings and values contained in the *tembang macapat* texts as clearly as possible so that students may be able to understand the messages intended by the writers of the *tembang macapat* (Setiyadi 2015a; Althof 2006; Aqib 2012; Ashraf et al 2013; Anwar 2016).

Besides studying the texts, the teachers also had another activitiy namely habituating students to sing the *tembang macapat* at the beginning or the end of learning activities. This habituation is intended to implant the values in students' deep heart. The activities are then followed by giving students home works namely memorizing the *macapat* and lyrics intended to make the values contained in the *tembang* directly implanted in the students' deep heart. The memorization of the lyrics is needed to implant noble attitudes or behaviors as stated in the lyrics automatically. Singing the *tembang* may be done at the beginning or at the end or just at the beginning or at the end of each learning activities in learning Javanese language. It is intended to habituate and to implant values in the *tembang* to the students. By singing the *macapat*, the contents of the values in the lyrics are expected to be implanted in the students' deep heart. Therefore, the values of character education contained in the lyrics of the *tembang macapat* may automatically are absorbed into students' deep heart and become their daily behaviors.

The model of implanting character education by singing the *tembang macapat* imitates the one that has been made by the ancestors of Javanese people in the past. They implanted character education through activities of *nembang macapat* (singing the *macapat*) where in the *macapat*, expressions containing character education may be found. Expression such as *guyuprukun, manungalingkawulagusti, lakuutama, narima, ajadumeh, adigang, adigung, adiguna, sudanenhawalannepsu, tapa brata, wongsatriya*, and the like exist in *the tembang macapat*. By singing the macapat every day, at last the young generation of Javanese ethnic group may absorb and understand the values of the character education. Then, they will be able to inherit the noble values integrated in the *tembang macapat* automatically in their daily life. By imitating the model, it is expected that the five main values of the character education that will be grown in the students' deep heart may be naturally implanted what happens in the Javanese ethnic group.

4 CONCLUSION

From the descriptions above it can be concluded that the *tembang macapat* texts can be used as learning materials for character education at secondary schools. The sources of the character education may be taken from the *tembang macapat* texts of *SeratWulangreh*, *SeratWedhatama* and *Tripama*.

The syllabus has been arranged according to the local content of the Indonesian 2013 curriculum. Then, the lesson plan has also been made in accordance with those found in each school area. The teaching materials for character education may be integrated in the Javanese language learning in line with the 2013 curriculum. The integration may be done through the basic competence containing the study of *piwulang* in the *tembang macapat* texts in line with the curriculum, syllabus and lesson plan. Besides integrating the *tembang macapat* in line with steps above, the integration may also made by singing the *tembang* either at the beginning, at the end of learning activities or both. Moreover, homework in the form of memorizing the lyrics of the *tembang macapat* may also be given. By memorizing and singing the *tembang macapat*, the implanting of values of the character education may be quickly absorbed in the students' deep heart.

REFERENCES

Althof, W., & Berkowitz, M.W. 2006. Moral education and character education: their relationship and roles in citizenship education. *Journal of Moral Education* 35(4): 495–518.
Anwar, C. 2016. Patterns of character education of primary school students, *Mediterranean Journal of Social Sciences* 7(2): 156–166.
Agboola, A., & Tsai, K.C. 2012. Bring character education into classroom. *European Journal of Education Research* 3(2): 163–170.

Aqib, Z. 2012. *Pendidikan Karakter di Sekolah: Membangun Karakter dan Kepribadian Anak [Character education in school: Building children character and personality]*. Bandung: Yrama Widya.

Ashraf, S., et al. 2013. Students' Preferences for the Teachers' Characteristics and Traits in Character Building of Students with Special Needs. *Mediterranean Journal of Social Sciences* 4(4): 423–430.

Banicki, K. 2017. The character–personality distinction: An historical, conceptual, and functional investigation. *Theory & Psychology* 27(1): 50–68.

Collin, R. 2012. Genre in discourse, discourse ingenre: anew approach to the Study of Literate Practice. *Journal of Literacy Research* 44(1): 76–96.

Cornelius, O.O., & Greg, E. 2013. Federal Character Principles, Nation Building and National Integration in Nigeria: Issues and Options. *Mediterranean Journal of Social Sciences* 4(16): 33–40.

Creswell, J.W. 2014. *Research Design: Qualitative, Quantitative, and Mixed Methods Approaches*. Newbury Park, CA: SAGE Publications.

Dinas Pendidikan Jawa Tengah. 2014a. *Kurikulum 2013: Muatan Lokal Bahasa Jawa SMA/SMALB/MA/SMK/MAK [Indonesian national curriculum: Javanese language as a local content for high school]*. Semarang: Author.

Dinas Pendidikan Jawa Tengah. 2014b. *Kurikulum 2013: Muatan Lokal Bahasa Jawa SMP/SMPLB/MTs [Indonesian national curriculum: Javanese language as a local content for middle school]*. Semarang: Author.

Flick, U. 2009. *Qualitatif Research*. London: SAGE Publications, Ltd. Retrieved from https://books.google.co.id/books?id.

Kemendikbud. 2013. *Bahasa Indonesia Ekspresi Diri dan Akademi: Buku Guru [Indonesian language: Self expression and academic]*. Jakarta: Kemendikbud.

Kemendikbud. 2017a. *Konsep dan Pedoman Penguatan Pendidikan Karakter Tingkat Sekolah Dasar dan Tingkat Sekolah Menengah [Concepts and guidelines for strengthening character education at Primary and Secondary School level]*. Jakarta: Kemendikbud.

Kemendikbud. 2017b. *Modul Pelatihan Penguatan Pendidikan Karakter bagi Guru [Module of strengthening character education training for teacher]*. Jakarta: Kemendikbud.

Kemendiknas. 2011a. *Panduan Pelaksanaan Pendidikan Karakter di SMP [Guidelines of character education implementation in middle school]*. Jakarta.

Kemendiknas. 2011b. *Pengembangan Pendidikan Budayadan Karakter Bangsa: Pedoman Sekolah [Development of Cultural Education and Nation Character: School Guidance]*. Jakarta: Puskur.

Mardikarini, S. & Suwarjo, S. 2016. Analisis muatan nilai-nilai karakter pada buku teks kurikulum 2013 pegangan guru dan pegangan siswa [Analysis of character value in teacher and student text book with 2013 curriculum]. *JurnalPendidikanKarakter* 6(2): 261–274.

Merino, M., et al. 2017. Narrative discourse in the construction of Mapuche ethnic identity in contexts of displacement. *Discourse & Sosciety* 28(1): 60–80.

Okeke, C., & Drake, M. 2014. Teacher as role model: the South African position on the character of the teacher. *Mediterranean Journal of Social Sciences* 5(20): 1728–1737.

Patton, M.Q. 2015. *Qualitative Research and Evaluation Methods*. USA: SAGE Publication, Inc. Retrieved from https://books.google.co.id/books?id.

Pino, M. 2016. Delivering criticism through anecdotes in interaction. *Discourse Study* 18(6) 695–715.

Pujianto, D., et al. 2014. A process-genre approach to teaching writtingreport texts to senior high school student. *Indonesian Journal of Applied Linguistics* 4(1): 99–110.

Saj, H.E., & Sarraf, C.M. 2013. Discourse and social values in Oprah Winfrey show hosting Queen Rania of Jordan. *Mediterranean Journal of Social Sciences* 4(11), 30–33.

Setiyadi, D.B.P. 2012. *Kajian Wacana Tembang macapat Kajian Wacana Tembang macapat Struktur, Fungsi, Makna, Sasmita, Sistem Kognisi, dan Kearifan Lokal Etnik Jawa [Discourse Discourse Tembang macapat Tesis Discourse Tembang macapat Structure, Function, Meaning, Sasmita, Cognition System, and Local Ethnic Wisdom of Java]*. Yogyakarta: Media Perkasa.

Setiyadi, D.B.P. 2012a. *Wulangreh sebagai sumber kearifan lokal etnik Jawa yang mengandung nilai-nilai Pendidikan karakter: sebuah kajian wacana [Wulangreh as Javanese etnic local wisdom source that contain character building value]. Proceeding Seminar Nasional Bahasa Ibu "Bahasa Ibu/ Daerah sebagai Sumber Kearifan Bangsa", Universitas Sebelas Maret, 20 April 2012*.

Setiyadi, D.B.P. 2013. Discourse analysis of *SeratKalatidha*: javanesecognition system and local wisdom". *Asian Journal of Social Sciences and Humanities* 2(4): 292–300.

Setiyadi, D.B.P. 2014. Pendidikan karakter bangsa berbasis teks cerita sejarah bangsa [Nation character building based on the history of nation text]. *Proceeding Seminar Internasional "Membangun Citra Indonesia di Mata InternasionalMelalui Bahasa dan Sastra Indonesia" dalamrangka PIBSI XXXVI; Prodi PBSI, FKIP, Universitas Ahmad Dahlan Yogyakarta, 11–12 Oktober 2014*.

Setiyadi, D.B.P. 2015. Teksastra Indonesia sebagai sarana membangun karakter siswa atau generasi muda penerus bangsa [Indonesian literature text as means of building student character. *Proceeding Seminar Nasional AsosiasiDosen Bahasa dan Sastra Indonesia (ADOBSI), 2015.*

Setiyadi, D.B.P. 2015a. *Baud Basa Jawa.* Klaten: Intan Pariwara.

Setiyadi, D.B.P., & Haryono, H. 2016. Character education model based on texts of *Macapat*song on primary and secondary school students in Central Java". *Paper in International Seminar "Character Building" 24 Maret 2016, in HuachiewChalermprakiet University Thailand.*

Setiyadi, D.B.P., & Wiyono, S. 2017. Thematic multikulturaltexts: astudy on building tolerance". *Proceeding10th International Conference "Revisiting English Teaching, Literature, And Translation In The Borderless World: My World, Your World, Whose World?".*

Silay, N. 2014. A study of moral education and its relationship with moral education. *Mediterranean Journal of Social Sciences* 5(2): 353–358.

Widyahening, C.E.T., & Wardhani, N. E., 2016. Literary Works and Character Education, *International Journal of Language and Literature* 4(1): 176–180.

Wodak, R. 2011. Complex text: Analysing, understanding, explaining and interpreting meanings. *Discourse Studies* 13(5): 623–633.

Do ethics education and religious environment mitigate creative accounting?

R.C. Sari
Universitas Negeri Yogyakarta, Indonesia

M. Sholihin
Universitas Gadjah Mada, Indonesia

D. Ratmono
Universitas Diponegoro, Indonesia

ABSTRACT: There are increasing volumes of corporate scandals and a decline in ethical reasoning of accountants. The credibility crisis of accountants arouses the awareness of educators concerning the importance of instilling ethical values in the context of education, because ethics is the most effective way to limit the scope of creative accounting. A university learning environment is critical to the development of ethical values. The social norm theory implies that religious institutions will shape people's behaviors. Otherwise, the theory of the sacred canopy argues that religion no longer affects the aspects of life due to the increased level of modern society materialism. The purposes of this study are: first, to examine the effect of ethics education on ethical perception of creative accounting; second, to test the differences of ethical perceptions of undergraduate students in religious and public universities; third, to examine the effects of individual's ethical ideology on ethical perceptions. The data collection method involved a questionnaire to a total of 225 respondents from two religious-based universities and one public university. SEM PLS was used to test the hypothesis. The results show that business ethics education did not influence ethical perceptions on creative accounting. However, there is a significant difference in students' ethical perceptions between the public university and religious-based universities. Individuals' ethical ideology has an impact on ethical perceptions. Educational implication and suggestion for future research are proposed.

1 INTRODUCTION

The increased global corporate scandals and decline in ethical reasoning of accountants over a 15-year period, made the public question the ethical behavior of accountants and auditors (Bean & Bernardi 2005). One possible explanation of the decline in ethical values of accountants and auditors, among others, could be the lack of effectiveness of the ethical training currently being provided for future practitioners. Some researchers argue that ethics education is not effective (Conroy & Emerson 2004). Another possibility is the increased pressure on accountants to produce reporting that is appropriate for their company's best interests (Zeff 2005), which encourages creative accounting.

Creative accounting is designed to prepare financial reports to achieve financial performance targets, resulting in a misrepresentation of corporate performances (Balaciu & Pop 2008). Accountants take advantage of the flexibility of accounting standards by making discretionary accounting to move reported earnings toward desired goals. It seems that creative accounting does not explicitly violate or only slightly deviates from the standards. But from a user's perspective, this practice is unethical because everyone who uses earning information generated from creative accounting tends to misinterpret and misdirect in decision making.

Ethical judgments on creative accounting vary, most accountants consider it is ethical and some others consider it is unethical (Bruns & Merchant 1990). While from the perspective of students as future professional candidates, there is no difference in the intention of doing creative accounting between students in public universities and religion (Sari & Sukirno 2015). The current study extends this line of research by investigating the effects of business ethics education and religiosity environments on ethical perceptions of creative accounting practices. The assessment of students' ethical perceptions of the practice measures students' ability to recognize and understand ethical issues.

We also test whether the internal factors of individual ethical ideologies influence ethical perceptions on creative accounting. Previous studies have extensively examined the influences of religions on personal behaviors (Lehrer 2004, Rawwas et al. 2006) but the extent to which religious environments have effects on ethical perceptions of creative accounting of students has limited attention.

Examining the factors that influence the perceptions of ethical students will provide useful insights to educators and regulators.

2 REVIEW LITERATURE AND HYPOTHESIS DEVELOPMENT

2.1 *Ethics education and ethical perception on creative accounting*

In general, creative accounting is unethical and causing negative consequence (Badertscher 2011, Cheung et al. 2006). But ethical judgement on creative accounting varies, most of the accountants and managers said ethically (Bruns & Merchant 1990). This raises the question whether or to what extent business ethics education has been effective in internalized ethical values, thereby reducing the intention of doing creative accounting in the future.

There are differing opinions about trainability of ethical decision making. Some argue that ethics cannot be taught because of character development being formed at college age (Cragg 1997). One ethical business class or series of classes in college is not sufficient to change the character that was built at the beginning of one's life. It is reasonable to argue that it is difficult to transmit the value of ethics in a business course because of the cruel nature of our focus on outcome and competitive economy. Conroy and Emerson (2004) provide evidence that business ethics education does not have an effect on ethical perception. Mayhew & Murphy (2009) suggest that ethics education does not result in internalized ethical value.

However, many professional organizations and researchers believed that some components of ethical decision making can be taught. They argue that with proper implementation, an ethics curriculum can be designed for effective learning (Sims 2002). Some empirical evidence showed that teaching ethics in the classroom would impact on improvement in ethical sensitivity, moral reasoning and even ethical behavior (Sims 2002). There are conflicting evidence about the effectiveness of ethics education in internalized ethical value. Therefore we propose the following hypothesis in the form of null hypothesis.

H_{null1}: There is no difference in the ethical perception of creative accounting ethics before and after ethics education.

2.2 *Religiosity and ethical perception of creative accounting*

There are two schools of thought about the influence of religion on ethical behavior: First, according to social norm theory, social norms influence behavior because individuals prefer to conform to their peer group (Kohlberg 1984) and avoid punishment for opposing standards, values and beliefs that are considered acceptable (Sunstein 1996). Religiosity as a social norm. Religion is a key social mechanism for controlling behavior and belief (Kennedy & Lawton 1998). Assuming all else is equal, the more salient a person's religious identity, the more the behavior increases in accordance with the expectation of his/her religion.

Weaver & Agle (2002) stated that religion influences business ethics. Longenecker et al. (2004) found that business managers and professionals believe that religious values are less likely to approve unethical behavior. Conroy & Emerson (2004) found religiosity is associated with lower acceptance of accounting manipulation. McGuire et al. (2012) found that firms in religious areas are less likely to engage in financial reporting irregularities. They argue that religiosity reduced acceptance of unethical business practices. Du (2014) found that religion can mitigate tunneling. The findings are consistent with the view that religion has an important influence on corporate behavior and can be a set of social norms and/or alternative mechanisms for reducing unethical tunneling behavior

Second, the theory of sacred canopy developed by Berger (1967). According to this theory, religious legitimations are constructed and maintained by human activity, yet when this aspect of their existence is forgotten, they also inherently carry with them the danger of alienation (Kline 2000).

Alienation is associated with the problems caused by rapid social change, such as industrialization and urbanization, which has broken down traditional relationships among individuals and groups. The Sacred Canopy theory argues that religions have lost their influence in many aspects of life due to the increasing materialism of our modern society (Berger 1967, Gorski 2000). People have become very pragmatic and are putting aside religious values.

Several studies support the theory of sacred canopy, among others: Rawwas et al. (2006) found that religion is not an impediment to academic dishonesty. McGuire et al. (2012) found positive association between religiosity and real activity manipulation.

Social norms theory predicts that the influence of religious norms increased in an environment with a strong religious social norms such as religious-based university (Kennedy & Lawton 1998). However, with an increasing materialism of our modern society and high pressure to meet targets, people have become very pragmatic in putting aside religious values as predicted by The Sacred Canopy theory.

Based on this discussion, we propose the following hypothesis:

H2: There is a difference in ethical perceptions of creative accounting between students at public universities and religious universities

2.3 Ethical orientation on ethical perception of creative accounting

Individual's ethical orientation (relativism vs idealism) impact on business decision making such as earnings management decision (Greenfield 2007). Individual ethical orientation is a continuum with relativism at one end and idealism at the other (Forsyth, 1980). Relativism is an individual concern for a universal set of rules or standards. Idealism focuses on human welfare. Individuals who lean toward idealism were less likely to manipulate earnings as it will cause harm to others, so they tend to avoid it (Forsyth 1982). On the other hand those who tend be more relativistic consider the circumstance first rather than potential harm a decision might cause. These individuals also tend to judge the decision more leniently and judge earnings management more ethically than the idealists do (Elias 2002). Individuals who tend to be more relativistic consider the circumstances first rather than the potential losses that may be incurred by the decision.

Greenfiled et al. (2008) found relativistic individuals are more likely to engage in creative accounting and idealistic individuals are less likely to engage in creative accounting practices. Based on this discussion, we propose the following hypothesis:

H3a: there are negative relationships between relativism and ethical perception on creative accounting. In particular, relativistic individuals are more likely to perceive that creative accounting is ethical.

H3b: there are positive relationships between idealism and ethical perception on creative accounting. In particular, idealistic individuals are more likely to perceive that creative accounting is unethical.

3 RESEARCH METHOD

3.1 Participants and procedure

Business students from two religious-based universities and one public university in Indonesia participated in this study. Table 1 shows the sample characteristics.

Participants are currently enrolled in an ethical business course. Pre-test and post-test field experimental design was used to examine effectiveness of business ethics education.

Table 1. Participant characteristics.

	Public university		Religious-based university	
	Mean	SD	Mean	SD
Age	20.6	0.78	20.5	0.86
GPA	3.57	0.16	3.47	0.21
Percent in each category				
	Male	Female	Male	Female
Number of samples observed	35	64	46	80

3.2 *Variable measurement*

Ethical Perception of creative accounting. We used a modified Burns & Merchant (1989) questionnaire to measure ethical perception of creative accounting. The respondents were required to rate each question on a five-point Likert scale ranging from 1 to 5 as follows: 1: ethical practice; 2: questionable practice; 3: minor practice; 4: serious infraction; 5; totally unethical.

Ethical Orientation. We used the Ethical Position Questionnaire (EPQ), developed by Forsyth (1980) to measure students' ethical orientation.

Ethics Education is measured by a dummy variable equal to one after students finish an ethical business course and 0 before. Type of university is measured by dummy variable, 0 for a public university and 1 for a religious based university. Gender is measured by variable dummy: 0 female, 1 male. Our model tests the influence of ethics education, religiosity and ethical orientation on ethical perception of creative accounting using the following form:

$$\text{Ethical Perception on Creative Accounting} = B_0 + \beta_1 \text{Ethics Education} + \beta_2 \text{Type of University} + \beta_3 \text{EthicsOrientation} + \beta_4 \text{Gender} + e$$

To test the hypotheses, a structural equation modeling with partial least squares (SEM-PLS) approach was employed because this model includes second order latent variables also reflective of measurement (ethic orientation).

4 RESULT

Hypothesis testing used a structural equation modeling with partial least squares (SEM-PLS).

4.1 *Measurement model analysis*

The measurement analysis indicated adequate discriminant validity and construct validity.

4.2 *Model fit*

We evaluated ten goodness-of-fit indices for the structural equation model used in this study. Each of these tests are considered a good fit.

4.3 *Test of hypothesis*

We tested the hypothesis based on their respective coefficients path value in the structural equation model and present those results in Table 2.

Hypothesis 1 predicts that there is no difference in the perception of creative accounting ethics before and after ethics education. The coefficient path of ethics education is not significant (Table 2), H1 is supported.

Hypothesis 2 predicted there is a difference in ethical perceptions of creative accounting between students at public universities and religious universities. The negative and significant path coefficient (−0.17) indicate that students at religious-based universities consider creative

Table 2. PLS results (Path coefficient, t-statistics, and R.

| Variable | Path to | |
	Ethical perception on creative accounting	Ethics orientation
Ethics education	0.02	
Type of university	0.17***	
Ethics orientation	0.16***	
Gender	0.03	
Idealism		0.664***
Relativism		0.664***

*** p < 0.01; **p < 0.05;*p < 0.10

accounting practices as acceptable practices compared to public university students. This result supports the theory of sacred canopy that the value of religion has faded due to the increasing materialism of our modern society.

Hypothesis 3a predicts relativistic individuals are more likely to perceive that creative accounting is ethical, suggesting the negative relationship between relativism and ethical perception on earnings management. The path coefficient is significant but in the opposite sign (0.664 × 0.16). An explanation of the difference in results will be discussed in the discussion section.

Hypothesis 3b predicts that idealistic individuals are more likely to perceive that creative accounting is unethical, suggesting the positive relationship between idealism and ethical perception on earnings management. The positive and significant path coefficient (0.664 × 0.16) supports hypothesis 3b. These results are consistent with Greenfield et al. (2007) who found that idealistic ethical orientation will be less likely to engage in earnings management behavior. The path coefficient for gender is not significant, indicating there is no difference in the ethical perception of male and female students.

5 CONCLUSSION, IMPLICATION AND LIMITATION

The flexibility of accounting standards is one of the drivers of creative accounting (Largay 2002). In general, creative accounting is unethical and causes negative consequences. The result indicates that there is no ethical difference in perception on creative accounting before and after a business ethics course. The results of this study are consistent with research by Conroy & Emerson (2004) and Mayhew & Murphy (2009), suggesting that ethics education does not result in an internalized ethical value.

Results of testing hypothesis 2, indicate that student's in religious-based universities perceive that creative accounting is more ethically acceptable than the student's in public universities. This phenomenon supports the theory sacred canopy which contends that religious values have faded due to high materialism. The results of this study also support McGuire et al. (2012) that the manager in religious areas also manipulates earnings using real activity manipulation due to high capital market pressure to achieve profit targets.

Relativism and idealism individuals view creative accounting as ethically unacceptable. These results are not consistent with Greenfield et al. (2007), the possible explanation because this study focused on students' ethical perceptions while Greenfield et al. (2007) focused on student's intentions.

Overall, the result may be helpful to academician, practitioners and accounting researchers. Improvements are needed in teaching business ethics in order to instill ethical values in prospective practitioners. According to Park (1998), the potential harm in business ethics education is that conducted in the area of cognitive reasoning. Most business ethics textbooks contain case studies of ethical scandals, remaining within boundaries of cognitive reasoning. Park (1998) proposed learning using experiential and practical levels of learning necessary for ethical reasoning, moral sentiments and ethical praxis.

Also, improvements should be made to the organizational infrastructure such as perform-ance evaluation and reward not only in the short term oriented. Short term orientation can escalate competition and deprive organizational members of ethical perspective (Park 1998) and religious value (Rawwas, et al. 2006).

The result of this study should be interpreted within the context of the following limita-tions. One such limitation relates to the inability to control the difference of teaching meth-ods or quality of the lecturers, although the field study provides a real assessment of the effectiveness of learning business ethics that occurs at the university.

REFERENCES

Badertscher, B.A. 2011. Overvaluation and the choice of alternative earnings management mechanisms, *The Accounting Review American Accounting Association* 86(5): 1491–1518.

Badertscher, B.A. et. al. 2009. Earnings management and the predictive ability of accruals with respect to future cash flows. *Working Paper,* University of Iowa, Northwestern University.

Balaciu, D. & Pop, C.M. 2008. Is creative accounting a form of manipulation? *Analele Universitatii din Oradea, seria Stiinte Economice, Sectiunea: Finante, Banci si Contabilitate* III(Tom XVII): 35–940.

Bean, D. & R. Bernardi. 2005. Accounting ethics courses: A professional necessity. *The CPA Journal* 75(12): 64–65.

Berger, P.L. 1967. *The sacred canopy: Elements of a sociological theory of religion.* New York, NY: Anchor Books Edition.

Bruns, W.J. & Merchant, K.A. 1990. The dangerous morality of managing earnings. *Management Accounting* 72(2): 22–25.

Cheung, Y., Rau, P.R, & Stouraitis, A. 2006. Tunneling, propping and expropriation: Evidence from connected party transactions in Hongkong. *Journal of financial economic* 82 (2): 343–386.

Conroy, S.J. & Emerson, T.L.N. 2004. Business ethics and religion: Religiosity as a predictor of ethical awareness among students. *Journal of Business Ethics* 50(4): 383–396.

Cragg, W. 1997. Teaching business ethics: the role of ethics in business and in business education. *Jour-nal of Business Ethics* 16(3): 231–245.

Du, X.Q. 2014. Does religion mitigate tunneling? Evidence from Chinese Buddheism. *Journal of Busi-ness Ethics* 125(2): 299–327.

Elias, R. 2002. Determinants of earnings management ethics among accountants. *Journal of Business Ethics* 40(1): 33–45.

Forsyth, D. 1982. Judging the morality of business practices: The influence of personal moral philoso-phies. *Journal of Business Ethics* 11(5): 461–470.

Greenfield, Jr. 2007. The effect of ethical orientation and professional commitment on earnings man-agement behavior. *Journal of Business Ethics* 83(3): 419–434.

Kennedy, E.J. & Lawton, L. 1998. Religiousness and business ethics. *Journal of Business Ethics* 17(2): 163–175.

Kohlberg, L. 1984. *Essays on moral development Vol II: The psychology of moral development.* San Fran-cisco, CA: Harper & Row.

Kline, T.C. 1981. *The philosophy of moral development: Moral stages and the idea of justice.* San Fran-cisco, CA: Harper & Row.

Largay, J. 2002. Lessons from Enron. *Accounting Horizons* 16(2): 154–160.

Longenecker, J.G., McKinney J.A., & Moore, C.W. 1989. Ethics in small business. *Journal of Small Busi-ness Management* 27(1): 27–31.

McGuire, S.T., Omer, T.C., & Sharp, N.Y. 2012. The impact of religion on financial reporting irregulari-ties. *The Accounting Review* 87(2): 645–673.

Park, H. 1998. Can business ethics be taught? A new model of business ethics education. *Journal of Business Ethics* 17(9): 965–977.

Rawwas, M.Y.A, Swaidan, Z., & Al-Khatib, J. 2006. Does religion matter? A comparison study of the ethical beliefs of marketing students of religious and secular universities in Japan. *Journal of Business Ethics* 65(1): 69–86.

Sari, R.C. & Sukirno, S. 2015. Creative accounting: do character education and religion matter? *Finance and Banking Journal* 17(Juni): 1–15.

Sims, R.R. 2002. Business ethics teaching for effective learning. *Teaching Business Ethics* 6(4): 393–410.

Sunstein, C.R. 1996. Social Norms and Social Rules. *Columbia Law Review* 96(4): 903–968.

Weaver, G.R. & Agle, B.R. 2002. Religiosity and ethical behavior in organizations: A symbolic interac-tionist perspective. *The Academy of Management Review* 27(1): 77–97.

Zeff, S.A. 2005. The evolution of U.S. GAAP: The political forces behind professional standards. *CPA Journal* 75(2): 19–29.

Enhancing the character of students' responsibility through context-based chemistry learning in vocational high school

A. Wiyarsi, H. Pratomo & E. Priyambodo
Universitas Negeri Yogyakara, Indonesia

ABSTRACT: Enhancement of students' character has become a significant topic in chemistry education today. Responsibility is one of the important pillars of character. Chemistry learning that provides a variety of learning experiences with the optimization of student involvement facilitates the development of the character of student responsibilities. Descriptive studies have been conducted to explore the character of students' responsibility through the implementation of context-based chemistry learning. The samples of this research are students in an automotive program of vocational education. The results showed that the character of students' responsibility developed during electrochemical and petroleum subject matters. The analysis showed that most students already have a good character of responsibility. There are four indicators including response, earnestness, acceptance and execution. The character indicator with the best response was that of acceptance of responsibility. The poorest was the task execution indicator. The implementation of context-based learning opens the minds of vocational students that chemistry is important in supporting vocational competencies. This fosters responsibility for completing good chemistry learning tasks.

1 INTRODUCTION

Active learning pedagogies that facilitate students' engagement play an important role in the vocational education system. Competence based education, as a new orientation of vocational education, has led to fundamental changes in developing active learning. These changes include the scope of courses offered, the content, goals, forms of instruction and coaching roles (Bruijin & Leeman, 2011; Biemans et al., 2009). Meanwhile, context-based learning has become a current trend in chemistry education to further the interest of students in chemistry and to increase understanding of the concepts studied (Ilhan et al., 2016).

Context-based chemistry learning is developed with regards to phenomena, technical applications and their relevance to students' life/work, so that it is adequately applied in vocational learning. Hence, the emphasis of chemistry concept related to student expertise is required on chemistry learning For example, on learning about petroleum in automotive engineering programs. It will be better if the learning emphasized on the characteristics and using of fuels. The discussion of the other petroleum fractions does not need to be deepened. This is necessary because the wrong notion that chemistry is not relevant to the engineering discipline decreases interest and motivation of vocational students in chemistry learning (Madhuri et al., 2012). In addition to increasing interest in learning chemistry, context-based chemistry learning also supports self-regulation skills (Bruijin & Leeman, 2011). Students' responsibility can then be better developed. It increase attitude and achievement students also (Rahdiyanta et al., 2017).

Enhancing students' character has also become a significant topic in chemistry education today. Responsibility is one of the important pillars of character (Lickona, 1999). The value of responsibility is needed to develop a healthy soul, concern for interpersonal relationships and social life. Teachers are required to provide education to ethically build students' characters and can position themselves as responsible sections of society. Context-based learning is the suggested learning approach to developing students' responsibility (Chowdhury, 2016a). There, students can handle various moral and ethical issues in society, take responsibility and build good character.

In previous studies, it was not identified that context-based learning was emphasized to enhance students' responsibility. Thus, it is important that chemistry teachers in vocational high schools enhance students' responsibility through context-based chemistry learning.

2 LITERATURE REVIEW

Context-based approaches are approaches adopted in chemistry learning where the integration of contexts and applications of chemistry are used as the starting point for the development of scientific ideas (Bennett et al., 2007; Ilhan et al., 2016). In vocational learning, context covers a mixture of aspects and includes the nature of the vocational subject, the learning setting, specification of students' qualifications and students' learning styles (Faraday et al., 2011). An important element of a context-based learning environment is active learning (Parchmann et al., 2006). In this learning, students are required to have a sense of ownership of the subject and are responsible for their own learning.

In context-based learning, using contexts to increase students' need-to-know, creating everyday life situations and doing in-class activities play a great role in the learning process (Ültay & Çalık, 2012). Previous studies have shown that context-based chemistry learning facilitates students to make connections with their own experiences, giving students personal responsibility for their learning, improving their motivation, contributing to attitude development and having students tackle problems together (Bennett et al., 2005; Vos et al., 2010; İlhan et al., 2015, 2016). From another perspective, context-based learning enhances students' responsibility and builds good character (Chowdhury, 2016a). Thus, it is important that educators emphasize character education to develop scientific attitudes, personality and leadership of students through context-based chemistry learning. In this study, the context is the integration of chemistry content with the content of vocational subjects.

Character education is essential for building a moral society, and it is the conscious effort to cultivate virtue. The psychological components of character education encompass the cognitive, affective and behavioral aspects of morality such as moral knowing, moral feeling and moral action (Lickona, 1999). Anderson (2000) stated that character is defined as moral excellence and firmness, while integrity refers to a firm adherence to a code of moral values.

Good character consists of the values that represent good human qualities such as wisdom, honesty, kindness, self discipline, responsibility, self-reliance, perseverance, leadership, tolerance, happiness and respect (Lickona, 1999; Weber & Ruch, 2012; Shoshani & Slone, 2013; Sanderse, 2013; Walker et al., 2015; Chowdhury, 2016a). Lickona (1991) stated that two main moral values are respect and responsibility. It is further said that responsibility is the ability to respond or answer. Reigosa and Aleixander (2007) states that the one form of students' responsibilities are to carry out the tasks in their learning. Students with high responsibilities characters will try to complete the task well. Responsibilities are oriented toward others, giving attention and emphasizing the positive obligation to protect one another.

Teaching strategies involving students in group work and discussions on current issues, project assessments, group work evaluations, observation techniques, interviews, pretest, posttest, anecdotal records and audio- visual evaluations are suggested in learning that promotes students' character (Chowdhury, 2016b). According to the results of previous research, it is stated that implementation of character learning indicates a potential improvement in academic achievement and an array of positive behaviors (Park & Peterson, 2006; Weber & Ruch, 2012; Snyder et al., 2012). Chemistry learning that facilitates the development of student responsibility characters will ultimately have a positive impact on improving the quality of learning.

3 RESEARCH METHOD

A descriptive research design was used in this study to describe the development of character responsibilities of vocational high school students. The samples were 62 students in a vocational education automotive program in Indonesia, namely SMK N 2 Yogyakarta. The samples were determined using a cluster sampling technique. All students were male. There were two classes, one class of students of XI grade learning electrochemistry and the other one students of X grade learning petroleum.

Context-based chemistry learning was implemented in three meetings for each electrochemistry and petroleum subject matter. Students learned in small groups of 4–5 people. Chemistry learning for each meeting was conducted in four stages according to a science, technology and society approach. The learning stages were initiation, concept formation, understanding of concept and consolidation of concept. In the initiation stage, students analyzed problems of electrochemistry and petroleum in the automotive field. For example, students analyzed the differences in petroleum products in Indonesia from a chemistry perspective. Then, they identified alternative solutions to solve the problems from various learning resources

Data collection was conducted using a responsibility observation sheet. There were four indicators measured including response, earnestness, acceptance and execution (Lickona, 1999; Reigosa & Aleixandre, 2007; Weber & Ruch, 2012; Shoshani & Slone, 2013). The instrument was judged by chemistry learning experts to ensure its accuracy. Data collection was conducted during the learning process.

Descriptive analysis was used to determine the category of students' responsibility character. Students' responsibility character consists of five criteria, namely, very good, good, enough, poor and very poor.

4 RESULTS

4.1 Students' responsibility character in each meeting

The first result of the study is the pattern of character development of students' responsibilities from first, second and third meetings. Based on Figure 1, the score of students' responsibility tends to increase both in electrochemistry and petroleum learning.

The content of chemistry learning in this study was adapted to the automotive vocational context. In the petroleum subject matter, the first meeting discussed petroleum fractionation and the second meeting discussed gasoline and diesel fuel. In the last meeting, the students did the practicum to determine the condensed numbers of lubricants. The score of students' responsibility in the petroleum subject matter increased from 3.07 to 3.23.

The first material in the electrochemical class was the voltaic cell, then the second meeting discussed corrosion. In the third meeting, students carried out the electroplating practicum. All material provided was adapted to the automotive vocational context. The results of the analysis showed that the character of students' responsibility increased from meeting 1, 2 and 3.

4.2 Category of students' responsibility character

This section describes the results of character analysis of responsibilities at the end of context-based chemistry learning implementation. The result showed that the category of

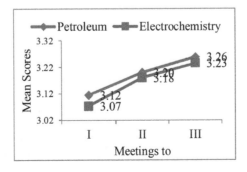

Figure 1. Trend of students' responsibility.

315

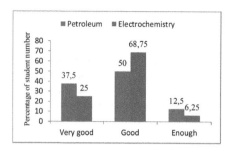

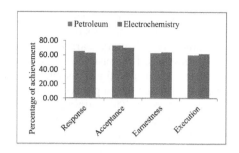

Figure 2. Distribution of students' responsibility.

Figure 3. Profile of students' responsibility in each indicator.

students' responsibilities in learning of electrochemistry and petroleum are good. The mean score of responsibility in petroleum learning (13,03) is better than in electrochemistry learning (12,93).

According to Figure 2, the distribution of students' responsibility categories can be known. In the class with petroleum learning, half of the students have good responsibility character, 37,5% of students have very good responsibility character and the others (12,5%) in enough category. There are slight differences in the electrochemical class. The largest percentage remained in the good category but with a larger number (68.75%). The percentage of students in the very good category was 25% and the students with responsibility character in the enough category was relatively small (6.25%).

4.3 *Description of students' responsibility character in each indicator*

Furthermore, each indicator in the responsibility character that observed was analyzed to find out more about which character has developed well or not. The result of data analysis for the indicator of responsibility is based on the percentage of scores obtained compared with the ideal score. Based on Figure 3, it can be concluded that there is no indicator of responsibility character that achieves a percentage of achievement of more than 75% compared to the ideal score. This means that the character of responsibility in petroleum and electrochemical learning is still in the development stage.

The first indicator is response. The percentage of ideality of student character achievement is 65.2% for petroleum learning and 63.2% for electrochemistry learning. The second indicator is the acceptance character with 73% idealization for petroleum learning and 70% for electrochemistry learning. This achievement is the highest compared to other indicators of responsibility. Meanwhile, for the third indicator (earnestness), the percentage of ideal achieved for petroleum and electrochemistry learning is 62.6% and 63.8%, respectively. The last indicator with the lowest percentage of ideality is execution. The percentage of ideality is 60% for petroleum learning and 61.8% for electrochemistry learning.

5 DISCUSSION

The research findings showed that the implementation of context-based chemistry learning in petroleum and electrochemistry classes is able to develop the character of automotive vocational students' responsibilities. The four character indicators developed during the learning stage took place with each characteristic. The first stage of learning was the initiation. The students analyzed cases related to the automotive field and compiled questions based on the results of the analysis. Students worked in groups so that the response character would appear when students were enthusiastic to hasten interaction with the team to complete the task. The acceptance character was particularly visible from the students' cheer while discussing the stage of concept understanding and concept formation. Character of earnestness especially arose when students presented the results of group work at the stage of concept formation. Likewise when they complete the tasks at the stage of consolidation concept. The execution character was shown by the performance of the students which developed in all the learning stages. The results of previous research strengthened this finding that the application of context-based chemistry learning opens the perspective of students to see the benefits of studying chemistry related to life and work fields. It has a positive impact on the development of student character including responsibility and perseverance (Snyder et al., 2012; Weber & Ruch, 2012; Chowdhury, 2013).

The character of student responsibility at the end of petroleum learning is better than in electrochemistry learning. This can be understood from the standpoint of chemistry content characteristics. Petroleum material has a closer relevance to the automotive field especially in studying fuel systems. Conceptually, the electrochemical content is slightly more complicated and abstract than the petroleum content. Chemistry involves different terminologies, structures and calculations. The learning of these elements, for many students, may cause different levels of difficulties, so chemistry is too broad for them to learn in a short time.

Two characters that developed better were response and acceptance. Students of automotive vocational schools have a good response to the chemistry learning developed in accordance with their vocational context. This good response was shown by the students who, as soon as possible, involved themselves in learning. This attitude is closely related to student interest because the chemistry content taught is relevant to the needs.

Recent studies have shown that the transfer of chemistry to the engineering education context and its material retention is facilitated when the concept is presented in a familiar and related context (Huettel et al., 2013). A context for the chemistry learning in this study is developed with regard to phenomena that are relevant to students' field competence. This is in line with a previous study about implementation of the contextual learning approach in engineering education to increase students' attitudes by strengthening their motivation and interest, and thus promoting meaningful learning (Kukliansky & Rozenes, 2015). Because the students have a good interest, the character of acceptance develops more. This was indicated when students did the task happily.

The character of earnestness will emerge and develop after being embedded with the feeling of being happy with something. This is the case with automotive vocational students in context-based petroleum and electrochemistry learning. With a high interest, students are moved to be serious in carrying out what is assigned. Earnestness is one form of responsibility character. The character of sincerity is also motivated by the knowledge that what is learned is needed in developing the competence of skills. Chemical learning on previous occasions that emphasized theoretical aspects was less attractive to vocational students. Automotive vocational students can see the relevance between petroleum and electrochemistry learning with their expertise in context-based learning.

6 CONCLUSION

Implementation of context-based chemistry learning in petroleum and electrochemistry provides student involvement and facilitates the development of the character of student responsibilities. Most students already have a good responsibility character. Among the four indicators of responsibilities, acceptance has the highest score. The lowest is the indicator of task execution. The implementation of context-based learning fosters responsibility for completing good chemistry learning tasks.

REFERENCES

Anderson, D.R. (2000). Character education: Who is responsible? *Journal of Instructional Psychology, 27*, 139–150.
Bennett, J., Lubben, F., & Hogarth, S. (2007). Bringing science to life: A synthesis of the research evidence on the effects of context-based and STS approaches to science teaching. *Science Education, 91*(3), 347–370.

Biemans, H., Wesselink, R., Gulikers, J., Schaafsma, S., Verstegen, J., & Mulder, M. (2009). Towards competence-based VET: Dealing with the pitfalls. *Journal of Vocational Education & Training, 61*, 267–286.

Bruijin, E.D. & Leeman, Y. (2011). Authentic and self-directed learning in vocational education: Challenges to vocational educators. *Teaching and Teacher Education, 27*, 694–702.

Chowdhury, M.A. (2013). Incorporating a soap industry case study to motivate and engage students in the chemistry of daily life. *Journal of Chemical Education, 90*(7), 866–872.

Chowdhury, M.A. (2016a). Emphasizing morals, values, ethics, and character education in science education and science teaching. *The Malaysian Online Journal of Educational Science, 4*(2), 1–16

Chowdhury, M.A. (2016b). The integration of science technology society/science technology society environment and socio-scientific issues for effective science education and science teaching. *Electronic Journal of Science Education, 20*(5), 20–37.

Faraday, S. Overton, C., & Cooper, S. (2011). *Effective teaching and learning in vocational education*. London, UK: LSN.

Huettel, L.G., Gustafson, M.R., Nadeau, J.C., Schaad, D., Barger, M.M., & Garcia, L.L. (2013). A grand challenge-based framework for contextual learning in engineering. *Proceedings of the 120th ASEE Annual Conference & Exposition*, Atlanta, 23–26 June 2013.

Ilhan, N., Dogan, Y., & Cicek, O. (2015). Preservice science teachers' context based teaching practices in "special teaching methods" course. *Bartin University Journal of Faculty of Education, 4*(2), 666–681.

Ilhan, N., Yildirim, A., & Yilmaz, S.S. (2016). The effect of context-based chemical equilibrium on grade 11students' learning, motivation and constructivist learning environment. *International Journal of Environmental & Science Education, 11*(9), 3117–3137.

Jegede, O.J. & Aikenhead, G.S. (1999). Transcending cultural borders: Implications for science teaching. *Journal for Science & Technology Education, 17*(1), 45–66.

Kukliansky, I. & Rozenes, S. (2015). The contextual learning approach in engineering education. *Proceedings of the 1st International Conference on Higher Education Advances*, Valencia, 24–26 June 2015.

Lickona, T. (1991). *Educating for character. How our schools can teach respect and responsibility*. New York, NY: Bantam Books.

Lickona, T. (1999). Character education: Seven crucial issues. *Action in Teacher Education, 20*(4), 77–84.

Madhuri, G.V., Kantamreddi, V.S.S.N., & Goteti, L.N.S.P. (2012). Promoting higher order thinking skills using inquiry-based learning. *European Journal of Engineering Education, 37*(2), 117–123.

Parchmann, I., Grasel, C., Bear, A., Nentwig, P., Demuth, R., & Ralle, B.. (2006). "Chemie im Kontext": A symbiotic implementation of a context-based teaching and learning approach. *International Journal of Science Education, 28*(9), 1041–1062.

Park, N. & Peterson. (2006). Moral competence and character strengths among adolescents: The development and validation of the values in action inventory of strengths for youth. *Journal of Adolescence, 29*(6), 891–909.

Rahdiyanta, D., Hargiyarto, P., & Asnawi. (2017). Characters-Based Collaborative Learning Model: Its Impact on Students Attitude and Achievement. *Jurnal Pendidikan Teknologi dan Kejuruan, 23*(3), 227–234.

Reigosa, C. & Aleixandre, M.P.J. (2007). Scaffolded problem-solving in the physics and chemistry laboratory: Difficulties hindering students' assumption of responsibility. *International Journal of Science Education, 29*(3), 307–329.

Sanderse, W. (2013). The meaning of role modelling in moral and character. *Research in Science & Technological Education, 17*(1), 45–66.

Shoshani, A. & & Slone, M. (2013). Middle school transition from the strengths perspective: Young adolescents' character strengths, subjective well-being, and school adjustment. *Journal of Happiness Studies, 14*(4), 1163–1181.

Snyder, F.J., Vuchinich, S., Acock, A., Washburn, I.J, & Flay, B.R. (2012). Improving elementary school quality through the use of a social-emotional and character development program: A matched-pair, cluster-randomized, controlled trial in Hawaii. *Journal of School Health, 8*(1), 11–20.

Ültay, N. & Calik, M. (2012). A thematic review of studies into effectiveness of context-based chemistry curricula. *Journal of Science Education and Technology, 21*(6), 293–302.

Vos, M., Taconis, R., Jochems, W., & Pilot, A. (2010). Teachers implementing context-based teaching materials: A framework for case-analysis in chemistry. *Chemistry Education Research and Practice, 11*, 193–206.

Walker, D.I, Roberts, M.P., & Kristjansson, K. (2015). Towards a new era of character education in theory and in practice. *Educational Review, 67*(1), 79–96.

Weber, M. & W. Ruch. (2012). The role of a good character in 12-year-old school children: Do character strengths matter in the classroom? *Child Indicators Research, 5*(2), 317–334.

The embedded values of the electronic textbook of Indonesia for primary school students

T. Subekti & K. Saddhono
Universitas Sebelas Maret, Indonesia

K.L. Merina
The University of Auckland, New Zealand

ABSTRACT: The electronic textbook (known as *Buku Sekolah Elektronik*/BSE) is a school textbook that can be accessed and downloaded from the Indonesian government's official website. The BSE consists of all the nationalism values of the country's generation. It is a learning material for school subjects within the levels of primary and secondary education. The BSE, as school teaching material, has a vital role in forming students' character. This is in line with the current situation where teaching materials need to be investigated in terms of content, especially nationalism values. This study aimed at explaining the nationalism character values contained within the BSE on the school subject of the Indonesian language for the primary school level. The methods used included a literature review and content analysis. The literature review was conducted to acquire the appropriate book through a detailed and accurate search of all BSEs. Meanwhile, the content analysis was performed through analyzing the nationalism character values contained in the BSE for the school subject of the Indonesian language, which covers (1) religiosity, (2) honesty, (3) tolerance, (4) discipline, (5) hard work, (6) creativity, (7) self-reliance, (8) democracy, (9) curiosity, (10) national spirit, (11) nationalism, (12) respect toward achievement, (13) friendliness/communicativeness value, (14) peace loving, (15) literacy/love of reading, (16) caring for the environment, (17) social care and (18) responsibility. The results of the content analysis of the Indonesian language for the BSE for the primary school level show that the nationalism character values contained in the BSE has not been fully and proportionally covered. Furthermore, the content was still dominated by the aspect of knowledge, without an emphasis on comprehension, value internationalization and their implementation.

1   INTRODUCTION

Character education is a topic that has been widely under the discussion of various countries and it has reached the climax in 2006–2010 (Arthur, 2014). Several conferences and seminars have been held connected to this topic (Marshall, 2001). Historically, Dewey believed that moral education and character development could not be separated from the school curriculum (Arroyo & Selig, 2004). Strategically planning to ensure character development is part of the mission and challenges for students to aspire to live by high standards and ethics (Ackerknecht, 2005). Character education is an effort to strengthen the sterling values owned by human beings during their learning to be better and wise persons (Kristjánsson, 2014). The essence of character education in the context of Indonesian education is the sterling values extracted from cultural values for the purpose of constructing youth personality. Character education is organized and applied systematically to help learners to cope with the human attitude values in relation to God, the own self, other people, society and nation, which are embedded in their thoughts, attitudes, feelings, utterances and behavior based on the religion, laws, manners, cultures and custom norms. This education not only provides knowledge but also good living norms and ethics (Awbrey, 2004). It provides habits in thinking and behavior which can help individuals live and work together as a family, society and country and to enable them make responsible decisions. In other words, character education offers smart thinking for learners who can solve all complex problems in their lives with thoughtful solutions (Larson, 2005). A comprehensive character development program has a positive effect on moral reasoning development, academic achievement and behavior of students (Larson, 2005). It also helps student affairs leaders worldwide learn how college can more effectively inspire students to lead ethical and civic-minded lives (Schwartz, 2007). Student experience should provide not only intellectual and social development, but also the development of commendable human qualities, which is commonly referred to as character development (Yoos, 2007).

Many studies discuss the implementation of character education for young learners as Arthur (2014) did in England, proving that character education is important in developing the students' characters starting from their first learning. One priority of character education is nationalism values. The Indonesian government agrees to implement character education in the recent curriculum which involves 18 items; religion, honesty, tolerance, discipline, hard work, creativity, self-reliance, democracy, curiosity, national spirit, nationalism, respect toward achievement, friendliness/communicativeness value, peace loving, literacy/love of reading, caring for the environment, social care and responsibility.

The implementation of character education requires a number of teachers who can provide character values in their learning materials for students (Cooke, 2014), the existence of outcome assessment of the students' characters (Pawelski, 2003) and the existence of learning materials covering nationalism character values. Learning materials can be textbooks or modules. The use of textbooks has been prioritized during the teaching and learning processes. One effort of the government to help society access and get textbooks is through providing electronic textbooks. The electronic textbooks are textbooks which are in the form of soft files and accessed through the Internet by all students in every part of the world.

The electronic textbooks (namely Buku Sekolah Elektronik/BSE) used still need to be studied to make sure that the nationalism character values are embedded in them. The content analysis was done to get the data on the implementation of nationalism character education in the BSE. The data gathered in this study is used as a benchmark of the quality of the book used in elementary schools.

2   LITERATURE REVIEW

Lickona (1991) defined character as a reliable inner disposition to respond to a situation in a morally good way, with a conscious effort to help people to understand, notice and conduct ethical values. He added that character education is the deliberate effort to cultivate virtue, that is, objectively good human qualities that are good for the individual person and good for the whole of society (Lickona, 2004). Suyanto (2010) defined character as ways of thinking and behaving which become the characteristics of individuals to live and work together in the scope of family, society, nation and country. The term character involves the characters of soul and morals which differs people from others by their habits and characters. Effective character education occurs in both the home and the school and requires parents and teachers who model good moral character (Ganiere, 2007). Parental care and parental control could affect adolescents' character strengths in the areas of authenticity, bravery, perseverance, kindness, love, social intelligence, fairness and self-regulation (Ngai, 2015). Changes in the environment will take place during a child's growth and development (Tinnfält, 2015). The instruction a child receives in the home lays the foundation for his or her individual future life (Ganiere, 2007). Character education is essential in building a moral society (Balas, 2006). Moreover, character education has a significant impact on the teacher's own character development (Hauer, 2000). Characters are measurable activities of somebody to respond to the surrounding conditions in good and respectable ways. Nationalism is commonly viewed as loyalty and love of nation/country. John Kane claims that "...Nationalism is an ideology that stresses allegiance to one's nation as a major political virtue and national preservation and self-determination as prime political imperatives...." (Kane, 2014). Based on the definition of character and nationalism, it can be concluded that nationalism characters are measurable activities of somebody in responding to the society condition of his/her country in good and wise ways. Individual responses are the acts of loving the country and self-defense of foreign threats.

The 18 nationalism values referred to as *Pusat Kurikulum Departemen Pendidikan Nasional* to be analyzed in BSE are: (1) religious is a behavior and attitude which exhibits obedience in implementing religious values, being tolerant toward the implementation of other religions' prayer activities and living in harmony with other different religious disciplines; (2) honesty is the behavior aimed at developing people trustworthy in their utterances, acts and works (Kristjánsson, 2014); (3) tolerance is the act and behavior of appreciating differences in religions, tribes, ethnicity, opinions, attitudes and behavior; (4) discipline is the act of order and obedience to the conditions and rules; (5) hard work is the serious effort to solve problems in learning and finishing tasks; (6) creativity is the act of thinking and doing something to get new ways or new finding; (7) self-reliance is the act of being independent in finishing tasks; (8) democracy is the way of thinking, behaving, and acting which views the equality of rights and obligations of him/herself and others (Ross, 2014); (9) curiosity means the behavior and attitudes of being curious in understanding something learned, seen and heard; (10) national spirit is the way of thinking, behaving, and having a concept which views nation and country as the most important aspects of all; (11) nationalism is the way of thinking ad behaving which show loyalty, concern and appreciation toward language, physical environment, social, culture, economy and national politics; (12) appreciation is the act of self-boosting to do beneficial things for society and respecting the success of others; (13) friendliness and communicativeness are behaviors showing the love to talk, mingle and work together with others; (14) peace loving is the attitude, utterances and acts that make other people happy and comfortable with themselves; (15) literacy/love to read is the act of providing time to share what she/he has read which has positive values for her/himself; (16) caring for society is the attitude of always avoiding destruction of the natural environment, and developing efforts to repair damage to the natural environment; (17) social awareness is the attitude of being eager to help others and society who are in need; (18) responsibility is the act of conducting all tasks and responsibilities for themselves, society, environment (nature, social and culture), country and God.

3 RESEARCH METHOD

This study is a content analysis study which measures the nationalism character values embedded in the electronic textbooks for the third level of elementary school. Roth (2017) elaborates that the main step of conducting this study is through analyzing the first part of the texts systematically. He also adds that analyzing the textbooks is pragmatism.

4 RESULTS

The electronic textbook for the third level of elementary schools, BSE, has implemented the national character values. There are 12 out of 18 characters embedded on the textbook. The twelve-character values are religious, tolerance, discipline, hard work, self-reliance, democratic, nationalism, love reading, friendliness and communicativeness, environment awareness, social awareness and responsibility. On the other hand, there are six characters which have not been implemented in the textbook. These are honesty, creativity, curiosity, nationalism spirit, achievement appreciation and peace loving. The result of the study is described in Figure 1.

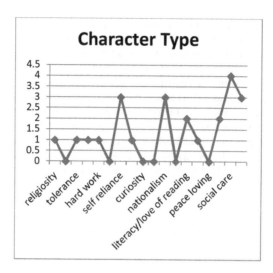

Figure 1.   Description of national character in BSE.

5   DISCUSSION

Based on the content analysis done for the third level textbook for elementary schools published by *Pusat Perbukuan Departemen Pendidikan Nasional*, the conclusions are as follows:

5.1   *Religious character value*

Being religious is the attitude of obedience and implementing all religious values, being tolerant toward other religions' fellows and living peacefully with others who have different religions. It is the level of the quality of behavior in everyday life. People who have high religious values will be able to behave based on their position in society, they are wise because the essence of religiosity is the honor of other majority cultures and religious culture of a certain country (Cush, 2014). Today, there are various religious communities both in real and virtual life. A study conducted by Alessandra Vitullo informed that the role of the Internet is very dominant in increasing the number of communities and online activities (Campbell, 2016). Only one religious character value can be found in the electronic textbook on page 36. On that page, there is a story about a family that lives in harmony. The sentence "*Sore harinya setelah salat magrib, keluarga Pak Abas sudah siap di depan meja makan*" in English: "In the Afternoon, after praying magrib, Mr. Abas's Familly have already in the dining room". It presents the religious value for the students. This sentence is a quotation from a text entitled "*Kegiatan Keluarga Pak Abas*".

5.2   *Honesty character value*

This character value is the act of being her/himself and being trustworthy in their utterances and behaviors. In fact, this value cannot be found in the BSE.

5.3   *Tolerance character value*

Being tolerant is the attitude of honoring differences in terms of religion, tribes, ethnicity, opinions and attitudes. The value of tolerance in the electronic textbooks can be found only on page nine. It says that it is important to respect other opinions during discussions regarding class leader elections.

5.4 *Discipline character value*

This character value shows the act of obedience to all rules and manners. In the electronic textbooks, on page 94, it says that every student has to organize his/her bed by his/herself after getting up.

5.5 *Hard work character value*

The hard work value is the attitude of being keen to do something. In the electronic textbooks, on page 34, it can be found that to get extra money for the family, *Pak Aba*s takes care of chickens in the backyard. This is taken from a story entitled *"Kegiatan Keluarga Pak Abas"*.

5.6 *Creative character values*

Creative character values involves the act of thinking and doing something to produce ways or new findings based on what is already held. Creative values are needed by students to face various events. Creativities can make students easier to solve all the problems with the best solution. It is lamentable that the electronic textbook does not have the creative character values.

5.7 *Self-reliance character values*

The self-reliance character value is the attitude of the students to be more independent in finishing tasks. In the electronic textbooks, there are three items for the students to do the tasks by themselves. Those are on pages 12, 13 and 25.

5.8 *Democracy character values*

This value is the way of thinking and behaving to make equal judgment on rights and obligations. In the BSE, there is a democracy character value on page nine. It is about the process of class leader.

5.9 *Curiosity character values*

Being curious means that the students can show their attitude to deepen and enlarge their knowledge to allow them easily understand the learning materials. Based on what the researcher has done, there is no curiosity character value in the BSE.

5.10 *Nationalism spirit character value*

This character value is the act of thinking, behaving and having insight which places the nation and country above all. In the electronic textbook, there is no nationalism character value.

5.11 *Nationalism character value*

This value is the way of thinking to retaining the nation and country from the enemy. There are three items found in the electronic textbook on the implementation of the nationalism character value. Those are on pages 7, 8 and 116. On page seven, there is a story about a flag ceremony every Monday in the school yard. On page eight, the text is about the school contest which celebrates the national education day. Lastly, on page 116, the text is about art performance. An illustration is also drawn by the writer in the form of a picture.

5.12 *Appreciation*

Appreciating other achievement is the attitude of supporting oneself to produce something beneficial for society and approving the success of others. In the electronic textbook there is no such value.

5.13 *Literacy/love to read*

This value is the habit of spending time reading which consists of positive moral values for the readers. Reading is a process of understanding words and combining the words to create sentences and paragraphs. In the BSE, there are two items on the love of reading character. On page 35, it is said that "*Winda mempunyai kegemaran membaca buku pelajaran, buku pengetahuan umum, buku cerita, koran, majalah, dan sebagainya*". In English: "Winda has a hobby reading a book, newspaper, and magazine". It shows that the text contains love to read character value. On page 38 it is noted that "*Pak Abas menunggui mereka belajar sambil membaca buku atau koran*". In English: "Mr Abbas waiting for them by reading a book or newspaper". These are taken from a text entitled "*Kegiatan Keluarga Pak Abas*".

5.14 *Friendliness and communicativeness character values*

The friendliness and communicativeness character values are in the forms of activities done in groups. Teamwork can be done in various ways but all of the activities are for achieving one goal. In the BSE, there is a friendliness and communicativeness character value on page 35. It says that "*Jika mengalami kesulitan, mereka menanyakan kepada ayah*". In English: "If they find difficulty, they will ask to their father". This is taken from a text entitled "*Kegiatan Keluarga Pak Abas*".

5.15 *Peace loving character value*

The peace loving character value is to support oneself to create something useful for society and approving and respecting others' successes. Unluckily, there is no such value in the electronic textbook.

5.16 *Environment awareness character value*

The environment awareness character value is the attitude of avoiding environmental destruction and developing efforts to repair natural destruction. In the electronic textbooks, two items of the same character value can be found. One is on page five which says "*Setiap hari Jumat sekolah kami mengadakan Jumat bersih*", in English: "Every Friday we always clean our school", taken from a text entitled "*Mengharumkan Sekolah*". The other one is on page 102, "*Dalam rangka menjaga kebersihan sekolah, akan diadakan kerja bakti di lingkungan sekolah*", in English: "We clean the school to keep our healthy", taken from a text entitled "*Pengumuman*".

5.17 *Social awareness character values*

Human beings must need other people to live their lives since every human is a social being. According to *Pusat Kurikulum Departemen Pendidikan Nasional*, social awareness is the attitude of helping others who are in need. There are four items about the implementation of this character value in the electronic textbook. First, on page 34, it is stated that "*Angga rajin membantu ayahnya memberi makan ayam-ayam dan membersihkan kandang setelah salat subuh*", in English: "Angga always help his father to feed the chicken and clean its cage", taken from a text entitled "*Kegiatan Keluarga Pak Abas*". The second, on page 38, it is stated that "*Winda rajin membantu pekerjaan kedua orang tuanya dirumah*", in English: "Winda always help her parents at home", taken from a text entitled "*Gemar Membaca*". The third, is on page 77 where it is written that "*Ibu sedang memberikan uang kepada orang yang membutuhkan*", in English: "Mother is giving some money to people in need". On the same page, there is an illustration of the event written. Lastly, on page 82, it is stated that "*Anak yang berbakti pasti senang membantu ibu*", in English: "A dutiful child would love to help mother", quoted from traditional poetry.

5.18 *Responsibility character value*

The responsibility character value is the attitude of bringing about tasks and obligations that must be done toward oneself, society, the environment (nature, social and culture), country and God. In the electronic textbook, there are three items found concerning this character value. First, on page two, it is stated that Arif's mother asked him to buy mercurochrome. This statement is taken from a text entitled "*Petunjuk Denah Suatu Tempat*". Second, on page three, it is written that "*Dina disuruh ayah untuk kerumah Pak Budi mengantar undangan rapat desa*", In English: "Dina asked by her father to deliver an invitation to Mr. Budi". Taken from a text entitled "*Membuat Denah Sesuai Petunjuk*". Lastly, on page seven, "*Merpaun adalah anak yang rajin. Setiap ada PR dari guru, dia selalu mengerjakan dengan baik*", in English: "Merpaun is a diligent student. He always do the exercise correctly", was presented.

6 CONCLUSION

It can be concluded that the electronic textbook for the third level of elementary schools, BSE, has implemented the national character values. There are 12 out of 18 characters embedded in the textbook. The 12 character values are religious, tolerance, discipline, hard work, self-reliance, democratic, nationalism, love of reading, friendliness and communicativeness, environment awareness, social awareness and responsibility. On the other hand, there are six characters which have not been implemented in the textbook. Those are honesty, creativity, curiosity, nationalism spirit, achievement appreciation and peace loving. Moreover, the content of the electronic textbook is dominated by knowledge and information. It has not reached the aspect of internalization and implementation.

REFERENCES

Ackerknecht, S.M. (2005). The cobleskill creed, guiding principles supporting community and character. *Journal of College and Character, 6*(2), 1–18. doi: 10.2202/1940–1639.1414.
Arroyo, A. & Selig, G. (2004). Differential assessment and development of character. *Journal of College and Character, 5*(7), 1–13. doi: 10.2202/1940–1639.1392.
Arthur, J. (2014). Foundations of character: Methodological aspects of a study of character development in three- to six-year-old children with a focus on sharing behaviours. *European Early Childhood Education Research Journal, 22*(1), 105–122. doi: 10.1080/1350293X.2012.707413.
Awbrey, S.M. (2004). Rekindling meaning in undergraduate education. *Journal of College and Character, 4*(3), 1–11. doi: 10.2202/1940–1639.1346.
Balas, G.R. (2006). The lessons of Anapra: International service learning and character education. *Journal of College and Character, 7*(7). 1–10. doi: 10.2202/1940–1639.1243.
Campbell, H.A. (2016). Assessing changes in the study of religious communities in digital religion studies. *Church, Communication and Culture, 1*(1), 73–89. doi: 10.1080/23753234.2016.1181301.
Cooke, S. (2014). Virtue, practical wisdom and character in teaching. *British Journal of Educational Studies, 62*(2), 91–110. doi: 10.1080/00071005.2014.929632.
Cush, D. (2014). Autonomy, identity, community and society: Balancing the aims and purposes of religious education. *British Journal of Religious Education, 36*(2), 119–122. doi: 10.1080/01416200.2014.884819.
Ganiere, C. (2007). Like produces like: John Heyl Vincent and his 19th century theory of character education. *Journal of College and Character, 8*(4), 1–10. doi: 10.2202/1940–1639.1612.
Hauer, J. (2000). The impact of character education on Russian teachers: An informal survey, spotlight on Josephine Hauer. *Journal of College and Character, 1*(5), 1–6. doi: 10.2202/1940–1639.1282.
Kane, J. (2014). The encyclopedia of political thought. *Reference List: Electronic Sources (web publications)*. Retrieved from http://hdl.handle.net/10072/66677.

Kristjánsson, K. (2014). Character and moral psychology. *Journal of Moral Education, 43*(4), 539–54. doi: 10.1080/03057240.2014.923133.

Larson, C.A. (2005). An examination of the effectiveness of a collegiate character education program. *Journal of College and Character, 6*(6), 1–17. doi: 10.2202/1940–1639.1468.

Lickona, T. (1991). *Educating for character: How our school can teach respect and responsibility.* New York, NY: Bantam Books.

Lickona, T. (2004). *Character matters: How to help our children develop good judgement, integrity, and other essential virtues.* New York, NY: Simon and Schuster Inc.

Marshall, J. (2001). Character education in preservice education: One institution's response. *Journal of College and Character, 2*(9), 1–7. doi: 10.2202/1940–1639.1300.

Ngai, S.S. (2015). Parental bonding and character strengths among Chinese adolescents in Hong Kong. *International Journal of Adolescence and Youth, 20*(3), 317–333. doi: 10.1080/02673843.2015.1007879.

Pawelski, J.O. (2003). The promise of positive psychology for the assessment of character. *Journal of College and Character, 4(2),* 1–8. doi: 10.2202/1940–1639.1361.

Ross, H. (2014). Citizenship & democracy in further and adult education. *Journal of Moral Education, 43*(4), 534–536. doi: 10.1080/03057240.2014.938437.

Roth, D. (2017). Morphemic analysis as imagined by developmental reading textbooks: A content analysis of a textbook corpus. *Journal of College Reading and Learning, 47*(1), 26–44. doi: 10.1080/10790195.2016.1218807.

Schwartz, A. (2007). Reflections on "character development" and the Journal of College and Character. *Journal of College and Character, 9*(1), 1–8. doi: 10.2202/1940–1639.1098.

Suyanto, S. (2010). *Aktualisasi pendidikan karakter* [The actualisation of character education]. Jakarta: Direktorat Jenderal Manajemen Pendidikan Dasar Menengah.

Tinnfält, A. (2015). What characterises a good family? Giving voice to adolescents. *International Journal of Adolescence and Youth, 20*(4), 429–441. doi: 10.1080/02673843.2015.1018283.

Yoos, C.J. (2007). "il faut cultiver notre jardin." cultivating the college garden of character: A process of character development at liberal arts colleges. *Journal of College and Character, 9*(2), 1–10. doi: 10.2202/1940–1639.1128.

The use of visual cues instruction in school to develop children's discipline and self-reliance

A. Listiana & A.I. Pratiwi
Universitas Pendidikan Indonesia, Indonesia

ABSTRACT: This study aimed to determine how the use of visual cues instruction could build discipline and independence in Bandung's kindergartens. A case study method was used, with data collection from teachers, principals and students. The instruments of data collection were observation records, teacher notes and interviews. The data were analyzed continuously during data processing and research, so that finally the study could draw a conclusion. The results show that visual cues instruction had a positive impact on children's discipline and self-reliance. Therefore, the use of visual cues instruction is recommended for use in schools.

1 INTRODUCTION

Discipline and self-reliance are important parts of character value that must be owned by every individual, as discipline and independence have a very big influence in private, community and state life. Discipline and self-reliance will be more effective if taught in the early childhood stage, as this allows children to learn to be accountable until they grow up (Holden, 2002). Discipline and self-reliance have a close relationship. The result of discipline teaching is to embed the value of self-reliance (Siswanti & Lestari, 2012). In addition, Schaefer (1996), Rimm (as cited in Romhmaniah et al., 2016), Howard (1996), Papalia et al. (as cited in Straus & Fauchier, 2007), Arthur and Thorne (1998) and Alwi et al. (as cited in Wiyani, 2012) noted that discipline and self-reliance are important for children because they influence the development of self-control and self-direction that is where children can direct and control themselves without external influence to adjust to the social environment. Discipline and self-reliance affect child self-esteem, development, aggression and behavioral problems (Socolar et al., 2007).

One of the institutions that are considered effective in developing children's discipline and self-reliance is school. The importance of character education in schools is emphasized by Law Number 20 of 2003 on the National Education System (UUSPN), article 3. In addition, it is also consistent with the objectives and principles of early childhood development of children's personality or life skills (Kasmadi, 2013; Saputri, 2016).

One of the strategies that can be used to build the characters of discipline and self-reliance is through the installation of visual cues instruction in the school environment. Visual cues instruction is a form of support for the creation of a supportive environment. In addition, visual cues instruction is one of the visual supports in learning that can help children to know and to control behavior and rules of what they should do (Blagojevic et al., 2011; Fox et al., 2010). Therefore, this study examined and analyzed the use of visual cues instruction in school to develop the attitude of children's discipline and self-reliance.

2 LITERATURE REVIEW

Visual cues instruction is a learning strategy that involves using pictographic and written language as instructional supports in the context of structured learning (Quill, 1995).

Quill (1995) argued that pictographic writing and cues support children's understanding of verbal and social cues, making it easier for children to learn, communicate, interact and develop self-control.

In addition, visual cues also provide an effective way for children to consistently perform daily routines (Sailor et al., 2009), learn basic skills to help themselves and understand spoken language (Norfolk Country Council Early Years & Childcare, 2015). Visual cues instruction has also been used in Theaching Pyramid Model (TPM) of the Curriculum 2013 program and its contents can provide behavior change to a more positive, especially in the behavior of discipline and manners (Maryadi et al., 2016). However, it should be noted that visual cues in schools should not be excessively given, especially in the marks affixed or hung on the wall. Therefore, it is necessary to plan and adjust the visual signs that are really needed in accordance with the location and the needs of the children.

Previous research has shown that visual cues instruction can be used at preschool levels (Krantz & McClannahan, 1998; Ganz & Flores, 2008a) and elementary school level (Sarokoff et al., 2001; Ganz & Flores, 2008a). Visual cues instruction is also capable of improving children's social and communication skills (Krantz & McClannahan, 1993; Sarokoff et al., 2001; Ganz et al., 2008b), helping children to communicate with adults (Krantz & McClannahan, 1998) and helping children answer questions (Charlop-Christy & Kelso, 2003) and comment (Ganz et al., 2008b).

3 RESEARCH METHOD

This research used a qualitative descriptive method. The subjects involved were two teachers, one principal and seven students in the kindergarten of Bunda Balita. The data were collected through observation, teacher and principal interviews, and daily diaries written by the teachers. The process of observation was done on two occasions namely, before the installation of visual cues and during the process of teaching and learning activities with the visual cues that were installed. Interviews and diaries were collected from teachers after three months. There were also types of visual cues used such as names and photos embedded in shoe racks, bag lockers, toothbrush glasses, chairs in the dining room and a circular sign in the circle time area.

4 RESULTS

The results of this research are based on interviews, observations, teachers' diaries about the use of visual cues instruction in the school environment to develop the attitude of discipline and children's self-reliance.

Before the use of visual cues, the children had difficulty obeying the rules. They were not neat when sitting on the carpet, sometimes confused when storing goods, saving shoes arbitrarily or irregular in front of the door because there was no available shelf so that they needed the help of teachers or parents. The following data were disclosed by the teachers before using visual cues instruction:

> Before using the carpet, the children sat to form an irregular circle and the teachers had difficulty encouraging the children to sit neatly, were constantly calling a child's name while asking them to sit down. (Observation notes, March 21, 2016)

However, after the use of visual cues, there appeared to be a change in the children's discipline and self-reliance for the better. Based on the data, it was revealed that the change in discipline and self-reliance was evident in the daily routine of the children such as keeping the shoes on the shelves and bags in their lockers, without the help of others and orderly following the rules of learning in the classroom.

The use of visual signs that affect the change in discipline of children include carpet circle, photos with names on shoe lockers and chairs, and glasses for toothbrushes. The following data about the visual sign instructions affects the child's disciplinary attitude. The notes on the existence of changes in attitude and self-discipline in children using visual cues are as follows:

> When the class started, the children were sleeping on the carpet. The teacher immediately instructed *'duduk di karpet'* (sit on the carpet) and the children neatly sat on the carpet without the need to call or arrange one by one. The children are more orderly in every school activity because the learning tools were supported such as the carpet to sit on, so the teacher easily instructed and the children can immediately sit in circle form on the carpet. (Observation notes, April 25, 2016)

The Figure 1 is an illustration of the use of visual cues with a circle on the carpet that helps children to sit during the circle time activity. The shape of the carpet used by children is a circle. With the shape, the children will sit in a circle so that every child can see the teacher clearly. The size of the carpet is also adapted to the number of children, not too small but also not too big. The shape of this carpet circle is as shown in Figure 2.

The carpet is stored in the study room, which is free from indoor play equipment and is an adequate size. In addition to the carpet circle, other visual cues also contribute to the change of discipline and self-reliance of children. These are the use of photos and names on the shoe racks, locker bags and chairs in the dining room. The data that support the condition are as follows:

> The children are the most affected (discipline and self-reliance) by the photos because the shoes are shuffled and the bags have no photos of their own. This shows

Figure 1. The circle form on the carpet.

Figure 2. Circle seats.

that the photos gives effect to the children. Sitting in the dining area, the children can follow (teacher instructions to sit) 'oh that's my chair', they can sit there, where the sandals are given a name and a photo. They are also happy to be more orderly. They feel that their property is recognized. (Interview G1, June 9, 2016)

By using the photo, a child does not need to be assisted by a parent or teacher. The children just look at each photo and they store the goods properly. Teachers just pay attention. Like this morning, "A" (name of a child) comes to the teacher and says hello. Then A squatted in front of the shelf while opening the shoes one by one from right to left without the help of the teacher. Then A knocks on the door and says 'assalamualaikum' and points to the locker on the photo to store the bag. (Observation notes, April 4, 2016)

Figure 3 shows the use of photos and names on the shoe rack.

Figure 3.   Photo and nametag on the shoe rack.

Figure 4.   'A' when opening and saving shoes on the shelf.

The picture is an illustration of the use of visual cues. It contains shelves that have the children's photos and names taped on. By using photos on some school facilities, the children have no difficulty to identify the place that belongs to them, so that their discipline is encouraged by the media.

Visual cues instruction is used to create a conducive environment for children to learn. The goal is to help children know what is to be done and to provide the means that can develop appropriate behavior consistently and continuously (Listiana & Rachmawati, 2016).

> The use of some environmental arrangements such as photographs and carpets helps children to be orderly in every school activities, and teachers do not seem to need to repeat and remind each child. (Observer notes, April 29, 2016)

5 DISCUSSION

Based on the findings, the use of visual cues instruction can provide a positive impact on changing the attitude toward discipline and self-reliance of children. Discipline and self-reliance are the habits of children that they must have in daily life. This is in accordance with the work of Sailor et al. (2009) in which visual cues were used to effectively train the child's consistency in performing daily routines. In practice, the visual cues used in the study included images, symbols and photographs as instructional supports in every child's playground (Hawkins & Clayton, 2011).

In addition, through visual cues children can easily translate abstract written rules into concrete behaviors that they can see and understand. Examples of the findings that childrens can sit neatly and regularly form a circle, they can understand through the carpet in the circle form. The quote that supports this is as follows:

> When the class started, the children were sleeping on the carpet. The teacher immediately instructed *'duduk di karpet'* (sit on the carpet) and the children neatly sat on the carpet without the need to call or arrange one by one. (Observation notes, April 25, 2016)

Visual cues can also help to translate an instruction through the appropriate media. This is based on the stage of child development. Piaget reveals that childhood is at a concrete pre-operational stage (Santrock, 2007).

Visual cues instruction makes it easier for children to learn and understand the behavior or instructions in accordance with the rules of the school. Furthermore, Hemmeter et al. (2006), Fox et al. (2003), Burchinal et al. (2002), Hunter and Hemmeter (2009), Curtis et al. (as cited in Hamalik, 2008) and RoI (2014) say that it affects changes in children's attitudes and helps to overcome behavior problems. The condition occurs because the use of visual signals is one of the visual supports in learning that can help children to know and control the behavior and rules of what should be done (Blagojevic et al., 2011; Fox et al., 2010). Research has shown that visual signs can improve children's social and communication skills (Krantz & McClannahan, 1993; Sarokoff et al., 2001; Ganz et al., 2008b).

This visual strategy is better than verbal instruction to children with the absence of visual media, when applied to early childhood. This is also supported by the results of research conducted by the British Audio Visual Association (as cited in Zaman & Hernawan, 2005) which explains that information for children is obtained 75% through visual senses (visual), 13% through the sense of hearing and 6% through the sense of touch.

6 CONCLUSION

Based on the discussion, it can be concluded that the use of visual cues instruction in the school environment in every area or place of children's activities can have a positive impact on changes in discipline and self-directed attitude in children. Visual cues contain rules by using symbols, images, photos and writings that can help children understand the rules and make it easier for them to perform simple activities without the help of others. The use of visual cues instruction proves to help the child to know and control behavior and rules of what should be done.

Based on the results of the research, the researcher recommends that schools use visual cues instruction. Teachers also can combine strategies to improve self-reliance and discipline of children by using visual cues instruction.

7 ACKNOWLEDGMENT

The researcher acknowledges the Directorate of Research and Community Service of the Directorate General for Research Strengthening and Development of the Ministry of Research, Technology and Higher Education who have funded the research in accordance with the Decree of the Director General of Research and Development Number 30/EKTP/2017 dated April 3, 2017 on "Higher Education PTNBH Budget Year of 2017".

REFERENCES

Arthur, D. & Thorne, S. (1998). Professional self-concept of nurses: A comparative study of four strata of students of nursing in a Canadian university. *Nurse Education Today, 18*(5), 380–388.

Blagojevic, B., Logue, M., Bennett-Armistead, S., Taylor, B., & Neal, E. (2011). Take a look! Visual support for learning. *Teaching Young Children, 4*(5), 10–14.

Burchinal, M.R., Peidener-Feinberg, E., Pianta, R., & Howes, C. (2002). Development of academic skills from preschool through second grade: Family and classroom predictors of developmental trajectories. *Journal of School Psychology, 40*(5), 415–436.

Charlop-Christy, M.H. & Kelso, S.E. (2003). Teaching children with autism conversational speech using a cue card or written script program. *Education and Treatment of Children, 26*(2), 108–27.

Fox, L., Carta, J., Strain, P.S., Dunlap, G., & Hemmeter, M.L. (2010). Response to intervention and the pyramid model. *Infants & Young Children, 23*(1), 3–13.

Fox, L., Dunlap, G., Hemmeter, M.L., Joseph, G.E., & Strain, P.S. (2003). The teaching pyramid: A model for supporting social competence and preventing challenging behavior in young children. *Young Children, 58*(4), 48–52.

Ganz, J.B. & Flores, M.M. (2008a). Effects of the use of visual strategies in play groups for children with autism spectrum disorders and their peers. *Journal of Autism and Developmental Disorders, 38*(9), 26–40.

Ganz, J.B., Kaylor, M., Bourgeois, B., & Hadden, K. (2008b). The impact of social scripts and visual cues on verbal communication in three children with autism spectrum disorders. *Focus on Autism and Other Developmental Disabilities, 23*, 79–94.

Hamalik, O. (2008). *Perencanaan Pembelajaran berdasarkan Pendekatan iSstem.* Cet. VII [Learning planning based on system approach 7th ed.]. Jakarta, Indonesia: Bumi Akasara.

Hawkins, H. & Clayton, C. (2011). *Using visual cues to improve classroom instruction for young children with developmental delays.* Atlanta, GA: Assistive Technology Dept. Fulton County School.

Hemmeter, M.L., Ostrosky, M.M., & Fox, L. (2006). Social and emotional foundations for early learning: A conceptual model for intervention. *School Psychology Review, 35*(4), 583.

Holden, G.W. (2002). Perspectives on the effects of corporal punishment: Comment on Gershoff 2002. *Psychological Bulletin, 128*(4), 590–595.

Howard, B.J. (1996). Advising parents on discipline: What works. *Paediatrics, 98,* 809–815.

Hunter, A. & Hemmeter, M.L. (2009). Addressing challenging behavior in infants and toddlers. The Center on the Social and Emotional Foundation for Early Learning. Zero to three. Retrieved from https://perspectives.waimh.org/2009/06/15/addressing-challenging-behavior-in-infants-and-toddlers-the-center-on-the-social-and-emotional-foundations-for-early-learning/

Kasmadi, K. (2013). *Membangun Soft Skill Anak-anak Hebat* [Building great children's skill soft]. Bandung: Alfabeta.

Krantz, P.J. & McClannahan, L.E. (1993). Teaching children with autism to initiate to peers: Effects of a script-fading procedure. *Journal of Applied Behavior Analysis, 26*(1), 121–32.

Krantz, P.J. & McClannahan, L.E. (1998). Social interaction skills for children with autism: A script-fading procedure for beginning readers. *Journal of Applied Behavior Analysis, 31*(2), 191–202.

Listiana, A. & Rachmawati, Y. (2016). The teacher's perception on TPM-Kurtilas implementation in Amal Keluarga Kindergarten-Bandung. *Advances in Social Science, Education and Humanities Research (ASSEHR), 58*, 1–6.

Maryadi, B., Listiana, A., Rachmawati, Y., & Zaman, B. (2016). The TPM-Kurtilas implementation program of children's behavioral changes in kindergarten. *Advances in Social Science, Education and Humanities Research (ASSEHR), 58*, 105–111. doi: 10.2991/icece-16.2017.17

Norfolk Country Council Early Years & Childcare. (2015). *Visual cues*. Retrieved from https://www.norfolk.gov.uk/-/media/norfolk/downloads/children-and-families/childcare/visual-cues.pdf?la=en

Quill, K.A. (1995). Visually cued instruction for children with autism and pervasive developmental disorders. *Focus on Autistic Behavior, 10*(3), 10–20.

Rohmaniah, N., Tegeh, I.M., & Magta, M. (2016). Penerapan teknik modifikasi perilak token economy untuk meningkatkan kedisiplinan anak usia dini [Application of token economy behavior modification techniques to improve the discipline of early childhood]. *E-Journal PAUD, 4*(2), 1–11.

RoI. (2003). Undang-Undang RI 2003 No. 20, Sistem Pendidikan Nasional (UUSPN) [Law RI 2003 No. 20, National Education System (UUSPN)]. Jakarta, Indonesia: Republik of Indonesia.

RoI. (2014). Peraturan Mentri Pendidikan dan Kebudayaan Republik Indonesia 2014 No. 146, kurikulum 2013 PAUD. [Regulation of the Minister of Education and Culture of the Republic of Indonesia 2014 146, curriculum 2013 PAUD]. Jakarta, Indonesia: Ministry of Education and Culture, Republik of Indonesia.

Sailor, W., Dunlap, G., Sugai, G. & Horner, R. (2009). *Handbook of positive behavior support*. New York, NY: Springer.

Santrock, J.W. (2007). *Life-span development: Perkembangan Masa Hidup (edisikelima) (Development of lifespan 5th ed.)*. (Penerj. Achmad Chusairi, Juda Damanik; Ed. Herman Sinaga, Yati Sumiharti). Jakarta, Indonesia: Erlangga.

Saputri, A.T. (2016). *Penanaman Nilai Kemandirian dan Kedisiplinan bagi Anak Usia Dini Siswa TK B di Kelompok BermainMutiara Hati Purwokerto* [Cultivation of independence and discipline value for early childhood students of Kindergarten B in the Playing GroupMoney of Purwokerto]. (Skripsi). Fakultas Tarbiyah dan Ilmu Keguruan Institut Tinggi Agama Islam Negeri, Purwokerto.

Sarokoff, R.A., Taylor, B.A., & Paulson, C.L. (2001). Teaching children with autism to engage in conversational exchanges: Script fading with embedded textual stimuli. *Journal of Applied Behavior Analysis, 34*(1), 81–84.

Schaefer, C. (1996). *Cara Efektif Mendidik dan Mendisiplinkan Anak* [How to effectively educate and discipline children]. Jakarta: Mitra Utama.

Siswanti, I. & Lestari, S. (2012). *Panduan Bagi Guru dan Orang tua Pembelajaran atraktif dan 100 permainan kreatif untuk PAUD* [Guides for teachers and parents attractive learning and 100 creative games for early childhood]. Yogyakarta, Indonesia: CV Andi Offset.

Socolar, S. et al. (2007). A longitudinal study of parental discipline of young children. *Southern Medical Journal, 100*(5), 473–477.

Straus, M.A. & Fauchier, A. (2007). *Manual for the dimensions of discipline inventory*. Durham, UK: Family Research Laboratory, University of New Hampshire.

Wiyani, N.A. (2012). Format PAUD Konsep, Karakteristik & Implementasi Pendidikan Anak Usia Dini [PAUD format the concept, characteristics & implementation of early childhood education]. Yogyakarta, Indonesia: Ar-Ruzz Media.

Zaman, B. & Hernawan, A.H. (2005). *Media dan Sumber Belajar TK* [Media and source of learning kindergarten]. Jakarta, Indonesia: Universitas terbuka.

How undergraduate students of mathematics education perform microteaching with the topic of the incircle and the area of a triangle for grade VIII

M. Marsigit, H. Retnawati & R.K. Ningrum
Universitas Negeri Yogyakarta, Indonesia

ABSTRACT: This study aimed to describe how undergraduate students of a mathematics education study program performed microteaching with the topic of the incircle and the area of a triangle for VIII graders. The study involved descriptive qualitative research. The population included all sixth semester undergraduate students of a mathematics education study program, in the academic year 2016/2017, with students being randomly selected. The data were collected through observation, interview and documentation, and descriptively analyzed. The results showed that through a lesson plan guiding process, revision, and teaching practice using guided inquiry learning based on a scientific model, undergraduate students could actively construct their knowledge of the topic under study.

1 INTRODUCTION

A teacher is one of the main components in the efforts to improve Indonesian educational quality and has a main role in the educational process (Saban & Coklar, 2013). In this case, a teacher does not only serve as the designer and the executor of learning activities but also as the designer and executor of training and guidance activities, research, development and program management school activities (Universitas Negeri Yogyakarta, 2011). Consequently, in order to support the teacher's role, which is very complex, a teacher's competencies should be continuously developed.

Based on Law Number 14, Year 2005 (RoI, 2005a) and Government Regulation Number 19, Year 2005 (RoI, 2005b), all teaching staff must have some competencies, namely pedagogical, personal, professional and social competencies. The teachers who have all of these competencies will be good facilitators in the classroom. Some competencies that are important for supporting classroom learning activities are the ability to develop lesson plans, organizing learning groups, arranging and conditioning classrooms (Baştürk, 2016), facilitating classroom discussions and the ability to deal with multiple situations or conditions that might appear within classrooms (Baştürk, 2016). These capabilities are necessary for the teachers in order to support their important roles as students' guides in constructing and developing their knowledge (Saban & Coklar, 2013).

To reach competence and professionalism, teachers should have well-planned preparation. This preparation starts from the educational degree for the teacher candidates. Undergraduate students should be supported with multiple skills and competencies. One of the activities for undergraduate students' preparation that might be performed is to establish their basic teaching capabilities both theoretically and practically. This activity might be trained through microteaching activities (Universitas Negeri Yogyakarta, 2011). In these activities, undergraduate students should be able to show teaching ability comprehensively.

Some studies have shown that the main problem in implementing learning activities for both teachers and undergraduate students is the imbalance between theory and practice.

Many teachers or undergraduate students who master the theories of how to teach are unable to implement these theories into their learning activities (Karckay & Sanli, 2015; Baştürk, 2016). Microteaching can be one of the main alternatives to solve this problem. Not only the theories of how to teach, but also the opportunity to plan and implement the learning strategies that undergraduate students have understood are provided (Saban & Coklar, 2013). Undergraduate students are divided into small groups with less than 20 people in a group and are given opportunities to perform teaching–learning activities of about 5–20 minutes (Saban & Coklar, 2013). The environment in the microteaching activities is deliberately created in order to facilitate undergraduate students to perform and get feedback and assessment of their performance using the result of reflection to develop their teaching ability (Ralph, 2014).

Through microteaching, it is expected that undergraduate students' emotional intelligence and interpersonal capabilities might be well-developed (Campos-Sánchez et al., 2013). Thereby, undergraduate students have enough time to prepare themselves well before teaching in classrooms. Concerned with the implementation of learning activities, some teachers had not understood about the assessment system in the implementation of Curriculum 2013 (the Indonesian national curriculum) so they had difficulties in developing instruments for it (Retnawati et al., 2016). Microteaching could train undergraduate students' capabilities in setting the learning time effectively and in deciding the assessment model that they will implement later in the classroom (Saban & Coklar, 2013), so this problem can be solved.

In addition, undergraduate students can identify and attain information about teaching skills and knowledge, and mutually criticize and give suggestions about their peers' performance (Amobi & Irwin, 2009; Ralph, 2014). For the undergraduate students, the feedback provided by both their peers and their academic supervisors might help them to: 1) see their strengths and weaknesses in teaching so that they have opportunities to evaluate and develop their teaching capabilities (Fernández & Robinson, 2006; Kpanja, 2001); and 2) decrease the negative impacts that might occur during the actual learning activities such as low self-confidence, noise and depression during learning activities (Arsal, 2015; Fernández, 2005; Kpanja, 2001; Peker, 2009; Şen 2010). Undergraduate students who perform microteaching activities have minor difficulties during their teaching assignment in comparison to those who do not perform microteaching activities. Therefore, microteaching should be attended by all undergraduate students to equip them with various skills which are necessary in designing and implementing learning activities.

Related to learning activities, it will be effective if students can be provided with opportunities to contemplate what they have been learning (Bruning et al., 2011). This matter is in accordance with the concept of constructivism learning. For the students of higher educational degrees, they should have been able to regard mathematics as something interesting and useful and as a medium for displaying their creativity (National Council of Teachers of Mathematics, 2000). Certainly, the teachers should have to create a learning process that could facilitates students to construct their knowledge and develop their creativity.

One of the topics in mathematics for VIII grade students is the incircle and the area of a triangle. Many constructivism-based learning models might be implemented for teaching this topic. Through constructivism-based learning models, students are directed to understand how to determine the incircle and the area of a triangle. The models of learning are selected based on the characteristics of the topic and the characteristics of the student. One of the models of teaching, which is appropriate to the characteristics of this topic, is guided inquiry based on a scientific method. This model could facilitate students to construct their knowledge actively (Rochani, 2016), find their understanding about the topic of learning (Hidayati, 2017) and motivate the student to solve their problem actively (Setianingsih, 2016). It will be meaningful learning (Siregar, 2015). This article will describe the studies that related to how undergraduate students perform microteaching with the topic of the incircle and the area of a triangle. It will describe how undergraduate students used a constructivism-based learning model, namely guided inquiry based on a scientific method in microteaching activities.

2 METHOD

This study involved qualitative descriptive research. This article describes how the undergraduate students prepared microteaching activities using guided inquiry based on a scientific model with the topic of the incircle and the area of a triangle for VIII grade students. The population of this study were all sixth semester undergraduate students of a mathematics education study program in Yogyakarta State University, Indonesia. From this class, students were randomly selected as participants.

Research data were collected by observation, interview and documentation. Observation was used to identify how undergraduate students performed their activities during microteaching, especially their abilities to design the learning process, explain the learning materials, implement the learning models, implement the learning methods and implement learning media. Interviews were used to clarify the results of observation and identify the difficulties or the obstacles that undergraduate students perceived when designing the learning process, implementing learning models, implementing learning methods or implementing learning media. Documentation was used to get pictures of the implementation of microteaching activities.

Descriptive analysis was employed to identify the preparation and the implementation of undergraduate students' activities in microteaching. The procedures of qualitative data analysis were reducing data, presenting data and drawing conclusions. The description of this study refers to the stages of microteaching activities as follows: (1) Supervision 1, including the process of consulting learning materials, models, methods and media that the teacher-candidate university students would implement in the microteaching activities with their academic supervisor; (2) Learning Set Development, including the process of developing the syllabus, lesson plans, student worksheets and learning media that would be implemented; (3) Supervision 2, including the process of presenting the learning sets that were developed with the academic supervisor; (4) Revision, including the process of revising the developmental results based on the feedback and the suggestions that were provided by the academic supervisor; and (5) Teaching Practice.

3 RESULTS AND DISCUSSION

3.1 *Results and description of the study*

The data were collected by observation, interview and documentation. The results of this study were described based on the phase of microteaching activities that undergraduate students attended, including (1) Supervision 1; (2) Learning Set Development; (3) Supervision 2; (4) Revision; and (5) Teaching Practice. A description of each stage is provided in the following sections.

3.1.1 *Supervision I*
Before entering Supervision 1 stage, undergraduate students prepared the preliminary microteaching activities. They must do a literature study first, to select some alternative topic of learning that would be employed in the teaching practice. Additionally, undergraduate students must select the model and learning media which is appropriate to the topic selected and review the fitness between the characteristic topic of learning with the model and media. After that, undergraduate students performed guidance and consultation activities with the academic supervisor. These activities were later regarded as the Supervision 1 stage.

In the Supervision 1 stage, undergraduate students consulted the fitness between the topic of learning with the model, method and learning media that they had reviewed and selected previously. After that, undergraduate students selected one design to implement in their teaching practice in the last stage of microteaching. Preservice teachers must select the model/method which is relevant to the implementation of Curriculum 2013.

The sample of this study chose guided inquiry based on a scientific method to teach the topic of the incircle and the area of a triangle. In the design of learning, the implementation of the guided inquiry learning model was elaborated into learning steps using scientific methods, namely observing, raising questions, gathering information, associating and communicating, so the students could construct their knowledge about how to determine the incircle and the area of a triangle by themselves within the learning activities.

3.1.2 *Learning set development*

In this stage, the topic and model of learning that had been selected and approved in the Supervision 1stage become the foundation of this stage. Undergraduate students developed a syllabus detailing the topic of learning. After the syllabus development was completed, they developed a lesson plan and student worksheet.

Undergraduate students designed a lesson plan according to the characteristics of guided inquiry based on a scientific method. Guided inquiry activities were integrated into the activities that referred to the steps of the scientific method, namely observing, raising questions, gathering information, associating and communicating. In each step, teacher and student activities were detailed.

In the apperception phase, undergraduate students prepared a quiz which is contained with the question to determine the characteristics of rectangles and parallelograms. Each student must complete the quiz to recall their knowledge of that topic. In core activities, students were divided into small groups as discussion groups. Through discussion activities, the students were expected to be able to discover how to determine the incircle and the area of a triangle. These activities would be facilitated by the student worksheets. The teacher guided the students to construct their knowledge during the learning activities.

Undergraduate students also designed a student worksheet to support the learning process. It was administered as the guideline for the students' learning activities within the discussion process. Student worksheets were designed according to the characteristics of guided inquiry based on a scientific method. The student worksheets provide illustrations and guiding questions to assist the students to construct their knowledge about how to determine the incircle and the area of a triangle. The students were directed to find the concept of the incircle and the area of a triangle using the concept of rectangles and parallelograms. Then, they were asked to identify the relationship between the incircle and the area of a parallelogram and that of a triangle. After discussing and completing the worksheet, the students were expected to be able to conclude how to determine the incircle and the area of a triangle. Each activity in the student worksheet was designed based on the learning steps in the scientific methods. Undergraduate students also prepared learning media that the students would use during the learning activities.

3.1.3 *Supervision 2*

In this stage, undergraduate students presented the results of the development learning kit to their academic supervisor. This stage aimed to evaluate the results of the learning kit that undergraduate students had developed. In this stage, undergraduate students with the academic supervisor evaluated the fitness between the syllabus, the lesson plan and the student worksheet with the characteristics of the learning models. In this stage, undergraduate students and the academic advisor also evaluated the readability of the worksheet.

Overall, the learning kit that undergraduate students developed belonged to the "Good" category. The learning activities designed in the lesson plan were in accordance with the learning characteristics of the guided inquiry learning model, while the learning stages referred to the scientific methods. Similarly, the student worksheet and learning media were also in accordance with the characteristics and might be employed in the learning activities. There were only several notes for minor revision that the academic advisor provided for improving the lesson plan and the student worksheet.

3.1.4 *Revision of the developed product*

In this stage, undergraduate students revised the learning kit based on the feedback and the suggestions that had been provided by the academic advisor. The learning kit that resulted from the revision would be employed in teaching practice.

3.1.5 *Teaching practice*

After the learning kit had been revised, undergraduate students performed their teaching practice. Teaching practice took place for around 30 minutes with the assumption that it was one learning period. In teaching practice, one student served as the teacher while the others served as the students. After undergraduate students finished performing their teaching practice, the academic advisor guided all preservice teachers to reflect on the activities. Every undergraduate student in the class was provided an opportunity to comment on their friend's performance as a teacher and the comment might be provided as questions, suggestions or feedback. Then, the academic advisor provided feedback and suggestions in relation to the overall learning activities within the teaching practice.

In the lesson plan, it was written that the learning activities started with the teachers who gave greetings and apperception, explained the learning objectives and provided learning motivation. However, after explaining the learning objectives, undergraduate students forgot to provide learning motivation; instead, they directly gave a quiz. During the clarification at the end of the activities, this incident occurred because the preservice student was nervous and had low self-confidence because he had to act as a teacher in front of his own friends and academic advisor. This was a matter that caused the preservice teacher to have low concentration and miss that part in the learning activities.

In the apperception phase, it was written on the lesson plan that the time allocated for the students to complete the quiz was three minutes. However, in the implementation, the students took more than three minutes to complete the quiz. After the students had completed the quiz, the teacher invited them to discuss the results of their quiz. After that, the teacher divided the students into small groups which consisted of three students. Then, the teacher distributed the student worksheet and learning media to each group. The students had discussion activities with their peers in a group to complete the worksheet. During the discussion activities, the teacher went around the classroom in order to see the process of the group discussion and guided the students who had difficulties by illustrating the examples and by providing triggering activities.

In the observation step, students read their worksheet. By using the learning media, students discussed the problems in the worksheet in their group. Then, the students wrote the results of the observation step into the available columns in the worksheet. After the observation step had been conducted, the students wrote the questions in their worksheet about the relation between the triangle and parallelogram that had been illustrated in the worksheet within the question raising step. These questions would be used by the students as their starting point in the information gathering step. In this step, the students discussed and looked for information from multiple learning sources namely books, the Internet or teachers.

The next step was the associating step. The students processed the information that they had attained in order to answer the questions related to the relationship between a parallelogram and a triangle to determine the incircle and the area of a triangle. After answering all questions in this step, students wrote their conclusions. During the discussion activities, the teacher went around the classroom to guide the students who had difficulties in completing the worksheet that had been distributed.

In the conclusion writing step, the time that was allocated for the implementation of the learning activities was not in accordance with what had been set in the lesson plan. The time that had been allotted for the students' discussion activities was still insufficient and there were some groups who could not complete their activities. It caused a time addition for discussion activities. Time addition caused a change in time allotment for subsequent activities. Consequently, the time for presenting the discussion results in front of the class and for the reflection activities should be decreased.

The last step in these learning activities was the communicating step. The students presented the results of their discussion and the conclusion of their group in front of the class. The students from the other group were provided with an opportunity to respond or provide different answers. In this step, the students also had a class discussion session. The teacher here served as a moderator.

After the class discussion session had been completed, the teacher guided the students to perform their reflection and provided reinforcement of the concept of the incircle and the area of a triangle. Then, the teacher assigned the students' homework as a means of reinforcing their understanding of the topic. Finally, in closing activities, the teacher informed the next topic that would be studied for the next meeting so that the students might study this first. Then the teacher asked the students to end the learning activities by praying together.

At the end of the teaching practice, the academic advisor gave opportunities for the other undergraduate students to provide their comments, questions, suggestions or feedback in relation to the performance of the undergraduate students who served as the teacher. Several suggestions were provided by these students and one of these suggestions was that the teacher still had to learn how to put the class into a conducive situation. Sometimes the teacher focused too much on one group, so the other groups potentially started to be noisy. However, in terms of learning material mastery and delivery, the teacher had been quite good. The use of learning media in the form of student worksheets had also been quite useful for the students. Overall, the learning process using a scientific method-based guided inquiry was effective in activating the students to construct their own knowledge.

After the discussion and the evaluation had been done, the academic advisor provided feedback in the form of overall conclusions and suggestions. Based on the results of the evaluation, the main problem for undergraduate students in performing learning activities was the occurrence of nervousness and the lack of self-confidence during the teaching activities. That main problem might trigger undergraduate students to frequently look at the lesson plan in order to make sure that they would not miss the learning stages again. In relation to the peer feedback, undergraduate students still had difficulties in organizing the classroom. They also had difficulties in using the learning time allocation effectively because the teaching duration in the teaching practice had been relatively shorter than the actual time allocation.

In general, the results of the observation showed that undergraduate students had been able to design and implement the learning process using guided inquiry based on a scientific method in teaching the incircle and the area of a triangle. This matter was apparent from the learning kit that was designed and its implementation in the classroom. Almost all learning activities implemented were in accordance with the design although there were some learning activities whose implementation demanded additional time.

Moreover, the undergraduate student's mastery of the learning material was good. They were able to explain the materials well and clearly to the students both in the discussion and reflection activities, though once in a while they still looked at the material summary. The feedback and suggestions provided by the academic advisor and the peers might be of benefit in order to improve the performance in implementing the learning activities designed.

3.2   Discussion

Microteaching facilitated undergraduate students to practice the theory that they have studied in a real learning process. This is in accordance with the opinion of Baştürk, (2006) and Karckay and Sanli (2015) who stated that microteaching becomes one of the alternatives for undergraduate students to combine the theory that they have studied and practice. Through the stages in microteaching, including consultational activities on designing the learning kit that consists of the syllabus, lesson plan and student worksheet along with the academic advisor's supervision, undergraduate students have opportunities to design the learning activities according to the theory that they have studied and practice the theory in a learning activity.

Evaluation activities with peers and the academic advisor are beneficial for undergraduate students both directly and indirectly. From feedback and suggestions from peers and the academic advisor, undergraduate students can witness their skills and improve them. Then, microteaching activities can also be a means for improving the lesson plan and the student worksheet that has been designed. These activities help undergraduate students to identify their strengths and weaknesses and develop their teaching skills (Arsal, 2015; Fernández, 2005; Kpanja, 2001; Peker, 2009; Şen, 2010). Similarly, Ralph (2014) stated that the feedback and reflection results that have been provided in microteaching activities might be used for improving undergraduate students' teaching skills.

Based on the results of the study, it is apparent that the obstacle that undergraduate students still have to deal with in performing their teaching practice is the self-mastery of nervousness and lack of confidence. In order to overcome these problems, the students can be provided with more opportunities to perform their teaching practice. The more the undergraduate students perform their teaching practice, the more negative impacts will decrease (Kpanja, 2001; Peker, 2009; Şen, 2010; Arsal, 2015). In addition, with the increasing activities within the teaching practice, the problems that have been related to the capabilities of organizing classrooms and allocating time might be minimized (Saban & Coklar, 2013).

In relation to the competencies that an educator should have, a teacher is required to have several teaching staff competencies that include pedagogical, personal, professional and social competencies in accordance with the statement in Law Number 14, Year 2005 (RoI, 2005a) and Government Regulation Number 19, Year 2005 (RoI, 2005b). One of the concrete forms of pedagogic competencies that a teacher should have is the skill in designing and implementing good learning processes, which departs from the learning model that will be implemented in the classroom.

In this study, undergraduate students selected guided inquiry based on a scientific method in teaching the incircle and the area of a triangle to VIII grade students. Based on the results of the model design and the implementation in the classroom, it might be concluded that the learning process using guided inquiry based on a scientific method is able to facilitate effective learning because this model provides opportunities for the students to contemplate what they have been studying (Bruning et al., 2011), the active teaching and learning can improve students' higher order thinking skill (Apino & Retnawati, 2017) and turn the mathematics learning into something interesting, useful and beneficial in displaying their creativity (National Council of Teachers Mathematics, 2000). Learning activities with that model can facilitate students to discover their own knowledge. It can have a positive effect on students' cognitive abilities and achievement (Hidayati, 2017; Rochani, 2016).

In this model, students have the opportunity to investigate and solve the problem actively. This activity supports the process of constructing or modifying their knowledge (Setianingsih, 2016). Therefore, it will be meaningful learning. By strengthening students' mathematical understanding through meaningful learning, student difficulties with mathematics can be overcome (Retnawati et al., 2017) and students' achievement will increase (Siregar, 2015). The main characteristics of the guided inquiry model (Garton, 2005) are questioning (raising questions at the beginning of the learning process), student engagement (engaging the students actively in the learning process where teachers serve as a motivator and facilitator), cooperative interaction (inter-student interaction in group discussions), performance evaluation (artwork exhibition) and variety of resources (using various learning sources); these characteristics might facilitate and activate the students to construct their own knowledge. In addition, the scientific model prioritizes the meaningful learning process, the curiosity, the creativity exercise and students' independence as well as training students' deductive and inductive thinking skills in the process of drawing conclusions (RoI, 2013).

A learning process will be effective if the learning situation is conducive, interesting and fun. This is the basis of selecting various learning models. With that consideration in mind, the learning process will be effective and the materials that have been delivered will be well understood by the students (Mulyasa, 2007). The selection of a guided discovery learning model has been appropriate for teaching the incircle and the area of a triangle. Through this model, the students not only retrieve learning materials from the teachers but also can actively construct their knowledge through the discovery activities that have been designed under the learning steps and the student worksheet. Then, the possibility that the students will feel bored might be minimized because they keep being active during the learning process (Faturrahman & Sutikno, 2007). It might be concluded that guided inquiry based on a scientific method is effective in facilitating students to actively construct their own knowledge.

The use of media in learning activities also has an important role. The learning media can be the means for assisting the students in retrieving and understanding the concept that they have been studying. Several studies have shown that learning media eases the students in understanding the learning materials. In addition, interesting learning media might improve students' learning motivation (Faturrahman & Sutikno, 2007). It is important for teachers to improve learning motivation of students, because a lack of learning motivation can cause difficulties for students to understand mathematics (Retnawati et al., 2017). Another kind of media used in teaching to improve students' motivation is computer utilization, both in teaching activities and testing (Retnawati, 2015).

4 CONCLUSIONS

Based on the results and the discussion of this study, it can be concluded that microteaching was effective in facilitating undergraduate students to implement the learning theory that they have been studying. Through all stages of microteaching, undergraduate students had complete opportunities to design their learning activity according to the theory that they have been learning and perform it in learning activity simulation. Using guided inquiry based on a scientific method, undergraduate students were able to implement a learning process that facilitated the students to actively construct their own knowledge about the topic of the incircle and the area of a triangle.

REFERENCES

Amobi, F. & Irwin, L. (2009). Implementing on-campus microteaching to elicit pre-service teachers' reflection on teaching actions: Fresh perspectives on an established practice. *Journal of the Scholarship of Teaching and Learning, 9*(1), 27–34.
Ananthakrishnan, N. (1993). Microteaching as a vehicle of teacher training: It's advantages and disadvantages. *Journal of Postgraduate Medicine, 39*(3), 142–3.
Apino, E. & Retnawati, H. (2017). Developing instructional design to improve mathematical higher order thinking skills of students. *Journal of Physics: Conference Series, 812*, 1–7. doi: 10.1088/1742–6596/755/1/011001
Arsal, Z. (2015). The effects of microteaching on the critical thinking dispositions of pre-service teachers. *Australian Journal of Teacher Education, 40*(3), 139–153.
Baştürk, S. (2016). Investigating the effectiveness of microteaching in mathematics of primary pre-service teachers. *Journal of Education and Training Studies, 4*(5), 239–249.
Bruning, R. H., Schraw, G. J., & Norby, M. M. (2011). *Cognitive psychology and instruction* (5thed). Boston, MA: Pearson.
Campos-Sánchez, A., Sánchez-Quevedo, M.C., Crespo-Ferrer, P.V., García-López, J.M., & Alaminos, M. (2013). Microteaching as a self-learning tool: Students' perceptions in the preparation and exposition of a micro-lesson in a tissue engineering course. *Journal of Technology and Science Education, 3*(2), 66–72.
RoI. (2005a). *Undang-Undang Nomor 14 Tahun 2005 tentang Guru dan Dosen [Law Number 14 Year 2005 about Teacher and Lecture]*. Jakarta, Indonesia: Department of National Education.
RoI. (2005b). *Peraturan Pemerintah Nomor 19 Tahun 2005 tentang Standar Nasional Pendidikan [Government Regulation Number 19 Year 2005 about National Education Standards]*. Jakarta, Indonesia: Departement of National Education.
Faturrahman, P. & Sutikno, M.S. (2007). *Belajar mengajar (Learning and teaching)*. Bandung, Indonesia: Refika Aditama.
Fernández, M.L. (2005). Learning through microteaching lesson study in teacher preparation. *Action in Teacher Education, 26*(4), 37–47.
Fernández, M.L. & Robinson, M. (2006). Prospective teachers' perspectives on microteaching lesson study. *Education, 127*(2), 203–215.
Garton, J. (2005). *Inquiry-based learning*. Willard, MO: Technology Integration Academy.
Hidayati, R. (2017). Keefektifan setting TPS dalam pendekatan discovery learning dan problem-based learning pada pembelajaran materi lingkaran SMP [The effectiveness of TPS settings in discovery learning and problem-based learning approaches in junior high school learning materials]. *Jurnal Riset Pendidikan Matematika, 4*(1), 78–86.

Karckay, A.T. & Sanli, S. (2015). The effect of micro teaching application on the preservice teachers' teacher competency levels World Conference on Educational Sciences 2009 teachers' teacher competency levels. *Procedia Social and Behavioral Science, 1*(12), 844–847. doi: 10.1016/j.sbspro.2009.01.151

Kpanja, E. (2001). A study of the effects of video tape recording in microteaching training. *British Journal of Educational Technology, 32*(4), 483–486. doi: 10.1111/1467-8535.00215

RoI. (2013). *Peraturan Menteri Pendidikan dan Kebudayaan Nasional Nomor 81 Tahun 2013 tentang Implementasi Kurikulum [Regulation of the Minister of National Education and Culture No. 81 of 2013 on Curriculum Implementation].* Jakarta, Indonesia: Ministry of Education and Culture, Republik of Indonesia.

Mulyasa, E. (2007). *Menjadi guru profesional, menciptakan pembelajaran yang kreatif dan* menyenangkan [Become a professional teacher, creating creative and fun learning]. Bandung, Indonesia: Rosda Karya.

National Council of Teachers Mathematics. (2000). *Principles and standards for school mathematics.* Reston, VA: The National Council of Teachers of Mathematics, Inc.

Peker, M. (2009). The use of expanded microteaching for reducing pre- service teachers' teaching anxiety about mathematics. *Scientific Research and Essay, 4*(3), 872–880.

Ralph, E.G. (2014). The effectiveness of microteaching: Five years' findings. *International Journal of Humanities Social Sciences and Education, 1*(7), 17–28.

Retnawati, H. (2015). The comparison of accuracy scores on the paper and pencil testing vs. computer-based testing. *The Turkish Online Journal of Educational Technology, 14*(4), 135–142.

Retnawati, H., Hadi, S., & Nugraha, A.C. (2016). Vocational high school teachers' difficulties in implementing the assessment in Curriculum 2013 in Yogyakarta Province of Indonesia. *International Journal of Instruction, 9*(1), 33–48.

Retnawati, H, Kartowagiran, B., Arlinwibowo, J., & Sulistyaningsih, E. (2017). Why are the mathematics national examination items difficult and what is teachers' strategy to overcome it? *International Journal of Instruction, 10*(3), 257–276. doi: 10.12973/iji.2017.10317a

Rochani, S. (2016). Keefektifan pembelajaran matematika berbasis masalah dan penemuan terbimbing ditinjau dari hasil belajar kognitif kemampaun berpikir kreatif [The effectiveness of problem-based learning mathematics and guided discovery is reviewed from the cognitive learning outcomes of creative thinking]. *Jurnal Riset Pendidikan Matematika, 3*(2), 273–283.

Saban, A. & Coklar, A.N. (2013). Pre-service teachers' opinions about the micro-teaching method in teaching practise classes. *The Turkish Online Journal of Educational Technology, 12*(2), 234–240.

Şen, A.İ. (2010). Effect of peer teaching and microteaching on teaching skills of pre-service physics teachers. *Education and Science, 35*(155), 78–88.

Setianingsih, H. (2016). Keefektifan *problem solving* dan *guided inquiry* dalam setting TAI ditinjau dari prestasi belajar, kemampuan berpikir krtis dan kedisiplinan diri [The effectiveness of problem solving and guided inquiry in the TAI setting is viewed from the learning achievement, the ability to think critically and self-discipline]. *Jurnal Riset Pendidikan Matematika, 3*(2), 221–233.

Siregar, N.C & Marsigit, M. (2015). Pengaruh pendekatan *discovery* yang menekankan aspek analogi terhadap prestasi belajar, kemampaun penalaran, kecerdasan emosional spiritual [The influence of discovery approaches that emphasize aspects of analogy to learning achievement, reasonability, spiritual emotional intelligence]. *Jurnal Riset Pendidikan Matematika, 2*(2), 224–234.

Universitas Negeri Yogyakarta. (2011). *Panduan pengajaran mikro [Micro teaching guidelines].* Yogyakarta, Indonesia: UNY.

Mastering 21st-century skills through humanistic mathematics learning

D.B. Widjajanti
Universitas Negeri Yogyakarta, Indonesia

ABSTRACT: Twenty-first century skills can be developed through good education, including the learning of mathematics in class. One of the recommended mathematics learning approaches is humanistic mathematics learning. In principle, humanistic mathematics learning is a learning approach that aims to treat students as human beings, stating that human beings can learn, can discover things, can solve problems, can work together, and can appreciate the beauty and usefulness of mathematics. Such humanistic learning is not easy to implement. Teachers must really get to know their students, prepare learning materials, and plan learning scenarios in detail. Humanistic learning has the potential to build harmonious teacher–student and student–student relationships. With such relationships, students will be more enthusiastic in the math lesson. Mathematics classrooms in which every student enthusiastically attends lessons give room for teachers to develop students' curiosity, critical and creative thinking skills, problem-solving skills, collaborative skills, and positive attitudes toward mathematics. These characteristics are a person's main capital in being open to change, and such willingness is necessary to master the skills needed for the 21st century.

1 INTRODUCTION

Trilling and Fadel (2009) wrote a book that really inspires and reminds us all as teachers to give thorough attention to the things students need to be successful in living in the 21st century. Their book, entitled "21st Century Skills: Learning for Life in Our Times", is well known and helps many, especially teachers, to review what they have planned and done in their classes. According to Trilling and Fadel, three key skills are needed to live in the 21st century, namely: (1) learning and innovation skills; (2) digital literacy skills; (3) career and life skills.

Having these three key skills will help students meet their needs to work and live successfully in the 21st century. The main focuses for learning and innovation skills are: (1) critical thinking and problem—solving; (2) communication; (3) collaboration; (4) creativity and innovation. The main focuses for digitally literacy skills are: (1) information literacy; (2) media literacy; (3) Information and Communication Technology (ICT) literacy. Lastly, the main focuses for career and life skills are: (1) flexibility and adaptability; (2) initiative and self-direction; (3) social and cross-cultural interaction; (4) productivity and accountability; (5) leadership and responsibility (Trilling & Fadel, 2009).

These 21st-century skills can be developed through good education. The problem is what kind of education is required for students to have all these skills. Jerald (2009) states that some of the things that have changed in the 21st century have resulted in changes in the skills needed. Such changes include automation, globalization, workplace change, demographic change, and personal risk and responsibility. These changes make some knowledge and skills more necessary in the 21st century. So, what should be improved? The first thing to be improved is the quality of education.

Hampson et al. (2017) presented ten ideas for 21st-century education, namely: (1) open up lessons; (2) think outside the classroom box; (3) get personal; (4) tap into students' digital expertise; (5) get real with projects; (6) expect (and help) students to be teachers; (7) help (and expect) teachers to be students; (8) measure what matters; (9) work with families, not just children; (10) empower the student. These ten ideas for 21st-century education are realistic enough to contribute to improving student skills. How to implement them in class, especially in math classes?

The interesting thing about these ten ideas is the need for teachers to be less rigid about rules while teaching, helping students to become teachers, and putting themselves in the same shoes as a student. In the math class, the teacher should give to the students the opportunity to choose the problem to be solved first; choosing how to solve the problem, and choosing how to represent data, for example, will improve student skills in critical thinking and problem-solving. These skills become the basis for the development of other skills, given the rapid changes that occur, both in everyday life and in the world of work, as a result of the rapid development of communication and information technology. According to Kivunja (2015), having critical thinking and problem-solving skills makes someone more open-minded, which helps them to be sensitive and caring for others, society, the environment and the world as a whole.

Engendering critical thinking and problem-solving skills is one of the objectives of giving mathematics lessons in schools. The results of some research show a positive impact from the use of problem-based learning in developing critical thinking and problem-solving skills (Roh, 2003; Masek & Yamin, 2011; Happy & Widjajanti, 2014; El-Shaer & Gaber, 2014; Birgili, 2015; Nugraha & Mahmudi, 2015).

Although problem-based learning positively impacts critical and creative thinking skills, there is still one psychological problem that disrupts student achievement in learning mathematics. The obstacle is the still excessive student anxiety toward mathematics, known as math-anxiety. There are several research results that conclude that math-anxiety has a negative correlation with learning achievement in mathematics (Karimi & Venkatesan, 2009; Rubinsten & Tannock, 2010; Khatoon & Mahmood, 2010).

The approach to mathematics learning that is recommended for implementation in mathematics classes such that students do not feel anxious about their math lesson is humanistic mathematics learning. Students who learn mathematics without anxiety can potentially master the many skills necessary to live and work in the 21st century. However, because implementing humanist learning is not easy for teachers, they need to understand in advance what is meant by humanistic education and humanistic mathematics learning.

Here we study the benefits of humanistic mathematics learning in helping students to master 21st-century skills. This study is important in giving mathematics teachers and prospective mathematics teachers a clear picture of how the approach should help students master 21st-century skills through mathematics lessons. If teachers have a good enough understanding of what is meant by humanistic mathematics learning, then it is more likely that they will be willing to implement it.

2 METHOD OF RESEARCH

This research uses a descriptive meta-analysis method. Several analyses of the existing literature were assessed to derive conclusions about the advantages of humanistic education. Then, inferences are derived from the results of the analysis as to what and how to implement humanistic mathematics learning, and what teachers and students can gain from mathematics learning in a humanist class.

3 MAIN DISCUSSION

3.1 *Humanistic education*

Everyone is unique. Thus, every student in a single class will be slightly different. They vary in various aspects, including their type of intelligence. Students with different types of intelligence tend to have different interests. It is important to understand that education should pay attention to the diversity of learners and appreciate the uniqueness of each so that the outcome of their education is optimized. Some research results show the advantage of learning based on the theory of multiple intelligences (Xie & Lin, 2009; Baş & Beyhan, 2010; Hanafin, 2014; Çelik, 2015). Zucca-Scott (2010) stated that "education without a true appreciation for the uniqueness of each and every individual is an empty endeavor." Thus, an education system is required that can support self-development for each student in all of their potential and uniqueness. Humanistic education is one such option.

The concept of humanism in education is not a new concept. The ideas introduced by Maslow, Sartre, Schiller, Schulz, Erickson, Roger, and so on, have influenced the social and educational framework for a long time (Thakur, 2014). In principle, the humanistic approach to education is based on a more holistic and humanistic perspective on students (Khatib et al., 2013). According to Lei (2007), a humanistic approach emphasizes the importance of the inner world of the learner and places the individual's thoughts, emotions and feelings at the forefront of all human development.

There are some very interesting principles of humanistic education. In summary, among the main premises underlying 'humanistic' education, Moskovitz enumerates the following: (1) the principal purpose of education is to provide learning and an environment that facilitate the achievement of the full potential of students; (2) personal growth as well as cognitive growth is a responsibility of the school and, therefore, education should deal with both dimensions of humans—the cognitive or intellectual, and the affective or emotional; (3) for learning to be significant, feelings must be recognized and put to use; (4) significant learning is discovered for oneself; (5) human beings want to actualize their potential; (6) having healthy relationships with other classmates is more conducive to learning; (7) learning about oneself is a motivating factor in learning; (8) increasing one's self-esteem is a motivating factor in learning (Khatib et al., 2013).

Moskovitz stated that the principles of humanistic education basically remind educators to pay attention to the human facets of students (Khatib et al., 2013). That is, every student has the potential, emotions, feelings, and also the motivation that influences their learning outcomes. Schools and teachers should be able to improve students' learning motivation and self-esteem if they expect each student to realize their potential.

Another principle of humanistic education that needs to be underlined is the importance of healthy/harmonious relationships, between both teachers and teachers, and teachers and students, as well as between students and students. Such healthy/harmonious relationships provide opportunities for the growth of a conducive learning environment that supports the achievement of defined learning objectives. Teachers play an important role in realizing such relationships. Zucca-Scott (2010) argued that the humanistic approach relies on the teachers' ability to truly reinvigorate the "know thyself" motto, even if it means that we need to rethink schooling as a whole.

The research results of Xiao (2016) strengthen the evidence for humanistic education. Xiao has done research on students in China. The results showed that during the implementation of humanistic education in college, the qualification rate of students' physical fitness test is maintained at 90%. Xiao's recommendation is that "college should pay attention to the university sports humanities education, fully mobilize the enthusiasm and creativity of the students, and promote coordinated development of physical exercise and social adaptation".

3.2 Humanistic mathematics learning

The essence of humanistic education is education that humanizes human beings. This approach to humanistic education refers to the principle that every student is unique and has the potential to learn. Teachers and schools must appreciate students' uniqueness and help each student to achieve their optimal performance. There are potential, feelings, values, and emotions of students that the teacher should pay attention to during the learning process.

Similarly, in learning mathematics, learning with a humanistic approach is called humanistic mathematics learning. Humanistic mathematics learning is a philosophy of teaching and learning which attempts to explore the human side of mathematical thought and to guide students to discover the beauty of mathematics (Tennant, 2014). So, what should be prepared and implemented by mathematics teachers so that the lesson is humane?

Based on the results of his research, Cibulskaite (2013) has formulated three main guidelines/directions to implement humanistic mathematics learning. All three are related to how the mathematics teacher should involve the humanist side of the students in learning, through the material, the method of teaching, and the creation of a harmonious academic atmosphere.

The three directions are summarized as follows: (1) introduce students to humanistic aspects (e.g. sensitivity, openness, dignity, and responsibility) through the teaching materials; (2) improve the quality of the mathematics learning methodology so as to create classroom conditions that enable each student to succeed in learning, to learn from each other, to communicate, to work together, and to work creatively and actively, individually and in teams; (3) create an environment and learning atmosphere based on respect, mutual respect, sincerity, trust, honesty, responsibility, and sincerity. These are three important things that teachers need to do in order that their learning can be called humanistic.

Given the presence of students who are afraid or anxious about their math lessons, the task of mathematics teachers is further augmented by having to reduce the anxiety of these students. Thus, teachers must also recognize the character of each student in order to motivate students in a timely, appropriate way, with appropriate targets. The students' expressions, smiles, movements, eye gaze, and body language are things that the teacher should recognize.

With respect to subject matter, mathematics should be introduced to students in a way that will reduce their anxiety. Math teachers are required to implement learning that makes every student feel comfortable, happy, and appreciate the need for math, but at the same time also feel challenged to learn mathematics (Widjajanti, 2011). It is important to provide challenges to students because the presence of a challenge will stimulate students to think harder. However, teachers must be careful with the challenges that they give. The challenge must not be too complicated, so that it can still be carried out by the students; otherwise, it would make the students anxious in math class, and disturb the learning spirit.

Haglund (2004) states that there are ten characteristics of humanistic mathematics learning, namely: (1) placing students in the position of inquirer, not just a receptor of facts and procedures; (2) allowing students to help each other understand a problem and its solution more deeply; (3) learning several ways to solve problems, not just an algebraic approach; (4) including historical background that shows mathematics as a human endeavor; (5) using interesting problems and open-ended questions, not just exercises; (6) using a variety of assessment techniques, not just judging a student on his/her ability to carry out memorized procedures; (7) developing an understanding and appreciation of some of the great mathematical ideas that have shaped our history and culture; (8) helping students see mathematics as the study of patterns, including aspects such as beauty and creativity; (9) helping students develop attitudes of self-reliance, independence and curiosity; (10) teaching courses at the university level from a 21st-century perspective, so students have a grasp of the mathematics that is being used today in science, business, economics, engineering, and so on.

On the basis of these characteristics, it is acknowledged that mathematics learning will take place in humanistic terms when it is able to treat students as human beings, such that human beings can learn, can discover things, can solve problems, can work together, and can appreciate the beauty and usefulness of mathematics. Of course, such humanistic learning is not easy. The teachers should really get to know their students, prepare the learning materials, and plan learning scenarios in detail (Widjajanti, 2011).

Because students can learn and work together with one another, the teacher should provide an opportunity for this; for example, by giving group assignments and requiring students to discuss things. Because each student has the potential to discover something for themselves, the math lesson should not be entirely in the form of ready-made materials. There are certain parts that could be handed over to the students to "find something". This will encourage students to develop their critical, creative, and innovative thinking skills.

Because each student has the ability to solve problems, though the speed and level may vary, it is the duty of teachers to provide troubleshooting problems that can be used as a means for students to learn about problem-solving. The opportunity to solve problems independently will improve the students' skills in understanding the problem, while the opportunity to solve problems as a group will help improve communication and cooperation skills.

Because every student is capable of appreciating the beauty and usefulness of mathematics, the math teacher must frequently show its beauty and usefulness to the students, for example as part of the motivation at the beginning of the lesson. For this purpose, teachers can be assisted by technology. This will make students more motivated in learning math and more technologically literate.

To help math teachers implement humanistic mathematics learning, today there are many sources of learning available to them. The *Journal of Humanistic Mathematics* is one such. In this journal, there are articles, features about the world of mathematics, and a poetry folder, among others. One article in the journal, entitled "Stop Ruining Math!" (Steinig, 2016) is very interesting to consider. This article delves into the many common reasons why math is ruined for so many students. To make math enjoyable for everyone, some of the solutions include societal shifts, things that math teachers can do in the classroom, how parents can shift their attitudes toward math, thus creating a healthier home culture around math, and, lastly, courses by which students can change the way they participate in mathematics classes to get the most out of them.

Another article, entitled "Every minute of your life has been interesting" by D'Agostino (2017), can also inspire mathematics teachers to find ideas on how to show mathematical links to everyday life. Some of the songs and poems contained in the journal can also help math teachers show the beauty of mathematics to the students.

In principle, math teachers must find a variety of ways to make students learn math with enthusiasm, because enthusiasm is a source of energy for students in not giving up easily. The characteristic of not easily giving up is also required by students to live and work successfully in the 21st century. To make students comfortable learning mathematics, it is recommended that teachers of mathematics always: (1) demonstrate the beauty of mathematics; (2) highlight the usefulness of mathematics; (3) stress that mathematics can be learned by anyone and that learning mathematics is fun.

Taking note of the various reviews, there are simply formulated learning steps that should be adopted by the mathematics teacher such that their class is a humanist class: (1) start the learning process by showing the beauty and usefulness of mathematics; (2) motivate students with the idea that math is easy and can be learned by anyone; (3) give the students the task of finding something (concepts, formulas, or relationships); (4) allow students to do their work individually or in groups; (5) give students/groups a chance to communicate what has been found; (6) give additional explanations for each student to fully understand; (7) assign tasks to students to solve challenging or open-ended problems; (8) ensure that each assignment is collected from the students, corrected and rewarded; (9) celebrate learning by reading poetry, singing songs, watching movies/video, or learning outside the classroom; (10) be kind to the class, control emotions, and pay attention to troubled students; (11) provide a learning scaffold to students that is precisely targeted, timely, and appropriate.

4 CONCLUSION

The essence of humanistic education is education that humanizes human beings. The ten ideas for 21st-century education of Hampson et al. (2017) are in line with this approach. The principles of humanistic education presented by Moskovitz (cited in Khatib et al., 2013) basically remind educators to pay attention to the human aspect of the students. That is, every student has potential, emotions, feelings, and also motivation that influences their learning outcomes. Therefore, implementing a humanistic approach in every class, including math classes, will help students master 21st-century skills.

Humanistic mathematics learning is a learning approach that takes place with regard for the human face of every student. This approach considers that every student: (1) can learn and work together with each other; (2) has the potential to discover things; (3) has the ability to solve problems; (4) can appreciate the beauty and usefulness of mathematics. Humanistic learning has the potential to build harmonious teacher–student and student–student relationships.

With harmonious relationships, students will be more enthusiastic in mathematics lessons. Mathematics classrooms in which every student enthusiastically attends lessons will give room for teachers to develop student curiosity, critical and creative thinking skills, problem-solving skills, collaborative skills, and positive attitudes toward mathematics. These characteristics are a person's main capital in being open to change, and such willingness is necessary to master the skills needed for the 21st century.

REFERENCES

Baş, G. & Beyhan, O. (2010). Effects of multiple intelligences supported project-based learning on students' achievement levels and attitudes towards English lesson. *International Electronic Journal of Elementary Education, 2*(3), 365–385.

Birgili, B. (2015). Creative and critical thinking skills in problem-based learning environments. *Journal of Gifted Education and Creativity, 2*(2), 71–80.

Çelik, S. (2015). Managing the classes by using multiple intelligence instruction. *Journal of Education, 4*(1), 25–29.

Cibulskaite, N. (2013). The humanization of mathematics education. *Procedia - Social and Behavioral Sciences, 83*(2013), 134–139.

D'Agostino, S. (2017). Every minute of your life has been interesting. *Journal of Humanistic Mathematics, 7*(1), 117–118.

El-Shaer, A. & Gaber, H. (2014). Impact of problem-based learning on students' critical thinking dispositions, knowledge acquisition and retention. *Journal of Education and Practice, 5*(14), 75–85.

Haglund, R. (2004). Using humanistic content and teaching methods to motivate students and counteract negative perceptions of mathematics. *Humanistic Mathematics Network Journal, 27,* 4.

Hampson, M., Patton, A. & Shanks, L. (2017). *10 ideas for 21st century education.* London, UK: Innovation Unit. Retrieved from https://www.innovationunit.org/wp-content/uploads/2017/04/10-Ideas-for-21st-Century-Education.pdf

Hanafin, J. (2014). Multiple intelligences theory, action research, and teacher professional development: The Irish MI Project. *Australian Journal of Teacher Education, 39*(4), 126–141.

Happy, N. & Widjajanti, D.B. (2014). Keefektifan PBL ditinjau dari kemampuan berfikir kritis dan kreatif matematis, serta self-esteem siswa SMP [The effectiveness of PBL on mathematical critical and creative thinking skills, and self-esteem of Junior High School students]. *Jurnal Riset Pendidikan Matematika, 1*(1), 48–57.

Jerald, C.D. (2009). *Defining 21st century education.* Alexandria, VA: The Center for Public Education. Retrieved from http://www.centerforpubliceducation.org/Learn-About/21st-Century/Defining-a-21-st-Century-Education-Full-Report-PDF.pdf

Karimi, A. & Venkatesan, S. (2009). Mathematics anxiety, mathematics performance and academic hardiness in high school students. *International Journal of Educational Science, 1*(1), 33–37.

Khatib, M., Sarem, S.N. & Hamidi, H. (2013). Humanistic education: Concerns, implications and applications. *Journal of Language Teaching and Research, 4*(1), 45–51.

Khatoon, T. & Mahmood, S. (2010). Mathematics anxiety among secondary school students in India and its relationship to achievement in mathematics. *European Journal of Social Sciences, 16*(1), 75–86.

Kivunja, C. (2015). Teaching students to learn and to work well with 21st century skills: Unpacking the career and life skills domain of the new learning paradigm. *International Journal of Higher Education, 4*(1), 1–11. doi:10.5430/ijhe.v4n1p1

Lei, Q. (2007). EFL teacher's factors and students affect. *US-China Education Review, 4*(3), 60–67.

Masek, A. & Yamin, S. (2011). The effect of problem based learning on critical thinking ability: A theoretical and empirical review. *International Review of Social Sciences and Humanities, 2*(1), 215–221.

Nugraha, T.S. & Mahmudi, A. (2015). Keefektifan pembelajaran berbasis masalah dan problem posing ditinjau dari kemampuan berfikir logis dan kritis [The effectiveness of problem-based learning and problem posing in term of the logical and critical thinking ability]. *Jurnal Riset Pendidikan Matematika, 2*(1), 107–120.

Roh, K.H. (2003). Problem-based learning in mathematics. In *ERIC Digest* (ERIC identifier: ED482725). Retrieved from https://files.eric.ed.gov/fulltext/ED482725.pdf

Rubinsten, O. & Tannock, R. (2010). Mathematics anxiety in children with developmental dyscalculia. *Behavioral and Brain Functions*, *6*, 46.

Steinig, R.M. (2016). Stop ruining math! Reasons and remedies for the maladies of mathematics education. *Journal of Humanistic Mathematics*, *6*(2), 128–147. doi:10.5642/jhummath.201602.10

Tennant, R. (2014). Interdisciplinary teaching strategies in the world of humanistic mathematics. *Visual Mathematics*, *4*(4). Retrieved from http://www.mi.sanu.ac.rs/vismath/tennant1/index.html

Thakur, G.K. (2014). Humanism education and management of teaching-learning. *EPRA International Journal of Economic and Business*, *2*(12), 137–140.

Trilling, B. & Fadel, C. (2009). *21st century skills: learning for life in our times*. San Francisco, CA: John Wiley & Sons.

Widjajanti, D.B. (2011). Managing students' math-anxiety through humanistic mathematics education. In *Proceedings of International Seminar and Fourth National Conference on Mathematics Education, 2011: "Building the Nation Character through Humanistic Mathematics Education"* (pp. 777–786). Yogyakarta, Indonesia: Department of Mathematics Education, Yogyakarta State University. Retrieved from http://eprints.uny.ac.id/1873/1/P%20-%2074.pdf

Xiao, H. (2016). Study on humanistic education of college physical teaching: Educational innovation based on web Survey. *International Journal of Database Theory and Application*, *9*(5), 149–158.

Xie, J. & Lin, R. (2009). Research on multiple intelligences teaching and assessment. *Asian Journal of Management and Humanity Sciences*, *4*(2–3), 106–124.

Zucca-Scott, L. (2010). Know thyself: the importance of humanism in education. *International Education*, *40*(1), 32–38.

Factors influencing students' performance in solving international mathematics tests

A. Hamidy & J. Jailani
Universitas Negeri Yogyakarta, Indonesia

ABSTRACT: This study was aimed to examine the effects of implemented curriculum by schools and national examination success towards students' performances in solving international mathematics tests. This study was carried out with 600 samples consisting of 300 8th graders and 300 9th graders of East Kalimantan students. The chosen samples represent different types of implemented curriculum (Curriculum 2006 and Curriculum 2013) and categories of mathematics national examination results in 2016 (high, moderate, and low). The study used 16 items PISA-like mathematics (Cronbach Alpha = 0.835) and 28 items TIMSS-like mathematics (Cronbach Alpha = 0.837). Then, the data was analyzed using two-ways analysis of variance. Our data analysis revealed that there were significant effects of implemented curriculum and national examination success toward students' performances in solving PISA-like mathematics tests. But, there was no significant interaction effect between implemented curriculum and national examination success toward students' performances in solving PISA-like mathematics tests. Also, there were significant effects of implemented curriculum, national examination success, and their interactions between them toward students' performance in solving TIMSS-like mathematics tests.

1 INTRODUCTION

Mathematics is one of the most important subjects for people to master. Besides the role as a core of the sciences, learning mathematics has also a role in 21st century character development, e.g. self-confidence (Pais 2013), creativity and critical thinking (Irfan 2017), and consistency (Satrianawati 2015). These importances are in line with schools' vision on mathematics stated by National Council of Teacher of Mathematics (2000, p. 5) that "in this changing world, those who understand and can do mathematics will have significantly enhanced opportunities and options for shaping their futures. Mathematical competence opens doors to productive futures. A lack of mathematical competence keeps those doors closed". Hence, improving mathematical competences as well as other competences is absolutely required to survive in this century. This can be realized by studying and observing mathematical education today, either in national or international scope.

Two international studies on mathematical performances are Programme for International Student Assessment (PISA) and Trends in International Mathematics and Science Study (TIMSS). PISA is a study whose purpose is on measuring literacy, mathematical, and scientific competences of 15 –year old children (OECD 2013). TIMSS is a study measuring mathematical and scientific achievements of 6th and 8th graders (Mullis et al. 2009). This research provides chances to countries to compare their national education quality to other countries. Hence, the countries can acknowledge their education quality based on an international comparison and learn other countries' educational systems as a reference to improve theirs. Such information will be of consideration in formulating educational policies. In Indonesia, the results of PISA and TIMSS have become the base of educational curriculum, changing from 2006 Curriculum to 2013 Curriculum (Kemendikbud 2012).

A country success in achieving high scores on either PISA or TIMSS study is affected by various factors either by schools or students' characteristics. Some research (Argina et al. 2017, Carnoy et al. 2016, Lam & Lau 2014, McConney & Perry 2010) argued that factors related to schools and students' characteristics and qualities contribute to students' performance. However, PISA and TIMSS study is still limited when investigated factors related to students' performance, especially in Indonesia. Therefore, research investigating effects of factors related to schools toward students' performance in solving PISA and TIMSS tests is required.

Scheerens (2002) explained that educational effectiveness related to students' performance can be explained by three main factors: a) educational resources such as teachers' allowances, teachers' qualifications, etc.; b) school managerial and organizational characteristics; and c) teaching-learning activities' effectiveness related to curriculum. Furthermore, Martin & Mullis (2006) hold an opinion that curriculum consists of three aspects: the intended curriculum, the implemented curriculum, and the achieved curriculum.

Indonesian schools have two implemented curriculum, which are 2006 Curriculum and 2013 Curriculum. Also, to fulfill national objective, evaluation on Indonesian students' mathematical competence which is implemented in the curriculum is conducted through National Examinations. Differences in the curriculum and students' success in the National Examination might cause any effect toward PISA and TIMSS results. Hence, investigations on effects of implemented curriculum and competencies achievement based on such curriculum are required to be make one consideration in formulating policies to improve qualities of education.

2 LITERATURE REVIEW

2.1 *PISA study*

PISA refers to Programme for International Student Assessment, an international research measuring students' achievement periodically over three years. This research is conducted by the Organization for Economic Co-operation and Development (OECD), a joint organization of country governments concerning matters on policy making as the joint goal (Wu 2010). PISA focuses their research on measuring reading, mathematics, and science literacy of 15 year old children (OECD 2013). The three competences examined by PISA are the most important competences for post-formal-education social life.

PISA sets their main goal in acknowledging 15 year old children's competences which might have significant roles in their future, especially for their working and social lives. Wu (2010) explained that PISA focuses their research mainly on answering questions, for instance, "What competence must be mastered by someone if he wants to constructively participate in the society?" In the working context, Pusztai & Bacskai (2015) explained that PISA aims to evaluate students' competence in implementing their knowledge and skills in their working lives and serve the results in order to be able to be compared to other countries. Wijaya et al. (2014) stated that PISA's goals are in step with mathematical learning's which is to provide insights of students' performance in implementing mathematics in any situation. Therefore, in the mathematical performance context, the PISA study has purposes on predicting students' performance to implement any mathematical concept to solve daily problems.

PISA test development frameworks measuring mathematical performance use the term 'literacy'. OECD (2013, p. 25) defined mathematical literacy as follows, "Mathematical literacy is an individual's capacity to formulate, employ, and interpret mathematics in a variety of contexts. It includes reasoning mathematically and using mathematical concepts, procedures, facts, and tools to describe, explain, and predict phenomena. It assists individuals to recognize the role that mathematics plays in the world and to make the well-founded judgments and decisions needed by constructive, engage, and reflective citizens." Based on such definition, the PISA' assessment focus is the accumulation of 15 year old children's mathematical knowledge which can be employed in any condition as well as reflect processes of formulation, employing concept, and interpretation of problem solving.

Mathematical literacy concepts are arranged based on three main domains linked to each other where students are regarded as a problem solver; namely process, content, and context domains. The domain process describes how an individual relates a problem to mathematical concepts to solve the problem (Stacey 2011). In mathematical literacy, such mathematizing and modelling processes are divided into three main processes to solve the problems, which are to formulate, employ, and interpret. Implementation of those processes (formulate, employ, and interpret) demands specific and basic mathematics competences in various forms and degrees. PISA assessment frameworks formulate seven mathematical competences suiting the three problem-solving processes in mathematical literacy, which are representation, mathematical tool use; communication; mathematization; strategic planning; reasoning and argumentation; and language and symbolic operation uses, both formally and technically (OECD 2013).

Those seven competences are not adequate for problems solved by those three processes, as the fact of the problem solving and interpretation will face many situations. Problems faced daily have the same characteristics as basic mathematical phenomena (de Lange 2006). Relying on mathematical essences in every phenomenon, OECD (2013) has formulated four mathematical contents on PISA, which are quantity, change and relationships, space and shape, and uncertainty and data.

Beside process and content domains, the PISA study also emphasizes on competences in implementing those two domains in any life context. The context domain of PISA study measure students' competences in comprehending essences of problems and deciding suitable strategies regarded to such contexts. This competence is urgently needed by this 21th-century generation when facing many complex problems and thus PISA' assessment uses mathematics tests containing any context to measure their mathematical literacy competences. OECD (2013) explained that contexts in PISA tests involve personal, occupational, societal, and scientific competences.

2.2 TIMSS study

Trends in International Mathematics and Science (TIMSS) is an international study measuring 6th and 8th graders' achievements in mathematics and science. This study is held in 4 years intervals under the International Association for Evaluation of Education (IEA) coordination (Yilmaz & Hanci 2015). IEA is an association consisting of various research organizations concerning education (Wu 2010).

TIMSS sets goals on providing any information about educational policies of the participant countries. Yilmaz & Hanci (2015) stated that TIMSS study collects comparative data about educational systems from various countries to give contribution toward educational development particularly in mathematics and science. Martin & Mullis (2006) also revealed the similar fact that TIMSS was designed to collect any information which is supposed to be able to improve mathematical and scientific learning in all over the world. Through collecting data in the form of students' achievement in mathematics and science, TIMSS contribute information about educational development trends periodically. Periodic monitoring on educational development of a certain country provides important information to develop educational policies, improve people accountabilities, and identify achievement improvement and decline.

The TIMSS curriculum model has three aspects, which are the intended curriculum, the implemented curriculum, and the achieved curriculum (Martin & Mullis 2006). The intended curriculum is mathematical and scientific knowledge that people want to be mastered by students and how educational systems manage to facilitate it. The implemented curriculum is learning activities in classes, who teaches in classes, and how he or she teaches the subject. The achieved curriculum is a competence achievement achieved by students from what they have learned and how they draw perspectives on it. The competence achievement is measured by TIMSS tests with certain perspectives.

Mathematical achievement assessments by TIMSS are designed by being based on a framework. The TIMSS framework does not completely suit the country's curriculum. The TIMSS framework is also designed by being based on future trends of mathematics (Wu 2010). The assessment frameworks are divided into two main domains, which are content and cognitive domains (Mullis et al. 2009). Content dimension is a material collection of mathematics assessed in TIMSS. The content domain of 4th graders and 8th graders is different, based on the broadness of materials that have been taught. For 8th graders, content domain of mathematics includes number, algebra, geometry, data and chance.

Students' achievements in mathematics through the TIMSS version are not only assessed by students' capabilities in mastering all content domains well but also students' skills in thinking. Mathematical tests developed by TIMSS demand students to think from the lower to the higher orders (Budiman & Jailani 2014). Tajudin & Chinnappan' study results (2016) showed a significant relationship between higher order thinking skills and performance in solving the TIMSS tests. Therefore, cognitive domain becomes an important aspect in developing the TIMSS assessment. The domain consists of three components, which are knowing, applying, and reasoning. The three components are required to solve mathematical tests.

3 METHODS

This study was an ex post-facto survey research to answer questions as follows: a) Is there any effect of implemented curriculum towards students' perfomance in solving PISA-like and TIMMS-like mathematics tests?, b) Is there any effect of national examination success towards students' performance in solving PISA-like and TIMMS-like mathematics tests?, and c) Is there any interaction effect between the implemented curriculum and national examination success towards students' performance in solving PISA-like and TIMMS-like mathematics?

Data was collected from 600 students of East Kalimantan Junior High School, consisting of 300 9th graders (for PISA) and 300 8th graders (for TIMSS) which were selected randomly. Those 600 students represented the school type based on the implemented curriculum (2006 Curriculum and 2013 Curriculum) and mathematics national examination results in 2016 (high, moderate and low). Distribution of samples based on curriculum type and national examination category which had been analyzed are shown in Table 1.

Data was collected using tests. The test instruments were PISA-like mathematics tests consisting of 16 items (Alpha Cronbach = 0.0835) and TIMSS-like mathematical tests consisting of 28 items (Alpha Cronbach = 0.837). The PISA-like mathematics tests were done by 9th graders; while the TIMSS-like mathematics tests were done by 8th graders.

Data acquired was in the form of scores then converted into correct answers percentages. Then, the percentages were analyzed descriptively and inferentially to test effects of the implemented curriculum and national examination success toward students' performance in solving mathematics tests. Statistical tests used were two-ways analysis of variance (Anova).

Table 1. Distribution of samples.

		9th grade (PISA)	8th grade (TIMSS)
National exam. categ.	High	100	100
	Moderate	100	100
	Low	100	100
Curriculum type	2006	153	151
	2013	147	149

4 RESULT

4.1 *Performance in solving the PISA-like mathematics tests*

Overall, the mean of correct answer percentage of PISA-like mathematics tests was 29.67% with the standard deviation of 17.39. This percentage was low since it was below the percentage of 50%. The highest correct answer percentage was 90%; while the lowest one was 5%. Students' mean and standard deviation of correct answer percentage based on the implemented curriculum and national examination success are shown in Table 2.

Based on two-ways analysis of variance (Anova) (Table 3), there was a significant effect of the implemented curriculum and national examination success towards students' performances in solving the PISA-like mathematics tests. However, the interaction effect between them was not significant towards students' performance in solving the PISA-like mathematics tests. Based on the implemented curriculum type, the mean of correct answer percentage of the PISA-like mathematics tests by students from schools which apply the 2013 Curriculum was significantly higher than students from schools which apply the 2006 Curriculum.

4.2 *Performance in solving TIMSS-like mathematics tests*

Overall, the mean of correct answer percentage of TIMSS-like mathematical tests was 27.76% with the standard deviation of 19.85. This percentage was low because it was below

Table 2. The mean and standard deviation of correct answer percentage of PISA-like mathematics test.

		C2006	C2013	Total
High	Mean	37.86	45.12	43.60
	Std. Dev	11.46	22.36	20.72
	Min	20	10	10
	Max	60	90	90
Moderate	Mean	21.61	20.97	21.35
	Std. Dev	8.27	9.37	8.69
	Min	10	5	5
	Max	45	45	45
Low	Mean	22.53	28.15	24.05
	Std. Dev	10.48	9.72	10.53
	Min	5	10	5
	Max	45	50	50
Total	Mean	24.28	35.27	29.67
	Std. Dev	11.18	20.67	17.39
	Min	5	5	5
	Max	60	90	90

Table 3. Two-ways Anova of the mean of correct answer percentage of PISA-like test based on implemented curriculum and national examination success.

Factor	Mean square	F	Sig.
Curriculum	984.77	4.87	0.028
National examination	8492.04	41.98	0.000
Curriculum * National examination	368.51	1.82	0.164

the percentage of 50%. The highest correct answer percentage was 96.77%; while the lowest one was 3.23%. The mean (standard deviation) of students' correct answer percentages based on the implemented curriculum and national examination success are shown in Table 4.

Based on Anova analysis result (Table 5), there were significant effects of the implemented curriculum, national examination success, and interactions between them toward students' performance in solving TIMSS-like mathematics tests. Based on the type of implemented curriculum, the mean of correct answer percentage of TIMSS-like mathematics tests by students from schools implementing the 2013 Curriculum was significantly higher than students from schools implementing the 2006 Curriculum. Post-hoc test of the effect of national examination success toward the correct answer percentage of TIMSS-like mathematics tests is shown by Table 6.

Table 4. The mean and standard deviation of correct answer percentages of TIMSS-like mathematics test.

		C2006	C2013	Total
High	Mean	38.34	48.82	46.09
	Std. Dev	12.35	24.25	22.21
	Min	19.35	6.45	6.45
	Max	67.74	96.77	96.77
Moderate	Mean	14.85	19.22	16.90
	Std. Dev	9.09	8.53	9.06
	Min	3.23	6.45	3.23
	Max	54.85	41.93	54.85
Low	Mean	15.86	31.68	20.29
	Std. Dev	5.32	10.13	9.96
	Min	6.45	12.90	6.45
	Max	29.03	54.84	54.84
Total	Mean	19.37	36.26	27.76
	Std. Dev	11.96	22.50	19.85
	Min	3.23	6.45	3.23
	Max	67.74	96.77	96.77

Table 5. Two-ways Anova of the mean of correct answer percentage of TIMSS-like test based on implemented curriculum and national examination success.

Factor	Mean square	F	Sig.
Curriculum	6638.96	32.99	0.000
National examination	15947.2	79.25	0.000
Curriculum * National examination	737.69	3.67	0.027

Table 6. Post-hoc test of the mean difference of correct answer percentage of TIMSS-like test based on national examination success.

National examination categ.	Mean diff.	Sig.
High–Moderate	29.19	0.000
High–Low	25.81	0.000
Moderate–Low	−3.39	0.277

Based on Table 6, the mean of correct answer percentages of TIMSS-like mathematics tests by students from high category schools was significantly higher than students from moderate and low category schools. While, the mean of correct answer percentages of TIMSS-like mathematics tests by students from moderate and low category schools was not significantly different.

5 DISCUSSION

The correct answer percentages of PISA-like mathematics tests by East Kalimantan students show their mathematical literacy. The literacy is related to students' competences in analyzing, reasoning, and communicating effectively to formulate, solve and interpret the problems in any situation (Saenz 2008). Study analysis shows that East Kalimantan students' mathematical literacy was low, that is the correct answer percentage was below 50%.

The correct answer percentage of TIMSS-like mathematics tests by East Kalimantan students shows their mathematical achievements. The achievements are related to students' competence achievement of what they have learned at schools (Martin & Mullis 2006). The study results show that East Kalimantan students' mathematical achievements was low, that is the correct answer percentage was under 50%.

The East Kalimantan students' low mathematical achievement and literacy showed that in general, most of them have not achieved mathematical competences that international people required. There might be various factors influencing this, either from students or schools. Those factors were the implemented curriculum by schools and mathematical competences that nationally achieved which are shown by national examination scores on mathematics.

Based on study results, it was found that factors of the implemented curriculum significantly influence students' performance in solving PISA-like and TIMSS-like mathematics tests. These indicate that mathematical competences formulated by PISA and TIMSS still have relation to the schools' curriculum, although OECD (2013) explained that the PISA mathematical literacy does not relate to the schools' curriculum, but it more emphasizes in problem solving abilities in real life contexts. Wu (2010) also explained that there is a relationship between students' achievements toward what they learn at schools.

Students' performance in solving PISA and TIMSS mathematics tests is related to Higher Order Thinking Skill (Budiman & Jailani 2014). Hence, HOTS in mathematics learning is important to be implemented. Some efforts are (1) engaging students in the activities of non-routine problem solving; (2) facilitate the students to develop the ability to analyze and evaluate (critical thinking) and the ability to create (creative thinking); and (3) encourage students to construct their own knowledge, so that learning becomes meaningful for students (Apino & Retnawati, 2017). Those efforts are much related to curriculum implementation. Besides, to implement effective learning, the schools should conduct the monitoring of curriculum implementation (Retnawati et al. 2016).

Students performances from schools that implement the 2013 Curriculum in solving PISA-like and TIMSS-like mathematics tests were significantly higher than students from schools that implement the 2006 Curriculum. This fact showed that the PISA-and-TIMSS mathematical competences have been accommodated better in the 2013 Curriculum. The differences between the 2006 Curriculum and 2013 Curriculum were emphasized by the learning and assessment approach (Retnawati et al. 2016). 2013 Curriculum emphasizes to students centered learning, while the 2006 Curriculum still uses teacher centered learning. Furthermore, 2006 Curriculum assesses students' performance using tests that emphasize cognitive and psychomotoric aspects. While, 2013 Curriculum is using authentic assessments which measure students' affective, cognitive and psychomotor domain based on learning process and product.

The curriculum changes in Indonesia from the 2006 to 2013 Curriculum is caused by Indonesian students ranking low in PISA and TIMSS (Kemendikbud 2012). It means, the aim of 2013 curriculum implementation is to improve students' mathematical performance in international tests. So, it makes sense that PISA and TIMSS mathematical competences have been accommodated better in the 2013 Curriculum. But, the implementation of 2013 Curriculum is still not adequate which causes students' performance in solving PISA and TIMSS tests to still be low. Wardhani & Rumiati (2011) and Ahyan et al. (2014) found that in learning practices, using tests whose characters are similar to PISA or TIMSS tests is still rare in Indonesia.

However, it assumed that the superiority of the 2013 Curriculum might caused of quality of school which implemented the 2013 Curriculum. Accreditation schools which implemented the 2013 Curriculum at first were A (high class schools) or international schools. So, the students' quality from schools which implemented 2013 Curriculum was very good.

Furthermore, study results also showed that curriculum competence achievement factors significantly influence students' performances in solving PISA-like and TIMSS-like mathematics tests. This finding was indicated by the significant effect of categories of national examination success toward the mean of correct answer percentages of PISA-like and TIMSS-like mathematics tests. This was caused by the efforts in developing mathematical competences of TIMSS and PISA have been started to be realized through inserting TIMSS-like and PISA-like problems in national examinations. Retnawati et al. (2017) explained that educational quality improvement was apparent from the addition of TIMSS or PISA test items, which have the international quality, into the national examination battery set package. Nizam explained that there are approximately 5–10% of 2016 National Examination problems that have PISA-like characteristics (Wiwoho 2015), yet the problems still belong to moderate and low levels (Zulkardi 2015).

Based on post-hoc test, it had been found that students' mathematical achievements from high category schools were significantly higher than students from moderate and low category schools. Moreover, students' mathematical achievements from moderate and low category schools were not significantly different. This finding showed that students with a high score in national examination mathematics gain bigger opportunities in having a high score too in solving TIMSS-like mathematics tests. Yilmaz & Hanci (2015) also explained that there is a close relationship between success in national examination and success in TIMSS.

Based on those two factors' interactions, it had been found that there was a significant effect of interaction between the implemented curriculum and national examination success towards performance in solving TIMSS-like mathematics tests. However, the interactions did not significantly effect toward performance in solving PISA-like mathematics tests. These findings clarified that the TIMSS study is aimed to evaluate educational curriculum effectiveness (Mullis et al. 2009) where the curriculum involves three aspects: the intended curriculum, the implemented curriculum, and the achieved curriculum. It is different from the PISA study which does not directly relate to the schools' curriculum although the curriculum always becomes a reference in deciding educational policies.

6   CONCLUSION

Based on the results, it was concluded that there were significance effects of implemented curriculum and national examination success towards students' performances in solving PISA-like mathematics tests. There was no significant interaction effect between implemented curriculum and national examination success toward students' performances in solving PISA-like mathematics tests. There were significant effects of implemented curriculum, national examination success, and their interactions between of them toward students' performances in solving TIMSS-like mathematics tests.

REFERENCES

Ahyan, et al. 2014. Developing mathematics problems based on PISA level. *IndoMS Journal Mathematics Education* 5(1): 47–56.
Apino, E. & Retnawati, H. 2017. Developing Instructional Design to Improve Mathematical Higher Order Thinking Skills of Students. *Journal of Physics: Conference Series* 812(1): 1–7.

Argina, et al. 2017. Indonesian PISA result: What factors and what should be fixed? *Proceedings Education and Language International Conference* 1(1).

Budiman, A. & J. Jailani. 2014. Pengembangan instrumen asesmen higher order thinking skill (HOTS) pada mata pelajaran matematika SMP kelas VIII semester 1 [Developing higher order thinking skill (HOTS) assessment for 8th grade junior high school mathematics on first semester]. *Jurnal Riset Pendidikan Matematika* 1(2): 139–151.

Carnoy, M., et al. 2016. Revisiting the relationship between international assessment outcomes and educational production: Evidence from a longitudinal PISA-TIMSS sample. *American Educational Research Journal* 20(10): 1–32. doi: 10.3102/000283121665318.

De Lange, J. 2006. Mathematical literacy for living from OECD-PISA perspective. *Tsukuba Journal of Educational Study in Mathematics* 25(1): 13–35.

Irfan, M. 2017. Role of learning mathematics in the character building. In *International Conference on Education* (pp. 599–604).

Kemendikbud. 2012. *Dokumen kurikulum 2013* [Document of 2013 Curriculum].

Lam, T.Y.P. & Lau, K.C. 2014. Examining factors affecting science achievement of Hong Kong in PISA 2006 using hierarchical linear modeling. *International Journal of Science Education* 36(15): 2463–2480.

Martin, M.O. & Mullis, I.V. 2006. TIMSS: Purpose and design. In S.J. Howie & T. Plomp (Eds.), *Contexts of learning mathematics and science—lessons learned from TIMSS* (pp. 17–30). New York, NY: Routledge.

McConney, A. & Perry, L.B. 2010. Science and mathematics achievement in Australia: The role of school socioeconomic composition in educational equity and effectiveness. *International Journal of Science and Mathematics Education* 8(3): 429–452.

Mullis, I.V.S., et al. 2009. *TIMSS 2011 assessment frameworks*. Chestnut Hill, MA: TIMSS & PIRLS International Study Center.

National Council of Teacher of Mathematics. 2000. *Principles and standards for school mathematics*. Reston, VA: NCTM, Inc.

Organization for Economic Co-operation and Development. 2013. *PISA 2012 assessment and analytical framework: mathematics, reading, science, problem solving and financial literacy*. Paris: OECD Publishing.

Pais, A. 2013. An ideology critique of the use-value of mathematics. *Educational Studies in Mathematics* 84(1): 15–34.

Pusztai, G. & Bacskai, K. 2015. Parochial schools and PISA effectiveness in three central European countries. *Social Analysis* 5(2): 145–161.

Retnawati, H., et al. 2016. Vocational high school teachers' difficulties in implementing the assessment in curriculum 2013 in Yogyakarta province of Indonesia. *International Journal of Instruction* 9(1): 33–48.

Retnawati, H., et al. 2017. Why are the mathematics national examination items difficult and what is teachers' strategy to overcome it? *International Journal of Instruction* 10(3): 257–276. doi: 10.12973/iji.2017.10317a.

Saenz, C. (2008). The role of contextual, conceptual and procedural knowledge in activating mathematical competencies (PISA). *Education Student Mathematics* 71(2): 123–143. doi: 10.1007/s10649-008-9167-8.

Satrianawati, S. 2015. Building character of student in the 21st century by learning mathematics. *Proceeding International Seminar Of Science Education*. Yogyakarta State University.

Scheerens, J. 2002. School self-evaluation: Origins, definition, approaches, methods and implementation. *School-based evaluation: An international perspective* 8(8): 35–69.

Stacey, K. 2011. The PISA view of mathematical literacy in Indonesia. *IndoMS Journal Mathematics Education* 2(2): 95–126.

Tajudin, N.M. & Chinnappan, M. 2016. The link between higher order thinking skills, representation and concepts in enhancing TIMSS task. *International Journal of Instruction* 9(2): 199–214. doi: 10.12973/iji.2016.9214a.

Wardhani, S. & Rumiati. 2011. *Instrumen penilaian hasil belajar matematika SMP: belajar dari PISA dan TIMSS* [Assessment instrument for junior high school: Learning from PISA and TIMSS]. Yogyakarta: PPPPTK Matematika.

Wijaya, A., et al. 2014. Difficulties in solving context-based PISA mathematics tasks: An analysis of students' errors. *The Mathematics Enthusiast* 11(3): 555–584.

Wiwoho, L.H. 2015. Mulai 2016, UN pakai sistem komputer [Start from 2016, national examination will use CAT]. *Kompas*. Retrieved from http://edukasi.kompas.com/read/2015/01/25/08000091/Mulai.2016.UN.Pakai.Sistem.Komputer

Wu, M. 2010. Comparing the similarities and differences of PISA 2003 and TIMSS, *OECD Education Working Papers* 32. doi: 10.1787/5 km4psnm13nx-en

Yilmaz, G.K. & Hanci, A. 2015. Examination of the 8th grade students' TIMSS mathematics success in terms of different variables. *International Journal of Mathematical Education in Science and Technology* 47(5): 674–695. doi: 10.1080/0020739X.2015.1102977.

Zulkardi. 2015. PISA's influence on thought and action in mathematics education. In Stacey, K. & Turner, R. (Eds.). *Assessing Mathematical Literacy* (286–290). New York, NY: Springer International Publishing Switzerland. doi: 10.1007/978-3-319-10121-7_15.

Cultural values-integrated mathematical learning model to develop HOTS and character values

H. Djidu & H. Retnawati
Universitas Negeri Yogyakarta, Indonesia

ABSTRACT: Higher-order thinking skills (HOTS) and character are two main competencies that students should master in order to deal with the challenges in the 21st century. The development of HOTS and character might be facilitated through mathematical learning process by designing the learning environment that supports both competencies. This study aimed at generating a cultural values-integrated mathematical learning model to develop HOTS and character values of students. The study was a design research. The learning model was designed by means of three main methods, namely preliminary study, development phase, and assessment phase. The data in the study was attained through test, observation, and interview. The results of the study show that the cultural values-integrated mathematical learning model has met the criteria of practicality in its implementation and effectiveness toward the development of students' HOTS and character. The character values of students' that have been developed through the implementation of the learning model are responsibility, tolerance, and positive attitudes toward mathematics learning process.

1 INTRODUCTION

The development of knowledge, technology, and information in the 21st century provides multiple easiness in the efforts to meet every need. On the other hand, the easiness triggers various problems in any life aspect, starting from the personal level to the macro level. Therefore, each individual should master multiple competencies in order to overcome the various problems that might appear. These competencies are namely knowledge, skills, and character (Alismail & McGuire 2015). The three competencies might not be considered as three separate aspects. The intersection among the three competencies should be focused toward and be integrated on the implementation of education (Bialik et al. 2015) so that the outcome that will be attained is individuals who are ready to overcome the challenges of the 21st century.

Skills are related to the capabilities of an individual in integrating the knowledge that he or she has possessed in order to solve the problems that the individual encounters. One of the skills is higher-order thinking skills (HOTS) related to the capabilities of an individual in solving or accomplishing problems, tasks, or jobs in a new context that he or she has not ever met before (Mainali 2012, Thompson 2008). These skills do not only need understanding toward concepts or theories, but also capabilities to create connections, categories, manipulations, and applications toward facts/concepts into new situations (Thomas & Thorne 2009). These skills are often associated to the top three levels in Bloom's taxonomy, namely analyze, evaluate, and create (Ramos et al. 2013). HOTS is important to be trained among students because the development of HOTS will ease the students in dealing with challenges from multiple life dimensions (Tan & Halili 2015) and they will solve problems effectively (Brookhart 2010) so that they can improve their life quality (Limbach & Waugh 2010).

Character is something that has been associated to an individual's attitude, and it has been the form of religious teaching or the results of interaction with parents, colleagues, and community. An individual who has good character will be able to differentiate the good and the bad of an action (Lumpkin 2008). In the same time, character in education has a close relationship to the competencies of attitudes, as mentioned by Krathwohl that character is at the top level of affection taxonomy (Paper & Thomas 2004).

Due to this important character, education should also be directed toward developing it. Although the character might be trained and be developed (Heckman & Kautz 2013), the development of character takes a relatively long time (Zuchdi et al. 2015). As a result, the development of the students' character should be conducted continuously and integratively through the educational process that students experience. Lickona, one of the experts who has been famous in character education, explained that character is built upon three basic concepts: moral knowing, moral feeling, and moral action (Lickona 1991). This statement implies that in order to develop the students' character, the students should understand, contemplate, and habituate the performance of actions of character. So, to foster students who have a positive character, we need to infuse character values during the learning process. Character values that focused on in this study were responsibility, tolerance, and positive attitudes toward mathematics learning process.

How the mathematical learning process is conducted depends on the learning model that will be implemented. Different learning models or teaching styles that a teacher implements influences the students' skills and character development (Pinxten 2016). A learning model has a step by step procedure (syntax), learning environment (social system and principles of reaction), supporting system, and instructional (direct) effect and nurturing (indirect) effect (Joyce et al. 2009). In relation to this explanation, the development of the students' HOTS and character values as the nurturing effect of the learning process will be manifested if the learning environment supports the development of both competencies.

Both the HOTS and the character values of students cannot be taught directly through the mathematics learning process because they are the nurturing effect of the learning process. HOTS might be trained by providing opportunities toward the students in order to be active in the learning process (Djidu & Jailani 2016, Limbach & Waugh 2010), and engage students to solve real-world problems (Rahmawati & Suryanto 2014, Riadi & Retnawati 2014, Susanto & Retnawati 2016). Beside that, character values can be developed by integrating cultural values in the learning process (Hasanah 2016). Cultural values that can be used in the learning process were local languages, or local objects, or local activities. In addition, both the students' HOTS and the character values might be developed through discussion or collaboration activities (Dhunnoo & Adiapen 2013, Goethals 2013) in solving problems.

Based on the description above, a learning model that integrates cultural values in learning activities is needed. Therefore, this study aimed at generating a cultural values-integrated mathematical learning model in order to develop the students' HOTS and character values.

2 METHOD

The study was a design research with the objective of generating a cultural values-integrated mathematical learning model in order to develop HOTS and character values of students. The learning model was designed by means of three main stages, namely preliminary study, development phase, and assessment phase (Plomp 2013). The preliminary study was conducted in order to identify the mathematical learning process conditions and the problems that had been encountered. Then, the development phase was conducted in order to attain the learning model design. Last, but not the least, the assessment phase was conducted during and after the model implementation in order to measure the quality of the learning model that had been designed and its impact toward HOTS and characters values of students.

The data in the study was gathered through test, observation, and interview. The test that had been administered was intended to measure the students' HOTS. Then, the observation was conducted by two observers in order to attain information regarding the students' behaviors during the learning process. Next, the interview was conducted at the end of the overall learning process in order to identify the students' opinions regarding the learning process implementation.

3 RESULT AND DISCUSSION

3.1 *Learning model that has been designed*

The mathematical learning model that has been designed consists of several main components, namely syntax, social system, principles of reaction, support system, and learning impacts. Before having been implemented in the classroom, three experts and two practitioners asked a judgment toward the learning model in order to attain information about validity evidence. The result showed that two experts and two practitioners judged that the learning model was qualified to be implemented, while one of them judged that it was qualified with some revision.

After the validation process had been completed, the learning model that had been designed was considered valid. Meanwhile, learning syntax refers to the operational steps in implementing the learning model and these steps consist of: (1) presenting the problems; (2) organizing the students to study; (3) identifying and formulating the problems; (4) investigating and solving the problems; (5) presenting the problem solution; and (6) evaluating and drawing conclusions.

The social system of the learning model urges that the students should be active in the learning process, should be learning independently and collaboratively, and should have freedom in selecting the problem-solving strategies. In line with the social system that makes use of collaborative pattern, the teachers' roles in the principles of reaction are facilitator, mediator, motivator, and evaluator. Moreover, in order that the learning model can be implemented practically and effectively, a support system that includes an instructional book containing the directions on the model implementation and the sets of problems that will be implemented in the learning process, the lesson plan, and the student's worksheet is designed. In general, the characteristics of the learning model that has been designed might be viewed in Fig. 1.

The learning activities that are reflected in the step by step procedure (syntax) have the objective of training the students' HOTS. It is in accordance to the results of previous studies that to be effective in developing the students' HOTS can be pursued by revealing that presenting problems, being active in problem-solving efforts, and presenting results in front of the class are proven (Apino & Retnawati 2017, Jailani & Retnawati 2016, Maharaj & Wagh 2016, Mokhtar et al. 2010, Rajagukguk & Simanjuntak 2015). The relationship between the students' activities and the students' HOTS that have been trained through the learning process might be viewed in Table 1.

Furthermore, in order to support the development of the students' character values, the researchers apply cultural values as a controlling tool or as a norm that governs the pattern of interaction between the students and the teachers within the learning process. These cultural values also serve as the teachers' guidelines in responding to every students' action, question, or response. The cultural values that have been applied are *sara pata anguna*, which consist of fourth main pillars namely *pomae-maeka*, *pomaa-maasiaka*, *popia-piara*, and

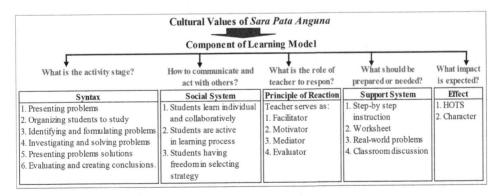

Figure 1. The mathematics learning model that has been designed.

Table 1. The activities that train higher-order thinking skills of students.

Syntax	Students' activities
1. Presenting problems	a. Observing
	b. Asking questions
2. Organizing students to learn	a. Observing
	b. Asking questions
3. Identifying and formulating problems	a. Identifying and writing important information
4. Investigating and solving problems	a. Planning problem-solving procedures
	b. Selecting appropriate strategies
	c. Creating ideas
	d. Creating assumptions
	e. Making patterns
	f. Modifying concepts
	g. Evaluating process and results that have been attained
5. Presenting problem solutions	a. Communicating mathematical ideas
	b. Evaluating ideas
6. Evaluating and drawing conclusions	a. Assessing a statement, an assumption, and a mathematical process
	b. Interpreting solutions that have been attained
	c. Drawing conclusions

Table 2. Cultural values of *Sara Pata Anguna* in the mathematical learning process.

Cultural values	Contained values	Implementation in the learning process
Pomae-maeka	1. Responsibility	1. Unwilling to violate others' rights
Pomaa-maasiaka	2. Social/environmental care	2. Performing communication by means of proper language
Popia-piara	3. Collaboration	3. Displaying equal treatment to all people
Poangka-angkataka	4. Friendliness/ communication	4. Pursuing mutually constructing cooperation and collaboration
	5. Tolerance	5. Giving assistance to people who have difficulties
		6. Positioning teachers as the parents for their students
		7. Embracing and appreciating differences

poangka-angkataka. Sara pata anguna is a social system that has been applied in the society of Buton, the City of Baubau, in the Province of Southeast Sulawesi. *Sara pata anguna* was chosen for two reasons. First, it contains morals values in the framework to maintain relationships with humans and God. These values are: responsibility, social/environment care and tolerance. Second, it contains the value of collaboration, friendliness/communication suggesting collaboration with each other in order to solve problems.

The inter-students' interaction in the learning process occurred during the group discussion. In the discussion, they had the opportunities to mutually collaborate, to mutually defend their opinion, to mutually ask questions, to mutually provide assistance, to mutually respond, and to mutually achieve agreement in selecting the appropriate strategy for solving the problems. The teacher's role in this was guiding, directing, and controlling the dynamics in the discussions. The implementation of the cultural values is presented in Table 2.

3.2 *Implementation of learning model*

The learning model was implemented for eight meetings in 11th grade of senior high school students. The results of the observation showed that most of the learning activities that were planned have been implemented successfully. In addition, the students' activities during the learning process has been more dominant than the teachers' activities. The implementation

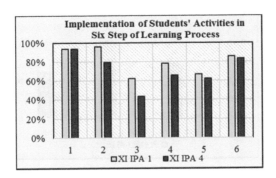

Figure 2.   The percentage of students' activities.

of this learning process shows that the learning model that was designed has met the criteria of practicality.

After the fourth and the fifth meeting, the learning process had been well implemented and the percentage of its implementation had been above 80.00%. Moreover, the students' activities during the learning process had been well implemented as well in accordance to the plan. However, the students' activities during the third learning steps had not been maximal (Fig. 2). The third stage was identifying and formulating problems. In this stage, the students should be able to write important information from the problems individually and then the results of the individual identification should be discussed with the fellow group members.

Observers noted that from the first to the third meeting there were still some students who did not perform these activities because they were hoping for the answers from their peers who were more intelligent. This finding is also strengthened by the results of the interview toward the students who admitted that there are still some friends who do not always participate in the problem identification. This result was also supported by Novferma (2016). The lack of understanding of the teachers in implementing learning activities also caused a number of learning activities to not be implemented. This is supported by the results of the study Jailani & Retnawati (2016), and Retnawati et al. (2016).

3.3   The development of students' higher-order thinking skills

The results of analysis toward the pre-test data showed that the students' HOTS had still been low (Fig. 3). The low HOTS were shown by the students' low average HOTS score, namely 27.75, with maximum score 60.00. The results of the analysis also showed that the score of 91.55% of students was less than 60.00, and there had only been 8.45% students whose score was equal to 60.00 and above.

On the other hand, the results of the post-test showed that the students' average HOTS score had improved to 57.61 with maximum score 80.00. Although the average score had not fallen into the "High" category, the students who had low scores showed significant improvement and they achieved better results. 69.01% of students at least attained 60.00, while 21.13% students attained 50.00. The remaining 9.86% students attained a score below 40.00.

Although the students' capabilities in the pre-test had been very low, the results of the post-test showed that the students' capabilities in all HOTS aspects had improved. Fig. 3 showed that the skills of students in all HOTS aspects had improved in terms of all aspects under measurement. These findings indicated that the learning process implementation provides quite significant impact toward the development of the students' HOTS.

From the three aspects of HOTS measured, it can be seen that the aspects of analysis and synthesis have a lower value than the evaluation aspect. This is allegedly because students are not accustomed to doing mathematical questions presented in narrative form, but in numerical form. This result also supported by Retnawati et al. (2017), Tias & Wutsqa (2015).

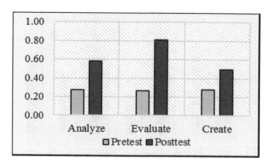

Figure 3.   The development of the students' higher-order thinking skills.

3.4   *The development of students' characters values*

The results of observation during the learning process implementation provided informa-tion regarding the students' characters that appeared. The students' characters that appeared were the capabilities of performing collaboration, mutual respect, and responsibility. The development of these characters was the impact of implementing the social system and the principles of teachers' reaction based on the cultural values. In addition, the results of the analysis on the students' attitude toward the mathematical learning process showed that most of the students displayed positive appreciation toward the learning process.

Meanwhile, the reduction of interview data provided the results that most of the stu-dents had a positive attitude toward the learning process. In general, there were four themes, namely: (1) the students were happy with the mathematical learning activities; (2) the discus-sion activities assisted the mathematical learning process; (3) the activities of presenting the problem-solving results improved the students' self-confidence, responsibility, and encour-agement in sharing their opinions; and (4) the learning process became easier with the assist-ance of students' worksheet.

The development of HOTS and character values of students as the researchers explain above shows that the expected learning process has been achieved. The use of integrated cultural values in the discussion, collaboration, and presentation activities proves to pro-vide positive impact toward the development of the HOTS and character values of students. These findings are in line with those of other studies which report that the implementation of cultural values, group investigation and collaboration in learning activities has positive influence toward the development of the students' character (Alexon & Sukmadinata 2010), and the development of students' character is in line with the development of the students' cognitive capabilities (Laili 2016). Although there were some students who had difficulties, their difficulties might be overcome by asking for the assistance of the students with better academic performance. Therefore, the learning model has met the criteria of effectiveness.

4   CONCLUSIONS

The cultural values-integrated mathematical learning model that has been designed meets the criteria of both practicality and effectiveness. It shown by the good implementation of the mathematical model so that the model is able to develop the students' higher-order think-ing skills and their characters values such as responsibility, tolerance, and positive attitude toward mathematical learning.

The diversity of local culture, especially in Indonesia, should be noticed by teachers when designing learning activities. The cultural values-integrated learning model that has been designed in this study proves to provide positive impact toward the development of the HOTS and characters values of students. In order to develop HOTS and character values of students through the learning process, teachers can use, adapt, or develop this learning model in another context.

REFERENCES

Alexon & Sukmadinata, N.S. 2010. Pengembangan model pembelajaran terpadu berbasis budaya untuk meningkatkan apresiasi siswa terhadap budaya lokal [The developing learning model based on culture to improve student's apreciation for local culture]. *Cakrawala Pendidikan* 29(2): 189–203.

Alismail, H.A. & McGuire, P. 2015. 21st century standards and curriculum: current research and practice. *Journal of Education and Practice* 6(6): 150–155. Retrieved from http://files.eric.ed.gov/fulltext/EJ1083656.pdf

Apino, E. & Retnawati, H. 2017. Developing instructional design to improve mathematical higher order thinking skills of students. *Journal of Physics: Conference Series* 812 (1): 1–7.

Bialik, M., et al. 2015. Education for the 21st Century: What Should Students Learn. *Center for Curriculum Redesign* 3(4): 415–420.

Brookhart, S.M. 2010. *How to assess higher-order thinking skills in your classroom*. Alexandria, VA: ASCD.

Dhunnoo, S. & Adiapen, V. 2013. Value-based education and teacher education in mauritius: Analysing the pertinence of value-based education at school to reconstruct society. *Purusharta* 6(1): 123–135.

Djidu, H. & Jailani. 2016. Aktivitas pembelajaran matematika yang dapat melatih kemampuan berpikir tingkat tinggi siswa [Mathematics learning activities which could examine HOTS]. In A.W. Kurniasih, B.E. Susilo, & M. Kharis (Eds.), *Prosiding Seminar Nasional Matematika X* 367–376. Semarang: Fakultas Matematika dan Ilmu Pengetahuan Alam.

Goethals, P.L. 2013. *The pursuit of higher-order thinking in the mathematics classroom*. Retrieved from http://www.westpoint.edu/cfe/Literature/Goethals_13.pdf

Hasanah, A. 2016. Al-Qiyam al-Thaqāfīyah wa Ṭabi' al-Sha'b li Jayl al-Muslimīn al-Shubbān al-Indūnīsīyīn. *Studia Islamika* 22(3): 535–562.

Heckman, J. & Kautz, T. 2013. *Fostering and measuring skills: interventions that improve character and cognitions*. Cambridge, MA: National Bureau of Economic Research.

Jailani & Retnawati, H. 2016. The challenges of junior high school mathematic teachers in implementing the problem-based learning for improving the higher-order thinking skills. *The Online Journal of Conseling and Education* 5(3): 1–13.

Joyce, B., et al. 2009. *Models of teaching 8th ed*. Upper Saddle River, NJ: Pearson Education. Inc.

Laili, H. 2016. Keefektifan pembelajaran dengan pendekatan CTL dan PBL ditinjau dari motivasi dan prestasi belajar matematika [Effectiveness CTL and PBL regarded motivation and student's achievement]. *PYTHAGORAS: Jurnal Pendidikan Matematika* 11(1): 25–34.

Lickona, T. 1991. *Educating for character how our school can teach respect and responsibility*. New York, NY: Bantam Books.

Limbach, B. & Waugh, W. 2010. Developing higher level thinking. *Journal of Instructional Pedagogies* 9. Retrieved from https://aabri.com/manuscripts/09423.pdf

Lumpkin, A. 2008. Teachers as role models teaching character and moral virtues. *Journal of Physical Education, Recreation & Dance)* 79(2): 45–49.

Maharaj, A. & Wagh, V. 2016. Formulating tasks to develop HOTS for first-year calculus based on Brookhart abilities. *South African Journal of Science* 112(11/12): 1–6.

Mainali, B.P. 2012. Higher order thinking in education. *Academic Voices: A Multidisciplinary Journal*, 2(1): 5–10.

Mokhtar, M.Z., et al. 2010. Problem-based learning in calculus course: perception, engagement and performance. In *Latest Trends on Engineering Education*: 21–25.

Novferma. 2016. Analisis kesulitan dan self-efficacy siswa SMP dalam pemecahan masalah matematika berbentuk soal cerita [Analysis of student's difficulty and self-efficacy in junior high school on arithmatics problem solving]. *Jurnal Riset Pendidikan Matematika* 3(1): 76–87. Retrieved from https://journal.uny.ac.id/index.php/jrpm/article/view/10403/8137

Paper, W. & Thomas, K. 2004. Learning taxonomies in the cognitive, affective, and psychomotor domains. *Rocky Mountain Alchemy*. Retrieved from http://www.rockymountainalchemy.com/whitePapers/rma-wp-learning-taxonomies.pdf

Pinxten, R. 2016. Multimathemacy: anthropology and mathematics education. *Multimathemacy: anthropology and mathematics education*. Retrieved from Cham: Springer International Publishing. https://doi.org/10.1007/978-3-319-26255-0

Plomp, T. 2013. Educational design research: an introduction. In T. Plomp & N. Nieveen (Eds.), *Educational design research*. Enschede: Netherlands Institute for Curriculum Development (SLO). Retrieved from http://international.slo.nl/publications/edr/

Rahmawati, U. & Suryanto. 2014. Pengembangan model pembelajaran matematika berbasis masalah untuk siswa SMP [The developing mathematics problem based learning model for students in junior high school]. *Jurnal Riset Pendidikan Matematika* 1(1): 88–97. Rajagukguk, W. & Simanjuntak, E. 2015. Problem-based mathematics teaching kits integrated with ICT to improve students' critical thinking ability in junior high schools in Medan. *Jurnal Cakrawala Pendidikan* 3(3): 347–355.

Ramos, J.L.S., et al. 2013. Higher order thinking skills and academic performance in physics of college students : a regression analysis. *International Journal of Innovative Interdisciplinary Research* (4): 48–60.

Retnawati, H., et al. 2016. Vocational High School Teachers' Difficulties in Implementing the Assessment in Curriculum 2013 in Yogyakarta Province of Indonesia. *International Journal of Instruction* 9(1):33–48.

Retnawati, H., et al. 2017. Why are the mathematics national examination items difficult and what is teachers' strategy to overcome it? *International Journal of Instruction* 10(3):257–276.

Riadi, A. & Retnawati, H. 2014. Pengembangan perangkat pembelajaran untuk meningkatkan HOTS pada kompetensi bangun ruang sisi datar [The developing learning media to improve HOTS for polyhedron]. *PYTHAGORAS: Jurnal Pendidikan Matematika* 9(2): 126–135.

Stedje, L.B. 2010. Nuts and bolts of character education. *A Literature Review*. Retrieved from www. characterfirst.com%5CnIntroduction

Susanto, E. & Retnawati, H. 2016. Perangkat pembelajaran matematika bercirikan PBL untuk mengembangkan HOTS siswa SMA [Mathematics learning media using PBL to improve student's HOTS in senior high school]. *Jurnal Riset Pendidikan Matematika* 3(2): 189–197.

Tan, S.Y. & Halili, S.H. 2015. Effective teaching of higher-order thinking (HOT) in education. *The Online Journal of Distance Education and E-Learning* 3(2): 41–47.

Thomas, A. & Thorne, G. 2009. How to increase higher order thinking. Retrieved from http://www.readingrockets.org/article/how-increase-higher-order-thinking

Thompson, T. 2008. Mathematics teachers' interpretation of higher-order thinking in Bloom ' s taxonomy. *International Electronic Journal of Mathematics Education* 3(2): 1–14.

Tias, A.A.W. & Wutsqa, D.U. 2015. Analisis kesulitan siswa SMA dalam pemecahan masalah matematika kelas XII IPA di kota Yogyakarta [Analysis of of student's difficulty in senior high school]. *Jurnal Riset Pendidikan Matematika* 2(1): 28–39.

Zuchdi, D., et al. 2015. *Pendidikan karakter: konsep dasar dan implementasi di perguruan tinggi* [Character education: the fundamental concept and implementation in the collage]. Yogyakarta: UNY Press.

Instilling character values through a local wisdom-based school culture: An Indonesian case study

S. Madya & I. Ishartiwi
Universitas Negeri Yogyakarta, Indonesia

ABSTRACT: This single-site case study examined the use of a local wisdom-based school culture in instilling character values in a senior secondary school in Yogyakarta Special Territory, Indonesia. Data were collected through observations, in-depth interviews, a survey, and document analysis. The findings indicated that the school culture is characterized by: (1) a clear vision and policy communicated in the local genre; (2) interactions among school members in local and national languages; (3) opportunities for students to (a) engage themselves in various types of learning activities inside and outside the classroom, (b) learn to cooperate, compete, share with others, and organize activities, and (c) express ideas in local, national, and international languages; (4) a clean, safe and orderly environment. This culture seems conducive to instilling the values of religiosity, tolerance, discipline, hard work, mutual respect, creativity, independence, democracy, national spirit, love for one's own culture, the pursuit of achievement, environmental concern, solidarity, responsibility, honesty, curiosity, friendship, and communicativeness.

1 INTRODUCTION

The tension between the global and the local is one of the seven tensions to be overcome in the 21st century (Delors, 1996). That is, "people need gradually to become world citizens without losing their roots and while continuing to play an active part in the life of their nation and their local community" (ibid., p. 15). We have observed the increasing tension from the people's complaint that foreign values have flooded our culture. Another tension is between competition on the one hand and the demand for equal opportunity on the other. To pursue higher quality people have to compete, but at the same time everybody needs the same opportunity to perform. When competition is emphasized, only selected people will enjoy opportunity and its related resources to fully develop their potential to excel compared to others, who are deprived of the same opportunity due to limtied resources. By contrast, when everybody has the same opportunity, the more people will not reach the highest level of their achievement potential. To overcome this tension Delors (1996) suggests the reconciliation of the following three values: competition for motivation, cooperation for building strength, and solidarity for unity.

Education is a strategic means to overcome these tensions. We agree with Cheng (2002) that globalization and localization are dynamically and interactively related. In this case, local culture, as long as it is open, will be enriched with global values as long as they are open and adaptive. In multicultural countries such as Indonesia, of the four scenarios of localization and globalization in education, the scenario of both highly localized and globalized education seems to be most acceptable.

In this globalization era, Indonesian schools are also encouraged to make efforts to help their students to develop triple citizenship, that is, at local, national and global levels. With such citizenship, an Indonesian is expected to think globally, but act locally, which means that he/she should have a global perspective and be able to act fruitfully to contribute to the solution of problems faced in the place where he/she lives locally and nationally, depending on the capacity of the person.

In response to this, Yogyakarta Special Territory, one of the regional governments in Indonesia, has developed a culture-based education regulated by Regional Regulation No. 5/2012 on the provision and management of education. This has been followed by schools through their efforts to develop local wisdom-based education that emphasizes Indonesian multicultural character building, while developing students' nationalist spirit and global perspective. How has this been implemented? Very little information, if any at all, is available.

1.1 School culture and school climate

We believe that the development of students' character is very much influenced by what is happening in the school under the principal's leadership, the surrounding environment and its local wisdom, and the school community members' perceptions of the school's effectiveness. School culture influences the ways people think, feel, and act (Peterson & Deal, 1998).

School culture is defined as follows:

> ... everything that goes on in schools: how staff dress, what they talk about, their willingness to change, the practice of instruction and the emphasis given student and faculty learning ... Culture is the underground stream of norms, values, beliefs, traditions and rituals that has built up over time as people work together, solve problems and confront challenges (Peterson & Deal, 1998, p. 28).

From his literature review, Peterson (2002) has identified negative (toxic) and positive school cultures, which should be placed in a continuum. This means a school can have more positive than negative features. It follows that the more positive features a school has, the more successful it will be, and the more negative features a school has, the less successful it will be.

However, in our observation, the notion of positive features in Indonesia may become negative when they are concerned with sensitive issues such as ethnicity, religion, race, and intercommunity interaction. Being religious, for example, is positive when it supports the effort to build a harmonious social life among members of different religions. This is often observable in public schools in which all students are encouraged to carry out their own religious teachings while being socially considerate, without a majority-minority dichotomy. In such a situation, the intensive practice of different religions contributes to the quality of social life, reflecting a positive school culture. However, if the practice of religion is limited to the religion embraced by most school members, then being religious can become negative because it may create disharmony in the social life of the school. This is because students that embrace other religions will feel uneasy andd disregarded. This will certainly influence student achievement.

Peterson (2002) cites the results of several research studies—conducted by Stein (1998), Lambert (1998), Fullan (2001), DuFour and Eaker (1998), and Hord (1998) – to conclude that the positive culture of schools possesses: (a) a widely shared sense of purpose and values; (b) norms of continuous learning and improvement; (c) a commitment to and sense of responsibility for the learning of all students; (d) collaborative, collegial relationships; (e) opportunities for staff reflection, collective inquiry, and sharing personal practice. In addition, successful schools often have a common professional language, communal stories of success, extensive opportunities for quality professional development, and ceremonies that celebrate improvement, collaboration, and learning (Peterson & Deal, 1998).

By contrast, the negative or toxic culture is characterized by "toxic" norms and values that hinder growth and learning (Peterson, 2002). It has been observed that the negative culture: (a) lacks a clear sense of purpose; (b) has norms that reinforce inertia; (c) blames students for lack of progress; (d) discourages collaboration; (e) often has actively hostile relations among staff. With such a culture, these schools are not healthy for staff or students. Students cannot learn effectively and staff members' professionalism does not improve.

School culture (shared norms) results in school climate (perceptions). School climate may be defined as the quality and character of school life shaped by students', parents', and school personnel's patterns of life within the school, reflecting its norms, purpose, values, interpersonal relationships, teaching and learning practices, and organizational structure. School climate factors include one's perception of personal safety, interpersonal relationships, teaching, learning, and the external environment (Jones et al., 2008).

From the literature on school climate (e.g. Jones et al., 2008; MacNeil et al., 2009), it can be inferred that a conducive school climate is necessary for students' growth and development. A conducive school climate is reflected in the school members' positive perceptions of: (1) social environment, which includes (a) personal relationships among staff and students, (b) respect for differences, (c) emotional welfare and sense of safety, (d) student involvement, (e) school–family collaboration, and (f) community partnership; (2) physical environment, including (a) building conditions, (b) physical safety, (c) school general protocols, and (d) class management; (3) behavioral environment, expectations, and support, including (a) physical and mental welfare, (b) prevention services and intervention, and (c) behavioral accountability (discipline and interventional responses).

From all the points presented above it can be inferred that both school culture and school climate can influence the school community, consisting of personnel, students, and parents. The school culture may then be developed intentionally under the lea-dership of the school principal with the support of the whole school community in the existing environment with its sociocultural aspects, especially community knowledge and local wisdom.

1.2 *Community knowledge and local wisdom*

The literature on school leadership talks very little about community knowledge and local wisdom. After searching, we found one work on community knowledge and some on local wisdom, and all by Asian writers. This implies that community knowledge and local wisdom matter in education in Asia but may not do so in other parts of the world.

Community knowledge can be categorized into: (a) knowledge to maintain the community, that is, knowledge of the community's history, important stories, main values, culture, traditions, regulations, and important teachings; (b) knowledge to obtain income, that is, knowledge of jobs, religion, training to improve members' quality and health; (c) knowledge to build harmony, that is, knowledge obtained from people's life enjoyment or general knowledge, and found in coffee shops, retail shops, or meeting places like halls and sport venues, which include storytelling, lullabies, harvest songs, daily stories and general news (Mungmachon, 2012).

Local wisdom is basic knowledge obtained from living harmoniously with nature. It is concerned with a community's accumulated culture, passed from generation to generation. This wisdom, which may be concrete or abstract, is gained from real-life experiences or truths revealed through real life. Nakorntap (Mungmachon, 2012, p. 176) states that in Thailand, wisdom integrates body, mind and environment, emphasizes respect for older people and their life experiences and, moreover, values morals above materialistic things. In our experiences, this local Thai wisdom is similar to Javanese wisdom.

In his study, Bauto (2016) found that traditional activities, representing local wisdom, contained a variety of values that could be grouped into religious values, social values, and political values. Another study, by Alfitri and Hambali (2013), found that local wisdom can be an effective tool for conflict resolution in the community. Meanwhile, Agung (2015) found that a local wisdom-based social learning model can be a learning source. At the university level, Suryadi and Kusnendi (2016) found that the Sundanese local-wisdom values of *silih asih, silih asah* and *silih asuh* increased actualization of scientific, educative and religious behavior.

By contrast, instilling character values through foreign language texts is not easy. Faridi (2014) found difficulties in instilling character values through English narrative stories. It can be assumed that local narratives with local-wisdom values are more conducive to instilling character values than foreign ones.

1.3 School–family cooperation

Ki Hajar Dewantara, the father of Indonesian education, and also the first Minister of Education in Indonesia, contends that schools, families and communities are the triple centers of education and should work together to ensure the quality education of children (Haryanto 2011). This has been practiced widely in Indonesia and in many other countries (Kim et al., 2012; Van Roekel, 2008; Webb & Healy, 2012). Research studies reviewed by NCSE (2005), as well as a study by Rahman (2014), showed that school–family cooperation or partnership influenced students' achievement and character development.

1.4 School physical environment

A school's physical environment is the condition of the school buildings, including their design, and the surrounding areas, including yards and sports fields. A good physical environment is one that is safe, clean, green and orderly.

The review of research studies by Koroye (2016) found that student academic achievement was influenced by building condition (well-equipped, decent buildings), and designs with friendly and agreeable entrance areas, supervised places for students, and public spaces that fostered a sense of community, with particular attention on the use of color. Koroye (2016) also found the influence of a beautiful environment on student achievement to be significant. Suleman and Hussain (2014) found that classroom physical environment had a significant effect on secondary school students' academic achievement. All of these findings imply that the school physical environment should be given due attention. It should be noted, however, that for Indonesian situations, especially in areas in which most of the people are Muslims, the physical environment should emphasize cleanliness and order—two of the values constituting Islamic beliefs—though not luxury. In addition, schools in Indonesia usually have a place for worship, that is, a mosque or *musolla* (a sort of mini mosque) for Muslims, and a prayer room for non-Muslims. The school physical environment should then reflect the local wisdom of the school community, subject to the leadership of the principal.

1.5 School leadership

The effectiveness of a school's culture very much depends on the effectiveness of the school principal. As suggested by studies of how principals operate in handling high rates of change, an effective principal is observed to carry out the following six functions: (1) providing and communicating a vision; (2) providing encouragement and recognition; (3) obtaining resources; (4) adapting standard operational procedures; (5) monitoring improvement efforts; (6) handling disturbances (Heller & Firestone, 1995, cited in Kruse & Louis, 2009). With such functions, a school principal can make efforts to shape a school culture that is conducive to learning, both affectively and cognitively.

The development of school culture within the existing environment, with its local wisdom, depends on the school leadership and school management—two necessarily different elements in a school. School management is focused on maintaining the existing organizational arrangements in an effective and efficient manner, and on running the school. By contrast, leadership activities involve efforts to influence other people to achieve a new desired end; they often involve change initiatives designed to achieve new or existing objectives. Spillane and Diamond (2007) state that, in particular, school leadership involves transforming the existing ways, stirring up school and classroom routines, although "managerial imperatives" often dominate school leaders' work. Spillane (2006, pp. 11–12) defines school leadership as referring to:

> ... *activities tied to the core work of the organization that are designed by organizational members to influence the motivation, knowledge, affect, or practices of other organizational members or that are understood by organizational members as intended to influence their motivation, knowledge, affect, or practices.*

1.6 School effectiveness

From the literature on school effectiveness, Saleem et al. (2012) list the indicators of school effectiveness, which fall into two categories: (1) the positive school culture, featuring (a) instructional leadership, (b) a clear vision and mission, (c) safe and orderly environment, (d) high expectations of students' achievement, (e) continuous assessment of student learning, (f) opportunities and time to do tasks, and (g) positive school–family relationships; (2) improvements of (a) students' various learning achievements, (b) students' attitudes to their school and to themselves, (c) teachers' attitudes to the school and students as learners; and (d) parents'/communities' attitudes to the school. Using teachers and school administrators in Pakistan as respondents, Saleem et al. (2012) found the 18 determinants gave rise to a model of school effectiveness comprising four factors: school environment, professionalism, management, and quality.

1.7 Purposes of the study

The literature reviewed above pays very little attention, if any at all, to the school culture which is developed according to local wisdom to facilitate the instillment of both globalized and localized character values. Our single-site case study aimed to fulfill the following purposes: (1) to unearth the characterisitcs of a local wisdom-based school culture as a setting for character education; (2) to discover the leadership policy used to instill character values; (3) to describe the strategies used to implement this policy; (4) to describe the school activities in a local wisdom-based school culture; (5) to uncover the character values instilled through this local wisdom-based school culture.

2 RESEARCH METHODS

To address the declared aims, a qualitative approach was applied to the study of a state senior secondary school in Sleman, Yogyakarta, as a school effective in instilling character values. This school is located in a suburban area about 15 kilometers from the city. Its surrounding environment is green with tall trees and not far from it there spreads a large paddy field. All of this creates a peaceful atmosphere. Entering the school through the main iron gate, we find a relatively large front yard with green Manila grass spread around its outskirts, together with pretty plants such as garden croton, *Acalypha siamensis*, orchids, and *Sansevieria*. In addition, we can see some tall leafy trees providing shade during the day when the sun is bright and hot. The two-storied school building is rectangulary laid out, with the administrative building in front and the classroom buildings on the right, left and rear sides, with a large middle yard paved with conblocks for various purposes, such as the flag ceremony and sports (basketball and volleyball). Tall shade trees also stand along the side of the middle yard. Here and there are garbage bins as well as some washbasins. Pot plants are also found on the school porch. Approaching the rear part of this area, there is a bicycle parking lot bordered by a wall full of pictures of national heroes. Further back is a small fish pool. In short, the school environment is peaceful, orderly, green and clean.

This school is guest-friendly and informative. It has a reception desk at which someone on duty will welcome visitors, who can immediately see the cups won through various student competitions. Statements of the school vision and mission are displayed on a wall outside the administrative building. This wall is also used to display school profile data and visitors can readily grasp what the school is like by reading these materials.

On some walls outside the classrooms are displayed Javanese proverbs, for example, *Sepi ing pamrih, rame ing gawe*, which highlights the importance of hard, diligent work without undesirable personal vested interest. Similarly, *Sapa nandur bakal ngundhuh*, which means that whoever does something will enjoy the yield accordingly, and *Sing bisa rumangsa, aja rumangsa bisa*, which means that one is expected to recognize honestly one's own capability and not overestimate it. These proverbs are used to stimulate the instillment of associated values.

Our study was conducted in five stages: (1) conducting a preliminary study, consisting of (a) analyzing relevant data and information about the school from different sources, (b) portraying the conditions and describing the policy of the school as an early picture of the school culture, (c) determining the focus of the study; (2) collecting information from the principal, teachers, students, administrative staff, and parents concerning their views of the school's ways of instilling character values in students through a local wisdom-based school culture; (3) conducting observations inside and outside the classroom, and in-depth interviews with school members to obtain information on their awareness and participation in building a local wisdom-based culture; (4) mapping the school's activities in the local wisdom-based culture to instill the desired character values; (5) inferring the character values successfully instilled.

3 FINDINGS AND DISCUSSION

The observations revealed that the school: (1) has a physical environment conducive to learning—clean, green, safe and orderly; (2) has attracted students from the middle-class economy; (3) is supported by qualified teachers; (4) has fairly adequate sports and art facilities and equipment; (5) has not been equipped with a Javanese gamelan, which are the main local musical instruments—doing so might have supported the efforts to improve student learning.

Efforts from 2012 to 2015 to pursue achievements in terms of student learning have resulted in a gradual improvement in academic, sports and arts achievements at the district, provincial and national levels. This might be related to leadership in creating a school culture by applying a participatory approach with local wisdom as a basis.

The interview with the principal revealed that the school policy is to create a local wisdom-based school culture to facilitate the development of the students' moral-spiritual, intellectual, sociocultural, aesthetic and autonomous potentials, with a love for their local culture, awareness of the need for national spirit, and great interest in the international outlook. For all of this, it is regarded as necessary to apply the following leadership principles (expressed in Javanese): *Ing Ngarsa Sung Tuladha, Ing Madya Mangun Karsa*, and *Tut Wuri Handayani* (role model, building motivation, and creativity). The principal stated his belief that in such a culture, in which learning takes place within the context of the learners' experiences, the instillment of character values would be effective.

Role model behavior was claimed by the principal to be effective in instilling values such as punctuality, religiosity, tolerance, cleanliness, openness and friendliness. An example is his model of consistently practicing his religion by praying in the school mosque, together with teachers and students of the same religion. Because the mosque was too small to accommodate the large number of Muslim students during noon prayers, many students were late for their afternoon class. To solve this problem, he invited parents and community members to contribute to the enlargement of the mosque so that students could say their noon prayers without feeling afraid of being late for their afternoon classes.

One strong policy is that students have equal opportunity to practice their religious teachings. This policy has been applauded by parents and members of the surrounding community. In this case, non-Muslim students are provided with a prayer room. It should be noted that they share the room with other religious groups because the number in each group is too small for a room each. In short, everyone is encouraged and facilitated to practice their religions with social tolerance being emphasized. Therefore, being religious is a positive feature of this school.

To instill the value of cleanliness as part of the principal's policy, he also practiced what he had advised others to do. For example, in relation to the policy of keeping the school environment clean, if he saw pieces of paper scattered on the floor, he picked them up and students seeing him doing so immediately joined him. To his observation, they continued paying attention to cleanliness afterwards.

Another important finding is the approach to keeping the school free from vandalism by incorporating community wisdom about vandalism, which is a persistent problem in the area. Based on his observation that communities also wanted to get rid of vandalism, the principal invited parents and representatives of the surrounding community to discuss ways of solving the problem. This resulted in an agreement that solving the problem is a joint responsibility. When a student was caught redhanded committing vandalism, his father was invited to the school to clean the wall involved and his son had to watch him doing so. This resulted in the son's improved behavior. Another action was a joint declaration of anti-vandalism by all school members, parents and surrounding community members. A preventative measure was taken by providing students with the freedom to draw whatever they wanted on the large wall near the parking lot. They have drawn pictures of national heroes and heroines; thus enhancing students' nationalist spirit. In short, incorporating community wisdom about vandalism is an example of an 'informal curriculum' for character education that can be extremely effective.

Another finding of the use of local-wisdom values to introduce change was related to a new uniform policy. After a long discussion with the teachers, it was decided that on the school anniversary date (the 22nd day of every month) teachers and other staff members must wear traditional Javanese costumes. Many parents who heard of this appealed to the principal to the effect that that their children should also wear Javanese costumes. The principal immediately agreed. Now, everyone has to wear Javanese costumes on these dates.

The local wisdom-based culture seems to be conducive to student learning in all areas. The school experienced a gradual improvement in academic, sports and arts achievements at district, provincial and national levels between 2012 and 2015. This might also be due to the teachers' positive attitudes to professional development and the school's high expectations of student achievements in all areas. While previous studies found a significant influence of physical environment on academic achievement (Koroye, 2016), and a relationship between character education and academic learning (Benninga et al., 2003), the present study indicated more areas of achievement although a further study is necessary to measure this influence.

Concerning shared leadership, the principal said that the vice-principal for curriculum affairs is to focus on running the teaching and learning programs, while he is to concentrate on non-academic affairs, but both should be directed to the achievement of the school's vision and mission. Shared leadership is believed to be more effective because the leaders concerned all have their own focus in leading yet aim for the same goal (Spillane & Diamond, 2007). In addition, the school has a high expectation of students' achievements too.

Compared with the findings of a previous study cited by Kruse and Louis (2009), the finding of the present study is of strength in local wisdom and shared leadership. In Yogyakarta Special Territory, in which education is to be culturally based, the finding of the present study has more practical significance for school development.

Other data revealed the activities conducted in the school as follows: (1) punctual teaching and learning activities both inside and outside the classroom; (2) students' involvement in sports and music groups to develop their talents for competition purposes; (3) students' practices of (a) their religious teachings, that is, for Muslims, daily joint prayers, regular Qur'anic recitation, a regular Islamic discussion hosted in turns by student representatives in their homes and also attended by teachers, holding ceremonies to commemorate important religious days, annual Islamic sacrifice rituals, fundraising/charity for the needy, and, for Christians, celebrating Christmas and commemorating religious holidays, and (b) shaking hands with teachers while bowing and kissing the teachers' hands to show respect; (4) practice of the local-wisdom values, including (a) mutual assistance through cooperative/group tasks, (b) making *batik* for the school uniform, (c) wearing Javanese traditional costumes on the school anniversary date (see above), (d) using appropriate and accurate Javanese for a flag ceremony and daily social communication, and (e) giving a student a chance to make a short speech using Javanese in the flag ceremony; (5) a form of service learning on certain occasions; (6) facilitating the development of students' international sense by giving them a chance to express ideas using foreign languages, either English, Japanese, or Mandarin, in the Monday flag ceremony; (7) using appropriate and accurate Javanese in social interactions accompanied by the S5 (*senyum/salam/sapa/sopan/santun* or smile/greet/say hello/be polite/be humble and respectful). However, not all students have had the opportunity to talk directly to teachers in social situations using good Javanese.

From this list of activities, the instillment in students of the following values might be inferred: discipline (from activity 1); competition and achievements (from activities in 2); religiosity, cooperation, unity, critical thinking, gratitude, generosity, solidarity, and respect (activities in 3); cooperation, independence, creativity, love for one's own culture, politeness, decency, and communicativeness, (activities in 4); social responsibility, coordination, empathy, solidarity, citizenship and civic engagement, social responsibility, and caring (activities in 5); international perspective (from activity 6); love and pride for own culture (from activity 7). Compared with previous studies, this case study is more comprehensive, although further study is necessary to provide measures.

The above practices imply an early triple (local, national, and global) orientation, which is recom-mended in Indonesian education and also an early effort to overcome the globalization–localization tension previously mentioned. The school has also made an early effort to overcome the competition–equality tension by practicing the ideas suggested by Delors (1996) to reconcile three values: competition, cooperation and solidarity. Competition and cooperation for the benefits of the community are encouraged in Islam, and so is the value of solidarity, which is observed through various charitable actions.

As already discussed, a previous study, by Wahyuni et al. (2015), found that religious orientation was significantly related to anti-corruption behavior. Our study showed that the school developed students' religiosity in a more comprehensive way, incorporating the values of cleanliness, punctuality, discipline, solidarity, tolerance, and cooperation, all of which might have a powerful preventative value.

The practices above have been made possible through the following strategies: (1) providing the facilities and equipment necessary for (a) developing musical, literary, and fine art talents, and (b) developing enterpreneurship through crafts, batik painting, and raising chickens, rabbits, and fish; (2) establishing relationships with relevant partners; (3) enhancing the practice of local-wisdom values through a dialog about local wisdom, as reflected in Javanese dress, Javanese customs, and local cuisine, together with ways of eating.

From the school culture the following values can be identified: (1) Yogyakarta cultural values are observed in the practices of decency, politeness, piety, respect and humility, self-confidence, customs, responsibility, natural conservation, discipline, wisdom, and love and care for others; (2) participatory values are observed in the practices of (a) involving students in sociocultural and civic activities, various student competitions (cooking, traditional fashion show, traditional songs, Javanese speech, traditional food decoration), (b) facilitating grade 12 students to ask for the consent and prayers of their parents, teachers, and younger fellow students for their success in national exams, and (c) simultaneously involving parents, teachers, and younger fellow students in giving moral support to grade 12 students taking the national exam; (3) cooperative and collaborative values are reflected in the practices of cooperation, communication and collaboration with relevant stakeholders (district educational officers, the school committee, parents, and surrounding communities).

Such observations were relatively consistent with the teachers' perceptions as revealed through a questionnaire. Of the 29 teachers completing the questionnaire, four teachers (13.79%) perceived the efforts to instill desired values through the local wisdom-based culture as very successful, 15 teachers (51.12%) as successful, and seven teachers (24.14%) as fairly successful. Although no teachers chose the option of 'unsuccessful', three teachers (10.35%) did not give any response.

Other data on the importance of values were col-ected through a four-scale questionnaire. Results of data analysis revealed that the school community members (teachers, administrative staff, students) agreed or strongly agreed about the importance of: teacher–student mutual respect; the character role models of the principal, teachers and administrative staff; student engagement/involvement in making decisions on character education; enhancement of students' creativity; good communication among members of the school community. These are consistent with the findings in relation to social interactions presented above.

The findings in this case study provide support to the findings of a study by Benninga et al. (2003) that schools addressing the character education of their students in a serious, well-planned manner tend to have higher academic achievement scores.

4 CONCLUSIONS AND SUGGESTIONS

Our findings give rise to the following conclusions: (1) character values can be instilled in a local wisdom-based school culture under the leadership of a principal who complies with the existing regulations, understands the local wisdom, has both managerial and leadership skills, and values school–parent–community partnerships through relevant activities involving the whole school membership; (2) local-wisdom values might provide an appropriate context for instilling character values; (3) students might be readier to observe values when they are involved in related decision-making and have ample opportunity to observe the values.

These conclusions lead to the following sug-gestion: to ensure participation and reflection of the whole school community in carrying out character education, the local wisdom-based school culture for character education should be developed conti-nuously through cycles of actions of role modeling, multi-way communication, participatory decision-making, and students' involvement in observing the values.

REFERENCES

Agung, L.S. (2015). The development of local wisdom-based social science learning model with Bengawan Solo as the learning source. *American International Journal of Social Science*, 4(4), 51–58.

Alfitri & Hambali. (2013). Integration of national character education and social conflict resolution through traditional culture: A case study in South Sumatra, Indonesia. *Asian Social Science*, 9(12), 125–135.

Anderson, G., Jr. (2013). Religion and morality in Ghana: A reflection. *Global Journal of Arts Humanities and Social Sciences*, 1(3), 162–170.

Bauto, L.M. (2016). Socio-cultural values as community local wisdom *katoba muna* in the development of learning mat-erials social studies and history. *HISTORIA: International Journal of History Education*, 14(2), 195–218.

Benninga, J.S., Berkowitz, M.W., Kuehn, P. & Smith, K. (2003). The relationship of character education implementation and academic achievement in elementary schools. *Journal of Research in Character Education*, 1(1), 19–32.

Billig, S.H. & Jesse, D. (2008). Using service-learning to promote character education in a large urban district. *Journal of Research in Character Education*, 6(1), 21–34.

Cheng, Y.C. (2002). *Fostering local knowledge and wisdom in globalized education: Multiple theories.* Keynote speech at the 8th International Conference on Globalization and Localization Enmeshed: Searching for a Balance in Education, 18–21 November 2002, Bangkok, Thailand. Bangkok, Thailand: Chulalongkorn University. Retrieved from https://home.ied.edu.hk/~yccheng/doc/speeches/18-21nov02.pdf

Delors, J. (1996). *Learning: The treasure within.* Paris, France: UNESCO Publishing.

Faridi, A. (2014). The difficulties of English teahers in instilling character building through narrative stories at elementary schools in Central Java, Indonesia. *International Journal of Contemporary Applied Sciences*, 1(2), 68–82.

Haryanto. (2011). Pendidikan karakter menurut Ki Hadjar Dewantara [Ki Hajar Dewantara's concept of character education]. *Cakrawala Pendidikan*, Edisi Khusus Dies Natalis UNY, 15–27. Retrieved from http://lppmp.uny.ac.id/sites/lppmp.uny.ac.id/files/02%20Haryanto%20KTP.pdf

Jones, M., Yonezawa, S., Mehan, H. & McClure, L. (2008). *School climate and student achievement.* Davis, CA: UC Davis School of Education Center for Applied Policy in Education. Retrieved from https://education.ucdavis.edu/sites/main/files/Yonezawa_Paper_WEB.pdf

Kim, E.M., Coutts, M.J., Holmes, S.R., Sheridan, S.M., Ransom, K.A., Sjuts, T.M. & Rispoli, K.M. (2012). *Parent involvement and family-school partnerships: Examining the content, processes, and outcomes of structural versus relationship-based approaches.* Lincoln, NE: Nebraska Center for Research on Children, Youth, Families and Schools. Retrieved from https://files.eric.ed.gov/fulltext/ED537851.pdf.

Koroye, T. (2016). The influence of school physical environment on secondary school students' academic performance in Bayelsa State. *Asia Journal of Educational Research, 4*(2), 1–15.

Kruse, S.D. & Louis, K.S. (2009). *Building strong school cultures: A guide to leading change.* Thousand Oaks, CA: Corwin Press.

MacNeil, A.J., Prater, D.L. & Busch, S. (2009). The effects of school culture and climate on student achievement. *International Journal of Leadership and Education: Theory and Practice, 12*(1), 73–84.

Mungmachon, R. (2012). Knowledge and local wisdom: Community treasure. *International Journal of Humanities and Social Science, 2*(13), 174–181.

National Center for School Engagement (NCSE). (2005). *What research says about school-family-community partnerships.* Denver, CO: National Center for School Engagement. Retrieved from http://www.ndpc-sd.org/documents/2012ITS/family_school_community_partnerships.pdf

Peterson, K.D. (2002). Positive or negative? *Journal of Staff Development, 23*(3), 10–15.

Peterson, K.D. & Deal, T.E. (1998). Realizing a positive school climate. *Educational Leadership, 56*(1), 28–30.

Rahman, B. (2014). Kemitraan orang tua dengan sekolah dan pengaruhnya terhadap hasil belajar siswa [Partnership between parents and schools and its impact on student's achievement]. *Jurnal Pendidikan Progresif, 4*(2), 129–138.

Saleem, F., Naseem, Z., Ibrahim, K., Hussain, A., & Azeem, M. (2012). Determinants of School Effectiveness: A study at Punjab level. *International journal of humanities and social science, 2*(14), 242–251.

Spillane, J.P. (2006). *Distributed leadership.* San Francisco, CA: Jossey-Bass.

Spillane, J.P. & Diamond, J.B. (2007). Taking a distributed perspective. In J.P. Spillane & J.B. Diamond (Eds.), *Distributed leadership in practice* (pp. 1–15). New York, NY: Teachers College Columbia University.

Suleman, Q. & Hussain, I. (2014). Effects of classroom physical environment on the academic achievement scores of secondary school students in Kohat Division, Pakistan. *International Journal of Learning & Development, 4*(1), 71–82.

Suryadi, E. & Kusnendi. (2016). The influence of local wisdom on the actualisation of educative, scientific and religious behaviour on an academic environment in a university. *American Journal of Applied Sciences, 13*(4), 467–476.

Van Roekel, D. (2008). *Parent, family, community involvement in education.* Washington, DC: National Education Association. Retrieved from http://www.nea.org/assets/docs/PB11_ParentInvolvement08.pdf

Wahyuni, Z.I., Adriani, Y. & Nihayah, Z. (2015). The relationship between religious orientation, moral integrity, personality, organizational climate and anti corruption intentions in Indonesia. *International Journal of Social Science and Humanity, 5*(10), 860–864.

Webb, M.A. & Healy, J. (2012). *Community schools: Working in partnership to support children, young people and families.* Belfast, UK: Barnardo's Northern Ireland. Retrieved from http://www.barnardos.org.uk/14457_prm_pp_briefing_no15.pdf.

Developing an Android application as a medium for mathematics learning and character forming: Needs assessment

N. Fitrokhoerani & H. Retnawati
Universitas Negeri Yogyakarta, Indonesia

ABSTRACT: This study was a preliminary study in the development of Android smartphone applications based on a problem-solving approach as a medium for mathematics learning and the development of hard work and curiosity characteristics of tenth grade high school students. The research used a survey-based approach and data was collected through literature review, interviews, observation, and focus group discussion. Subjects for the interviews included ten students and three math teachers in Karawang, Indonesia. Data was analyzed qualitatively. The findings suggested that: (1) the use of Android-based smartphones can support mathematics learning in developing student characteristics of hard work and curiosity; (2) most teachers and students have not utilized Android applications as a learning medium; (3) there is limited application to mathematics learning of Android-based smartphones; (4) a problem-solving approach using integrated media will help students work harder and improve their curiosity. These findings provide a basis for consideration in the development of an Android application as a medium in mathematics learning that also increases curiosity and hard work in tenth grade high school students.

1 INTRODUCTION

Mathematics is a form of knowledge that has an important role in the development of technology and other disciplines, and can also be found in other contexts of human life. The importance of the role of mathematics in human life means that it must be studied, especially in formal institutions such as schools. The aim of mathematics learning activities in schools is not only to understand math concepts, but also to help learners become mathematically literate so they can understand the mathematics that exists in the problems of life, both now and in the future (De Lange, 1999). In Indonesia, mathematics learning is not only aimed at developing students' mathematical skills but also student attitudes, such as hard work and curiosity (RoI, 2016).

Hard work is an attitude to doing something seriously to get optimal results (Maharani, 2014). Furthermore, curiosity is an emotion that is associated with naturally inquisitive behavior such as exploration, investigation, and learning (Mustari, 2011). Hard work and curiosity are an important part of the learning process, and are among the 18 values of Indonesian character. In order to develop hard work and curiosity in mathematics learning, teachers can use a learning approach such as problem-solving, including the use of mathematical problems. This approach requires teachers to assist students in learning to solve problems through hands-on learning experiences (Jacobsen et al., 2009).

However, the reality shows that many students are not familiar with problems in mathematics learning, especially for complicated issues. This is supported by Jaelani and Retnawati (2016), and Latifah and Widjajanti (2017), who claim that many students quickly become desperate in the face of mathematical problems and tend to give up. Therefore, in order to improve students' interest in solving the problems presented in the mathematics lesson, the problems can be presented using a learning medium.

The use of media in mathematics learning is an important and constitutive part because such media can be one way of creating a more interesting learning experience (Villareal & Borba, 2010). Latuheru (1988) states that the use of media in the learning process can make learning more appropriate and efficient, so that the quality of education can be improved. Futhermore, Hamalik (cited in Arsyad, 2010) argues that the use of learning media in teaching and learning can generate new desires and interests, generate motivation and stimulation in learning activities, and even exert psychological influence on students.

Today, there are a lot of media that can be used as an alternative by teachers in learning, one of them being applications on smartphones. Williams and Pence (2011) state that the use of smartphone applications in learning can facilitate the visualization and sharing of knowledge, making it more interesting and easier to understand. Supporting this statement, Shanmugapriya and Tamilarasi (2011), and Martono and Nurhayati (2014) added that learning by using smartphones can make learning more flexible because it can be implemented anytime, anywhere, and under any circumstances.

Nowadays, there is a wide choice of smartphones, which use a variety of operating systems. One of the most commonly used operating systems is Android. This system has a higher efficiency than many other smartphone operating systems such as Symbian or Windows Mobile (Gandhewar & Sheikh, 2010; Hanafi & Samsudin, 2012). Furthermore, Martono and Nurhayati (2014) revealed that many students are satisfied with the use of smartphones with an Android operating system in learning activities. Not only that, but Android is an open source operating system, which can help teachers to serve learning content to students.

The advantages associated with the Android operating system on smartphones seems to have created its own interest for developers. This is evident from the data presented on the AppBrain website page. As of September 2016, it was noted that the number of Android applications developed and published on the Google Play Store had reached over 2,400,000. However, math learning was supported by only a few of these applications, such as Quipper, Brainly, Math Formulas Free, and Math Tricks.

This study was conducted with the purpose of determining the needs of teachers and students for the existence of media that can assist the learning activities of mathematics. In addition, the results of this study will be used as a foundation for the development of Android applications based on problem-solving as a medium for learning mathematics that will develop student characteristics such as hard work and curiosity.

2 LITERATURE REVIEW

The literature review in this study included study of the theory of problem-solving approaches, study of learning media theory, study of Android application theory, and study of relevant research results. The results of the literature review activities are as follows.

2.1 Problem-solving approaches

Problem-solving has different definitions according to how it is perceived. VanGundy (2005) and Schunk (2012) defined problem-solving as a process of seeking and understanding how to do something different to achieve a goal when a person does not have an automatic solution. Problem-solving can involve the mastery, stability, and usefulness of a system of production, which is a network of condition–action circuits, conditions being a series of circumstances that activate the system, while actions are a series of activities that occur in the system as a consequence (Schunk, 2012).

In learning activities, problem-solving is defined as an approach that requires teachers to help learners solve problems through hands-on learning experiences (Jacobsen et al., 2009). Furthermore, Taplin (2007) explains that there are several characteristics of problem-solving approaches, namely: (1) the existence of interaction between students as well as between teachers and learners; (2) the existence of (mathematical) dialog and consensus among learners; (3) the provision by the teacher of sufficient information about the problem, with students clarifying, interpreting, and trying to construct a solution; (4) teachers receive a clear answer and do not have to evaluate it themselves; (5) teachers guide, train, and enquire with insightful questions and share in the problem-solving process; (6) teachers know when to intervene and when to stand back and let students find their own way; (7) students are encouraged to generalize rules and concepts, a central process in mathematics.

Problem-solving is an essential part of mathematics learning. This is because students thereby become skilled at selecting relevant information, analyzing it, and, ultimately, re-examining the outcome; intellectual gratification will arise from within, which is an intrinsic gift for students; students' intellectual potential is improved and students learn how to make discoveries through the process (Hudojo, 2003). On this basis, it can be seen that the problem-solving approach can encourage students to work hard to solve the problems given. Hard work is a behavior that shows genuine efforts to overcome learning barriers and complete tasks as well as possible (Narwanti, 2011).

In addition, the characteristic of problem-solving can also develop students' curiosity. Curiosity is the desire to learn something to get new information or knowledge. Learning is not just knowing but exploring to discover more, and to give meaning to what is gained in the learning process (Schmitt & Lahroodi, 2008). Curiosity occurs when a person perceives a difference or conflict between what is believed (known) and the reality (Loewenstein in Matheson & Spranger, 2001). Curiosity is usually characterized by feedback that has revealed inconsistency in a person's base knowledge, so they are motivated to understand what they do not yet know (Matheson & Spranger, 2001). Curiosity can be indicated by four aspects, namely: (1) the desire to learn; (2) the desire to investigate; (3) the desire to obtain new information or knowledge; (4) the desire to solve problems (Latifah & Widjajanti, 2017).

2.2 Learning media

Learning media are tools used or provided by teachers in learning activities. Such tools serve as communication tools that can convey messages, instructions, or information from sources to students (Moreno, 2006; Muhson, 2010). With media, the learning can stimulate students' attention, interest, thoughts, and feelings in relation to the learning activities (Daryanto, 2010). In addition, the use of different media can also help students gain different knowledge. (Villareal & Borba, 2010). Moreover, learning media can be categorized into three types, namely, visual media, audio media, and audiovisual media (Muhson, 2010; Arsyad, 2010). A computer-assisted learning media can improve the competence of students, indicated by students' enthusiasm and actively involved in the learning process. (Kurniawan et al., 2017)

The use of media in learning has several benefits, namely: (1) clarification of abstract concepts, so the verbalism can be reduced; (2) generation of students' desire, interest, motivation, and positive attitude toward learning; (3) learning becomes more interesting and more interactive, because it allows direct interaction between students and learning resources; (4) confers uniformity in observation and experience, so students share the same perceptions; (5) allows students to study independently because it provides consistent learning information that can be repeated or stored according to need; (6) more effective in improving the achievement of student competence and the quality of learning interaction especially in using a web based e-learning media (Muhson, 2010; Daryanto, 2010; Arsyad, 2010; Jaedun, 2007).

2.3 Android applications

Android is a software stack for mobile devices that includes an open source operating system, middleware, and main applications. It is supported by Google Corporation, the world's leading search engine company (Gandhewar & Sheikh, 2010). The Android operating system is now a project run through the Open Handset Alliance, which includes more than 30 ICT companies, and is freely distributed and used by any vendor. Android is also a very complete platform for both operating systems, and applications and development tools. The Android application market has very high support from the world's open source community (Pocatilu, 2010; Safaat, 2015).

Android is one of the operating systems that uses the Linux 2.6 framework (Team EMS, 2015), and phones based on the Android platform have become an indispensable communication tool for many people, especially in the younger segment of the population. One of the main reasons for the widespread adoption of Android in the mobile market is that mobile applications developed through Android development technologies are more efficient and effective than other technologies, such as Windows Mobile or Symbian operating systems, and produce fast, user-friendly and interesting applications (Hanafi & Samsudin, 2012).

As explained above, the openness of the Android operating system allows anyone to develop applications for the platform. An Android application is an application written in the Java software language, so the base programming of Android applications is conducted in Java. However, during the development of an application, Java is not exclusively used because XML and Apache Ant are also used (Team EMS, 2015). Android applications consist of several features, including an application framework, Virtual Machine (VM), integrated browser, optimized graphics, and SQLite data handling (Katysovas, 2008).

The number of Android applications developed and published on the Google Play Store is growing, and this includes the education category, as shown in Table 1.

The number of Android applications for the education category is also similar to the increase in Android apps that can be used in math, as illustrated in Figure 1. However, most such applications are included in the gaming category and there are only a few applications that can help in mathematics learning activities, especially for high school students.

2.4 Study of relevant research

Research about Android applications has started to be conducted. However, there have not been many studies that specialize in research on the use of Android applications in mathematics learning activities, especially in terms of developing student characteristics of hard work and curiosity. Nonetheless, based on several journal articles on the use of Android applications in learning, the use of an Android application as a learning medium is not unfeasible.

Table 1. Number of Android applications in August and September 2016 (Source: https://www.appbrain.com).

Category	Date	
	August 2016	September 2016
Education	186,686	189,297
Lifestyle	178,957	182,632
Entertainment	165,297	167,544
Business	155,464	157,706
Personalization	140,263	141,880

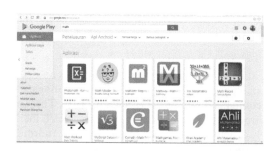

Figure 1. Android applications for mathematics.

In fact, students like it because they can learn the material anywhere, anytime, and under any circumstances. With the use of Android applications, learning can be more interesting and easier to understand. It is hoped that by integrating a problem-solving approach into Android applications, a student's hard work and curiosity in relation to mathematics can also be developed.

3 METHOD

The focus of this study is an analysis of student needs in mathematics learning that will be used as a foundation for the development of learning media in mathematics for high school students. This research used a survey-based approach and was conducted in Kabupaten Karawang, West Java, Indonesia. Data collection was conducted as follows: (1) review of literature for additional information about the media that can be used in mathematics learning for high school students; (2) observations were made to determine the learning conditions and available resources; (3) interviews were conducted with several respondents to identify a topic and problems in mathematics learning that might be addressed by the development of learning media; (4) a Focus Group Discussion (FGD) was conducted with experts in mathematics. The collected data was analyzed using a qualitative analysis method.

4 RESULTS

A preliminary study on a framework of need assessment was conducted to identify information about the needs associated with learning media for mathematics for high school students. Data was collected as described above. The results of this study serve as a basis for determining subsequent research steps. In general, the preliminary study results can be summarized as follows.

4.1 *Observations*

The observations were carried out on tenth graders majoring in science (Senior High School 1 Karawang). Based on these observations, it was found that the learning activities of mathematics in schools still tend to be conventional. The activity began with an explanation of the material by the teacher and was followed by issuing sample questions. After this, students were assigned to work on exercises in a mathematics textbook. During the learning activities, some students were very serious in following every activity, but there were also a few students who were not seriously following every learning activity.

4.2 *Interviews with mathematics teachers*

This activity was conducted using unstructured interviews with three mathematics teachers from three different high schools in Kabupaten Karawang. The purpose of this activity was to obtain an overview of the activities of mathematics learning at the schools. Based on the results of these interviews, it was again found that the learning activities of mathematics in schools still tend to be conventional. Learning begins with teacher explanations and examples, followed by practice questions. Teachers are still using simple practice questions and rarely use practice questions that involve mathematical problems.

In the learning activities, teachers usually only use the available textbook and have not used other media as a tool to support the activities. Not only are visual media, such as models of space, not used but even technology-based media such as PowerPoint have not been used. Nevertheless, the teachers stated that they have been using smartphones in their daily life as a tool of communication. However, they have never used them as a tool to support mathematics learning in the classroom. This is due to their lack of knowledge of learning applications that can be used with smartphones. In addition, according to the narratives of two teachers, the smartphone was never used as a medium in learning activities because students are not allowed to bring smartphones to school. In contrast to this, the mathematics teacher at High School 1 Karawang stated that their school has, indeed, allowed students to bring their smartphones, but their use is still very minimal: smartphones are usually only used in mathematics learning as a calculation tool.

4.3 *Interview with students*

This activity was conducted with tenth graders majoring in science (namely SMAN 1 Karawang, Indonesia). According to the results of the student interviews, the students are rarely given mathematics questions that contain non-routine mathematical problems. Many of them stated that they have an Android-based smartphone but more often use it as a communication tool rather than as a medium for learning. In terms of school learning activities, students revealed that the smartphone was used several times in certain learning activities. Such use is usually intended to help students obtain information from Internet that is not in the textbook. However, they said that in mathematics learning, smartphones are rarely used other than as a counting tool.

4.4 *Focus group discussion*

A focus group discussion was held on "The use of smartphones in mathematics learning" in October 2016 at the *Universitas Negeri Yogyakarta* Graduate School. The participants were two mathematics experts and 18 postgraduate students of the mathematics education program. Based on this discussion, the following output was produced:

– A problem-solving approach can provide an opportunity for students to develop their hard work and curiosity;
– The usage of smartphones in learning can have a good impact if used properly;
– The usage of smartphones in learning cannot be implemented in every school because of school rules and the expense of Android smartphones;
– There are not many Android applications that can be used in mathematics learning;
– There are a number of developer constraints in creating Android applications for use in mathematics learning.

5 CONCLUSION

Based on the research results, it can be concluded that: (1) the use of Android-based smartphones can support mathematics learning in developing student characteristics of hard work and curiosity; (2) most teachers and students have not utilized Android applications as a learning medium; (3) there is limited application to mathematics learning of Android-based smartphones; (4) a problem-solving approach using integrated media will help students work harder and improve their curiosity.

On the basis of these results, an Android application based on problem-solving should be developed as a flexible learning medium that increases the hard work and curiosity of students.

REFERENCES

Arsyad, A. (201005). *Media* pembelajaran [Instructional media]. Jakarta, Indonesia: RajaGrafindo Persada.
Daryanto. (2010). *Media pembelajaran: Peranannya sangat penting dalam mencapai tujuan pembelajaran* [Learning media: The role is very important in achieving learning objectives]. Yogyakarta, Indonesia: Gava Media.

De Lange, J. (1999). *Framework for classroom assessment in mathematics*. Utrecht, The Netherlands: Freudenthal Institute.

Elliot, S.N., Kratochwill, T.R. & Littlefield, J.Cet al. (2000). *Educational psychology: Effective teaching, effective learning* (3rd ed..). New York, NY: McGraw-Hill.

Gandhewar, N. & Sheikh, R. (20101). Google Aandroid: An emerging software platform for mobile devices. *International Journal of Computer Science and Engineering, NCICT 2010 Special Issue*, 12–17.

Hanafi, H.F. & Samsudin, K. (2012). Mobile Learning Environment System (MLES): The case of Aandroid-based learning application on undergraduates' learning. *International Journal of Advanced Computer Science and Applications, 3*(3), 63–66.

Hudojo, H. (2003). *Pengembangan kurikulum dan pembelajaran matematika* [Development of curriculum development and mathematics learning]. Malang, Indonesia: JICA.

Jacobsen, D.A. et al., Eggen, P. & Kauchak, D. (2009). *Methods for teaching: Promoting student learning in K-12 classrooms.* (A. Fawaid & K. Anam, Trans.). Yogyakarta, Indonesia: Pustaka Pelajar. (Original work published 2009 by Pearson Education).

Jaedun, A. (2007). Rancang Bangun dan Implementasi Web Based Learning untuk Meningkatkan Pencapaian Kompetensi Mahasiswa Bidang Aplikasi Komputer melalui E-Learning UNY [Design and Implementation of Web Based Learning to Improve Student Competency Achievement on Computer Applications through E-Learning UNY]. *Jurnal Pendidikan Teknologi dan Kejuruan, 16*(2), 187–208.

Jaelani & Retnawati, H. (2016). The challenges of junior high school mathematic teachers in implementing the problem-based learning for improving the higher-order thinking skills. *The Online Journal of Counseling and Education, 5*(3), 1–13.

Katysovas, T. (2008). *A first look at Google Android.* Retrieved from http://everythingcomputerscience.com/books/ android.pdf

Kurniawan, W., Budijono, A., & Susanti, N. (2017). Developing Computer Assisted Media of Pneumatic System Learning Oriented to Industrial Demands. *Jurnal Pendidikan Teknologi dan Kejuruan, 23*(3), 304–309.

Latifah, U.H. & Widjajanti, D.B. (2017). Pengembangan bahan ajar statistika dan peluang berbasis multiple intelligences berorientasi pada prestasi, pemecahan masalah, dan rasa ingin tahu [Developing statistics and probability teaching material based on multiple intelligences and oriented to the achievement, problem- solving, and curiosity]. *Jurnal Riset Pendidikan Matematika, 4*(2), 176–185.

Latuheru, J.D. (1988). *Media pembelajaran dalam proses belajar-mengajar masa kini* [Learning media in the process of teaching and learning today]. Jakarta, Indonesia: Depdikbud.

Maharani, H.R. (2014). Nilai-nilai karakter dalam pembelajaran project based learning materi statistika SMP [Values of character in learning with project- based learning in statisctical material for jyunior high school]. *Jurnal Pendidikan Matematika FKIP Unissula, 2*(2), 199–217.

Martono, K.T. & Nurhayati, O.D. (2014). Implementation of Aandroid based mobile learning application as a flexible learning media. *International Journal of Computer Science Issues, 3*(1), 168–174.

Matheson, D. & Spranger, K. (2001). Content analysis of the use of fantasy, challenge, and curiosity in school based nutrition education programs. *Journal of Nutrition Education, 33*(1), 10–16.

Moreno, R. (2006). Does the modality principle hold for different media? A test of the method-affects-learning hypothesis. *Journal of Computer Assisted Learning, 22*,: 149–158.

Muhson, A. (2010). Pengembangan media pembelajaran berbasis teknologi informasi [Development of learning media based information technology]. *Jurnal Pendidikan Akuntansi Indonesia, 8* (2),: 1–10.

Mustari, M. (2011). *Nilai Karakter [Character Value]*. Yogyakarta, Indonesia: LaksBang PRESSindo.

Narwanti, S. (2011). *Pendidikan karakter: Pengintegrasian 18 nilai pembentuk karakter dalam mata pelajaran* [Character education: Integration of 18 forming 18 character-forming values in subjects]. Yogyakarta, Indonesia: Familia.

RoI. (2016). *Peraturan Menteri Pendidikan Nasional Tahun 2016 No. 21, Standar Isi Pendidikan Dasar dan Menengah* [Regulation of the Minister of National Education Year 2016 No. 21, Content Standards of Elementary and High Education]. Jakarta, Indonesia: The Ministri of National Education, Republic of Indonesia.

Pitafi, A.L. & Farooq, M. (2012). Measurement of scientific attitude of secondary school students in Pakistan. *Academic Research International Journal, 2*(2), 379–392.

Pocatilu, P. (2010). Developing mobile learning applications for Aandroid using web services. *Informatica Economica, 14*(3), 106–115.

Safaat, N. (2015). *Android: Pemrograman aplikasi mobile smartphone dan tablet PC berbasis android* [Android: Programming mobile smartphone and PC tablet applications based on Android]. Bandung, Indonesia: Informatika.

Santoso, F.G.I. (2011). Mengasah kemampuan berpikir kreatif dan rasa ingin tahu melalui pembelajaran matematika dengan berbasis masalah: Suatu kajian teoritis [Honing the ability of creative thinking and curiosity through math-based learning: A theoretical study]. In *Proceedings of National Seminar on Mathematics and Mathematics Education, 3 December 2011, Presented in National Seminar on Mathematics and Mathematics Education, Yogyakarta State University, Indonesia* (pp. 230–240). Retrieved from http://eprints.uny.ac.id/7376/1/p-21.pdf

Schmitt, F. & Lahroodi, R. (2008). The epistemic value of couriositycuriosity. *Educational Theory*, *58*(2), 125–148.

Schunk, D.H. (2012). *Learning theories: A an educational perspective* (E. Hamdiah & R. Fajar, Trans.). Yogyakarta, Indonesia: Pustaka Pelajar. (Original work published 2012 by Pearson Education).

Shanmugapriya, M. & Tamilarasi, A. (2011). Designing an m-learning application for a ubiquitous learning environment in the Aandroid based mobile devices using web service. *Indian Journal of Computer Science and Engineering*, *2*(1),: 22–30.

Taplin, M. (2007). Mathematics through problem solving. *Math Goodies*. Retrieved 2 August 2016, from http://www.mathgoodies.com/articles/problem_solving.html.

Team EMS. (2015). *Pemrograman android dalam sehari* [Programming Aandroid in a day]. Jakarta, Indonesia: Elex Media Komputindo.

VanGundy, A. (2005). *101 Activities for teaching creativity and problems solving*. California San Francisco, CA: John Wiley & Sons.

Villareal, M.E. & Borba, M.C. (2010). Collectives of human-with-media in mathematics education: Nnotebooks, blackboards, calculators, computers and … notebooks throughout 100 years of ICMI. *ZDM Mathematics Education*, *42*, 49–62.

Williams, A.J. & Pence, H.E. (2011). Smart phones, a powerful tool in the cChemistry classroom., *Journal of Chemical Education*, *88*, 683–686.

The teaching of morality through musical elements based on learning models of the Netherlands and Indonesia

K.S. Astuti & A. Widyantoro
Universitas Negeri Yogyakarta, Indonesia

ABSTRACT: This article is based on a two-year study entitled "Developing the music thematic teaching model based on teaching implementation in the Netherlands", in which one of the objectives is to encourage children's character formation using the elements of music. The research used an experimental design, and the data collection methods were literature review and observation in both the Netherlands and Indonesia. The data was analyzed using a *t*-test. The findings of the research are: 1) music teaching in both the Netherlands and Indonesia used elements of music to build children's characters; 2) the rhythmic element can shape self-discipline; 3) changes from major to minor scale elements can develop sensitivity; 4) changes in dynamics and tempo can enhance tolerance; 5) both inward-facing and canon singing can improve students' ability to focus; 6) a *t*-test analysis showed that these musical elements are able to influence students' behavior with a significance level of 0.05%.

1   INTRODUCTION

Wang (2015) reported that in Indonesia, particularly in Yogyakarta and Central Java, there could be found children's songs or what are usually called *dolanan* in Javanese. Some educational character values could be found in these songs, which have existed for a long time: in the 1930s, Hans Overbeck identified at least 690 such songs.

With regards to the new Indonesian curriculum, Curriculum 2013, which makes character education compulsory, these songs could be used both as teaching materials and media for building students' character.

Similar songs can also be found in other areas. The power of these songs lies not only in their melody, which is easy to listen to, but also in their lyrics, which are full of advice in a language that is easy to understand.

Local songs containing advice that is more valuable can be found in the *Saman* performance from Aceh. This art uses vocals and handclaps to express its meaning. In the lyrics, poems that contain advice for life are found. Desyandri (2015) found that songs in Minangkabau (West Sumatra) also contain the essence of moral values.

Astuti et al. (2017) found that in the Netherlands, such as in The Royal Conservatorium Den Haag, Cals College in Nieuwegein, Montessori Lyceum Herman Jordan in Zeist, and Da Costa School, Hoograven, there is integrated music teaching that can be used by students to solve behavioral problems in daily life. It can be said that both Indonesia and the Netherlands use music as a medium to cultivate morality.

The basic elements of music are rhythm, melody, and harmony. How the elements of music can be used to cultivate morality in students, and how effective such a learning model is in developing musical ability and building students' morality is the focus of this research.

Morris (2000) states that morality is a set of norms, ideas, and dispositions that governs conduct and thought and which claims authority. Norms making up common morality must satisfy the current condition.

Morality cultivates character. Hasan (2010) states that character is one's disposition or personality, formed through the process of internalization of virtues and used as the foundation of one's perspective, thought, behavior, and action.

Dewantara (1977) argues that music can train listening sensitivity, which leads to more sensitive feeling and behavior. This practice can balance language and behavior, which are inseparable.

Moreno et al. (2009) found that musical training influences linguistic abilities in eight-year-old children. Kushartanti (2004) explains that the two sides of the brain are connected by a corpus callosum switch system which is very complicated with 300 million active nerve cells. This system balances messages in holistic pictures with concrete and logical messages. This relates to Rudolf Steiner's theory (Dewantara, 1977), which argues that music—in this case, rhythm—can facilitate physical work, support the brain's working, improve moral behavior, and invigorate spiritual power.

In their study, Wilson and McCrary (1996) conclude that music teaching and learning change the behavior of students. Meanwhile, the different characteristics of music teaching have different behavioral impacts. Astuti (2011) found that learners who play traditional music are calmer and more obedient, more readily accepting referrals from people who are more mature. Cassity (1976) proves that music not only influences children, but it makes a contribution to the group cohesiveness and interpersonal relationships of adult psychiatric patients. The music learning process cultivates behavior. Additionally, Brendell (1996) found some off-task students in the initial group rehearsal activities and hence the group might not accomplish the task.

Based on these descriptions, it can be said that music contains elements that lead to more sensitive feelings. Furthermore, the proposed hypothesis is that musical elements are effective in cultivating morality.

2 METHOD

The research method used a pretest/post-test experimental design. The dependent variable of the research is morality; the independent variables are rhythm, melody, and harmony.

The data was collected using documentation, observations, and interviews. In Indonesia, the study was conducted at several high schools in Yogyakarta (namely, SMP Negeri 2 Temanggung, SMP Muhammadiyah Wonosari and SMA Negeri Depok). In the Netherlands, this study was conducted at Cals College Junior High School, Nieuwegein, Herman Jordan Lyceum Montessori School in Zeist, and Conservatorium Den Haag. These schools were selected on the basis of observations that had been made in previous studies, which showed that the schools had been very conducive for the development of teaching–learning innovations.

Students are taught music by focusing on three music elements, which are rhythm, melody, and harmony. During the teaching process, the researcher conducted an observation of students' behavioral changes. If behavioral changes were found during the teaching and learning process, it could be said that musical elements could influence students' behavior and vice versa.

There is one continuous dependent variable and one independent variable. According to Pallant (2007), a paired-sample t-test can be used with the same population in these circumstances.

3 RESULTS AND DISCUSSION

In this section, the teaching and learning process as well as the changes it brought to students' behavioral changes are presented. The teaching and learning process focused on rhythm, melody and harmony, and behavioral changes involve qualitative and quantitative data.

3.1 Teaching of rhythm

The teaching of rhythm was done through two activities. In the first one, students learned rhythm by singing and moving in time with the song. In the second, students used rhythmic musical instruments in accordance with the songs.

The first teaching was conducted in the 5–6 age group in Conservatorium Den Haag. The procedure was as follows: a) teacher and students sang a local song that students already knew; b) teacher and students sang the song off by heart while walking 'on the spot'; c) teacher asked appointed students to sing part of the song more loudly; d) teacher appointed students to take turns in singing; e) the appointed students sang part of the song loudly.

The second rhythm learning was conducted in SMP Negeri 2 Temanggung, The teaching method used was adapted from the Netherlands, where it was developed by Suzan Lutke and Christianne Nieuw Meijer: both are lecturers at Hogeschool voor de Kunsten, Utrecht. The procedure was as follows: a) rhythmic musical instruments were distributed to the learners, who were also allowed to use the instruments they had brought; b) teachers and learners discussed and agreed musical symbols; c) learners played music in accordance with the agreed symbols; d) the teacher played music and each student played a musical instrument to accompany the song *Radetzky March* according to the agreed symbols.

3.2 Teaching of melody

The teaching of melody was done in Conservatorium Den Haag on the basis developed by Elleke Bijsterveld to increase students' sensitivity to tone. Two teaching methods were adopted. In the first, students were trained to differentiate between a major scale and a minor scale. In the second, students were asked to vary the scale.

The first approach was conducted in early childhood music education. The parents were involved in the learning process, so that they could help their children when studying at home. Extracurricular learning of music was held on Saturdays, when parents were not at work. The purpose was to develop sensitivity to major and minor scales because these scales contrast with each other in terms of their nuance. The major scale provides a happy nuance while the minor scale gives a sad nuance. The procedure was as follows: a) the teacher and students sang a famous Netherlands folk song, *Lazy Boy*, which contains both a major and a minor scale, while walking 'on the spot' and sang while forming a circle (facing into the circle); b) the song was played repeatedly with increasing variation in modulation; c) when the scale changed from major to minor, students changed their position (to face out of the circle).

The second method was developed by Jos Overmars et al. (2012) in Cals College, Nieuwegein. The steps were as follows: a) the teacher asked students to sing a minor scale; b) the teacher asked students to improvise up and down the minor scale but with different students using different rhythms. This method was implemented in Indonesia at two junior high schools, namely SMP Negeri 2 Temanggung and SMP Muhammadiyah Wonosari, and in modified form at a senior high school, namely SMA 1 Depok Sleman, Yogyakarta.

3.3 Teaching of harmony

The teaching of harmony was conducted in Cals College, Nieuwegein, Utrecht. The teacher played a chord progression and students were asked to improvise the melody based on those chords. This teaching was modified and implemented in SMA 1 Depok Sleman. The chord progression used was Am–Dm–G–C–F–Dm–E–Am. These chords were played by the teacher continuously. In response to the chords, each student was asked to sing an improvised melody, with each student in the classroom producing a different melody.

Teaching harmony was done by asking students to work in pairs to create a melody by forming question and answer sentences. The first student created an interrogative sentence and the second student answered this question with an improvised melody.

3.4 Behavioral changes of students

3.4.1 Students' behavioral changes while learning rhythm

Students' behavioral changes in Conservatorium Den Haag during the learning of rhythm using the first method showed that kindergarten-age students could better control their emotions when they sang while walking 'on the spot'. Students that were usually 'chaotic' by the end of the lesson became quiet when following the rhythm. In addition, with the technique of singing by heart (from memory), students could learn to concentrate.

At the start of the lesson, students tended to follow their emotion by playing the instruments as loudly as possible so that no rhythmic sound could be heard. However, by the end of the lesson, students could control themselves and were able to play the instruments well, and could also play them by paying attention to the volume of other instruments.

Teaching rhythm was also done by playing a piano together (ensemble). Students had to continuously play the melody played by the teacher or other students. At the outset, students played loudly before the teacher advised them by giving an example. Eventually, students understood that they should play the instrument dynamically and smoothly, not loudly. By the end of the lesson, all students could play the music with the same dynamics and tempo.

It could safely be said that learning rhythm influenced students' ability to control their emotions such that they could play music harmonically with the right dynamic and tempo and with other students. Students that played ensemble music well showed that they could control their emotions and had good competency. Xiao et al. (2014) demonstrated the effectiveness of learning rhythm by walking in the course of piano teaching.

3.4.2 Students' behavioral changes during teaching of melody

Students' behavioral changes in Cals College in Nieuwegein, SMA 1 Depok Sleman, SMP Muhammadiyah Wonosari, and SMP Negeri 2 Temanggung showed that students put hard work into finding a melody that had not been played by other students. Meanwhile, the melody teaching conducted in extracurricular music for students of Conservatorium Den Haag showed that at the beginning of the lesson only a few students changed their position (to face out of the circle) when the song changed into a minor scale, but by the end of the lesson, all students were changing their position when the scale changed.

3.4.3 Students' behavioral changes during the learning of harmony

The teaching of harmony improvisation in Cals College in Nieuwegein, SMA 1 Depok Sleman in Yogyakarta, and SMP Muhammadiyah Wonosari showed that at the beginning of the lesson, students were still hesitant in creating a melody in accordance with the harmony of the instruments. However, by the end of the lesson, students were more confident in creating new melodies in the form of either a question or an answer.

3.4.4 Quantitative data of behavioral changes

Students' behavioral changes during the music learning processes that emphasized rhythm, melody and harmony are as shown in Table 1.

Table 1. Descriptive statistics of students' behavior.

	Pretest	Post-test
N	20	20
Minimum	70	80
Maximum	90	95
Mean	76	87.5
Standard deviation	5.98	5.96

Table 2. The output of paired-sample *t*-test of students' behavior between pretest and post-test.

Paired samples test

| | | Paired differences | | | 95% confidence interval of the difference | | | | Sig. |
		Mean	Std. deviation	Std. error of mean	Lower	Upper	*t*	df	(2-tailed)
Pair 1	Behavioral ability	80.25	7.94	1.26	77.71	82.79	63.938	39	.000

Based on analysis of a paired-sample *t*-test, the results are as shown in Table 2, which shows a *t*-value of 63.93 with a significance of 0.00. Therefore, there are significantly different behaviors between pretest and post-test. The mean of behavioral ability in the post-test is higher than in the pretest. Thus, it could be said that learning musical rhythm, melody and harmony can change students' behavior.

The data implicitly shows that this kind of teaching can be widely applied, because almost all students can improve their behavior. Learning rhythm by singing while walking 'on the spot' controls the tempo of the song so that it becomes more stable. This movement also controls the students' behavior so that they do not move around.

Differentiating between major and minor scales in a song is a teaching that could develop student sensitivity. It is hoped that with such abilities, students would be more sensitive to things happening in their surroundings so that they could adapt better to their social or natural environment.

Learning melody by varying melody gave students experience of being creative, which should stimulate them to be more creative. In improvising by creating question and answer sentences with particular chord progressions, students experience free creativity while still adapting in the context of a group. This is consistent with Gardner (2004), who said that developing interpersonal intelligence can be done by giving students the opportunity to conduct activities based on children's capacities, such as learning to pay attention to others, and inviting or influencing others to do something.

4 CONCLUSIONS

The conclusions of this research are:

1. Teaching elements of music in Indonesia or the Netherlands uses these elements to change students' behavior. Teaching rhythm by singing while walking 'on the spot' can make students better able to concentrate and to control their emotions. All students could play the music with the correct dynamic and tempo; teaching melody with major or minor scales can make students more sensitive to their environment; teaching harmony by creating question and answer sentences in accordance with chord progressions can train students' ability to cooperate with other students.
2. Music teaching in Indonesia and in the Netherlands are similar, that is, the use of well-known folk songs to cultivate morality and encourage musical ability at the same time.
3. Musical elements can have a significant effect on cultivating behaviors.

The implication of the results of this research is that, to build students' morality, students could learn music, particularly with elements such as rhythm, melody, and harmony as the focus of lessons. By implementing such an approach, it not only focuses on musical ability but also on the development of students' behavior.

REFERENCES

Astuti, K.S. (2011). *Developing model for teaching and learning music in public school based on comparative study between Indonesia and the Netherlands.* Jakarta, Indonesia: DP2M. Retrieved from http://eprints.uny.ac.id/26122/

Astuti, K.S., Widyantoro, A., Boel, C. & Berendsen, M. (2017). Developing an integrated music teaching model in Indonesia based on the Dutch music teaching model as the implementation of the 2013 curriculum. *Researchers World: Journal of Arts, Science, and Commerce, 8*(2), 115–124.

Brendell, J.K. (1996). Time use, rehearsal activity, and student off-task behavior during the initial minutes of high school choral rehearsals. *Journal of Research in Music Education, 44*(1), 6–14.

Cassity, M.D. (1976). The influence of a music therapy activity upon peer acceptance, group cohesiveness, and interpersonal relationships of adult psychiatric patients. *Journal of Music Therapy, 13*(2), 66–76.

Desyandri, D. (2015). Nilai-nilai edukatif lagu-lagu Minang untuk membangun karakter peserta didik [Educational values of Minang's songs in building the character of learners]. *Jurnal Pembangunan Pendidikan: Fondasi dan Aplikasi, 3*(2), 126–141.

Dewantara, K.H. (1977). *Pendidikan* [*Education*]. Yogyakarta, Indonesia: Majelis Luhur Persatuan Taman Siswa.

Gardner, H. (2004). Workshop: Tapping into multiple intelligences. *Concept to Classroom.* New York, NY: Educational Broadcasting Corporation/WNET. Retrieved from http://www.thirteen.org/edonline/concept2class/mi/index.html

Hasan, S.H. (Ed.). (2010). *Pengembangan pendidikan budaya dan karakter bangsa* [*Developing cultural education and national character*]. Jakarta, Indonesia: Puskur Balitbang. Retrieved from http://gurupembaharu.com/home/wp-content/uploads/downloads/2011/11/Panduan-Penerapan-Pendidikan-Karakter-Bangsa.pdf

Kushartanti, W. (2004). *Optimalisasi Otak dalam Sistem Pendidikan Berperadaban* [*Brain optimization in the civilized education system*]. Foundation Day speech at Universitas Negeri Yogyakarta. Yogyakarta, Indonesia: Universitas Negeri Yogyakarta. Retrieved from http://staffnew.uny.ac.id/upload/131405898/penelitian/Optimalisasi+Otak+Dalam+Sistem+Pendidikan+Berperadaban.pdf

Moreno, S., Marques, C., Santos, A., Santos, M., Castro, S.L. & Besson, M. (2009). Musical training influences linguistic abilities in 8-year-old children: More evidence for brain plasticity. *Cerebral Cortex, 19*(3), 712–723.

Morris, C.W. (2000). Morals, manners, and law. *The Journal of Value Inquiry, 34*(1), 45–59. doi:10.1023/A:1004790123233

Overmars, J., van de Putte, R., Tempelaar, W., Timmer, M., de Vriend, H. & Zweers, M. (2012). *Intro* (Vol. 1). Amersfoort, The Netherlands: ThiemeMeulenhoff. Retrieved from https://issuu.com/thiememeulenhoff/docs/intro_onderbouw_english_edition_hav

Pallant, J. (2007). *SPSS survival manual: A step by step guide to data analysis using SPSS for Windows* (3rd ed.). New York, NY: McGraw-Hill

Wang, J.-C. (2015). Games unplugged! Dolanan Anak, Traditional Javanese children's singing games in the 21st-century general music classroom. *General Music Today, 28*(2), 5–12.

Wilson, B. & McCrary, J. (1996). The effect of instruction on music educators' attitudes toward students with disabilities. *Journal of Research in Music Education, 44*(1), 26–33.

Xiao, X., Tome, B. & Ishii, H. (2014). Andante: Walking figures on the piano keyboard to visualize musical motion. In *NIME '14, 30 June–03 July 2014, Goldsmiths, University of London, UK* (pp. 629–632). Retrieved from https://www.media.mit.edu/publications/andante-walking-figures-on-the-piano-keyboard-to-visualize-musical-motion/

A model for development of extracurricular activity-based civic intelligence in primary schools

M. Masrukhi
Universitas Negeri Semarang, Indonesia

ABSTRACT: The purpose of this study is to identify an effective model for development of extracurricular activity-based civic intelligence in primary schools. The design of this study is based on research and development with the research subjects, namely the primary schools of Semarang Municipality in Java, Indonesia. The research technique is the Delphi technique. The findings of the study were as follows: first, the civic intelligence of primary school students was low; second, all primary schools implemented adequate extracurricular activities; third, a model for development of extracurricular activity-based civic intelligence in primary schools can be implemented by democratic methods such as search efforts and joint activities, modeling, simulation, lived experience, and value clarification.

1 INTRODUCTION

Civic education is one of the mandatory subjects that must be studied by students in Indonesia from elementary school to high school. The decision for the subject of civic education to be compulsory has been long-standing, but it has not shown any real success. This can be seen from the response of students in Indonesia who are not very interested in this lesson, whereas the noble goal of this subject is the development of active participation as citizens (Zuriah, 2011), which represents the peak of the intelligent and good citizen. To achieve this level, of course, there are competences that must be exhibited by citizens. These include knowledge of citizenship, citizenship skills, and disposition of citizenship. These three competencies are essential indicators of whether a citizen can be regarded as a good and intelligent citizen. In order to achieve these goals, efforts should be made, although some may be not easy. Ineffective civic education lessons tend to be old-fashioned and ignored. This learning should be reconstructed such that student learning interest increases and citizenship competence is eventually achieved. The ultimate goal of the subject of civic education to create a good and intelligent citizen cannot be negotiable if the State of Indonesia wants its citizens to compete in the global era. The formation of citizenship knowledge, skills, and disposition, which are the three elements of the indicators of good and intelligent citizens, are the central points that the subject of civic education must address (Patrick & Hoge, 1991; Dynneson, 1992; Sears, 1994; UNDP Democratic Governance Group, 2004; Branson & Quigley, 1999; Print & Smith, 2000; Kirlin, 2002; Zaff et al., 2003; Solhaug, 2006; Youniss, 2009; Blais & Vincent, 2011).

To develop the nation's intellectual life is one of the goals of the Indonesian nation listed in the fourth paragraph of the Preamble of the State Constitution of 1945. This objective is then realized through education conducted in schools where civic education becomes one of the vanguards by which it happens. More specifically, Wahab and Sapriya (2011) stated that the goals of civic education are the creation of good, intellectual, emotional, social and spiritual citizens with a sense of pride and responsibility, able to participate in the life of the community and the state, and to foster a sense of citizenship and patriotism. The progress of a nation can be seen from the quality of its citizens. To fulfill the mandate of the State

Constitution, the contribution of the subject of civic education is awaited by the Indonesian nation.

The mandate of civic education subjects to form qualified citizens is, of course, not easy to achieve because it takes quite a long time to make it happen. For this reason, by looking at the condition of education in the 21st century, where today's human beings are in the era of the fourth industrial revolution, it is deemed necessary to reconstruct civic education learning. This reconstructed learning technique is expected to make learning more meaningful so that the characteristics desired of citizens can be achieved. One approach to reconstructing civic education learning is to integrate it with the activities of the school, either intra-curricular or extracurricular. The learning of civic education needs to be integrated with the student's real life so that students can capture real examples they find in their lives (Cogan, 1998).

In order to form good citizens, the basic consideration is the need to develop civic intelligence. Civic intelligence is the citizens' ability to play a proactive role as citizens in the governance of complex life on the basis of a normative national identity. If the intelligence of nationality thrives in a person, it will will create a model citizen. Thus the key to the formation of good citizens is civic intelligence.

According to Winataputra and Saripudin (2012), there are seven skills that must be developed to establish civic intelligence: civic knowledge, civic disposition, civic skills, civic confidence, civic commitment, civic competence and civic culture. These skills must be integrated harmoniously in the activities of thinking, behaving and acting as Indonesian citizens according to civic values. These skills are combined in the psychiatric processes of the students, according to the civic values.

Thus, the origin of civic intelligence comes from the values inside the field of affection of the citizens themselves. Fraenkel (2007) explains that this affection field includes ideals and goals adopted or expressed, aspirations expressed, the attitude shown or revealed, the feeling favored, and the actions undertaken, as well as any worries revealed.

However, the reality is that in the classroom learning process, civic education places more emphasis on the importance of citizenship without involving positive social and cultural contributions to society and the life of the nation. Consequently, many citizens who have understood the concepts, attitudes, norms and values, the relationship between citizens and the state, the rights and obligations of citizens, and early education to defend the state, are still limited to a low level of rote theoretical knowledge. Meanwhile, the understanding and insight, attitude, confidence, commitment, and behavior of society, nation and state are still far from a reflection of the character of good citizens that could be relied upon for the benefit of the nation.

This reality happens because the knowledge and understanding of the values of citizenship obtained by learners are not powerful. Less powerful learning means that it is less meaningful, less integrated with the real lives of students, less value-based, less challenging, and less actively engages learners in learning (NCSS, 2000).

Another relevant aspect of the renewal of the concept of citizenship for educational purposes in schools is the formulation of civic education as part of the democratization process in Indonesia. In this context, civic intelligence is directed to achieve through democracy, in democracy, and for democracy (Winataputra & Saripudin, 2012).

To implement the aspects described above is very difficult to do simply through curricular activities in the classroom because during in-class activities related to the internalization of the values of civic intelligence, implementation is still limited to the provision of knowledge and understanding (with minimal changes in the structure of cognition), and does not engage at affective or psychomotor levels. Furthermore, knowledge is limited to the theoretical, and to a low level. The imbalance of curricular activities that have been arranged and designed in the education field has led to efforts to develop the potential of learners, is still far below expectations.

Within the framework of educational activities, extracurricular activities are activities and counselling services that are conducted outside school hours to help develop students' potentials to the fullest. According to the National Education Minister Regulation No. 39, Year 2008, on Fostering Students, the extracurricular program is an integral part of the learning process that places emphasis on meeting the needs of students. The implementation of these extracurricular activities is tailored to the needs, potential, talents, and interests of students, as well as the availability of schools' resources. Hence, the flexibility of extracurricular activities is very high.

The elementary schools of Semarang are very dynamic in implementing their extracurricular activities. Based on a previous interview that I conducted with the Head of the Education Department of Semarang, the majority (80%) of primary schools in the city, both public and private, conduct extracurricular activities every Friday and Saturday afternoon. There are, on average, four types of extracurricular activity, including sports, art, religion, and, in particular, scouting.

Further, the Head of the Education Department of Semarang said that because there are public and private universities in Semarang, there is collaboration between the universities and schools, mainly for conducting extracurricular activities.

Student participation in training or coaching extracurricular activities in primary schools of Semarang is also encouraged by university student affairs policies. Universities set such policies in order to elaborate the activities of each Student Activity Unit (known as *Unit Kegiatan Mahasiswa* or UKM). They are required to have units built, both at schools and in the community, according to the provisions of the UKM for the unit activities.

This condition means that extracurricular activities in the primary schools in Semarang are relatively dynamic. However, the question is whether, in practice, these dynamic extracurricular activities have achieved the mission of developing students' civic intelligence. In this light, the question can be resolved through research and development, namely: (1) What is the real condition of civic intelligence in primary schools? (2) What is the design of the ideal model for development of extracurricular activity-based civic intelligence in primary schools.

2 METHOD

This study used a qualitative approach, where the analysis started from the empirical facts obtained in the field for subsequent abstraction and drawing of conclusions. The respondents of this research were a number of principals and teachers of primary schools in Semarang City. The determination of the respondents was done by purposive sampling and snowball sampling. The research informants were the heads of education offices at both city and subdistrict levels. Data were collected through participatory observation techniques, in-depth interviews, and documentation studies. Furthermore, in conducting data analysis, an interactive analysis model was used, namely, the interactions among three components: data reduction, data presentation, and data verification. These activities were carried out during the data collection process. If deemed less stable toward the conclusion (by virtue of verification data) because of possible weaknesses in reducing and presenting the data, a further investigation of the field notes was then conducted.

3 RESULTS AND DISCUSSION

In characterizing students' civic intelligence, there are seven targeted aspects, namely civic knowledge, civic disposition, civic skills, civic confidence, civic commitment, civic competence, and civic culture.

The results show that the presence of civic intelligence in primary school students of Semarang is generally at low levels. This is indicated by the average score of questionnaires in civic intelligence of 5,465, as compared to the ideal score of 12,250, which represents a percentage coefficient of 44.61%, and positions it in the 'low' category in relation to the parameters set.

In terms of individual aspects of civic intelligence, the highest percentage coefficient relates to civic knowledge at 52.8%, which places it in the 'medium' category. This is followed by civic disposition with a coefficient of 49.1%, which is in the 'low' category. After this is civic

Table 1. Profile of Civic Intelligence (CI).

No.	Aspects of CI	Percentage	Category
1	Civic knowledge	52.8%	Medium
2	Civic disposition	49.1%	Low
3	Civic skills	48.3%	Low
4	Civic confidence	42.8%	Low
5	Civic commitment	46.8%	Low
6	Civic competence	35.7%	Low
7	Civic culture	36.6%	Low

skills with a percentage coefficient of 48.3%, civic commitment at 46.8%, civic confidence at 42.8%, civic culture at 36.6%, and, finally, civic competence with a percentage coefficient of 35.7%, all of which fall into the 'low' category. Table 1 shows the profile of civic intelligence in Semarang primary school students for each of these aspects.

The findings show that the intelligence profile of citizenship in primary school children is still low. All but one of the seven aspects of civic intelligence are in the 'low' category, the exception being civic knowledge, which is in the 'medium' category.

The low civic intelligence of the primary school students is associated with the school system of counseling or coaching. Coaching responsibility is not only at the level of curricular activities in the classroom through related subjects (e.g. Civics), but also through extracurricular activities outside school hours. This is because civic intelligence is among the soft skills that must be possessed by students, especially in primary schools. In accordance with its level, soft skills coaching is the domain of extracurricular activities.

In all primary schools targeted, the research finds that the schools' extracurricular activities are adequate. This means that all surveyed schools are conducting an average of four to five kinds of extracurricular activities. Schools that perform four types of extracurricular activities account for 70%, while 30% of them carry out five types of extracurricular activities.

The adequate implementation of extracurricular activities in the primary schools of Semarang is due to the presence of both state and private colleges in the area. The principals of some primary schools said that their schools are near the campus of Semarang State University. This leads to a mutual benefit between university students and the schools and improves the quality of education in the primary schools. The same thing was expressed by the principal of a primary school in Sambiroto, who also said that the existence of Muhammadiyah University of Semarang near the school makes for better extracurricular activities because the school's students receive guidance directly from the university. In general, the university students need to practice activities such as training, in accordance with their expertise. Meanwhile, the school is greatly in need of trainers to develop extracurricular activities for the students.

Therefore, schools in Semarang, including Islamic primary schools, invite students who are commonly activists to become trainers of extracurricular activities. Sometimes, the students themselves come to the schools without being asked to conduct extracurricular activities as an implementation of the programs of student organizations on their campus.

In general, there are four to five types of extracurricular activities organized in primary schools and madrasahs in Semarang. The selection of the type of activities organized is adjusted to the needs of each school. However, in all surveyed schools four basic extracurricular activities are indicated, that is, scouting, art, sport, and religion.

All primary schools and Islamic primary schools organize scouting as an extracurricular activity. Besides being commonly practiced, organizing scouting as an extracurricular activity is also required to meet the implementation of the Indonesian Curriculum of 2013, which has been gradually implemented since 2014 at all levels of schooling.

Following the implementation of the curriculum, extracurricular scouting activity became obligatory in schools. This provision has brought consequences for every school in preparing people who possess sufficient competence as trainers. At the very least, such trainers must have been certified as proficient in the basic course.

In terms of other types of extracurricular activity, each school has its own selection according to the school's needs and policies. Besides scouting activities, the schools select types of activities that include arts, sports, religion, and leadership, and the findings showed that all surveyed schools (100%) organized extracurricular activities involving these activities.

In the field of art, 65% of schools choose to organize extracurricular activities such as dance, while 20% of schools choose to pursue music, and 10% choose visual art. The interesting thing is that for these artistic extracurricular activities, the schools work together with the students to manage them because there are no teachers specializing in arts in the primary schools surveyed.

In sport, there are various choices of extracurricular activity. Seven forms of sports are practiced, of which the most common is futsal (35%), followed by karate (20%), swimming (15%), indoor soccer (10%), table tennis (10%), badminton (5%) and volleyball (5%).

In the field of religious activities, all schools surveyed choose the activity of reading the Qur'an, as well as the guidance of daily worship (*fiqh yaumiyyah*). This reflects the fact that students in the primary schools of Semarang are mostly Muslims (94%). This religious activity is directly handled by teachers of religion, in cooperation with the activists of Islamic organizations.

All surveyed schools and madrasahs performed the extracurricular activities every Friday and Saturday, outside school hours, specifically between 15:00 and 17:00.

In order to guide the learners in determining the options for extracurricular activity (besides the obligatory scouting), schools seek parents' opinions and approval for their children's options, for which they provide registration forms. The purpose of this common understanding includes establishment of the search-related interests of learners, safety in the implementation of activities, and any funds that must be invested by parents.

In managing extracurricular activities in the participating primary schools, schools cooperate effectively with public and private campuses, particularly in the field of student affairs. This is done by having written requests to student affairs, so that extracurricular activities held in schools get trainers' assistance from state and private universities in Semarang. The demand is in accordance with the student affairs program, one aspect of which is to bring all Student Activity Units of public and private campuses closer to the schools around the campus. Every unit is required to conduct training activities in schools, according to the characteristics of the activity.

Extracurricular activities in primary schools and Islamic primary schools of Semarang run in accordance with the schedules set by the school. The enthusiasm of the students is high, indicated by the students' presence and activeness in following the activities of their own choosing.

Although extracurricular activities run well, with the exception of scouting activities, extracurricular activities such as art, sports, and religion do not have standard guidelines set out. There have not been stages of activities in terms of planning, organizing, implementing, and evaluating. Most student activists who train extracurricular activities in schools do not have a vision of coaching values for each activity. According to them, the orientation of the activities provides the necessary skills and expertise to the participants such that they master the basics and techniques. Thus, the framework has not yet been structured, the target of interest for each activity has not been identified, nor is there any system around either the theoretical material provided or field activities, the methods and media used, or the targeted values of each activity.

The planning of the activity that will be implemented at each session is still carried out only on a partial basis, planning in the moment, and showing no signs of overall development. The researcher interviewed Imam from Volleyball unit. Imam was also coaches at Sekaran Primary School. He said that the target of the coaching is that the participants can play volleyball skillfully. In each activity session, the participants (students) are invited to play volleyball while being given direction by a team builder. This activity takes place repeatedly so that when the study was conducted, it was entering its third year in Sekaran Primary School.

The fact above shows that development of a model for civic intelligence based on extracurricular activities in primary schools can be done through a combination of models that are both integrated within the learning process and outside it. Planting values through formal teaching is integrated with outside learning activities. This model can be implemented either in cooperation with a team of teachers or with people outside the school.

Further, the method for delivering such a model should use democratic means, with joint search efforts, joint activities, exemplary, direct lived experience or simulation, and value clarification.

4   CONCLUSION

The results of the discussion can be summarized as follows. First, in exposing them to civic intelligence, primary school students are serving as research subjects, and are targeted in terms of seven aspects, namely, civic knowledge, civic disposition, civic skills, civic confidence, civic commitment, civic competence, and civic culture. The study finds that the civic intelligence of primary school students is generally at low levels. This is indicated by the score returned though civic intelligence questionnaires of 5,465, compared to the ideal score of 12,250, which if presented in percentage coefficient terms equates to 44.61%. If this coefficient is calibrated against standardized parameters, it is obviously at a low level.

Second, all primary schools implement adequate extracurricular activities. This means that all schools surveyed carry out an average of four or five kinds of extracurricular activities. Schools that implement four types of extracurricular activity account for 70% of the total, while 30% of schools carry out five types of extracurricular activity.

Third, the development of a model for extracurricular activity-based civic intelligence in primary schools can be done through democratic methods such as search efforts and joint activities, modeling, simulation, lived experience, and value clarification. In principle, all the methods involve all aspects of cognitive, affective, psychomotor and social intelligence. Therefore, understanding the concept, the introduction of context, reactions and actions become important parts of the whole development model for extracurricular activity-based civic intelligence in primary schools.

The ideal model of civic education development in elementary students is through systemic reconstruction of extracurricular activities, starting with planning, and then implementation and evaluation, adopting the development of civic intelligence as its spirit. This can be implemented through partnership with the nearest universities.

REFERENCES

Blais, A. & Vincent, S. (2011). Personality traits, political attitudes and the propensity to vote. *European Journal of Political Research*, 50(3), 395–417.
Borg, W.R. & Gall, M.D. (1983). *Educational research: An introduction* (4th ed.). New York, NY: Longman.
Branson, M.S. & Quigley, C.N. (1999). *The role of civic education* (An Education Policy Task Force position paper from The Communitarian Network). Washington, DC: Institute for Communitarian Policy Studies, George Washington University. Retrieved from https://www2.gwu.edu/~ccps/pop_civ.html
Cogan, J.E. (1998). *Multidimensional Citizenship Education*. Illinois, IL: Scott & Co Publication.
Dynneson, T. (1992). What's hot and what's not in effective citizenship instruction. *The Social Studies*, 83(5), 197–200.
Fraenkel. (2007). *Developing the civic society: The role of civic education*. Colorado, CO: Englewood Cliffs, Inco.
Freire, P. (2000). *Pedagogy of the oppressed*. New York, NY: Continuum.

Graham, D. (2002). *Citizenship for the 21st century: An international perspective on education.* London, UK: Kogan Page.

Kirlin, M. (2002). Understanding how organizations affect the civic engagement of adolescent participants. *New Directions for Philanthropic Fundraising, 2002*(38), 19–35.

NCSS. (2000). *National standards for social studies teachers* (Vol. 1). Washington, DC: National Council for Social Studies.

Patrick, J. & Hoge, J. (1991). Teaching government, civics, and the law. In J. Shaver (Ed.), *Handbook of research on social studies teaching and learning.* New York, NY: Macmillan.

Print, M. & Smith, A. (2000). Teaching civic education for a civil, democratic society. *Asia Pacific Education Review, 1*(1), 101–109.

Sears, A. (1994). Social studies as citizenship education in English Canada: A review of research. *Theory and Research in Social Education, 22*(1), 6–43.

Solhaug, T. (2006). Knowledge and self-efficacy as predictors of political participation and civic attitudes: With relevance for educational practice. *Policy Futures in Education, 4*(3), 265–278.

UNDP Democratic Governance Group. (2004). *Civic education: Practical guidance note.* New York, NY: United Nations Development Programme.

Wahab, A.A. & Sapriya. (2011). *Teori dan Landasan Pendidikan Kewarganegaraan* [Theory and Foundation of Civic Education]. Bandung, Indonesia: Alfabeta.

Winataputra & Saripudin, U. (2012). Jati diri pendidikan kewarganegaraan sebagai wahana sistemik pendidikan demokrasi: Suatu kajian konseptual dalam konteks pendidikan IPS [The identity of civic education as a systematic vehicle of democratic education: A conceptual study in the context of social science education]. *Jurnal Pendidikan Program Pascasarjana, 1*(1), 39–75.

Youniss, J. (2009). Why we need to learn more about youth civic engagement. *Social Forces, 88*(2), 971–976.

Zaff, J.F., Moore, K.A., Papillo, A.R. & Williams, S. (2003). Implications of extracurricular activity participation during adolescence on positive outcomes. *Journal of Adolescent Research, 18*(6), 599–630.

Zuriah, N. (2011). Model pengembangan pendidikan kewarganegaraan multikultural berbasis kearifan lokal dalam fenomena sosial pasca reformasi di perguruan tinggi [Development model of multicultural civic education based on local wisdom in social phenomenon in post-reform higher education]. *Jurnal Penelitian Pendidikan, 11*(2), 75–86.

Character Education for 21st Century Global Citizens – Retnowati et al. (Eds)
© *2019 Taylor & Francis Group, London, ISBN 978-1-138-09922-7*

Individual student planning: Counselors' strategies for development of academic success in middle-school students

A.R. Kumara & C.P. Bhakti
Universitas Ahmad Dahlan, Indonesia

B. Astuti & Suwardjo
Universitas Negeri Yogyakarta, Indonesia

ABSTRACT: Students' academic success includes aspects of academic achievement, the achievement of learning outcomes, the acquisition of skills and competencies, satisfaction, and persistence, and determines post-college further education. The purpose of guidance and counseling in schools is to play a role in the optimal development of student potential. Guidance and counseling use multiple services to develop student academic success. Individual student planning services support learners in formulating and conducting activities related to planning their futures according to an understanding of their strengths and weaknesses, as well as an understanding of the opportunities available in their environment. In terms of the development of student academic success in middle school, counselors facilitate students by helping them understand their potential through interventions such as individual appraisals, helping students to interpret their self-potential test results, and providing information on further study and work on the basis of student potential. Individual planning service strategies can be implemented through classroom guidance, in both group and individual settings, and may involve other relevant stakeholders.

1 INTRODUCTION

Middle school represents a major transition in a student's academic career. For most students, it means changing schools, adjusting to a longer school day, changing teachers according to course content, and meeting the demands of more complex assignments that require independent learning and critical thinking skills. Given these challenges, the fact that many students require additional support to enjoy academic success in middle school is not surprising (Johnson & Smith, 2008).

It is unsurprising that researchers hesitate to define what constitutes student success. The term has been applied with increasing frequency as a catch-all phrase encompassing numerous student outcomes. The term 'academic success' is only slightly narrower, with the nuanced descriptor 'academic' intended to limit the term's application to the attainment of outcomes specific to educational experiences (York et al., 2015).

The 21st century presents an array of challenges and opportunities for school counselors to renew their practice and respond to the climate of school reform. Voices from the profession have called for a shift in the role of the professional school counselor from that of service provider to one of promoting optimal achievement for all students (Dahir & Stone, 2012).

The American School Counselor Association (ASCA, 2012) has advocated that school counselors establish their identity and clearly articulate and define the role that school counseling programs play in promoting student achievement and educational success. New-vision school counselors work intentionally with the express purpose of reducing the environmental and institutional barriers that impede student academic success. School counselors are challenged to demonstrate accountability, document effectiveness, and promote the contributions of school counseling to the educational agenda (Dahir & Stone, 2012), and they are in a unique position to exert a powerful influence. The contributions of school counseling can support every student's progress through school to help each emerge more capable and more prepared than ever before to meet the challenging and changing demands of the new millennium.

Brown and Trusty (2005) suggest that the school counselor's primary mission is the improvement of academic achievement. The research literature regarding school counselors' efforts in relation to academic achievement concludes that there is little support for the supposition that comprehensive school counseling programs improve achievement. Conversely, there is a growing body of evidence that suggests that school counselors can use strategic interventions to improve academic achievement.

Within the systemic framework of comprehensive school counseling programs, middle school counselors have a vital role in collaborating with other educators to promote the academic development of early adolescents. Research pertinent to contemporary middle-school counseling in this developmental domain is summarized by Sink (2005), especially as it may relate to program activities and interventions; implications and recommendations for best practice are included.

Gysbers and Henderson (2012) argued that the successful implementation of a comprehensive guidance and counseling program is supported by the implementation of four components: (1) basic services, (2) responsive services, (3) individual planning services, (4) system support. In order to establish a comprehensive guidance and counseling model, the Indonesian government issued Ministry of Education and Culture Regulation no. 111 of 2014 on Guidance and Counseling in Basic Education and Secondary Education (RoI, 2014). This regulation refers to the presence in the program components of basic services, responsive services, individual planning services and specialization, and system support. In line with the implementation of the 2013 curriculum, it places emphasis on individual planning services and specialization, with a particular focus on specialization services.

2 METHOD

2.1 Basic concepts of individual planning services

The term individual planning is sometimes confusing for those investigating school counseling programs. One might logically assume that because the word 'individual' appears in the title, the associated services are delivered to students on an individual basis. As related to this program element, the word individual actually means that the counselor uses whatever methods are most appropriate for helping individuals make plans about their future (VanZandt & Hayslip, 2001). School counselors want to help young people make careful decisions by exploring all their options, using many sources of information and identifying probable outcomes. That goal guides this program element. Some examples of the approaches counselors can take to accomplish this include individual counseling and group counseling of students, consultation with parents and teachers, coordination of community resources, classroom guidance focused on careers education, and the development and maintenance of careers information centers. In these ways, counselors and others assist all students in the development of career life plans consistent with their personal/social, academic, and career goals (Gysbers & Henderson, 2012). Gysbers and Henderson categorize these interventions as individual appraisal, advisement, and placement and follow-up.

ASCA (2012) states that individual student planning consists of ongoing systemic activities designed to help students establish personal goals and develop future plans, such as individual learning plans and graduation plans. School counselors use these activities to help all students plan, monitor and manage their own learning as well as to achieve academic, career and personal/social competencies aligned with the school counseling core curriculum.

According to the Ministry of Education and Culture (RoI, 2008), individual planning is defined as an aid to learners in formulating and conducting activities related to planning of their future based on an understanding of their personal advantages and disadvantages, as well as an understanding of the opportunities available in the environment. In-depth understanding of the counselee with all their characteristics, interpretation of their assessment results, and the provision of accurate information in accordance with the opportunities and potential of the counselee are needed so that the counselee is able to choose and make the right decisions in optimally developing their potential, including gifted and special needs counselees. This is in line with the opinions expressed by Gysbers and Henderson (2012): individual planning is a systematic activity designed to help learners understand and take action to develop plans for their future.

2.2 *Purpose of individual planning services*

According to RoI (2008), individual planning aims to assist the counselee in: (1) having an understanding of themselves and their environment; (2) being able to formulate the goals, planning, or management of their own development, whether involving personal, social, learning or career aspects; (3) being able to perform activities based on the understanding, goals, and plans that have been formulated. The purpose of individual planning can also be formulated as an effort to facilitate the counselee in planning, monitoring, and managing their own educational, career, and social development plans. The contents of individual planning services are things that the counselee needs to understand, specifically:

a. Preparation for further education, career planning, and development of sociocultural skills, based on their own knowledge, and information about school, the world of work, and their community.
b. Analysis of their strengths and weaknesses in relation to the achievement of their goals.
c. Measures for the level of achievement in their goals.
d. Decision-making that reflects their planning.

2.3 *Focus on individual planning development*

According to RoI (2008), the focus of individual planning services is closely related to the development of academic, career, and social-private aspects. In detail, the scope of the focus includes, among others: (1) the development of academic aspects, including the utilization of learning skills, conducting advanced education electives or choice of majors, choosing appropriate courses or additional lessons, and understanding the value of lifelong learning; (2) careers, including the exploration of career opportunities and work practices, and understanding the need for positive work habits; (3) socio-personality aspects, including the development of a positive concept of self and effective social skills.

2.4 *Focus individual planning activity at each level of education*

According to Cobia and Henderson (2003), the focus of individual service planning activities is illustrated by the levels and categories of service interventions shown in Table 1.

2.5 *Individual planning service strategy*

Schwellie-Giddis and Kobylarz (cited in Cobia & Henderson, 2003) include and expand on the interventions identified by Gysbers and Henderson (2012) to help students gather, analyze, synthesize, and organize information related to their future, and which can be modified for different age levels:

a. Outreach. An approach used to alert all students to the information and services available.
b. Classroom instruction. Curriculum activities delivered by teachers and counselors in large group activities; integrating career concepts into academic instruction makes material meaningful to students.
c. Counseling. Focusing on helping students in an individual or small group forum to explore personal issues that relate to their plans for the future. Students examine ways to apply

Table 1. Focus of individual planning service activities in each level of education.

	Appraisal	Advisement	Placement	Career
K–3	Special needs	As needed for special concerns	Special needs	Awareness of work/self; exposure to concepts
4–5	Special needs Achievement Aptitude Intelligence Interests	Educational planning; middle-school curriculum	Middle-school choices	Awareness of work/self; choices; exposure
6–8	Special needs Achievement Intelligence Interests	Educational planning; high-school curriculum	High-school choices; work experience/ service learning possibilities	Exploration of self/careers; set preliminary goals
9–12	Special needs Achievement Aptitude Intelligence Interests Personality Entrance exams	Educational planning/ career planning; post-secondary preparations	Continuing high-school choices; work experience/service learning possibilities/ post-secondary options	Continuing and deeper exploration of self/ careers; set goals; implement plans

information and skills they have learned to the development of their individualized educational and career plans.

d. Assessment. Assessment includes the administration and interpretation of both formal and informal measures and gives students a clearer understanding of their skills, abilities, interests, achievements, and needs.

e. Career information. Resources that provide current and unbiased information to students about occupations, educational programs, post-secondary training, the military, and employment opportunities.

f. Career information delivery system. In some states, a computer-based career information delivery system includes comprehensive, accurate, and current information about occupations and education/training opportunities.

g. Work experience. Students have chances to participate in actual work settings.

h. Placement. Resources and assistance are provided to help students make a successful transition from high-school to employment, post-secondary education, military service, or other options.

i. Consultation. Counselors give direct assistance to teachers, administrators, parents, and others who interact with students to help them better understand career development and strategies for supporting it.

j. Referral. For students who have barriers that may inhibit career development, school counselors recognize the problems and make appropriate referrals.

k. Follow-up. Counselors maintain long-term contact with students as they move through their school years and beyond

3 DISCUSSION

Interventions by individual planning services in the development of academic success can be through a variety of strategies. One of the most effective interventions for developing academic success is group intervention, as illustrated by the research of Webb and Brigman (2007) into the Student Success Skills (SSS) small group intervention developed for school counselors targeting academic outcomes. The SSS program is based on extensive reviews of research about the skills that students need to be successful. Studies supporting program effectiveness were briefly reviewed and showed consistent patterns of improved academic achievement outcomes for group treatment of students. The structured group format and intervention are described in detail, together with effective school counselor group work practice.

Shi and Steen (2012) reinforced this view with their research on the Achieving Success Everyday (ASE) group model, which is used to promote self-esteem and academic performance in students of English as a Second Language (ESL). The findings from the preliminary data indicated that the participants' self-esteem was improved by participation in the group. There was no significant improvement in the total Grade Point Average (GPA) of the participants, although 75% of them made modest improvements in GPA. Research by Webb et al. (2005) reported that a counseling approach led to improvements in the academic and social competence of elementary and middle-school students.

In addition to group intervention, the development of academic success can be done through individual interventions. The meta-analysis of Casey (cited in Dimmitt et al., 2007) reported that student planning is implemented through application of several strategies. In appraisal, school counselors work with students to analyze and evaluate their abilities, interests, skills and achievement. Test information and other data are the basis for helping students develop immediate and long-range plans. In advisement, school counselors help students make decisions for future plans based on academic, career and personal/social data.

School counselors work with stakeholders, both inside and outside the school, as part of the comprehensive school counseling program. Through school, family and community collaboration, school counselors can access a vast array of support for student achievement and development that cannot be achieved by an individual, or a school, alone.

School counselors collaborate in many ways. Within the school, school counselors build effective teams by encouraging collaboration among students, teachers, administrators and school staff to work toward the common goals of equity, access and academic success for every student. Outside of school, school counselors create effective working relationships with parents, community members and community agencies, tapping into resources that may not be available at the school. By understanding and appreciating the contributions made by others in educating all children, school counselors build a sense of community, which serves as a platform to create an environment that encourages success for every student.

4 CONCLUSION

School counselors have established their identity and clearly articulate and define the role that school counseling programs play in promoting student achievement and educational success. School counselors work with the express purpose of reducing the environmental and institutional barriers that impede student academic success. The purpose of guidance and counseling in schools is to play a role in the optimal development of student potential. Guidance and counseling use multiple services to develop student academic success. Individual student planning services aid learners in formulating and conducting activities related to planning of their future according to an understanding of their strengths and weaknesses, as well as an understanding of the opportunities available in their environment. In terms of development of student academic success in middle school, counselors facilitate students by helping them to understand their potential with interventions such as individual appraisal, helping students interpret their self-potential test results, and providing information on further study and work according to student potential. Individual planning service strategies can be implemented through classroom guidance, in both group and individual settings, and may involve other relevant stakeholders.

REFERENCES

ASCA. (2012). *ASCA national model: A framework for school counseling programs*. Alexandria, VA: American School Counselor Association.
Brown, D. & Trusty, J. (2005). School counselors, comprehensive school counseling programs, and academic achievement: Are school counselors promising more than they can deliver? *Professional School Counseling, 9*(1), 1–8.

Cobia, D.C. & Henderson, D.A. (2003). *Handbook of school counseling*. Upper Saddle River, NJ: Prentice Hall.

Dahir, C.A. & Stone, C.B. (2012). *The transformed school counselor* (2nd ed.). Belmont, CA: Brooks/Cole, Cengage Learning.

Dimmitt, C., Carey, J.C. & Hatch, T. (2007). *Evidence-based school counseling: Making a difference with data-driven practices*. Thousand Oaks, CA: Corwin Press.

Gysbers, N.C. & Henderson, P. (2012). *Developing and managing your school guidance and counseling program* (5th ed.). Alexandria, VA: American Counseling Association.

Johnson, E.S. & Smith, L. (2008). Implementation of response to intervention at middle school: Challenges and potential benefits. *Teaching Exceptional Children, 40*(3), 46–52.

RoI. (2003). *Undang-Undang Nomor 20 Tahun 2003 tentang Sistem Pendidikan Nasional [Law Number 20 Year 2003 regarding National Education System]*. Jakarta, Indonesia: Department of Law and Legislation, Republic of Indonesia.

RoI. (2008). *Penataan pendidikan profesional konselor dan layanan bimbingan dan konseling dalam jalur pendidikan formal [Structuring professional education counselors and guidance and counseling services in the formal education path]*. Jakarta, Indonesia: Ministry of Education and Culture.

RoI. (2008). *Peraturan Menteri Pendidikan Nasional Nomor 27 Tahun 2008 tentang Standar Kualifikasi Akademik dan Kompetensi Konselor [Regulation of the Minister of National Education Number 27 Year 2008 regarding Academic Qualification Standards and Counselor Competencies]*. Jakarta, Indonesia: Ministry of Education and Culture, Republic of Indonesia.

RoI. (2014). *Peraturan Menteri Pendidikan dan Kebudayaan Nomor 111 Tahun 2014 tentang Bimbingan dan Konseling Pada Pendidikan Dasar dan Menengah [Regulation of the Minister of Education and Culture No. 111 of 2014 on Guidance and Counseling in Primary and Secondary Education]*. Jakarta, Indonesia: Ministry of Education and Culture, Republic of Indonesia.

Shi, Q. & Steen, S. (2012). Using the Achieving Success Everyday (ASE) group model to promote self-esteem and academic achievement for English as a Second Language (ESL) students. *Professional School Counseling, 16*(1), 63–70.

Sink, C. (2005). Fostering academic development and learning: Implications and recommendations for middle school counselors. *Professional School Counseling, 9*(2), 128–135.

VanZandt, Z. & Hayslip, J. (2001). *Developing your school counseling program: A handbook for systemic planning*. Belmont, CA: Wadsworth/Thomson Learning.

Webb, L. & Brigman, G.A. (2007). Student success skills: A structured group intervention for school counselors. *The Journal for Specialists in Group Work, 32*(2), 190–201.

Webb, L.D., Brigman, G.A. & Campbell, C. (2005). Linking school counselors and student success: A replication of the Student Success Skills approach targeting the academic and social competence of students. *Professional School Counseling, 8*(5), 407–413.

York, T.T., Gibson, C. & Rankin, S. (2015). Defining and measuring academic success. *Practical Assessment, Research & Evaluation, 20*(5), 1–20.

Character Education for 21st Century Global Citizens – Retnowati et al. (Eds)
© 2019 Taylor & Francis Group, London, ISBN 978-1-138-09922-7

Integration of Javanese characters and chemistry concepts on developing TOKIJO as chemical literacy learning media

S. Purtadi, D. Dina & R.L.P. Sari
Universitas Negeri Yogyakarta, Indonesia

ABSTRACT: Character education in chemistry teaching and learning is still a big problem for teachers. Character education is seen as education in general, not integrated in every subject. Character education would be great if it was contextual. The use of culture elements well known to students will make character education more acceptable. TOKIJO are imaginary figures created to represent the integration of cultural elements and chemistry concepts. This article reports on the preliminary research process aimed at developing TOKIJO and evaluating it. This article describes the results of need assessment and development of TOKIJO based on the character of the elements and its correspondence to the character of cultural elements, namely *wayang* (Javanese traditional puppets). The result of need assessment analysis reveals that *wayang* characters are well known but not yet developed as a character education deliverer that is integrated to chemistry concepts. The development of TOKIJO figures representing the properties of the elements that are featured on Javanese characters is done as the initial step of this research.

1 INTRODUCTION

The goals of science learning, including chemistry, have changed in recent years. The purpose of science learning today is more direct students to be ready to face the century-oriented science and technology (Guo 2007). Science learning no longer prepares the selected elite for a career in the scientific world, but is expected to increase scientific literacy for the students. In high school science education, the two important goals are that students can master basic science concepts, and develop student curiosity and inspire future generations of scientists (Bethel & Lieberman 2014).

Good science learning needs to be more grounded with rooted knowledge and practical local knowledge (Indigenous knowledge) (Ugwu & Diovu 2016). The inclusion of science, including chemistry in original knowledge—local wisdom will create a generation that is aware that science actually belongs to everyone and they need to learn and apply it. Chemistry in context is one component of chemical literacy (Shwartz et al. 2006). Therefore, the placement of science that is directly related to the knowledge of local wisdom is part of the process for scientific literacy.

The practices of Javanese life are also not separated from the local wisdom transmitted from generation to generation. This knowledge has not been widely expressed and used as a learning resource in chemistry learning. Abah et al. (2015) see that local wisdom knowledge needs to be integrated in science learning. Based on this, it is necessary to find a solution to integrate indigenous knowledge in chemistry learning that emphasizes chemical literacy. Local clarification in Javanese society needs to be studied and made as part of the acquisition of chemical concepts to make high school students more literate.

2 LITERATURE REVIEW

2.1 *Chemical literacy*

Scientific literacy is the knowledge and understanding of scientific concepts and processes required in personal decision making as well as in societal and cultural, and economic productivity (NSES 1996). Science literature implies that one can identify the scientific issues underlying national and local decisions and state positions that indicate a person has good scientific and technological information. A person who is said to be literate must be able to evaluate the quality of scientific information within the framework of the source and method used to express it. Furthermore, scientific literacy also refers to the capacity to convey and evaluate arguments based on facts and to apply the conclusions of the correct argument (NSES 1996).

The dimensions of scientific literacy need to be addressed as they relate to decisions about how to distribute lessons and subject matter in the classroom and how to respond individually to students who show a lack of understanding (Trowbridge et al.2000). The dimensions of scientific literacy according to Trowbridge et al. (2000) are: 1) the dimension of literacy, 2) nominal scientific literature, 3) functional scientific literacy, 4) conceptual and procedural scientific literature, and 5) multi-dimensional scientific literacy. According to Shwartz et al. (2005), the commonly used dimensions related to science literacy are, a) understanding of the nature of science—the norms and methods of science and the nature of scientific knowledge, b) the understanding of key concepts, principles, and scientific theories (knowledge of science content), c) understanding of how science and technology are interrelated, d) appreciation and understanding of the influence of science and technology in society, e) communication competencies in scientific contexts—reading, writing, and understanding of human knowledge; f) application of knowledge and reasoning skills in everyday life.

Based on this dimension, Shwartz et al. (2005) developed components of chemical literacy by stating that, a chemistry literate understands the basic idea in chemistry, namely: 1) common scientific ideas, 2) chemical characteristics, 3) chemistry in context, 4) High Order Thinking Skills, and 5) affective aspects.

This literacy has been recognized by many observers of education, but attention to the size of the literacy is still limited to PISA and TIMMS. It is necessary to develop scientific literacy assessment instruments and scientific inquiry (Wenning 2007).

2.2 Indigenous knowledge

Education is an integral part of social life. The social and cultural forces surround each individual to form an indigenous education (Ekwam et al. 2014). Furthermore, humans from prehistoric times can survive because they can learn from examples and experiences to adapt to their lives in the environment (Ekwam et al. 2014). This capability can be maintained from generation to generation in the form of indigenous knowledge *(Indigenous knowledge)*.

Knowledge of indigenous *(Indigenous knowledge),* or often called wisdom, refers to how knowledge is unique from a cultural perspective and owned by local people through the accumulation of experience, informal experiments, and deep understanding of the cultural environment (Abah et al. 2015). This can be a system of technology, social, economic, and philosophy, learning, and government.

2.3 Wayang (Javanese Traditional Puppet) and TOKIJO

Wayang play story is a representation about nature and character of a in the world that reflects the nature and character of a typical human (Dwiandiyanta et al. 2012). The story itself comes from India mythology (Kusumanugraha et al. 2011), so the main character of *wayang* comes from these stories. In Javanese culture, the appearance of *wayang* has a pattern standard that symbolizes the trait of the character. The pattern standards comprise of pattern of eye, nose, eye brow, mouth, shoulder, and skin color. Every pattern has its own meaning.

TOKIJO comes from Tokoh Kimia Jowo (Javanese Chemistry Character), and is a learning media created to teach chemistry concepts integrated with Javanese characters TOKIJO is developed from the characteristics of *wayang* and the correspondence to the chemistry elements.

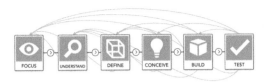

Figure 1.   The design process consists of 6 iterative phases: focus, understand, define, conceive, build and test.

3   METHODS

3.1   *Design based research*

This article presented a part of research developing TOKIJO as a media to teach chemical literacy. The whole research is a Design Based Research method by following steps developed by Easterday et al. (2014) as shown in Figure 1.

This article will describe two steps of six steps, namely define the goal, and conceptualize the solution steps.

To define the goal is a step done by analyzing the theory about the chemical characteristics and knowledge of Java, and analysis of the needs of chemistry teachers about indigenous knowledge-based character learning. The results of this analysis are further used to conceptualize and develop the desired TOKIJO character.

3.2   *Need assessment*

Need assessment is conducted to capture the needs and opinions of teachers about the characteristics of TOKIJO to be developed. Need assessment is done by giving a questionnaire to 30 high school teachers. The questionnaire was then analyzed and used as a reference in developing TOKIJO.

3.3   *Developing the TOKIJO*

TOKIJO is developed based on needs analysis and theoretical analysis on the integration between characterizing character of Java with the characteristics of chemistry elements. Each selected element is analyzed to obtain prominent characters that can be visualized. These characters are then designed in the form of two-dimensional images. Characters in the form of two-dimensional images are then assessed by experts to know the suitability of the characters of Java and the truth of the concept of chemistry.

4   RESULTS AND DISCUSSION

4.1   *Need assessment*

Based on the results of the questionnaire, 25% of chemistry teachers are S2 educated and are entirely from ethnic Java. The development of this media does not want to highlight the characters of Java and leave the other ethnic, but the characters of Java are chosen because *wayang* is known in various cultures, not just by Javanese people. This Javanese *wayang* takes place as a pilot project which can later be developed in other cultural products.Therefore, it is expected that the respondents who come from Java will better understand these Javanese *wayang* characters.

Character learning is understood differently by respondents. Although all respondents know that character teaching emphasizes the need for good character building, only 20% of respondents can accurately select the definition of character learning as desired by the Ministry of Education and Culture. All respondents stated that their schools have applied character learning, among others, by conducting honesty canteens and increasing the intensity of religious activities. Furthermore, in character study, it is better to see how students do not cheat in tests (honest nature), and do their own work (responsible) on time (discipline). This kind of context appears in every subject. There is no chemical characteristic. Within this framework, such chemistry learning is attached only by character learning, rather than compounding.

The cultural association with chemical learning that can be shown by respondents is "Giving examples of the use of elements and chemical compounds in traditional societies" (76.67%) and "Explaining local Javanese wisdom related to the concept of chemistry" (13.33%). Meanwhile, cultural products that are most likely to be thought of as a form of media that can be used to teach character in learning chemistry are a puppet, batik, and song. Puppet was chosen because of the characters of *wayang* figures who are considered to have exemplary as a reflection of the nature of Javanese society and also as a legitimate cultural heritage. Batik is the pride of the nation of Indonesia. Today, the introduction of chemistry through batik is the reason for teachers to choose batik as a medium of character learning in chemistry lessons. Meanwhile, songs that have been memorized by students can be used in memorizing chemistry concepts.

Focusing more on the figure seen by respondents to teach chemistry, the use of figures has been widely known by respondents in teaching chemistry concepts. Its use is in the form of comics (still images), with the appearance of multimedia, and so forth. Puppet figures have not been widely known as a figure that can be used as a conveyor of chemical concepts. Even in conveying character study in chemistry lessons, respondents prefer chemist characters (53.33%) than puppet characters (26.67%).

Based on this need assessment, it can be seen that the puppets are considered to give a humanist figures as an example of character learning for students. However, when associated with the study of chemistry, the use of *wayang* (puppet) figures has plastered not been seen able to convey two sides at once, namely the character and the concept of chemistry. In chemistry teaching and learning, it is necessary to look for the peculiarities of the chemical character itself. Therefore, the next steps is to combine the puppet potential that has been recognized by respondents being able to deliver moral message (good character) and to introduce chemistry concepts.

4.2 *The Developoment of TOKIJO*

The characters in *wayang* are very carefully described in terms of their properties. The character of each character can be seen from various faces and body shapes, as well as attributes of clothing worn. The character depiction of this character can be seen from the shape of the eyes, nose, eyebrows, lips, skin color, shoulder inclination, body shape, and so on. TOKIJO will initially be developed based on this standard.

Each Tokijo figure is built on the physical and chemical properties of the elements in question. Based on these traits, the face and body of TOKIJO are determined. This determination includes its gender determination.

However, to build the figure from the beginning by following this grip is more difficult, especially as the resulting figure aspect should be easily recognized by the students. Based on these considerations, the FGD discussion result decided that the TOKIJO figure is derived from the puppet characters. This is intended to simplify the process of sketches. This derivation is not meant to copy all traits from the character but is intended to facilitate the developer's imagination.

The result of this process is the determination of the elements to be constructed in the form of figures and puppets who are the inspiration for the figures. There are just 30 TOKIJO figures sketched. Here are two examples of them (Figures 2 and 3).

Cobalt is corresponded to Anila character (Figure 2). Anila is a monkey puppet from Ramayana that has a blue body color like cobalt, is shiny, wears silvery-white metal accessories. Anila has a violent character, and his eyes, nose, eyebrows, lips portray a fierce character. Similarly to the character from Anila, the TOKIJO figure was developed with silver-gray wire and magnets attached to the body. This is because of the cobalt metal is the strongest

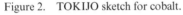

Figure 2.   TOKIJO sketch for cobalt.

Figure 3.   TOKIJO sketch for calcium.

magnet in the world and can retain magnets at high temperatures, making them reddish around the magnet. Cobalt metal is a hard but fragile metal that makes Anila's body cracked.

Calcium is corresponded to the Gatotkaca character (Figure 3). Calcium is shiny and silvery when it is first cut, but it can be oxidized with air to a gray color so that the Gatotkaca puppet character is depicted as having gray leather. Calcium is a good element for bone and tooth formation, so it is described that Gatotkaca that has strong muscles and bones. The famous calcium is contained in milk so the Gatotkaca puppets are made carrying milk. Calcium is a mild element to be described as a floating Gatotkaca. One of Gatotkaca's legs is tied to limestone, since calcium in nature is found in limestone.

In the sketch, Anila's characterization may be completely different from Anila's appearance in *wayang*. In this sketch it only takes the prominent features of the character that inspires it. Likewise with the character of TOKIJO figure for calcium. The inspirational figure used is Gatotkaca who is known for his strong bones.

The next step of our research was changing TOKIJO sketches so that they can be used as a media in chemistry learning in various ways. First, Tokijos are formed as cards that provide information about the elements and are organized into the periodic table of the elements. This will make the periodic table more interesting and increase students' curiosity. The TOKIJO visualization will make it easier for studies to remember chemical properties and usages of the elements. Secondly, TOKIJOs are used as an image design printed on t-shirts and other media. By sticking to the media that is easy to see and has unique impression, TOKIJOs will increase the chemical curiosity of the students and general people as well. This design should certainly be supplemented with other useful information. Third, TOKIJOs are formed into three dimensional figures. These three-dimensional figures can be used as a character of character education in chemistry learning, and also can be used as a unique souvenir that will increase interest in chemistry and curiosity both in students and others.

5   CONCLUSION

Character learning in learning in general can be done by displaying characters that are considered to have an exemplary nature. Puppets are known by the Java community to provide valuable lessons and give examples of a good attitude directly. Although *wayang* has potential as a medium in teaching character, this potential has not been seen in chemistry learning. Based on need assessment, puppets can be developed further in the development of character learning media in chemistry lessons. Its development is to utilize the characters that have been known in puppet characters and modify their properties in accordance with the physical and chemical properties of the elements and their compounds. There are 30 figures derived from *wayang* figures with chemical attributes. The TOKIJO figure is then ready to be reviewed and implemented for further evaluation

REFERENCES

Abah, J. et al. 2015. Prospect of integrating African indigenous knowledge systems into the teaching of sciences in Africa. *American Journal of Educational Research* 3(6): 668–673.

Bethel, C.M. & Lieberman, R.L. 2014 Protein structure and function: an interdisciplinary multimedia-based guided-inquiry education module for the high school science classroom. *J. Chem. Educ.* 91(1): 52–55.

Çelik, S. 2014. Chemical literacy levels of science and mathematics teacher candidates. *Australian Journal of Teacher Education*, 39(1): 1–15.

Dwiandiyanta, B.Y. et al. 2012. New shadow modeling approach of wayang kulit. International *Journal of Advanced Science and Technology* 43: 95–104.

Easterday, M.W. et al. 2014. Design-based research process: problems, phases, and applications. *Proceedings of International Conference of the Learning Sciences, ICLS*, 1(January), 317–324.

Ekwam, L. et al. 2014. Influence of community indigenous knowledge of science on students' performance in chemistry in secondary schools of Samburu County, Kenya. *Asian Journal of Management Sciences & Education* 3(4): 19–44.

Kusumanugraha, S. et al. 2011. An analysis of Indonesian traditional "*wayang kulit*" puppet 3D shapes based on their roles in the story. Second International Conference on Culture and Computing, *Proceeding*, 147–148.

NGSS Lead States. 2013. *Next generation science standards: For states, by states. Volume 1: the standards—arranged by disciplinary core ideas and by topics*. Washington, D.C.: The National Academies Press.

NSES. 1996. *National science education standard*. Washington, D.C.: The National Academies Press.

Shwartz, Y. et al. 2006. Chemical literacy: what does this mean to scientists and school teachers?. *Journal of Chemical Education* 83(10): 1557–1561.

Trowbridge, L.W. et al. 2000. *Teaching secondary science school science: strategy for developing scientific literacy*. New Jersey, N.J. Prentice Hall. Inc.

Ugwu, A.N. & Diovu, C.I. 2016. Integration of indigenous knowledge and practices into chemistry teaching and students' academic achievement. *International Journal of Academic Research and Reflection* 4(4): 22–30.

Wenning, C.J. 2006. Assessing nature of science literacy as one component of scientific literacy. *J. Phys. Tchr. Educ. Online* 3(4): 3–14.

A needs assessment for development of a learning set to improve students' higher-order thinking skills and self-confidence

E. Kurnianingsih & H. Retnawati
Universitas Negeri Yogyakarta, Indonesia

ABSTRACT: This research was a needs assessment to develop learning sets for improving students' Higher-Order Thinking Skills (HOTS) and self-confidence in junior high school. Higher-order thinking skills are a student's ability to connect new information with information stored in their memory and to rearrange and/or extend this information to achieve a purpose or find a possible answer in a complicated situation. This ability includes deciding what to believe or what to do, creating a new idea, new object, or an artistic expression, making predictions, and solving non-routine problems. This research was descriptive research by means of a qualitative approach. Data was collected through observations, interviews, and focus group discussion, and the subjects were two university lectures, 18 university students, ten eighth-grade students and two mathematics teachers from the same junior high schools in Yogyakarta, Indonesia. Data was analyzed qualitatively. The results of the study were as follows: first, students had difficulties in solving HOTS problems because they were unwilling to solve problems in long and complicated forms. They also did not have the self-confidence to solve such problems. Second, the problems that students are used to practicing are routine problems. Third, the lesson plans and student worksheets that teachers had were not HOTS-oriented. However, lesson plans and student worksheets are the kind of learning set that has the most important role in determining the orientation of a lesson. Based on these results, it is necessary to develop learning sets (lesson plans and student worksheets) to improve students' HOTS and self-confidence that use a problem-solving approach, ideally in the cooperative setting of a group investigation.

1 INTRODUCTION

Education is one of the most important tools for people in handling the challenges of the era of globalization. The 21st century is a dynamic era. People are required to be active, open-minded, and competitive. Therefore, education in the 21st century must facilitate people to learn well so they can survive life in this competitive era. The flows of information and technological developments also run very fast, and people should, therefore, have some competencies such that they can handle them judiciously.

McKenzie (2005) explained that the Information Age demands that people have additional skills to remain competitive. Thus, workers need: (1) **information technology skills**, which means the ability to access information and manipulate it using a variety of digital tools; (2) **information literacy skills**, which means the ability to evaluate information for validity and reliability through a variety of critical thinking strategies; (3) **problem-solving skills**, or the ability to generate efficient and effective solutions that meet the needs of the marketplace; (4) **collaboration skills**, or the ability to interact with colleagues, even in geographically disparate locations, to complete complex tasks; (5) **flexibility**, which is the ability to adapt and adjust to ideas as new information becomes available; (6) **creativity**, or the ability to present information and ideas in novel or unique ways in the marketplace.

These skills are abilities that cannot just be acquired through low-level thinking skills; these skills are acquired through more complex thinking skills or what we usually call Higher-Order Thinking Skills (HOTS). High-level thinking skills, including critical thinking skills, creative thinking skills, and problem-solving skills (Brookhart, 2010) are indispensable for students so that they can solve problems that might be found in their everyday lives. Lewis and Smith (1993) define higher-order thinking as occurring when a person takes new information and, together with information stored in memory, interrelates and/or rearranges and extends it to achieve a purpose or find possible answers in perplexing situations. A variety of purposes can be achieved through such higher-order thinking, which would include deciding what to believe, deciding what to do, creating a new idea, a new object, or an artistic expression, making a prediction, and solving a non-routine problem. Forster (2004) explained that in today's world it is necessary, but not sufficient, for students to achieve minimal competence in areas such as reading, writing and numeracy. Beyond the achievement of minimum competence, students also need to develop skills such as critical literacy, critical numeracy and cross-curricular competencies.

Given the importance of high-level thinking skills in the 21st century, learning processes in schools should be HOTS-oriented. Since the advent of Curriculum 2013, education in Indonesia has been oriented to these capabilities. In the Indonesian National Curriculum (Permendikbud No. 20, 2016), it is explained that graduates of high schools are expected to have competence in skill dimensions such as thinking and acting creatively, productively, critically, independently, collaboratively and communicatively through a scientific approach, as studied in education units and other sources. The capabilities to think critically and creatively form part of the higher-order thinking skills.

In fact, these skills have not been taught in schools. Jailani and Retnawati (2016) stated that HOTS-oriented learning has not been implemented well by math teachers, especially those in Yogyakarta, Indonesia. There were obstacles that made implementation of HOTS-oriented learning difficult to apply. These constraints stem from both teacher factors and student factors. Constraints associated with teacher factors include teachers' difficulties in understanding HOTS, teachers having difficulty in developing HOTS-oriented learning tools, and teachers having difficulty in finding test items to measure HOTS that are written in Bahasa Indonesia (Indonesian). This led to the HOTS-oriented learning set being limited. Therefore, it is necessary to develop a HOTS-oriented learning set that can facilitate teachers in implementing HOTS-oriented learning in schools. The learning tools that are targeted are lesson plans and student worksheets because these are the tools most needed by teachers to support the success of the classroom learning process.

The speed of information flow and the rapid development of technology in the 21st century must be balanced with a strong human character. People who do not have good character will tend to be destabilized by the uncertainty that enters their lives. With good character, people can respond to this globalization age with more wisdom so that they can make good decisions and do not succumb to negative things so easily. Character can be invested in primarily through formal education. One characteristic that has to be developed is self-confidence. Hannula et al. (2004) argued that "self-confidence is another variable that seems to be an important predictor for future development. A pupil's self-confidence predicts largely the development of self-confidence in the future, but also the development of success orientation and achievement." Students' self-confidence can be developed through learning in the classroom. Jurdak (2009) explained that "student self-confidence in learning mathematics is primarily formed as a result of students' interactions with the math teacher and with classroom peers during math instruction."

Based on the above, it is necessary to develop a learning set that not only develops students' HOTS but also their self-confidence.

2 LITERATURE REVIEW

2.1 *HOTS*

Thinking is one of the most important processes in a human body. Moseley et al. (2004) explained that the word 'thinking' is usually used to mean a consciously goal-directed process, such as remembering, forming concepts, planning what to do and say, imagining situations, reasoning, solving problems, considering opinions, making decisions and judgments, and generating new perspectives. 'Thinking skill' means expertness, practical ability or facility in the process or processes of thinking (processes that occur spontaneously or naturally, or which are acquired through learning and practice). Meanwhile, according to Presseisen (1985), thinking is generally assumed to be a cognitive process, a mental act by which knowledge is acquired. From these two definitions, we can conclude that thinking is a cognitive process such as remembering, forming concepts, planning what to do and say, imagining situations, reasoning, solving problems, considering opinions, making decisions and judgments, and generating new perspectives where the goal is to get new knowledge.

Bloom's well-known thinking taxonomy consists of six categories which are hierarchically ordered from the simplest to the most complex as follows: (1) knowledge, (2) comprehension, (3) application, (4) analysis, (5) synthesis, and (6) evaluation. Krathwohl (2002) subsequently revised this taxonomy thus: (1) remember—retrieving relevant knowledge from long-term memory; (2) understand—determining the meaning of instructional messages, including oral, written, and graphic communication; (3) apply—carrying out or using a procedure in a given situation; (4) analyze—breaking material into its constituent parts and detecting how the parts relate to one another and to an overall structure or purpose; (5) evaluate—making judgments based on criteria and standards; (6) create—putting elements together to form a novel, coherent whole or make an original product. The last three of these—analyze, evaluate, and create—are categorized as higher-order thinking skills.

Thinking can also be classified into four levels. Of these, the first level of thinking is *recall*, which includes those skills that are almost automatic or reflexive. The second level of thinking is *basic*, which includes the understanding and recognition of mathematical concepts such as addition, subtraction and so on, as well as their application in problems. The third level is *critical* thinking, which is thinking that examines, relates, and evaluates all aspects of a situation or problem, and includes gathering, organizing, remembering and analyzing information. Critical thinking also includes the ability to read with understanding and to distinguish between extraneous and necessary material. It also means being able to draw proper conclusions from a given set of data and being able to determine inconsistencies and contradictions in a data set. The critical thinking as an aspect of transferable skills can be identified based on levels of awareness between male and female students (Baser et al., 2017). The fourth level is *creative* thinking, which is thinking that is original and reflective and that produces a complex product. It includes synthesizing ideas, generating new ideas, and determining their effectiveness, and also includes the ability to make decisions. It often involves the generation of some new end product. The third and fourth levels here are classified as HOTS.

Cohen suggests at least four different complex thinking processes (Presseisen, 1985). These are: (1) *problem-solving* – using basic thinking processes to solve a known or defined difficulty; (2) *decision-making* – using basic thinking processes to choose the best response among several options; (3) *critical thinking* – using basic thinking processes to analyze arguments and generate insights into particular meanings and interpretations; (4) *creative thinking* – using basic thinking processes to develop or invent novel, aesthetic, and constructive ideas or products, related to precepts as well as concepts, and stressing the intuitive aspects of thinking as much as the rational ones.

From these statements, we can conclude that higher-order thinking skills consist of critical thinking and creative thinking, while problem-solving is recognized as the basis of HOTS. Based on the explanations of critical thinking and creative thinking, we can see that critical thinking is equivalent to *analyze* and *evaluate* in the revised Bloom taxonomy, while creative thinking is equivalent to the *create* category. Thus, we can summarize the indicators of HOTS as shown in Table 1.

Resnick (1992) explained that we cannot define HOTS exactly, but we can recognize them when they occur from some of their key features. Consider the following: (1) higher-order thinking is non-algorithmic; (2) it tends to be complex; (3) it often yields multiple solutions, each with costs and benefits, rather than unique solutions; (4) it involves nuanced judgment and interpretation; (5) it involves the application of multiple criteria, which sometimes conflict with one another; (6) it often involves uncertainty; (7) it involves self-regulation of the

Table 1. Indicators of HOTS.

Element of HOTS	HOTS indicator	Sub-indicators of HOTS
Critical thinking	Analyze	Differentiating
		Organizing
		Attributing
	Evaluate	Checking
		Critiquing
Creative thinking	Create	Generating
		Planning
		Producing

thinking process—we do not recognize higher-order thinking in an individual when someone else "calls the plays" at every step; (8) higher-order thinking involves imposing meaning, and finding structure in apparent disorder; (9) higher-order thinking is effortful—there is considerable mental work involved in the kinds of elaborations and judgments required.

2.2 Self-confidence

Self-confidence seems to be an important predictor of students' future development. Experts have provided various definitions and features of self-confidence. Goel and Aggarwal (2012) define self-confidence as a personality trait that is a composite of thoughts and feelings, hard work and hope, fear and awe, a person's views on what they are, what they are up to, what they will become, and their attitudes in relation to the values in which they believe.

Schunk (2012) defines self-confidence as the trust in oneself to be able to deliver results, achieve goals, or perform tasks competently. Srivastava (2013) adds that a confident person will have a positive view of themselves and the situation being experienced. Such a person also believes in their own abilities for realistic reasons, and will be able to do what they want, plan and expect. Yoder and Proctor (1988) say that a child who has self-confidence has the ability to (a) be firm, without being aggressive, (b) be firm in belief, even when others stand up against him, (c) be easygoing, (d) remain with a job until it is finished and is sufficient to ensure that the best is good enough, (e) accept defeat and rejection calmly, and rebound rapidly and enthusiastically, (f) work well with others as a "team" member, (g) hold a leadership role without hesitation at the right time, and (h) expect to be a leader, at least on several occasions.

JIST (2006) mentions that people who have confidence can finish almost any work undertaken, and keep trying (not hesitating to try) even if they fail. In addition, people who have confidence know their advantages and disadvantages. They always think positively of failure; failure does not bring them down but gives them the spirit to do better in the future. The teacher can use the training within Industry Strategy for improving of the students' self-confidence especially in conducting the practices matter (Marini, 2016).

Indicators of self-confidence include being independent, communicating easily with others, daring to accept new tasks/challenges, and being able to express emotion reasonably (Adywibowo, 2010). Thus, self-confidence is the attitude of someone who believes in themselves and their abilities, has an optimistic attitude, and dares to accept the challenges given.

2.3 Learning sets

As facilitators, teachers have to help students to construct their own knowledge. One of the methods is by preparing lesson plans and student worksheets that suit the needs of students. This is explained in Permendikbud No. 22 (RoI, 2016), which obliges every educator in the educational unit to develop complete and systematic lesson plans so that learning takes place interactively, is inspiring, fun, challenging and efficient, motivates learners to participate actively, provides enough space for initiative and creativity, and is based on the talents, interests, and physical and psychological development of learners. One of the elements in the lesson plan is the student worksheet. Therefore, teachers should be able to develop lesson plans and student worksheets that match the needs of students.

Learning sets are part of lesson planning, which includes the preparation of learning implementation plans and preparation of media and learning resources, learning appraisal tools, and learning plans. Learning plans should be developed in detail by referral to the syllabus, textbooks and teacher manuals. The components of learning plans listed in Permendikbud No. 103 (RoI, 2014) include: (1) school/madrasah identity, subject, and class/semester; (2) time allocation; (3) core competencies, basic competencies, and indicators of competency achievement; (4) learning materials; (5) learning activities; (6) assessment; (7) media/tools, materials, and learning resources.

Student worksheets are worksheets that have to be completed by students (Depdiknas, 2008). Prastowo (2012) states that student worksheets are printed material in the form of sheets of paper containing materials, summaries, and instructions on the implementation of learning tasks that must be done by students, and refer to the basic competencies that must be achieved. As a minimum, student worksheets consist of: 1) a title, 2) the basic competency to be achieved, 3) completion time, 4) equipment/materials needed to complete the task, 5) brief information, 6) work steps, 7) tasks to be performed, and 8) reports to be worked on (Depdiknas, 2008).

3 RESEARCH METHOD

The research methods that are used in this research consist of observation, interviews, and Focus Group Discussion (FGD).

3.1 *Observation*

The purpose of the observation is to evaluate the habits of students in the classroom and their higher-order thinking skills. Observations were made during classroom learning.

3.2 *Interviews*

Interviews were conducted with two math teachers from two state junior high schools in Sleman, Indonesia. One teacher was from a grade A school and the other from a grade B school. Interviews were also conducted with ten eighth-grade students from the two schools: five from the grade A school and five from the grade B school.

3.3 *Focus group discussion*

The focus group discussion was held in October 2016 and involved two lecturers majoring in mathematics education and 18 students of the mathematics graduate program of *Universitas Negeri Yogyakarta* (UNY).

4 RESULTS

4.1 *Results of observation*

The observations were conducted in March 2017 in two schools: one school in the high category and one in the middle category. From the observations in one class of each school, the results were not very different.

During the lesson, teachers explained the material orally. After the teacher had finished their explanation, students were given an exercise to complete. Teachers gave the students two choices: they could complete it on their own or discuss it with a friend that sat near them. Most students chose to discuss with friends who sat next to or behind them. Then, at the end of the learning exercise, the answers were discussed. The questions given were routine problems that did not require much strategy to solve them.

4.2 Results of interviews with teachers

Based on interviews with the teachers, we obtained information that students prefer learning in which they are presented with material orally by the teacher and are then given exercise questions. Teachers only occasionally conducted group discussions using discovery-based student worksheets. Teachers rarely used LCD projectors in the classroom because they were considered to reduce the focus of learners. The learning set they use is one adapted from the 2013 curriculum but is not specifically HOTS-oriented. Teachers agreed that students' HOTS should be developed through learning in the classroom.

4.3 Results of interviews with students

During student interviews, students were given HOTS questions and asked to review them. From the students' answers, we concluded that nine out of ten students did not like story-based problems. Even before trying to do them, they said that such problems were difficult and complicated. They preferred 'ordinary' questions that do not require much strategy to solve. When asked to work on a problem, they tend to be confused by having to determine what strategy to use and which formula to choose, and would immediately ask the teachers, "*What should I do first? What formula should be used? Can we just do the computation?*" They said that they did not have the self-confidence to solve such problems. They also explained that their teachers would not give them such difficult questions during classroom exercises or daily tests. They prefer questions for which the answers can be directly calculated. In solving the problems, they would not conduct any analysis of what is known and what is being asked.

4.4 Results of FGD

HOTS must be possessed by students. Therefore, learning processes in classrooms should be integrated with HOTS. However, to integrate HOTS with learning activities is not easy because of the lack of time allowed to teachers when they have so much material that must be delivered. In addition, teachers themselves also have difficulty in understanding HOTS. Therefore, it is difficult to create a HOTS-oriented learning set. In addition, students who are familiar with routine questions find it difficult to work on problems that should be solved not only through calculations but also through a process of analysis, evaluation and creativity. It is, however, not impossible to develop students' HOTS through learning. What is needed is to familiarize students with HOTS questions. This can be done through a problem-solving approach. In implementing a problem-solving approach, learning activity takes a real problem as its starting point. To make it easier for students to exchange ideas, the learning can be integrated with cooperative learning. One suitable form of cooperative learning is Group Investigation (GI). Thus, we need to develop a HOTS-oriented learning set that is based on a problem-solving approach in a GI setting.

Based on the results of the observations, interviews and FGD described above, we can conclude that it is necessary to develop a learning set in accordance with learning approaches that can improve students' thinking skills and self-confidence. The learning set should enable students to be more active in discussions so they will not be dependent on teacher explanations. A problem-solving approach is appropriate to develop such a learning set, and can provide an opportunity for students to solve non-routine problems, thereby developing students' thinking abilities. Because junior high school students are able to consider many point of views at once, learning would be better if conducted using cooperative learning. Cooperative learning can provide opportunities for students to discuss, consider and exchange ideas so that their thinking and self-confidence is developed. In this case, the appropriate cooperative learning is group investigation. Therefore, the specifications of the necessary learning sets are:

a. Consist of a problem as the starting point of the learning process.
b. Learning steps correspond to the steps of a problem-solving approach in the setting of a group investigation, that is: 1) determine the topic, 2) form appropriate groups, 3) plan

the learning procedure, 4) problem presentation, 5) group investigation, 6) presentation, 7) evaluation, and 8) generalization.

c. In the student worksheet, there should be an answer sheet in the form of problem-solving steps that can guide students to solve the problem, based on the problem-solving steps of Pólya (1945), that is, understand the problem, devise a plan, carry out the plan, and then reflect back.

d. The student worksheet should also have a conclusion feature that can facilitate students in analyzing patterns such that they can perform a generalization and acquire new knowledge.

e. There should be sections that take the form of clues to help develop students' self-confidence.

f. There should be exercises in the form of non-routine problems that can help students to develop their thinking skills.

g. There should be a reflection column that can be used to monitor student difficulties, so the teacher can consider things that can be used to make improvements for the next cycle of learning.

5 DISCUSSION

Based on our observations, interviews, and focus group discussion we concluded that teachers agree that students' HOTS must be developed through learning. However, HOTS-oriented learning sets are limited, and their development is difficult because of teachers' lack of knowledge about HOTS. In lessons, teachers often deliver learning materials orally. They facilitate student discussion during exercise questions for which the answers are routine and can be obtained through calculation, and this situation causes students to have difficulties in working on higher-level questions. When they face problem-solving questions, especially story-based problems, students feel pessimistic even before they have tried to address them. In other words, they did not have the self-confidence to solve such problems, thinking them too difficult and complicated. They were also reluctant to read and analyze the problem before starting to solve it. Therefore, students tend to be confused when trying to determine their problem-solving strategies, and could not determine which formula should be used. For example, when they were given a problem concerning the volume of a cube, they did not know whether they should use a formula for the volume of a cube or for the surface area of a cube. They preferred to work on problems that they could directly compute. This indicates that, in addition to HOTS, teachers must also instill a self-confident character in students and thus it is necessary to develop learning sets that are oriented to both HOTS and student self-confidence.

The development of HOTS-oriented learning sets can be done through a problem-solving approach. Wiederhold (cited in Widodo & Kadarwati, 2013) states that the problem-solving model is seen as a learning model that can improve students' ability in higher-level thinking (HOTS). Problem-solving models involving higher-order thinking will bring students the experience of using knowledge and skills they have mastered to the application of non-routine problem-solving, the discovery of patterns of problem-solving, and good communication skills, so that the meaningfulness of their learning will be more pronounced. Schoen and Charles (2003) explained that teaching mathematics through problem-solving involves presenting students with complex, open tasks, often in a "real world" context, with the intention of encouraging the development of deep conceptual understanding through engagement in mathematical thinking, reasoning, and problem-solving. This means that in learning activities based on problem-solving we should use problems that are complex, open tasks with a "real world" context as their starting point. The problem-solving learning steps are as follows: (1) the teacher presents the problem to the students, (2) the students discuss the problem in the group, (3) the students present the results of their discussion, and (4) the students, with the guidance of teachers, generalize the material from the problem they had discussed and solved (Lester & Mau, 1993).

The use of a problem-solving approach in learning is more effective if combined with cooperative learning (Shanti & Abadi, 2015). This is because cooperative learning provides an opportunity for students to work together on a task, and they must coordinate their abilities to complete the task (Arend, 2012) so that through discussion and exchange of ideas, tasks can be solved in accordance with the learning objectives. Cooperative learning can also improve self-confidence. Killen (2009) reported that cooperative learning can boost students' confidence and self-esteem because it allows all students to experience learning success. A suitable form of cooperative learning that can be combined with a problem-solving approach to develop HOTS and self-confidence is group investigation. Slavin (2010) explains that group investigation is appropriate for integrated study projects that deal with such matters as mastery, analysis, and synthesis of information relating to multi-faceted problem-solving attempts. This indicates that group investigation is suitable for developing HOTS. In addition, in group investigation, students in the group are asked to conduct an in-depth investigation of a particular topic and in the investigation process they are expected to use all the skills and resources available to solve the problem. This is in accordance with the problem-solving approach in which students are required to exercise all of their capabilities to solve non-routine problems. From the above description, it can be concluded that a problem-solving approach with a cooperative group investigation setting is suitable for the development of a HOTS-oriented learning set.

6 CONCLUSION

From the discussion above we can conclude that:

a. Students had difficulties in solving HOTS problems because they were unwilling to solve problems in long and complicated forms. They also lacked the self-confidence to solve these problems.
b. The problems that students usually practice upon were routine problems.
c. The lesson plans and student worksheets that teachers had were not HOTS-oriented even though lesson plans and student worksheets represent the learning sets that have the most important role in determining the orientation of a lesson.
d. It is necessary to develop learning sets (lesson plans and student worksheets) that improve students' HOTS and self-confidence.
e. A problem-solving approach in the cooperative setting of a group investigation should be chosen as an appropriate approach to develop lesson plans and student worksheets that are oriented toward development of HOTS and self-confidence in students.
f. The specifications for learning sets based on a problem-solving approach in the cooperative setting of a group investigation are as follows: (1) consist of a problem as the starting point of the learning process; (2) learning steps correspond to the steps of a problem-solving approach in the setting of a group investigation; (3) in the student worksheet, there should be an answer sheet in the form of problem-solving steps that can guide students to solve the problem; (4) the student worksheet should also have a conclusion feature that can facilitate students in analyzing patterns such that they can perform a generalization and acquire new knowledge; (5) there should be sections that take the form of clues to help develop students' self-confidence; (6) there should be exercises in the form of non-routine problems that can help students to develop their thinking skills; (7) there should be a reflection column that can be used to monitor student difficulties, so the teacher can consider things that can be used to make improvements for the next cycle of learning.

REFERENCES

Adywibowo, L.P. (2010). Memperkuat kepercayaan diri melalui percakapan referensial [Reinforcinge the confidence by using referential conversation]. *Jurnal Pendidikan Penabur*, 15(9),: 37–49.
Arend, R.I. (2012). *Learning to teach*. New York, NY: McGraw-Hill.

Baser, J., Hasan, A., Asha'ri, A., & Khairudin, M. (2017). A Study on the Transferable Skills of the Engineering Students at Universiti Tun Hussein Onn Malaysia. *Jurnal Pendidikan Teknologi dan Kejuruan, 23*(3), 257–264.

Brookhart, S.M. (2010). *How to assess higher order thinking skills in your classroom*. Alexandria, VA: Association for Supervision and Curriculum DevelopmentASCD.

Clark, D. (1999, June 5). Bloom's Taxonomy of learning domains. *Big Dog & Little Dog's Performance Juxtaposition*. Retrieved from http://www.nwlink.com/~donclark/hrd/bloom.html

Depdiknas. (2008). *Panduan pengembangan materi pembelajaran dan standar sarana dan prasarana sekolah menengah kejuruan, madrasah aliyah, sma/ma/smk/mak* [*The A gGuide of to developing materials, method and infrastructure of learning for vocational, Aliyah and senior high school*]. Jakarta, Indonesia: BP. Mitra Usaha Indonesia.

Forster, F. (2004). Higher order thinking skills. *Research Developments, 11*(1), 1–6.

Goel, M. & Aggarwal, P. (2012). A comparative study of self-confidencet of single child and child with sibling. *International Journal of Research in Social Sciences, 2*(3), 89–98.

Hannula, M.S., Maijala, H. & Pehkonen, E. (2004). Development of understanding and self confidence in mathematics; Ggrades 5–8. *Proceedings of the 28th Conference of the International Group for the Psychology of Mathematics Education, 3*, 17–24. Retrieved from http://www.emis.de/proceedings/PME28/RR/RR162Hannulula.pdf.

Jailani, J. & Retnawati, H. (2016). The challenges of junior high school mathematic teachers in implementing the problem- based learning for improving the higher-order thinking skills. *The Online Journal of Counseling and Education, 5*(3), 1–13.

JIST Live. (2006). *Young person's caracter education handbook*. Indianapolis, IN: JIST Publishing, Inc.

Jurdak, M. (2009). *Toward equity in quality in mathematics education*. New York, NY: Springer.

Killen, R. (2009). *Effective teaching strategies: Llessons from research and practice* (5th ed.). South Melbourne, Australia: Cengage Learning.

Krathwohl, D.R. (2002). A revision of Bloom's Taxonomy: Aan overview. *-Theory iIInto Practice, 41*(4), 212–218. *College of Education*, The Ohio State University Learning Domains or Bloom's Taxonomy: The Three Types of Learning.

Lester, F.K., Jr., F.K. & Mau, S.T. (1993). Teaching mathematics via problem solving: A course for prospective elementary teachers. *For the Learning of Mathematics, 13*(2), 8–11.

Lewis, A. & Smith, D. (1993). Defining higher order thinking. *Theory Iinto Practice, 32*(3), 131–137.

Marini, C. (2016). Strategi Training within Industry sebagai Upaya Peningkatan Kepercayaan Diri Siswa pada Mata Pelajaran Pengolahan Makanan Kontinental [Strategy Training within Industry as an Efforts in Improving the Student's Self-Confidence in the subject of Continental Cuisine]. *Jurnal Pendidikan Teknologi dan Kejuruan, 22*(4), 410–423.

McKenzie, M. (2005). *Multiple intelligences and instructional technology*. Washington, DC: Iinternational Ssociety on for Ttechnology in Eeducation.

Moseley, D., et alBaumfield, V., Higgins, S., Lin, M., Miller, J., Newton, D., ... Gregson, M. (2004). *Thinking skill frameworks for post-16 learners: Aan evaluation*. TrowbridgeLondon, WiltshireUK: Learning and Skills Research Centre.

Pólya, G. (1945). *How to solve it*. Princeton, NJ: Princeton University Press.

Prastowo, A. (2012). *Panduan kreatif membuat bahan ajar inovatif* [*The guide of to designing the innovative learning material creatively*]. Yogyakarta, Indonesia: Diva Press.

Presseisen, B.Z. (1985). Thinking skills: Mmeanings and models. In Arthur A.L. Costa (Ed.s), *Developing minds: A resource book for teaching thinking* (pp. 43–48). Alexandria, VA: Association for Supervision and Curriculum DevelopmentASCD.

Resnick, L.B. (1992). *Education and learning to think*. Washington, DC: National Academy Press.

RoI. (2014). *Peraturan menteri pendidikan dan kebudayaan Republik Indonesia No. 103, 2014, tentang pembelajaran pada pendidikan dasar dan pendidikan menengah* [Regulation of the Minister of Education and Culture Republic of Indonesia number 103 of 2014 concerning learning for primary school and middle school]. Jakarta, Indonesia: Ministry of Education and Culture, Republic of Indonesia.

RoI. (2016). *Peraturan menteri pendidikan dan kebudayaan Republik Indonesia No. 22, 2016, tentang standar proses pendidikan dasar dan menengah* [Regulation of the Minister of Education and Culture Republic of Indonesia number 22 of 2016 concerning process standards for primary school and middle school]. Jakarta, Indonesia: Ministry of Education and Culture, Republic of Indonesia.

Schoen, H.L., et al. & Charles, R.I. (Eds.). (2003). *Teaching mathematics through problem solving: Ggrades 6–12*. Reston, VirginiaVA: National Council of Teachers of MathematicsNCTM.

Schunk, D.H. (2012). *Learning theory* (6th ed.). Boston, MA: Pearson.

Shanti, W.N. & Abadi, A.M. (2015). Keefektifan pendekatan problem solving dan problem posing dengan setting kooperatif dalam pembelajaran matematika [Effectiveness of problem solving and problem posing approach by using cooperative setting in mathematics learning]. *Jurnal Riset Pendidikan Matematika, 2*(1), 121–134.

Slavin, R.E. (2010). *Cooperative learning*. Bandung, Indonesia: Nusa Media.

Srivastava, S.K. (2013). To study the effect of academic achievement on the level of self-confidencet. *Journal of Psychosocial Research, 8*(1), 41–51.

Widodo, T. & Kadarwati, S. (2013). Higher order thinking berbasis pemecahan masalah untuk meningkatkan hasil belajar berorientasi pembentukan karakter siswa [Higher order thinking based on problem solving to improve learning achievement by using student's character building]. *Cakrawala Pendidikan, 5*(1), 161–171.

Yoder, J. & Proctor, W. (1988). *The self-confident child*. New York, NY: Fact on File Publication.

Students' character learning through internalization of character values from *wayang* figures

B. Nurgiyantoro
Universitas Negeri Yogyakarta, Indonesia

ABSTRACT: *Wayang* is a cultural object recognized in Indonesia as one of character building since it encompasses glorious value and plays an important role in human civilization. It was also recognized by UNESCO as a Masterpiece of Oral and Intangible Heritage of Humanity in 2003. The content of *wayang's* value aspect can be seen in the characterization of the characters, stories and other good elements which refer to students' character learning. *Wayang* stories provide many characterizations which can be used as a learning source, such as the loyal attitude when we give a promise. The promise, which is expressed to others, must be fulfilled whatever the risk. Data resources were derived from the stories and characters of *Sumantri*, *Kumbakarno*, *Karna* and *Bhisma*. The data were collected through discourse analysis and analyzed using descriptive qualitative techniques. The four *wayang* characters which were investigated hold onto something although life is insecure. The internalization of character learning in a *wayang* story can be learned through telling a story, role play and rewriting the story which must be continued with reflection and affective questions.

1 INTRODUCTION

Wayang performance are epic stories based on *Mahabharata* and *Ramayana* which come from India. After entering Java, the *wayang* stories were edited to the old Javanese language by adding and adapting, corresponding to the development era. Thus, it has gone through a long history. The fact that there are many people enjoying the stories currently, shows the value of the *wayang* stories to society. This fact is also acknowledged on an international level, such as the UNESCO (United Nations Educational, Scientific and Cultural Organization). In 2003, UNESCO declared the Indonesian *wayang* a *Masterpiece of Oral and Intangible Heritage of Humanity* (Wibisono, 2009; Sudjarwo et al., 2010). UNESCO also instructed that the inheritance of the *wayang* culture must be maintained and preserved. Characters' self-actualization and functions of actualization of puppet characters are also shown intensively in Indonesian fiction of the twenty-first century (Nurgiyantoro & Efendi, 2017).

Essentially, the *wayang* stories, either *pakem Mahabharata* or *Ramayana*, tell the heroic stories of the good characters who faced and devastated the evil characters. It can be seen that through some aspects, such as plot, characters and the implicit values, the *wayang* stories can become one reference, life principle and value finding. *Wayang* is not only one form of popular traditional art, but also a local wisdom resource. Substantially, *wayang* values are related to the problems in a human's life which is regarded as an individual, social and religious life. Those matters are suitable to teach to students through the stories until there is an internalization process of values in their soul.

Currently, character learning is based on the importance of education in developing students as people who have character, culture and can create morals as a part of life's principles. A student's character becomes the guarantee of the life-long building of a nation. We can adopt the example of education in Japan, which places the traditional culture as an important resource of moral value and puts values and morals at the core of educational activity. This must inevitably focus on what happens in elementary and junior high schools, for until very recently in Japan, moral education was provided exclusively to those between the ages of 6 and 15 years (Nakayama, 2015).

Thus, education must become an integral part in the learning process which is done consciously and planned. Davison and Lewis (2005) also stated that character education is a specific approach to morals and values education which is consistently linked with citizenship education. Education must be able to motivate students to behave based on moral principles. Character is the set psychological characteristics that motivates and enables an individual to function as a competent moral agent (Nucci & Narvaes, 2008). Bajovic et al. (2009) said that character is a socio-moral competency that incorporates moral action, moral values, moral personality, moral emotions, moral reasoning, moral identity and foundational characteristics.

Character education deals with teaching students to develop the ability to decide how to behave in an appropriate manner in various social situations with the purpose of developing individuals who are capable of understanding moral values and who choose to do the right thing (Almerico, 2014). However, character learning values which are learned need to be identified. RoI (2010) stated there are 18 character values included in the character learning guidelines in schools, such as religiousness, honesty, tolerance, discipline, hard work, creativity, independence, democracy, curiosity, nationalistic enthusiasm, nationalism, appreciation of achievement, friendliness/communicativeness, peacefulness, love to read, care about the environment, care about the social condition and responsibility.

However, this does not mean that there are only 18 values which require the teacher's attention in the learning process. One example, the character value to fulfill a promise, is important and is stated in the Al-Qur'an (23:8) as: "Those who are faithfully true to their trusts and to their covenants". In the *wayang* stories, this value is also emphasized in the following study. Almerico (2014) also stated that the content of programs typically align with the core principles and values of generosity, kindness, honesty, tolerance, trust, integrity, loyalty, fairness, freedom, equality, and respect of and for diversity.

One way to bring character education into a crowded curriculum, is to make it part of the literacy program by embedding character lessons in reading and language arts instruction through the vehicle of high quality literature, either traditional or modern. Thus, it is necessary to observe and choose the literature work which can facilitate the purpose of character learning in school, one of which is literature work based on the *wayang* stories.

2 DATA RESOURCES

The data sources in this study were the *wayang* stories of *pakem Mahabharata* and *Ramayana* which are focused on certain characters, life plots and character learning values, especially the loyal attitude when we give a promise. *Wayang* characters which have character values are the focus in this study, including *Sumantri, Kumbakarna, Karna* and *Bhisma*. These four characters were chosen since they are role models of how to behave in Javanese society, as is shown in *Serat Tripama* (Mangkunagara IV). The data sources of those characters were books of *wayang* stories which were published in the Javanese language or Indonesian language. The characters of *Sumantri, Karna* and *Kumbakarna* were also based on *Serat Tripama*.

The books of *wayang* stories that became the references are *Silsilah Wayang Purwa Mawa Carita, jilid I—VII* (Padmosoekotjo, 1992), *Rupa & Karakter Wayang Purwa* (Sudjarwo et al., 2010) and *Ensiklopedi Wayang* (Dwiyanto et al., 2010). The comparison sources refer to the story books by an Indian author, such as *Mahabharata* (Rajagopalachari, 2008).

The data were collected through a literature review using a discourse analysis technique with the knowledge of the world instrument (Brown & Yule, 1983). The data collection was focused on aspects of characterization, life plot and character learning values. The results of the data were analyzed by using a descriptive qualitative technique, especially making inference on giving meaning and deriving a conclusion.

3 FINDINGS AND DISCUSSION

The results of the study focused on the character value identification of the four *wayang* characters *Sumantri, Kumbakarna, Karna* and *Bhisma*. The brief stories and characterizations are described as follows:

Sumantri (Suwanda). Brief story: He was a son of *Resi Suwandagni*, who had high ability and power. He served *Prabu Arjuna Sasrabahu* in *Maespati*, became the chief minister, and changed his name to *Suwanda*. With the help of his brother, *Sukasrana*, the dwarf giant, he could fulfill the queen's request. However, his brother was killed intuitively, he felt sorry and then he focused his attention to serve the king. Finally, he was killed by the cruel King of *Alengka, Dasamuka*, when he helped his king. The dominant characteristics are fulfilling a promise, being loyal to the king, being honest and hating evil.

Kumbakarna. Brief story: He was the son of *Begawan Wisrawa* and also the brother of *Dasamuka*, King of *Alengka*. Physically, he was a giant, but had a knight-heart which hated evil. When there was the Great War in *Alengka*, the war between *Alengka* and *Pancawati*, *Dasamuka* asked him for help. He knew that *Dasamuka* was guilty since he kidnapped Rama's wife Sinta. Rama is the king of *Pancawati*. *Kumbakarna* refused to help. However, finally he decided to go to war with the purpose of fighting for the nation, not to fight for his cruel brother. He died on the battlefield. The dominant characteristics are loyalty to the nation, strength to fulfil the promise to fight for the nation, being honest and hating evil.

Karna (Basukarna). Brief story: He was the son of God (*Batara Surya* and *Dewi Kunti*). He had five step brothers, including *Pandawa*, who is known for agood character. He served the King of *Kurawa, Prabu Duryudana* who had an evil character; he had a promise to fight for him since he had an honorable position. He loved his five brothers. However, when the *Bharatayuda* War happened between the *Pandawa* and *Kurawa* families, he served *Kurawa* and died on the battlefield. He sacrificed his life intentionally in order to destroy the evil in this world and give the victory to *Pandawa*. The dominant characteristics are fulfilling a promise, loyalty to the king, loving the good brothers, willingness to sacrifice and hating evil.

Bhisma. Brief story: He had a young name, *Dewabrata*; he was the son of *Prabu Santanu*, King of *Astina*, with the fairy, *Dewi Gangga*. When he was an adult, he became the crown prince. After a long time of being a widower, his father wanted to marry again, but he failed since his future wife, *Dewi Durgandini*, wanted her future son to become king. *Bhisma* convinced his future step-mother by making an oath that he would never become the king and marry. In fact, his step brothers died before they had children. He still refused to become the king and also refused to get married. *Bhisma* loved his grandchildren, the *Pandawa*, since they had a good character. Yet, when the *Bharatayuda* War happened, he took the side of *Kurawa*, for the sake of fighting for the nation. He died on the battlefield. The dominant characteristics are fulfilling a promise, being loyal to the nation, being honest, being fair and hating evil.

In conclusion, there is a similarity between the four characters above, such as fulfilling a promise, loyalty to the nation, honesty and hating evil. Those four moral values are the character values which are suitable for students to learn in order to integrate character learning in schools. Frankly, those four characters have more than four kinds of moral values; however, those are the dominant values. In addition, the learning of character values need to be conducted by focusing on a certain value.

3.1 *The role model of the four wayang characters*

The four *wayang* characters above often become the reference point as role models of how to behave in Javanese society since their characters are good role models. Why are they chosen? As suggested by the Mangkunagara IV, the three characters: *Sumantri, Kumbakarna* and *Karna*, are role models of good characters. However, in fact, *Bhisma*, who has a similarity with the other characters, shows his excess. It can be seen from *Bhisma* life story, his characteristic in fulfilling the promise is more inspiring. Therefore, *Bhisma's* behavior is also suitable as a reference of a role model.

The first role model of *Sumantri* can be indicated through three things, *guna, kaya* and *purun. Guna* and *kaya* mean the ability to solve the problem, winning the battle. *Purun* is *Sumantri's* model of braveness to face the evil King, *Dasamuka*, who always had a bad plan to King Rama. In loyalty to the king, he fights until death. *Sumantri* gives all the ability and braveness; he also sacrifices his body and soul as the manifestation of fulfilling the promise of loyalty to the king. *Sumantri* succeeded in gaining a high position because he was helped by his brother, *Sukasrana*, the dwarf giant. However, he killed his brother intuitively. Philosophically, *Sukasrana* is a symbol of worldly passion. *Sumantri* who killed *Sukasrana*, is a symbol of one who can defeat his passion. Frankly, when human is life in this world, the passion must not be killed, but it should be managed (Sudjarwo et al., 2010).

The second role model is *Kumbakarna* who is a giant-shaped prior hero. He is an honest character who emphasizes good and refuses bad things because of what has been done by his brother, the King of *Alengka, Dasamuka*. He refuses to fight against *Rama* since he knows that his brother is guilty. However, he has a promise to be loyal to the nation; he decides to go to the war with the purpose of fighting for the nation, not to fight for his cruel brother. He died on the battlefield. This is the main role value of *Kumbakarna,* such as heading toward the principle and promising to fight for the nation, not to fight for his cruel brother, although he died.

The third role model is *Karna,* who is the oldest brother of the *Pandawa* family. As a knight, he has a promise to be loyal to the king who gives him an honorable position. He has a principle as a knight, he cannot break the promise. Then, when the *Bharatayuda* War happened, he serves *Kurawa* although, he knows that the wrong side is King of *Kurawa,* who gives him a position; and the right side is his brothers, the *Pandawa*. Then, there is a hard fight in his heart to make a decision. He decides to serve *Kurawa* and dies on the battlefield in order to destroy evil and give the victory to *Pandawa* who is the right side. So, the principle of fulfilling the promise is defensible although the life is in danger (Sudjarwo et al., 2010).

The fourth role model who fulfills the promise is *Bhisma*. He is not included in *Serat Tripama*, but his loyalty principle in a promise is more inspiring than the other three characters. The young *Bhisma* is already chosen as a crown prince. But, when his father wants to marry again and his future wife requests that their future son must become king, he appears as the good son of his father. He makes two oaths, he gives the throne to his future brother and he swears to be unmarried in his life, so he cannot have a descendant who may be able to take the throne. Because of those oaths, the God named him, *Bhisma,* meaning "the very afraid"'. His long life sees him live until the *Bharatayuda* War. The *Kurawa family* is the wrong side, he prefers to take a side for *Kurawa*, for the sake of fighting for the nation. He died on the battlefield while fulfilling his promises.

There is a different story context between these four characters, each with their problems, conflicts and intensity, but all characters are strongly heading toward their principles until death. The characters' attitude and behavior are good examples of attitude and behavior in character learning through classical literature which is recognized as local wisdom. Literature is certain to become a means of character learning's purpose (Suryaman, 2013). However, the literary work must be chosen and focused on the good one. Literary works which contain character education is called as literary works which can be to reanimate local literature, which expresses habit in religiosity, customs in a certain ethnical, and patterns of behavior and the other habit which expresses diversity (Widyahening & Wardhani, 2016). The *wayang* stories fulfill those requirements.

Character learning by using the literature can help students to internalize the character values. Literature could tell about the life model of a character. Literature is culture in an act. Students will learn the characters' attitude and behavior concretely, either verbally or non-verbally, in life in a concrete context. This becomes the reason why many experts took the example of a story when it delivers certain moral values so it can have a concrete context. It is also suggested by Agboola and Tsai (2012), that the easiest way to promote character education is by using literature studies, since the stories serve as role models that connect experiences and morals.

In fact, other traditional stories besides the *wayang* stories, also contain more character learning values, such as Classic Fairy Tales. This can be seen from a study conducted by

Rosidah (2013) about the original and Indonesian translated version of the Classic Fairy Tales. The result of the study showed that those stories contain some types of values, which can be classified in the values which are related to self-development, the relationship between human and God, others and the environment. Similar to this study, Normawati (2015) observed the character learning values in the text book of the *Bahasa Indonesia* subject (Indonesian language for middle school) in the city of Yogyakarta, Indonesia. Therefore, there are many choices of literary work which can be used as a learning media of character value.

3.2 Internalization of character values

The common belief of character education is from psychological and philosophical perspectives that virtues can be taught and learned through the proper pedagogy (Agboola & Tsai, 2012). Through those ways, it is hoped that character values which are learned can be internalized in students themselves, so those values internalize in the form of moral action in daily life.

The moral values of the *wayang* stories are delivered through the character's attitude, the moral action. The final objective of literature learning is more about the affective formation. The problem is how it occurs and is internalized in the students themselves. Almerico (2014) states that the intent of the full study, presented in part here, is integrating the teaching of character with research-based literacy instruction through children's literature. The first thing to do should focus on developing the instruction. A certain value should be chosen. So, what is the character? One example is fulfilling a promise like what has been done by the *wayang* characters above. The definite thing is that character education can become an everyday opportunity (Agboola & Tsai, 2012).

The strategy of character education is described by Romanowski (2005) as: 1) the involvement of the teachers in program planning, what value messages are being sent to our students explicitly and implicitly; (2) the curriculum must be relevant to the lives of students and challenge students intellectually, emotionally and socially; (3) the pedagogical strategies used must engage students in relevant class discussions; (4) administration should support and give enough space for teachers to exercise flexible pedagogy in specific character traits; (5) schools should consider ways to foster student leadership and involvement in character education. Teachers must be role models of behaving appropriately with a focus on the character value being learned. A role model is an important aspect in character education, such as religious, discipline, nationalism, peaceful and others, which can become examples to students in elementary school (Prasetyo & Marzuki, 2016).

The chosen learning strategy must be suitable to the learning context. There are many strategies which can be chosen to internalize moral values including: (1) Telling a story. The teacher tells a story. In general, children love to hear a story, making this is a good strategy to choose; (2) Reading a story. Choose a good text which is appropriate with the character value which becomes the learning focus. After that, some points related to the story and students' affective response can be stated; (3) The book is read by the teacher. This is done when the students cannot read by themselves. This reading is ended by an affective question focused on character values; (4) See and hear a story. This activity should end with reflection and affective questions focused on character values; (5) Role play. Students are asked to play a certain role which represents a certain character. This activity is also ended with reflection and affective questions focused on character values; (6) Writing practice or rewriting the story which has been read, retelling it to others or students creating original work which has a certain moral value. This activity ranks as a high one, although it can give experience, itself, which has a deep influence.

4 CONCLUSION

The *wayang* stories are one local wisdom source which is rich in moral values. A few characters in the *wayang* stories become role models on how to behave in everyday life. Some examples are the loyal attitude when we have a promise, loyalty to the nation, being honest, loving to do good things and on the other hand, hating to do bad things. Those moral values are suitable to teach to students through character learning internalization in schools. Thus, character values must be concretely included in the curriculum, be integral with the learning activity, showed in daily behavior. Character values can be learned through telling a story, reading, role play and rewriting the story which must be continued with reflection and affective questions.

REFERENCES

Agboola, A. & Tsai, K.C. (2012). Bring character education into classroom. *European Journal of Educational Research, 1*(2), 163–170.

Almerico, G.M. (2014). Building character through literacy with children's literature. *Research in Higher Education Journal, 26*(October), 1–13.

Bajovic, M., Rizzo, K. & Engemann, J. (2009). Character education re-conceptualized for practical implementation. *Canadian Journal of Educational Administration and Policy, 92* (March), 1–23.

Brown, G. & Yule, G. (1983). *Discourse analysis.* Cambridge, UK: Cambridge University Press.

Davison, A.J. & Lewis, M. (2005). *Professionals values and practice achieving the standards for QTS.* London, UK: Routledge.

Dwiyanto, D., Susantin, S & Widyowati, W. (2010). *Ensiklopedi Wayang [Puppet encyclopedia].* Yogyakarta, Indonesia: Media Abadi.

RoI. (2010). *Pengembangan Pendidikan Budaya dan Karakter Bangsa [Development of cultural education and nation character].* Jakarta, Indonesia: Kementerian Pendidikan Nasional.

Mangkunagara IV, K.G.P.A.A. *Serat Tripama, Dhandhangnggula [Letter of Tripama Dhandahangnggula].* Retrieved from https://tanahmemerah. wordpress.com/kasastraan/naskah-kuno/serat-tripama/.

Nakayama, O. (2015). Actualizing moral education in Japan's tertiary sector: Reitaku university's response to today's challenges. *Journal of Character Education, 11*(1), 39–50.

Normawati, N. (2015). Nilai pendidikan karakter dalam buku teks pelajaran bahasa Indonesia SMP di DIY [The character education value in the Indonesian literature book of junior high school in Yogyakarta]. *Jurnal Pendidikan Karakter, 5*(1), 48–69. doi: 10.21831/jpk.v0i1.8612.

Nucci, L. & Narvaes, D. (2008). *Handbook of moral and character education.* New York, NY: Routledge.

Nurgiyantoro, B. & Efendi, A. (2017). Re-actualization of puppet characters in modern Indonesian fictions of the 21st century. *3L: The Southeast Asian Journal of English Language Studies, 23*(2), 141–153. doi: 10.17576/3L-2017-2302-11.

Padmosoekotjo, S. (1992). *Silsilah Wayang Purwa Mawa Carita, jilid I-VII [Puppet lineage of Purwa Mawa Carita volume I-VII].* Surabaya, Indonesia: Citra Jaya Murti.

Prasetyo, D. & Marzuki, M. (2016). Pembinaan karakter melalui keteladanan guru pendidikan kewarganegaraan di sekolah Islam Al Ashar Yogyakarta [Character building through exemplary teachers of civic education in private school. *Jurnal Pendidikan Karakter, 6*(2), 215–230. doi: 10.21831/jpk.v6i2.12052.

Rajagopalachari, C. (2008). *Mahabharata, sebuah roman epik pencerah jiwa manusia [Mahabharata, an epic roman illmuniating human soul].* Yogyakarta, Indonesia: IRCiSoD.

Romanowski, M.H. (2005). Through the eyes of teachers: High school teachers' experiences with character education. *American Secondary Education, 34*(1), 6–23.

Rosidah, A.A. (2013). Pendidikan karakter pada classic fairy tales [Character education in classic fairy tales]. *Jurnal Pendidikan Karakter, 3*(3), 250–263. doi: 10.21831/jpk.v0i3.2748.

Sudjarwo, H.S., et al. (2010). *Rupa & Karakter Wayang Purwa. [The form and character of Purwa puppet].* Jakarta, Indonesia: Kaki langit Kencana Prenada Media Group.

Suryaman, M. (2013). Pengembangan model buku ajar sejarah sastra indonesia modern berperspektif gender [Developing of learning resource for hystory of modern Indonesian literature in gender perspective]. *Litera, 12*(1): 106–118. doi: 10.21831/ltr.v12i01.1333.

Wibisono, S. (2009). *Wayang, Karya Agung Dunia [Puppet, the world greatest work].* Retrieved from http://www.Sastra-Indonesia.com/2009/12/Wayang,KaryaAgungDunia/.

Widyahening, E.T. & Wardhani, N.E. (2016). Literary works and character education. *International Journal of Language and Literature, 4*(1), 176–180.

Developing young children's characters using project-based learning

O. Setiasih, E. Syaodih & N.F. Romadona
Universitas Pendidikan Indonesia, Indonesia

ABSTRACT: It is important to start character education as early as possible. Young children aged 0–6 years old experience rapid growth and development of their brain. They can absorb enormous information without any ethical filtering. Thus, teaching moral values become very crucial in early childhood education to assist children growing with characters highly expected by society. The success of character education depends on teacher's pedagogical skills and well-developed appropriateness of the learning method used in the pedagogical process. One of learning approach that is effective to build children's character is *project-based learning approach*. This approach opens up the opportunities for children to develop their own curiosity, creativity, confidence, acceptance, teamwork ethics and cooperation, respect, and other important social skills. The present paper writing uses a literature study examining and analyzing textbooks and journal articles on character education and project-based learning approach. This paper serves as a reference for teachers of early childhood education to apply a *project-based learning approach* to develop young children's character.

1 INTRODUCTION

Character proves to be one of the developmental aspects imperative to be stimulated in early age since in this phase children in their early age are very sensitive in terms of receiving and absorbing various stimulations as they appear in their environment. These children in their early age manage to absorb various information of its both sides: the good one and the less useful one. This leads to an indication that early age moral education brings with it fruitful responses to begin character education for children. The key factor being, referring to developing nation's character with quality, is starting the right earliest possible time. Character education developed on the right track results in children growing with characters involving personal qualities applicable to the children's daily practices in life. The success of inseminating moral values in young generation in an early stage is a strategic effort leading to a dependable outstanding generation in the future. On the contrary, failure of building character among children in their early age triggers personality formed with a problem once the growing phases take place.

Character development right at the early age plays a pivotal role in making it grow in the right direction and finally leads to solving the existing social problems.

Thus, character education for the children in their early age proves to be unavoidable. Towards the end, accordingly, starting character education as early as possible will result in individuals with strong characters making it possible for the individuals to be people with good deeds, honest, disciplined, independent, responsible, skilled, creative, emphatic, and eventually be good democratic citizens. Findings of some research indicate that there is a strong correlation between the success of character education and academic achievement with children's pro-social attitude. It is imperative to generate some interesting and conducive atmosphere on the institution's part so as to allow children to have a well-supported effective teaching-learning process (Kementerian Pendidikan dan Kebudayaan 2012a).

The Government of Indonesia through Curriculum and Collection Research and Development Board has issued Guidelines on Performing Character Education Program. At the early childhood educational level, the guidelines are put into an operational set of Guidelines of Character Education for Early Childhood Education (Kementerian Pendidikan dan

Kebudayaan 2012a). Efforts made denote that the Government of Indonesia pays a very good attention and care to the importance of performing a character education for students of all levels of education. Experts like Lickona (1996) and Nodding (1984) recommend that character education be part of school curriculum. The teachers at institutions of early childhood education, they play an important key role to especially stimulate the development of a good character building on the children through an appropriate approach according to the development principles and ways of the children learn. There is a number of approaches a teacher of early childhood education can decide to apply. Among others are role modeling, habituating, and applying a learning model constructed through a systematic program. Another approach worth applying is *project-based learning approach* which is considered appropriate.

2 METHODOLOGY

This study applied scientific literature study method. Some literature used as sources in this study, among 1) research journal review, 2) reference book, and 3) Indonesian Republic government policies (law and regulations). Literature searched manually by visiting some libraries as well as visiting places that are sources of information and also searching online using *characters education, project-based learning*, and *young children* as keywords. Data analysis method used in this study is comparison analysis between theory.

3 RESULT AND DISCUSSION

3.1 *Character education for young children*

There exists quite a number of definitions of character education by experts in the field. McClellan et al. (1999) puts forward that character education is a very important teaching of ethical values with its aim to encourage good attitudes during learning in class. The US Department of Education Character Education Pilot Project defines character education as *an inclusive term encompassing all aspects of how a school, related social institution, and parents can support the positive character development of children and adults"*. Character education is not only addressed to children but also to those categorized as adults. Another definition comes from Shield (2011) saying that there are four universal definitions termed as a sub-category or character dimension comprising "intellectual", "moral", "civic", and "performance. " One account on a good character as stated is "a disposition to seek the good and right".

Character education is referred as an education involving an act of disseminating knowledge, a sense of loving to do good deeds, and making good attitudes well-generated leading to a pattern or some degree of habituation. As confirmed by The Indonesia Directorate General of Early Childhood Education in the year of 2012, character education is an endeavour of disseminating good values and wisdom to God, oneself, peer individuals, environment, and nation in order to become man of moral.

Character education for young children deals with various dimensions including the values of honesty, patience, loyalty, respectfulness, belief, and responsibility. Character education comprises basic values of their being good. Referring to the Guidelines on Character Education for Early Childhood Education (Kementerian Pendidikan dan Kebudayaan 2012a), values worth disseminating to young children consist of four aspects, namely (1) Spiritual aspect, (2) Personal aspect/personality, (3) Social aspect, and (4) Environmental aspect. The four aspects are then elaborated into basic values to be introduced to young children and need to be internalized into attitudes of daily life as (1) Love to God, the Almighty, (2) Honesty, (3) Discipline, (4) Tolerance and Peace Loving, (5) Self confidence, (6) Independence, (7) Helping each other, collaboration, and working in a team, (8) Respectfulness and being civilized, (9) Responsibility, (10) Willingness to work hard, (11) Leadership and justice, (12) Being creative, (13) Being low profiled, (14) Caring to environment, and (15) Great respect to nation and country.

These great values of character can be put into daily actions in life with the help and care from adults living around and, especially, parents and teachers. Nodding (1984) echoes that

"The greatest obligation of educators inside and outside of formal schooling is to nurture the ethical ideals of those with whom they come in contact". Other than Nodding, Porter et al. (2010) says that the implementation of character education can be done by creating an educated community with care. Another, Olweus (1993) confirms that character education in schools derives from the role model by the adults. Playing the role models as reflected by the adults includes the way of producing words when communicating, attitudes and taking actions and all those kinds of good deeds since children will be more than likely imitate and identify the behavior as demonstrated by adults.

3.2 Project-based learning

3.2.1 Definition of project based learning

As has been explained that disseminating the values of character to children in their early age can be implemented through various ways which are among others role modeling, habituation, and repeating doing good deeds in daily life. Other than the above-mentioned ways, educators of young children have to be creative in looking for various alternatives to approach learning in applying the character education in such a way that children will find it interesting to get involved. The project was initiated by John Dewey of the USA (1920), which was then made popular by his colleague, William Kilpatrick using a term *project method* (Katz & Chard 2003). In its development, the project has been influenced by Dewey's idea and implemented in schools as a model of learning. Katz & Chard (1993) say that *project* is a study which is vast in nature on topics usually conducted by a group of children either in a small group or a big one but sometimes it is conducted by children as individuals. Its implementation is executed according to the time allocated, interest, the children's capability, and the educational institution's capability as well. A study in a project is an investigation of a specific topic attracting children's interest so that they get involved actively in the course of activity. Project to children in their early age proves to be more effective if it is implemented in small groups rather than in individual groups or big groups. Part of the reason is that in a smaller group, children will interact easily and will express their feeling more confidently when sharing, and they will learn to conduct tolerance towards others.

Henry (1995) puts forward six criteria as a definition of working on a project for young children, namely

The students:
a. (usually) Select the project topic
b. Locate their own source material
c. Present an end product (usually a report and often for assessment)
d. Conduct an independent piece of work (though there are also group projects)
   The project:
e. lasts over an extended period, and
   The teacher:
f. Assumes as the consultant

The above criteria denote that the *project* is a children-based approach of learning since the beginning until the end of an activity while the teacher's role is more to a facilitator and a consultant.

3.2.2 The aim of project-based learning

The aim of project-based learning as proposed by Katz & Chard (1991) comprises the followings: (a) acquisition skill including social competence, communicative competence, and academic skill; (b) development of disposition consisting of development of interest, effort, mastery, challenge seeking, and social disposition; (c) development of feeling.

The young children develop knowledge and skill through first-hand experiences with physical and social environments. Knowledge and skills the children gain will be meaningful if it interests them and is relevant to their needs. Skills are a conduct identifiable with ease and can be conducted in a relatively short time.

Disposition is a tendency of children to respond a certain situation using certain ways. Disposition as it is expected to develop in children is a positive disposition such as a feeling of curiosity, a deep involvement in doing a work, cooperating in a teamwork, creativity, feeling tough in facing troubles, a spirit to solve problems, generosity, responsibility, initiatives, being creative, and willing to help others.

A feeling is an emotion or a subjective attitude expressed positively or negatively such as a feeling of being accepted, confidence, or worries. Teachers and parents, in general have the same expectation from their children as a feeling of being accepted, a feeling of happiness, confidence, and other feelings of the same kind. Some children have feelings of being negative, therefore, the teachers should feel challenged and take steps to modify learning method to generate the feelings of being positive.

Disseminating and developing character value to children does not only transfer knowledge about a matter of being good or being bad but also, more importantly, generate consciousness and apply good values in daily life. There are five steps necessary to take to disseminate character to children of early age, namely: (1) introducing the way to knowing the good to children; (2) giving children a room for thinking the good in that they are introduced to what is good and what is not and why; (3) asking children to feel good and do good things to gain the feeling of the good; (4) asking children to act on the right track and act the good; (5) practising doing good things in daily life to take the chance of habituating the good (Kementerian Pendidikan dan Kebudayaan 2012b).

Character values are worth disseminating to children in their early age since this phase of growing determines the degree of absorbing various information in a fast natural way. Educators (teachers and parents) would benefit from this phase of development by giving stimulation the right way through various kinds of approaches appropriate to children's characteristic development and the children's way of learning.

Project is a learning approach making it possible for the teachers to develop character values of children in their early age. How the *project* works enable children to enhance their social competency. Through activities done by children in their groups, working in a team as a practice is developed, a skill to initiate a good start is generated, and developing an attitude to work together in a team while keeping a relationship with especially fellow students among the children in a good balance is enhanced. Teachers play a key role in creating a social atmosphere in class leading to a conducive condition so as to ensure that all children become part of a social system. They will find the values in appreciating each other, helping each other, sharing good things leading to a possibility. This fact is in line with Morefield's (1999) school of thought that teachers can take part as a facilitator in class and teach social values to children to learn how to appreciate others, protecting individual's right, and appreciate minority groups.

Learning at an early age by children should enable them to strengthen a positive disposition, such as a feeling of curiosity, creativity, cooperation, generosity, being humorous, independence, being creative, and willing to help with pleasure. A *Project* is a learning approach viewed as to have characteristics to develop positive disposition and emotion in children. This feeling is learnable when children interact with others in a group. Katz and Chard (2000) say "....*including projects in the daily work of the whole group alleviate the pressure on all children to succeed at the same task at the same time*".

Previous research on project-based learning indicates that project-based learning has many benefits for children of early age. Research findings by Brown and Campione (1996) show that project-based learning develops a skill of thinking at a high level such as problem-solving, designing, and self-monitoring. Other than that, children as students involved in a project group of work are able to develop conceptual ideas in a cross-learning situation of various kinds (Brown & Campione 1996, Scardamalia & Bereiter 1991). Katz's (1994) research finding shows that students at Regio Emilia project-based school in Italy show a feeling of curiosity, ability to reflect senses, and a sense of caring for other people's views. Other finding shows that a project work enables to develop students' feelings about themselves in that they are able to learn to give emotional response accurately to the success or the failure they experience and others' less successful accomplishments. This self-evaluation reflects an admittance on one's self-superiority and limited ability as well (Mitchell et al. 2008).

Results from the present study as mentioned above relate to an influence of project-based learning on children's characters. The findings show that project-based learning is applicable and thus serve as principles of developing positive characteristics for children in their early age.

4 CONCLUSION

Characters are one of the aspects of development worth investigating further. Children in their early age are the right subject for the investigation since in this time of development children are at their sensitiveness in terms of receiving stimulation.

Educators have a very important role to develop character values of young children through various approaches including role modeling, habituating, repeating good deeds in daily life and integrating the values into various learning strategies. Project-based learning through a small group is a learning strategy considered as having a potential to accommodate character development in young children.

REFERENCES

Brown, A.L. & Campione, J.C. 1996. Psychological theory and the design of innovative learning environments: On procedures, principles, and systems. In L. Schauble & R. Glaser (Eds.), *Innovations in learning: New environments for education*. Hillsdale, NJ: Lawrence Erlbaum Associates.

Freeman, G.G. 2014. The implementation of character education and children literature to teach bullying characteristics and prevention strategies to preschool children: An Action Research Project. *Early Childhood Education Journal* 42(5): 305–316.

Henry, J. 1995. *Teaching through project: open & distance learning series*. London: Kogan Page.

Katz, L.G. & Chard, S.C. 1993. The project approach. In J.L. Roopnarine & J.E. Johnson (Eds.). *Approaches to early childhood education* (2nd ed., pp. 209–222). New York: Macmillan.

Katz, L.G. & Chard, S.C. 2000. *Engaging children's minds: The project approach*. Norwood, NJ: Ablex.

Kementerian Pendidikan dan Kebudayaan. 2012a. Guidelines of Character Education for Early Childhood Education. Jakarta: Author.

Kementerian Pendidikan dan Kebudayaan. 2012b. Guidelines on curriculum 2013 implementation for early childhood education. Jakarta: Author.

Lickona, T. 1996. Eleven principles of effective character education. *Journal of Moral Education* 25(1): 93–100.

McClellan, et al. (1999). Assessing Young Children's Social Competence. ERIC Digest. Retrieved from https://www.ericdigests.org/2001-4/assessing.html.

Mitchell, S. et al. 2009. The negotiated project approach: Project based learning without leaving the standards behind. *Early Childhood Education Journal* 36(2): 339–346.

Morefield, J. 1999. *The Classroom-A Community of Learners*. Retrieved from http//education.alberta.ca/media/307119/o.pdf.

Noddings, N. 1984. *Caring: A feminine approach to ethic and moral education*. Berkeley, CA: University of Calivornia Press.

Olweus, D. 1993. *Bullying at school: What we know and what we can do*. Cambridge, MA: Blackwell.

Porter, W. et al. 2010. Bully-proofing your elementary school: Creating a caring community. In S.R. Jimerson, S.M. Swearer, & D.L. Espelage (eds.), *Handbook of bullying in school: An international perspective*. New York, NY: Routledge.

Sascha, M. et al. 2008. The negotiated project approach: Project based learning without leaving the standards behind. *Early Childhood Education Journal* 36(4): 339–346.

Scardamalia, M. & Bereiter, C. 1991. Higher levels of agency for children in knowledge building: A challenge for the design of new knowledge media. *Journal of the Learning Sciences* 1(1): 37–68.

Shields, D.L. 2011. *Character as the aim of education*. SAGE Journals 92(8): 48–53.

U.S Department of Education, Office of Safe and Drug-Free School, Character Education and Civic Engagement Technical Assistance Center. 2008. *Partnerships in character education, State Pilot Projects*, 1995–2001: Lesson learned, Washington, DC.

Cultivating character in junior high school students through the subject of Islamic religious education

M. Marzuki
Universitas Negeri Yogyakarta, Indonesia

ABSTRACT: Character education can be conducted through various methods, one being the provision of adequate textbooks to support the smooth process of learning in the classroom. This study aims to review the payload of the textbook *Pendidikan Agama Islam dan Budi Pekerti* (Islamic Religious Education and Character) in developing the character of junior high school students in Indonesia in the context of the guidelines for the implementation of the Indonesian national curriculum of 2013. This research is a content analysis in relation to the curriculum, and the sample is the Student Book of *Pendidikan Agama Islam dan Budi Pekerti* for Class VII. Data collection involved documentation techniques supported by interviews. Data analysis used content analysis techniques with steps of data collection, sampling (unit of analysis), data recording, data reduction, results description, and conclusion drawing. The research concluded that the textbook *Pendidikan Agama Islam dan Budi Pekerti* for junior high school already contains material that can develop the character of junior high school students, and many of the character values sought already grow and flourish among junior high school students as a result of this book.

1   INTRODUCTION

1.1   *Background*

Education plays a very big role in national development. The direction and purpose of education programs in a country will align with the direction and purpose of the state and nation and a state's success in implementing development depends on the success of its education program.

Experts, such as Lapsley (2008), have concluded that the moral formation of children is one of the foundational goals of formal education, and there has been increasing recognition that neighborhoods and communities play critical roles in inducting children into the moral and civic norms that govern human social life. Saeed (2007) has argued that education acts as a key to changing attitudes: in the case of Islam, it was the Islamic world view, with its positive views or biases, prejudices, and suspicions of other systems, that colored the attitudes that believers adopted.

Tilaar (2009) stated that an independent Indonesian was an Indonesian who can realize their personality or morals in an Indonesian nation based on Indonesian culture. In this context, national education is a process of Indonesian human liberation. As an independent nation, Indonesian people should not be swept along by the flow of globalization or just stand idly by and be ignorant of the big changes in everyday life, but must be aware of being persons who have an identity as an Indonesian nation. The occurrence of changes in national education policy can be assessed through policy analysis, that is, analysis that makes generalizations or presents information in such a way as to improve the basis on which policy makers evaluate their decisions (Assegaf, 2005).

Within a period of one decade (2003–2013), there have been various national policies on education, as outlined in various pieces of national legislation on the subject, ranging from laws, government regulations, and regulations of the Ministry of National Education, to decrees of the Ministry of National Education. Through these various legislative products we can analyze national policies related to education in Indonesia from a variety of aspects or areas of emerging policy, as well as the background that accompanies their emergence. In 2013 the government of Indonesia imposed a new curriculum, the 2013 curriculum. Up until 2016, the government of Indonesia had gradually introduced this curriculum through various efforts, such as the issuing of various laws and regulations that supported the smooth implementation of this curriculum to various facilities and infrastructure, and created the implementation manuals and textbooks to be used in the learning process of the 2013 curriculum for both students and for teachers.

In the 2013 curriculum the minimum materials and competencies to be achieved were established for each level of education, ranging from elementary school (*Sekolah Dasar*/SD or *Madrasah Ibtidaiyah*/MI) to high school (senior high school – *Sekolah Menengah Atas*/ SMA or *Madrasah Aliyah*/MA; vocational high school – *Sekolah Menengah Kejuruan*/SMK). The competencies to be achieved at each level include spiritual attitudes, social attitudes, knowledge and skills. The two initial competencies (spiritual and social attitudes) are closely related to character values. Thus, the 2013 curriculum is very concerned with the issue of character development in students.

One of the supports in the implementation of the 2013 curriculum is a special textbook prepared by the government for all subjects at all levels of education from elementary school to high school. The existence of textbooks is crucial in the implementation of the 2013 curriculum, so the government made it policy that this textbook must be used by students and teachers in the 2013 curriculum. The textbook had been prepared by the government, although other textbooks may also be used to support and complement it. The success of curriculum enforcement is not only based on textbooks, but is also supported by other factors, especially teachers. Teachers play an important role in the success of the classroom learning process. Saeed (2007) confirmed that teachers have several roles, including the facilitation of critical discussion of an issue, the exploration of links between the issue and society, and not seeking one correct answer but exploring all possible aspects of the associated problems.

1.2 *Research questions*

This research will focus on the study of this textbook of Islamic Religious Education and Character (*Pendidikan Agama Islam dan Budi Pekerti*) in junior high school (or middle school: *Sekolah Menengah Pertama*/SMP) and whether the textbook can foster the character of the student through its content and messages. It will examine the character values contained in *Pendidikan Agama Islam* (PAI) *dan Budi Pekerti* in junior high school and how they are presented. These questions will be examined through content analysis research in the hope that the results will be beneficial to the parties concerned with the implementation of the 2013 curriculum and to activists for character education.

Based on this background, the research questions can be formulated as follows. The first question is, what are the contents of the government textbook of Islamic Religious Education and Character that can foster the desired characteristics of junior high school students in Indonesia? The second question concerns how the implementation of the 2013 curriculum is able to foster the character of junior high school students in Indonesia through this textbook of Islamic Religious Education and Character.

1.3 *Meaning of character education*

Education, which is an agent of change, should be able to improve the character of the nation. In other words, education should be able to carry out the mission of character building, so that learners can participate in the fulfillment of national development in the future without losing these character values.

To underpin the study of these issues, it is necessary to understand some important concepts related to character education. Thomas Lickona (1991) described character as "A reliable inner disposition to respond to situations in a morally good way." He added: "Character so conceived has three interrelated parts: moral knowing, moral feeling, and moral behavior."

Thus, character refers to a set of knowledge (cognitives), attitudes, and motivations, as well as behaviors and skills, and character education consists of three main elements: knowing the good, loving the good, and doing the good (Lickona, 1991; Alberta, 2005). Character education does not just teach what is right and what is wrong; character education inculcates habituation about the good so that students understand good, are able to feel it, and want to do it. The concept of character education has, at times, been understood as being interchangeable with the concept of moral education (Qoyyimah, 2016; Hill, 1991; Lickona, 1996), and character education thus involves the same mission as moral education.

Frye et al. (2002, p. 2) defined character education as "A national movement creating schools that foster ethical, responsible, and caring young people by modeling and teaching good character through an emphasis on universal values that we all share." Thus, character education should be a national movement for the cultivation of values of noble character through learning and modeling.

1.4 *Character building of students in school*

Noble character needs to be habituated, and the realization of noble character as the ultimate goal of an educational process is highly coveted by every institution that organizes such processes. The culture that exists in institutions, whether schools, colleges, or others, plays an important role in building noble character among the academic community employed or taught therein.

To realize noble character in the life of every person, the habituaton of noble character is necessary. This is achieved in schools or educational institutions through the provision of subjects of moral education, ethics education, or character education. Lately in Indonesia, this mission has been carried out through two main subjects, namely Religious Education and Civic Education. Following the Act of Republic of Indonesia No. 20 of 2003 on a National Education System, and its reinforcement by Government Regulation of Republic of Indonesia No. 19 of 2005 on National Education Standards (RoI, 2005), the government decreed that each subject group be implemented holistically such that the learning of each subject group influences the understanding and/or appreciation of the learner (RoI, 2005, article 6, para. 4). Fostering the character of students in schools refers to the various efforts undertaken by a school in order to form the character of students, and the establishment of an appropriate school culture is now encouraged in schools.

1.5 *Islamic religious education in school*

Religious education is an integral part of the implementation of education conducted in formal educational institutions in Indonesia, and is also part of national education. In article 3 of Act No. 20 of 2003 on a National Education System (RoI, 2003), it was reiterated that the national educational function was to develop the ability and form the character and civilization of a dignified nation in order to develop the nation's life. The purpose of Indonesian national education is to develop the potential of learners to become human beings who believe in and are respectful of God, have noble character, are healthy, knowledgeable, capable, creative, independent, and who become democratic and responsible citizens.

As part of Indonesian national education, religious education has a very important and strategic role in the realization of its functions and objectives. Religious education, especially Islamic religious education, has a huge role and responsibility in realizing national education goals, especially when it comes to preparing learners to understand the teachings of religion, the variety of knowledge learned, and its implementation in everyday life. Theologically, Islamic education has a philosophy that is different from any other education, because Islamic education bases its fundamental principles on the Qur'an and the Hadith. Islamic education is as old as Islam itself and its history shows conceptual, ideological, and structural diversification (Niyozov & Memon, 2011). Through Islamic education, the basis of Islam (Islamic theology, Islamic law, and Islamic morals) can be transferred to other education, including science, humanism, and laws of nature. Amjad Hussain (2010) said that "Islamic education for Muslims is just an element of the wider Islamic theology where primacy of science and autonomous human and secular reality are non-existent". There are basic principles in Islamic education that are contained in the term *tarbiyah* (Qur'an Surah 17:24), which indicates that Islamic education involves nurturing and caring for the child. *Ta'lim* (Q.S. 96:4–5) explicitly indicates that one of the purposes of Islamic education is to impart knowledge, and *ta'dib* fully demonstrates the importance of the three parts of human existence that Islam upholds: the mind, the body, and the soul. As the prophet Muhammad said, "My Lord educated me, and so made my education most excellent" (Hussain, 2010; Hussain, 2004; Cook, 1999; Halstead, 2004; Pohl, 2009).

Realizing the goals of religious education as described above is not easy. Many things must be considered, from the materials, management, methodology, facilities and infrastructure, to educators and learners. As one of the subjects in school, Islamic Religious Education and Character should strive to follow developments and demands of the times so as to carry out the general national education mission and Islamic education, in particular.

A curriculum for Islamic education in Indonesia should be arranged better than before. If implemented well in the educational system, this curriculum helps to produce learners who are spiritually, physically, intellectually and emotionally strong and balanced, making for a more dynamic and progressive generation. Thus, the curriculum of Islamic education needs to be planned well, requiring comprehensive preparation while paying careful attention to details and aspects of practicality (Lubis et al., 2010).

2 METHOD

2.1 Types of research

In terms of approach, this research is qualitative descriptive research based on library research (Zed, 2004). In terms of analytical techniques, this research is based on content analysis, which is a form of research conducted with the aim of exploring the contents or the meaning of symbolic messages in a book or other work (Krippendorff, 2004).

2.2 Population and sample

The population for this research is all of the policies concerned with the implementation of the 2013 curriculum in Indonesia. The sample for the research is the textbook of Islamic Religious Education and Character (*PAI dan Budi Pekerti*) in relation to junior high school (SMP). Not all of the contents of the manuscripts will be reviewed, but only what relates to the effort to develop the character of junior high school students. Specifically, the research sample is the Student Book for Junior High School, Class VII of *PAI dan Budi Pekerti* (RoI, 2014).

2.3 Data collection techniques

The data collection technique used in this research is a documentation technique, given that all of the manuscripts used as research data have been documented either through print media (in book form) or electronic media (stored in files on internet). Another technique used is that of library research, because the data is also captured in books stored in the library. To complete the research data, the researcher also used interviewing techniques with people who were specifically associated with the writing of these texts, as well as with the users of the book.

2.4 Data analysis

The collected data was then selected and analyzed by qualitative analysis based on content analysis. The technique used was a cognitive map technique that describes the location of several concepts and the nature of the relationship between them (Krippendorff, 2004).

3 FINDINGS AND DISCUSSION

3.1 *Developing the character of junior high school students through PAI dan Budi Pekerti textbook*

As it relates to the specificities of the 2013 curriculum, which emphasizes character education, the textbook *PAI dan Budi Pekerti* adequately meets the curriculum requirements, although there are still some concerns. The authors of the book seek to incorporate the basic principles of character education within, for example, in the presentation of the material in each chapter.

Using a methodical approach, the authors consistently begin by inviting the reader (learners) to observe the various phenomena of human life through a limited number of images (two to four). Before inviting additional observations, the authors invite reflection tailored to the messages contained in each chapter or study topic. With the invitation to reflect upon and observe the pictures, the authors hope to encourage character in the reader (learners) in accordance with the messages conveyed. Following on from this observation and contemplation, the writers then present description of the material to completion.

Based on the pictures that form the objects of observation, the messages to learners to grow and develop certain characteristics are not very clear. The pictures are more directed at familiarization with the topics to be discussed, rather than leading to certain characteristics in the reader. There are some clear pictures of messages about noble character, but only a few. This is understandable given the limitation of places to explore the wider picture until the messages about the desired character have been clearly illustrated.

In each part of the description (subsection) of the existing topics, the authors begin by inviting the learners to exhibit noble character. This can be seen in the title and introduction of each chapter or subsection, where there are many sentences in the form of a call-to-action or statement that contain an exhortation to display a given characteristic, such as "*Mari Berperilaku Jujur*" ("Let's behave honestly"), "*Mari Berperilaku Amanah*" ("Let's be trustworthy"), and "*Mari Berperilaku Istiqamah*" ("Let's be steadfast"), in addition to other titles that are not invitations. Titles like this have large significance in the growth of the character of the learners who read them. Before the learner reads the detailed content of the textbook, the idea has been placed in their mind that they are invited to behave commendably, as described in the titles. Although the description of the material may indicate expected characteristics, not all such characteristics are clearly illustrated in the material.

The character values revealed in the book are many. Almost all of the material presented reflects important character values that the book is attempting to induce in the everyday behavior of learners. Comparing these with the main character values developed by the Ministry of Education and Culture, almost all of the latter values (18 of them) are contained in the *PAI dan Budi Pekerti* textbook, although the values are not always expressed in the same terms. However, if the number of character values addressed in *PAI dan Budi Pekerti* were identified and counted, the number is greater than those identified by the Ministry of Education and Culture.

Another aspect of *PAI dan Budi Pekerti* associated with the development of the character of learners is the model or method used in presenting the material. If we examine the content of the material presented in each chapter and subsection of the textbook, we can identify two models or modes of character education, namely inclusion (advice and motivation) and modeling (Kirschenbaum, 1995; Zuchdi, 2012).

3.2 *Content of PAI dan Budi Pekerti textbook*

The character education content in *PAI dan Budi Pekerti* is also quite clearly visible, although not yet entirely clear and perfect. Clarity in character education in *PAI dan Budi Pekerti* can be seen in its content in terms of five aspects of Islamic Studies, namely *Akidah* (Islamic Theology), *Akhlak* (Islamic Morality), *Syariah* (Islamic Law), *Sejarah Islam* (Islamic History), and *Al-Quran-Hadis* (Quran—Tradition). The authors of *PAI dan Budi Pekerti* seem to have tried to follow the principles of good book writing and book development in the context of the 2013 curriculum, which emphasizes strengthening character education.

From the interview with one of the authors of *PAI dan Budi Pekerti*, Muhammad Ahsan (16 October 2016), it appears that the book is written in accordance with the principles of character education emphasized in the 2013 curriculum. He asserted that the contents of character education in *PAI dan Budi Pekerti* are adequate, in terms of the characteristics of both spiritual attitudes and social attitudes. Presentations in *PAI dan Budi Pekerti* have also been pursued in accordance with the principles of true learning that can foster the characters of the students themselves.

Dr. Muhammad Kosim, Mag., one of the national instructors (interview, 31 October 2016), asserted that the contents of *PAI dan Budi Pekerti* are in line with the principles developed in the 2013 curriculum in terms of developing the character of learners. He declared that the content of the book had been loaded with adequate character education in each chapter, especially at the beginning of the subsection "*Renungkanlah*" ("Meditate"), and at the end that gave rise to the "*Cerita*" ("Stories") that were exemplary in character. According to Kirschenbaum (1995), one of the most effective methods for cultivating the character of students is inculcation and facilitation, for example, by providing motivation and assignment to behave with certain characteristics. It is also part of the revitalization of religious education in schools (Marzuki, 2013).

4   CONCLUSION

From the research findings described above and the discussion, the conclusion is that the textbook of Islamic Education and Character (*PAI dan Budi Pekerti*) for junior high school (SMP) already contains materials that can develop the character of junior high school students. It seeks to develop and grow many character values among junior high school students, such as honesty, discipline, obedience, sympathy, respect for others, and so on.

The presentation of the material in this book has implemented recognized models or methods of character education, especially those of inculcation and modeling. This is in line with the principles developed in the 2013 curriculum. Nevertheless, there are still some weaknesses in *PAI dan Budi Pekerti* that need to be refined, for example, in the presentation of facilitation models and the development of special skills related to *PAI dan Budi Pekerti*, and in motivating students to gain character through evaluation and assignment.

REFERENCES

Alberta. (2005). *The heart of the matter: Character and citizenship education in Alberta schools*. Edmonton, Canada: Alberta Education.

Assegaf, A.R. (2005). *Politik pendidikan nasional: Pergeseran kebijakan pendidikan agama Islam dari proklamasi ke reformasi* [National education policy: Shifting the policy of Islamic education from proclamation to reform]. Yogyakarta, Indonesia: Kurnia Kalam.

Cook, B.J. (1999). Islamic versus western conceptions of education: Reflections on Egypt. *International Review of Education, 45*(3/4), 339–357. doi:10.1023_a-1003808525407.

Frye, M., Lee, A.R., LeGette, H., Mitchell, M., Turner, G. & Vincent, P.F. (Eds.). (2002). *Character education: Informational handbook and guide for support and implementation of the Student Citizen Act of 2001*. Raleigh, NC: Public Schools of North Carolina.

Halstead, M. (2004). An Islamic concept of education. *Comparative Education, 40*(4), 517–529. doi:10.1080/0305006042000284510.

Hill, B.V. (1991). *Values education in Australian schools*. Hawthorn, Australia: Australian Council for Educational Research.

Hussain, A. (2004). Islamic education: Why is there a need for it? *Journal of Beliefs & Values: Studies in Religion & Education, 25*(3), 317–323. doi:10.1080/1361767042000306130.

Hussain, A. (2010). Islamic education in the west: Theoretical foundations and practical implications. In K. Engebretson, M. de Souza, G. Durka & L. Gearon (Eds.), *International handbook of interreligious education* (pp. 235–248). New York, NY: Springer. doi:10.1007/978-1-4020-9260-2_15.

Kirschenbaum, H. (1995). *100 ways to enhance values and morality in schools and youth settings*. Boston, MA: Allyn & Bacon.

Krippendorff, K. (2004). *Content analysis: An introduction to its methodology* (2nd ed.). London, UK: Sage.

Lapsley, D.K. (2008). Moral self-identity as the aim of education. In L.P. Nucci & D. Narvaez (Eds.), *Handbook of moral and character education* (pp. 30–52). London, UK: Routledge.

Lickona, T. (1991). *Educating for character: How our school can teach respect and responsibility*. New York, NY: Bantam Books.

Lickona, T. (1996). Eleven principles of effective character education. *Journal of Moral Education, 25*(1), 93–100. doi:10.1080/0305724960250110.

Lubis, M.A., Yunus, M.M., Embi, M.A., Sulaiman, S. & Mahamod, Z. (2010). Systematic steps in teaching and learning Islamic education in the classroom. *Procedia – Social and Behavioral Sciences, 7*, 665–670. doi:10.1016/j.sbspro.2010.10.090.

Marzuki, M. (2013). *Revitalisasi pendidikan agama di sekolah dalam pembangunan karakter bangsa di masa depan* [Revitalization of religious education in schools in character building of nation in the future]. *Jurnal Pendidikan Karakte, 3*(1), 64–76.

Niyozov, S. & Memon, N. (2011). Islamic education and Islamization: Evolution of themes, continuities and new directions. *Journal of Muslim Minority Affairs, 31*(1), 5–30. doi:10.1080/13602004.2011.55 6886.

Pohl, F. (2009). Interreligious harmony and peacebuilding in Indonesian Islamic education. In C.J. Montiel & N.M. Noor (Eds.), *Peace psychology in Asia* (pp. 147–160). New York, NY: Springer. doi:10.1007/978-1-4419-0143-9_8.

Qoyyimah, U. (2016). Inculcating character education through EFL teaching in Indonesian state schools. *Pedagogies: An International Journal, 11*(2), 109–126. doi:10.1080/1554480X.2016.1165618.

RoI. (2003). *Undang-Undang Republik Indonesia Nomor 20 Tahun 2003 tentang Sistem Pendidikan Nasional* [Law of the Republic of Indonesia Number 20 Year 2003 regarding National Education System]. Jakarta, Indonesia: Ministry of Education and Culture, Republic of Indonesia.

RoI. (2005). *Peraturan Pemerintah Republik Indonesia Nomor 19 Tahun 2005 tentang Standar Nasional Pendidikan* [Government Regulation of the Republic of Indonesia Number 19 Year 2005 regarding National Standards of Education]. Jakarta, Indonesia: Ministry of Education and Culture, Republic of Indonesia.

RoI. (2010). *Kebijakan nasional pembangunan karakter bangsa tahun 2010–2025* [National policy for nation character building in 2010–2025]. Jakarta, Indonesia: Puskur Balitbang Kemdiknas, Republic of Indonesia.

RoI. (2014). *Pendidikan Agama Islam dan Budi Pekerti: Buku Siswa untuk SMP Kelas VII* [Islamic religious education and character: Seventh grader students' textbook]. Jakarta, Indonesia: Ministry of Education and Culture, Republic of Indonesia.

Saeed, A. (2007). Towards religious tolerance through reform in Islamic education: The case of the state institute of Islamic studies of Indonesia. *Indonesia and the Malay World, 27*(79), 177–191. doi:10.1080/13639819908729941.

Tilaar, H.A.R. (2009). *Kekuasaan dan pendidikan: Manajemen pendidikan nasional dalam pusaran kekuasaan* [Powers and education: Management of national education in power whirls]. Jakarta, Indonesia: Rineka Cipta.

Zed, M. (2004). *Metode penelitian kepustakaan* [Biblical research methods]. Jakarta, Indonesia: Yayasan Obor Indonesia.

Zuchdi, D. (Eds). (2012). *Pendidikan karakter konsep dasar dan implementasi di perguruan tinggi* [Character education: Basic concepts and implementation in higher education]. Yogyakarta, Indonesia: UNY Press.

Revitalization of Banyumas traditional games as media for instilling character

S. Yulisetiani
Universitas Sebelas Maret, Indonesia

T. Trianton
Universitas Muhammadiyah Purwokerto, Indonesia

ABSTRACT: Traditional games are part of a culture that can help to form a more noble character. Such games have flourished in Banyumas, a city in Central Java, Indonesia. They have distinctive characteristics and play a role in creating a more egalitarian society. However, their practice has largely been displaced by modern games. Therefore, traditional games that can foster positive character in children at an early age require a revitalization effort. Both parents and teachers have a role to play in this. This research takes the form of a qualitative descriptive study; the aims are to inventory and revitalize the diversity of traditional games as character-implanting media in schools. The research data are traditional games and content analysis is employed. The results are analytic descriptions of traditional games that can be implemented through language learning in elementary school. These traditional games can foster a sense of solidarity, honesty, ability, unity, skill, and courage, and also have a role in shaping a more egalitarian society.

1   INTRODUCTION

One effort that can be made to instill the values of national character in learners is to introduce traditional games. Indonesian has many *adiluhung* (worthy) cultural heritages; one of these is the traditional game. However, today the existence of traditional games has been displaced by the presence of games based on modern and digital technologies. Therefore, it is important to undertake the revitalization of traditional games as one of the efforts in realizing the cultivation of national character.

A game is a non-formal activity that can manifest skills. Each game has rules as its basis. The rules in the game usually contain activity signs, which must be obeyed by every player. One of the rules is about permissible and impermissible treatments. Rules in a game can usually be changed and adapted to the situation. The goal is to create an interesting game plot (Alessi & Trollip, 2001).

Traditional games can form naturalistic, kinesthetic, and intrapersonal intelligence in children. They are a form of folklore. The traditional game is an activity that is governed by rules of the game that are an inheritance from the previous generation by humans (children) in order to elicit enjoyment (Danandjaja, 1987; Tzeng & Huang, 2010). Traditional games can be used as a good means of developing children's education, and they contain moral and cultural values of the community. These games often give priority to creativity values so can be used as learning media.

Traditional games can instill life attitudes and skills, such as values of cooperation, togetherness, discipline, and honesty. The value of cooperation can be obtained through games conducted in groups. Meanwhile, the value of togetherness can grow through both individual games and group games. Usually, each player will put the joy and happiness together. Then the value of discipline and honesty in the game is realized through agreement to obey the rules.

Traditional games have flourished in Banyumas, a city in Central Java, Indonesia. The traditional games have distinctive characteristics and play a part in manifesting an egalitarian society. However, today they have replaced by modern games that are often not in accordance with the identity of the Banyumas people.

The traditional games play a role in forming children's character. However, in practice, the traditional games have been abandoned and children have switched to modern games, such as PlayStation, online games and others. Therefore, the traditional games need to be revitalized, for the preservation of culture, as well as in an effort to nurture positive characteristics in children naturally.

2 RESEARCH METHODS

This research is conducted using an analytical descriptive method to reveal the diversity of traditional games that can be implemented through language learning in elementary school. The primary data sources of this research are traditional games that have flourished in the Banyumas, which are *Bedoran, dut-dut Kiradut, Sliring Genting, ancak-ancak Alis* and *cublek-cublek Suweng*. The secondary data sources are previous journals, books and relevant research. The data collection techniques are observation, interview and documentation study. The data analysis technique used is content analysis, which is a strategy for capturing messages in text as well as cultural products. The analysis is conducted using a three-stage interactive analysis model: data reduction, data presentation and conclusion drawing (Miles & Huberman, 2007; Creswell, 2012; Endraswara, 2008).

3 DISCUSSION

3.1 *Banyumas traditional games*

Traditional games are often identified as one of the characteristics or markers of local culture. In the Banyumas region, traditional games have special characteristics. The games have flourished rapidly, and have had a role in manifesting egalitarian society. Egalitarian refers to actions based on an open attitude to issues without differentiation according to rank, position, or other distinguishing attributes, and provides a solid foundation for growing democracy.

In Banyumas culture, an egalitarian attitude manifests through some typical characteristics, namely *cablaka, sabar lan nrima* and *ksatria* (*nobility*). *Cablaka* is a sincere open attitude with careful consideration of what is spoken spontaneously with plain, straightforward language. This open attitude in Banyumas people is mainly reflected in the use of the Javanese dialect of *Banyumasan*. This is an identity of Banyumas people that cannot be replaced. *Sabar lan nrima* is the attitude of patient acceptance of all the events that have been experienced in daily life for what they are. This characteristic has become an identity of Banyumas society in coping with the problems of daily life. Banyumas people have a noble soul: honest, good personality and tolerant, harmonious, a help to others, and prioritizing the common interest. This character manifests in the tradition of *gotong royong*, which is a form of Banyumas cultural communication that indicates high empathy and sympathy (Priyadi, 2003, 2007; Trianton, 2013; Trianton et al., 2016).

Traditional games in Banyumas are generally similar to traditional games in other parts of Indonesia. Traditional games are one of the fastest-growing forms of folklore in the archipelago. Some traditional games have been developed in Banyumas and implemented in language learning: *Bedoran, dut-dut Kiradut, Sliring Genting, ancak-ancak Alis* and *cublek-cublek Suweng*. In learning activities, teachers can modify the five games by inserting relevant learning materials for discussion during play.

The traditional game of *Bedoran* is an individual game with no limit on the number of players; more players will enliven the game. The game is very suitable to be played by all of the students in a class. The children can start the game with a lottery activity (*hompimpah*) to determine the guard or keeper.

In *Bedoran*, one child plays as guard player and many children act as *bedoran*. The guard player is responsible for chasing *bedoran* until they can link hands successfully. The chased *bedoran* can take refuge in fellow *bedoran* to avoid the chasing guard, and the guard should not attack the *bedoran* that have sought mutual shelter. Once captured, a *bedoran* becomes the next guard player. Teachers can modify this game in terms of the theme of play. Thus, the captured *bedoran* has to create a question for the guard according to this theme. Only if the guard player can answer the question, can they swap roles and become *bedoran*. At the end of the game, the teacher can ask the students to recount the game chronology based on the experiences they have gained during play.

Dut-dut Kiradut is a traditional game that introduces children to the customary visit. In the game, one child acts as a homeowner, another child serves as a guest, and the other children act as sweet yams and wasps owned by the homeowner. The homeowner offers sweet yams to the guest and gives an opportunity to choose between them. The homeowner also cautions against wasps around the sweet yams. The guest must cleverly choose yams in order not to pick those with wasps. The sweet yams selected belong to the guest but if the guest mistakenly chooses one with wasps, then the wasps will chase the guest. The game ends if the guest is caught by the wasps. The teacher can modify the game by describing and narrating things around the playground as discussion points during the game.

Sliring Genting is a traditional game that teaches a cautious attitude toward strangers. One child acts as a chicken owner. Another child acts as a robber or an evil alien while the other children act as chickens ready to be stolen by the robber. The chickens line up behind their owner and the robber tries to catch the chicken at the back. The teacher can modify the game by applying a rule that the chicken caught by the robber becomes their property if the robber is successful in telling an interesting story when requested by the child who acts as the chicken.

Ancak-ancak Alis is another very popular game, which is known in other areas as the dragon snake game. At the beginning of the game, two players of equal height are selected to become leaders. The two place their palms together to form an entrance that will be passed through by the other players. Meanwhile, the other players join hands to form a long snake. When a song is sung, the snake players run through the portal. When the song finishes the two leaders catch the player passing through at that point and ask them a question, usually about agriculture. However, to develop students' knowledge, the teacher can determine other relevant learning themes.

Cublek-cublek Suweng is a traditional game that is accompanied by traditional song. The losing child has to close their eyes and bow down and the palms of the other players are placed on their back. One child, who acts as a guide, moves a coin from one palm to the next accompanied by a song. When the song stops, the hands are clasped and the guard must guess who holds the coin. If they guess correctly they must tell a story to entertain their friends. In this game the teacher can choose a story theme that is relevant to the desired learning.

3.2 *Value character building*

The government has implemented various strategies in pursuit of character education. One such strategy is character education at the educational unit level. The values of character formation in schools are instilled through the operational program of these units, and implementation of character education is done in every aspect, especially in teaching and learning activities. Character education is integrated into classroom learning, and teachers play a very important role.

Character education is imbued with more meaning than moral education, because it is not only about teaching what is right and what is wrong, but also instills habituation in relation to good actions, so that learners acquire an understanding (the cognitive domain) about what is good and bad, are able to feel and live (the affective domain) good values, and routinely perform them (the psychomotor domain) (Suroso, 2012).

There are two aspects to character education, namely behavior and personality. Aspects of behavior determine how a person behaves in terms of good or bad, while the personality aspects concern how a person is characterized when exhibiting a personality that is in accordance with moral values.

Thus, there are values, abilities and inner mechanisms that can be learned and can be taught by teachers, known as mega skills, which include confidence, motivation, effort, responsibility, initiative, strong will, affection, cooperation, logical thinking, problem-solving skills, and concentration (Rich, 1998).

Moral values must be totally integrated into daily classroom life and teachers have a central role in shaping the character of students. In this case, the teacher must be able to develop innovative and creative learning that can stimulate the development of student character (Bryk, 1988; Goodlad, 1992; Hansen, 1993).

Six characteristics are described in *The Six Pillars of Character*, which is produced by the Character Counts Coalition (a project of The Joseph Institute of Ethics) (Character Counts!, 2017). The six are: (a) trustworthiness, the characteristic that makes a person act with integrity, honesty, and loyalty; (b) fairness, the characteristic that gives a person an open mind and they do not take advantage of others; (c) caring, the characteristic that gives a person a caring attitude and concern for others, as well as the social conditions of the surrounding environment; (d) respect, the characteristic that causes a person always to appreciate and respect others; (e) citizenship, the characteristic that makes a person aware of laws and regulations, as well as caring for the natural environment; (f) responsibility, the characteristic that makes a person responsible, disciplined, and always doing things to their best ability.

Instilling character through classroom learning is one of the solutions. The teacher should play an active role in instilling these character values implicitly when choosing a topic, responding to student questions, stimulating students to seek the truth, and determining classroom activities (Campbell, 2003; Hansen, 1993; Tom, 1984).

Traditional games make an effective contribution to character formation in learning through manipulative and locomotor skills. These games can have a positive effect on improving character education in schools (Akbari et al., 2009).

Some of the character values that can be conveyed through traditional games are: (a) teaching to share with fellow friends, as the games require students to interact directly with their playmates; (b) training every player to be sporting in every game played and able to accept defeat; (c) training every player not to be easily discouraged; (d) creative thinking— every player will think creatively in every game.

3.3 *The revitalization of traditional games*

Each traditional game can influence the development of children's intelligence: intellectual, spiritual and emotional. The games teach cooperation, tolerance, honesty and other positive attitudes. This is what distinguishes traditional games from the modern games that have been developed in recent times. Therefore, in order that the existence of traditional games will not be obliterated by the presence of modern games, it is necessary to preserve them.

The preservation of traditional games as one form of the cultural wealth of the archipelago should be undertaken seriously, and can be achieved through government oversight and community support.

In elementary schools, traditional games can be implemented in the subject of Bahasa Indonesia for basic competence in speaking skills. Speaking skills are skills in expressing ideas and responding to other people's ideas verbally through considerations such as a mastery of grammar, social status and social situation, and mental condition in developing the topic of conversation according to the needs of the listener or the interlocutor. Therefore, speaking skills and speaking materials require sufficient knowledge (Hughes, 2003; Widdowson, 2004; Luoma, 2009; Nurgiyantoro, 2010; Fulcher, 2014).

The implementation of traditional games in learning activities can create a positive impression and experience in learners, as well as knowledge. In the learning process, play activities can support the apperception carried out before the core activities of learning take place.

Thus, teachers first give opportunities to students to play. After the students have gained direct experience from these play activities, the teacher gives students the opportunity to talk through the storytelling activities. Through language learning via the competence of speaking skills, especially storytelling skills, traditional games can bring students closer to the traditional arts, and grow the senses of solidarity, honesty, ability, unity, integrity, skill and courage.

4 CONCLUSION

The Banyumas region is part of Java, Indonesia. In this region, traditional games have distinctive characteristics, and these games have a role in manifesting the egalitarian society of Banyumas. The efforts to revitalize traditional games are important in terms of cultural preservation, as well as in the effort to develop positive character in children naturally. The traditional games of *Bedoran, dut-dut Kiradut, Sliring Genting, ancak-ancak Alis* and *cublek-cublek Suweng* can be integrated into language learning in elementary school, especially for learning speaking skills through the activity of storytelling. These traditional games bring learners closer to the traditional arts, and grow the senses of solidarity, honesty, ability, unity, integrity, skill and courage.

REFERENCES

Akbari, H., Abdoli, B., Shafizadeh, M., Khalaji, H., Hajihosseini, S. & Ziaee, V. (2009). The effect of traditional games in fundamental motor skill development in 7–9 year old boys. *Iranian Journal of Pediatrics, 19*(2), 123–129.

Alessi, S.M. & Trollip, S.R. (2001). *Multimedia for learning, methods and development* (3rd ed.). Boston, MA: Allyn & Bacon.

Bryk, A.S. (1988). Musings on the moral life of schools. *American Journal of Education, 96*(2), 256–290.

Campbell, E. (2003). *The ethical teacher.* Philadelphia, PA: Open University Press.

Character Counts! (2017). The Six Pillars of Character. *charactercounts.org.* Retrieved from https://charactercounts.org/program-overview/six-pillars/.

Creswell, J.W. (2012). *Research design: Qualitative, quantitative, and mixed methods approaches.* Los Angeles, CA: Sage.

Danandjaja, J. (1987). *Folklore Indonesia.* Jakarta, Indonesia: Gramedia.

Endraswara, S. (2008). *Metodologi penelitian sastra* [*Literary research methodology*]. Yogyakarta, Indonesia: Medpress.

Fulcher, G. (2014). *Testing second language speaking.* New York, NY: Routledge.

Goodlad, J. (1992). The moral dimensions of schooling and teacher education. *Journal of Moral Education 21*(2): 87–98.

Hansen, D.T. (1993). From role to person: The moral layeredness of classroom teaching. *American Educational Research Journal, 30*(4), 651–674.

Hughes, A. (2003). *Testing for language teachers.* Cambridge, UK: Cambridge University Press.

Luoma, S. (2009). *Assessing speaking.* Cambridge, UK: Cambridge University Press.

Miles, M.B. & Huberman, A.M. (2007). *Analisis data kualitatif: Buku sumber tentang metode-metode baru* [*The analysis of qualitative data: The resource of new methods*]. Jakarta, Indonesia: Universitas Indonesia Press.

Nurgiyantoro, B. (2010). *Penilaian pembelajaran bahasa: Berbasis kompetensi* [*Assessment of language learning: Based on competency*]. Yogyakarta, Indonesia: BPFE.

Priyadi, S. (2003). Beberapa karakter orang banyumas [The character of Banyumas people]. *Jurnal Bahasa dan Seni, 31*(1), 14–37.

Priyadi, S. (2007). Cablaka sebagai inti model karakter manusia Banyumas [Cablaka as the role model of the character of Banyumas people]. *Jurnal Diksi, 14*(1), 11–18.

Rich, D. (1998). *Mega skills: Building children's achievement for the information age.* Boston, MA: Houghton Mifflin.

Suroso. (2012). Menggali nilai-nilai luhur budaya membangun karakter bangsa [Unearthing cultural values builds national character]. Paper presented at an international seminar on Language,

Literature, and Culture of Nusantara, 16 February 2012, University of Muhammadiyah, Bengkulu, Indonesia. Retrieved from http://staffnew.uny.ac.id/upload/131572386/penelitian/Menggali+Potensi+ Pendidikan+karakter+Bangsa.pdf.

Tom, A. (1984). *Teaching as a moral craft*. New York, NY: Longman.

Trianton, T. (2013). *Identitas wong Banyumas* [*The identity of Banyumas people*]. Yogyakarta, Indonesia: Graha Ilmu.

Trianton, T., Suwandi, S., Waluyo, H.J. & Saddhono, K. (2016). Ethics values as the portrayal of Banyumas local wisdoms in the novels of Ahmad Tohari. *International Journal of Languages' Education and Teaching*, 4(3), 306–319.

Tzeng, S.-K. & Huang, C.-F. (2010). A study on the interactive "Hopscotch" game for the children using computer music techniques. *The International Journal of Multimedia and its Applications*, 2(2), 32–44.

Widdowson, H.G. (2004). *Teaching language as communication*. Oxford, UK: Oxford University Press.

Character Education for 21st Century Global Citizens – Retnowati et al. (Eds)
© 2019 Taylor & Francis Group, London, ISBN 978-1-138-09922-7

Role models in language acquisition and character education

S. Sugirin
Universitas Negeri Yogyakarta, Indonesia

ABSTRACT: The success of language acquisition is dependent on *comprehensible input in a low affective filter* (Krashen 2009). This input can be provided by a role model who is fluent in speech and understands the condition of the language acquirer. This understanding makes the language acquirer feel comfortable so that the acquisition process may run smoothly. This acquisition principle matches the condition demanded in character education as a nurturing character will be well facilitated if there is a reliable role model in the family, education institution and community. The consistency of the role model's words, attitudes, and actions will be clearly perceived, easily understood, and effortlessly internalized by the targeted groups so that without being instructed, they will acquire and voluntarily practice it in their life. It is the intention of this paper to suggest the application of the principle in character education.

1 INTRODUCTION

Character education is nothing new anywhere in the world. Long before Thomas Lickona published his monumental work in 1991, the Indonesian schools already practiced moral education. As time went by, it was included in the religious education and *Pancasila* (the Indonesian Five Principles), and citizenship education. While the religious education focuses on the development of noble morality, the *Pancasila* and citizenship education aim at developing nationalism. With varying modes of delivery, such an effort of nurturing morality and nationalism has been common all over the world (USA, Europe, and Asian countries). There have also been studies on integrating character education in the teaching of school courses such as mathematics, English, sports, etc. However, character education is seen as a standalone discipline focusing on moral nurturance, as if it did not have any relation to language discipline such as language acquisition.

While language acquisition needs a role model to provide comprehensible input (Krashen 2009, Gass & Selinker 2008), Sugirin (2010) and Khisbiyah (2010) assert the demand for role models to develop the young generation into whole persons.

Similarly, Tong & Christodoulou (2017) suggest the importance of positive role models for children and youth in character education. This indicates that both character education and language acquisition rely on the roles of the role models. However, there has not been any study investigating these roles. It is the intention of this paper to explore the possible compatibility between the roles of the role models in character education and language acquisition.

To achieve the goal, this paper will be organized as follows: 1) The first section presents some theories of language acquisition and their potentials for application in character education, 2) The second highlights the importance of role models in both language acquisition and character education, 3) The third illustrates the compatibility between the roles of the role models in language acquisition and character education, and 4) the fourth provides an example of monitoring the practice for maximizing success.

2 THEORIES OF LANGUAGE ACQUISITION AND THEIR POTENTIAL FOR APPLICATION IN CHARACTER EDUCATION

In order to avoid confusion it is worthwhile presenting the distinction between language acquisition and language learning. Krashen (2009 p. 10) asserts that *acquisition* is a process similar to, if not identical with, the way children develop ability in their first language. Hence, language acquisition is a subconscious process; language acquirers are unaware that they acquire the language; they just realize that they use the language for the purpose of communication. In contrast, *learning* refers to consciously obtaining knowledge about the target language, consciously knowing the language rules, and having the ability to talk about those rules.

Krashen (2009) further explains that first language acquisition usually refers to how infants acquire their native language, while second language acquisition refers to how both children and adults acquire another language other than their mother tongue. Acquisition is a subconscious process while learning is a conscious process (Krashen 2009, Williams & Leung 2011). Children blessed with complete five senses will automatically acquire their mother tongue. What they need is *comprehensible input in a low affective filter*. *Comprehensible input* is input that can be easily understood as it is "roughly-tuned" to the child's current level of linguistic competence (Krashen 2009). The input should be provided in *a low affective filter*, in a situation in which the child feels comfortable, with little or no anxiety.

Krashen (2013) even claims two amazing facts about language acquisition, "First, it is effortless; it involves no energy, no work. All that is necessary is to understand messages. Second, language acquisition is involuntary. Given comprehensible input, you must acquire—you have no choice." This means that language does not have to be taught but used for interaction in life. Without undermining the importance of input, Brown (2007) reminds Krashen that input is not the only causative variable in the second language (L2) acquisition; the role of the learners' engagement in the process is also important. Through interaction with other people at work and in the community as well as exposure to the communication and social media, without intentionally learning, children and adults also further develop their language mastery. As abstract thinking has developed in adults, learning may have a greater role than acquisition. Adults' experience in using the mother tongue and understanding of language rules will aid in their learning and acquisition process.

3 THE IMPORTANCE OF ROLE MODELS IN LANGUAGE ACQUISITION AND CHARACTER EDUCATION

One of the requirements for language acquisition to occur is the availability of *comprehensible input* (Krashen 2009) which can only be provided by role models qualified in the target language. As the input must be *in a low affective filter*, in a situation in which the acquirer feels comfortable, a special technique is required. In the context of a child learning the mother tongue, a mother or a caretaker often speaks slowly, clearly, and repeatedly to a child so as to make the utterance (a new word, someone's name, etc.) comprehensible. Knowing that the child may have difficulty pronouncing a certain sound, the consonant /r/, for example, the mother will accept whatever sound the child produces. When the child has difficulty pronouncing the sound /r/ such as saying /lɔtI/ (loti) instead of /rɔtI/ (roti) meaning "bread," the mother accepts it and even uses the same utterance /lɔtI/ not /rɔtI/ to give the child the feeling of being accepted before the child is able to pronounce the word correctly. This mother has not only provided *comprehensible input* but also provided it *in a low affective filter,* in which the child does not have to be concerned with language accuracy. The mother's permissiveness is intended to build the child's self-confidence, to lower their *affective filter* (the feeling of discomfort which has the potential of blocking the acquisition to happen).

A good role model must be able to adjust himself or herself with the language acquirers' condition, understand the problems in comprehending the linguistic input as well as in producing the language correctly. The mother and the caretaker do not realize that they have played an important role in the child's language development as they have applied the language use called *caretaker speech* (Krashen 2002). Step by step, along with the child's language development, they will change the pronunciation into the normal pronunciation standard.

In character education, a good educator or role model will undoubtedly understand the problems faced by the learners in understanding and applying character education traits or values expected to become part of their life. A good educator will set a good example of the implementation of the values but he/she will be tolerant towards the learners' shortcomings. Some form of dispensation is therefore given, for example, to children under ten years of age to fast during Ramadan for three hours, a half day, or two thirds of the day instead of a full day.

Another example, a strategy in shaping the character of *generosity* does not require the explanation of the meaning of the word "generous." It is sufficient to ask children to share the food with their siblings, give some sweets to their peers, give some money to the beggars, and put some coins into the alms box, etc. These techniques will have long-lasting impressions on the children and will grow the seeds of care and whole-heartedness in sharing what they have with needy children. At some point in time, by their adolescence, these children will understand that they have practiced the character of *generosity*. However, suppose the parent gives the child a dollar bill to put into the alms box and the child replaces it with a quarter coin and keeps the dollar, the parent must accept the decision on the level of generosity the child considers just.

In the first language acquisition process, children directly hear the language used by their parents or other family members in their daily communication. Children gain the knowledge and ability to use the language holistically including pronunciation, grammar, vocabulary, and the social contexts or situations in which the language is used. This enriches the understanding of the language use as it is not only based on the language utterance but also the condition of the environment, the speaker's facial expressions, the tone accompanying the speech, etc. Children will also slowly learn the differences in meaning of the same expression spoken in different tones. For instance, the expression "Good!" can have different meanings or interpretations when it is spoken in different modes. This expression can mean a compliment, a satire, or a warning. Children will also learn variations in vocabulary use when a dialog involves an elder person, a peer, and a younger person, especially in a culture where age, positions, and relations determine the kind of expressions used. Understanding the interrelation between language and culture is also one of the targeted aspects in character education. The different language use in communicating with an elder person, a peer, and a younger person is what the learners need to model from the role models (e.g. parents and teachers). This is part of the culture, the asset of the nation that we have to preserve (Nugiyantoro 2011).

In the language acquisition process, the figures serving as reference models for the children are those directly interacting with them in daily communication. The children's language will develop in accordance with the examples of the language used by the parents, siblings, caretakers, housemaids, and peers in the neighborhood. When they come to their school age, the role models will shift or increase due to their contact with their teachers and new peers at school. If the family, the neighbors, the community and the school provide them with the ideal language input, they will also gain the ideal language mastery.

The same mode seems to apply to character education. Analogous with Krashen's theory (2013), the nurturance of character education values does not have to be carried out in formal instructions but through their application in the daily life. Formalities usually end in an assessment which generates high scores but low or even zero in implementation. In this regard, Pentcheva & Sopov (2003 p. 52) point out, *"One demonstrates one's ability to swim not by answering questions about swimming but by performing the act."* Hence, character nurturance should be contextual and holistic, following the real life application of the values in the family, school, and community.

Pentcheva & Sopov's (2003) claim may be considered extreme because language teaching experts accept the importance of comprehensible input and the natural processes in language acquisition, but they also recognize the benefit of conscious language teaching and learning, particularly in the environment where the target language is not used in the learners' daily life (Gilakjani & Ahmadi 2011). They further suggest that without conscious and purposeful teaching and learning processes or the provision of artificial environment of language use such as video or television programs, the learners will not be able to access the language input. Besides, the teaching and learning process in the form of *awareness raising*, such as *grammatical awareness raising*, has proved to contribute positively to the language acquisition, especially for adult learners.

In the context of character education, if formal lectures are needed, the content should reflect real life contexts; there is no need for image building slogans if they do not match the conduct of the role models that can be directly observed by the learners. Wooden (2009) reminds us, "Be more concerned with your character than your reputation, because your character is what you really are, while your reputation is merely what others think you are."

In this era of information technology, image building often becomes a priority for individuals, organizations, or institutions; however, it must be based on reality. Institutional slogans which are not founded on commitment for realization will disadvantage the institution itself in the long run. The slogan of "Say NO to corruption" popularized by some public figures and celebrities has tragically sent many of the role players to jail. One by one they are trapped by the Act to which they have contributed in the bill drafting and sanctioning. Therefore, educators should not easily coin a slogan when there is no guarantee that it will be realized, or if there is no clear sanction for those acting against the slogan.

The past failure in nurturing the noble values of *Pancasila* (the Indonesian Five Basic Principles) through the P4 (*Pancasila* nurturing) program to the young generations was, among others, due to the scarcity or even the absence of the noble values within the public figures or officials who were expected to be role models setting examples for the laymen. The P4 lectures at university level consisted of textual analysis of the materials with which the participants were familiar, as they had already participated in the same program over and over in the junior and senior high schools. Furthermore, the available real life examples were the noble value violations by public figures or those who were expected to be role models. As a result, during the question and answer sessions, the instructors were often confronted with the participants' non-academic questions which were difficult to answer. Both the participants and the instructors actually knew the problems and the answers because they centered round the discrepancy between the program content and the day-to-day living examples which started to find exposure in the mass media. It was not uncommon (that time and even lately) for traffic violators to settle their problems through negotiated bribery (Ganjarsetia 2011). There could be many other examples out there.

Educators do not expect the failure in the P4 program will be repeated in character education or other related programs. Therefore, the strategic position of the role models should be maximally utilized in day-to-day living in the family, school, and community. The strategic position of the role models should become the main pillar of character education. Our young generations need role models of noble value application, not highly packaged programs but away from practical applications of the targeted values.

In addition, the language acquisition theory reminds us that "optimal input is interesting and/or relevant" (Krashen 2009 p. 66). Therefore, the language elements the learners are expected to acquire should also be presented in an interesting mode, in a relaxed and natural situation, without any constraint or disagreeable attitude, *in a low affective filter* (Krashen 2009) so that the learners pay attention to and unconsciously acquire them. Therefore, for children as well as adults, language is often presented through the media of songs or lullabies.

A similar mode may also be needed for character education. On the one hand, the character traits to be passed on should be relevant to the children's needs, on the other, the values should be presented in an enjoyable manner so that they can be easily understood and internalized. For instance, the obligation to love their parents, their siblings, and other people, is presented in a song "Satu-satu aku sayang Ibu ...." (First, I love mommy) while for higher education students, the song "Heal the World" by Michael Jackson, for instance, may arouse the care for others and the living environment. Ebit G. Ade's songs of the environmental theme and religious chants of Kyai Kanjeng are meant to socialize the message to the public through enjoyable lyrics and melody so that they can be easily understood, remembered, and implemented in day-to-day living.

Table 1. The compatibility of the roles of role models in LA and CE.

Aspect	Language acquisition	Character education
Function	Parent, caretaker, sibling, teacher	Parent, caretaker, sibling, teacher, manager, leader, public figures
Input	Comprehensible language input	Comprehensible targeted character traits
Activity	Involving learners' in language use	Involving learners in practicing the targeted character traits
Constraint prevention	Lower affective filter (e.g. caretaker's speech)	Tolerating learners' performance at early stages
Method	Humanizing teaching technique	Humanizing nurturing technique

4 THE COMPATIBILITY BETWEEN THE ROLE MODELS IN LANGUAGE ACQUISITION AND CHARACTER EDUCATION

Upon examining the discussion in the previous sections, it becomes clear that role models in language acquisition must be present to provide comprehensible input in an anxiety-free situation. Similarly, character education also requires role models who set good examples representing the targeted character traits. This may sound extreme but, while recognizing the benefit of lecturing, modeling should be the key approach to character education as it provides direct exposure to conditions facilitating the acquisition of the nurtured values (Sugirin 2011). This is in line with Lickona's statement (Nucci & Narvaez 2008) that without the direct involvement of cognition, affection, and real actions, character education will not be effective. What makes the acquisition of the targeted values easy is the consistency of the role models who insure the oneness of their words, attitudes and actions. As Nucci & Narvaez (2008 p. 115) point out, if we expect our young generations to have noble character, while the precondition of the noble character is a good role model, then the parents, the caretakers, and the teachers themselves must set examples for the good character traits. This is in accordance with the policy of the Peraturan Mentri Pendidikan Nasional Indonesia (2010 p. 8) which emphasizes the need for consistent role modeling and habituation in creating a conducive environment to facilitate character building. In order to facilitate acquisition, consistency is also needed in language acquisition, particularly in input provision to avoid learners' confusion.

The points discussed in this section and the previous ones have shown that there are aspects in the language acquisition (LA) which are compatible with those in character education (CE). The compatibility can be illustrated in the following table.

5 MAXIMIZING SUCCESS THROUGH MONITORING THE PRACTICE

Language acquisition theories suggest that language acquirers will master the target language maximally if they are not in the defensive position. Stevick (Krashen 2009 p. 73) reminds us that the defensive position will happen if the acquirers feel they are being tested or their weaknesses being revealed. This is in line with the technique of error correction. The best way to treat learners' errors is by giving them the correct models of the language use (Brown 2007) not by pointing out errors. The more the learners are exposed to the correct examples, the more chance they will internalize the correct examples observed, and the greater possibility they will use them in line with the context of use given.

Similar phenomenon will most probably apply to character education. Adolescents and university freshmen are usually sensitive towards issues related their prestige and self-identity. Comparing the work ethos of the older generations with that of the present one can be a source of defensiveness on the part of the students. They were born and brought up in an era with the living conditions different from the older generations. It is unfair to expect them to think and live the way the parents and the previous generations did in their youth. Understanding on the part of parents and educators on the nature of the lives of the young, the school and university students, which are different from theirs, will make them feel comfortable interacting and communicating with the older generations. Parents and educators do not need to show the weaknesses of the young generation as they need only good models of language, attitudes, and actions reflecting character education values such as honesty, discipline, fairness, respect for others, autonomy, responsibility, etc. Should showing what the older generations have achieved be needed, it could be done sparingly without any pretention whatsoever.

In this regard, Margalit (Nixon 2008 p. 131) reminds us that *"... a civilized society is one whose members do not humiliate one another."* Although someone has done something wrong, he or she does not deserve humiliation. Feeling guilty is painful already, while being humiliated turns the pain into an offence. The best thing to do is in setting good examples. If students are exposed to correct models of truths and noble values, they will slowly but surely leave the wrong doings and model the truths and values observed in their interaction and communication with the elders.

It is further stated in the theory of language acquisition that in order to encourage learners to develop comprehensible input consistently, Krashen (2009) suggests providing tools to help students obtain more input. Chances to get maximum input will happen if the language acquirers are provided with opportunities to communicate using the target language in various contexts which enable the teachers to spot language deviations in their communication. Through this way, chances are open for the teachers to provide correct and appropriate language models so that the language acquirers will use their *"monitor"* (comparing the language rules with their language use) so that they can make efforts to correct their performance errors.

A similar mode can be implemented in character education. To find out whether the target groups have developed the targeted character traits, they can be given a role to play or a task to complete. Observation on the individual or group performance will reflect whether the intended values have been implemented. What they need next is positive reinforcement in the form of compliment, appreciation, and encouragement to improve the performance. When the individuals or the groups have not reflected the intended values, the role models can clarify the targeted traits and provide more examples on various opportunities and settings with the expectation that they develop deeper awareness of the values to implement. This is in line with the suggestion that input frequency affects the success of language acquisition (Ellis & Collins 2009).

Koellhoffer (2009) suggests that people of good character know when they need to reveal all the truths and when to save some for themselves. In language teaching and learning, teachers understand all the errors the learners have made, but they do not have to reveal them all. Without referring to whoever has made the errors, they can show the crucial ones for the purpose of awareness raising and classroom remedy. Many teachers may be impatient using this mode of error treatment, but good role models always believe that providing correct models is more effective than showing errors. Showing errors may humiliate the learners, create defensive attitudes, and make them hate the teachers; setting good examples will motivate them to monitor their own performance. This will make the learners feel safe, comfortable, protected and tolerated, which will, in turn, make them grow the feeling of respect and loyalty towards the role models, develop commitment to actions, and model them for their interactions with peers and their future learners.

6 CONCLUSION

From the issues and discussions above, it can be concluded that principles of the theory of language acquisition can be applied in character education. Among the principles is the importance of *comprehensible input* (in language and character traits). The intended value can be easily understood as they can observe, feel, and experience its application by the teachers or the role models in day-to-day living, not only through verbal lectures or empty slogans. Because the role models display tolerance, learners have strong commitment for implementation and the rate of refusal is low (*low affective filter*). All of these can only happen if the role models reflect the targeted values in their words, attitudes and actions.

REFERENCES

Brown, H. D. 2007. *Principles of language learning and teaching (5th ed.)*. New York: Addison Wesley Longman.

Ellis, N. & Collins, L. 2009. Input and second language acquisition: the roles of frequency, form, and function. *The Modern Language Journal,* 93(3) doi: 0026-7902/09/329-335.

Ganjarsetia, G. G. 2011. Rahasia (umum) bebas tilang (The (public) secret of exempt in trespassing). *Kompasiana*, 18 Juni 2011, retrieved from http://www.kompasiana.com/ge2pro/rahasia-umum-bebas tilang_5500e 535a 333117c6 f5125a3.

Gass, S. M. & Selinker, L. 2008. *Second language acquisition: An introductory course*. Third edition. New York: Routledge.

Gilakjani, A. P. & Ahmadi, S. M. 2011. Role of consciousness in second language acquisition. *Theory and practice in language studies*, 1(5): 435-442.

Koellhoffer, T. T. 2009. *Character education: Being fair and honest.* New York: Infobase Publishing.

Krashen, S. D. 1981. *Second language acquisition and second language learning.* New York: Pergamon Press Inc.

Krashen, S. D. 2009. *Principles and practice in second language acquisition.* First internet edition. Pergamon Press Inc.

Krashen, S. D. 2013. *Second language acquisition theory, applications, and some conjectures.* Cambridge: Cambridge University Press.

Lee, S. Y. 2005. How robust is in-class sustained silent reading? *Studies in English Language and Literature, 15, 65–76.*

Lickona, T. 1991. *Educating for character: How our schools can teach respect and responsibility*. New York: Bantam Books.

National Council of Teachers of English. 2006. *The role of English teachers in educating English language learners (ELLs)*. Urbana, IL: NCTE.

Nixon, J. 2008. *Towards the virtuous university: the moral bases of academic practice.* New York: Routledge.

Nucci, L. P. & Narvaez, D. (Eds.). 2008. *Handbook of moral and character education.* New York: Taylor & Francis.

Nuh, M. 2017. *Pendidikan karakter untuk kemandirian bangsa menuju kejayaan Indonesia 2045* [*Character education for the nation's independence to the glory of Indonesia 2045*]. Retrieved from https://www.uny.ac.id/fokus-kita/prof-dr-ir-kh-mohammad-nuh-dea.

Nurgiyantoro, B. 2011. Wayang dan pengembangan karakter bangsa [The puppet and development character of nation]. *Jurnal Pendidikan Karakter XXX (1):* 18-34.

Pentcheva, M. & Sopov, T. 2003. *Whole language, whole person. A handbook of language teaching methodology*. Viseu: Passagem Editores.

Pionnah, J. 2008. Free voluntary reading and the acquisition of grammar by adult ESL students. *International Journal of Foreign Language Teaching, 4(1), 20–24.*

Sugirin. 2010. Affective domain development: reality and expectation. *Cakrawala Pendidikan* XXIX(3): 267-279.

Williams, J. & Leung, J. 2011. *Unconscious language learning.* retrieved from http://www.cam.ac.uk/research/news/ unconscious-language-learning.

Wooden, J. 2009. *Ten best quotes*. Retrivied from https://id.pinterest.com/pin/28006828913247951/.

Tim Pendidikan Karakter. 2010. *Desain induk pendidikan karakter 2010–2025* [Design of character education education 2010–2025]. Jakarta: Kementerian Pendidikan Nasional.

Yayah Khisbiyah. 2010. Karakter bangsa terbentuk dari karakter warganya [*National character is formed from the character of its citizens*]. *Suara Muhammadiyah,* XCV (18): 11.

Developing social care and hard work characteristics through a service learning program: A case study of a natural resource management course

I.Y. Listyarini & L.D. Handoyo
Universitas Sanata Dharma, Indonesia

ABSTRACT: Currently, the characteristics of social care and hard work among students are in decline. One method that can be used to develop such characteristics in students is Service Learning (SL). SL is a learning method that combines lecture materials with the problems that exist in the community: students apply the knowledge they have acquired in lectures to help solve problems in the community. This method is employed in a Natural Resource Management (NRM) course. This study aims to describe the development of the hard work and social care characteristics through the implementation of SL in this course. This research uses qualitative and quantitative methods, with 76 student participants from the NRM course as subjects. Data is collected using observation, questionnaire, and documentation techniques. The results show that through the application of SL activities in NRM courses, social care and hard work characteristics can be developed in students. This is because students are dealing directly with the problems that exist in the community and they can solve them by direct application of the knowledge they have gained in the classroom.

1 INTRODUCTION

In the 21st century, the rapid development of technology and information has brought many changes in all areas, including education. This development has had positive impacts such as easier access to information from anywhere, anytime, and the development of digital media that increasingly support learning. However, there have also been negative impacts, such as a decline in character of the younger generation, which is evident in the declining spirit of hard work among students. This is indicated by the existence of an instant culture when performing tasks, the emergence of copy-paste culture, a dependence on the internet, and a declining interest in reading among students.

In addition to the declining characteristic of hard work, another characteristic that has also started to decline is social attitude. In the current era, often referred to as the digital era, almost everyone is never without a gadget or smartphone. With smartphones, everyone can access virtually all of the information that exists in any part of the world, and is able to communicate with anyone without distance limitations through the use of social media. The variety of social media makes almost everyone more interested in cyberspace. They spend a lot of time with their smartphones instead of interacting with their immediate family, friends, and surroundings. This causes social attitudes toward each other and the surrounding environment to degrade.

This declining character, especially among the next generation in the form of students, has attracted serious attention from the Indonesian government. Through the Ministry of Education and Culture, it has formulated 18 character values for the nation that will be integrated into learning activities at all levels of education, from early childhood to higher education. The 18 national character values include religious values, honesty, tolerance, discipline, hard work, creativity, independence, democracy, curiosity, the spirit of nationality, love of country, appreciation of achievement, friendship, peace, reading, social care, and responsibility (Listyarti, 2012).

At Sanata Dharma University, a private university in Indonesia, the biology education study program, in particular, has been trying to develop the character of students through academic and non-academic activities. One of the compulsory subjects in the program for a second-semester student is Natural Resource Management (NRM), and one of the learning methods developed in this course is Service Learning (SL). Service learning is a learning method that combines learning materials in the classroom with service activities in the community. With service learning, students can directly apply the science or concept they learn in the classroom to help solve problems that exist in the community. Student character will be developed through all the stages of service learning, which are referred to as IPARD, namely Investigation, Planning and Preparation, Action, Reflection, and Demonstration. From the lectures on NRM, students are expected to be able to master the concepts, principles and application of biological knowledge in the fields of food, health, environment (biological) and biological resources. In NRM lectures, SL is applied to Participatory Rural Appraisal (PRA) materials.

2 STATE OF THE ART

2.1 *Service learning*

The learning method is one of the important aspects of achieving learning objectives. Service learning is a method of learning that has been developed in America since the 1970s (Kezar & Rhoads, 2001). Service learning is experiential learning, in which students experience and are directly involved in activities devoted to human and community needs (Jacoby, 2015), and adheres to Dewey's opinion that a person learns from their experience (Speck & Hoppe, 2004). Service learning provides students with experience through interaction with the community. In this case, service learning integrates community service activities with learning concepts and is reinforced by reflection to enrich the learning experience, caring for the community and encouraging community engagement. This is in line with the definition provided by The Community Service Act of 1990 (Jacoby, 2015):

> *A method by which students or participants can learn and develop themselves through active participation in an organized service activity that is conducted based on the problems encountered in the community; in its implementation there is coordination between primary schools, secondary schools, higher education institutions, or service programs to communities with the communities to be served; integrated in the academic curriculum or the educational component as well as allowing time for students or participants to reflect on the service experience to the community.*

Service learning can also be categorized as contextual learning, where students learn directly from the community. Community context is important as a learning resource in service learning. Armed with the theories they have learned in the classroom, students try to apply the theories to solve problems in society. Service learning enables a learning process in which students use their academic understanding and abilities in various contexts inside and outside the school to solve real-world problems (Curry et al., 2012). Contextual learning also makes learning more meaningful and memorable in students' minds (Johnson, 2014). Thus, students can prepare themselves to be citizens who have a share in solving community problems using their acquired knowledge.

Youth Service America (2011) describes steps in the implementation of service learning, summarized as IPARD, and illustrated in Figure 1. In the Investigation stage students are invited to analyze the context of the problems in the community. In the Planning and Preparation stage, students create a plan of what activities will be done to overcome the problems that exist in the community. In the third stage, Action, students carry out this activity plan directly within the community. At the Reflection stage, students look back at the experience of their action and learn to evaluate the activities that have been conducted. The last stage is Demonstration, where students convey and share the results of their activities and their evaluations (Cahyani et al., 2012).

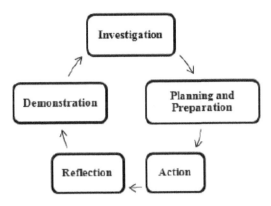

Figure 1. Schematic stages in service learning.

A meta-analysis by Warren (2012) describes the benefits of service learning in improving learning outcomes. This is in line with Weiler's research, which states that there are significant differences in learning achievement between students involved in service learning and those who are not (Weiler, et al., 1998). In addition to the cognitive domain, service learning also has an impact in the affective domain, such as increasing self-esteem and social responsibility, developing tolerant behavior toward cultural differences, developing student leadership capacity, increasing motivation to learn, and developing characteristics such as creativity, responsibility, hard work, and communication in students (Warren, 2012; Kezar & Rhoads, 2001; Handoyo, 2014).

2.2 Social care characteristic

The characteristic of social care is one of the values of humanity (Soenarko & Mujiwati, 2015). The term "social care" is obviously formed from two words: according to the online Indonesian Dictionary, *Kamus Besar Bahasa Indonesia* (KBBI, 2017), "care" means to heed or pay attention to something; "social" means (1) with respect to society or (2) paying attention to the public interest (aid, charity, etc.). On the basis of these meanings, the term "social care" can be interpreted as attentiveness to something happening in society.

Being a caring person means being able to see what others need and doing something to help them. Raatma (2014) states that caring people choose not to be selfish or hurt others. Nowadays, social care has a broad scope, not only caring for oneself and one's immediate family, but also for the surrounding environment, society and the wider world. Soenarko and Mujiwati (2015) state that social care is not just a feeling in the heart, but involves real action: "When seeing people who are victims of disaster or suffering, directly or on television, then saying 'pity', it has not really touched the essence of social concern if it is not followed by an action."

The characteristic of social care involves not only knowing whether something is right or wrong, but being willing to take some action in response. This social caring characteristic is, in fact, a sensitivity of attention that will ultimately lead to an attitude of empathy toward the distress of others. Among others, the characteristic of social care can be seen in three indicators: (1) ability to share in the suffering of others; (2) willingness to give help to alleviate the suffering of others; (3) the willingness to make sacrifices in providing any kind of help in response to the suffering of others (Soenarko & Mujiwati, 2015). The indicators of social care developed in this research were a sense of empathy to the problems of society, and an ability to provide solutions to those problems.

2.3 Hard work characteristic

Hard work is one of the characteristics identified in the character education of Indonesia. According to Listyarti (2012) and Mustari (2011), hard work is a behavior that demonstrates genuine efforts to overcome barriers to learning and tasks, and to complete tasks as well as possible. This definition is consistent with the opinion of Kesuma et al. (2011), which states that hard work is a continuous effort to complete a task. In this case, working hard does not mean completing the task thoroughly and then stopping, but is more directed to a larger vision that must be achieved for the good of people and the environment.

In terms of learning, hard work can be interpreted as a behavior that shows genuine efforts in overcoming barriers to learning and tasks, and complete the task as well as possible (Listyarti, 2012; Mustari, 2011). The hard work indicator used in this research is not succumbing to despair in performing the given tasks.

3 RESEARCH METHODOLOGY

This study uses a research method that combines the qualitative and the quantitative. Qualitative data were obtained from an initial questionnaire and student reflection, while quantitative data was obtained from this initial questionnaire and the final student questionnaire about student perceptions of the development of social caring and hardworking characteristics. The initial questionnaire took the form of an open questionnaire to explore the characters of students before service learning, and the final questionnaire was a closed questionnaire about the perceptions of students in relation to their characters.

The subjects of this research were 76 students participating in the NRM course. Service learning was conducted in the sub-village of Turgo, Pakem sub-district, Sleman regency, Yogyakarta, in May 2017. The target community was Turgo villagers who owned livestock in the form of cattle and goats.

4 RESULTS AND DISCUSSION

4.1 *Implementation of service learning for Natural Resource Management course*

Service learning in NRM courses was conducted in the second semester of 2016/2017. A total of 76 students were divided into 16 groups with a total of four to five students per group. Students undertook service learning, especially in relation to Participatory Rural Appraisal material. In this material, students were invited to provide assistance to the community according to the PRA concept in terms of management of the natural resources that exist around it.

The student groups carried out service learning according to the five previously described IPARD stages, as detailed below.

4.1.1 *Investigation*
The Investigation phase was conducted on 20 May 2017 in Turgo village. At this stage of the process, the student groups carried out observations in the residents' houses to identify the potential and constraints in the management of natural resources, especially in the management of livestock and cattle. The 16 groups of students were distributed among 16 residents who became resource persons.

From the observation, it was found that the Turgo community has the potential in cattle and goats for milk and animal waste. Until now, the milk produced has been sold in raw form to a cooperative at a relatively low price. The selling price for the milk does not cover the cost to the farmers of their care of the livestock, such as feed, medicines, or vitamins. According to residents, cow's milk has previously been processed into instant ginger milk but now such processing has stopped. There needs to be an effort to assist the community in overcoming obstacles in terms of milk management as a natural society resource. The data obtained at the Investigation stage provides the material for planning of the next activity.

4.1.2 *Planning and preparation*
The results of the students' observations are, as indicated above, the basis for the Planning and Preparation stage, which is held in the classroom at a regular time. In this stage, the data collected by the students during the observation provided the data for the class as a whole.

From the collected data, students jointly formulated what activities can be done in conjunction with the community to overcome obstacles in the management of natural resources. Theories that students have acquired during lectures form a basis for the activities they will carry out. From the discussion, it was agreed that the students would focus their assistance activities on cattle and goat farmers, especially in the case of processing milk into more valuable products such as ice cream and yogurt. The choice of processed products is determined by the availability of tools in the community, and students donated two sets of ice-cream-making machines and yogurt makers to community groups as an incentive.

4.1.3 *Action*

The Action stage, in the form of a Participatory Rural Appraisal implementation, was held on 25 May 2017 at the home of Mr. Musimim. This activity was attended by citizens who were members of the cattle and goat breeder groups. At this stage, the student group, together with the associated lecturers, presented the details of their earlier observations. Afterwards, the students started the PRA in conjunction with the community in terms of planning future activities with the input and the provision of tools by the students. PRA is principally a mentoring of, by, and for society. The students act as facilitators for community discussions, and discuss matters related to milk processing plans for yogurt and ice cream with the community.

4.1.4 *Reflection*

In the Reflection stage, students reflect on their experiences during the preceding Investigation, Planning and Preparation, and Action stages. This reflection is guided by questions such as "How did you feel when you first got the task of doing service learning?", "What can you learn from the community in relation to the PRA activities you pursued?" and "What intentions arise in your personal life after the experience of your service learning activities in the community?"

From the students' reports, it was found that most students were afraid, worried, hesitant, and/or curious when they first received the service learning task. Some students felt happy about the forthcoming service learning task because they were pleased to be out of the classroom and gaining more understanding of the course through direct practice in the community.

On the basis of the reflections captured, the service learning activity increases the intention in students to directly apply the science or theory they have acquired in class to the community, as exemplified in the following quotations:

> *Intentions that arise in service learning activities are to take advantage of the ability that has been obtained in the classroom to be applied into the community. The way it is applied is not only done alone, but it can invite communities to utilize the existing natural resources by producing a local product so as not to rely on products from abroad.* (Angela Ivanka Novitasari)

> *More active in socialization activities in the community as well as developing the potentials that exist in society.* (Ghina Salsabila)

> *The intention that came to me after doing the activity is that I want to do a similar activity in my hometown, Manado; there are no cows but many plants like nutmeg and cloves; I want to manage the existing plants and create a new activity that can help the people in my village.* (Maria Ni Luh Rosariana)

Students also feel happy when undertaking service learning, as expressed in the following quotation:

> *My impression when doing the observation is very happy. Because what I get in the lectures I can apply in the community; one of them is that I can directly analyze what resources exist in Dusun Turgo and what are the obstacles faced by citizens directly.* (Refika Yeisa Mukswadini)

In general, students acquire learning from the service learning activities performed. They learn from the community not in terms of cognitive development, but more in relation to the character and behavior of the communities they observed during the activities. Students can also learn from the community, for example, in terms of the hard work of citizens in working and caring for livestock, understanding of the potential of surrounding natural resources, spirit in overcoming the problems of livestock management, and the struggle in life for citizens facing obstacles every day.

Students' understanding of PRA materials is also reinforced by the application of service learning. PRA is an approach and method that allows communities to jointly analyse life issues in order to formulate real plans and policy. In this case, the community plays a direct role as planners, executors, and supervisors of the activities undertaken. In simple terms, the PRA principle is from, by and for the community. In the NRM lectures, students are invited to apply the PRA principle in the Turgo village community. In this case, the student acts as a facilitator in service learning activities conducted directly in the community to explore the potential and problems of natural resource management in Turgo, here in relation to cattle and goats. From the results of these observations, students jointly formulate with the community what activities can be done. The results of student discussions are shared with the community in routine monthly meetings. In the meetings, the students present the observations made and offer alternative activities that can be done within the community to overcome the problems they face. The students and the community discuss these activities together. Through all of these stages, students learn to act as facilitators in PRA activities in the community. From this practice, students really experience the PRA process so that their understanding of PRA material and classroom theory is far more profound.

4.1.5 *Demonstration*
In the Demonstration stage, students document their service learning experience in a personalized report. These reports are used as data in understanding the character development of students as a consequence of undertaking service learning.

4.2 *Development of social care characteristic*

The development of students' social care characteristic was seen in the results of the initial and final questionnaires administered. From the initial questionnaire, it was found that 50% of students feel they have done something for the community. This relates to the activities associated with the observation of water quality in the Boyong river in the village. In the activity, the students observed the condition and the quality of water and then presented the results to the people in the watershed as inputs for future river management. The results from the final questionnaire are shown in Figure 2.

The social care characteristic is seen in terms of two aspects, that is, student involvement and empathy. From the questionnaires it was found that 12.5% of the students reported being very actively involved in offering ideas to help solve problems in the community, with a further 67.1% reporting active involvement. However, 17.8% of the students reported far less involvement, and 2.6% felt uninvolved in contributing ideas. In terms of empathy, 43.4% of

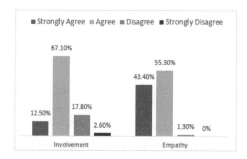

Figure 2.   Increased student involvement and empathy after implementation of service learning.

464

the students reported feeling very empathetic, with a further 55.3% feeling moderately empathetic, and just 1.3% not feeling empathetic to the problems that exist in society.

From the data, it can be seen that after implementing service learning, most students have an increased sense of social care in relation to the problems existing in the community. This sense of social care arises from the students' interactions with the problems that exist in society. Through direct observation, students can observe the real conditions in society, along with the problems faced. This improves students' sense of empathy and gives them the desire to help the community with new ideas and direct engagement. Students have a desire to help people using the knowledge and skills they have acquired in their course.

The social care characteristic is basically an attitude that heeds (shows concern about) something that happens in society. Being a caring person means being able to see what others need and doing something to help such people. Someone who has social concern not only knows about whether something is right or wrong, but shows a willingness to do something to help people in need (Soenarko & Mujiwati, 2015). This is seen in the students after they have carried out service learning. Besides the empathetic attitude exhibited, it also appears in the creative ideas the students offer; for example, the processing of cow's milk into products such as ice cream and yogurt.

In addition to the questionnaire results, students' social caring character also appears in their reflections, as seen in the following quotations: "I feel sad because there are various problems that arise in the citizens" (Ghina Salsabila); "The intention that comes to me is a hard work effort ... In addition, I also want to see high caring in the community so as to seek solutions when faced with obstacles" (Isidorus Purnama Jaya).

4.3   *Development of hard work characteristic*

In this research, the hard work characteristic focused on the ability not to easily succumb to despair in carrying out the given tasks. The results of the preliminary questionnaire indicate that during the course of NRM, hard work is needed, especially when doing tasks, making reports, carrying out observations, and undertaking the exam.

The final questionnaire data is shown in Figure 3. The results showed that 23.7% of students felt that the existence of challenges such as lecturing tasks did not decrease their spirits at all in terms of implementing service learning, and a further 58.5% of students felt other duties did not lower their spirits. By contrast, 16.45% of students reported that many duties lowered their spirits, and 1.3% of students reported that many tasks in implementing service learning greatly lowered their spirits.

Kesuma et al. (2011) state that hard work is the ongoing effort in seeing a task through to completion. Student participants of the NRM course showed the characteristic of hard work during the implementation of service learning. On the basis of the results of the final questionnaire, most (82.2%) students felt that tasks from other courses did not lower their spirits in implementing service learning. The observation also shows that students can carry out the service learning tasks well and in accordance with the schedule that has been determined. Busyness and other tasks did not cause students to neglect the service learning tasks. Students can manage their time effectively so that service learning activities can be done. This shows that students already possess the hard work characteristic, which must continue to be developed.

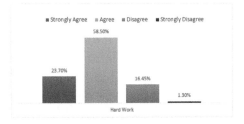

Figure 3.   Hard work characteristic of students after implementation of service learning.

5 CONCLUSION

From the results it can be concluded that service learning can develop students' social care and hard work characteristics. The characteristic of social care is visible in students' ability to show empathy, and in the efforts by students to offer their ideas and involvement in giving assistance to society through Participatory Rural Appraisal in relation to the problem of milk processing. The hard work characteristic appears in the efforts made by students to implement and complete service learning activities effectively, in accordance with a predetermined schedule.

REFERENCES

Cahyani, V.A., Santosa, S. & Indrowati, M. (2012). Pengaruh penerapan service learning terhadap hasil belajar biologi siswa kelas XI SMA Negeri 1 Boyolali tahun pelajaran 2011/2012 [The influence of service learning toward biology learning achievement of XI degree students at SMA Negeri 1 Boyolali in academic year 2011/2012]. In *Biologi, Sains, Lingkungan, dan Pembelajarannya dalam Upaya Peningkatan Daya Saing Bangsa, Seminar Nasional IX Pendidikan Biologi FKIP UNS [Biology, science, environment, and learning in efforts to increase the competitiveness of the nation, National Biology Education Seminar IX, FKIP, UNS]* (pp. 76–83). Retrieved from http://eprints.uns.ac.id/12281/1/1027-2406-1-SM.pdf.

Curry, K.W., Wilson, E., Flowers, J.L. & Farin, C.E. (2012). Scientific basis vs. contextualized teaching and learning: The effect on the achievement of postsecondary students. *Journal of Agricultural Education, 53*(1), 57–66.

Furco, A. & Root, S. (2010). Research demonstrates the value of service learning. *The Phi Delta Kappan, 91*(5), 16–20.

Handoyo, L.D. (2014). Menumbuh kembangkan karakter mahasiswa melalui service learning program di mata kuliah ilmu gizi dan kesehatan [Growing student character through service learning program in nutrition and health sciences course]. *Jurnal Kependidikan Widya Dharma, 25*(2), 183–200.

Hudson, C.C. & Whisler, V.R. (2008). Contextual teaching and learning for practitioners. *Systemics, Cybernetics and Informatics, 6*(4), 54–58.

Jacoby, B. (2015). *Service-learning essentials: Questions, answers, and lessons learned.* San Francisco, CA: Jossey-Bass.

Johnson, E.B. (2014). *Contextual teaching and learning: Menjadikan kegiatan belajar-mengajar mengasyikkan dan bermakna [Contextual teaching and learning: Making learning activities fun and meaningful].* Bandung, Indonesia: Penerbit Kaifa.

KBBI. (2017). *Kamus Besar Bahasa Indonesia [Indonesian dictionary].* Retrieved from http://kbbi.web.id/sosial.

Kesuma, D., et al. (2011). *Pendidikan karakter: Kajian teori dan praktik di sekolah [Character education: Theory and practice at school].* Bandung, Indonesia: Remaja Rosdakarya.

Kezar, A. & Rhoads, R.A. (2001). The dynamic tensions of service learning in higher education: A philosophical perspective. *The Journal of Higher Education, 72*(2), 148–171.

Listyarti, R. (2012). *Pendidikan karakter dalam metode aktif, inovatif, dan kreatif [Character education in active, innovative, and creative methods].* Jakarta, Indonesia: Erlangga.

Mustari, M. (2011). *Nilai karakter: Refleksi untuk pendidikan karakter [Character value: Reflection for character education].* Yogyakarta, Indonesia: Laksbang Pressindo.

Raatma, L. (2014). *Caring.* Ann Arbor, MI: Cherry Lake.

Soenarko, B. & Mujiwati, E.S. (2015). Peningkatan nilai kepedulian sosial melalui modifikasi model pembelajaran konsiderasi pada mahasiswa tingkat I program studi PGSD FKIP Universitas Nusantara PGRI Kediri [Increasing the value of social concern through modification of learning model consideration for level I students of PGSD study program, FKIP, Universitas Nusantara PGRI Kediri]. *Efektor, 26*(April), 33–47.

Speck, B.W. & Hoppe, S.L. (2004). *Service-learning: History, theory, and issues.* Westport, CT: Praeger.

Warren, J.L. (2012). Does service-learning increase student learning? A meta-analysis. *Michigan Journal of Community Service Learning, 18*(2), 56–61.

Weiler, D., LaGoy, A., Crane, E., and Rovner, A. (1998). *An Evaluation of K-12 Service-Learning in California: Phase II Final Report.* Emeryville, CA: RPP International with The Search Institute.

Youth Service America. (2011). *Semester of Service Strategy Guide Revised for 2011.* Washington, WA: YSA Committed.

Design of a learning environment for children's basic character development

R. Mariyana, A. Listiana & B. Zaman
Universitas Pendidikan Indonesia, Indonesia

ABSTRACT: The objective of this research is to obtain an overview of the impact of a learning environment structuring on children's basic character development using the teaching pyramid model based on the Indonesian Curriculum 2013. The approach used in this research was the Research and Development (R & D) method of Borg and Gall (1983). The research was conducted in Bandung City, Indonesia, using a descriptive qualitative approach. Data collection techniques were questionnaires, observations, documentation, and literature studies. Data analysis was done qualitatively. The results show that the design of a conducive and integrated learning environment can support children's basic character development through the application of the teaching pyramid model based on the 2013 curriculum.

1   INTRODUCTION

Kindergarten-age is a sensitive period. Children start to be aware of efforts on the development of their potential. This sensitive period is the maturation period of physical and psychological functions that are ready to respond to the stimulation provided by the environment. This is the time to lay the early foundation for developing physical, cognitive, language, emotional, social, self-concept, discipline, self-reliance, artistic, moral and religious values. One of the objectives of applying the Indonesian Curriculum 2013 is to develop social and spiritual aspects as the basic characteristics of the child. The curriculum also implies that elementary school children should receive a greater share of social behavior aspect development (80%) than knowledge aspect development (20%).

The teaching pyramid model is a model developed by Fox et al. (2003) aimed at improving children's social competence. It contains four stages of activity, namely building positive relationships, creating a conducive environment, teaching emotional social skills and providing individualized services. These four activities are grouped into three strategies, namely preventive, development, and curative. The teaching pyramid model is adapted and developed, in order to become a model that can develop children's basic character tailored to the Indonesian Curriculum 2013.

The learning environment design is one of the key factors of success in building the ability and behavior necessary for developing a child's basic character. The implication is that the environment for children should be prioritized, especially the indoor environment (Mariyana et al., 2010). The environment and indoor spaces should be intimate and entertaining, reducing the transition from the home atmosphere to the early school setting. This involves complementing the physical environment with soft furnishings, within small and quiet rooms. As Beckley (2012) states, the best room arrangement eases the transition by serving all children. Furthermore, an integrated outdoor learning environment is also one of the ways in which teachers can encourage children's activities by stimulating their curiosity, inquiry and exploration, and it provides a number of sensual experiences for children to encourage them to use all their senses in a safe way (Beckley, 2012).

2 LITERATURE REVIEW

Every child is sensitive to receiving any stimulus as long as their psychological and physical functions are ready for the stimulus provided by the environment. A learning environment can increase various developmental stages of children optimally if it is designed carefully and properly. Thus, specific planning and environment selection should be considered seriously. On the other hand, the learning environment is one of the key factors in developing children's capability and behavior. Thus, the implication is that children's learning environment should be prioritized.

Derived from the words "environment" and "learning", learning environment could be interpreted as place or condition that could influence human behavior. From this explanation, it can be inferred that changing the environment could have a relatively permanent impact on behavior. It means that a stronger environmental influence leads to greater predicted change in the subject. It proves how significant environmental influence can be on children's learning behavior. That is why it will be very good for teachers to show environmental influence in the development of individuals, especially children (Mariyana et al., 2010).

The indoor environment is very important for children. It should be familiar, comforting, and it should ease the transitions from home care to early year settings. This may involve applying the physical environment with soft furnishing within small rooms and quiet spaces. The best setting might ease transition by catering for all children (Beckley, 2012).

In order to get ideal classroom, we have to pay attention to its setting. An indoor playing room for children is commonly a large square room with some dividers that separate one area from another. In every corner of the room, there is usually a storage box for certain items that could be used in the activities.

3 RESEARCH METHOD

The approach used in this research was the Research and Development (R & D) method of Borg and Gall (1983) because the aim was to develop a model that can improve the social and spiritual characters of children as regulated in the 2013 Early Childhood Education curriculum. The development of the model was carried out by adapting the teaching pyramid model developed by Fox et al. (2003), while the R & D approach is a research method employed in ten steps, including: (1) research and information collecting; (2) planning; (3) development of preliminary form of product; (4) preliminary field testing; (5) main product revision; (6) main field testing; (7) operational product revision; (8) operational field testing; (9) final product revision; (10) dissemination and implementation. The ten steps can simply be categorized into four stages of activity, namely: (1) preliminary study; (2) model planning and development; (3) trial and revision; (4) model validation.

4 RESULT

The results show that the design of the learning environment arrangement is effective in developing children's basic characters using the teaching pyramid model based on the Indonesian Curriculum 2013. Hence, the learning environment and classroom within the school environment have to be designed and organized properly. The considerations for setting up the learning environment are:

1. Providing a reward board that contains the name of every child.
2. Drawing a circle on the carpet for the child to sit in during circle time.
3. Drawing a line to help children line up.
4. Writing the name and putting a photo of the child on each of their chairs.
5. Writing the name and putting a photo of the child on their locker.
6. Writing the name and putting a photo of the child on their backpack storage.
7. Writing the name and putting a photo of the child on their shoe storage.

Figure 1.   The teacher and the children posting a list of rules together.

Figure 2.   Learning equipment arrangement based on children's names.

Example arrangements of the learning environment are shown in Figures 1 and 2.

The implementation of providing social and spiritual learning is designed into sections (teaching, displaying posters, and providing the means and media that enable children to be healthy, confident, independent, orderly and disciplined to the school's daily rules; teaching healthy living, honesty, courtesy and faith (*aqidah*), and timely worship; teaching creative and aesthetic attitudes through activities; teaching adjustments and self-control). The activities provide spiritual social teaching, such as teaching and displaying posters about:

1. Being tidy and clean at school.
2. How to enter a room.
3. How to store shoes correctly.
4. How to store backpacks correctly.
5. How to learn in the classroom.
6. Ordinance using indoor gaming tools.
7. How to use outdoor gaming tools.

5   DISCUSSION

Character education is a process that involves instilling knowledge, love and behavior that will become habit. Character education cannot be separated from the basic values that are considered beneficial for early childhood education. It is very important that these values are to introduced, exemplified and internalized into the children's behavior. Instilling character values in children at an early age encompasses the components of knowledge, awareness or willingness, and the actions to implement those values toward God, self, others and the environment in order to become perfect human beings. Character development is taught, mastered, and implemented by children in their everyday life through the introduction of norms or values and the internalization of actions in real life every day.

The teaching pyramid model is a model developed by Fox et al. (2003) that is aimed at developing children's social competence. Based on the author's study of the model, the following discussion is presented. The teaching pyramid model is specifically designed to improve social competence and prevent deviant behavior of children aged two to five years old. The model is a comprehensive model because it uses various interventions, including giving examples by teachers as role models, designing a conducive environment, teaching social and emotional skills, and providing individualized services. In addition, the activities of the model are carried out systematically or gradually from the first level, to the second level and to the third level. Each level has the following activities and objectives. The first level builds positive relationships and designs a supportive environment as preventive acts, intended for all children. The second level teaches social skills and emotional control; the activities include developmental activities aimed at children who are likely to have social and emotional development difficulties. The third level is curative in the form of providing individual services for children who, after given the first and second level, have not shown the expected behavior.

Research conducted by Listiana (2011) showed that the teaching pyramid model is effective in improving social competence. This means that children who get exposure to the teaching pyramid model show a better behavior in sharing, helping, empathizing and controlling emotion. The results also found that generally, children's behavior in sharing, helping, empathizing and controlling emotion is still at a moderate level.

The results of the research of Feil et al. (2015) illustrated the adaptation of pre-school intervention programs as the first steps of early intervention. Participants, staff, and professionals chose appropriate interventions to address the emergence of antisocial behavior and externalization behavioral disorders prior to school entry (Feil et al., 2015).

The teaching pyramid model to support social-emotional competence in infants and teenagers (Fox et al., 2003) is a positive intervention and supports behavior that early educators can use to promote the youth's social and emotional development, and to prevent and overcome destructive behavior.

Discussions related to the arrangement of a conducive learning environment have an impact on the development of children's basic character. Mariyana et al. (2010) advise that some drawings and learning symbols are displayed in every corner of the room to describe various activities that occur. For example, reading is usually done in a learning area called "the language area," and drawing activity in "the art area". The principle is that these activities are carried out in areas already designed according to their needs (Mariyana et al., 2010). In this case, the arrangement of the learning environments, both indoor and outdoor, are designed and arranged in such a way and given the rules and related important symbols and usage procedures in order to improve the children's basic character, especially related to discipline. Each design of the learning environment is placed in accordance with the purpose and children's needs to facilitate the development of their basic character.

The strategies in the teaching pyramid model are applied using following techniques:

5.1 *Establish positive relationships with children, family and colleagues*

Establishing positive relationships with children is achieved by carrying out activities that enable children to have a positive self-concept, to have self-confidence, and to encourage them to develop a sense of security. The activities include always giving a positive response, being friendly, gentle, and being a role model for children.

A positive relationship between the children and the teacher is needed by the children in order to improve positive behavior. In this case, Webster-Stratton (in Joseph & Strain, 2010) states that with a relationship built on a sense of trust, understanding, and caring, the children will display a cooperative and energetic behavior. Positive relationships also provide important protection from the likelihood of misbehavior, because children who feel well will behave well. Corsini (2002) states that children who have a sense of trust in the teacher, will take the teacher's speech and behavior as examples, hence it is easy for teachers to direct the child's behavior.

5.2 *Designing a supportive environment*

Creating a supportive environment means that the teacher creates a learning environment. It lets the children know what they should do and provides a way for them to consistently and continuously develop appropriate behavior, engage in activities and obtain feedback. An orderly and calm condition is needed to reach the learning objectives. The Washington State Office of the Superintendent of Public Instruction (OSPI) indicates a supportive learning environment as one of the nine characteristics of a successful school criterion (McClellan and Katz, 2001).

5.3 *Using social and emotional teaching strategies*

Social-emotional teaching is concerned with learning about difficult behaviors such as communicating emotions effectively, controlling anger, and tackling social problems. Social-emotional teaching becomes very important; most children do not have many problem-solving skills because they are still involved in parallel games, which means that they generally do not have much experience to solve conflicts (Beckley, 2012). Social skills that should be taught to children are learning to wait for their turn, and helping them to understand and resolve conflict (Fettig & Barton, 2014).

5.4 *Provide individual intervention*

Individual intervention is aimed at children who constantly display misbehavior. Individual intervention is planned based on an understanding of students' behavior. Children's behavior is observed, analyzed and it is then planned how the development and intervention will be carried out to overcome the particular behavior.

In line with the results of the aforementioned research, Fettig and Barton (2014) applied intervention-based functional assessment, and discovered that this intervention successfully reduced challenging behavior and increased appropriate behavior in children.

6 CONCLUSION

Based on the implementation model, the following results are obtained: (1) the design of the effective learning environment can improve the basic character of children; (2) the implementation of the teaching pyramid model influences teachers' understanding and ability; the teachers' and principal's opinion and responses are generally positive as they stated that the teaching pyramid model is beneficial and has a positive impact on changes in children's behavior; (3) there is a change of attitude and behavior of children after the teaching pyramid model implementation, especially on the characteristics of discipline and independence.

Children's character development has become the main focus of current education, especially based on the curriculum of 2013. This is due to the current condition of learners who are considered to not have the ability to show politeness and virtuous behavior as part of their basic character. In general, the use of the teaching pyramid model as children's basic character development model based on the 2013 curriculum aims at obtaining a model or character development strategy that can be implemented in a planned, systematic, sustainable and comprehensive manner. Moreover, the teaching pyramid model has four development strategies: (1) building positive relationships with children and families; (2) creating a supportive environment; (3) using character education teaching; (4) providing intensive individual services. The focus of the character development refers to the 2013 curriculum, in which the development of social and spiritual characters serves as the basic character.

ACKNOWLEDGMENTS

Gratitude is extended to the Directorate of Research and Community Service, and the General Directorate for Research Reinforcement and Development of the Ministry of Research, Technology and Higher Education who have funded this research in accordance with the Decree of General Director of Research Reinforcement and Development No. 30/EKTP/2017 of 3 April 2017 on "Higher Education Research Funding Receiver of State Owned University in Fiscal Year 2017".

REFERENCES

Beckley, P. (2012). *Learning in early childhood.* London, UK: Sage.

Borg, W.R. & Gall, M.D. (1983). *Educational research: An introduction* (5th ed.). New York, NY: Longman.

Corsini, R.J. (2002). *The dictionary of psychology.* New York, NY: Routledge.

Feil, E.G., Frey, A., Walker, H.M., Small, J.W., Seeley, J.R., Golly, A. & Forness, S.R. (2015). The efficacy of a home-school intervention for preschoolers with challenging behaviors: A randomized controlled trial of preschool first step to success. *Journal of Early Intervention, 36*(3), 1–20. doi:10.1177/1053815114566090

Fettig, A. & Barton, E.E. (2014). Parent implementation of function-based intervention to reduce children's challenging behavior: A literature review. *Early Childhood Special Education, 34*(1), 49–61. doi:10.1177/0271121413513037

Fox, L., Dunlap, G., Hemmeter, M.L., Joseph, G.E. & Strain, P. (2003). The teaching pyramid: A model for supporting social competence and preventing challenging behavior in young children. *Young Children, 58*(4), 48–52.

Joseph, G.E. & Strain, P.S. (2010). *Promoting children's success: Building relationships and creating supportive environments* Nashville, TN: The Center on the Social and Emotional Foundations for Early Learning. Retrieved from http://csefel.vanderbilt.edu/modules/module1/script.pdf

Listiana, A. (2011). *Layanan Bimbingan Dengan Menggunakan Model Pembelajaran Piramid Untuk Meningkatkan Kompetensi Sosial Anak: Studi Kuasi Eksperimen pada anak-anak TK di Kecamatan Sukasari Bandung Tahun Ajaran 2010–2011* [Guidance Services Using Pyramid Learning Models to Improve Child Social Competence: Quasi-Experimental Studies of Children on kindergarden in Sukasari, Bandung Year 2010–2011]. Program Pasca Sarjana Universitas Pendidikan Indonesia.

Mariyana, R., Nugraha, A. & Rachmawati, Y. (2010). *Pengelolaan lingkungan belajar* [Managing learning environment]. Jakarta, Indonesia: Prenada Media.

McClellan, D.E. & Katz, L.G. (2001). *Assessing young children's social competence.* Goodyear, AZ: At Health, Inc. Retrieved from https://athealth.com/topics/assessing-young-childrens-social-competence-2/.

The effect of science fair integrated with project-based learning on creativity and communication skills

I.D. Tantri & N. Aznam
Universitas Negeri Yogyakarta, Indonesia

ABSTRACT: This research aims to reveal the effect of science fair integrated with project-based learning on students' creativity and communication skills. This research used quasi-experimental design with pre-test-post-test non-equivalent control group design. The population in this research were all grade eight students of a private junior high school in Yogyakarta, Indonesia (namely SMP 3 Muhammadiyah). Samples of this research were class VIII F students as experimental group and class VIII E students as control group. Experimental group implemented science fair integrated with project-based learning, while control group implemented conventional learning. Samples were selected by means of cluster sampling technique. The data was collected through test and non-test. Data of creativities skill was collected by using test and the data communication skills were collected by using observation sheet. The data was analyzed using multivariate analysis of variance (manova). The results indicated that science fair integrated with project-based learning has significant effect on creativity and communication skills based on manova with sig. $0.000 < 0.05$.

1 INTRODUCTION

Education is an attempt to create a learning atmosphere and learning process to develop the students' competence that includes knowledge, skills, and attitudes. These competencies must be developed in accordance with the demands of the times. Students are expected to have some complex competencies to deal with various problems faced today and in the future. One of the problems facing Indonesian society is the use of food additives that are not safe for health. The results of Syah et al. (2015) showed that the problem of food and snacks at school was dominated by chemical contamination problems due to the misuse of hazardous chemicals in food. Therefore, the problem of school children's snacks needs to be fixed.

The problem of school children's snacks can be solved by optimizing science learning, especially on "Additive Substance" theme. The theme can be used to increase students' understanding of the harmful use of forbidden food additives while stimulating students' creativity in creating solutions and communicating the solutions. But, the tendency in the field showed that science learning was still focused on the cognitive aspect. The learning content presented was limited to the content in the teachers' books and students' books, as well as the content of the national examination. This caused teachers to be less motivated to innovate learning activities and use various learning models. As a result, the students' creativity and communication skills were still low. It could be known based on the teacher's assessment of the student products, the experimental report and the presentation.

Student's creativity and communication skills can be improved using appropriate learning methods. There are several learning methods used in the science learning. One method of science learning is science fair. Science fair is an activity that facilitates students to exhibit and communicate their project results. According to Grote (1995), science fair enhanced student's enthusiasm for science and provided experience in communication skills. In addition, Barry & Kanematsu (2005) research showed that science fair provided an opportunity for students to develop creative thinking skills. In line with these two studies, Smith's (2013) research results also showed that science fair could be used to develop language skills, writing skills, and higher order thinking. Based on the these results, it can be concluded that science fair activities can be used as an alternative solution to improve the creativity and communication skills of students.

Science fair activities will be implemented optimally if integrated with an appropriate learning model. Bencze & Bowen (2009) stated that science fair is the activities of students exhibiting and discussing project summaries. Therefore, the appropriate learning model for applying science fair activities is a project-based learning model. This is also reinforced by the results of Mihardi, Harahap, and Sani (2013) research that indicated that student's creativity increased after receiving project-based learning. Meanwhile, according to the Ministry of Education (2013), project-based learning can also improve student's communication skills. Thus, the science fair integrated with project based learning has potential to be applied to improve student's creativity and communication skills. Therefore, it needs research to test the effect of science fair integrated with project based learning on students' creativity and communication skills.

2 LITERATURE REVIEW

2.1 Science learning

Science learning is inseparable from the nature and characteristics of science. According to Carin & Sund (1989), science is a system for understanding nature through collecting data from the results of observation and controlled experiments. In line with the statement, Hewitt et al. (2007) mentioned that science is more than a knowledge, but it is also a method to explore nature. More specifically, Collette & Chiappetta (1994) defined science as a way of thinking in understanding nature, a way of investigating to reveal a phenomenon, and a body of knowledge resulting from the inquiry process. From some of these statements, it can be concluded that science is a knowledge gained by understanding and exploring the universe through the collection of data from the results of observation and experimental activities.

Associated with that meaning, science contains three essential elements. These three elements are processes or methods, products, and attitudes (Carin & Sund 1989). Processes element is a method that includes investigation and observation of a problem to produce science product. The process elements or methods will be related to the scientific attitude when scientists are doing investigation and observation. Therefore, science learning should present concepts as products, and also series methods or processes to develop scientific attitudes.

Science learning will be accomplished optimally if presented contextually and implemented using appropriate models and instructional media. Learning model used should be able to facilitate students to carry out the process in order to produce science products and develop scientific attitudes. Examples of appropriate learning models are problem-based learning, project-based learning, and discovery learning (Ministry of Education 2014). Therefore, teachers are expected to organize science learning with appropriate learning models.

2.2 Food additive substances

Food additive substances is one of the themes in science learning that fits closely with the students' daily life. According to Effendi (2012), food additives are substances that are added to the food in the process of production, packaging or storage with a specific purpose. Based on the source, the additive is distinguished into natural and synthetic additives. The use of synthetic additives is more frequently found. This is because synthetic materials have advantages, which are more concentrated, more stable, and cheaper. However, some additives are carcinogens that can stimulate the occurrence of cancer in animals and humans (Padmaningrum 2009). Given the dangers of food additives abuse, society needs to be educated. This can be done by involving students through science learning on the theme food additive substances. Through science learning, students are expected to create solutions to avoid the danger of food additives abuse. Thus, society can avoid the danger of food additives abuse.

2.3 Creativity

Creativity has been defined in several definitions by some experts. According to Woolfolk (2007), creativity relates to the process of linking the stored knowledge with the problems faced to produce something new. In line with that statement, Santrock (2011) defined creativity as the ability to think about something in an extraordinary way and contains a unique solution to a problem. Referring to some of these statements, creativity can be interpreted as an ability to create something new as a solution to problems through flexible and practical thinking by combining or synthesizing ideas.

2.4 Communication skills

Cook & Weaving (2012) defined communication skills as an ability to express and interpret concepts, thoughts, feelings, facts and opinions either through oral or written (listening, speaking, reading and writing). According to Rezba et al. (1995), the forms of communication can be oral descriptions, body language, numbers, written language, music, data tables, images, charts, concept maps, models, graphs, symbols and maps. Thus, it can be concluded that the communication skills are done to disseminate or share an idea orally or non-verbally.

2.5 Science fair

Science fair is one form of activity in science learning. Science fair is an activity that allow students to deeply study problems and communicate their findings (Collette & Chiappetta 1994). While Bencze & Bowen (2009) mentioned that science fair is the activity of students exhibiting and discussing project summaries that have been done. Referring to these opinions, science fair in this study is defined as an activity that facilitates students to exhibit and communicate project results or findings.

Science fair in science learning has a good effect on the students' skills. The results of Smith's (2013) study showed that science fair can be used to develop language skills, writing skills, and higher order thinking. In addition, projects undertaken outside of school require students to communicate the results of their project activities to all school members through display boards and other materials. Such activities can facilitate students developing communication skills of students (Rillero 2011). Thus, it can be concluded that science fair activities can be used to improve the student's communication skills.

In addition to improving communication skills, science fairs enhance the students' creativity. According to Collette & Chiappetta (1994), the categories covered by the science fair project assessment are creativity, procedure of investigation, understanding concepts, display quality, and oral presentation. Based on these studies, science fair can improve creativity and communication skills. Thus, science fair is an appropriate learning activity to be used in facilitating students to develop creativity and communication skills.

2.6 Project based learning

Science fair activities will be implemented optimally if integrated with the appropriate learning model. Bencze & Bowen (2009) stated that science fair is the activity of students exhibiting and discussing project summary. Therefore, the appropriate learning model for applying science fair activities is a project-based learning model. This is also reinforced by the results of Mihardi et al. (2013) research that showed that students' creativity improved after project-based learning. Meanwhile, according to the Ministry of Education (2013), project-based learning can also improve the students' communication skills.

Project-based learning is a learning model that uses problems as a first step in collecting and integrating new knowledge based on experience in real activity (Ministry of Education 2013). Meanwhile, Patton (2012) defined project-based learning as a learning that puts students to design, plan, and execute an expanded project to produce publications such as products, publications, or presentations. In line with these opinions, Klein et al. (2009) suggested that project-based learning is a learning strategy that provides students with the opportunity to acquire knowledge independently and demonstrate their new knowledge through different types of presentations. The syntax of project based learning according to Ministry of Education (2013) can be described as follows.

1. Start with the essential question
2. Design a plan for the project
3. Create a schedule
4. Monitor the students and the progress
5. Assess
6. Evaluate the experience.

From some of these opinions, project-based learning can be defined as a learning model that allows students to design and implement projects related to real issues so as to produce products or publications in order to gain knowledge.

2.7 Science fair integrated with project based learning

Based on the definition of science fair, it can be known that science fair is closely related to the project. In other words, science fair can be applied in science learning using project-based learning. Therefore, this study uses science fair term integrated project-based learning to state the activities of students planning and executing the project and communicate the results of the project in the fair held in science lesson.

Through its stages, science fair integrated with project-based learning model has potential to be applied to enhanced students' creativity and communication skills.

3 RESEARCH METHOD

3.1 Types of research

This research was a quasi-experiment research using non-equivalent control group design done by giving certain treatment to experimental group and providing control group as comparison.

3.2 Time and place

The research was done in the semester of the year 2016/2017 (August 2016–December 2016) in a private junior high school (SMP Muhammadiyah 3 Yogyakarta).

3.3 Target/research subject

The population in this study was eight grades. The subject of the samples were selected by cluster sampling. The sample were the students of class VIII E and VIII F. The control group was class VIII E and the experimental group was class VIII F.

3.4 Procedure

This research was a quasi-experimental research using non-equivalent control group design done by giving certain treatment to experimental group and providing control group as comparison. The research design can be shown in Figure 1.

Where O1: pre-test of experimental group; O_2: post-test measurement of experimental group; O_3: pre-test measurement of the control group; O_4: post-test measurement of the control group; X_1: treatment of experimental group; and X_2: treatment of control groups. In this study, X_1 symbolized the science fair integrated with project-based learning and X_2 was a symbol of conventional learning.

	Pretest Measure	Treatment	Posttest Measure
Experimental group	O_1	X_1	O_2
Control group	O_3	X_2	O_4

Figure 1. Research design.
Source: (Johnson & Christensen 2000).

3.5 Data, instruments, and data collection

Data was obtained through both test and non-test techniques. Test instruments were used to measure creativity, while non-test instruments were used to measure communication skills.

3.6 Data analysis technique

3.6.1 Creativity test, report assessment sheet, and presentation assessment sheet
The technique of analyzing the creativity test, the report assessment sheet, and the presentation assessment sheet was done by summing the score obtained by students from all points of all aspects.

3.6.2 Hypothesis testing
This study was an experimental quasi-experiment with experimental group and the control group had an unqualified chance on the variable that can affect the dependent variable. The variable that may affect was the pre-test score (initial ability). Therefore, hypothesis testing was done through post-test analysis with pre-test as covariate. Thus, a pre-test analysis was necessary before determining the hypothesis test.

3.6.3 Student ability analysis before treatment
To ensure students' initial ability in the two classes did not differ significantly and was not a covariate variable, multivariate variance analysis (manova) was performed on the pre-test score of creativity and first measurement of communication skills. The analysis was done through a manova test.

3.6.4 Student ability analysis after treatment
Based on the initial ability analysis of students, the initial ability of both groups of samples did not differ significantly (not to be a covariate variable). Therefore, hypothesis testing can be done by using multivariate analysis of variance (manova).

4 RESULTS

4.1 Student ability analysis before treatment

Before testing the research hypothesis, testing is required on the students' ability before treatment. To ensure that the initial ability of the students did not differ significantly, a manova test was performed on the pre-test score of creativity and communication skills at meeting 2.

The manova test was performed to analyze the pre-test score of students on aspects of creativity and skills. The manova test results are listed in Table 1.

Based on the results of Multivariate Tests, it is found that the significance value in Wilks' Lambda is 0.830. Therefore the significance value > 0.05 then Ho accepted. In other words, the initial ability of both groups of samples did not have a significant difference. Based on these results, it can be concluded that the initial ability of students did not meet the requirements to be used as a covariate variable.

4.2 *Student ability analysis after treatment*

Having confirmed that the initial ability of the two sample groups did not differ significantly, the hypothesis tested the students' ability after the treatment was performed manova. The manova test was performed to analyze the students' post-test score on creativity and skill aspects.

Based on the analysis result, Ho is rejected because F = 22,32, sig value 0.000 < 0.05. Thus, there is a significant effect of implementing science fair integrated project-based learning towards the students' creativity and communication skills.

5 DISCUSSION

Science fair is an activity in science learning that allowsstudents to deeply study problems and communicate their findings (Collette & Chiappetta 1994). In order to produce an invention that will be communicated, students do the learning with appropriate learning model. Learning model that can facilitate students to produce an invention is project based learning. Project based learning is a learning model that facilitates students planning and executing of a project.

Science fair integrated project-based learning applied to the treatment class affects the creativity and communication skills of students. The effect of science fair integrated with project-based learning on both aspects was proven through statistical hypothesis testing. Based on the results of data analysis through manova test, obtained sig value < 0.05 so Ho is rejected. Thus, it can be concluded that science fair integrated project-based learning significantly influences the students' creativity and communication skills.

The result was in line with research conducted by Mihardi et al. (2013) which indicated that the students' creativity increases after receiving project-based learning. The research of Kusumaningrum & Djukri (2016) also showed that the learning based on project-based learning model was effectively used to improve the students' science and creativity skills especially in the aspect of creative thinking ability. In addition, ChanLin (2008) research results showed that project-based learning can improve the skills of reporting practicum activities as a form of written communication. In addition to project-based learning, several studies related to science fair also show the same thing. Barry & Kanematsu's (2005) research showed that science fair provided an opportunity for students to develop creative thinking skills when designing and creating displays and posters. In line with this, Smith's research (2013) also showed that science fair can be used to develop language skills, writing skills, and higher order thinking.

The effect of science fair integrated with project based learning on students' creativity and communication skills could not be separated from the learning stages. Initial steps in the form of initial questioning could stimulate students to generate ideas for solving the problem. The first stage triggers the fluency as students are asked to generate as many ideas as possible. In addition to fluency, students are stimulated to develop originality. Originality could be developed because students were asked to generate ideas that are unique or different from existing troubleshooting ideas. The development of creativity in generating ideas was strengthened at the next stages.

At the planning stage, students with a group of friends carry out brainstorming to sort and select ideas that will be implemented and embodied in project activities. In the opinion of Woolfolk (2009), creativity development can be done through brainstorming activities. Planning stages packed through brainstorming activities can facilitate students to develop flexibility and elaboration sub-aspect. Through brainstorming activities, students are trained to think from different points of view so as to stimulate flexibility. From the ideas generated by the group, new ideas will be developed that will be developed into elaboration or ideas. These elaboration sub-projects will continue to be used throughout the project implementation process.

After carrying out the project activities, students make a written report in the form of a display board. This stage trains students to develop written communication skills. In addition to written communication, students can develop creativity as it is required to produce a unique and exciting display board. After the project deadline, students presented the display board and demonstrated the project's results in science fair activities. Through presentations in science fair, students were trained to develop oral communication skills. From a series of learning activities undertaken through the science fair integrated project-based learning, students' competencies on the aspects of creativity and communication skills can be improved. In other words, the results showed that giving treatment gives a significant influence on the students' creativity and communication skills.

6 CONCLUSION

Based on the results of research and discussion, it can be concluded that the science fair integrated project-based learning significantly influence the students' creativity and communication skills. The first stage triggers the students' fluency and originality on generating troubleshooting ideas. At the planning stage, students' with groups of friends carry out brainstorming to sort and select ideas that will be implemented in project activities. Planning stages through brainstorming activities can facilitate students to develop flexibility and elaboration. These elaborations will continue to be used throughout the project. After carrying out the project activities, students make a written report in the form of a display board. This stage facilitates students to develop written communication skills. After the project deadline, students presented the display board and demonstrated the project's results in science fair activities. Through presentations in science fair, students were trained to develop oral communication skills.

These results can be used as a consideration for teachers to apply to other themes, especially on themes that require creativity and communication skills. However, teachers need to do orientation and motivation in advance so that students learn to be stable.

REFERENCES

Barry, D.M. & Kanematsu, H. 2005. *Science Fair Competition Generates Excitement and Promotes Creative Thinking in Japan.* Japan: Gendai Tosho.

Bencze, J.L. & Bowen, G.M. 2009. A national science fair: exhibiting support for the knowledge economy. *International Journal of Science Education* 31(18): 2459–2483.

Carin, A.A. & Sund, R.B. 1989. *Teaching Science Through Discovery.* Columbus: Merrill Publishing Company.

ChanLin, L.J. 2008. Technology integration applied to project-based learning in science. *Innovations in Education and Teaching International* 45(1): 55–65.

Collette, A.T. & Chiappetta, E. L. 1994. *Science Instruction in the Middle and Secondary School.* USA: Macmillan Publishing.

Cook, R. & Weaving, H. 2012. *Key Competence Development in School Education in Europe.* Brussels: European Schoolnet.

Effendi, S. 2012. *Teknologi Pengolahan dan Pengawetan Pangan* [Food processing and preservation technology]. Bandung: Alfabeta.

Grote, M.G. 1995. Teacher Opinions Concerning Science Projects and Science Fairs. *Ohio Journal of Science* 95(4): 274–277.

Hewitt, P.G., et al. 2007. *Conceptual Integrated Science.* San Francisco: Pearson Education.

Johnson, B. & Christensen, L. 2000. *Educational Research: Quantitative and Qualitative Approaches.* Massachusetts: Pearson Education Company.

Klein, J.I., et al. 2009. *Project-Based Learning: Inspiring Middle School Students to Engage in Deep and Active Learning.* New York: Division of Teaching and Learning Office of Curriculum, Standards, and Academic Engagement.

Kusumaningrum, S. & Djukri, D. 2016. Pengembangkan perangkat pembelajaran model project based l earning (PjBL) untuk meningkatkan keterampilan proses sains dan kreativitas [Develop a learning based project model learning kit (PjBL) to improve the skills of science and creativity processes]. *Jurnal Inovasi Pendidikan IPA* 2(2): 241–251.

Mihardi, M., et al. 2013. The Effect of Project Based Learning Model with KWL Worksheet on Student Creative Thinking Process in Physics Problems. *Journal of Education and Practice* 4(25): 188–200.

Ministry of Education. 2013. *Modul Pelatihan Implementasi Kurikulum 2013* [Module of training curriculum implementation 2013]. Jakarta: Kementerian Pendidikan dan Kebudayaan.

Padmaningrum, R.T. 2009. Bahan aditif dalam makanan [Additives in food]. *Pendidikan dan Pelatihan Kesalahan Konsep dalam Materi IPA Terpadu.* Yogyakarta: FMIPA UNY.

Patton, A. 2012. *Work That Matters The teacher's Guide to Project-Based Learning.* San Diego: Paul Hamlyn Foundation.

Peraturan Menteri Pendidikan Nasional Tahun 2014 Nomor 103, Pembelajaran pada Pendidikan Dasar dan Menengah [Regulation of the Minister of National Education 2014 Number 103, Learning on Primary and Secondary Education].

Rezba, R.J., et al. 1995. *Learning and Assessing Science Process Skill.* USA: Kendall/Hunt.

Rillero, P. 2011. A Standards-Based Science Fair. *Science and Children* 48(8): 32–36.

Santrock, J.W. 2011. *Educational Psychology.* New York: McGraw-Hill.

Syah, D. 2015. Akar masalah keamanan pangan jajanan anak sekolah: studi kasus pada bakso, makanan ringan, dan mi 1 [Roots of problem of food consumed by school children safety: case study on meatball, snack, and noodle]. *Jurnal Mutu Pangan* 2(1): 18–25.

Smith, V.L. 2013. *Science Fair: Is It Worth the Work? A Qualitative Study on Deaf Students' Perceptions and Experiences Regarding.* Mississippi: University of Southern Mississippi.

Woolfolk, A. 2007. *Educational Psychology.* Boston: Pearson Education.

Character Education for 21st Century Global Citizens – Retnowati et al. (Eds)
© *2019 Taylor & Francis Group, London, ISBN 978-1-138-09922-7*

Water safety as preventive action and for child character behavior development in aquatic learning

E. Susanto
Universitas Negeri Yogyakarta, Indonesia

ABSTRACT: Pool accidents can happen to everyone, whether they can swim or not. A frequent accident in swimming pools is drowning and this is one of the biggest risks in aquatic activity. For this reason, this research is conducted to develop a water safety model and to develop character behavior in primary school aquatic learning. To achieve these targets, research is conducted through five research and development stages: (1) product analysis; (2) developing the initial product; (3) expert validation; (4) field trials; (5) product revisions. The correlation test was performed by inter-rater test. Data analysis was performed using a multi-factor ANOVA General Linear Model to test two variables of ordinal type and scale with normal/parametric distribution. The instrument validity test was tested against ten subjects/students. The data of the observed validity test showed a high degree of correlation (r = 0.961; p = 0.001; valid), and a reliability test between raters using a correlation coefficient test between classes indicated reliability (r = 0.985). The results showed the creation of a product in the shape of a guidance manual for 28 forms of water safety activity. These 28 water safety activities have been tested on a small scale to measure the successful implementation of the model. The contents of the water safety manuals include names of activities, drawings, purposes, equipment used, water depth, site arrangements, participants, and implementation guidelines.

1 INTRODUCTION

Drowning can be overcome by using a minimum salvage standard owned by each person. Some cases illustrate that drowning incidents are caused by oversight, negligence of swimmers, tools, inadequate facilities, and most importantly because of failure in first aid for handling emergency cases of accidents in the water. Based on cardiovascular physiology, when submerged for five minutes there is a high risk of death. If first aid is applied quickly and precisely it is helpful for salvation (Meaney & Culka, 2005). Currently, not many places can teach how children, in particular, can learn about security and safety in the water. The Physical Education (PE) program is one of the functions that has the capability to teach children about safety behavior around the pool (Sawyer, 2000). Children can be introduced to the basic rules about personal water safety while swimming or during activities outside the classroom. An accident in the pool can happen to everyone, whether they can swim or not. One type of accident that often occurs in the pool is drowning and this is one of the biggest risks in aquatic activities.

Several other risks that may occur include water injury, cramps, respiratory problems, headaches, and fainting. In addition there is also a safety risk to the health of student participation in learning aquatics. Health risks include hypothermia, water poisoning, skin irritation, eye irritation, and possible spread of infectious diseases. These incidents are a serious problem that could threaten the health and safety of students. Reducing the possibility of drowning or other types of water injury is a shared responsibility among PE teachers, parents, and lifeguards. However, they need to equip themselves with knowledge of the security and rescue capability as a prudent action (Graver, 2003). Lessons in the curriculum of physical education, scope or aquatic material remain. This suggests that the role of PE is very important, as it gives students the opportunity to be directly involved in a variety of learning experiences through physical activity, play and sport (Bucher & Wuest, 1995). The experience of motion obtained by students in PE is an important contributor to the increase in enrollment, and at the same time is an important contributor to welfare and health throughout life (Siedentop, 1991; Ratliffe & Ratliffe, 1994).

In line with the scope of the subject matter of Physical Education, Sport and Health (known as *PJOK*) in elementary school, there are seven main elements: (1) Games and Sports; (2) Development Activities; (3) Test yourself/Gymnastics; (4) Rhythmic Activities; (5) Aquatics (water activity); (6) Outdoor Education; (7) Health. Aquatics involves activities carried out in the pool, such as water games, water safety, swimming styles, and development aspects of the relevant knowledge and values contained therein (RoI, 2004). This suggests that the role of physical education is very important, giving students the opportunity to be directly involved in a variety of learning experiences through physical activity, play and sport. The experience of motion obtained by students in physical education is an important contributor to the increase in enrollment, and at the same time is an important contributor to the welfare and health throughout life (Siedentop, 1991: Ratliffe & Ratliffe, 1994). The aquatic learning process of elementary schools cannot be separated from the development of children's potential in three domains: basic psychomotor skills, basic attitude, and basic understanding (Langendorfer & Bruya, 1995; Dougherty, 1990; Graver, 2003). The learning approach at elementary school level is conducted by referring to an activity program that has been prepared so that all the basic behaviors and abilities that exist in the child can be developed as much as possible (Sawyer et al., 2001; Clement, 1997). There are several guidelines for aquatic learning with a games-based approach, including *Wet games: A fun approach to teaching swimming and water safety* (Meaney & Culka, 2005), and *Water fun: 116 fitness and swimming activities for all ages* (Lees, 2007), as well as some other sources of aquatic learning guides.

On the one hand, learning aquatics is now increasing in demand broadly at the elementary level. Teachers and students have begun to take an interest in the aquatic program because this program creates a sense of fun and social atmosphere for child development, builds confidence, and improves physical fitness health. Aquatic learning can also reduce the delinquency or reduce the level of activity of the child due to an energy distribution for positive activities. In Indonesia, there is often news about a child who had an accident in the pool. However, the event drowned in the pool until death, has not been handled properly by the pool attendant. Our preliminary survey showed a 40% incidence of drowning (near-drowning), occurring because of the absence of a Standard Operational Procedure (SOP) for safety in the water. About 70% of victims drowned in public pools. Several public swimming pools do not have an accident emergency response SOP.

2 METHOD

The research was conducted through research and development (Borg & Gall, 1983) with five stages: (1) product analysis; (2) developing the initial product; (3) expert validation; (4) field trials; (5) product revisions. The correlation test was performed by inter-rater test. The data was analyzed using a multi-factor ANOVA (Analysis of Variance) General Linear Model (Thorndike, 1982) to test two variables of ordinal type and scale with normal/parametric distribution. The instrument of validity test was tested on ten subjects/students. The data of the observed validity test showed a high degree of correlation ($r = 0.961$; $P = 0.001$; valid), and the reliability test between raters using a correlation coefficient test between classes indicated reliability ($r = 0.994$). The study was conducted in the aquatic primary school learning class. The subjects of the study were elementary school students following aquatic learning. The trial sample was ten students. The data collection instrument was observation guidelines. Observations are used to obtain information from experts to advise about safety activities in the water. The expert comes from aquatic learning. In order to obtain data, an instrument test is performed to determine the validity and reliability. The validity and reliability of the instrument was tested using the inter-rater correlation test and data analysis using a multi-factor ANOVA model (Thorndike, 1982).

3 RESULTS

3.1 *Data requirement analysis*

Data requirement analysis needs analysis in the preparation of research products was required to develop and explore the problems of water safety activities for students. This activity was done by analyzing the process of learning in the field, making observations, and conducting literature studies/reviews. The products produced were (1) draft forms of water safety activities for students, and (2) validity and reliability of the instrument.

3.2 *Expert initial draft validation*

The design of water safety activities that be tested on a small scale and validated by experts. To validate the product being produced, the researcher involved two aquatic learning experts as well as an expert lecturer in the field of sports education and physical education. Validation is done by providing the initial product draft, with a dissertation evaluation sheet for the expert. The result of the evaluation was in the form of value for quality aspect using a Likert scale from 1 to 4. The data obtained from questionnaires by experts was a guideline to state whether the products of water safety activity forms can be used for small-scale and large-scale trials. Table 1 represents a draft of the forms of water safety activity that were identified.

3.3 *Small-scale trial*

After the product of the forms of water safety activity was validated by experts and revised, the product was tested on the students. This trial was conducted on ten students. This trial aims to identify problems such as weaknesses, deficiencies, and product effectiveness. Observations made by the raters are one indicator to determine the effectiveness of the product. Observations by the raters were performed during the trial. Based on the observations, 28 forms of water safety activity and instrument validity values were found.

3.4 *Test validity of test instruments*

The number of subjects or students used in small-scale trials was a total of ten students. The correlation test was done by inter-rater test. The data was analyzed using a multi-factor ANOVA model (Thorndike, 1982) to test two variables of ordinal type and scale with normal/parametric distribution. The data of the observation validity test showed a high degree of correlation with an average of 0.995.

Table 1. List of water safety activities identified (using local Indonesian terms).

No.	Name of activity	No.	Name of activity
1	Entries/Exits	15	Ular Naga
2	Sculling	16	Komando
3	Strokes	17	Menolong
4	Survival/PFD	18	Mengirim Benda
5	Underwater Skill	19	Menyelam
6	Rescue Skill	20	Penyelam
7	Bola dan Simpai	21	Perahu Naga
8	Kincir Bola	22	Selancar Air
9	Gelap Total	23	Penyelam Estafet
10	Ambil Koin	24	Pegang Rambut
11	Bugi-Wugi	25	Melarikan Diri
12	Pindah Benda	26	Kaki Pencari
13	Menjejak Air	27	Cincin Jatuh
14	Ujung Lingkaran	28	Injakan Kaki

Table 2. Level of validity of water safety.

No.	Score	Correlation coefficient	P	Status
1	Rater 1	r = 0.999	0.001	Valid
2	Rater 2	r = 0.996	0.001	Valid
3	Rater 3	r = 0.991	0.001	Valid

Therefore, based on the statistical calculation of the validity of the instrument test, it has been shown that there is a high level of positive relationship, so the instrument is valid and can be used for data collection on a wide scale. Based on the results of factor analysis, it can be confirmed that the instrument has a good construct, meaning that the instrument can be used to measure symptoms in accordance with the recommendations. (Sugiyono, 2003).

3.5 Test instrument reliability test

When testing reliability between raters, the kappa statistic is used when there are two raters, while an Intra-class Correlation Coefficient (ICC) is used when more than two people are rating (Widhiarso, 2006). This research uses three raters and employs a correlation coefficient between classes. The Intra-class Correlation Coefficient shows the comparison between variations as measured by the overall variation. Based on the calculation of reliability statistics of small-scale testing of the instrument, it has been shown that there is a high reliability value between raters of 0.985, so the instrument is considered reliable and can be used for data retrieval.

4 DISCUSSION

Based on the results of frequency distribution in ten students, it has been shown that: (1) according to Rater 1, there are four students (40%) included in the good category, there are five students (50%) who belong to medium category, and there is one student (10%) which is low category; (2) according to Rater 2, there are five students (50%) who are categorized good, there are five students (50%) who belong to medium category, and there are zero students (0%) which belong to low category. In psychomotor factor, there are ten students with the following details: (1) there are four students (40%) who are in good category; (2) five students (50%) who are of medium category; (3) there is one student (10%) who belong to low category. Based on the steps of research and development, then obtained the final product in the form of water safety activities. The indicator of the success of this product is the similarity of perception between the raters, in the form of the assessment sheet, and the observation of all subjects tested in the research. Based on field trials of ten students with similar characteristics, the same results were obtained. Thus the tested product can be applied to groups of students with similar characteristics.

This study has some limitations, among others, elementary school children who still love to play with friends so that the learning process is disrupted. In the learning process there are several types of water safety at the advanced level, so that it is not uncommon that it is done in a deep pool/pond. The research subjects were relatively difficult to find considering that only a few elementary schools run regular aquatic activities. The use of public swimming pools as research sites that are shared with general visitors caused the research process to be slightly unreliable.

5 CONCLUSION

The results showed that aquatic learning in primary schools has not handled the aspects of safety in the water. Thus, there are draft water safety activities and 28 water safety activities have been compiled. Other forms of water safety activity among others: the name of the activity, the image, goals, equipment used, the depth of the water, the place settings, the number of participants, the potential of the developed swimming skills, and the water safety purposes intended. The test validity value was 0.995 and the test reliability value was 0.985.

ACKNOWLEDGMENTS

We thank the physical education teachers who have taken the time to discuss the development of water safety from: *SD* (Primary School) *Muh Suronatan, SD Tumbuh, SD Samirono, SD Budi Mulia Dua Sedayu*, and *SD Muh Gendol Tempel*.

REFERENCES

Borg, W.R. & Gall, M.D. (1983). *Educational research: An introduction* (5th ed.). New York, NY: Longman.

Bucher, C.A & Wuest, D.A. (1995). *Foundations of physical education & sport* (12th ed.). New York, NY: Mosby Year Book.

Clement, A. (1997). *Legal responsibility in aquatics*. Aurora, OH: Sport and Law.

Dougherty, N.J. (1990). Risk management in aquatics. *Journal of Physical Education, Recreation & Dance*, *61*(5), 46–48.

Graver, D. (2003). *Aquatic rescue and safety: How to recognize, respond to, and prevent water-related injuries*. Champaign, IL: Human Kinetics.

Langendorfer, J.S. & Bruya, L.D. (1995). *Aquatic readiness. Developing water* competence *in young children*. Champaign, IL: Human Kinetics.

Lees, T. (2007). *Water fun: 116 fitness and swimming* for *all ages*. Champaign, IL: Human Kinetics.

Meaney, P. & Culka, S. (2005). *Wet games: A fun approach to teaching swimming and water safety*. Clifton Hill: Robert Andersen & Associates.

Ratliffe, T. & Ratliffe, L.M. (1994). *Teaching children fitness: Becoming a master teacher*. Champaign, IL: Human Kinetics.

RoI. (2004). *Competency standards for physical education elementary school*. Jakarta, Indonesia: Department of Education, Republic of Indonesia.

Sawyer, T.H. (2000). Aquatic facility safety and responsibility. *Journal of Physical Education, Recreation & Dance*, *71*(5), 6–7.

Sawyer, T.H., Spengler, J.O. & Connaughton, D. (2001). Planning for emergencies in aquatics. *Journal of Physical Education, Recreation & Dance*, *72*(3), 12–13.

Siedentop, D. (1991). *Developing teaching skills in physical education* (3rd ed.). Mountain View, CA: Mayfield.

Sugiyono, S. (2003). *Business research methods*. Bandung, Indonesia: Alfabeta.

Thorndike, R.L. (1982). *Applied psychometrics*. Boston, MA: Houghton Mifflin.

Widhiarso, W. (2006). *SPSS untuk psikologi* [SPSS for psychology]. Retrieved from. http://widhiarso.staff.ugm.ac.id/files/bab_2_estimasi_reliabilitas_via_spss.pdf.w

Character Education for 21st Century Global Citizens – Retnowati et al. (Eds)
© 2019 Taylor & Francis Group, London, ISBN 978-1-138-09922-7

The implementation of integrity values in the elementary schools: Problems and potential solutions

S. Rahmadonna & Suyantiningsih
Universitas Negeri Yogyakarta, Indonesia

R.A. Wibowo
Universitas Gadjah Mada, Indonesia

ABSTRACT: This preliminary research evaluates the implementation of delivering integrity values from the teachers towards their students in several elementary schools in Yogyakarta province, Indonesia. This study adopted the descriptive quantitative method. The subjects of this research were 33 teachers in eight schools. The data was acquired by document study, observation, interviews and questionnaires. This research found that although almost 85% of teachers who participated in this research claimed that they understood about integrity values, only one-fifth of them could elaborate it comprehensively. It indicates that they have been facing obstacles to deliver integrity values toward their students. Therefore, this research attempts to explain why that problem could exist and to discuss the potential solution to cope with the matter.

1 INTRODUCTION

Indonesia has shown a consistent trend of reducing levels of corruption since President Soeharto fell in 1998. As it may be known, Soeharto was the Indonesian dictator during 1965 to 1998. He centralized power, suppressed opposition parties (Bunte & Ufen 2009), and restricted freedom of expression (Robertson-Snape 1999). He granted business monopolies to his relatives and cronies and tolerated widespread corruption among his men in exchange for their loyalty (Robertson-Snape 1999).

According to the Corruption Perception Index (CPI) 2016, Indonesia obtained a score of 37. This index is prepared by an international non-government organisation, Transparency International (TI). It has the range point from 0 to 100 with 0 being perceived as 'highly corrupt' while 100 is considered as 'very clean' from corruption. With that score, Indonesia is ranked 90 out of 176 countries (TI 2016). In other words, obtaining this score is unsatisfactory, and it indicates that Indonesia is still classified in the top 50% of countries suffering from corruption.

As a fundamental problem which has to be addressed seriously by multi stakeholders, the Corruption Eradication Commission or *Komisi Pemberantasan Korupsi* (KPK) has cooperated with educational institutions to internalise the anti-corruption value to the students. This initiative was started in October 2008 when the KPK handed in the module to the Indonesian Minister of National Education to officially launch a module on anti-corruption.

These two institutions believe that education can be a promising alternative to building the character of the younger generation. It is expected that through education they can be responsible men or women with integrity and hold firm to morals. This program has been implemented in some schools; however, it remains unknown whether or not this program has been optimally implemented. Therefore, this research aims at assessing and evaluating the implementation of internalisation of integrity values in some elementary schools in Yogyakarta, Indonesia.

According to the United Nations Office on Drugs and Crime (UNODC), corruption is a complex social, political, and economic phenomenon occurring in all countries. It is considered that corruption undermines democratic institutions, triggers unstable political situation, and retards the economic development.

The gloomy situation above is in line with the data compiled by the KPK. Since its establishment, the Commission has been tackling 411 corruption cases with bribery and corruption in public procurement being the two most frequent types (Wibowo 2015). If the corruption cases prosecuted by the attorney general are also counted, the number will be increased.

At this point, it is relevant to discuss the terminology of corruption. According to UNODC (2004), there was no fixed definition towards corruption. The United Nations Covenant on Against Corruption and Indonesian Law on Eradication of the Criminal Act of Corruption are also silent on the matter. However, there is a global common understanding on interpreting corruption as "the misuse of a public or private position for direct or indirect personal gain" (UNODC 2004).

Referring to a publication prepared by the KPK (2006), corruption in Indonesia can be classified into seven classifications: (i) unlawful acts which may be detrimental to the finances of the state; (ii) bribery; (iii) embezzlement; (iv) extortion; (v) fraud: (vi) conflict of interest in public procurement; and (vii) personal gratification civil servant is not permitted to receive any gift whenever it shall be assumed given due to his/her position as public officials and against his/her obligation or task. The gift will be classified as a bribe except if the recipient reports the gratification.

It is believed that integrity is the essential element to curb corruption. It is indicated by the development of the national integrity system initiated by Transparency International; Public Sector Integrity Reviewers prepared by the Organisation for Economic Cooperation and Development (OECD), etc. The KPK itself has realised the urgency to nurture the integrity of society. According to a document named titled (translation) the road map of KPK to curb corruption 2011–2023, one of the KPK's grand strategies to achieve a national integrity system is through education (2011).

Realising the pivotal position of education to nurture integrity, the student should get used to promoting integrity. It is conducted to internalise and to create what is called 'habitus'. The internalisation and habitus can prevent someone in making mistakes. Regarding this, Haryatmoko explained Bourdieu related to integrity (Kompas 04/06/2013) as follows, "public integrity must be developed and trained, because it is not something that will occur spontaneously. Instead, integrity can be achieved as the output from training, education, and habitus which are directed to the public ethics. A leader who has been growing up from an area which concerns to the community welfare may have public integrity because his involvement is the habitus (In Indonesia: *pembatinan*) processes of public ethics".

The habitus processes will be more effective whenever these are conducted in childhood. An educational institution such as a school may take an important role to embed good character values. A school is a place where a child spends the most (or at least considerable) time. Thereby, to achieve the goal of educating on anti-corruption, Hujair (2009) explained that educational processes should be oriented to moral action. The processes shall not stop merely in the competence but should embrace on creating the will and habitus to perform an action in accordance with good values. In other words, a school should be the place for raising awareness especially to nurture integrity values.

Lickona (1991) argued that three stages of sustainable coaching are needed to educate the moral of a child to ensure him conducting a moral action. It is started with moral knowing, then moral feeling, and finally moral action. These three shall be developed/coached holistically. Whenever one has not been coached until the moral action stage, it will be easy for him to neglect the values. The educational institution, therefore, shall support the nurturing processes of the students and shall ensure the coaching process has reached the moral action stage. By doing so, the habitus will be created, and the student will act according to good values in the future.

3 THE CHARACTERISTICS OF THE ELEMENTARY SCHOOL STUDENTS

According to Piaget's theory of development, a student in elementary school is in the phase of operational concrete, meaning the student is starting to think logically relating to a concrete object. The student in this stage can conduct direct manipulation to an observed object. One of the general characteristics of the concrete operational structure is the balance between affirmation and its opponents.

Margaret (2013) concluded that the student would be able to conduct reasoning logically and it will be limited to direct manipulation to the object. Pons and Harris (Margaret 2013) also explained that this stage is accompanied by standard conceptualization processes. The learning process at this stage should be designed to involve the student actively so that the learning process would be meaningful. The learning process can exercise the medium; however, the characteristic of the student in the operational concrete should be regarded. Indeed, a good medium is indicated by whether or not it can help the student in opening or broadening their logical reasoning by learning involvement.

4 FINDINGS

This research has conducted observations, interviews and focus group discussion (FGD) to assess the implementation of integrity values in the learning processes.

Observation was being carried out with observing activities of the elementary school students, teachers, and the headmaster. The observation is focused on the behaviour of the students; it aims at monitoring whether or not they have manifested the integrity values. The implementation should be based on their internal awareness and should not be caused by external pressure. The behaviour, however, may be promoted or conditioned by the school environment. Echoing the point mentioned by Simmons et al. (2016), wellbeing at school is having relationships based on equality and respect both with friends and teachers.

The research team visited numerous schools and invited 10 school teachers in Yogyakarta province; one of those 10 schools was Samirono Elementary School which will be discussed further as follows. The researchers found that this school is interesting to be discussed because we found that students in Samirono Elementary School have good characteristics that are defined as integrity values.

Based on the direct communication with students at Samirono Elementary School, there was a student who showed higher sensitivity than other students. He offered his help to bring the bag of the authors. The authors thought that his action was related to one of the integrity values; caring. His action was also in accordance with one of the basic learning principles; learning processes occur due to the condition which can influence (and can be affected by) the learning processes itself (Illeris 2009).

The observation also found that only one school has obtained socialisation on anti-corruption education from the KPK. That school has been appointed as the partner of the Commission. Nonetheless, other schools have not known about that program, including Samirono Elementary School. Therefore, it can be said that the implemented values performed by the student in Samirono Elementary School are not influenced by the socialisation from KPK.

After interviewing with the teachers and headmasters, it can be concluded that they have not understood appropriately about education on anti-corruption. The main problem is that the teachers are not familiar with the integrity values as the core elements in educating on corruption. Besides, many teachers have difficulties in integrating the anti-corruption values to the learning activities and supporting activities at the school.

The teachers explained that the learning processes are more emphasised on character values; they delivered repetitively the message (containing the moral issues) so that the students will get used to the moral issues (and it is hoped that they will act morally). They admitted that this strategy is quite effective in internalising the character. Indeed, this strategy is coherent with the research finding conducted by Looijenga et al (2015) that iteration during the design process is an essential element; spontaneous playing behaviour of students indicates that iteration fits in a natural way of learning.

The authors also conducted FGD to get the conformity of the observation and interview which have been conducted above. Thirty-three elementary school teachers in Yogyakarta were invited to attend the FGD. The FGD was focused on discussion and sharing opinions pertaining to education on anti-corruption. In here, one of the interesting statement made by a teacher (Ms PM) was that:

> "Our school has understood about the education on anti-corruption because the KPK has chosen us to be one of their partners. However, honestly, we merely know about corruption and maladministration. We are still unaware of integrity values and how to embody or manifest these values into school actual activities."

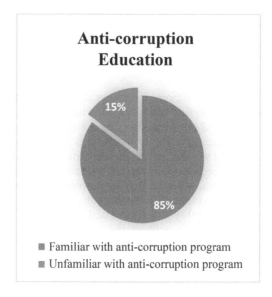

Figure 1.   The familiarity of the teachers related to the issue of education of anti-corruption.

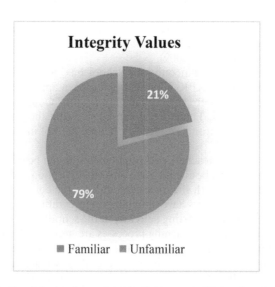

Figure 2.   The familiarity of the teachers related to the concept of integrity values.

490

Furthermore, another teacher (Mr WD), admitted different information than teacher Ms PM;

> "Our school has never obtained the socialisation on the education of anti-corruption. We have just heard about the matter from this discussion. Nonetheless, after we were informed about the integrity values, we believe that we have been trying to implement the messages to the students, especially when we teach civic education."

The final agenda of the FGD was requesting the teachers to fill out the questionnaire. It aimed to re-clarify the preliminary findings and the FGD discussion which had been conducted. The result of the questionnaire showed that 84.84% of the teachers admitted that they understood the issue of education of anti-corruption. They knew about the issue from the newspaper and electronic media. Only 15.16% of them were not aware of the issue on this matter. Nonetheless, whenever these teachers were asked about their familiarity concerning the integrity values, the crucial element of the educating anti-corruption, it was found that merely 21% of them could explain the concept of the integrity values. The rest admitted that they have no idea about the concept of these values.

Observation result, interview, and FGD indicated that the program initiated by the KPK – education of anti-corruption for childhood – has not been effectively implemented. The schools need further assistance to understand the concept and to apply the integrity values in the learning processes corruption which can be integrated with learning subjects at the schools, therefore, the media can be used in the learning process in schools.

5 CONCLUSION

According to the elaboration above, it can be concluded that the KPK has been socialising and calling the schools to internalise the integrity values in the schools. Nevertheless, the implementation has not been satisfactory. Only a small percentage of elementary schools in Yogyakarta have obtained the socialisation from the KPK. The finding confirms the matter that 85 percent of the respondent (the teachers) are well informed about the anti-corruption program initiated by the KPK, but only 21% of those understand the integrity values. It is surprising since the integrity values are the primary values in educating anti-corruption.

This finding indicates that further research should be conducted to determine how the integrity values should be internalised in schools. Furthermore, this result recommends the Indonesian government to (a) intensify the socialization regarding to anti-corruption and integrity values; (b) develop a learning strategy which can internalize integrity values; (c) develop media education pertaining to anti-.

REFERENCES

Bunte, M. & Ufen, A. 2009. The new order and its legacy: reflections on democratization in Indonesia. In M. Bunte, & A. Ufen, *Democratization in post-Soeharto Indonesia* p. 5–25. New York: Routledge.
Gredler, Margaret E. 2013. *Learning and Instruction: Teori dan Aplikasi* [Learning instruction: theory and application]. Jakarta: Kencana Prenada Media Grup.
Hujair AH. Sanaky. 2009. *Pendidikan Anti Korupsi* [Education on anti-corruption]. Retrieved from: http://sanaky.staff.uii.ac.id.
Illeris, Knud. 2009. Contemporary theories of learning: learning theories in their own words. New York: Routledge.
Komisi Pemberantasan Korupsi (KPK). 2006. *Memahami Untuk Membasmi: Buku Saku Untuk Memahami Tindak Pidana Korupsi* [Understanding to Eradicating: Handbook for Understanding Corruption as Crime Actions]. Jakarta: KPK.
Komisi Pemberantasan Korupsi (KPK). 2011. *Road Map KPK dalam Pemberantasan Korupsi di Indonesia Tahun 2011–2023* [KPK's Road Map to Eradicate Corruption in Indonesia 2011–2023]. Retrieved from: http://acch.kpk.go.id/.
Kompas. 2013. *Mencari Pemimpin Berintegritas Publik* [Finding A Leader Who Has Public Integrity]. Retrieved from: http://nasional.kompas.com/read.

Law Number 31/1999 concerning on the Eradication of Crime on Corruption.

Law Number 20/2001 of the revision of the Law number 31/1999 concerning on the Eradication of Crime on Corruption.

Lickona, Thomas. 1991. *Educating for character: how our schools can teach respect and responsibility.* New York: Bantam Books.

Looijenga, Annemarie et al. 2015. The effect of iteration on the design performance of primary school children. *International Journal of Technology and Design Education* 25(1): 1–23. Doi: 10.1007/s10798-014-9271-2.

Organisations for Economic and Co-operation and Development (OECD). 2016. *Public Sector Integrity Reviews.* Retrieved from: http://www.oecd.org/gov/ethics/.

Perdamean, Daulay & Malik, Abdul. 2015. *Pengembangan Model Pendidikan Anti Korupsi Melalui Komik di Sekolah Dasar* [Model Development for Anti-Corruption by the Medium of Comic Book], 22(1). http://journal.um.ac.id/.

Robertson-Snape, F. 1999. Corruption, collusion and nepotism in Indonesia. *Third World Quarterly* 20(3): 589–602.

Simmons, Catharine et al. 2015. Imagining an ideal school for wellbeing: locating student voice. *Journal of Educational Change* 16(2): 129–144. Doi 10. 1007/s10833-014-9239-8.

Transparency International Indonesia, Corruption perceptions index 2016. Retrieved from http://www.ti.or.id/index.php.

UNODC. 2004. *United Nations Handbook on Practical Anti-Corruption Measures for Prosecutors and Investigators.* Vienna: UNODC.

UNODC. 2016. *UNODC's action against corruption and economic crime.* Retrieved from: https://www.unodc.org/unodc/.

Wibowo, R.A. 2015. Mencegah korupsi di pengadaan barang dan Jasa: apa yang sudah dan yang masih harus dilakukan? [Preventing corruption in public procurement: what has been done and what should be done]. *Jurnal Integritas* 1(1): 37–60.

Intellectual character in the learning of musical interpretation as an enhancement of 21st-century skills

A.N. Machfauzia
Universitas Negeri Yogyakarta, Indonesia

ABSTRACT: This study aims to describe students' intellectual character in the learning of musical interpretation as an improvement of skills for the 21st century. A qualitative research method was adopted according to the phenomenology involved, and the research subjects were 15 students who took the course of Guitar 4 major instrument practice. The students were observed while playing the guitar, interviewed, and documentation was studied. The findings indicate that intellectual character makes a significant contribution to the learning of musical interpretation. Most students had not used structured thinking skills, which relates to a lack of critical thinking and open-mindedness, and lack of curiosity towards the knowledge behind the work of music being played. As an improvement to their creative and critical thinking skills, these thinking skills of open-mindedness, curiosity, and metacognition need to be developed and built into the learning of musical interpretation.

1 INTRODUCTION

Currently, a lot people are discussing character education, but few discuss intellectual character. Intellectual character is a collection of attributes or dispositions that distinguish a person as someone who is able to think clearly and effectively. It is, as discussed by Ritchhart (2002) that the intellectual character is an agglomeration of the habit of thinking, mindset, and the motivation that orient a person's activities. Intellectual character is a set of attributes or dispositions that distinguish a person as someone who is able to think clearly and effectively (Ritchhart, 2002). Furthermore, Ritchhart (2002) explains the concept of intellectual character that is associated with good thinking and productiveness, and not just related to intelligence alone. Specifically, intellectual character describes a set of properties that not only shape behavior, but also motivate intellectual behavior.

In terms of related intellectual character attributes, Jamie (2009) and Ritchhart (2002) explain that there are six essential attributes of the intellectual character, namely open-mindedness, curiosity, metacognition, seeking truth and understanding, strategic thinking, and skepticism. These character attributes are not only a habit of thinking, but a conviction that a more permanent learning is embedded in the way a person sees and interacts with the world. Proust (2013) adds that if a student has good character, they can generate good profits in the academic field, such as higher value, and this supports good intellectual character too. Thus everyone, including students, can improve their capacity to learn by building their own intellectual character, especially in learning musical interpretation.

As an improvement of 21st-century skills, for which one must think creatively and critically, students can build their thinking ability as intellectual character through learning musical interpretation. An intellectual character is very useful for a person in being a lifelong learner and this can be a personal quality. In order to be a lifelong learner, one must have a broad base of knowledge in one's field, both theoretically and practically (Baehr, 2006). This is important so that the students of the next generation can continue to improve their ability to think openly, critically and regulate their mindset, so that they can create morally good relationships with everyone in the world. These relationships can be improved indirectly through continuous involvement with music interpretation and the performance of music, as both include the same skills. Learning musical interpretation is an implicit skill in musical instrument practice. It is closely related to the performance of music, and a good performance of music cannot be separated from good interpretation. Developing interpretations of the music being played, according to Ford (2011), is one of the most important aspects of making music enjoyable for the players of the music. To be able to interpret the work of music well, students are required to learn the historical background of the work of music, which includes playing, listening, and analyzing its elements before playing the music. Furthermore, it requires the ability to think, which includes attributes of open-mindedness, curiosity, and metacognition to interpret a musical masterpiece well. Open-mindedness means that students can be open to the aspects that are needed for interpreting the work of music. Curiosity is about the meaning behind the work of the music being played. Meanwhile, metacognition is about how students organize awareness about their cognition. That is how students are aware of knowledge about the interpretation of music. It can be said that metacognition is a form of self-attribute (Proust, 2013).

In fact, there are still many students who play music who concentrate only on the ability to read music notation and the techniques of playing. However, to properly express the form of a piece, music students must employ the essentials of artistic interpretation to show listeners where the action crests and reposes, to communicate mood, and to let phrases breathe (Klickstein, 2009). All of this is sometimes astonishing (Blum, 1977). Therefore, it is worth exploring the intellectual character of music students in learning musical interpretation in terms of the improvement of 21st-century skills.

2 RESEARCH METHOD

A qualitative research method of descriptive phenomenology was used because it refers to the study of students' experiences in playing musical works and requires a description or interpretation of the meaning of phenomena experienced by students in the investigation (Diaz, 2015). Determination of the application of phenomenology is the existence of research problems that require an in-depth understanding of students' experience in playing musical works that are common in musical instrument practice, including learning music interpretation (Creswell, 2010; Diaz, 2015).

Related to this study, researchers used various data collection procedures based on a predetermined time (Creswell, 2010). In addition, it is described that the intellectual characteristics of the learning process of music interpretation based on what happens in the field (in the implementation of Guitar 4 major instrument practice) as a further study to identify flaws and weaknesses in the mindset of students in the learning of musical interpretation.

2.1 Research subjects (participants)

The participants in this study were students who took Guitar 4 major instrument practice. They consisted of 15 people, determined by purposive sampling. Furthermore, the students were divided into three groups, each consisting of five students.

In playing a piece of music, students were observed in relation to intellectual character traits that included open-mindedness, curiosity, and metacognition in learning musical interpretation.

2.2 Procedures

In the first stages, students who took the Guitar 4 major instrument practice course were grouped into three groups. Each group was asked to play music. They were observed and recorded through field notes. Observed aspects included the open-mindedness of the students to knowledge related to the work of music being played, the curiosity of the students about the historical background of the music, and the meaning contained in the work of music. Another aspect observed of students was metacognition, which is how effectively students organize their cognitive activities when playing a musical piece. In addition, students were recorded on video.

In the second phase, the students were interviewed, with questions relating to students' ability to interpret a piece of music being played, and the development of intellectual character, including open-mindedness, curiosity, and metacognition. In playing a piece of music, musical interpretation is required, and in learning musical interpretation there are three dimensions that need to be fulfilled by a music player: the knowledge dimension, the dimension of perception, and the dimensions of the musical experience.

The third stage of the research reviewed the validity of the data and provided an explanation of the research conducted, accompanied by evidence in the form of field notes, voice recordings of interviews, video recording observations, and photographs. In the next phase, for confirmability, researchers reviewed these records again, and conducted triangulation techniques to recheck the data that had been obtained from all the students who were taking Guitar 4 major instrument practice, including the results of individual (face to face) interviews, the results of observations captured on video, and the documentation.

Finally, the data collected was analyzed using an interactive model (Miles & Huberman, 1994) consisting of reduction, data display, and conclusion/verification.

3   RESULTS

Based on the observations and interviews that have been obtained, it is known that there are three students (coded as IP, SDW, and AND) who do not yet have good thinking skills. Some students still think that in playing a piece of music the most important priority is the fluency in reading notation and skill in playing the guitar. In playing music, students are also required to be able to interpret the work of the music well. Therefore, before playing a musical piece, students need to analyze the music, and before analyzing, the students are required to listen to the music being played. From these three students it can be seen that they have not been able to develop patterns of thought (intellectual character) in learning interpretation; the students still have narrow thinking that involves merely reading notation and playing the guitar smoothly.

In contrast to these three students, the thinking skills (mindset) of two other students (coded as RHM and IHN) were better and more developed. This is evidenced from the interviews, which explain that before playing the music, the students have read the history of the composer. Furthermore, they are looking to understand the background of the creation of works/songs, as well as the meaning of the songs' titles. In addition, they look and listen to the song through YouTube. From the results of these interviews, it appears that these students have an intellectual character (an ability to think or a mindset) that is more developed (better thinking) when compared with the three previous students. Such abilities can be said to be the basis of good intellectual character and are possessed by these two students. They think this would determine their guitar performance in the future.

Meanwhile, four students, namely YFN, IRD, FRM, and DMS, also have better thinking skills than IP, AND, and SDW in terms of interpreting the work of music being played. This is evidenced by their analyzing the song first, before playing the music. Two other students, AL and GN, are still concentrating on the ability to read the music. Conditions like this certainly do not provide an opportunity to think openly about the knowledge contained in the work itself because the students are using only their viewing perspective, not their thinking. In addition, these students are less accustomed to listening to works of music played by others. Such students tend to think they can only play music through skill on the guitar. This was expressed by YFN who explained that there is more focus on guitar-playing skills. This shows that YFN prioritizes the playing of the musical instrument, without examining the other aspects involved in the musical scores being played. In fact, Cook (2013) highlights that a knowledge of music is where meaning is generated in real time through the process of musical performance. Therefore, in interpreting musical works students need to interpret what is contained in the work of music. Thus, thinking skills are needed in order to interpret a musical work. Such thinking skills include open-mindedness to the dimensions in the interpretation of music, curiosity, and metacognition. These attributes of intellectual character can be established and developed in the classroom if lecturer's faculties allow for it.

Similarly to RHM and IHN, students AL, IN, and ML tend to have an intellectual character with good curiosity and metacognitive attributes. This is evidenced by their activities in first reviewing and analyzing the work of music that will be played. In addition, these three students also studied the background of the music's creation. After that, the students listened to the piece of music, but played by others. This is evidenced from the interviews in which the students said that are almost always listening to musical works played over the internet. After these activities had been performed, the three students began to play the guitar. This shows that these three students can develop their thinking skills using internet technology.

Meanwhile, the other student participants (DO, LE, FI) were not much different from IP, SDW, and AND. They tended to prioritize the ability to read musical notation, and guitar-playing skills. They seemed to be lacking in their thinking skills. Seeing these situations, the lecturer can condition the class of Guitar 4 major instrument practice to regularly engage in activities that develop their thinking skills, so that students can improve their skills for the 21st century.

4 DISCUSSION

To be able to interpret a work of music well, students need to learn the background, related to both the composer and the time of the creation of the music, the meaning of the music, as well as listening to it and analyzing it. This process of interpretation is necessary in the playing of a work of music because, in music, interpretation is a process of translating meaning to the symbols and elements of music contained in a score, as this is the closest thing to supporting music (Machfauzia, 2014). The same thing is also said by Kitelinger (2010); that the meaning of interpretation is in the music, namely "finding implied meaning in the written symbols".

In music, interpretation is closely related to its performance. Therefore, in performing a work of music, students must not only be skilled in playing guitar, but also need the ability to interpret the work of music that is being played. To be able to achieve this requires good thinking skills.

Based on the results obtained in this research, it has been shown that in playing music on the guitar, most students still focus on the ability to read notation and having fluency in playing the guitar. This is not enough to present the work of music well. It can be said that students' ability to read musical notation still has obstacles, so this also needs to be overcome in order to improve their thinking skills (intellectual character) as well as improving skills for the 21st century in general.

Basically, to build students' thinking skills in learning the interpretation of music through open-mindedness, curiosity, and metacognition, lecturers must use effective means. One of these is to develop creative thinking skills by brainstorming about music-related works that have been heard by students, either through technology (e.g. the Internet), or through their peers. With the use of such methods, students can be expected to: (1) create useful ideas, especially in learning musical interpretation; (2) develop, implement and communicate those ideas to others effectively, more openly, and more responsively to different perspectives; (3) cooperate with their group; (4) see failure as an opportunity to learn better, and understand that developing and improving thinking and innovating skills is a long-term process (Trilling & Fadel, 2009). Thus, if this can be carried out on a regular basis, it is expected that music students can develop their intellectual character and support the improvement of skills for the 21st century.

5 CONCLUSION

Intellectual character contributes to the learning of musical interpretation. In learning the interpretation of music, students are asked to listen to a piece of music (western art music), analyze it, and discuss the musical elements that have been heard. Listening to a work of western art music requires high-order thinking skills, because students also analyze the elements of music contained within it. This is reinforced by the results of research carried out by Johnson (2006), who wrote that there are some advantages in listening to music if done seriously and with full concentration. These advantages include providing musical experience and improving high-order thinking skills, because in listening to music students need to analyze, synthesize, and make judgments about what they hear. As an improvement in 21st-century skills, namely creative thinking, critical thinking and problem-solving skills, thinking skills that include open-mindedness, curiosity, and metacognition need to be developed in music students and built especially in the learning of musical interpretation. In playing music, most students have not used their structured thinking skills, which relates to a lack of critical thinking and open-mindedness, and lack of curiosity towards the knowledge behind the work of music being played. Students must not just concentrate on reading musical notation, but must as Pablo Casals said (Blum, 1977), ensure that every note sings.

REFERENCES

Baehr, J. (2006). Character, reliability, and virtue epistemology. *The Philosophical Quarterly, 56*(223), 193–212.

Blum, D. (1977). *Casals and the art of interpretation*. Berkeley, CA: University of California Press.

Cook, N. (2013). *Beyond the score: Music as performance*. Oxford, UK: Oxford University Press.

Creswell, J.W. (2010). *Research design: Qualitative, quantitative, and mixed methods approaches*. Newbury Park, CA: Sage.

Diaz, P.M. (2015). Phenomenology in educational qualitative research: philosophy as science or philosophical science? *International Journal of Educational Excellence, 1*(2), 101–110.

Ford, M. (2011). Marimba: An interpretation. Nashville, TN: Innovative Percussion. Retrieved from http://www.innovativepercussion.com/docs/documents/405/MarimbaAnInterpretation.pdf.

Jamie. (2009, May 29). Building intellectual character: 6 attributes that will make you a better learner. *Self Made Scholar*. Retrieved from http://selfmadescholar.com/b/2009/05/29/building-intellectual-character/.

Johnson, D.C. (2006). Music listening and critical thinking. *International Journal of the Humanities, 2*(2), 1163–1164.

Kitelinger, S. (2010). Musical performance for the instrumental conductor. Paper presented at California Music Educators Association, Southern Border Section Conference 2010.

Klickstein, G. (2009). *The musician's way: A guide to practice, performance, and wellness*. New York, NY: Oxford University Press.

Machfauzia, A.N. (2014). *Interpretasi Musik dalam Pembelajaran Praktik Instrumen di SMK Negeri 2 Kasihan Bantul* [Interpretation of music in practice instruments learning at Vocational High School 2 Kasihan Bantul] (Unpublished doctoral dissertation, Technology and Vocational Education, Yogyakarta State University, Indonesia).

Miles, M.B. & Huberman, A.M. (1994). *Qualitative data analysis*. Thousand Oaks, CA: Sage.

Proust, J. (2013). *The philosophy of metacognition: Mental agency and self-awareness*. Oxford, UK: Oxford University Press.

Ritchhart, R. (2002). *Intellectual character: What it is, why it matters, and how to get it*. San Francisco, CA: John Wiley & Sons.

Trilling, B. & Fadel, C. (2009). *21st century skills: Learning for life in our times*. San Francisco, CA: Jossey-Bass.

Character Education for 21st Century Global Citizens – Retnowati et al. (Eds)
© *2019 Taylor & Francis Group, London, ISBN 978-1-138-09922-7*

Fostering students' character of patriotism and critical thinking skills

R. Yo
Global Sevilla Puri Indah School, Jakarta, Indonesia

N. Sudibjo & A. Santoso
Universitas Pelita Harapan, Jakarta, Indonesia

ABSTRACT: Patriotism and critical thinking are two significant factors in preparing the young generations in facing globalization in order to build our nation. However, the negative impacts of globalization either diminish people' patriotism or escalate chauvinism. As a result, critical thinking serves as an important filter to reduce those negative impacts. The purpose of this study was to analyze the implementations of the Creative and Productive Learning (CPL) model to enhance the students' patriotism and critical thinking skills in their Civics lessons. The action research design, performed among the seven Grade 8 students at Global Sevilla Puri Indah School, a private high school in Jakarta, Indonesia was carried out in the classroom in three cycles. The qualitative data were collected through observations, interviews, field notes, video recordings, photos, and school documentations. The study obviously suggests that the CLP model has enhanced the sense of patriotism and critical thinking skills among the students involved. In the CLP model, both the teacher and students are more engaged in reaching their learning objectives. More importantly, the teacher, as a facilitator and motivator, may perform an important role in stimulating the teaching-learning interactions while the students may take more ownership in exploring the learning materials presented in their lessons.

1 INTRODUCTION

The young generations are the future of our nation and it lies in the hands of those who will lead and develop this nation later on. According to Budimansyah (2010), the young generations should be prepared to participate well in the community. Thus, the nurturing process to be smart and good citizens is the main key. In the Indonesia National Education System law in 2003, it is stated that the national education aims to develop the students' potentials to be become citizens who have faith in God and good morals, as well as those who are healthy, knowledgeable, creative, independent, democratic, and responsible. In Indonesia, since its independence day, schools have been given responsibility to nurture the national insights and nationalism through Civics education, initially introduced with different names. It began formally with a lesson called Civics or Citizenship in the Senior High School curriculum in 1962 and continued with Civics Education in the standardized contents in 2006 (Budimansyah 2010). One of the learning objectives of Civics Education which is stated in the 2006 standardized documents is to educate the students to think critically, objectively, and creatively in dealing with issues of citizenship.

Based on the observations of the researcher, the students at Global Sevilla still did not possess the habits to ask questions critically. In general, when the teacher asked the students to write reflective questions in their journals at the end of the lesson by asking them "What questions do you still want to ask?". In most cases, the students would respond "Is it okay if I don't have any questions?" Civics Education, which is an important lesson to enhance the sense of patriotism, was still received by the students as an abstract lesson with difficult terminologies. The source for learning was merely the handbook which was not adequate to help the students to understand the teaching materials and to reach the learning objectives. The other factor was the application of both the national and international curricula, which limited the time allocation for that lesson or sometimes it was even replaced by any other subjects. Since 2015, as a teacher for the Civics education, the researcher has been motivated to improve the learning processes of that subject.

As stated by Wardani (2014) in her lecture, the principles behind the Creative and Productive Learning (CPL) model are active participation, constructivism, collaborative and cooperative learning, and creative learning. The model has apparently been able to challenge the students' critical thinking.

With the above background, the researcher was motivated to enhance the learning practice quality in Civics Education so that it can enhance the patriotism and critical thinking of the students. Accordingly, four research objectives were formulated as follows:

1. To describe the implementation processes of the CPL model as a tool to enhance the patriotism and critical thinking of the Grade 8 students at Global Sevilla, Puri Indah.
2. To analyze how the processes of the CPL model in enhancing the critical thinking of the Grade 8 students at Global Sevilla, Puri Indah.
3. To analyze how the processes of the CPL model in enhancing the patriotism of the Grade 8 students at Global Sevilla, Puri Indah.
4. To evaluate the obstacles in the implementations of the CPL model in enhancing the patriotism and critical thinking of Grade 8 students at Global Sevilla, Puri Indah.

2 LITERATURE REVIEW

In this section, some relevant concepts used in this paper are discussed.

2.1 *Creative and Productive Learning (CPL) model*

According to Wardani (2014), the CPL model is a learning method that provides the students with the opportunities to produce creative work as a result of re-creation of their understandings of the concepts and main topics learned. The steps are orientation, exploration, interpretation, re-creation, and evaluation.

2.2 *Critical thinking skills*

According to Robertson & Szostak (1996), critical thinking is a skill that will help children to become effective participants in the society. Table 1 describes the expectations of applying the critical thinking skills.

Table 1. Critical thinking skills rubrics.

Aspects to be assessed	3 (High)	2 (Fair)	1 (Low)
Can ask relevant questions as a way to understand the lessons thoroughly	> 3 questions	2–3 questions	1 question
Can ask higher-order thinking questions	3 questions	2 questions	1 question
Can use more than 1 source and state the sources	Always use more than one source	Only use one source and state it	Use more than one source but not state them
Can give relevant and logical explainations	Very logical and relevant explanations	Moderately logical and relevant explanations	Moderately logical but not relevant

2.3 Attitudes of love towards the motherland

Schulz et al. (2008) indicated that patriotism is defined as the loving or concern and good support to the motherland, which allows someone to take action to support their country. Aman (2011, as cited in Ismawati & Suyanto 2015) stated that patriotism is a key factor in building a nation. A nation with people who love their nation will bring their nation to its development. However, with the nation in which the people do not love their country, they will face the risk of bankruptcy. Table 2 summarizes the indicators of the attitudes of love towards the motherland, which were synthesized from the literature.

2.4 Previous research

The following is the list of three research projects on the Creative and Productive Learning (CPL) model, critical thinking, and patriotism:

First, Budiningsih & Rahmadona (2010) conducted a research project on forty-nine students. According to their research, the CPL model increased two aspects on their samples, namely: the students' social abilities and the students' creativities and productivities. For the students' social abilities, 33,33% resulted in the "Very Good" category, 58,97% were in the "Good" category and 7,69% were in the "Medium" category. For the students' creativities and productivities, the research results showed that 35,89% were "Very Good", 53,84% were "Good" and 10,25% were "Medium".

Second, Robertson & Szostak (1996) in their research, entitled "Using dialogues to develop critical thinking skills: A practical approach", conducted a two-step approach model which included class discussions to develop and apply critical thinking skills. During the first step, Robertson & Szostak (1996) chose four to five students randomly. The purpose was to evaluate their written dialogues that were related to biases, interpretations, evidence and errors while the students were reasoning. In the second step, the students were divided into three groups of nine students. In the first group, the students played a different role and job. The roles included analysing thinking errors, identifying errors in thinking, reasoning, evaluating ethical implications and analysing roles that encouraged or restrained the group discussions. The second group became the observers whose role was to evaluate the processes and contents of the discussions which involved the nine students in the first group. The third group's role was to observe the thinking biases and errors, the applications of the reasoning skills

Table 2. Rubrics for the attitudes of love towards the motherland.

Indicators	Questions
A sense of loving towards people with different ethnics, religions, races and groups that live in Indonesia.	Have the learning processes enhanced your caring attitude to your friends so that you don't discriminate their etnics, religions, races, and groups in Indonesia? Yes or No. Explain.
A sense of pride towards the struggles and achievements of this nation.	Have the learning processes enhanced your pride of the struggles and achievements of this nation?
A sense of belonging towards the identities, facilities, and resources of this country.	Have the learning processes enhanced your sense of belonging to maintain the identities, facilities, and resources of this country? Yes or No. Explain.
A sense of appreciation or respect to your nation.	Have the learning processes enhanced your sense of appreciation or respect to your nation? Yes or No. Explain.
Being loyal to defend the country by studying harder in order to participate in the future development of the nation.	Have the learning processes enhanced your motivation to study harder so that you can participate in developing this country in the future? Yes or No. Explain.

*Assessment Criteria: Yes = 1, Yes and No = ½, No = 0.

and the implications of the ethical aspects of the discussions. The research showed that the two-step approach model was effective to get the students involved actively in the discussions where they were given opportunities to criticize, observe the group discussions, apply critical thinking skills provided with guidance. There was not only any guarantee that the students would apply the same critical thinking skills in the same situations again but it also opened up an opportunity for them to apply the skills in various situations.

Finally, Ersner-Hershfield et al. (2010) conducted a research project that involved twenty-five female and fifteen male American university students in 2010. The participants followed the measurement of the political attitudes, which was grouped randomly in a condition called "*Counter*", which consisted of ten questions used to measure ten patriotism attitudes with a scale of seven. The results showed that the patriotism attitudes could be developed through counterfactually reflecting on where people come from or their origins. It indicates that thinking about the non-existence of their countries will lead to the growth of their commitments to their countries without questioning their conditions.

3 METHODOLOGY

In this section, the research methodology which was used in the research is presented.

3.1 *Research design*

The Classroom Action Research (CAR) emerges from the teachers' self-awareness that the practices that they do in the classrooms, bearing the problems to be dealt with and fixed (Wardani 2007). Researchers improved the practice of education in terms of the learning processes and the teacher's performances in Grade 8 at Sevilla Puri Indah Global School. The CAR model used in the study was from the Kemmis & McTaggart's (1986, as cited in Arikunto, 2006) model as illustrated in the spiral model (Fig. 1). This includes planning, implementation (action), observation and reflection. During the research, the improvement occurred gradually and continuously in the three cycles. The activities in the Cycle 1 can be repeated in Cycle 2, which can then be repeated again in Cycle 3. This cycle repetitions can be done until the researchers assume that the results are satisfactorily and; therefore, the research can be completed.

3.2 *Place, time and research subjects*

The classroom action research was conducted at Sevilla Puri Indah Global School at Jalan Raya Kembangan Blok JJ No 1A, West Jakarta by observing the learning proc-

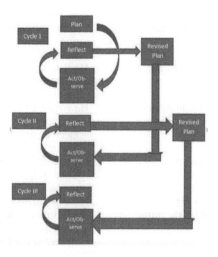

Figure 1. The CAR spiral model of action research.

esses of the Civics lessons conducted by the teacher, who was the researcher herself. The research was conducted between April and May 2016. The subjects of the research were seven students of Grade 8. The ages of the students in the class ranged from thirteen to fourteen years.

3.3 *Techniques and instruments of data collection*

The research procedure, adopted from Wardani (2014), is described as follows:

1. Problem Identification

The teacher found two main problems with the students' critical thinking skills, especially in the questions raised during the Civics lessons and the students' low interests in the Civics lessons.

2. Analysis and problem formulation

The teacher investigated the causes of the problems by interviewing the previous Civics teachers and some students. Furthermore, the teacher held a discussion with the principal to find the alternative solutions for the problems.

3. Improvement planning

At this stage, the plans for the corrective actions, appropriate topics of the learning materials, lesson plans, student worksheets, assessment sheets, observation sheets and the other instruments required to analyze the data were formulated.

3.4 *Actions and observations*

The actions implemented were adjusted to the action settings specified in the lesson plan. The assessment of the students' learning processes was carried out from the beginning to the end of the learning processes. The observations and actions were implemented simultaneously to observe the students' activities in the learning processes. The observations were conducted to identify, record, note, and collect the data from each indicator of the students' performances in the learning processes.

3.5 *Reflections*

A reflection is an activity of the synthesis-analysis, interpretations and explanations of all data or information collected from the conducted action research. The data collected were analyzed and interpreted to find the results of the actions taken.

3.6 *Data analysis*

The data of the students' patriotism and critical thinking skills attitudes were analyzed using the following steps:

a. Giving a score for each assessment criterion observed based on the scoring sheets developed
b. Summing up the scores of all assessment criteria
c. Determining the values based on the scores of each indicator using the following formula:

$$\text{Score} = \frac{Obtained\ Score}{Maximum\ Score} \times 100$$

After that, the calculated results were compared to the classification table of the value category of the students' critical thinking skills. The classification level of the students' critical thinking skills and patriotism attitudes is presented in Table 3.

The classification level of the patriotism attitudes is presented in Table 4.

Table 3. Classification level of the students' critical thinking skills.

No.	Total score	Value	Conclusion
1.	4–6	40–59	Low
2.	7–10	60–79	Fair
3.	11–12	80–100	High

Table 4. Classification of the growth of patriotism.

No.	Total score	Value (%)	Conclusion
1.	4–5	80–100	High patriotism attitude
2.	2–3	40–79	Fair patriotism attitude
3.	≤ 1	0–39	Low patriotism attitude

4 RESULTS AND DISCUSSIONS

In this section, the steps to implement the CPL Model are described in detail.

4.1 *Cycle 1*

Action planning
Cycle 1 was carried out in three meetings of 40 minutes each on 18, 19 and 22 April 2016. The third meeting was an additional meeting for individual presentations and the completion of a written interview about the patriotism attitudes. The learning objectives of the Civics lessons in Cycle 1 were understanding the history of democracy in the world and explaining the core meanings of democracy.

Actions and observations
Tables 5 and 6 provide the assessment results of the seven participants on their critical thinking skills and patriotism after Cycle 1. Five of the seven participants show high enhancement in their critical thinking skills after the three meetings in Cycle 1. Meanwhile, five of the seven participants show high enhancement in their sense of patriotism.

Reflections
The CPL model consisted of the stages of orientation, exploration, interpretation, re-creation and evaluation. It needed longer time so that the students could also present the results. Because of the limited time, the teacher decided to choose the students who did not participate in the competitions and performances as research subjects. The self-study using the Internet encouraged the students to seek extensive references. To provide the sufficient Internet access, the teacher had to conduct the research in the computer laboratory.

4.2 *Cycle 2*

Action planning
Cycle 2 consisted of two meetings on 25 and 26 April 2016 with a 40-minute duration in each meeting. The learning objective of the Civics lessons in Cycle 2 was the importance of democracy in the life of the community, nation, and state.

Actions and observations
Tables 7 and 8 describe the assessment results of the seven participants on their critical thinking skills and patriotism after Cycle 2. Four of the seven participants show high enhancement in their critical thinking skills after the two meetings in Cycle 2. At the same time, four of the seven participants show high enhancement in their sense of patriotism.

Table 5. Assessment results on critical thinking skills rubrics in cycle 1.

Name	Seeking for as many explanations as possible	High order questions	Source and writing	Logical and relevant reasons	Score	Description
PRS	3	3	3	2	11	High
ZLK	2	2	3	3	10	Fair
MCL	2	3	2	3	10	Fair
VLR	3	3	2	3	11	High
ASH	3	2	2	3	10	High
ATS	2	3	3	3	11	High
EVN	3	3	3	3	12	High

Table 6. Assessment results of written questionnaire on patriotism attitudes in cycle 1.

Name	Sense of love/care	Pride	Sense of belonging	Respect	Loyalty	Score	Description
PRS	0	1	1	1	1	4	High
ZLK	0	1	0	0	1	2	Fair
MCL	1	1	1	1	1	5	High
VLR	1	1	1	1	1	5	High
ASH	0	1	0	1	1	3	Fair
ATS	1	1	1	1	1	5	High
EVN	0	1	1	1/2	1/2	3	High

Table 7. Assessment results on critical thinking skills rubrics in cycle 2.

Name	Seeking for as many explanations as possible	High order questions	Source and writing	Logical and relevant reasons	Score	Description
PRS	3	3	3	3	12	High
ZLK	2	2	3	3	10	Fair
MCL	2	3	2	3	10	Fair
VLR	3	3	3	3	12	High
ASH	2	3	2	2	11	Fair
ATS	2	3	3	3	11	High
EVN	3	3	3	3	12	High

Table 8. Assessment results of written questionnaire on patriotism attitudes in cycle 2.

Name	Sense of love/care	Pride	Sense of belonging	Respect	Loyalty	Score	Description
PRS	1	1	1	1	1	5	High
ZLK	0	1	0	1	1	3	Fair
MCL	1	1	1	1	1	5	High
VLR	0	1	1	1	1	4	High
ASH	1	1	0	1	0	3	Fair
ATS	1	1	1	1	1	5	High
EVN	1	0	1	0	0	2	Fair

Reflections

In the self-study in Cycle 2, selecting learning materials based on the learning objectives was more relevant to the students. The students had a very positive impression when learning that the 2014 US election success was praised by the US President, Barack Obama. The achievements of the

Indonesian young generations who successfully watched over the presidential election via www.kawalpemilu.org also inspired the students. As explained by Yuen & Bryam (2007), the teachers could enhance the patriotism of the students if the teachers helped the students develop an emotional interest in the learning materials in addition to adopting the rational approach. During the self-study, the students demonstrated high enthusiasm and focus because the questions were appropriate with the students' interests. Although the students experienced some challenges in reading the Indonesian texts, the students showed great willingness to read more than one source.

4.3 Cycle 3

Action planning
Cycle 3 consisted of two meetings on 2 and 3 May 2016. The learning objective of the Civics lessons in Cycle 3 was the application of democratic values in the life of democracy in the community, nation, and state.

Actions and observations
Tables 9 and 10 indicate the assessment results of the seven participants on their critical thinking skills and patriotism after Cycle 3. Four of the seven participants show high enhancement in their critical thinking skills after the three meetings in Cycle 3, while six of the seven participants show high enhancement in their sense of patriotism.

The first meeting was delayed and; therefore, the students were such in a hurry when they worked on their questionnaire on critical thinking. Cycle 3 can be regarded as the culmination of the learning processes where the students became more aware of the applications of democracy in the school community (from the other students as well as their school leader) and from the leaders of this country.

Reflections
In a self-study of Cycle 3, some students initially hesitated during the interviews with the school leader, but were encouraged by the teacher to do it. The CPL model in Cycle 3 went

Table 9. Assessment results on critical thinking skills rubrics in cycle 3.

Name	Seeking for as many explanations as possible	High order questions	Source and writing	Logical and relevant reasons	Score	Description
PRS	2	1	3	3	9	Fair
ZLK	2	3	2	3	10	Fair
MCL	3	3	3	3	12	High
VLR	2	3	3	3	11	High
ASH	2	3	3	3	11	High
ATS	2	1	2	3	8	Fair
EVN	3	3	3	3	12	High

Table 10. Assessment results of written questionnaire on patriotism attitudes in cycle 3.

Name	Sense of love/care	Pride	Sense of belonging	Respect	Loyalty	Score	Description
PRS	1	1	1	1	1	5	High
ZLK	1	1	1	1	1	5	High
MCL	1	1	1	1	1	5	High
VLR	1	1	1	0	1	4	High
ASH	0	1	0	1	0	2	Fair
ATS	1	1	1	1	1	5	High
EVN	1	1	1	1	1	5	High

well as expected. The students followed the explanations of the guest speaker who shared about her experiences as a leader as well as a parent who had to implement democracy well both at school and at home. This was revealed by some students in the questionnaire of love of the motherland. However, when asked to formulate the questions, some students seemed less enthusiastic than ever before. One of the students (ASH) revealed that the question had been asked and answered during the Question and Answer session. Then, the teacher directed ASH to further explore the answers provided by the guest speaker.

The developments of the students' critical thinking skills and love towards the motherland are depicted in Figures 2 and 3.

4.4 *Constraints in the applications of the CPL model*

The processes of applying the CPL model could run quite well according to the expectations of the researcher due to the support of many parties, notably: fellow teachers, the principal, and the Grade 8 students at **SMP** Global Sevilla Puri Indah. Some obstacles faced by the researcher were among others:

Constraints from the students
The tight schedule of this long study seemed to cause boredom to the students so that it was likely to influence the research results. In addition, the entire series of research cycles were carried out in the final hours of the school so that it was possible for the students to be in the conditions of being extremely tired.

Constraints from the teacher
The teacher who acted as a researcher and observer could not fully record all the dynamics of the learning activities so that she may have lost some important data during the executions of the activities which had to be analyzed further.

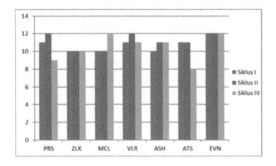

Figure 2.   Developments of critical thinking abilities.

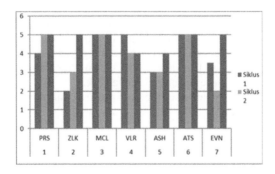

Figure 3.   Growth of love towards the motherland.

Constraints from the school
The Civics lessons were conducted during the last hours of the school and were often used for the students to prepare themselves for musical performances or in-and-out-of-school competitions.

4.5 Some solutions to the existing constraints

The teacher used relevant and actual learning materials and tried to approach the students who were getting burned out during the implementations of the research.

The researcher communicated with the teacher responsible for the competitions and performances well in advance, related to these research activities. In addition, the researcher also chose the times of the research implementations in a quieter week, which was after the Cambridge Checkpoint exam.

5 CONCLUSION AND RECOMMENDATIONS

In this section, the conclusion, recommendations, constraints and limitations, and further research are described.

5.1 Conclusion

The steps of implementing the CPL model could run well with the adjustments to the availability of time. The individual presentations in the classroom were limited due to the time constraints. The researcher provided individual evaluations to the students, both orally and in written form.

The development of the students' critical thinking skills through the CPL model showed some enhancement in the two students at the end of Cycle 3. Two students showed a stable growth. Three students showed a down-and-then-up graph. One of the students showed a declining graph in Cycle 3. The decline was related to the number of the questions asked by the student in Cycle 3. The student just wrote down a question that was personal enough. The student seemed motivated by the question and asked for more time to do the research.

The development of the students' attitudes towards the motherland in the Civics class through the CPL model was significant. Out of the seven students, four students indicated a high level of growth in their attitudes towards the motherland and three students experienced a moderate level of growth. The CPL model encouraged the teacher to explore creative and relevant learning materials. The teacher who acted as a facilitator and motivator needed to have the enthusiasm to deliver the learning materials. The conditions of such a teacher could encourage the students to be more passionate and they could also appreciate in a better way what the teacher was trying to do.

5.2 Recommendations

The Internet connection which was very slow at times could inhibit the students in performing their research. As a result, it is recommended that the school should attempt to improve the Internet speed to provide a more conducive teaching-learning atmosphere.

Due to the long duration of the Classroom Action Research (CAR), the teachers can assign the participants with more structured homework.

More participants may be included in the next research and more extra observers may be involved in collecting the data.

5.3 Constraints and limitations in the implementations of the CPL model

The constraints in the implementations of the CPL model and their solutions are as follows:

The tight research schedule was conducted in seven meetings might have caused some fatigue and boredom for the students. The researcher tried to create fun activities and interesting teaching materials, accompanying and encouraging the students during the learning activities. The teacher attempted to maintain the students' enthusiasm in their learning activities. The constraints could also be overcome because Grade 8 consisted of the students who had positive attitudes toward learning. The students showed a positive response when asked to give more time to fill out the questionnaire.

The CPL model in the Civics classes guided the students to read long enough Indonesian texts. This was a challenge for the students because the students were more familiar with the English texts. However, with the motivations and exercises conducted from Cycles 1 to 3, the students seemed to have begun to get used to the Indonesian texts and the complaints from the students were slowly diminishing.

The CPL model needed to be supported by the availability of various references. The Internet connections and availability of PC's were important to support the success stories of the research activities. These problems could be overcome because the students could have access to the computer laboratory, which was normally used by upper-level students as they had already taken their exams.

5.4 Further research

It is suggested that in the next research, the CPL model can be introduced and implemented to see if such a model may also contribute better results in the other disciplines (i.e. history and religion), which surely require a great deal of comprehension of values, concepts, actual problems as well as applications in real life.

REFERENCES

Arikunto, S. 2002. *Metodologi penelitian* [Research Method]. Jakarta: Rineka Cipta.
Budimansyah, D. 2010. *Tantangan globalisasi terhadap pembinaan wawasan kebangsaan dan cinta tanah air di sekolah* [Globalisation challenge to building nation perception and patriotism in school]. Retrieved 3 January, 2015, from http://jurnal.upi.edu/ educationist/view/308/tantangan-globalisasi-terhadap-pembinaan-wawasan-kebangsaan-dan-cinta-tanah-air-di-sekolah.html.
Budiningsih, C.A. & Rahmadona, S. 2010. *Model pembelajaran kreatif produktif untuk meningkatkan kemampuan sosial, kreativitas, dan produktivitas belajar mahasiswa TPFIP UNY* [Model of creative learning to enhance social ability, creaivity, dand productivity in learning of students TPFIP UNY]. Research Report, Yogyakarta: FIP UNY.
Ersner-Hershfield, H., Galinsky A. D., Kray, L. J., & King, B.G. 2010. Company, country, connections: counterfactual origins increase organizational commitment, patriotism, and social investment. *Journal of Psychological Science* 21(10): 1479–1486.
Fraenkel, J.R. & Wallen, N.E. 2008. *How to design and evaluate research in education*. New York: McGraw-Hill Companies.
Ismawati, Y.T., & Suyanto T. 2015. Peran guru pkn dalam membentuk sikap cinta tanah air siswa [Teachers role in building patriotism]. *Kajian Moral dan Kewarganegaraan* 2(3): 877–891.
Jannah, E. R. 2013. *An analysis of a teacher's questions in EFL classroom: A qualitative case study of English teachers in Public Junior High School Bandung.* http://repository.upi.edu/3202/1/S_ ING_0801193_TITLE.pdf Retrieved 7 December, 2014.
Peraturan Menteri Pendidikan Nasional Number 20 Year 2001. *Standar Penilaian Pendidikan* [Standard assessment education].
Robertson, J. F., & Szostak, S. 1996. Using dialogues to develop critical thinking skills: A practical approach. *Journal of Adolescent and Adult Literacy, 39*(7) 552–556.
Schulz, W., Fraillon, J., Ainley, J., Losito, D., Kerr, D. 2008. *International Civic and Citizenship Education Study: Assessment Framework*. Retrieved 10 February, 2016, from http://research.acer.edu.au/civics/10/.
Wardani, I.G.A.K. 2014. *Strategi Kognitif.* [Cognitive strategy]. Jakarta: Universitas Pelita Harapan..
Wardani, I.G.A.K. (2007. *Penelitian tindakan kelas* [Classroom action research]. Jakarta: Universitas Terbuka.
Yuen, T., & Bryam, M. 2007. National identity, patriotism and studying politics in schools: A case study in Hongkong. *Compare: A Journal of Comparative and International Education, 37(1)* 23–35.

Character Education for 21st Century Global Citizens – Retnowati et al. (Eds)
© *2019 Taylor & Francis Group, London, ISBN 978-1-138-09922-7*

Improving character education through industrial ethics using a model of private sector cooperation

M. Khairudin, K.I. Ismara & S. Soeharto
Universitas Negeri Yogyakarta, Indonesia

ABSTRACT: This study aims to identify the private sector's contribution to improving character education on the basis of industrial ethics in technical vocational education and training. This study uses a descriptive, qualitative approach to explore a cooperation program that was conducted between a vocational high school and industry. The subjects were 30 final-year students of a vocational high school sampled using a proportional purposive technique. Data was collected using questionnaires and analyzed qualitatively using descriptive statistics in the form of percentages. The findings of this study showed that: (1) most (93.75%) of the 30 students put work experience of industrial internship in the Very Good category; (2) based on the assessment of industry instructors, a majority (58%) of students have the competence to cooperate with others, are ready for working under pressure, and have good soft skills.

1 INTRODUCTION

The new Indonesian government policies for vocational education consist of adjusting the proportions between vocational high school (*Sekolah Menengah Kejuruan* or SMK) and general high school, which are 30% and 70% respectively. The government has a goal to reduce high unemployment rates among educated youth, pledging to reverse the current proportions of general high school to vocational high school by 2015 (RoI, 2006), which will have the impact of around five million students moving into SMKs. The outcomes for vocational high school graduates in Indonesia are presented by Chen (2009). Although the Indonesian target is probably not feasible, the ministry has frozen the construction of new public general high schools and converted selected general schools to vocational schools, despite scant evidence that vocational education improves labor market outcomes (Newhouse & Suryadarma, 2009). Worldwide empirical evidence on the merits of vocational education is mixed.

Furthermore, Horowitz and Schenzler (1999) noted that general graduates in Suriname have a higher salary in their country. Similarly Kahyarara and Teal (2008) analyzed several Tanzanian students and found that those who can continue their study into university have high salary in Tanzania. In contrast, El-Hamidi (2006) showed that vocational graduates receive a premium salary in Egypt. Similar finding is also found in Thailand (Moenjak & Worswick 2003).

The sustained economic growth of the past decade has brought several important production process changes in Indonesia and new skills are demanded at both enterprise and individual levels. The availability of skills for enterprises, as well as the relevance of their skills set for the individual, are crucial to ensure that: (a) the Indonesian private sectors can make the best of economic opportunities; (b) Indonesian workers can find decent jobs. The growth in sales of consumer goods has increased the requirement for localized skilled services, which are often provided by family-based industries and other micro-enterprises. In this sense, the Indonesian skills training system, together with the current economic growth, has improved the availability of labor market information.

The absence of discrimination can play a crucial role in ensuring: (a) that individuals and enterprises may adapt to economic changes, in line with the Indonesian pro-growth, pro-job, and pro-poor economic policies; (b) that the benefits of economic progress trickle down to all Indonesians (ILO, 2011).

Vocational high schools adhere to Dual Vocational Education Training (DVET) in order to produce graduates who have job readiness. DVET is one kind of education provision that combines professional expertise in systematic and synchronous education programs for schools and mastery programs obtained through direct industrial work internships (Rochmadi, 2016). It is directed to achieve a certain level of professional expertise.

Characteristics of vocational education include: (1) being geared to preparing students for entering employment (Marfu'ah et al., 2017); (2) being demand-driven (for the needs of the workforce); (3) having educational content that is focused on the knowledge acquisition, skills, attitudes, and values needed by the demand side (private sector); (4) assessment of student achievements according to demand performance indicator (private sector); (5) having a close relationship with the demand side (private sector), which is the key to successful vocational education; (6) being both responsive and adaptable to technological advancement with more emphasis on learning through hands-on experience.

Several principles in mutualism relationship between vocational education with the private sectors, among others, are as follows: (1) vocational education should be implemented as soon as possible (education-in-short); (2) the development of vocational education should be oriented to the kind of jobs that are needed in the field (orientation); (3) vocational education is arranged in such a way that students can join and exit educational institutions easily (free entry/exit); (4) development of vocational education should be open to interdisciplinary interaction (cross-discipline); (5) vocational education should be adjusted to market demand (demand-driven) and the market not adjusted to vocational education; (6) vocational education must dare to develop technologies that will evolve advanced technology.

The program proposed in this study is a cooperation program, which will be conducted by vocational high schools and industry. In the final semester, SMKs will join with industry to improve students' vocational skills. They will work in industries for one semester. Teachers will be collaboration partners between schools and industry instructors. This study is conducted in a public vocational high school in Yogyakarta. Yogyakarta is a heritage and education city that establishes a lot of vocational high schools. It means schools in Yogyakarta need more creativity and effort for sustained schools, especially SMKs. As only a few industries are established in Yogyakarta, this creates additional challenges for SMKs.

Some of the advantages to be gained by students or schools through the cooperation program are: (1) students experience the real work situation, not just seeing or observing; (2) students get paid reasonable care; (3) students gain work experience with a certificate of competence or other recognition; (4) students get employment opportunities in the same workplace after graduation; (5) schools obtain information about the real state of industrial processes; (6) schools get information about worker demand; (7) schools can save costs in in-house laboratory or workshop activities; (8) an interwoven synergistic relationship of symbiotic mutualism and other advantages can be gained by SMKs.

This paper expresses the need to develop the concepts of the cooperation program, which supposes that: (1) the process aims slightly higher than industrial internship; (2) the program should be conducted with the industrial partner; (3) the program is conducted by final-year vocational students in an industry that will treat them as employees of the company; (4) generally, the rights, obligations, and responsibilities of the students are at the same level as permanent employees.

2 COOPERATION PROGRAM BETWEEN SMK AND INDUSTRY

This study was a case study involving one SMK and several industries. The school was located in Yogyakarta and the industries were located in a number of places around Yogyakarta. The distance between the school and the industries for students' work experience was between 10 and 40 kilometers.

The participants of the cooperation program are 32 final-year students of the vocational high school who have several duties at the school, so the cooperation program should involve placement in low-risk, light jobs, or as assistants to a particular job. Conducting the program needs cooperation between the schools and the third sector or private sector. A person will take a placement if they will get gains from the cooperation. A person will not be likely to take a placement if it has no logic or is unsystematic; a highly rational explanation of the purpose is needed to avoid an unprofitable cooperation. Thus, industry will undoubtedly be interested in cooperation offered by SMKs. The cooperation program will create a mutual cooperation between industry and school. In one model, the company takes the initiative to find suitable school partners. The company offers specialized training that is relevant to vocational schools. The aims are to increase the company's production processes, and provide benefits for companies and schools.

In a second model, schools choose the partner companies and make a deal to provide the human resources training that is stated in their agreement. The company is involved in the school's management and provides assistance with funding, equipment, and practice for students. It also provides a number of specialist teachers. Schools establish special programs and training programs, develop a training curriculum to fulfill business objectives, and set the requisite courses. The school also contributes to the development of new technologies and products, and provides training and consultancy services.

The instruments of this study were two questionnaires, one for the students and the other for the industries. The student questionnaires consisted of 20 items that asked five categories of question: (1) students' understanding of the cooperation program; (2) maturity, competence and the ability to work; (3) ability to adapt/adjust to working environments; (4) willingness and abilities to cooperate with others; (5) critically and responsible attitude.

The industry instructors' questionnaires consisted of: (1) guiding process in industry; (2) instructor evaluation of students; (3) curriculum relevance to industrial technology.

The questionnaires were administered at the end of the cooperation program to each group. The responses were categorized as 4, 3, 2 and 1, representing Very Good, Good, Satisfactory and Inadequate, respectively. The data was also analyzed using descriptive statistics in the form of percentages.

3 FINDINGS AND DISCUSSION

With a cooperative relationship between school and company, there is expected to be more extensive opportunities for students to be accepted in the companies. In addition, after doing an industrial internship in a particular company, it is likely that after graduating the student will be directly drawn to or have an employment contract with that company. Therefore, vocational graduates will gain employment and will create a prosperous society through national development while reducing levels of unemployment. In order to improve their capabilities, students can also continue their study through college or polytechnic. Thus, knowledge and skills will improve, leading to better performance and professionalism.

Another advantage that will be achieved is a zero period of adaptation. In this collaboration model, schools can identify the knowledge and skills requirements for the company, and then make full use of company resources to provide appropriate training programs and produce graduates for the industry. This program also allows students to move directly into a position without strict requirements. This includes restructuring the program and curriculum, new training for teachers, and establishing field practice placements in schools and industry. Based on data collected for industrial work experience, the final-year students of the Department of Power Electrical Engineering in a public SMK in Yogyakarta (namely SMK N2) obtained lowest and highest scores of 75 and 106 respectively. This result has mean and median values of 93.3 and 88.0, respectively, and a standard deviation of 9.2.

To identify trends in the performance of work experience in industry for final-year students at SMK N2, Yogyakarta, performance was compared with the lowest and the highest ideal performance. These performances were calculated using four categories, consisting of Very Good, Good, Satisfactory, and Inadequate. A majority (93.75%) of the 30 students stated that work experience in industrial internship was in the Very Good category. The remaining two students (6.25%) placed it in the Good category. It is concluded, therefore, that work experience of industrial internship is categorized as Very Good from the students' perspective.

Industrial internship experience provides good quality for students and also for the school. This can be seen in: (1) final-year students at the Department of Power Electrical Engineering at SMK N2, Yogyakarta, agreeing that industrial internship objectives to generate job readiness of students after graduation were met; (2) the cooperation program participants said the guidance and placement of the cooperation program for industrial experiences was according to student capabilities and conditions—its impact is that students will be competent in their knowledge, skills and ability, which will support the readiness of students to work; (3) the cooperation program participants agreed that knowledge, skills and ability obtained during a cooperation program are in accordance with the industries' demands; (4) the cooperation program participants agreed that facilities used during the cooperation program had been fully equipped to meet the needs of students; (5) the cooperation program participants agreed that the industry instructors provided guidance and motivation about how to work correctly, helped solve problems, and gave encouragement, which will have an impact on students' readiness to work after graduating.

Industry instructors also gave feedback for the cooperation program. Based on the assessment of industry instructors, a majority of students (58%) showed the competence to cooperate with others, were ready to work under pressure, and had good soft skills. However, several students needed more improvement in working skills and attitude. It was noted that the vocational high school should make more preparations in collaboration with industry, and curriculum improvements are also needed in the SMK.

The main features of this collaborative model are work preparation, training by order, and advanced skills training. The school has managed to conduct an extensive study of labor market needs to identify what the market requires, and the specific knowledge and skills requirements of different occupations and jobs. Schools are able to design comprehensive and specialized programs to meet those needs. This has included the successful integration of theoretical and practical training plans, and implementation of approaches for training that recognize the needs of both students and companies. The schools should also be implementing strategies for developing a new kind of professional teacher to integrate the concept of industrial production and teaching. In addition to developing strategies and dual qualities in teachers, the school also has to implement the development of highly skilled teachers in specific competences.

Industry will gain advantages from the cooperation program, including: (1) obtaining cheap labor in the form of probationary employees; (2) getting employees who are observant; (3) needing no layoff terms; (4) having no need to provide severance; (5) obtaining potential students.

In this model, school collaboration aims to improve teaching quality and the provision of services to the local economy by contemplating and utilizing the experience from other countries to improve its own operations, and to build a network for international cooperation and exchange.

To sustain the cooperation program, several activities should be conducted: (1) immediately reorganize the system preparation and operation of a vocational curriculum; (2) immediately re-examine the implementation of a competence assessment and certification system; (3) implement the cooperation program for each department and industry according to their particular characteristics.

4 CONCLUSION

A cooperation program conducted between industry and a SMK has been presented. The cooperation program is designed to improve the existence of technical vocational education and training and to achieve readiness of students for work. The findings of this study showed that: (1) most (93.75%) of the 30 students put work experience of industrial internship in the Very Good category; (2) based on the assessment of industry instructors, a majority (58%) of students have the competence to cooperate with others, are ready for working under pressure, and have good soft skills.

REFERENCES

Chen, D. (2009). Vocational schooling, labor market outcomes, and college entry. *Policy Research Working Paper 4814*. Washington, DC: World Bank.

El-Hamidi, F. (2006). General or vocational schooling? Evidence on school choice, returns, and 'sheepskin' effects from Egypt 1998. *Journal of Policy Reform, 9*(2), 157–176.

Horowitz, A.W. & Schenzler, C. (1999). Returns to general, technical and vocational education in developing countries: recent evidence from Suriname. *Education Economics, 7*(1), 5–19.

ILO. (2011). *TVET centres in Indonesia: Pathway to revitalization*. Jakarta, Indonesia: International Labour Organization.

Kahyarara, G. & Teal, F. (2008). The returns to vocational training and academic education: evidence from Tanzania. *World Development, 36*(11), 2223–2242.

Marfu'ah, S., Djatmiko, I. W., and Khairudin, M. (2017). Learning goals achievement of a teacher in professional development. *Jurnal Pendidikan Teknologi dan Kejuruan, 23*(3), 295–303.

Moenjak, T. & Worswick, C. (2003). Vocational education in Thailand: A study of choice and returns. *Economics of Education Review 22*(1), 99–107.

Newhouse, D. & Suryadarma, D. (2009). The value of vocational education high school type and labor market outcomes in Indonesia. *Policy Research Working Paper*. Retrieved from https://openknowledge.worldbank.org/handle/10986/4229.

Rochmadi, S. (2016). Industry partnerships learning models for surveying and mapping of vocational high schools. *Jurnal Pendidikan Teknologi dan Kejuruan, 23*(2), 210–225.

RoI. (2006). *Rencana strategis departemen pendidikan national tahun 2005–2009* [*Strategic plan of the ministry of national education 2005–2009*]. Jakarta, Indonesia: Ministry of National Education, Republic of Indonesia.

Character Education for 21st Century Global Citizens – Retnowati et al. (Eds)
© 2019 Taylor & Francis Group, London, ISBN 978-1-138-09922-7

An exercise model to develop the biomotor ability of endurance in teenage martial arts athletes

N.A. Rahman & S. Siswantoyo
Universitas Negeri Yogyakarta, Indonesia

ABSTRACT: Endurance is the basic biomotor component that needs to be developed at the beginning of a workout. Today, some coaches do not recognize the importance of this basic biomotor ability in the development of other biomotor abilities. This study aimed to: (1) produce endurance exercise models for adolescent martial arts athletes; (2) investigate the effectiveness of the developed exercise models in improving the endurance of adolescent martial arts athletes. The study procedure employed steps adapted from the research and development model of Borg and Gall (1983), consisting of a preliminary study, development and expert validation, a small-scale tryout and revision, a large-scale tryout and revision, and creation of a final product. The results of the study were as follows: (1) the development produced a ten-exercise model in the form of a guidebook and DVD to improve the endurance of adolescent martial art athletes; (2) the result of the test of the effectiveness of this endurance exercise showed a significant t-test for pretest and post-test data with $p = 0.000$. Therefore, the exercises were effective in improving the endurance of teenage martial art athletes.

1 INTRODUCTION

In Indonesia, there is traditional martial arts are called pencak silat. This is an exercise which focused on skills of = physical defends by maintaining harmony of self with the environment and the natural. This exercise is believed to increase faith and piety to God Almighty (Paiman, 2010).

For a fighter to perform well in matches, whether in the competition category, solo category, dual category, or team category, they must have an excellent biomotor component. The biomotor components needed in martial arts by the fighter include strength, speed, power, flexibility, endurance and coordination. In addition, the psychological aspects of emotional mastery, motivation, intelligence and other related elements are needed in order to be a good fighter.

Physical trainers and good martial arts trainers will first provide training menus to adolescent fighters for the improvement of physical condition in the form of strength and endurance biomotor abilities, which will eventually impact the increase of other physical biomotor abilities that will lead to maximum achievement. In addition, this can reduce the occurrence of injury to the fighter when performing techniques and tactics from the simple to the complex (Nugroho, 2001).

This study focuses on the importance of giving the first practice schedule to the adolescent fighter, which is for the biomotor component of endurance. In teaching endurance to adolescent fighters, physical trainers and martial arts trainers aim to teach good components of endurance, so that the fighter is able to work longer, not experience fatigue too quickly, and make a faster recovery. Getting the fighter to the peak of achievement should start from the development of aerobic capability, anaerobic excitatory thresholds, and anaerobic exercises (Hariono, 2006).

Adolescence is the phase from childhood to adulthood, in both women and men, beginning and ending at different ages. In women it starts at age 10 and ends at the age of 18, while in men it is from the age of 12 to the age of 20. Basic motion and physical exercise learning for boys and girls increases with age and can be done with a variety of movements and exercises. The development of motion in the adult phase continues to run rapidly; along with increasing body size and increased physical ability, it also increases the capability of movement in adults.

Based on preliminary studies through interviews with martial art trainers, there is little knowledge about how to train physical exercise, especially endurance, although they can train based on their experiences of training while having been students themselves.

Physical trainers and martial arts trainers said that because of the limited facilities and infrastructure in the majority of traditional martial arts colleges, fostering and developing teenage fighter physique in terms of endurance is still hard, disciplined, and heavy in physical exercise. Moreover, most coaches still provide models of physical exercise and motion based on their belt level even though specialized endurance exercises should be based on the development and growth of a fighter. In addition, endurance training should take into account the principles of practice and FITT (Frequency, Intensity, Time, Type), as this will positively affect a child's growth and performance achievements. Given these issues, it is necessary to create an exercise model to develop endurance in adolescent fighters with specifications of motion technique and martial art characteristics that can be expected to help the process of physical development in adolescent fighters. A training model to develop the correct biomotor program for adolescents will help physical trainers and martial arts trainers in solving very basic problems so that the physical needs of adolescent fighters can be achieved optimally. In addition, it can produce outstanding martial arts athletes.

2 METHODS

This research method uses research and development to produce a particular product and test its effectiveness. This development research was conducted to produce an exercise model to develop the endurance ability of teenage martial arts athletes. The research was conducted over a two-month period.

The development procedure included collecting data and information, planning, developing start-up products, preliminary trials, major product drafting revisions, key field trials, revisions of operational product preparation, product trial runs, and final product revisions. The types of data obtained in this research and development were qualitative and quantitative. Qualitative data was derived from interviews with trainers and fighters, as well as input information from subject experts, while the quantitative data was obtained through the assessment of the exercise model at the time of the modeling exercises to develop the endurance of the adolescent fighter. The assessment was done by the coach. The instrument of data collection was an interview with guidance, and a value scale in the form of an assessment observation sheet aiming to gather information about the development of the exercise model to develop endurance.

Indicators to develop models of adolescent endurance exercises were derived according to empirical theory, empirical studies and discussions with experts on model development exercises for endurance. Data processing in this study was done using descriptive statistical analysis of quantitative and qualitative data in accordance with applicable rules. Quantitative descriptive analysis was performed to analyze the following data: (1) data from the assessments by subject experts; (2) small—and large-scale test data; (3) tests of effectiveness. After the percentages of eligibility were obtained, the product was deemed to be worthwhile for values above 75%. There were four feasibility categories according to the percentage values: Worthwhile; Less worthwhile; Not worthwhile; Very unfeasible. Qualitative descriptive analysis aimed to describe the results of observation, interview, and statistical results. The feasibility categorization in terms of percentages is shown in Table 1.

Table 1. Feasibility categorization.

Percentage range	Appropriateness
>75%–100%	Worthwhile
>50%–75%	Less worthwhile
>25%–50%	Not worthwhile
≤25%	Very unfeasible

3 RESULTS AND DISCUSSION

3.1 Results of initial product development

The results of the needs analysis conducted by observation, survey, and interviews with four trainers and several fighters in this study are as follows:

1. In accordance with the development of adolescent fighters, an exercise that needs to be developed and providing to them on the basis of their age is endurance training. This is because the initial foundation must be good for the development of the other biomotor abilities as well. In addition, in some models the exercises are created in the form of games because these affect a child's growth: frequent or continuous practice is boring because the training model is always the same due to the lack of variation.
2. The abilities of coaches in physical coaching should be very diverse when developing the physical abilities of fighters, especially teenagers, because many branches of science need to be mastered by the coach.
3. The condition of facilities and infrastructure in training sites is very limited. This is illustrated by the tools available at the two practice sites for developing adolescent endurance.

Based on the above requirements analysis, an exercise model is needed to develop adolescent endurance that can accommodate the needs in the field, especially for fighters developing their endurance according to training objectives. Thus, an exercise model for developing an endurance ability is prepared by taking into account the safety rules of adolescents, which must be clear, challenging and fun.

3.2 Initial draft

Initial draft exercise models to develop the endurance of teenage martial arts athletes are expected to: a) add new variations in endurance exercises during the training process to adolescents at the practice site; b) motivate trainers in utilizing existing facilities and infrastructure; c) assist trainers in developing the endurance of adolescents; d) stimulate the trainer's willingness to teach using modular infrastructure facilities.

3.3 Result of validation data

The validation of the draft model came from subject experts who were physicians, martial arts experts and coaches, and reported eligibility above 75%, while media- and product-focused linguists also reported eligibility above 75%, which means the draft exercise model for developing the endurance of adolescent fighters was suitable for a small-scale test.

3.4 Small-scale test results

Based on the results of the experiment, the draft of the small-scale exercise model shows the amount of value attributed to the "material" at 156 (21.4%), and the total value of the books and videos at 571 (78.54%), while the conformity assessment of the draft exercise model (material and product) by seven trainers declared 56 (7.7%) of the draft model exercises to be unsuitable, 507 (7.7%) of the exercises to be appropriate, and 164 (22.56%) draft model exercises were described as very suitable. Thus, in terms of the material, it can be concluded that the results of the implementation of the exercise model to develop the endurance of the adolescent martial arts athlete in small-scale trials can be said to have conformed with the objective, being easily and safely done by adolescents, easy to understand, and useful to trainers when providing endurance training to adolescents.

3.5 Large-scale test results

The experimental results of large-scale testing of the exercise model report the value obtained from the "material" as 765 (40.9%) and the total value of books and videos as 1106 (59.1%). In the assessment of conformity of the exercise model (material and product), 13 trainers declared 16 (0.8%) exercise models to be very unsuitable, 88 (4.7%) exercise models were declared unsuitable, 1011 (54%) training models were declared as fit, and 756 (40.4%) model exercises stated that the manual and video model of exercise could develop adolescents' endurance in large-scale trial in terms of the material, can be said to have conformity with the objectives of developing endurance, namely (1) the exercise model is easy to implement; (2) the exercise model can develop endurance; (3) the exercise model can motivate fighters, and is safe in execution with easily obtainable tools. It can be concluded from the assessment that the exercise model is appropriate for developing endurance in adolescent fighters.

3.6 Review of the final product

The final product made in this research and development is in the form of guidebooks and video execution of exercise models to develop endurance in adolescent fighters. In the process of drafting a model of an exercise product to develop adolescent endurance, the products that have been created are first validated by the experts to obtain their input before being tested. After that, the testing was carried out so that in the end it produced a model of an exercise product to develop the endurance of the adolescent fighter. The final product created in the form of "Model Exercise for Developing Endurance Ability of Teenage Warriors" is detailed in the manual and comes with DVDs that can be viewed in accordance with the model instructions. The final endurance training model consists of ten exercise models, circuits 1 to 10.

3.7 Result of end-product effectiveness test

The final product of the development is an exercise model for developing endurance in adolescents. Using the final product created, an effectiveness test was performed in MAN 2 Semarang, which involved ten fighters. Field testing of the final model of development was done through observation. The effectiveness test was conducted in 12 sessions of physical exercise.

Following conversion of the raw scores into values, pretest and post-test result data can be seen in Table 2.

Based on the results in Table 2 and Figure 1, it can be seen that the average test value at pretest was 36.31 and at post-test was 38.59. Thus, there is a pretest to post-test increase of 2.28. Tables 3 and 4 summarize the programs and the amount of endurance training included.

Endurance is a condition of the body that describes its ability to practice for a long time, without experiencing excessive fatigue after completing the exercise (Kardjono, 2008). If an athlete has good endurance ability, the cardiovascular, respiratory, and circulatory systems

Table 2. Pretest and post-test results.

Test	Pretest	Post-test	Δ
Balke endurance test	36.31	38.59	2.28

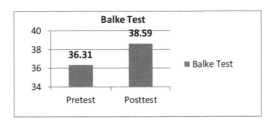

Figure 1. Effectiveness of the exercise to develop endurance biomotor ability.

Table 3. Endurance training program.

Type of exercise	Amount of exercise	Purpose of exercise
Endurance exercise	Intensity : 85–90% max Duration : 2–5 minutes Repetition : 6 rep. Set : 1 set T. interval : 1:1/2–3	Train the capabilities of aerobic and anaerobic energy systems

Table 4. Amount of endurance exercise.

Amount of exercise per week

Week 1		Week 2	
Monday, Wednesday, Friday		Monday, Wednesday, Friday	
Intensity	: 90% max	Intensity	: >95% max
Duration	: 30 seconds	Duration	: 25 seconds
Repetition	: 5 rep.	Repetition	: 8 rep.
Set	: 5 set	Set	: 4 set
T. interval	: 1:3	T. interval	: 1:3
T. recovery	: 2:1 minutes	T. recovery	: 2:1 minutes
Tempo	: Fast	Tempo	: Fast
Week 3		Week 4	
Monday, Wednesday, Friday		Monday, Wednesday, Friday	
Intensity	: >95% max	Intensity	: >95% max
Duration	: 20 seconds	Duration	: 25 seconds
Repetition	: 10 rep.	Repetition	: 10 rep.
Set	: 4 set	Set	: 4 set
T. interval	: 1:3	T. interval	: 1:3
T. recovery	: 2:1 minutes	T. recovery	: 2:1 minutes
Tempo	: Fast	Tempo	: Fast

work well so that the provision of energy during activity can take place smoothly. The advantage will be felt by athletes at the time of competing as they will be quicker to recover and will be able to work longer at high intensity.

Physical trainers and martial arts trainers must understand the predominance of the energy system used, as this can form the basis for consideration in determining an appropriate method of improvement. Based on simple observation, the predominant energy system in martial arts is adenosine triphosphate/creatine phosphate (ATP-CP; 73.75%), lactic acid/O_2 (LA-O_2; 16.25%), and O_2 (10%) (Hariono, 2005). Paying attention to this predominance will result in measurable exercise that ultimately produces athletes who can display the results of their exercise at the time of competition with a brilliant performance because of their excellent durability.

In a study (Helgerud et al., 2007), high-intensity aerobic interval training compared with long-range intensity and lactate threshold. give conclusion, that high-intensity aerobic interval training was significantly more effective than doing the same total work in terms of both lactate threshold or 70% of maximal heart rate (HR max) in increasing VO_2 max (maximal oxygen consumption). In other studies (Roziqin & Widodo, 2013), the results obtained for average endurance were 41.8 ml/kg/min in pretest and 44.4 ml/kg/min at post-test. This means that physical endurance training using a varied exercise model, as long as it remains in accordance with an amount that has been adapted to individual principles, overload principles, progressive principles, and careful and measurable variation principles, can positively impact endurance capabilities in athletes.

In our study, using a circuit training model with a high-intensity interval method is also consistent with the results of research indicating that regular aerobic exercise can increase VO_2 max by making the heart and respiratory systems more efficient, thereby channeling O_2 to more active muscles. The exercise muscles themselves become increasingly able to use the O_2 distributed to them (Seiler, 2010). Aerobic exercise is exercise that is carried out continuously in which the oxygen needs can be met by the body. The best exercise is that done between the ages of 18 and 20 years old because this is when the value of VO_2 max peaks. There is research that aims to determine the effect of aerobic exercise in increasing VO_2 max in adolescents aged 18–20 years through the form of running for 30 minutes four times a week for four weeks. VO_2 max is measured using the Cooper test. The results showed that such aerobic exercise increased VO_2 max in adolescents aged 18–20 years (Kumarudin, 2013). This is in accordance with research conducted into the provision of a circuit training model to develop endurance capabilities in adolescent fighters.

In a subsequent study, it was aimed to determine the effect of interval training and circuit training on aerobic endurance improvement, and to understand the difference between interval training and circuit training. From the research it was found that: (1) there is an influence of interval training and circuit training in increasing aerobic endurance, with aerobic endurance increased by interval training by 13%, and by circuit training by 8%; (2) interval training is a better influence than circuit training in increasing aerobic power (Khotimah, 2011).

This is highly consistent with research conducted by researchers who want to develop and improve endurance using an exercise model based on endurance training. The results obtained from a circuit training model with a high-intensity interval method show an increase in VO_2 max in teenage martial arts athletes. Indeed, VO_2 max is very influential in everyday life as it can increase physical activity, which is especially important for martial arts athletes (*pencak silat*) in achieving maximum performance. In order to improve VO_2 max it is necessary to practice careful, systematic and orderly exercise. This was seen in futsal athletes where research wanted to establish the increase in VO_2 max of futsal players through continuous running and circuit training. The results obtained showed an increase in VO_2 max in futsal athletes. However, it was found that circuit training was more effective in increasing VO_2 max than continuous running exercises (Masdar, 2017). This validates the development of a circuit training model that aims to improve and develop endurance and that can be used as an alternative exercise in supporting physical qualities of athletes, because research shows that circuit training can be a very significant training exercise for athletes.

This study also aims to prove the influence of circuit training on leg muscle strength and VO_2 max in adolescent boys. Based on the results of data analysis and discussion, it can be concluded that circuit training has an effect on increasing muscle strength of limbs in adolescent male students with significant value: less than 0.01. It also has an effect on increasing VO_2 max in teenage students with a significance value below 0.01. It is recommended that sports athletes use this training as an alternative for improving leg muscle strength and VO_2 max (Hariyanta et al., 2014). An effective circuit training model is used for defined practice purposes. A circuit training model can be used with the aim of improving physical quality. Physical qualities include increased strength, endurance, speed, and agility. It is seen above that circuit training can be used to increase leg muscle strength while increasing VO_2 max.

There are other studies that confirm that circuit training can improve VO$_2$ max. The results showed that there were significant improvements through circuit training of 22.36 and cross country of 33.54 in VO$_2$ max in extracurricular taekwondo students' (Ambarwati & Jubaedi, 2014). The interval of training can also provide a significant effect in improving physical fitness and VO$_2$ max (Syaifudin et al., 2015).

All of the above research results from both forms of endurance training show such training should be the focus at the start of exercise, which includes the stage of specialization. If this initial foundation is well established, it will also affect the development of other biomotor abilities so that fighters will be more efficient during both exercise and competition in applying energy to technique and movement, as well as also being better tactically and psychologically. In addition, it will support their sports achievements.

4 CONCLUSIONS

In the development of an exercise model to develop the biomotor ability of endurance in adolescent martial art athletes there are two conclusions.

First, a model exercise to develop the biomotor ability of endurance of adolescent martial arts fighters based on circuit training has been created in accordance with validated martial arts techniques. There is also a training model for how to execute the exercise model in the form of guidebooks and accompanying DVD videos to develop the biomotor abilities of strength and endurance in adolescents that is classified in good category.

Second, increased effectiveness of training to develop endurance is achieved with a significant endurance training model through exercises that emphasize the effectiveness of developing strength and endurance in adolescent combatants. Observation of testing the end product in the field through a pretest/post-test experimental method showed increased endurance from one test conducted with adolescent martial arts fighters. In the accompanying t-test, the strength and endurance increases were "significant" with p = 0.000. Thus, the exercise is declared effective for increasing the endurance of adolescent fighters. This ten-circuit exercise model can be used as an alternative exercise to improve endurance.

Observation data of end product test in field through a pretest-posttest experimental method showed endurance biomotor from one test conducted to adolescent material arts fighters experience improvement with "increase" information.

There is arranged model of exercise to develop endurance biomotor ability of adolescent martial art that is a model of circuit training and how to execute the model, a training model to be trained in accordance with the validated martial arts techniques as well as the compilation of guidebooks and accompanying DVD videos in developing the biomotor strength and endurance of adolescent teens in the "decent" category.

REFERENCES

Ambarwati, R.H. & Jubaedi, A. (2014). The influence of circuit training and cross-country exercises on VO$_2$ max. *Journal of Education, 2*(2), 99–110.

Borg, W.R. & Gall, M.D. (1983). *Educational research: An in-troduction* (5th ed.). New York, NY: Longman.

Hariono, A. (2005). Predominant energy system in martial arts category. *Sports Scientific Magazine, 11*(3), 427–440.

Hariono, A. (2006). *Method of physical training of martial arts*. Yogyakarta, Indonesia: Faculty of Sports Science, Universitas Negeri Yogyakarta.

Hariyanta, I.W.D., Parwata, I.G.L.A, Wahyuni, N.P.D.S. (2014). Pengaruh circuit training terhadap kekuatan otot tungkai dan VO$_2$ max [Effect of circuit training on leg muscle strength and VO$_2$ max]. *Jurnal Ilmu Keolahragaan I*, 1–11.

Helgerud, J., Høydal, K., Wang, E., Karlsen, T., Berg, P., Bjerkaas, M.,... Hoff, J. (2007). Aerobic high-intensity intervals improve VO$_2$ max more than moderate training. *Journal of the American College of Sports Medicine, 39*(4), 665–671. doi:10.1249/mss.0b013e318 0304570.

Kardjono, K. (2008). *Physical condition course module*. Bandung, Indonesia: UPI.

Khotimah, N. (2011). *Effect of interval and circuit training on aerobic endurance improvement* (Unpublished thesis, Universitas Negeri Yogyakarta, Indonesia).

Kumarudin, A. (2013). *The effect of aerobic exercise on maximal oxygen volume increase (v max) in adolescents aged 18–20 years*. Surakarta, Indonesia: Universitas Muhammadiyah Surakarta.

Masdar, R.I. (2017). *Effects of continuous running and circuit training on VO$_2$ max futsal players*. Surakarta, Indonesia: Universitas Muhammadiyah Surakarta.

Nugroho, A. (2001). *Handout of martial arts practice guidelines*. Yogyakarta, Indonesia: Faculty of Sports Science, Universitas Negeri Yogyakarta.

Paiman, P. (2010). Pengaruh gizi terhadap hasil belajar pencak silat [The influence of nutrition on the performance of martial arts sport]. *Proceedings of Seminar Nasional III, Universitas Negeri Yogyakarta, Indonesia* (pp. 104–110).

Roziqin, A.K. & Widodo, A. (2013). Effect of physical exercise models using ball against aerobic football players age 15–18 years. *Sports Health Journal, 1*(3), 53–56.

Seiler, S. (2010). What is the best practice for training intensity and duration distribution in endurance athletes? *International Journal of Sports Physiology and Performance, 5*, 276–291.

Syaifudin, A.W., Jubaedi, A., and Wiyono, W. (2015). Pengaruh interval training terhadap kebugaran jasmani dan VO$_2$ max [The influence of training interval on physical fitness and VO$_2$ max]. *Jurnal Penjaskesrek, 3*(1), 1–8.

Implementation of a work-based learning model to improve work attitudes and learning achievements of students in vocational education

D. Rahdiyanta & A. Asnawi
Universitas Negeri Yogyakarta, Indonesia

ABSTRACT: This research aims to study the impact of a work-based learning model of students' work attitudes and learning achievements in machining practice in vocational education. This experimental study was conducted in two manufacturing industries in Yogyakarta Special Region and in a machine workshop in the Mechanical Engineering Department, Faculty of Engineering, Universitas Negeri Yogyakarta. The research population included all students undertaking the course of Complex Machining Process, with a total number of 85 students. The research sample consisted of 32 students, determined by a purposive sampling technique. The experiment was conducted using post-test-only control design. The data collection techniques were questionnaires, observation sheets, documentation, and learning output assessment. Data was analyzed with descriptive analysis and different-test with significance level of 0.05. Before the analysis of test-different, first the normality and homogeneity of the research data were tested. The research results revealed that there are significant differences in students' work attitudes and learning achievements in the class using the work-based learning model (experimental class) and the class that did not use the work-based learning model (control class). The students' work attitudes and learning achievements in the experimental class were better than those in the control class.

1 INTRODUCTION

Vocational education, as part of the education system, plays a strategic role in the realization of a competent and professional workforce. Competent and professional workforces require the basic knowledge and ability to adapt to the demands and dynamics of ongoing developments. Thus, to face the intense competition of this global era requires not only workforces with hard competence but also, soft competence. The challenge for vocational colleges is therefore to integrate the two components, thereby giving students the ability to work and to develop in the future.

Along with advances in technology and the workplace dynamic, vocational education must be able to anticipate and face the challenges associated with the demand from employers for higher work competencies. One of the strategies that can be employed to address the challenges associated with the demand for increasing job competencies is to improve the quality of practical teaching in vocational colleges. The practical course is the core course and becomes the hallmark of vocational education. The development of practical teaching in vocational education must therefore be done continuously so that the quality of graduates meets the demands of employers. To ensure the outputs of vocational education are in line with the demands of the workplace, vocational education institutions should be able to cooperate with the business world and industries. Theories of experiential learning, contextual teaching and learning, and work-based learning become highly relevant in the delivery of vocational education.

In order to follow the above aspirations, it becomes the responsibility of vocational colleges to produce graduates with excellent academic competences and characters. Accordingly, it becomes an obligation to immediately implement character values in the learning process like methods of based learning (Rahdiyanta et al, 2017). One of the ways to implement character values in learning is to develop a Work-Based Learning (WBL) model that emphasizes the integration of aspects of academic competence (hard-skill) and aspects of the character (soft-skill) in practical courses in vocational colleges.

In order to improve the quality of vocational education, it is important to examine whether the implementation of a work-based learning model with integrated character education has an effect on the formation of students' work attitudes and learning achievements in the sector of machining process.

2 LITERATURE REVIEW

2.1 Work-Based Learning (WBL)

Work-Based Learning (WBL) is a learning approach exploiting the workplace to create workplace experiences that contribute to social, academic, and student career development. Workplace learning experiences are applied, refined, and expanded in teaching both on campus and at work. With WBL, students can develop the attitudes, knowledge, skill, insight, behavior, habits, and associations related to real-life work activities (Lynch & Harnish, 1998). Several forms or models of WBL include apprenticeship opportunities, career mentorship, cooperative work experience, credit for prior learning, internships, job shadowing, practicum, school-based entrepreneurship, service learning, teacher externship, tech-prep, vocational student organizations, volunteer service, and worksite field trips (Iowa Workforce Development, 2002).

In vocational education, the application of the WBL model to practical courses is highly appropriate. This is as expected, since practical courses are the hallmark of vocational education. Therefore, the development of practical teaching in vocational education must be done continuously so that the quality of graduates meets the workplace demands. To align the outputs of vocational education with the demands of the workplace, the operation of vocational education should be closely tied to the business world and industries.

2.2 Character education

Character education is not merely to teach what is right and what is wrong. Character education is an effort to internalize good habits (habituation) so that students are able to behave and act on the values that have become incorporated into their personalities. According to Lickona (1991), a good character education should involve moral knowing, moral feeling, and moral behavior. In Indonesia, character education is engaged as a foundation to realize the visions of national development, which are to create a noble, moral, ethical and civilized society based on the philosophy of Pancasila (RoI, 2010). Character education plays an essential role in overcoming current national problems, such as the shifting of ethical values in the life of the nation and state, the fading of the nation's cultural value awareness, the threat of national disintegration, and the weakening of the nation's independence. To strengthen the implementation of character education in educational units, 18 values have been identified derived from religion, Pancasila, culture, and national education goals; these are religiosity, honesty, tolerance, discipline, diligence, creativity, independence, democratism, curiosity, patriotism, nationalism, appreciation of achievement, being sociable/communicative, pacifism, having a culture of reading, environmental awareness, social awareness, and responsibility (RoI, 2011).

Even though the 18 values of national character building have been formulated, educational units can base their priorities on existing values they have developed. The implementation of national character values can be initiated from the essential and applicable ones, such as cleanliness, neatness, comfort, discipline, honesty, and politeness.

Teaching should focus on the teaching process and not simply the transfer of knowledge. The method of merely transferring knowledge is described by Hiltz (1998) as "the stage on the stage". This method is giving no chances for interactions and transactions with the learners. Therefore, the learning process has to provide the students with practices that stimulate critical thinking and social interaction. Teachers should also consider the cultivation of character values or soft-skills, such as cooperation, appreciating opinions, sense of belonging, sense of responsibility, honesty, and sacrifice.

In reality, teaching that encourages critical thinking and social interaction is still rarely seen. Consequently, it is undeniable that the development of aspects such as cooperation, respecting opinions, recognizing one's self and others are neglected in the learning process. Thus, it is necessary to take steps to improve our education process and system, especially the learning process, to put more emphasis on the development of cognitive, affective, and psychomotor dimensions in a balanced way.

2.3 *Vocational education*

According to Calhoun and Finch (1976), vocational education is an education program directly linked to one's preparation for entering the workforce or for supplementary training required in a career. Furthermore, according to Finch and Crunkilton (1979), vocational education is defined as an education that prepares students to work for their futures. Based on those opinions, it means that vocational education is needed to prepare students to be ready for work both within and outside their social communities or environments. Therefore, the main mission of educators and policy-makers is to prepare a strong foundation in the teaching and learning process for the mastery and application of academic skills as well as the necessary concepts needed to face the real workspace.

Wardiman (1998) states that a vocational college has the following characteristics: 1) directed to prepare learners entering the workplace; 2) based on a demand-driven model; 3) focused on the mastery of knowledge, skills, attitudes, and values required by the workplace; 4) assesses students' achievements based on hands-on or work performance; 5) has good linkage to the workforce as the success key of vocational education; 6) is responsive and anticipates technological advancements; 7) emphasizes learning through practice and hands-on experience, 8) requires cutting-edge facilities for practical work; 9) demands more investigation and operational costs than other standard educational institutions.

The implementation model of character-based WBL in this study is a development of the existing model of machining practice learning. The improvements are: 1) practice jobs that is making real products produced by partner industries; 2) existence of field practice activities in partner industries; 3) implementation of integrated learning, which is the involvement of partner industries in the process of providing guidance to students both in the forms of hard-skill and soft-skill.

From the above descriptions, it is the responsibility of education, especially vocational education, to create graduates who have both academic competencies and good character. Practical courses are the main courses that characterize vocational education. It is, therefore, a necessity to integrate character values into the practice learning process. One way of integrating the values is through developing a WBL learning model that is based on characters in vocational education.

In order to improve the quality of vocational colleges in Indonesia, it is essential to examine whether the teaching of machining practices with character-based WBL influences the establishment of students' attitudes and achievements in the machining process.

3   RESEARCH METHOD

The implementation of a character-based WBL model on a Complex Machining Process course was conducted using the experimental design of post-test-only control design. In this research, pre-test is no needed, because the result can be seen from the practice result work piece. There for the research design is presented in Figure 1.

$$\begin{array}{ccc} R & X & O2 \\ R & & O4 \end{array}$$

Figure 1.   Post-test-only control design.

Note: R = control class and experimental class took randomly; O2 = *post-test* of experimental class; O4 = post-test of control class.

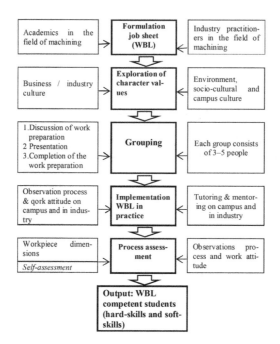

Figure 2.   The character-based WBL model.

This research was conducted in two manufacturing industries in Yogyakarta Special Region and in a machine workshop in the Mechanical Engineering Department, Faculty of Engineering, Yogyakarta State University. The study was conducted over three months.

The study population included all students undertaking the course of Complex Machining Process; a total of 85 students. The research sample included 32 students, determined using a purposive sampling technique. The sample of the study, which amounted to 32 students, was grouped into two classes, each of which amounted to 16 students in each of the experimental and control classes.

The data collection technique used observation sheets, documentation, and evaluation of learning outcomes. The validity of the research instrument was performed by expert judgment. The data of the research were then analyzed both qualitatively and quantitatively. To examine the effectiveness of the developed model compared to the prior model, a *t-test* was applied. Before being analyzed, the data passed through normality and homogeneity tests.

The design of the character-based WBL model is displayed in Figure 2.

4   RESULTS

This research was carried out for eight meetings. The first and second meetings were for the explanation and preparation of learning activities, while the third to eighth meetings formed the core of this research activity. Thus, starting from the third until the eighth meetings, aspects of work attitudes and student learning achievements needed to be considered carefully. In accordance with the characteristics of the machining process course, aspects of work attitudes to be observed were whether the students were honest, disciplined, diligent, conscientious, independent, hard-working, and caring. Additionally, learning achievement was also reflected by the completed job sheet during learning complex machining process, particularly components of *speed-reducer,* namely worm gear, worm screw shaft, and worm gear shaft.

4.1 *Description of data*

Data from the research activity on student attitudes in Complex Machining Process learning, both in the experimental class and the control class, can be seen in Table 1.

Table 1. Students' attitudes.

Aspects	Average number of students in the six meetings		Percentage (%)	
	T1	T2	T1	T2
Honest	13.0	8.8	0.81	0.74
Disciplined	15.0	10.0	0.94	0.83
Diligent	12.0	6.5	0.75	0.54
Conscientious	11.7	6.2	0.74	0.52
Independent	12.5	6.5	0.78	0.54
Hard-working	11.7	5.8	0.73	0.49
Caring	13.8	7.8	0.86	0.65
Total Average	12.81	7.38	0.81	0.63

Note: T1 = Experimental class, T2 = Control class.

Table 2. Students' learning achievements.

Students	The average score of three jobs	
	T1	T2
1	83.67	67.00
2	83.33	63.33
3	79.33	68.67
4	82.33	70.67
5	81.00	68.00
6	81.67	64.00
7	83.67	65.00
8	81.33	65.67
9	78.67	65.33
10	80.67	67.33
11	81.33	66.67
12	80.33	65.33
13	81.67	66.67
14	82.33	63.33
15	80.00	65.33
16	82.33	64.00
17	–	70.67
Total Average	82.31	65.28

Note: T1 = Experimental class, T2 = Control class.

Accordingly, in the eighth meeting, almost all students in the experimental class possessed all aspects of attitude expected. Furthermore, overall, seen from the third to the eighth meeting, more than 80% of students in the experimental class had the attitudes of discipline, honesty, and caring.

Data on the students' learning achievements were taken from the assessment of work pieces from three practices. the full data of the students' learning achievements in the Complex Machining Process course can be seen in Table 2.

4.2 *Testing analysis requirements*

The next stage was to test the analysis requirements in accordance with the type of analysis to be used, which was the *t-test*. An analysis requirement test is a test of normality and homogeneity. To test the data of the research results, whether or not it had a normal distribution, skewness and kurtosis ratio tests were employed. Data are considered to have a normal distribution if the value of skewness kurtosis ratios range from –2 to +2. From the result of the normality test, we concluded that the data distribution in both the experimental and control class followed a normal distribution. For the experimental class, the skewness ratio was –1.525 and the kurtosis ratio was (0.247), while for the control class, the skewness ratio was 0792 and the kurtosis ratio was –0.372.

The homogeneity of the research data was tested using Levene's statistic. The test result on student learning achievement obtained a significance of 0.159 based on a mean bigger than 0.05. Likewise, the result of data testing on student attitudes gained a significance of 0.156 based on a mean bigger than 0.05. Therefore, the test revealed that the data was homogeneous. According to the result of analysis requirements test, t-test technique can be used in this resereach.

4.3 *Data analysis*

In line with the research result, the average of students' learning achievements in the experimental class was 82.31, while in the control class it was 65.28. The t-test result showed that $t = 8.473$; $p = 0.000$. Thus, it was proven that there was a significant difference between student learning achievements in the experimental class and the control class. The learning achievements of the experimental class were better than that of the control class (X experimental = 82.31 > X control = 65,28).

The results of research on student learning activities indicate that 81% of students in the experimental class have a good working attitude, while in the control class this figure is only 63%. The t-test result showed that $t = 7.521$; $p = 0.000$. This means that there is a significant difference in the students' work attitudes between the experimental class and the control class.

5 DISCUSSION

Based on the implementation of character-based WBL model, it has been determined quantitatively that the learning model is able to combine aspects of attitudes or behaviors that build the character of the learner. This is reflected in the activities during the practical learning process seen during the model implementation process.

Important stages in the implementation of this model are the determination of the types of jobs that will be done in the practice on campus. Job practice should create real objects that match what is produced in the industry. Job practice in the form of project work done in groups, is appropriate to encouraging students to learn with passion.

At the exploration stage of work attitude, it shows that the character-based WBL learning model is very effective in exploring student awareness related to aspects of work attitude. In this stage, learners are required to convey their opinions related to aspects of work attitude that should be owned by a person, especially in implementing the practice learning processes. The purpose of the implementation is to instill students' awareness of the character aspect.

Thus, it is expected that students will more easily focus on applying the character aspects in the process of machining practice learning. This is confirmed by the observation of student activities during the learning process: students carry out the character aspects well. So, the character-based WBL learning model is indeed effective in integrating aspects of the character in the process of practice learning.

The composition stage of the Work Preparation Sheet also showed student activities that were very positive. In preparing the Work Preparation Sheet, students were required to carry out learning collaboratively. Implementing this collaborative learning means the students become accustomed to having the courage to deliver opinions, respect others' opinions, and cooperate well. This is in line with Hill & Hill (1993), which states that there are several merits gained through collaborative learning, such as: 1) promoting higher learning achievement; 2) providing deeper understanding; 3) experiencing fun learning; 4) developing leadership skills; 5) upgrading positive attitudes; 6) boosting self-esteem; 7) learning inclusively; 8) achieving a sense of belonging; 9) developing prospective skills. Therefore, students are required to be able to collaborate and respect their team-mates and others. Similarly, according to Qin, et al. (1995), there is extensive empirical evidence that collaborative and cooperative learning experiences can improve higher academic achievement more than individual learning and competitive learning experiences do.

Another stage in the process of character aspects integration is held during the process of evaluation of the product from the practice work result. Before assessed by the teacher, the workpiece was formerly assessed by the students themselves—self-assessment. In this process, the students were obliged to carry out self-measurement of their own workpieces, of which the assessment was filled in on the provided sheets. The data of the self-assessment result was then cross-checked by the teacher. Through this activity, it can reveal the level of student honesty, especially in carrying out the self-assessment.

Based on the results of implementation, it can be concluded that there is a difference in aspects of work attitude of students in the experimental class and those in the control class. This is revealed from the differences in the student activities during the learning process; the experimental class was far more active or better than the control class. Therefore, it could be said that the learning achievement results were in accord with the observation data. Based on the obtained data, in the experimental class, where the level of activity was better, the learning achievement was also much higher than in the control class. Some of these facts are in accordance with a study conducted by Berkowitz and Bier (2007), which states that there could be an increase of motivation in academic achievement when schools implement character education. The results of this study are also in line with the results of a research study conducted by Bailey and Merrit (1997), which concludes that the use of a character-based WBL model in vocational education has a positive influence on student achievement and learning motivation.

6 CONCLUSION

The research results revealed that there are significant differences in students' work attitudes and learning achievements in the class using the work-based learning model (experimental class) with the class that did not use the work-based learning model (control class). Students' work attitudes and learning achievements in the experimental class are better than those in the control class.

Implementation of a work-based learning model integrated with character education proved very effective in improving students' work attitudes and learning achievements in machining practice. This can be seen from the striking difference between the average score of students' work attitudes and learning achievements in the experimental and the control class. Therefore, this learning model needs to be implemented in practical learning for other areas of expertise in vocational education.

Implementation of a work-based learning model integrated with character values puts more emphasis on the activities of learners in the learning process. Therefore, the role of lecturer/instructor should be more focused on the process of supervision and guidance to learners.

REFERENCES

Bailey, T. & Meritt, D. (1997). Youth apprenticeship: A lesson from the U.S. experience. *Center Focus, 1*.

Berkowitz, M.W. & Bier, M.C. (2007). What works in character education. *Journal of Research in Character Education, 5*(1), 29–48.

Calhoun, C.C. & Finch, C.R. (1976). *Vocational educational: Concepts and operation*. Belmont, CA: Wadsworth.

Finch, C.R. & Crunkilton, J.R. (1979). *Curriculum development in vocational education*. Boston, MA: Allyn and Bacon Inc.

Hill, S. & Hill, T. (1993). *The collaborative classroom. A guide to cooperative learning*. Armadale, Australia: Eleanor Curtin Publishing.

Hiltz, S.R. (1998). *Collaborative learning in asynchronous learning networks: Building learning communities*. New Jersey, NJ: New Jersey Institute of Technology.

Iowa Workforce Development (2002). *Annual Learning*. Retrieved from https://www. iowaworkforcedevelopment.gov/2002-annual-report.

Lickona, T. (1991). *Educating for character: How our school can teach respect and responsibility*. New York, NY: Bantam Books.

Lynch, R.L. & Harnish, D. (1998). Preparing pre-service teachers education students to used work-based strategies to improve instruction. In Workplace and Beyond, *Contextual teaching and learning: Preparing teachers to enchance student success in the workplace and beyond* (pp. 127–158). Columbus: OH: ERIC Clearinghouse on Adult, Career, and Vocational Education.

Qin, Z, Johnson, D.W., and Johnson, R. T. (1995). Cooperative versus competitive efforts and problem solving. *Review of Educational Research, 65*(2), 129–143.

Rahdiyanta, D., Hargiyarto, P., & Asnawi. (2017). Characters-Based Collaborative Learning Model: Its Impact on Students Attitude and Achievement. *Jurnal Pendidikan Teknologi dan Kejuruan, 23*(3), 227–234.

RoI. (2010). *Buku induk pembangunan karakter* [Character building master book]. Jakarta, Indonesia: Curriculum Center.

RoI. (2011). *Pedoman pelaksanaan pendidikan karakter* [Character education implementation guideline]. Jakarta, Indonesia: Research and Development Center of Curriculum and Book Center.

Wardiman, J. (1998). *Pengembangan sumber daya manusia melalui SMK* [Development of human resources through Vocational High School]. Jakarta, Indonesia: PT Jayakarta Agung Offset.

Culture-based character education in a national character education course

D. Kumalasari
Universitas Negeri Yogyakarta, Indonesia

ABSTRACT: Character education in the History Education Study Program, Social Science Faculty, *Universitas Negeri Yogyakarta* (UNY) uses a monolithic approach, meaning that character education is presented as a stand-alone course and is not integrated into other relevant courses by the name of National Character Education. National Character Education courses held in the History Education Study Program are developed with activity-based learning, meaning that the learning process in order to plant, practice, and habituate these values is conducted through selected and designed activities, in addition to the presentation of concepts via lecture, question and answer, and other learning methods. Lectures are started by: (1) stipulation of the lecture contract building hope and commitment; (2) finding and determining prioritized words of wisdom, which are obtained from community figures of Yogyakarta, and placing them in classrooms and strategic places on campus; (3) students observing and relating examples of goodness in the surrounding environment (storytelling); (4) lecturers noting goodness in their environment and stating it; (5) developing an appreciative atmosphere; (6) improving the cleanliness and tidiness of the classroom; (7) starting the lecture by praying; (8) having concern for problems in society and the victims of disaster; (9) watching character education films together; (10) reflection—building habits of introspection or self-examination, and encouraging all class members to improve virtue and improve honesty to one's self.

1 INTRODUCTION

Post-independence education in Indonesia has experienced many changes and developments. Education that was previously colonial-centric, prioritizing the interests and needs of colonial government, changed its orientation and aimed to educate the nation under the mandate contained in the preamble of the constitution of the Republic of Indonesia in 1945. In Act No. 20 of 2003 on National Education System, Article 3 asserts that "National education functions to develop the ability and form the character and civilization of dignified nation in order to educate the nation's life, aims to develop the potential of learners to become human beings who believe and piety to God Almighty, noble, healthy, knowledgeable, capable, creative, independent, and become a democratic and responsible citizen". From this, it can be seen that national education holds a significant function—to build a whole human being who has good character values, in addition to having faith and devotion. Therefore, education becomes an agent of change which must hold national character building.

Nowadays, character education has become the priority of national policy. Government, through the Ministry of National Education, has proclaimed the implementation of character education for all education levels, from elementary school to higher education (Zuchdi, 2010). The development of character education that consists of moral and religious values is increasingly recognized as an urgent need, given that cognitive intelligence alone does not guarantee someone's success (Kneller, 1964).

Building a continuous balance between cognitive, affective, and psychomotor aspects is the most important. In Lickona's view (1991), education is a process involving efforts to develop three aspects of ones' life, which are life view, life attitude, and life skill. Education can be described as civilizing or as "enculturation", which is a process to enable someone to live in a certain culture. Later, Zamroni (2002) stated that education is a process that takes place in a certain culture. There are many values of culture and its orientation that can inhibit and encourage education. Also, there are many culture values that can be utilized consciously in the education process. Dewantara (1977b), also reminded that in overcoming this culture, it takes a guarded attitude in choosing which one is good to add to the glory of life and which is not profitable.

Character education in higher education, particularly in the History Learning Study Program, Social Science Faculty, *Universitas Negeri Yogyakarta* (UNY), necessary to instill self and personality concept and equip prospective teachers with learning insights through integrated character education. With regard to the integration of character education into history learning, history itself is a subject that includes historical contents as a study object for the students to learn and explore various historical stories of the process of the formation of NKRI (Negara Kesatuan Republik Indonesia/Unitary State of the Republic of Indonesia). The development of an appropriate character education integrated model in history learning can provide an alternative solution to restoring the spirit of nationalism, loving the homeland, fighting spirit, confidence, the spirit of unity and entity.

2 EDUCATION AND CULTURE

According to Dewantara (1964), culture is the result of human mind. Culture is the result of human efforts toward two strong influences, which are nature and times (nature and society). In culture, there is evidence of the glory of human life to overcome various obstacles and difficulties in life and living in order to reach safety and happiness, which is essentially orderly and peaceful. As a result of the struggle of human life that lives in the same character, same age, and same society, culture is always national (national) and manifests the nation's personality (national independence).

The purpose of the culture itself is actually to preserve and advance human life toward civilization. Culture in its existence sometimes experiences cult (worship), and often experiences "frozen life" in the sense that it does not develop (Dewantara, 1951a). Therefore, Dewantara reminded us that people have cultures that consider several important things such as: (1) maintaining culture with the aim to promote and adjust culture with every turn of nature and the times; (2) "exile" (isolation) of culture causing decline and death, thus there must be a connection between culture and society; (3) the renewal of culture also needs the connection with other cultures that can develop (promote perfect) or enrich (add) to our own culture; (4) putting in other cultures that are incompatible with the environment and time, even to the replacement of cultures that violate the demands of environmental and community conditions will always be harmful; (5) cultural progress must be a direct continuation of its own culture (*kontinuitet*), moving toward the unity of world culture (convergence), and continuing to have personality traits within the world humanitarian environment (*konsentrisitet*).

Indonesian culture, according to Dewantara (1977a), is originally a collection of regional cultures, therefore we must mobilize to become a unity of culture for all people. Local cultures in Indonesia are not impossible to incorporate into cultural unity because basically the areas in the archipelago have the same nature, history, society, and the times; hence according to Dewantara (1964), the unity of culture in Indonesia is only a matter of time. Efforts to build unity of culture are needed as every "peak culture" exists in all areas of Indonesia as its substance.

The meaning of cultural education according to Dewantara (1977a) is organizing the growth of best character, among mind, feeling, and will; those three must be educated.

For mind and intelligence, the value of education is clear—they are the main priority in our education system even now. However, the two that are often forgotten are the education of feeling and will.

3 EDUCATION AND DECENCY

Teaching activity, according to Dewantara (1964), does not generate difficulty, because there are many science books and a teacher usually has their own scientific knowledge. Otherwise, many teachers find it difficult when they try to educate because of lack of preparation and lack of books that can be used as a guide. Therefore, teaching manners and decency needs to be understood by the teachers first (Dewantara, 1951b).

Character teaching is related to the levels of mental development that exist in the lives of children, from childhood to adulthood. Otherwise, according to Dewantara (1977a), the content of teaching is conducted in comparison with the tradition of religious education (Islam), which has long been known by the methods of "*shari'ah, hakikat, tarikat, and ma'rifat*".

Shari'ah education is used for young children and can interpreted as habitual behavior and following general rules or habits. In shari'ah phase, children must get used to what is good, so teachers need to constantly reprimand them when they do something wrong, but at the same time always considering the nature of children, which is spontaneous.

The next phase is *hakikat,* which means "reality" or "truth", and which has the intent of giving understanding to children, so that they become "*insyaf*" and "conscious" about good and the opposite. The teaching of *hakikat* is used for children in the period of *balig* – that is, when the intellect or the power of reasoning develops.

The next phase is *tarikat*, which means "behavior", which is the act that we intentionally do with the intention of training ourselves to carry out various good, though it is difficult. This phase is an exercise for older children to compel and suppress or rule and control themselves. In religion, the activities of '*tarikat*' are done in the form of fasting, walking to distant places, reducing sleep and eating and suppressing various passions in general (Dewantara, 1951c). This is actually the core content contained in the character education. In the modern educational setting, the exercises are not only related to the intent of *kebatinan* but can also be manifested as art, sport, community and state exercises, starting with the scouting and youth movement, social movements and so on. The activities of these exercises are aimed to train young people to use all their abilities to help the public interest.

The last phase after *tarikat* is the *ma'rifat* method, used in character education for children who have become teenagers. Ma'rifat phase means "understand". This phase is a good time to inculcate in order for adults not to be hesitant, tossed around by circumstances they have never experienced before. They have to understand the existence of the relationship between the order of birth and inner peace, have enough practice and control of themselves, and put them in the lines of *shari'ah* and *hakikat* phases. They are considered to have understood all desires and possibilities. If they still make a wrong decision, at least they can think responsibly, so they will not be swayed by inner contradictions.

Ma'rifat phase is an education for teenagers whose age is approaching adulthood—those who are 17–20 years old. In this phase, they have entered the phase of "understandability", are used to do good and realize the means and goals. Character education given in this phase is in the form of science and knowledge which are rather deep and broad; they are introduced to the teaching of "ethics" – the law of decency—which deals also with components such as nationality, humanity, religion, philosophy, science, state, politics in its general nature, culture, customs and others. In that phase, it is also necessary to teach the character of morals listed in the lessons given at certain times and when possible to invite experts in the field to contribute to the teaching (Dewantara 1977b).

4 METHOD AND IMPLEMENTATION OF CULTURE-BASED CHARACTER EDUCATION

As a stand-alone course, the procedure and method of character education implementation in the History Education Study Program is similar to other courses. The procedures are:

a. Stipulation of lecture contracts agreed along with the student and then formulated in the course syllabus (attached);
b. Explanation of the system/method of lecturing to be conducted;
c. Explanation of tasks that will and should be done by students;
d. Explanation of the scoring system to be conducted;
e. Implementation of lectures as planned;
f. Implementation of evaluation.

The National Character Education course held in the History Education Study Program is developed on the basis of activities. It means that the learning process in order to plant, practice, and habituate these values is conducted through selected and designed activities, in addition to the presentation of concepts via lecture, question and answer, and other learning methods.

The implementation of activity-based National Character Education as stated above is conducted in accordance with the character education concept of Dewantara (1977a) by doing some adjustment and development. The activities include those discussed in the following sections.

4.1 *Stipulation of lecture contract by building hope and commitment*

The aims of this activity are to
a. Develop the personal responsibility of students to build good habits;
b. Develop introspective habits in order to make improvements;
c. Develop the power to control oneself;
d. Develop social intelligence.

Description of activity:
 i. At the beginning of the meeting, the lecture contracts between lecturers and students are discussed. In the course, the contract is discussed about the applicable provisions and becomes a reference for lectures, such as: the number of days, presence, class entry requirements, clothing, assignment execution, assessments and so forth.
 ii. Although what is discussed is already written in the college's book or academic guidance, the contract is made to build commitment from students, so they are involved and feel that the rules are organized and made together, for the common good.
 iii. In the initial meeting, students are also invited to discuss what characters will be developed during the lecture (one semester), and the desired things toward the class or class characteristics that make them proud to be class citizens (such as ethics/courtesy, environment, discipline, etc.). The characters developed are based on the inherent culture-based values of society. Each student is given an opportunity to submit their proposal. Once collected and directed by the team of lecturers, the characters that will be developed together in the classroom are agreed upon in accordance with the agreed cultural values. At this point, the lecturers and students are committed to work together to implement what has been agreed upon.
 iv. The profile of characterized human, expected class characteristic and mutual commitment are then made in the form of a frame and installed in the classroom as a driver, as well as a reminder of the things that should be achieved by all class members.

4.2 *Find and determine prioritized words of wisdom obtained from community figures*
    of Yogyakarta and put them in classrooms and strategic places on campus

The aims of this activity are to
a. Introduce the students to the local wisdom of Yogyakarta;
b. Inspire students to develop local wisdom and goodness based on common perceptions;
c. Recognize the importance of good character;
d. Develop leadership;
e. Develop social intelligence.

Description of activity:
  i. Each student is given the task to look for words of wisdom that can inspire and give
     motivation to anyone who reads them.
 ii. Each student enters and joins into groups of 3–5 people.
iii. Each group selects a number of words of wisdom that they like best and then compiles a
     list of these words.
 iv. Each group presents the words of wisdom they have chosen in front of the class and
     recounts the process they did in discovering and declaring which person said those words
     of wisdom (emphasized to take local figures of Yogyakarta to bring up the character of
     local culture).
  v. The group creates a beautiful or interesting format for displaying those words of wisdom
     and installing them in classrooms and other strategic places.
 vi. Classes compile books of words of wisdom.

4.3 *Students observe and state goodness in the surrounding environment (story telling)*

The aims of this activity are to:
a. Increase student awareness of good habits that exist around them;
b. Grow the desire to develop the observed good habits;
c. Grow the conviction that they can become a better person;
d. Train students to be caring and sharing.

Description of activity:
  i. At the beginning of each meeting, several students take turns in storytelling—telling a
     story of virtue, in which there are life values that can give inspiration to everyone who
     listens and can motivate them to become a better person.
 ii. The storytelling is followed by a discussion of the virtues contained in the story and of
     the virtues that can be developed by everyone in everyday life.
iii. At the end of the storytelling, the lecturer underlines the results of the discussion.

4.4 *Lecturers note goodness in their environment and state it*

The aims of this activity are to:
a. Increase attention to virtue;
b. Inspire students to do good things and become better people;
c. Increase the willingness of students to care and share.

Description of activity:
  i. Lecturers note good behavior seen in the community or on television, heard on the radio,
     or read in books or newspapers.
 ii. The lecturer briefly relates these virtues to the students in the class at the beginning of the
     lesson—either before or after the students' storytelling.

4.5 *Develop an appreciative atmosphere*

The aims of this activity are to:
a. Develop virtue through the power or strength owned by a person or group;
b. Develop optimism and confidence;
c. Develop good behavior to become a good habit.

Description of activity:
i. Lecturers pay attention to good behavior or good things done by the students, no matter how small.
ii. Lecturers then appreciate the goodness.
iii. This appreciation can be conveyed by thanking students for the good behavior or kind acts, giving personal praise or mentioning the goodness in front of others, or giving praise in front of others, or mentioning, or giving praise or awards in front of the public.

4.6 *Find idol figures/role models*

The aims of this activity are to:
a. Find or identify local figures with good characters who have been meritorious in the progress of society, the progress of the nation, or the progress of the world;
b. Contemplate the strong character or the kindness that is admired;
c. Inspire and cultivate motivation to emulate the kindness shown by the admirable figure;
d. Encourage the students to believe that there is no success without kindness.

Description of activity:
i. Each student is asked to identify a successful person as their idol.
ii. Students are asked to highlight the characteristic of the idol that is believed to produce success or asked to write briefly in a few sentences about the privilege of an admired character.
iii. One by one, the students are asked to publicly declare their chosen character.
iv. The lecturer compiles all the characteristics mentioned by the students that are believed to be the source of the idol's success.
v. The lecturer concludes by affirming that the idol's success is due to the good traits (good character) he has.

4.7 *Improve cleanliness and tidiness of the classroom*

The aims of this activity are to:
a. Build and strengthen the responsibility of students as good campus residents;
b. Build a clean-living habit;
c. Build a habit of living neatly;
d. Build and strengthen self-discipline.

Description of activity:
i. Lecturers at the beginning of each lecture always remind students to look around the classroom and observe whether it is clean and neat or not.
ii. All the class members take the garbage that is still in the classroom and throw it into the trash that is outside the classroom, then tidy up each seat.
iii. Lecturers give appreciation for the effort and creativity of students in improving the cleanliness and tidiness of the class.
iv. When the classrooms are clean and tidy, the lectures begin.

4.8 *Start the lecture by praying*

The aims of this activity are to:
a. Establish the habit of being grateful for all the mercy, gifts and grace given by God according to the cultural basis of our religious society;
b. Encourage all class members to improve good behavior and kind acts;
c. Increase faith and devotion to God.

Description of activity:
  i. After the classroom has been made clean and tidy and before the lecture begins, the lecturer gives the students the opportunity to offer their own willingness to lead the communal prayer.
 ii. Students who are willing to lead the prayer come to the front of the class then invite everyone to pray.

4.9 *Concerns for problems in society and disaster victims*

The aims of this activity are to:
a. Develop and strengthen empathy, sympathy and caring for others, especially for people who are experiencing difficulties or are victims of disaster;
b. Develop a sense of responsibility, honesty, teamwork and leadership skills;
c. Develop pride in making a contribution;
d. Train and develop the ability to manage activities.

Description of activity:
  i. Students are divided into several groups (each group containing up to 5 people) and assigned to undertake a humanitarian project. This project can be any kind of activity, using effective and efficient principles.
 ii. The project is carried out outside the learning activities in the classroom but is still completed in the same semester as the course is delivered.
iii. The students involved in the activity are asked to be responsible for all activities or donations obtained in writing, neatly and honestly.
 iv. Students are asked to communicate or declare all activities or donations gained, and those distributed or used, in front of the class and to all members of the community.
  v. Lecturers show appreciation for the attempts or efforts that have been made by the students and the results that have been achieved, in a forum that involves the public.

4.10 *Watching films together with a character education theme based on the values of Yogyakarta society local culture, the history of struggle and humanity*

The aims of this activity are to:
a. Inspire students or viewers to develop good characters based on Yogyakarta cultural values, such as moral or *unggah-ungguh*, respecting older people, and courtesy;
b. Develop students' insights into the history of the nation, the culture of the archipelago, the culture of nations, and human civilization;
c. Develop an appreciation of diversity;
d. Develop a love for the values of truth and justice, homeland and humanity;
e. Develop the spirit and sense of togetherness.

Description of activity:
  i. Movies are chosen that inspire or motivate the audience to develop good characters including virtues such as persistence, courage, honesty, caring, affection, justice, and leadership.
 ii. Lecturers and students watch these movies together in the classroom.

iii. Lecturers and students conduct a brief discussion of the characters from the characters in the film, or the theme of the movie they watched together.

4.11 *Working in groups*

The aims of this activity are to:
a. Develop the habits of sharing, appreciating, mutual support, mutual respect, and leadership potential;
b. Train students to be responsible.

Description of activity:
i. Lecturers plan the section of lecture materials that will be taught through the learning process in a team (cooperative learning).
ii. Groups are formed that will perform specific learning tasks.
iii. Each group member conveys or recounts the role and contribution to the group results.
iv. The results are presented to the group.

4.12 *Reflection*

The aims of this activity are to:
a. Build a habit of introspection or self-examination;
b. Encourage all class citizens—lecturers and students—to improve their virtues.
c. Increase honesty to one's self.

Description of activity:
At the end of the course, all members of the class community engage in introspection and write down:

i. The goodness they have done, to others and for their own good, during the semester.
ii. Morally questionable things they have done consciously or unconsciously.
iii. The things they will do in the future to add to their good characters and reduce the development of morally questionable actions.
iv. Everyone signs their introspection sheet.
v. Students are asked to give their impressions of character education during the course, the benefits they received and the things that are judged to be lacking, and their suggestions for future improvement of the course.

5 CONCLUSION

Character education in the History Education Study Program, Social Science Faculty, UNY uses a monolithic approach—this means that character education is presented as a stand-alone course and is not integrated into other relevant courses by the name of National Character Education. National Character Education courses are held for second semester students with two credits. Culture-based character education becomes the basis of the implementation of National Character Education courses. Cultural base means culture of society in Yogyakarta.

National Character Education courses held in the History Education Study Program are developed with activity-based means that learning process in order to plant, practice, and habituate these values is conducted through selected and designed activities, in addition to the presentation of concepts via lecture, question and answer, and other learning methods. Lectures are started by (1) stipulation of the lecture contract, building hope and commitment; (2) finding and determining prioritized words of wisdom, which are obtained from community figures of Yogyakarta, and placing them in classrooms and strategic places on campus; (3) students observing and relating examples of goodness in the surrounding environment (storytelling); (4) lecturers noting goodness in their environment and stating it; (5) developing an appreciative atmosphere; (6) improving the cleanliness and tidiness of the classroom; (7) starting the lecture by praying; (8) having concern for problems in society and victims of disaster; (9) watching films together with a character education theme based on the values of Yogyakarta society culture, the history of struggle and humanity; (10) reflection—building habits of introspection or self-examination and encouraging all class members to improve virtue and improve honesty to one's self.

REFERENCES

Dewantara, K.H. (1951a). Hal pendidikan; Diktat K.H.D. [Educational; Diktat K.H.D]. *Pusara, XIII*(3), 59–64.
Dewantara, K.H. (1951b). Sifat dan maksud pendidikan I [Nature and purpose of education I]. *Pusara, XIII*(4), 65–68.
Dewantara, K.H. (1951c). *Sifat dan maksud pendidikan II. Pusara* [Nature and purpose of education II]. *Pusara, XIII*(5), 66–69. Dewantara, K.H. (1964). *Kenang-kenangan promosi doktor honoris causa* [Doctorate monograph]. Yogyakarta, Indonesia: Majelis Luhur Taman Siswa.
Dewantara, K.H. (1977a). *Karya Ki Hadjar Dewantara, bagian pertama: Pendidikan* [The work of Ki Hadjar Dewantara, the first part: Education]. Yogyakarta, Indonesia: Majelis Luhur Persatuan Tamansiswa.
Dewantara, K.H. (1977b). *Karya Ki Hadjar Dewantara, bagian kedua: Kebudayaan* [The work of Ki Hadjar Dewantara, the second part: Culture]. Yogyakarta, Indonesia: Majelis Luhur Persatuan Tamansiswa.
Roi. (1945). *Undang-undang dasar negara republik Indonesia tahun 1945* [The constitution of the republic of indonesia]. Jakarta, Indonesia: Author.
Kneller, G.F. (1964). *Introduction to the philosophy of education.* New York, NY: Chichester, Brisbane, Toronto.
Lickona, T. (1991). *Educating for character: How our schools can teach respect and responsibility.* New York, NY: Bantam Books.
Zamroni, Z. (2002). *Paradigm of national education development in realizing the civilization of the nation. In, Education for the new Indonesian community.* Jakarta, Indonesia: Grasindo.
Zuchdi, D. (2010). *Humanization of education.* Jakarta, Indonesia: Bumi Aksara.

Developing the culture of quality in learning: A case study in Indonesia

A. Ghufron
Universitas Negeri Yogyakarta, Indonesia

ABSTRACT: Qualified learning has become universally desired. This type of learning is believed to be used to prepare the excellent Indonesian human resources. However, the problem is how to create effective learning according to the expectations of different groups. Various methods were taken by the nation, the people, and the country of Indonesia, and even the issue of quality was raised to be one of the educational development strategies. Based on the study of various views on improving the quality of learning, the activities of developing the culture of quality in learning is an interesting topic to review. Developing the culture of quality in learning can be done through developing the work culture in learning activities, formulating the focus of improving the quality of aspects of learning, motivating all parties to improve the quality of learning, observing the behavior of excellent performance in learning, and assessing the culture of excellence in learning.

1 INTRODUCTION

Learning is an essential activity for preparing students to be valuable human resources. The learning process involves the transmission and transformation of all learning experiences from the teacher to the student. At the end of the learning activities, students are expected to optimally develop their potential.

Ensuing learning activities for preparing the learners to be an excellent human resource is not too excessive if all interested parties in the field of learning (family, government, and society) are willing to improve the quality of learning at all levels, extents, and sorts of education. The process should involve continuous improvements in the quality of the learning activities.

In Indonesia, improving learning excellence at every level of education is a strategic step that needs to be followed if the nation wishes to be competitive in various fields of life in the global era. Why do we need to improve the quality of learning? One reason is that learning is concerned with the development of human resources. Tilaar (1998) says that education (learning) is part of the effort to improve human wellbeing and it is part of national development. The Indonesian National Education Department's strategic plan (National Education Department, 2005), states that education is a systematic process that serves to improve human dignity. In Indonesia, education is formulated to educate the nation so that every citizen has the capacity to actualize themself in their society.

Trianto (2010) suggests that learning is a two—way interaction between a teacher and a student, where both intense and targeted communication (transfer) leads to a pre-determined target. According to Sagala (2009), learning is teaching students to use the principle of education and learning theory as a major determinant of educational success. Learning is a two-way communication process; teaching is done by teachers as educators, while learning is done by students. While Suherman et al. (2003) state that learning is the process of education in the scope of schooling, so the meaning of the learning process is the interaction between a student with a school environment, such as a teacher, a source or facility, or a fellow student.

There are many methods that have been used by those who care about the field of learning to attempt to improve the quality of learning. These methods can be grouped into two categories—the structural way, and the cultural way. Perhaps, more structural ways may be more workable than cultural. How successful has this strategy been? Based on observation, methods to improve quality of learning have not yet been maximized. Implementing new methods to improve the quality of learning also tends to introduce new problems.

What efforts can be made to further improve the excellence of learning? One way that this can be achieved is through developing a culture of quality in learning. This method is relevant because it touches on habituation and action in improving the quality of everyday learning.

2 DISCUSSION

Three major topics are addressed in the discussion: the definition of learning quality; improving the quality of learning; and the development of a culture of quality in learning. These are three important areas to consider when addressing the problem of the ineffectiveness problem of the learning improvement movement in the learning activities.

2.1 *Definition of learning quality*

The meaning of quality can be studied from two perspectives, namely etymology and epistemology. Etymologically, the term of quality is known as "standard" of Quality means size, namely as a degree. Therefore, if one refers to a book as being of a high quality, then the person wants to say that the book has a good degree of content (substance) and coherent exposure.

Epistemologically, quality has several meanings—as many as the experts who develop the meaning of quality. Also, the meaning of quality is strongly influenced by the point of view of the developers. For example, Sudarwan (2007) defines the meaning of quality as the degree of excellence of a product or the work, both in the form of goods and services.

In the author's view, although the meanings of quality may vary, the focus of the study remains on the "degree, excellence, meaning, and significance" of an object or activity. Quality is always related to the term of "superior, degree, and benefit".

What about the meaning of qualified learning? By applying the aforementioned meanings, qualified learning can be defined as the degree or level of quality of all aspects of learning. Matthew (2016) states that "instructional quality" refers not to any measure of actions taken in the classroom (such as observations of class sessions), but rather to the full set of classroom interactions that affect student learning, including the ability of the instructor, the quality of instruction delivered by that instructor (including curriculum, teaching methods, etc.), and other classroom-level factors such as peer effects.

With this definition, the meaning of qualified learning encompasses all aspects of learning. Even if it is related to the concept of integrated quality management, learning is said to be excellent if there is a win-win situation—that is, all those involved in learning activities feel happy, are treated fairly, and are served well (Pulungan, 2001). In this context, the aspect of satisfaction becomes one of the criteria related to the qualified learning activity (to determine whether it is qualified or not).

Based on the description above, it can be inferred that qualified learning can be defined as a code that describes the learning process taking place as superior, adequate, and provides benefits for the development of the learners. This is not only examined by the learning outcomes, but also by the learning process.

2.2 *Improving the quality of learning*

To improve the quality of learning, it is necessary to know which aspects of education will be improved. The aspects of learning need to be upgraded first prior to the activities. Logically, we will not be able to determine and conduct activities to improve the quality of learning when the aspects of learning to be improved are not well identified. In addition, efforts to improve qualified learning should also consider the inter-relationship between aspects of learning that form an integrated system.

Improving the quality of learning is intended as an activity to improve learning performance in all its aspects. Aspects of learning may relate to input (raw, environmental, and instrumental), processes, and products (output and outcome). All of these aspects can be regarded as the scope of activities to improve the quality of learning through implementing various approaches and ways.

In Indonesia, improving the quality of learning is positioned as part of the four strategic issues of educational development. Improving the quality of education needs to be done immediately because the results are related to the students' learning abilities significantly. This is in line with the opinion expressed by Darling-Hammond (2003), who stated that "during the 1990s, a new policy hypothesis—that focusing on the quality of teaching would provide a high-leverage means for improving student achievement—began to gain currency".

There are several things that can be done to improve the quality of learning. In terms of its scope, improving the quality of learning can be done at a macro or micro scale. Based on the method or approach, improving the quality of learning can be achieved through structural or cultural approaches.

The question that should be addressed is, "Which learning quality improvement approach is most relevant?" To answer that question, the first thing to do is to recognize the aspects of learning that require improvement. The next step is determining the method of learning that will be implemented, in accordance with the aspects of learning to be improved.

For example, we can use the PDCA (plan, do, check, and act) approach to improve the quality of learning. When using this approach, the first thing to do is to develop a design related to the learning problem to be solved; this can be done by using various problem analysis techniques to discover the causes of the existing learning problems, then developing effective ways of solving the problems or identifying improvements that can be made. The second step is to take action in accordance with the procedure of learning approach that has been planned. The third step is to evaluate the results of any changes or the improvement of learning quality related to the actions taken. The final step is to follow up on any changes that occur and make adjustments if there has been no improvement in the quality of learning.

2.3 *Developing a culture of excellence in learning*

A culture of excellence can be defined by a circumstance or condition that indicates that most or even all people are already behaving, practicing, or accustoming themselves to performing their best. It cannot be separated from the meaning of culture as a tradition or a habit that has been developed.

With the above understanding, people often use the term "culture of excellence" as a goal to be realized in the future. Similarly, this term is often used to describe a situation in which the traditions and habits of excellent performance have developed in a particular institution or work unit.

Furthermore, if the meaning of a culture of excellence is brought into the context of learning activities, it can be interpreted as a condition or situation which indicates that some or even all parties involved in learning activities have internalized and familiarized themselves with excellent performance in organizing the learning activities. They carry out the learning activities not merely to fulfill the wishes of the leaders, but in theirself there has grown an awareness to always behave and make excellent performance in organizing the learning activities.

How do we develop a culture of excellence in learning? We realize that culture contains many aspects: artifacts, norms, values, and assumptions; thus, the development of the quality of learning must consider those aspects. Consequently, the first step should be to carry out a study of the essential aspects contained in the culture of excellence in learning, then setting a strategy for its development.

2.3.1 *Work culture in the learning activities*

The first step in developing a culture of excellence in learning is to document the learning culture that takes place in the school. This information-gathering step can focus on the artifacts, norms, values, and assumptions that develop in that school.

The results of this process are followed up with activities to translate and interpret the content of the artifacts, norms, values, and assumptions. Based on the translation and interpretation of this content, we can figure out the characteristics of the performance and the parties that involved in the learning activities in the school.

2.3.2 *Formulating the focus of improvement of the quality of learning*

The formulation of the focus of improvement of the quality of learning is informed by the results of the findings or performance profiles of the parties involved in the learning activities. Thus, the focus of improving the quality of learning does not cover all aspects of learning; it only covers certain aspects of the quality of learning.

In this fashion, the actions of improving the quality of learning become increasingly focused and the success rate becomes easily measurable. At this point, we are expected to behave excellently step by step toward a broader and more comprehensive stage in order to realize the quality of learning in all aspects of learning.

2.3.3 *Motivating all parties to improve the quality of learning*

This activity aims to invite, direct, and encourage all parties to always improve the quality of learning by using a variety of approaches. It is not meant to force excellence in organizing learning activities.

The key idea of the activity is "how to raise self-awareness of the parties to always behave excellently in organizing learning activities". Hence, it is necessary to use various techniques to motivate others, so that they consciously perform excellently in organizing learning activities.

2.3.4 *Observing excellent performance behaviors in learning*

It is necessary to observe the excellent performance behaviors in conducting learning activities. This observational activity is conducted to identify, recognize, and understand various behaviors that have excellent performance in the learning, both inside and outside the classroom.

The observations include looking at the various actions to improve the quality of learning, observing the habits that show the excellent performance in learning, and observing the positive impact of habits with excellent performance in learning.

2.3.5 *Assessing the culture of excellence in learning*

Assessment needs to be conducted to find out whether the excellently performed learning activity has become a tradition or habit for most or even all parties involved in the learning activities. This assessment focuses on aspects related to the culture of excellence of the learning such as artifacts, norms, values, and assumptions.

The form of activities may be in quantitative or qualitative form in accordance with the purpose of assessment. Quantitative assessment is used to assess aspects related to the numbers. Meanwhile, qualitative assessment is used to assess aspects related to characters and values.

3 CONCLUSION

The quality of learning is something that has long been everybody's concern. Nevertheless, the issues of the quality of learning have not been solved until recently. The government has tried to solve the problem of how to improve learning quality by using various methods, but the result is not yet as expected.

In order to participate in developing and improving the quality of learning, with the serious efforts and support from various parties, the activities of developing a culture of excellence in learning in the educational unit are necessary. There are various ways to develop a culture of excellence in learning, such as by describe the benefits of the culture in learning activities, formulating the focus of improving the quality of aspects of learning, motivating all parties to improve the quality of learning, observing the behavior of excellent performance in learning, and assessing the culture of excellence in learning.

REFERENCES

Darling-Hammond, L. (2003). *Building instructional quality: "Inside-out" and "outside-in" perspectives on San Diego's school reform* (Research Report). Seattle, WA: Center for the Study of Teaching and Policy University of Washington.

Matthew, C. M. (2016). Instructional quality and student learning in higher education: Evidence from developmental algebra courses. *The Journal of Higher Education, 87*(1), 84–114. doi:10.1080/002215 46.2016.11777395

National Education Department. (2005). *Rencana strategis departemen pendidikan nasional 2005–2009* [The national education ministry's strategic plan 2005–2009]. Jakarta, Indonesia: Pusat Informasi dan Humas Depdiknas.

Pulungan, I. (2001). *Manajemen mutu terpadu* [Integrated quality management]. Jakarta, Indonesia: Depdiknas P2UT—Ditjen Dikti.

Sagala, S. (2009). *Konsep dan makna pembelajaran* [The concept and meaning of learning]. Bandung, Indonesia: Alfabeta.

Sudarwan, D. (2007). *Visi baru manajemen sekolah* [New vision of school management]. Jakarta, Indonesia: Bumi Aksara.

Suherman, E., Turmudi, Suryadi, D., Herman, T., Suhendra, Prabawanto, S., Nurjanah, & Rohyati, A. (2003). *Strategi pembelajaran matematika kontemporer* [Contemporary mathematics learning strategy]. Bandung, Indonesia: JICA UPI.

Tilaar, H. A. R. (1998). *Beberapa agenda reformasi pendidikan nasional dalam perspektif abad 21* [Some of the national education reform agenda in the 21st century perspective]. Magelang, Indonesia: Tera Indonesia.

Trianto. (2010). *Mendesain model pembelajaran inovatif-progresif* [Designing innovative-progressive learning models]. Jakarta, Indonesia: Prenada Media Group.

Character education in Indonesian English as a Foreign Language (EFL) textbooks: Does it still exist?

S.D. Amalia
Universitas Muhammadiyah Surakarta, Indonesia

ABSTRACT: Prior to the focus on scientific approach in the recent implementation of the 2013 Curriculum, the Indonesian government drafted 18 values abstracted from *Pancasila* (the official philosophical foundation of Indonesia) and *UUD 1945* (the 1945 Constitution). The values, also known as National Character Building, were aimed to promote the betterment of Indonesians' characters, especially the younger generation. Since the scientific approach moved into the spotlight, a character education approach seems to be missing from both classroom activities and textbooks. As a reminder of how the 18 values of character education could have been an important component of the latest curriculum, alongside the scientific approach, my previous study regarding the depiction of these characters in English commercial textbooks showed that the characters can help learners understand themselves and express their identities, as stated in the philosophical and ideological foundations of Indonesia. However, the result also revealed that some authors of the textbooks did not fully understand the values, which is illustrated in some ambiguous samples. Based on my previous research, I investigate the representation of character education in other textbooks by using the same data collection and data analysis methods, and I offer some alternative learning activities of relevance. This paper elaborates the two aforementioned topics by providing comparisons to my previous study.

1   INTRODUCTION

Character education has long been favored as one of the most essential contents inserted in textbooks written by curriculum makers around the world, including the Indonesian government. Also known as national character building, character education is one of the main values inserted into many educational programs, either explicitly or as a hidden curriculum. Textbooks, published by either the government or commercial publishing companies, are commonly used as the media to transmit the values of character education.

As posted in Antara News (The Communications and Public Service Bureau of the Education and Cultural Ministry, 2017), the Minister of Education and Culture, Muhadjir Effendy, made a statement in his speech in 2017 National Education Day commemoration that "character building allows students to equip themselves with high competitive skills that the world needs in the 21st century." It implies that character education is one of the main issues to be inserted in the national curriculum, among other pivotal issues intended by the government.

Some values of character education had actually been inserted in the previous curriculum in Indonesia, such as in the KTSP (the Multi-tiered Education Curriculum) of 2006. It was galvanized when the government started drafting the 2013 Curriculum. Ever since the early draft of the latest curriculum was formulated, the government tried to promote the 18 values of national characters and wished to include them in every level of education, especially in the elementary and secondary levels.

The establishment of character education in the 2013 Curriculum was based mainly on *Pancasila* (the official philosophical foundation of Indonesia) and *UUD 1945* (the 1945

Constitution), the basic foundations of the nation. Hence, it is not an exaggeration to say that if Indonesians can embody the values of both Pancasila and UUD 1945 in their conducts, the intended national character proposed by the government can be eventually transformed into the national identity (Amalia, 2014). As previously mentioned, many nations aiming to encourage its citizens to embrace their national character have done many efforts to promote their national identity such as by inserting some values of national character into education. The significance of promoting national identity among citizens is highlighted by Woodward (2001), who emphasizes that identity "gives us a location in the world and presents the link between us and the society in which we live" and "gives us an idea of who we are and of how we relate to others and to the world in which we live". As for Indonesia, the form of character education that was based on *Pancasila* and *UUD 1945* is divided into 18 characters, namely:

No.	Character	Definition
1.	Religiosity	Showing obedience in every attitude and behavior when conducting any religious activity, being tolerant with others' religious activities, and living harmoniously with people with different religions and beliefs.
2.	Honesty	Being someone who can be trusted in every word, behavior and attitude.
3.	Tolerance	Being tolerant to others of different faith, tribe, race, opinions, or actions.
4.	Discipline	Committing to respect the laws and rules that prevail.
5.	Hard work	Indicating an earnest effort to overcome barriers in learning activity and completing tasks verily.
6.	Creativity	Generating ideas to produce something new or to modify some alternative ways from what they already have.
7.	Independence	Not being dependent on others when completing tasks.
8.	Democracy	Treating others by giving them equal rights and responsibilities.
9.	Curiosity	Demonstrating effort to know more about what is being learned, seen and heard.
10.	Sense of nationalism	Placing the interests of the nation above oneself.
11.	Patriotism	Showing loyalty, care and respect towards the language, environment, society, culture, economics, and politics of the nation.
12.	Achievement orientation	Striving to produce something advantageous for society and recognising as well as respecting others' success.
13.	Friendship or communication	Demonstrating willingness to be communicative, friendly, and cooperative to others.
14.	Love of peace	Behaving in a way that make others feel pleasant and safe toward our existence among them.
15.	Fondness for reading	Spending time reading a variety of worthy publications.
16.	Environmental awareness	Always seeking to prevent damage to the environment and developing efforts to repair environmental damage.
17.	Social care	Showing generosity to others who are in need.
18.	Responsibility	Carrying out duties and obligations sincerely for one's self, society, surroundings (natural, social and cultural), the nation, and God.

(Adopted from Puskurbuk, 2010).

In the early establishment of the 2013 Curriculum, the promotion of character education in Indonesia EFL textbooks was quite intense. This can be proven by the research I conducted to investigate the representation of national character building in two EFL textbooks published commercially in Indonesia, namely *Contextual English for grade twelve of senior high schools: physical sciences and social sciences majors*, by Bambang Sugeng and Noor Zaimah (2012), and *Bahasa Inggris: Berbasis pendidikan karakter bangsa* [Grounding on the national character education] for senior secondary school students in year ten, by

Otong Setiawan Djuharie (2012). Both books were written in 2012 and were based on the Indonesian 2013 Curriculum.

In that study, the data analyzed revealed that national character values were being represented more in *Contextual English* than in *Bahasa Inggris*. In *Contextual English*, the national character values depicted in either images or the text mostly referred to hard work, creativity and independence, whereas in *Bahasa Inggris* (English), they mostly dealt with friendship or communication, discipline, and environmental awareness. Additionally, there was an interconnection among the values of hard work, creativity, and independence found in *Contextual English*, from which the writers might aim "to highlight the values that are strongly related to profession and employment as well as entrepreneurship" (Amalia, 2014). This inclination was considered to be "the speculation of addressing a hidden curriculum on the current economical situation." On the other hand, it was found that the writer of *Bahasa Inggris* put more concern on depicting the national character values related to humanity, such as friendship, communicative skill, and environmental awareness.

Both books also seemed to represent *Bhineka Tunggal Ika* (Unity in Diversity), the Indonesian national slogan, by portraying values like tolerance, love of peace, friendship or communication, and social care. The values above are very likely to promote the multicultural education of Indonesia; as argued by Lie (2000, p. 82), "multicultural education is needed to foster peace, understanding, and respect." Furthermore, the values found in the two books were explicitly displayed in the books; the writers of the books put a note in each chapter so that it was apparent to the readers that every chapter contained particular values. This also helped me scrutinize the values in each chapter. However, some of the data analyzed were not consistent with some of the national character values because some of the texts or images shown in the textbook did not correspond to the value intended by the writers, instead those data were related to other values. For instance, there is information in particular chapter that the chapter contains the value of discipline; however, the image reflects the value of responsibility. This resulted in a significant finding that should be carefully examined by any writer who wants to write a textbook containing character education.

Above all, my previous study and many other similar studies have demonstrated that the government, as well as those dealing with education in Indonesia, such as publishing companies, have made a significant effort to promote the 2013 Curriculum through the insertion of character education into several aspects—in this case English as a Foreign Language (EFL) textbooks. Nevertheless, as the formulation of a scientific approach has become the spotlight of the latest curriculum, character education seems to be missing from both classroom activities and the textbooks being used in schools.

Based on the aforementioned rationale, this paper intends to present another study with a similar topic, specifically on whether the values of character education proposed by the Indonesian government still exist in EFL textbooks. This paper examines the representation of character education in selected books published by the government and provides some alternatives that can be applied by teachers to impose character education during teaching and learning activities.

2 METHODS

2.1 *Data source*

To create a comparison with my previous study, I decided to select textbooks published after 2012 but still based on the same curriculum: the 2013 Curriculum. Three senior high school English textbooks were chosen for this purpose. The first is entitled *Bahasa Inggris SMA/MA SMK/MAK Kelas X (10) Semester 1* (Widiati et al., 2014). The second is entitled *Bahasa Inggris SMA/MA SMK/MAK Kelas XI (11) Semester 1* (Bashir, 2014). The third is entitled *Bahasa Inggris SMA/MA SMK/MAK Kelas XII (12)* (Widiati, et al., 2015).

These books were chosen because the topics and kinds of materials in senior high school textbooks are more variable than in junior high books or even in elementary school books. Thus, content related to character education is more likely to be found in those textbooks than in junior high or even elementary school books. Moreover, the textbooks selected are all published by the Ministry of Education and Culture. Compared to my previous study which used commercially published textbooks, in this recent study textbooks published by the government are preferred, to see whether these books contain more values of character education.

2.2 Data collection and data analysis methods

By adopting the constructivist perspective, this study is a qualitative descriptive study, as the data collected are in the form of text and images containing character education values. Moreover, the data are not analyzed numerically with statistical methods as in a quantitative research but are analyzed through the content analysis procedure. Even though content analysis is often associated with quantitative research, as long as the data obtained are not analyzed statistically it is categorized as a qualitative study by nature. Thus, content analysis that is used qualitatively can help the researcher gain more meaningful results by interpreting the analyzed data based on the relevant theories. Amalia (2014) supports this by stating that "studying visual data by using content analysis can be done not only in a ruled-governed way but also through Interpretive analysis."

The methods of data collection and data analysis used in this study are the same methods as used in my previous study. They only differ in the data source, with different textbooks examined in each.

3 RESULTS

3.1 Representation of character education in books published by the government

The three textbooks investigated have shown different results in their representation of the 18 characters. Below are some samples of the data that were analyzed.

Bahasa Inggris SMA/MA SMK/MAK Kelas X (10) Semester 1, which is referred to as BI-X, contains several characters such as religiosity, which is mostly depicted through images showing people wearing a head scarf symbolizing a certain religion, and an image of a funeral ceremony that also symbolizes a type of funeral commonly practiced in a particular religion. There is also a passage of text that reminds the readers to praise God for His Greatness in creating beautiful nature, which indicates that Indonesians believe in God.

Tolerance is shown by portraying an Indonesian from Papua talking to another Indonesian from Java, whereas a sense of nationalism is represented by showing famous images of a fictional hero and tourist destinations in Indonesia.

Friendship or communication are depicted through passages of text describing how to compliment others and to express sympathy, an image of people shaking hands, and text about someone's best friend.

Environmental awareness is represented by promoting ecotourism, such as text about an orangutan sanctuary, an image of people hiking in a mountain, and a note discusses what students should do with their rubbish when visiting a tourist destination.

Responsibility is represented through a note that people should be responsible in their life.

BI-XI, which refers to *Bahasa Inggris SMA/MA SMK/MAK Kelas XI (11) Semester 1* represents religiosity by giving students a task to write their gratitude to God and by showing letters to God, while honesty is depicted in text about the importance of being honest.

Sense of nationalism is shown through an image stating that Indonesia is well known for its various kinds of traditional markets, while patriotism is represented through a text of Soekarno's speech (The first president of Indonesia) about the danger of colonialism. Friendship or communication is represented by text that teaches students how to respond to suggestions and offers, whereas social care is shown by text describing how to react to bullying and to stop it.

Meanwhile, *Bahasa Inggris SMA/MA SMK/MAK Kelas XII (12)*, which is referred as BI-XII, represents religiosity by showing a person with religious attire. Tolerance is shown through images of people with different physical characteristics, such as images of people with darker skin color. Discipline is represented by images showing a satirical message that people should obey traffic regulations in order to be safe, while sense of nationalism is represented in text about the Baduy tribe.

Friendship or communication is shown through text about a best friend. Environmental awareness is represented by text about the danger of tornados, tsunamis, and volcanic eruptions, and text describing how students can adopt environmental-friendly practices, whereas social care is shown through text about causes of teenage bullying.

3.2 *Learning activities as an alternative to teach character education*

Since the findings have shown that there are only a few representations of national character in the three books being analyzed, teachers, especially EFL teachers, can make use of several teaching and learning activities described below as an alternative—to convey the values of national character to their students. These activities can be utilized when the textbook being used does not contain enough representation of national character, as in the case of the three textbooks described above. The possible activities are:

a. The teacher can directly perform an example of an attitude reflecting one particular character in front of the students and explain it afterwards.
b. The teacher can find an interesting media item, such as a video from YouTube containing a certain character, and then ask the students to discuss it together.
c. The teacher can ask students to give some examples of characters that they have expressed or known and then discuss them together.
d. The teacher can source text from a recent newspaper or online news article containing a particular character and then ask the students to discuss it.
e. The teacher can always try to remind the students about the importance of embracing and practicing those national characters as Indonesian citizens.

In addition, there are many other activities that can be carried out by the teacher and the students to promote character education, as proposed by the Indonesian government. According to Sultoni (2016), extracurricular activities can also be used as the media for students to develop their national character. Moreover, as proposed by Kaimuddin (2014), it is important to support the government's effort in promoting character education by building a relationship between schools and other informal institutions in society. The key is that many educational practitioners should work together in order to help the government.

4 CONCLUSION

From the data being analyzed in this study, it can be concluded that there are not many values of character education inserted in the three EFL textbooks being scrutinized. BI-X indeed shows more representations of national character than the other two books. In terms of representing the characters of religiosity and friendship, BI-X shows more examples than the other two. On the other hand, BI-XII represents more examples of the character of environmental awareness. Depictions of the other characters cannot be found in the three textbooks.

The results above have shown that the three books published by the government do not really represent the 18 values of character education promoted in the 2013 Curriculum. In addition, most of the data found are inserted implicitly, without a note referring to a particular value like in my previous study, and the data are mostly in the form of images rather than text.

REFERENCES

Amalia, S. D. (2014, September). *Representations of national character building in Indonesian EFL textbooks: A qualitative study.* Paper presented at the The 3rd UAD Tefl International Conference, Universitas Ahmad Dahlan, Indonesia.

Bashir, M. (2014). *Bahasa Inggris SMA/MA SMK/MAK Kelas XI (11) semester 1 (English for Senior High School Grade XI 1st semester).* Jakarta, Indonesia: The Ministry of Education and Culture of Republic of Indonesia.

Djuharie, O. S. (2012). *Bahasa Inggris berbasis pendidikan karakter bangsa: untuk SMA/MA kelas X semester 1 & 2* [English based on national character building: For senior secondary schools grade X semester 1 & 2]. Bandung, Indonesia: Yrama Widya

Kaimuddin. (2014). Implementasi pendidikan karakter dalam kurikulum 2013 [Character education impelementation in curriculum 2013]. *Dinamika Ilmu, 14*(1), 47–64. doi:10.21093/di.vl4il.7

Lie, A. (2000). The Multicultural Curriculum: Education for Peace and Development. *Human Rights Education in Asian Schools, 6(1),* 81–102

Puskurbuk (The Centre for Curriculum and Textbooks). (2010). *Development of cultural and national character building: A school guide.* Jakarta, Indonesia: Ministry of Education and Culture of Republic of Indonesia.

Sugeng, B., & Zaimah, N. (2012). *Contextual English for grade XII of senior high schools: Physical sciences and social sciences majors.* Solo, Indonesia: Platinum—Tiga Serangkai

Sultoni, A. (2016). Pendidikan karakter dan kemajuan negara: Studi perbandingan lintas negara [Character education and development nation: Comparison study accross country]. *JOIES: Journal of Islamic Education Studies, 1*(1), 184–207.

Widiati, U. Rahmah, Z., & Furaidah (2014). *Bahasa Inggris SMA/MA SMK/ MAK Kelas X (10) semester 1* [English for Senior High School Grade X 1st semester]. Jakarta, Indonesia: The Ministry of Education and Culture of Republic of Indonesia.

Widiati, U. Rahmah, Z., & Furaidah. (2015). *Bahasa Inggris SMA/MA SMK/MAK Kelas XII (12)* [English for Senior High School Grade XII]. Jakarta, Indonesia: The Ministry of Education and Culture of Republic of Indonesia.

Woodward, K. (2001). *Identity and difference.* London, UK: Sage Publications.

Profile of student character on discipline behavior and cleanliness culture in higher education

M. Mutaqin

Universitas Negeri Yogyakarta, Indonesia

ABSTRACT: Students discipline and cleanliness culture were strongly supports the learning process quality in higher education. This study aims to determine profile of students character on discipline behavior and cleanliness culture. The study was conducted by survey approach, with the unit of analysis being Universitas Negeri Yogyakarta (UNY) students. Data collection was carried out using questionnaires and observations. Data was carried analyzed by using qualitative descriptive studies. The results showed that most of students (63.33%) were on good category of disciplined behavior. However, most of students (53.33%) were on less category of cleanliness culture.

1 INTRODUCTION

Character interpreted as a series of knowledge, attitudes and behaviors leads to commitment to doing well (Lickona, 1991). Furthermore, character education was a typical values of a person's personality formed from the results of internalization and used in attitude, and behavior in everyday life (Dharmawan, 2014). Nowadays, character educations are still relevant to be discussed because it is an important part of developing nation character. One of the nation's character education activities program in the micro context was through the daily life activities of campus community in higher education (RoI, 2010). In RoI (2011), it has been identified that there are 18 values of character that need to be implanted to students who come from Religion, Pancasila, Culture, and National Education Objectives. Two character values are disciplined behavior, and care for the environment (Zuchdi, 2012).

Discipline control is one of the school cultures needed to support successful learning (Setiyati, 2014). Awareness of discipline, interpreted as an attitude of obedience to the rules of the campus community in a college, is still one of the serious problems that requires more attention (Winataputra, 1997). Awareness of disciplinary behavior in the academic community in most universities is still low. In fact, discipline behavior present a lot of profit for ourselves such as comfortable feels. Conversely, undisciplined behavior creates a lot of discomfort for us and the others. Some examples of undisciplined behavior in the campus environment are ranging from orderly queuing, waste disposal, even to behaviors that can be harmful to self and surrounding communities. If this is so, then how can undisciplined behavior be changed into disciplined behavior? Disciplined behavior must be enhanced through awareness built by a person or institution. Disciplined behavior, growth the awareness of cleanliness living behavior in a person or an institution.

In general, everyone likes cleanliness. Environmental hygiene is the main factor of comfort in everyday life. A clean campus is the dream of every campus. With a clean, neat and beautiful campus, students, lecturers, and employees will be able to perform their tasks well and feel comfortable. Students can learn comfortably because they are supported by a pleasant atmosphere, and lecturers can also teach comfortably because the place is clean and tidy, so cleanliness supports the teaching and learning process. The cleanliness of the campus environment is a shared responsibility.

The culture of clean living is within each individual student concerned about personal cleanliness, appearance of dress and shoes, and how to care for the cleanliness of limbs. Furthermore, the culture of living clean of students in campus environment, both when students were in the inside and when outside the classroom. When students are outside the classroom in the campus environment, the culture of clean living example were throwing waste it should be disposed of in the trash that is available. In addition, the students should also take care of cleanliness when in a place of worship, in the library, in the cafeteria, and in sports venues, etc.

In practice, the behavior of discipline and cleanliness culture of students inside campus has not obtained the most accurate data. Is student discipline, both in the lecture and outside the class, good or not? Similarly, in terms of cleanliness, there is no empirical data to support the need for students to be clean in the campus environment. This research is therefore important to do.

2 METHODS

The type of research used in this study was survey research, with qualitative descriptive studies. In this study, the student data sources were scattered across several faculties. This research took at Universitas Negeri Yogyakarta (UNY). This research reveals profile of students character on discipline behavior and cleanliness culture. The methods of data collection were using questionnaires and observation techniques. The validity of the data was checked with a cross-check technique, so that only valid data is used for analysis, improving the accuracy of the research result. Data analysis in this research was carried out using qualitative descriptive analysis technique.

3 RESULTS

Discipline and cleanliness culture of students is organized by students as the result of learning process in the campus environment. Based on the results of this study, the profile UNY students in terms of the characters of disciplined behavior and cleanliness culture can be described. Five of the seven faculties in UNY are represented, namely Faculty of Engineering (FT), Faculty of Social Sciences (FIS), Faculty of Economics (FE), Faculty of Mathematics and Natural Sciences (FMIPA), and Faculty of Sport Science (FIK). The results of the study in five faculties based on the categorization of the implementation of character education on discipline behavior and cleanliness culture are presented below.

3.1 *Disciplined behavior and cleanliness of FT students*

The student discipline level in FT is viewed on the basis of disciplined behavior in the classroom, discipline in administration, and discipline in non-formal activities. Culture of cleanliness can be seen from personal hygiene and clean culture of students on campus. The culture of the school environment cleanliness students are grouped into two: the culture of cleanliness inside and outside the classroom. In detail the behavior of discipline and cleanliness of FT students can be seen in Table 1.

3.2 *Disciplined behavior and cleanliness of FIS students*

Discipline levels of FIS students are seen on the basis of disciplined behavior in lectures, discipline in administration, and discipline in non-formal activities. Culture of cleanliness can be seen from personal hygiene and clean culture of students on campus. The culture of the school environment cleanliness students are grouped into two: the culture of cleanliness inside and outside the classroom. In detail, the behavior of discipline and clean culture of FIS students can be seen in Table 2.

Table 1. Disciplined behavior and cleanliness of FT students.

No.	Disciplined behavior						Cleanliness						Category
	Course		Adminis-trative		Non-formal		Personal hygiene		In class		Outdoor		
	f	%	f	%	f	%	f	%	f	%	f	%	
1	0	0.00	2	5.00	3	7.50	1	2.50	3	7.50	3	7.50	Very good
2	20	50.00	24	60.00	16	40.00	19	47.50	16	40.00	17	42.50	Good
3	17	42.50	11	27.50	18	45.00	17	42.50	21	52.50	18	45.00	Less good
4	3	7.50	3	7.50	3	7.50	3	7.50	0	0.00	2	5.00	Not good
Sum	40	100	40	100	40	100	40	100	40	100	40	100	

*f = frequency.

Table 2. Disciplined behavior and cleanliness of FIS students.

No.	Disciplined behavior						Cleanliness						Category
	Course		Adminis-trative		Non-formal		Personal hygiene		In class		Outdoor		
	f	%	f	%	f	%	f	%	f	%	f	%	
1	4	13.79	0	0.00	3	10.34	1	3.45	1	3.45	0	0.00	Very good
2	12	41.38	16	55.17	11	37.93	16	55.17	16	55.17	17	58.62	Good
3	11	37.93	11	37.93	14	48.28	9	31.03	10	34.48	9	31.03	Less good
4	2	6.89	2	6.89	1	3.45	3	10.34	2	6.90	3	10.34	Not good
Sum	29	100	29	100	29	100	29	100	29	100	29	100	

*f = frequency.

3.3 *Disciplined behavior and cleanliness of FE students*

The student discipline level in FE is viewed on the basis of disciplinary behavior in the class-room, discipline in administration, and discipline in non-formal activities. Culture of cleanliness can be seen from personal hygiene and clean culture of students on campus. The culture of the school environment cleanliness students are grouped into two: the culture of cleanliness inside and outside the classroom. Detailed results can be seen in Table 3.

3.4 *Disciplined behavior and cleanliness of FMIPA students*

The student discipline level in FMIPA is viewed on the basis of disciplined behavior in the class room, discipline in administration, and discipline in non-formal activities. Culture of cleanliness can be seen from personal hygiene and clean culture of students on campus. The culture of the school environment cleanliness students are grouped into two: the culture of cleanliness inside and outside the classroom. Detailed results can be seen in Table 4

3.5 *Disciplined behavior and cleanliness of FIK students*

The student discipline level in FIK is viewed on the basis of disciplined behavior in the class-room, discipline in administration, and discipline in non-formal activities. Culture of cleanliness can be seen from personal hygiene and clean culture of students on campus. The culture of the school environment cleanliness students are grouped into two: the culture of cleanliness inside and outside the classroom. Detailed results on the behavior of discipline and net culture of FIK students can be seen in Table 5.

Table 3. Disciplined behavior and cleanliness of FE students.

No.	Disciplined behavior						Cleanliness						Category
	Course		Adminis-trative		Non-formal		Personal hygiene		In class		Outdoor		
	f	%	f	%	f	%	f	%	f	%	f	%	
1	1	3.70	0	0.00	1	3.70	0	0.00	3	11.11	0	0.00	Very good
2	14	51.85	18	66.67	14	51.85	13	48.15	11	40.74	16	59.26	Good
3	8	29.63	6	22.22	11	40.74	12	44.44	13	48.15	11	40.74	Less good
4	4	14.81	3	11.11	1	3.70	2	7.41	0	0.00	0	0.00	Not good
Sum	27	100	27	100	27	100	27	100	27	100	27	100	

*f = frequency.

Table 4. Disciplined behavior and cleanliness of FMIPA students.

No.	Disciplined behavior						Cleanliness						Category
	Course		Adminis-trative		Non-formal		Personal hygiene		In class		Outdoor		
	f	%	f	%	f	%	f	%	f	%	f	%	
1	0	0.00	0	0.00	2	5.88	0	0.00	2	5.88	3	8.82	Very good
2	21	61.76	20	58.82	17	50.00	17	50.00	14	41.18	14	41.18	Good
3	10	29.41	11	32.35	15	44.12	14	41.18	16	47.06	16	47.06	Less good
4	3	8.82	3	8.82	2	5.88	3	8.82	2	5.88	1	2.94	Not good
Sum	34	100	34	100	34	100	34	100	34	100	34	100	

*f = frequency.

Table 5. Disciplined behavior and cleanliness in FIK students.

No.	Disciplined behavior						Cleanliness						Category
	Course		Adminis-trative		Non-formal		Personal hygiene		In class		Outdoor		
	f	%	f	%	f	%	f	%	f	%	f	%	
1	4	13.33	0	0.00	2	6.67	5	16.67	1	3.33	3	10.00	Very good
2	9	30.00	19	63.33	15	50.00	11	36.67	13	43.33	12	40.00	Good
3	15	50.00	8	26.67	11	36.67	13	43.33	15	50.00	14	46.67	Less good
4	2	6.67	3	10.00	2	6.67	1	3.33	1	3.33	1	3.33	Not good
Tot	30	100	30	100	30	100	30	100	30	100	30	100	

*f = frequency.

4 DISCUSSION

This study focuses on student behavior in terms of discipline and cleanliness on campus. The value of the discipline in non-formal activities conducted in the campus environment. For The value of clean cultural behavior can be assessed from the behavior of clean life as individuals (personal hygiene). This aspect can be viewed based on cleanliness of the personal limbs, clean in dress, shoe and in appearance. Additionally this aspect can be reviewed based on student behavior when in a campus environment, whether in the explanation of the results of research on students from five faculties: FT, FIS, FE, FMIPA and FIK.

The level of student discipline in FT, in terms of disciplined behavior in lectures, discipline in administration, and discipline in non-formal activities, is grouped into four categories: very good, good, less good, and not good. Similarly, the level of the culture of cleanliness is grouped into four categories: very good, good, less good, and not good. culture of cleanliness is reviewed based on student behavior when in a campus environment, whether in the classroom or outside the classroom.

Based on the data analysis as described above, it can be concluded that student discipline in FT for lecture and administration activity including in less good category. The clean cultural behavior, especially in and outside the classroom is at less good category, but in terms of personal hygiene is included in good category.

For student FIS, it is assumed that in the course of lectures and administration is at good category, but for non-formal activity including in less good category. About the cleanliness cultural behavior in terms of personal hygiene and clean culture in the classroom is in the good category.

Students at FE have good category for lectures, administration and non-formal activities. Similarly, the behavior of cleanliness culture in terms of personal hygiene and clean cultural behavior inside the classroom also is in the good category. But in net cultural behavior outside the classroom, the students is in less good category.

As students in FE, the results of students in FMIPA are in the same category for lectures, administration, non-formal activities, and in term of personal hygiene. But in term of clean culture is in the less category.

Different results come from the students in FIK. According to data analysis, the terms of lecturing activities of students is in the less category. In terms of administrative and non-formal discipline are in the good category. The cleanliness cultural behavior is in the less category.

5 CONCLUSIONS

Based on the analysis, it can be concluded that the implementation of character education behavior discipline of students in the campus UNY is in the good category (63.33%). Meanwhile, for the behavior of cleanliness culture is in the less good category (53.33%).

6 RECOMMENDATIONS

Based on the results of this study, it is suggested that: character education in universities, including UNY in developing disciplined behavior and a culture of cleanliness, especially among students, should be given serious attention from all parties, including from the leaders in work units within UNY. Therefore, all academic community must have a commitment, which is willing to build and internalize character education to behave discipline and clean culture. It should be a habit of living on campus.

REFFERENCES

Dharmawan, N.S. (2014). *Implementasi pendidikan karakter bangsa pada mahasiswa di perguruan tinggi* [Implementation of national character education in university]. Paper presented at Character Education Training Lingkungan Kopertis Wilayah VIII.

Lickona, T. (1991). *Educating for character: How our school can teach respect and responsibility.* New York, NY: Bantam Books.

RoI. (2010). *Kebijakan nasional pembangunan karakter bangsa tahun 2010–2025* [National policy of character building in 2010–2025]. Jakarta, Indonesia: Curriculum center, The Ministry of National Education, Republic of Indonesia.

RoI. (2011). *Panduan Pelaksanaan Pendidikan Karakter* [Character Education Implementation Guidance]. Jakarta, Indonesia: The Ministry of National Education, Republic of Indonesia.

Setiyati, S. (2014). Pengaruh kepemimpinan kepala sekolah, motivasi kerja, dan budaya sekolah terhadap kinerja guru [The influence of headmaster leadership, work motivation, and school culture on teacher performance]. *Jurnal Pendidikan Teknologi dan Kejuruan, 22*(2), 200–207

Winataputra, U.S. (1997). *Strategi Belajar Mengajar* [Teaching learning strategy]. Jakarta: Universitas Terbuka.

Zuchdi, D. (2012). *Pendidikan karakter konsep dasar dan implementasi di perguruan tinggi* [Character education basic concept and implementation in university]. Yogyakarta, Indonesia: UNY Press.

Character Education for 21st Century Global Citizens – Retnowati et al. (Eds)
© *2019 Taylor & Francis Group, London, ISBN 978-1-138-09922-7*

The implementation of a local wisdom-based character education model in primary schools in Bantul, Yogyakarta

R. Rukiyati, L. Hendrowibowo & M. Murtamadji
Universitas Negeri Yogyakarta, Indonesia

ABSTRACT: The purpose of this study was to design a model of local wisdom-based character education in primary schools in Bantul, Yogyakarta, Indonesia. The method used was a research development model consisting of ten stages according to Borg and Gall (1989). In this research, the stage that has been reached is field test phase implementation. The subjects of the research were seven teachers of the fourth grade in primary schools in the sub-districts of Pajangan and Banguntapan, Bantul, Yogyakarta, Indonesia. The data was obtained by conducting Focus Group Discussions (FGD), observations, interviews, and documentation. The data was then analyzed using the quantitative and qualitative descriptions. The research concludes that a character education model based on local wisdom in primary schools in Bantul, Yogyakarta, Indonesia has been successfully implemented in the integrated learning by teachers of five primary schools. The character education model developed from local wisdom consists of the *dolanan* song (be careful, love science, cooperation, honesty, harmony, humility, caution, responsibility) and batik art (cooperation, perseverance, responsibility, care, cleanliness, creativity).

1 INTRODUCTION

Local wisdom is basic knowledge gained from living in balance with nature. It is related to the culture in the community in which it is accumulated and passed on. The wisdom comes from experiences or truth gained from life. It emphasizes respect for elders and their life experiences. Moreover, according to Nakorntrap, it values morals more than material things (Mungmachon, 2012).

As a multicultural nation, Indonesia has many traditions and values of local wisdom that need to be preserved and developed so that the character and characteristics of Indonesian human beings, with their various cultural values, do not just disappear from the effects of the negative influence of modern life that tends to emphasize materialism and freedom, almost without limit. Unfortunately, these values have not been seriously included in the national education improvement agenda. Thus, it needs creativity and innovation to integrate the values of local wisdom in the learning process in schools under the umbrella of the nation's character education.

Teachers are the spearheads of formal education (schools), and play a very important role. The teacher should be responsible for educating character to their students. They need to design the character education inserted in the learning process by integrating it into various local wisdoms that have been entrenched in society over the years. Humphreys, Post and Ellis (as cited in Lake, 1994, p. 1) state "An integrated study is one in which children broadly explore knowledge in various subjects related to certain aspects of their environment". They see links among the humanities, communication arts, natural sciences, mathematics, social studies, music, and art. Skills and knowledge are developed and applied in more than one area of study.

Yogyakarta and Central Java in Indonesia are known for their various local cultures that have their own values of wisdom. The forms are varied, including cooperation and caring activities (*gotong royong*), village discussion (*rembug desa*), songs, traditional Javanese dances, as well as traditional games.

Ryan and Bohin (1999) stated that character consists of three prior aspects, i.e., knowing the good, loving the good, and doing the good. In character education, goodness is often represented in good attitudes. Shea and Murphy (2009) concluded character education programmed by the school and stakeholders is considered as a magical experience. Osterman (Mingchu et al., 2007) found that students' experience of school embraces the norms of collaboration and cooperation and the core value of benefiting others. Goodenow (1993) and Wentzel (1998) found students experiencing caring relationships have more positive attitudes and are more motivated toward learning and involvement in school activities (Mingchu et al. 2007). The study conducted by Benninga et al. (2003), on the relationship between character education and school achievement in some primary schools in California, concluded that the schools where character education is taught seriously tend to have high academic achievement. Davidson et al. (2007) claimed there is the role of character in all school achievement, whether curricular or non-curricular activities.

With these considerations, it is necessary to conduct research that develops a thematic-integrative learning model of character education based on the values of local wisdom for primary schools.

The formulation of the research problem is "How to develop a thematic-integrative learning model of character education based on the values of local wisdom in primary schools in Bantul Yogyakarta."

2 RESEARCH METHOD

The research method is designed with a research and development approach. As stated by Borg and Gall (1989), there are ten research steps. In this study, seven steps have been achieved, namely field implementation tests involving five primary schools in Bantul, Yogyakarta, Indonesia.

This research was conducted using various data collection methods including conducting observation and written tests and distributing the questionnaires. Observation was used to obtain data about the learning practices in each school's classroom. Observation sheets and field notes were used as the instruments for data collection. The test was used to obtain the students' learning outcomes. The questionnaire was used to collect the data of the students' attitudes toward the embedded character values.

The data analysis method used is as follows. In the field testing phase, an evaluation of the data gained from the observation of the learning process was done. The data was analyzed with scores based on the rubric that had been prepared. Meanwhile, the evaluation of the students' attitudes was done using a scale in the questionnaire about student attitudes of attitude of students toward the learning which has been implemented. The analysis of attitude scale results was done by finding the mean score on the attitude scale scores with range 1–4. Then, the students' cognitive evaluation was done by multiple choice written tests of which the score range was 0–100.

3 RESULTS AND DISCUSSION

3.1 *Research results*

The field testing implementation was conducted in five schools involving ten teachers. Gradually the field implementation test began with the following activities.

Before the field testing was conducted by the primary school teachers, the researcher coordinated and socialized the guidebook, lesson plan, and played the learning recordings in the field testing stage in a workshop in Pajangan, Bantul on 31 August 2016.

During the workshop, it had been agreed that teachers would implement an integrative thematic learning of character education based on local wisdom in each school by first making a few changes to the lesson plans that had been prepared.

After the teachers prepared the lesson plans in accordance with the Curriculum 2013 format, they sent them to the researcher via email. With a few improvements and adjustments, they were sent back to the teachers to be ready for implementation in the lesson. Results of implementation follow.

3.1.1 *Results of implementation at Muhammadiyah Kalakijo Primary School*

During the practice, there were two teachers who implemented the teaching, namely Mrs. Andin and Mr. Ridwan. The students were divided into seven groups consisting of 2–3 children to make batik caps by using a natural stamp of the plant in the form of banana leaf pieces, starfruit, papaya leaf and other plants from the surrounding environment. The purpose was that they would learn how to make a batik stamp on a flower vase media in groups as well as practicing the values of cooperation, responsibility, tenacity and creativity.

The teacher gave an example of how to make batik stamps with the stamp tool using plants around the student environment. Of the nine groups, it appears that almost all students are concerned, except for two groups. The students seem to start negotiating on the value of cooperation when creating patterns. Teachers always checked the learning process of students in each group.

When the batik stamping activity was finished, the teachers did an assessment of the students' learning outcomes. They said that all batik vases produced were good, but only three of them were considered the most beautiful from the use of stamps, motif patterns and mix of colors. Then, the students were asked to write down their feelings when they made the batik. They expressed that they were happy to learn to make batik stamps using simple tools derived from nature. Furthermore, they also agreed the batik lesson taught them diligence, responsibility and cooperation with friends.

In general, character learning integrated in batik education with a batik stamp theme using the plant media ran as expected.

3.1.2 *Results of implementation at Krebet Primary School*

The practice of character learning with the integration of local content with the theme of the diversity of Indonesia combined Civics learning and *Dolanan Anak "Cublak-Cublak Suweng"*. Mrs.Tia, as the executing teacher, started the lesson by praying according to the teachings of Islam (as all students were Muslims) followed by the apperception material of questions and answers about traditional houses in Indonesia.

After that, she conveyed the purpose of the lesson and continued by inviting students to sing *"Cublak-cublak Suweng"*. After singing, the teacher gave clarification of the value or explanation about the game of *Cublak-cublak Suweng* as a game that taught harmony, unity and honesty. After giving the clarification, she asked the students to form groups and played

> *Cublak-cublak suweng/suwenge teng Grendel/*
> *Mambu ketundung gudel/Pak Lempong lela lelo/*
> *Sopo gelem ndelik'ake/Sir pong dele kopong/sirsir pong dele kopong.*

After playing for the third time, the teacher re-explained the meaning of the game and the values of the characters contained in it. She gave reinforcement and praised the students who could answer the questions related to the game of *Cublak-cublak Suweng* and its relation to the value of harmony, unity, and honesty.

After this reinforcement by the teacher, the students were asked to do the test on the learning outcomes and complete the attitude scale questionnaire about the *Cublak-cublak Suweng* game integrated with Civics. The test result of the study of the 20 students obtained an average score of 8.80, so as a whole it appeared that the learning outcomes had reached the targeted criteria score of 8.00). The affective learning results obtained from the questionnaire scale of attitudes are shown in Table 1. Table 1 indicated that the students had been positive toward honesty, respect, responsibility, and cooperation behavior. In the range of scores 1 to 4 (Strongly Disagree = 1, Disagree = 2, Agree = 3, Strongly Agree = 4), the students tended to be very amenable to the values of characters raised to be addressed.

Table 1. Student attitude result (N = 20).

No.	Behavior characteristic measured	Mean
1.	Honesty	3.7
2.	Appreciation and respect for others	3.9
3.	Responsiblity	3.65
4.	Work well together	3.8

Based on a whole series of learning process, it appears that the field implementation test had been carried out well, as planned, although there were still some shortcomings in the process.

3.1.3 *Results of implementation at Triwidadi Primary School*

Implementation at the participating public elementary school was conducted on Friday, October 14, 2016 with the theme: Piggy bank batik stamp with media from plants. The implementation in the school was nearly the same as that in the participating private elementary school in Kalakijo, but the students studied individually.

Ms. Marini (teacher 1) taught the students about batik stamps using medium from nature on the piggy banks that had been prepared as well as learning to be responsible, creative, and diligent. Mr. Irwan (teacher 2) acted as the observer and assessor during the learning process.

Ms. Marini conveyed the learning objectives and targets that the students must achieve. The students had been sitting quietly in their respective chairs. Then, the students started to work on batik stamping by using the stamps made from banana stem, papaya leaf bark, carrot and starfruit. They seemed enthusiastic about doing their work. The students directly responded to tasks assigned by teachers with enthusiasm. All were busy working with their own creativity. On average, the piggy banks could be completed within 30 minutes which indicated that the students were responsible enough to finish the work according to the time determined by the teacher. The observation scores by the researcher and teacher within the range of 1–4 for each character: creativity, responsibility, diligence and cooperation, are shown in Table 2. Based on Table 2, it can be concluded that the implementation of character learning has been able to train the students to realize the values of creativity, responsibility, diligence, and cooperation characters as the results were Good and Very Good.

3.1.4 *Results of implementation at Sendangsari Primary School*

Sendangsari Primary School is located on the edge of the Pajangan highway. The field implementation at the school was conducted by the fourth-grade teacher, Mrs. Nasia. The teaching was conducted on October 26, 2016 in accordance with the jointly designed lesson plan.

The practice of the character learning integrated in local content was done using the theme of *Indahnya Kebersamaan* (The Beauty of Togetherness). The integrated local wisdom was the "*Gundul-gundul Pacul*" game song. The character values instilled were honesty, self-discipline, responsibility, courtesy, caring and confidence in interacting with family, friends, teachers and society.

Before starting the class, the teacher had already arranged it dividing the students into six groups. Then, the teacher invited them to sing a song, "*Gundul-gundul Pacul*". After singing twice, the teacher asked students to discuss what the meaning of the song was. Then, the respective representatives from each group read out the results of their discussion and the students from other groups were asked to provide verbal feedback. When each group of students received feedback, they recorded it to improve their discussion results.

After the discussion, each group demonstrated the song using their creativity. Each group seemed to create various dances while singing the song:

Gundul-gundul Pacul cul Gembelengan
Nyunggi nyunggi wakul kul gembelengan
Wakul glimpang
Segane dadi sak latar

Table 2. Observation results of character values N = 20.

No.	Targeted character value	Score value	Result
1.	Creativity	3.45	Good
2.	Responsibility	3.25	Good
3.	Diligence	3.30	Good
4.	Cooperation	3.60	Very good

Aesthetically, it could be seen that the most expressive performers were groups two and four. The learning activities continued with clarification of the character's values contained in the song. Mrs. Nasia said that the song taught people not to be arrogant, but to be humble and careful in their behavior. Moreover, an arrogant leader will bring inconvenience for the citizens he leads. The teacher continued that one day the children would become leaders. Perhaps they would be a village head, chief of the sub-district or an official. Therefore, the teacher reminded them not to be arrogant, to remain humble and responsible, and to hold the trust.

Giving clarification of the values, the teacher assigned the students to take the formative test about the lessons of the character values of the song. It was revealed that all students could answer the question about the meaning of the song as well as the character values contained in the song correctly. The teacher closed the class by reflecting on the learning activities that had been implemented.

Based on the observation of the learning process and the analysis of the students' learning outcomes, it could be concluded that the field implementation of character education based on local wisdom had been successful implemented.

3.1.5 *Results of implementation at Kanisius Sorowajan Primary School*

Implementation of the learning was also carried out in Kanisius Primary School (Catholic primary school) in Bantul. The field implementation was conducted together with Ms. Sandra.

The integrated local wisdom was the *"Gundul-gundul Pacul"* dolanan song. The focus of the learning was in the social science lesson, sub-theme: The cultural diversity of my country. The character values embedded were honesty, discipline, responsibility, politeness, caring and confidence in interacting both as a leader and members of the community.

Mrs. Sandra started the class by exchanging greetings followed by an apperception in the form of questions and answers about regional cultural variety in Indonesia.

Mrs. Sandra started the class by exchanging greetings followed by apperception materials which were given in the form of questions and answers about the local culture in Indonesia.

After that, Mrs. Sandra invited them to sing the *"Gundul-gundul Pacul"* song. After singing, the teacher clarified the character values contained in the song. In a very interesting way, she explained about the values of characters in the song. *"Gundul"* referred to a hairless head. Hair was the crown. "Bald" was a metaphor for a leader without a crown. So, the leader could be an ordinary person who became a leader. *"Nyunggi wakul"* meant the leader must be willing to accept a mandate or task without being *"gembelengan"* (arrogant, ignorant) so that the people prosper and be prosperous. the people are prosperous and prosperous. If *"wakul gelimpang"* (the basket fell to the ground) and *"segane dadi sak latar"* (the rice scattered on the ground), then the people's trust would be lost because the leader did not hold the mandate to *"nyunggi wakul"* as well as possible.

Mrs. Sandra asked the children: "What values are there in the *Gundul-gundul Pacul*?"

It turned out that they were able to answer it easily, "Responsibility", "fairness", "firmness", "intelligence", "discipline", "no discriminating", and "honesty".

Mrs. Sandra then continued asking the reflective question, "Are you disciplined in obeying the rules?"

Some children answered, "Yes, Ma'am." The teacher asked again, "What if we disobey the rules?"

The children replied, "Exposed to punishment, Ma'am."

Furthermore, Mrs. Sandra related the values in the song to Kanisius' values, namely love, discipline, and honesty.

Then, Ms. Sandra invited the students to come out of the classroom to the *pendapa* (hall) to practice the songs and choreography of *"Gundul-gundul Pacul"* expressively. Each group seemed cheerfully singing and dancing to the rhythm of the song.

The next activity was the task of writing the students' reflections on the feelings and actions that would be performed after singing and dancing to the song. In general, they were happy. They said: "exciting", "happy to play happily with friends", "happy to sing and be creative", and other happy statements about the newly done learning.

It could be said that the implementation of character education based on local wisdom in this school had been successful implemented.

3.2 *Discussion*

It has been seen that the results achieved at this stage of learning practices that have been done can be implemented properly.

It has been seen that the results achieved in the teaching practice that has been done can be implemented well.

There were some methods of character education implemented by the teachers. They displayed themselves as people of good character, by preparing facilities in the form of materials needed from the both the school and the environment, as well as value skills in the form of working together in groups and discussing these. This is in line with Kirschenbaum's (1995) assertion that values or character education should be comprehensively carried out using four main methods: value-building, exemplification, facilitation, and skills of moral/character values. Likewise, Lickona (2012) reinforces that educating characters should be by example using direct instruction to form conscience and habits, giving the opportunity to practice virtue and encouraging spiritual development.

If the teachers are consistent in integrating the character education based on local wisdom in the subject matters, it is very possible for students with good character to be produced. Local wisdom as is stated by Adimiharja (Sadjim et al., 2015) becomes a common commitment as a form of social cohesiveness of harmonic complexity of the society.

4 CONCLUSIONS AND SUGGESTIONS

The research has successfully taught the students to know and practice the values of good character, especially the value of cooperation, creativity, humility, responsibility, diligence and to not be easily discouraged.

From the results, it could be concluded that the model of character education based on local wisdom in primary schools has passed the test and can be continued to the next stage of wider implementation.

Teachers could implement the character education based on local wisdom by integrating it into art and culture subjects, and social sciences with guidance books prepared by the researcher.

REFERENCES

Benninga, J.S., Berkowitz, M.W., Kuehn, P., and Smith, K. (2003). The relationship of character education implementation and academic achievement in primary schools. *Journal of Research in Character Education*, 1(1), 19–32.

Borg, R.W. & Gall, M.D. (1989). *Educational research: An Introduction* (Fifth Edition). London, UK: Longman.

Davidson, M., Lickona. T. & Khmelkov, V. (2007). Smart and good schools. *Education Week*. Retrieved March3, 2008 from http://www.edweek.org/ew/articles/2007

Kirschenbaum, H. (1995). 100 ways to enhance values and morality in schools and youth settings. Boston, MA: Allyn and Bacon.

Lake, K. (1994). *Integrated curriculum. school improvement research series*. Retrieved from http://educationnorthwest.org/sites/default/files/integrated-curriculum.pdf

Lickona, T. (2008). *Effective character education*. Retrieved February 20, 2008 from www.cortland.edu/character/articles/char_v.asp

Lickona, T. (2012). *Character matters: Character problems*. Jakarta, Indonesia: Bumi Aksara.

Mingchu, L., Huang, W., and Najjar, L. 2007. The relationship between perceptions of a Chinese high school's ethical climate and students' school performance. *Journal of Moral Education, 36*(1), 93–111.

Mungmachon, R. (2012). Knowledge and local wisdom: Community treasure. *International Journal of Humanities and Social Science, 2*(13), 174–181.

Ryan, K., & Bohin, K.E. (1999). *Building character in schools—practical ways to bring moral instruction to life*. San Francisco, SF: Jossey-Bass A Wiley Imprint.

Sadjim, U.M., Muhadjir, N. & Sudarsono, F.X. (2015). Revitalisasi nilai-nilai Bhinneka Tunggal Ika dan kearifan lokal berbasis learning society pasca konflik sosial di ternate [Revitalization of unity in diversity values and local wisdom—based learning society post-social conflict in Ternate]. *Jurnal Pembangunan Pendidikan: Fondasi dan Aplikas,i 4*(1), 79–91.

Shea, K. & Murphy, K.B. (2009). *A perfect match: Living values educational program and Aventura city of excellence school, USA. [Electronic version]*. Retrieved on January 15, 2010 from www.springer.com

The strengthening of multicultural values within civic learning: How do teachers make it work?

S. Samsuri & S. Suharno
Universitas Negeri Yogyakarta, Indonesia

ABSTRACT: This study aimed to elaborate the understanding and experience of civic education teachers when they implemented multicultural values in the *Bhinneka Tunggal Ika* (Unity in Diversity) concept, which is one of the main subjects of *Pancasila* and civic education (*Pendidikan Pancasila dan kewarganegaraan,* PPKn) in Indonesian schools. The strengthening of multicultural values becomes a necessity in the formation of civic competences. This research involved teachers of civic education in two junior high schools in Yogyakarta, Indonesia. They were trained to use the civic competence development model which was developed by the research team. The results found that, firstly, the teachers understood the academic content of developing citizenship competence in the curriculum year 2013, which is stated in the subject of PPKn in the schools. The model covered the aspect of civic attitude/disposition, knowledge and skill competences. Secondly, the values of nationalism and the teachings of local wisdom in multicultural societies have been implemented by teachers as the main subject of PPKn in schools with various innovations in their respective schools. Some problems exercising the model program are discussed.

1 INTRODUCTION

The diverse nationalism and statehood of Indonesia encourages the importance of developing citizenship competencies in the students' experiences based on national values and local wisdom developed through school. Following the findings of Suharno, et al. (2013a) studies on the model of peace building resolution in a multicultural society, civic competence should develop peaceful means for any conflicts that may arise. Similarly, the ways local communities such as Sampit and Poso resolve conflicts in multicultural communities by using local wisdom (Suharno et al., 2013b), should be used as a model of competence development of citizenship in the teaching of citizenship education (civics) in schools.

The study by Samsuri and Marzuki (2016) elaborated on the needs of character formation of the civic multicultural Madrasah Aliyah, especially in Yogyakarta. Its development based on the pillars of national according to the People's Consultative Assembly (*Majelis Permusyawaratan Rakyat, MPR*). That study shown the formation of the multicultural citizenship character seemed a number of good practices in school. In line with the intention of Miller-Lane, et al. (2007), the competence of multicultural citizenship can be developed through a variety of studies such as social studies learning which includes topics such as civic education in the United States. Thus, teachers are expected to prepare students to face the life of a global community in a multicultural society.

Maximally achieving components of citizenship competence in learning civics is expected to reduce the underestimate on strategic position of civics as one of the compulsory subjects from primary education to higher education. Along with the complexity of national and state life in Indonesia entering the reform era, the component of citizenship competence as an indicator of the success of civic curricular programs in schools needs to make national values and local wisdom contextually. Task of civics as the forming of good citizen character need civics teachers who are competent and skilled.

Competence development of citizenship based on national values and local wisdom in multicultural communities in schools is backed by a number of research results (Samsuri, 2010; Pawiroputro, et al., 2010; Samsuri, 2011; Suharno, et al., 2013a; Suharno, et al., 2013b; Samsuri & Marzuki, 2016). From the research findings, there are a number of important vertices in the roadmap to this study.

Competence citizenship became the principal aspect of the components of successful achievement of learning citizenship education in schools that cannot be separated from the paradigm shift of the curricular program in Curriculum 2006 (known as Unit Level Curriculum, *Kurikulum Tingkat Satuan Pendidikan*, KTSP). The 2006 curriculum also generally views the importance of competence as a differentiator from the previous curriculum, thus known as the Competency Based Curriculum (CBC). On the other hand, Curriculum 2013 has provided academic and pedagogical opportunities to establish citizenship competencies. The subjects of *Pancasila* and Citizenship Education (PPKn) as one of the compulsory subjects in elementary to secondary education as set forth in the curriculum document of 2013, have a role and strategically equip citizenship competence for the learners.

Components of civic competence introduced in civics (*Pendidikan Kewarganegaraan*, PKn) (Curriculum 2006) and PPKn (Curriculum 2013) are in line with the new paradigm of civic education that has developed since the end of the cold war in some democratic countries in the world. For the Indonesian context, acknowledged or not, the influence and role of Center for Civic Education (CCE), Calabassas, California, USA, introducing the concepts and components of citizenship competence is essential in civic education studies. Components of CCE's civic competence include: civic knowledge, civic skills and civic dispositions (Samsuri, 2011).

However, understanding of the importance of achieving the competence of citizenship by some teachers of PKn/PPKn still encounter obstacles. For example, in the research by Pawiroputro, et al. (2010) on the junior high school's civics teachers in Sleman, Yogyakarta, some complain of difficulty translating basic competencies civics lesson at secondary school level (Curriculum 2006). On the other hand, the civics teachers in Madrasah Aliyah throughout Special Region of Yogyakarta Province have optimized the achievement of civic competency especially through curricular program of civics subject with the national values based in the national pillars of the MPR version along with the religious values (Samsuri & Marzuki, 2016).

As explained in the preceding paragraph, this paper elaborates on a number of important components of citizenship competence in the study of civic education in multicultural societies. Theoretically, the components are spelled out from various perspectives of academic and pedagogical studies. In addition, the discussion will touch on the importance of funding the values of nationality and local wisdom as the content of the mastery of citizenship competence of learners with diverse backgrounds of life, especially the implementation of curriculum programs PKn/PPKn in both curricula (Curriculum 2006 and Curriculum 2013) in schools.

2 LITERATURE REVIEW

2.1 *Multicultural values based national pillars*

The concept of the "Pillars of Nationality" of Indonesia or fully known as the "Four Pillars of Nationality and State Life" (General Secretariat of the People's Consultative Assembly, 2012) was massively promoted by the People's Consultative Assembly from 2009–2014. The four pillars according to People's Consultative Assembly (MPR) include Pancasila, the 1945 Constitution of the Republic of Indonesia, the Unitary State of the Republic of Indonesia and *Bhinneka Tunggal Ika*. Although, there is a debate about the position of *Pancasila* which is one of the pillars because *Pancasila* is the foundation of the state, but MPR RI insists on appointing *Pancasila* as one of the four pillars of nation and state (Samsuri, 2012).

In practice, the pillars were promoted by the MPR through a number of television media, newspapers and online. In this research, the values of multicultural citizenship character developed following to the guidebook on the four pillars. According to the MPR, the reason for the importance of the four pillars is that the constitutional mandate that it embodies as the embodiment of the spirit of the Indonesian family. It has the responsibility to strengthen the fundamental values of national and state life. To that end, the MPR performs its tasks by providing an understanding of the noble values of the nation contained in *Pancasila*, the 1945 Constitution of the State of the Republic of Indonesia, the Unitary State of the Republic of Indonesia and *Bhinneka Tunggal Ika* to the community, known as the Socialization of the Four Pillars of Nation and State Life (Secretary General of MPR Republic of Indonesia, 2012, p. v).

2.2 *Multicultural values based local wisdom*

There are a number of explanations of "local wisdom". The term "local wisdom" is often paralleled as an English translation of the compound "local wisdom". This is often defined as "a local cultural treasure that contains a life policy, a way of life, that accommodates wisdom and living wisdom" (Suyatno, 2011). With this understanding, local wisdom can be a moral or ethical teaching that is taught from generation to generation through oral literature such as proverbs, folklore and manuscripts (Suyatno, 2011).

The special region of Yogyakarta in which the study was conducted also has a number of local wisdoms instituted through a regulation in the form of provincial level regulations. Article 2, paragraph (2), of the Provincial Regulation of the Special Territory of Yogyakarta No. 5, Year 2011 on the Management and Implementation of Cultural Based Education mentioned a number of noble values of culture as follows: honesty; modesty; order/discipline; decency; courtesy/politeness; patience; cooperation; tolerance; responsibility; justice; concern; confidence; self-control; integrity; hard work/tenacity /perseverance; accuracy; leadership and/or toughness (Special Region of Yogyakarta Province, 2011b).

The noble values as a form of local wisdom in the special region of Yogyakarta are used as "Yogyakarta Cultural Values". The values are mobilization of all resources (*golong gilig*), integrated (*sawiji*), the persistence and hard work (*greget*), accompanied by confidence in action (*sengguh*), persistent of any risk (*ora mingkuh*) (Article 1, point (2) of Provincial Regulation No. 4 of 2011 in Special Region of Yogyakarta Province, 2011a).

Thus, the local wisdom in this research is the cultural values of the Yogyakarta Special Region that are institutionalized both culturally and structurally. This local wisdom should be adapted in the formation of citizenship competency curricular programs in schools, whether in certain subjects such as PPKn/PKN or other subjects such as history, sociology of anthropology and religious education.

3 METHODS

This research study, funded by the Ministry of Research and Higher Education, in the Postgraduate Research Team (PTP) research scheme, uses a research and development model. In this study, the development is done by: (1) identifying the need of the civic competency development model based on the national values and local wisdom of the multicultural community in the schools throughout Yogyakarta; (2) photographing the experience of the civic competence development model based on the values of nationality and local wisdom of the multicultural community in a school in Yogyakarta; (3) designing a model of citizenship competency development based on the values of nationality and local wisdom of the multicultural society in Indonesia through school programs; and, (4) producing a civic competency development model based on the values of nationality and local wisdom of multicultural societies in Indonesia through schooling programs. Methods of data collection included focus group discussion (FGD) with a number of civics teachers and academics about the components of citizenship competence and experience of good development in schools.

Along with these development steps, research students, who are members of the PTP Grant, also implemented the results of learning civics in one or all components of citizenship competence submitted by a number of experts.

4 RESULTS

This research yielded quantitative and qualitative data. Quantitative data were derived from a survey on the need for developing citizenship competency in a FGD with teachers of PKn/PPKn, Teacher Consultative of Subject (*Musyawarah Guru Mata Pelajaran*, MGMP) in secondary schools (*Sekolah Menengah Pertama*, SMP), and senior high school (*Sekolah Menengah Atas*, SMA) and vocational schools (*Sekolah Menengah Kejuruan*, SMK). These data were developed as material for developing the civic competence development model based on the national values as contained in the Basic Competency Document on the Subject of Curriculum 2006 and Curriculum 2013 in secondary school to senior high school. The national values are derived from the four national consensus introduced by the MPR RI as "The Four Pillars of National and State Life", namely *Pancasila*, the 1945 Constitution of the Republic of Indonesia, *Bhinneka Tunggal Ika* and the Unitary State of the Republic of Indonesia. Components of civic competence include knowledge, attitudes and civic skills.

Qualitative data were in the form of verbatim data from identification of competency components of citizenship that are contained in the Basic Competency of the PPKn' Curriculum 2013 in secondary to senior high school. In addition, the experience of the teachers of PPKn in secondary to high school outlined the national values that are the subject matter in Curriculum 2013. The subject of PPKn is described as the "good practices" part of each teacher in establishing the citizenship competence of the learners.

The results of the survey and FGD seem that teachers have developed in such a way in the PPKn interpretation civic competence development component of the document Curriculum 2006 and Curriculum 2013. Contextualizing competence development aspects of national values developed from PPKn subject matter includes the study of *Pancasila*, the 1945 Constitution of the Republic of Indonesia, *Bhinneka Tunggal Ika* and the insight of the Unitary State of the Republic of Indonesia.

Specifically, teachers felt that the PPKn' Curriculum 2013 makes it easier for them to develop the competency component of citizenship and content of their studies with national values and local wisdom. The Curriculum 2013 confirms that the PPKn subject (together with religious education and character) has described in detail the components of its basic competence that aligns with the three civic domain competences, namely civic knowledge, civic disposition and civic skills.

Aspects of multicultural values developed from the pillars of nationality and local wisdom. They translated as contents of competence that must be achieved by students in civic learning. Starting from unity and national unity material to globalization materials, all equip the learners with national and state values in PKn' Curriculum 2006. In another side, PPKn's Curriculum 2013 the national values in the pillars combined with the values of local wisdom in Yogyakarta.

On the other hand, from the FGD results with the civics subject teachers, it was found that, so far, the development of citizenship competence was still limited to the translation of basic competencies (*Kompetensi Dasar, KD*) in each grade level. Development is done in two ways. First, following the standard process patterns covered in curriculum documents, syllabuses and lesson plans (*Rencana Pelaksanaan Pembelajaran*, RPP). Second, the development of citizenship competence in addition to translating the components of operational verbs in the KD and teaching materials, while also paying attention to the appropriate learning model to achieve certain competencies. In general, FGD participants were aware of the uniqueness of each KD when translated in RPP.

The values of local wisdom that became one of the aspects of control of citizenship competence in the study of civics and PPKn among others comes from Javanese culture, which is centered on the Sultanate of Yogyakarta. The civics teachers have done both direct and indirect teaching for values such as tolerance, respect for diversity and the nature of mutual love in different societies, put harmony and balance. The values of the local wisdom teachings in each school are mostly presented in the moral messages that exist in the character education sites.

The integration of national values and local wisdom in the development of citizenship competence by teachers has not been done entirely. Nevertheless, there are efforts from a number of teachers to integrate them, especially when presenting material on *Bhinneka Tunggal Ika* in the PPKn class of junior high school or senior high school.

The experiences of a number of teachers presented in the FGD on the development of citizenship competency in the curricular program of PPKn and PKN stated that the translation of competence is done by looking at two aspects. First, the content aspects of the study, or the content of the subject matter contained in the curriculum structure, contained in core competencies and basic competencies in Curriculum 2013, or competency standards and basic competencies in the 2006 Curriculum. Second, performance aspects that appear in verbs operational in the formulation of each basic competence.

The general tendency of civics teachers is to develop grounded study materials and references to textbooks published by the Ministry of Education and Culture of the Republic of Indonesia. The civics textbooks of Curriculum 2006 is primarily an electronic schoolbook (BSE) uploaded by the Center for Curriculum (*Puskurbuk*) Research and Development Agency (*Balitbang*), Kemdikbud. The books were written by various authors from various publishers, who had previously passed the feasibility criteria based on the guidance of the National Education Standards Agency (BSNP), then they were bought by Kemdikbud. In the observation of researchers, civics books by BSE are available widely for elementary school to senior high school. In addition to the schools a selection of complementary books/enrichment tools published by private publishers from various authors are also available.

What are the national values and local wisdoms developed by teachers to achieve the establishment of citizenship competencies for learners in schools? The explanation of civics teachers when in the FGD yielded various results. First, the national values contained in the subjects of PKn/PPKn must definitely contain the values contained in the *Pancasila*, both basic values, instrumental and praxis. The basic values of the *Pancasila* are universal such as the value of God, humanity, unity, people and justice (Ministry of Education and Culture Republic of Indonesia, 2013b, pp. 16–19; Ministry of Education and Culture Republic of Indonesia, 2013c, pp. 86–87) textually and contextually have been translated in such a way in each KD in each level of class, especially in the PPKn Curriculum 2013. The basic values of the *Pancasila* are derivative and emanative, reflected in the integrity of the 1945 Constitution of the Republic of Indonesia, motto of *Bhinneka Tunggal Ika* and the insight of the Unitary State of the Republic of Indonesia.

The subject of multicultural awareness built through Curriculum 2013, is strongly emphasized when discussing diversity in the realities of life, or the tolerant behavior of religious, race, cultural and gender diversity in the *Bhinneka Tunggal Ika* (Ministry of Education and Culture Republic of Indonesia, 2013a; Ministry of Education and Culture Republic of Indonesia, 2013c).

The ability to solve diversity problems such as ethnic, religious, racial and intergroup conflicts (*Suku, Agama, Ras* and *Antar-Golongan*, SARA) that arise in the community, from the very beginning is of great concern. This is understandable given that Indonesians are very diverse in their religious, ethnic, geographical and local culture backgrounds. With spirit of *Bhinneka Tunggal Ika*, the conflict-prone diversity in multicultural society can be solved peacefully. The civic skills associated with conflict resolution in multicultural societies through PPKn subjects is one of the main competencies that must be mastered by students in schools and outside schools as young citizens.

Nationality values that became the basis for the development of citizenship competence by the teachers of civics integrated in the character education program have been intensively promoted since 2010 by the Minister of National Education, Muhammad Nuh. The ministry office simplified the national values into 18 character values. The eighteen values include: religious; honesty; tolerance; discipline; hard work; creativity; independent; democracy; curiosity; spirit of nationality; love of the country; appreciation of achievement; friendly/communicative; love of peace; like to read; environmental care; social care; and responsibility (Curriculum Center of the Ministry of National Education, 2010). Administrativelydocumented, the character values by the PPKn/PKN teachers are inserted into their RPP. This is seen predominantly in the RPP format of the 2006 Curriculum after 2010.

5 CONCLUSION

Based on the presentation of research findings on the development of citizenship competence based on the values of nationality and local wisdom in the multicultural community in schools, which is part of the research report of PTP Grants 2016, it has been shown that:

First, the values of nationality and local wisdom in addition to the content of the study of civics/PPKn in schools can also be used as a goal of achieving the component of citizenship competence in the multicultural community in school.

Secondly, from the two curricula that became the basis for the development of citizenship competence in schools, Curriculum 2013 makes it very easy for teachers to determine the components of citizenship competency. In contrast, civics in the Curriculum 2006 has not explicitly grouped the components of its core competencies into three domains. The development of its review content faces a number of complexities, as reported in the study of Pawiroputro, et al. (2010).

The strengthening of the values of multicultural citizenship as found in the study of this paper, its attainment is determined by the teacher. Civics teachers must understand the substance of subject matters and component of the competency of civic education. They have play an important role as model, as well as perpetrators of a multicultural life that brings closer the national values and local wisdom comprehensively.

The ability of civics teachers to integrate national values and local wisdom contextually for learners is another indicator of how the strengthening of multicultural values success. Even in practice, the reinforcement of these values requires hard work in a challenging millennial era innovatively.

REFERENCES

Curriculum Center of the Ministry of National Education, (2010). *Pengembangan pendidikan budaya dan karakter bangsa* [Developing cultural and character education]. Jakarta, Indonesia: Pusat Kurikulum Badan Penelitian dan Pengembangan Kementerian Pendidikan Nasional.

Miller-Lane, J., Howard, T.C., & Halagao, P.E. (2007). Civic multicultural competence: searching for common ground in democratic education. *Theory & Research in Social Education, 35*(4), 551–573.

Ministry of Education and Culture Republic of Indonesia (2013a). *Pendidikan pancasila dan kewarganegaraan. Buku siswa kelas VII* [Pancasila and civic education. Students' book grade VII]. Jakarta, Indonesia: Kementerian Pendidikan dan Kebudayaan.

Ministry of Education and Culture Republic of Indonesia. (2013b). *Pendidikan pancasila dan kewarganegaraan. Buku siswa kelas VIII* [Pancasila and civic education. Students' book grade VIII]. Jakarta, Indonesia: Kementerian Pendidikan dan Kebudayaan.

Ministry of Education and Culture Republic of Indonesia. (2013c). *Pendidikan pancasila dan kewarganegaraan. Buku siswa kelas IX* [Pancasila and civic education. Students' book grade IX]. Jakarta, Indonesia: Kementerian Pendidikan dan Kebudayaan.

Pawiroputro, E., Samsuri, & dan Halili Samsuri & Halili. (2010). *Pengembangan materi mata pelajaran pendidikan kewarganegaraan SMP se-Kabupaten Sleman* [Development of civic education lesson in junior high school in Sleman]. Laporan Penelitian. Yogyakarta, Indonesia: FISE UNY.

Samsuri. (2011). Kebijakan pendidikan kewarganegaraan era reformasi di Indonesia [Civic education policy in reformation era in Indonesia]. *Cakrawala Pendidikan,*. Vol. *XXX,* (No. 2), pp. 267–281.

Samsuri. (2012). *Pendidikan karakter warga negara* [Character education of civics]. Surakarta, Indonesia: Pustaka Hanif.

Samsuri & Marzuki. (2016). Pembentukan karakter kewargaan multikultural dalam program kurikuler di madrasah aliyah se-Daerah Istimewa Yogyakarta. [Forming multicultural citizen character in curricular in Islamic high school in special region Yogyakarta] *Cakrawala Pendidikan,* February, *XXXV* (1), 25–31.

Secretary General of MPR Republic of Indonesia. (2012). *Empat pilar kehidupan berbangsa dan bernegara* [Four pillars of nation and state life]. Jakarta, Indonesia: Setjen MPR RI.

Special Region of Yogyakarta Province (2011a). Peraturan Daerah Provinsi Daerah Istimewa Yogyakarta No. 4 Tahun 2011 tentang Tata Nilai Budaya Yogyakarta [Local regulation of special region Yogyakarta on culture value of Yogyakarta].

Special Region of Yogyakarta Province. (2011b Peraturan Daerah Provinsi Daerah Istimewa Yogyakarta No. 5 Tahun 2011 tentang Pengelolaan dan Penyelenggaraan Pendidikan Berbasis Budaya [Local regulation of special region Yogyakarta on managing and establishing culture-based education].

Suharno, Samsuri, & Hendrastomo, G.(2013a). *Pengembangan model* peace-building teaching and learning: *intervensi pencegahan kekerasan melalui pendidikan formal* [Developing peace-building model of teaching and learning: Iintervention of preventing violence through formal education]. Laporan Tahunan Penelitian Hibah Bersaing Tahun I. Yogyakarta, Indonesia: LPPM UNY.

Suharno, Samsuri, & Nurhayati, I. (2013b). *Pengembangan model resolusi konflik untuk masyarakat multikultural di Poso dan Ambon (studi implementasi kebijakan resolusi konflik di Sampit, Kotim, Kalimantan Tengah)* [Developing conflict resolution model for multicultural society in Poso and Ambon]. Laporan Tahunan Penelitian Hibah Bersaing Tahun I. Yogyakarta, Indonesia: LPPM UNY.

Suyatno, S. (2011). *Revitalisasi kearifan lokal sebagai upaya penguatan identitas keIndonesiaan* [Revitalization of local wisdom as an effort to strengthen Indonesian identity], retrieved from http://badanbahasa.kemdikbud.go.id/lamanbahasa/artikel/1366 accessed October 28, 2016.

Character education through innovative literary learning using film-based teaching material

F. Nugrahani
Universitas Veteran Bangun Nusantara, Indonesia

A.I. Alma'ruf
Universitas Muhammadiyah Surakarta, Indonesia

ABSTRACT: The purposes of this study are to describe (1) the character values in film-based teaching materials, (2) the role of teachers in character education through innovative literary learning using film-based material, and (3) success indicators of character education through innovative literary learning. This study used a descriptive and qualitative method. Data were collected by document analysis, observation, and in-depth interviews. Data analysis was performed inductively with an interactive model. The results showed that: (1) The dominant character values contained in film-based literary materials are trustworthiness, caring, and citizenship; (2) The role of teachers in innovative literary learning is as managers, compilers of teaching materials, motivators, as well as models for students in finding their characters; (3) The success of character education in schools can be measured by the following indicators: (a) Promoting ethical values as the basic character; (b) Interpreting character in thought, feeling and behavior; (c) Supporting curriculum; (d) Encouraging the development of motivation; (e) Supporting character findings; and (f) Providing the opportunity to display good characters.

1   INTRODUCTION

Recently, Indonesian television has been broadcasting news about corruption events involving politicians and state officials. There is also news of other criminal acts, such as torture, rape, theft, robbery, and murder. All this news is a clue to low morality, as well as the manifestation of the failure of character education in Indonesia so far.

To overcome this, character education needs to be re-established, through formal and non-formal education in the family and community. Implementation of the program needs to get special attention and support. The program also needs to be disseminated through media to reach the public. The amelioration in the nation's character, is a crisis for its human resources which is the largest asset of a nation. Therefore, a person's character needs to be established, and developed through education.

Meanwhile, the study result of Muzaki (2016) showed that high academic achievement at school is not a guarantee of a student's success in real life. This happens because learning at schools usually focuses only on cognitive learning. Regarding the findings in that study, it is suggested that the educational concept, which prioritizes effective learning through character education is delivered with attention to the individual intelligence of each student, because according to Gardner (2004), clever means being the champion in one's own field. Thus, character education can be provided to students through enjoyable learning, for example through the appreciation of films that are produced based on literary works.

The ever-changing and changing technological environment creates an interconnected relationship in the world. Institutions and communities, including educators, are challenged to answer the call of globalization in various ways. One of them is awareness of technology for the purpose of education. The need to use technology in education is inevitable, because it is a necessity (Leynard, 2015). Therefore, the use of film as a media in character education is also a demand of the times.

Literary works as a reflection of socio-cultural circumstances must be inherited by the younger generation (Suryaman, 2010). Literature has great potential to influence the changing character of a person. Literature is an expression of language and art that is reflective and interactive, therefore it can be the catalyst for the emergence of changes in society, as well as the source of inspiration and motivation for the dissemination of life values, and as the agent for the development of society's cultural order to a more advanced civilization. Essentially, literature should be entertaining and useful.

A literary work is a social document, because it is an expressed life story as a picture of reality, despite the subjectivity of its author. This in in line with Abrams's (1981) theory that literary works can be seen from four points of view: (1) as an objective, autonomous work, regardless of any element; (2) as a mimetic work, an imitation of the universe and the rest; (3) as a pragmatic work, which benefits the reader; (4) as an expressive work, mirroring the experiences and thoughts of its creator. From this theory, it can be said that literary works have benefits for the audience, especially with regards to the values contained therein to enrich the insight of knowledge and experience in life.

Literary works, such as novels, contain characters such as the perpetrator who presents the culture of the society in setting the story. Through the characters and characterizations in the novel, the reader can understand various aspects of life, as well as learn the character and culture of a nation. Finally, considering the nature and function of such literature, it can be understood that literature is important for character education for students at school.

In this context, the character values contained in the literary works are presented to the students through learning by using literary film media, which is better known as ecranization. When it was realized that film can do more than provide simple entertainment, various theories and approaches were developed to help analyze films to understand how they create responses for viewers through plot presentation and story material, as well as the use of various formal narrative elements (such as characters, settings, repetition/variation, chronological structure, etc.) to convey meaning to audiences. With photographic techniques, camera movements, sound editing in relation to images, etc., movie viewers can feel the scene and interpret the beauty of the story in a literary work delivered through the film (Stirbetiu, 2001).

Film is a form of artistic expression, which tells the story of the character The easiest way to explain it to others is through the film, because the film is both entertaining and educating (Bohlin, 2005). When watching a movie, someone can feel emotions such as joy, anger, laughter, relaxation, love, imagination, or even boredom. The emotions that appear when watching the movie are often triggered or reinforced by the mood created by certain visual scenes, actors, and/or background music, which has a powerful effect on the mind and the senses of audiences (Berk, 2009).

In respond to the empirical phenomena about people's behavior that show poor character, teachers can make efforts to improve character through learning activities in schools. In this context, it is character education via literary learning. The manifestation of that role is by designing literary learning strategies that contribute to the formation of student character (Tasri, 2011). One of them is by presenting a film-based instructional material entitled *Laskar Pelangi (LP)* [*Rainbow Troops*] from a novel written by Andrea Hirata (2006). This article presents the values of the characters contained in that film, the role of teachers in character education and the success indicators of character education through innovative literary learning in schools.

2 RESEARCH METHOD

This research was conducted using a qualitative descriptive approach. Based on the characteristics of the method, this study belongs to the group of embedded case studies (Yin, 2000). The case studied was about character education through innovative literature learning with film-based teaching materials. The subjects of this study were teachers and students in character education through innovative literary learning. The research data was in the form of words and sentences that indicate the existence of: (1) the character values contained in the literary materials based on the *LP* film; (2) The role of teachers in innovative literature learning based on the *LP* film; and (3) The success of character education through innovative literary learning based on the *LP* film. The data sources of this research were: (1) Place and event, where the process of character education took place in a school using the literary learning process with film-based teaching materials; (2) The resources are the students and literature teachers in the school under study; and (3) Documentation of literature materials based on the *LP* film and reference books about education character. The research place was a senior high school in Sukoharjo, Indonesia, while the time of research was January to June 2017. The sample was taken by purposive sampling, that is part of the character education process through literary learning in school. Data was collected by document analysis, observation, and in-depth interviewing. Data analysis was performed since the data collection process that occurred in the field until obtained a solid conclusion in the verification through cycle process. The data in this study were analyzed inductively with interactive analysis model (Miles & Huberman, 2004).

3 RESULTS AND DISCUSSION

Data collected in this research were grouped into three: (1) data about dominant character values contained in film-based literary materials; (2) data about the role of teachers in character education through innovative literature learning; and (3) data about the success of character education through innovative literary learning in schools.

3.1 *The character values in film-based literary materials from the novel exploitation of LP*

LP is an educational film produced by Miles Films. The producer is Mira Lesmana and the director is Riri Reza. The scriptwriter is Salman Aristo. As for the players, these are famous Indonesian actors and actresses, such as Slamet Raharjo, Alex Komang, Mathias Muchus, Cut Mini, Ikranegara, and Ario Bayu. It also involves ten native children from Bangka Belitong. The film is an ecranization of a novel with the same title by Andrea Hirata.

Andrea Hirata is a young writer whose work receives exceptional reception from readers in Indonesia and the world community. His novel become the best-selling novel of the year (2007) and translated into many languages to be sold outside Indonesia. The success of the novel is the main reason why the film rights to the story were bought and turned into a very well known movie and watched by many people from various circles. The *LP* film is not only famous in Indonesia, but also in many foreign countries, as it has been screened at various film festivals such as the Asian American Film Festival in New York, Philadelphia, and Los Angeles (Nugrahani, 2016).

The *LP* film, which is an dramatization of the novel of the same title, provides many good character examples for students. The story presents the reality of the poor who live in the inland area of the country, which is far from prosperous. Nevertheless, the characters in the story remain happy, and in togetherness and solidarity they are passionate in living their lives and reaching their goals by being educated in the elementary school in their village which have minimal facilities. Therefore, this story really inspires others to struggle in life to achieve their goals. The character values in the *LP* film are trustworthiness, caring, and citizenship. Here are excerpts of the visualization and dialog.

Dialog 1: Trustworthiness. This scene illustrates the strong confidence of Lintang when following a careful race. It seems that the judges and other participants disparaged him because he came from a poor and backward village school. However, he was able to show that he was the smartest and became the champion. Here is the 105th minute dialog.

The jury: "the boy must be cheating, because we never see the child counting". Lintang: "I can explain it". The committee: "Let me read it out". Lintang: "No need, I still remember". (With Lintang confident answering the questions in front of the jury to prove that his answer is correct). Jury: "we made a mistake, this boy's answer is correct". (Everyone looked amazed and applauded Lintang).

Dialog 2: Caring. The scene with the following dialog illustrates a strong sense of concern of Muslimah (a teacher in the Muhammadiyah elementary school) for the education and future of her students in the underdeveloped school, without the support of surrounding community. Below is the 38th minute dialog.

Bahri: "I got an offer to teach in Bangka elementary school". Muslimah: "So, you can leave Muhammadiyah. Our duty is heavy, our students are just few, but we have an obligation to educate the children who cannot afford the education". Bahri: "Mus, no one wants to send their children here. Mus, they think their son is better to be a heaver to support their family".

Dialog 3: The sense of nationality (citizenship). The scene with the following dialog illustrates the strong sense of nationality of the students in their spirit and togetherness to enliven the carnival to commemorate the Independence Day of Indonesia. Even without facilities, and obviously different from other schools, these children remain passionate and creative to produce costumes for their team, which

Figure 1. Trustworthiness.

Figure 2. Caring.

Figure 3. Citizenship.

eventually attract the attention of the audience in the event due to their rough appearance. This is the 55.30th minute dialog.

Mahar: "I know what we should display in the carnival. Open your shirt, open your shirt…". Friends: "What are you doing, Mahar?" (While undressing and put make up on his friends for the carnival event commemorating Indonesia's Independence Day on 17 August).

3.2 *The role of teachers in character education through innovative literary learning*

This paper is specifically related to the learning materials that have a prominent position in achieving goals, and learning media as a means of conveying the message. If learning is viewed as a process of communication, then learning is the process of delivering messages from sources through certain media to the recipient of the message (Riyadi, 2014). Messages in the form of learning materials in the curriculum, message sources, media, and students as the recipient of the message, are components in the communication process.

Various studies on media use in learning concluded that the process and the outcomes will be better if the students learn using media assistance. Therefore, the use of media is highly recommended in improving the quality of learning, including the use of cinema for learning literature in schools.

Using innovative literature learning with cinema media, students can easily understand the material because the message is delivered through visual images. Through cinema, and the transformation of literary works into films, learning can be carried out with a strategy that pleases students (Dirga, 2016). There is a fundamental difference in the processes of writing novels and making films, because filmmakers must be able to arrange words that are able to give the audience imagination. In a novel, text is used as the medium, whereas in film, a visual form or image is used to convey the author's message (Istadiyantha and Wati, 2015). To keep the author's message for the audience, the filmmaker should avoid changes in the story, including characters, plots, settings, or themes.

A film-based on literary novel is an alternative to teaching materials as well as a medium for innovative literature learning. It overcomes the problem of unavailability of text and time in massive quantities. Film, as the exploitation of a novel, can be a bridge to help in literary learning, so that the learning can be more effective, efficient and pleasant (Febriana et al., 2014). The lesson should be continued by reading the original literary texts at a later time.

By their respective characters, the behavior of a human being is influenced by their past experiences The background of family and the environment where a person grows is what shapes his or her character. The character will give color to his/her behavior in life in the present and future. In this case, the role of the teacher is as a motivator, as well as a model for students in finding their characters.

3.3 *Success indicators of character education through innovative literature learning*

The process of character formation involves knowledge, feelings, and actions, coherently and comprehensively. Therefore, the process involves many components, including community consensus of the karmic organization, stakeholders, comprehensive curriculum preparation, adult readiness as a model, and the availability of activities to implement good character values in schools. If it refers to the formulation of Character Education Partnership as cited by Schaps et al (1996), character education can be measured through eleven indicators (Samsuri and Marzuki, 2016). Among those suitable to measure the success of character education through literary learning in schools are: (1) Promoting ethical values as the basis for character formation; (2) Showing the way of thoughts, feelings and behaviors depicting a character; (3); Academic curriculum encouraging the development of student character; (4) Encouraging the development of student motivation; and (5); Providing opportunities for students to display good character. Through these indicators, the formation of student characters in schools can be monitored by the teacher as the person in charge.

4 CONCLUSION

The adaptation of the *LP* novel into a film inspires much appreciation of the spirit of life, and achieving the ideals of the future, even in poverty. The characters in the film can be examples for the audience, especially about trustworthiness, caring, and citizenship. The authors convey the message that all people must be able to live their lives, if they have belief and confidence. In addition, caring for others can make our lives happy. Sharing with others is what makes success in life. The sense of nationality (citizenship) needs to be nurtured so people have self-esteem and pride in their identity. This sense of nationality is important for students as it can be the basis for the awareness of the importance of unity to defend their homeland.

The character values conveyed by the authors are important for the student in the process of character building. Discovering character values through literature is certainly more fun for students, because literature is a work of art. The teacher can act as a leader by provides the material, methods, and media learning, as well as its assessor. The teacher must become a model for good character, as well as a motivator and a controller of student behavior who is ready to help students in finding their characters.

The process of character education in schools involves many components, including principals as policy makers, teachers as models, a supportive curriculum, school communities who have manners, and positive activities to implement good character values at school. For its success, it should periodically be evaluated with several indicators. (1) The ability of teachers to promote ethical values through literature; Teachers' abilities to motivate students to discover their character; and provide opportunities for students to display good character. (2) Students' abilities to show their way of thinking, feelings and behavior that characterize their good characters; and (3) the goodwill of the school in preparing an academic curriculum that supports the development of student character. Through these indicators, the success of character formation through literature learning in schools can be monitored.

Through these indicators, the success of character building through literary learning in schools can be monitored.

REFERENCES

Abrams, M.H. (1981). *A glossary of literary terms*. New York, NY: Holt, Rinehart & Winston Inc.
Berk, R.A. (2009). Multimedia teaching with video clips: TV, movies, YouTube, and mtvU in the college classroom. *International Journal of Technology in Teaching and Learning, 5*(1): 1–21. doi: 10.15640/jehd.v5n2a18
Bohlin, K.E. (2005). *Teaching character education through literature*. New York, NY: Routledge Falmer.
Dirga, R. (2016). Inovasi pembelajaran sastra pada Mata Pelajaran Bahasa Jepang di SMK (Innovation of literature learning in Japanese language subjects at SMK. Cendekia). *Language and Cultural Studies Center, 10*(1), 101–108.
Febriana, N., Thahar, H.E., and Ermanto. (2014). *Nilai-nilai pendidikan karakter dalam novel rantau satu muara karya ahmad fuadi: tinjauan sosiologi sastra*[Character education values in novel *rantau satu muara* by Ahmad Fuadi: A review of literature sociology The values of character education in a single overseas novel by Ahmad faudi: a review of the sociology of literature]. *Literature and Learning Journal, 2*(3), 92–107.
Gardner, H.E. (2004). *Frames of mind: the theory of multiple intelligences* (2nd ed.). New York, NY: Basic books.
Hartana, P.P.E., Rasna, I., W., and Wisudariani, N., M., R. (2014). Penggunaan media film untuk meningkatkan keterampilan menulis cerpen siswa kelas X2 di SMA Negeri 1 Tampaksiring [The use of film media to improve the skills of writing short stories of class X2 students of Senior High School 1 Tampaksiring]. *E-Journal Universitas Pendidikan Ganesha Education of Indonesian, Language and Literature, 2*(1), 1–12.
Hirata, A. (2006). *Laskar pelangi: Rainbow Troops*. Yogyakarta, Indonesia: Bentang Pustaka.
Istadiyantha & Wati, R. 2015. *Ekranisasi sebagai wahana adaptasi dari karya sastra ke film* [ecranization as adaptation from literature media to adaptation film media from literary work to film]. *Journal of Literature and Culture, 33*(1), 1–19.

Leynard L.G. (2015). Filipino film-based instructional plan for pre-service education students. *International journal of e—education, e-business, e-management, and e-learning 6*(1), 56–70. doi: 110.17706/ijeeee.2016.6.1.56–70

Miles, M.B. & Huberman, A.M. (1994). *Qualitative data analysis: a sourcebook of new methods.* Beverly Hills, United States of America: Sage Publication.

Muzaki F.I. (2016). Character education in the last song: A conceptual review. *Journal of Education and Human Development, 5*(2), 149–154. doi: 10.15640/jehd.v5n2a18

Nugrahani, F. (2014). Laskar pelangi novel by Andrea Hirata as a creative industry and educative media (A review of sociologi literature). *Proceeding Seminar Antarbangsa Kesusastraan Asia Tenggara (SAKAT) Brunai Darussalam,* 16–26.

Riyadi, S. (2014). Penggunaan film adaptasi sebagai media pengajaran sastra [The use of adaptation film as a literary teaching media]. *Journal of Language and Literature Universitas Indonesia, 14*(2), 241–251.

Samsuri, S., and Marzuki, M. (2016). *Pembentukan karakter kewargaan multikultural dalam program kurikuler di Madrasah Aliyah se-DIY* [The establishment of multicultural citizenship character in curriculum program at Madrasah Aliyah in Yogyakarta Special Region]. *Journal Cakrawala Pendidikan, XXXV*(1), 24–32.

Lickona, T., Schaps, L., and Lewis, C. (1996). Eleven principles of effective character education. *Journal of Moral Education, 25*(1), 93–100.

Stirbetiu. M. (2001). *Literature and film adaptation theory.* Ovidius, Romania: University of Constanța.

Suryaman, M. (2010). Pendidikan karakter melalui pembelajaran sastra [Character education through literary learning]. *Cakrawala Pendidikan, XXIX,* Edisi khusus Dies Natalis UNY), 112–126.

Tasri, L. (2011). Pengembangan bahan ajar berbasis web [Material learning development based on web]. *MEDTEK Technical Faculty of UNM, 3*(2), 1–8.

Yin, R.K. (2000). *Case study research: design and methods.* Jakarta, Indonesia: Raja Grafindo Persada.

Assessing student's character development
(Values acquisition assessment)

Development of the psychosocial skills scale and its relationship with the negative emotional states of elementary school children

S. Nopembri
Universitas Negeri Yogyakarta, Indonesia

Y. Sugiyama
Kyushu University, Fukuoka, Japan

ABSTRACT: The current studies aimed to develop a Psychosocial Skills Scale (PSS) and examine the relationship between psychosocial skills and the negative emotional states of elementary school children. The first study involved four experts in the educational and psychological fields and 745 fourth- to sixth-grade children at nine elementary schools. The second study involved 810 fourth- to sixth-grade children at 15 elementary schools. The scale development process (DeVellis, 2003) was conducted to develop the PSS in the first study. In the second study, the students completed the Depression Anxiety Stress Scale (DASS) and the valid version of PSS. Exploratory and confirmatory factors, multiple correlations, and Cronbach's coefficient (Alpha) analysis were used in the first study and Pearson correlation analysis was used in the second study. The PSS with four subscale structures (stress coping, communication, social awareness, and problem-solving skills) was validated as reliable which was indicated by a good fit in construct validity, internal validity, and internal consistency/ reliability. These results provide some support for using the scale to measure children's psychosocial skills in Yogyakarta, Indonesia. Furthermore, in the second study, the Pearson correlation analysis suggested that the relationship between negative emotional state and psychosocial skills is fragile and there tended to be no connection between them.

1   INTRODUCTION

1.1   *Background*

Psychological and social problems have been found to arise in daily life, making it difficult for children to avoid these problems. It is our responsibility that educational programs should be based on problems that children may face in their life and imparted through several activities that support them (Tasgin, 2011; Yigiter, 2013). Therefore, many psychosocial skills need to be developed by children to face their daily life problems. Our study found that the children in the fourth- to sixth-grades had good psychosocial skills, but required some improvement in their stress coping, communication, social awareness, and problem-solving skills based on teacher perception (Nopembri et al., 2013). Therefore, in this study, we explored some children's psychosocial skills including stress coping, communication, social awareness, and problem-solving skills.

There are some reasons to explore these specific psychosocial skills of children. Children need to have stress coping skills to build coping and rapid recovery skills, coziness and sustainability (Kar, 2009), and to promote and maintain their physical and psychological well-being (Wagner et al., 1999; Kadhiravan & Kumar, 2012). Communication skills are required for a mutual transfer of feelings and thoughts (Aydin, 2015), to express him/herself in relation to others (Erdogan & Bayraktar, 2014), and to communicate effectively (Hollander et al., 2003). Social awareness is needed to empathize mentally and emotionally with others from diverse backgrounds and cultures, to understand social and ethical norms for behavior, and to recognize family, school, and community resources and supports (Smith, 2006; Cavo-jová et al., 2012; Collaborative for Academic, Social, Emotional Learning, 2015). It is neces-sary for children to have problem-solving skills because these are essential for every part of students' future lives, both personal and social (Gorucu, 2016), and for identifying effective solutions to the problem (Thompson et al., 2013).

An instrument to measure the psychosocial skills of children for these specific skills is needed. This scale should pay attention to validity and reliability so that it can be used for data collection. Therefore, we strived to develop a scale based on semantic differential atti-tudes toward statistics, especially validation using factor analysis. The psychosocial Skill Scale (PSS) consisted of four subcategories of stress coping, communication, social awareness, and problem-solving skills. The scale was considered an indirect measure because no one can directly observe psychosocial skills. The development of psychosocial skills scale followed the guideline process recommended by DeVellis (2003).

Furthermore, we investigated the correlation between psychosocial skills and the nega-tive emotional state of children. We assumed that negative emotional states have an inverse relationship with the psychosocial level of children. It is in line with some previous studies that show the existence of these links. The relationship between depression levels and some psychosocial aspects of children has been investigated in several studies (Yasin & Dzulkifli, 2010; Becker-Weidman et al., 2010; Moghaddam et al., 2012; Tully et al., 2016). The level of anxiety of the children is also indicated to be related to some of the psychosocial skills investigated (Henley, 2005; Almeida et al., 2011; Jellesma, 2013). Likewise, the stress level seen there has a relationship with psychosocial aspects (Chou et al., 2011; Park et al., 2015, Karademir & Taşçi, 2015).

1.2 *Purposes of the study*

This consists of first and second studies. The purpose of the first study is to develop an instrument to measure children's psychosocial skills and the second study aims to examine the relationship between psychosocial skills and the negative emotional states of children.

2 METHOD

2.1 *Participants*

Educational and psychological experts were involved in reviewing the items of scale in the first study. The educational experts included classroom, PE and sport elementary school teachers and an educational researcher, while the psychological expert is in the field of social psychol-ogy. A total of 745 children in the fourth to sixth-grades from nine elementary schools: three schools located near Merapi volcano (a disaster area); two schools based in Yogyakarta city (an urban area); and four schools located in the Sleman district (a suburban area), as shown in Table 1, participated in the study.

A total of 810 fourth to sixth-grade students (440 girls and 370 boys) from 15 elementary schools participated in the second study. The children's ages ranged from 7 to 15 years old (ages: Mean = 10.3, SD = 1.09). Participants in this study are described in Table 2.

Table 1. The characteristics of children participating in the first study.

School area	Sex		Age		Grade		
	F	M	Mean	SD	4th	5th	6th
Disaster	67	91	10.3	1.08	52	50	56
Urban	92	94	10.5	1.58	59	58	69
Suburban	204	197	10.6	1.13	138	123	140

Note: F = Female, M = Male, SD = Standard Deviation.

Table 2. School and children participating in the second study.

| Schools | Children in grade | | | N |
	4th	5th	6th	
1	12	15	17	44
2	24	23	15	62
3	16	17	17	50
4	15	9	12	36
5	31	23	20	74
6	11	13	19	43
7	9	11	11	31
8	11	6	15	32
9	14	14	22	50
10	24	19	15	58
11	27	20	26	73
12	18	22	18	58
13	21	31	18	70
14	16	16	18	50
15	33	27	19	79
				810

2.2 Procedure

The development of the Psychosocial Skills Scale (PSS) in the first study was based on specific guidelines for scale development (DeVellis, 2003): (1) determine clearly what is to be
measured; (2) generate an item pool; (3) determine a format for measurement; (4) have the
item pool reviewed by experts; (5) include scale validation items; (6) administer the items to
a development sample; (7) evaluate the items; and (8) complete the final version of the scale.
In the second study, the children completed the Depression Anxiety Stress Scales (DASS 42)
and the final version of the PSS.

2.3 Data collection

The pilot version of PSS included four self-reporting subscales designed to measure coping
with stress, communication, social awareness, and problem-solving skills among the children.
Each of the four subscales contained ten items. Respondents rated each item on a four-point
scale according to their circumstances, with response options ranging from; not according to
me (0), less suited to me (1), moderately according to me (2), and completely according to
me (3).

Based on the results of the first study, the final version of the PSS was used to measure
coping with stress, communication, social awareness, and problem-solving skills among the
children in the second study. The negative emotional state of the children was measured
using the DASS 42 (Lovibond & Lovibond, 1995). The 42-item self-reporting questionnaire
consisted of three subscales; each scale consisted of 14 items, divided further into subscales
of 2–5 items with similar content. Respondents were asked to use a four-point severity scale
to rate the extent to which they experienced each symptom over the previous week.

2.4 Data analysis

All analyses were performed using SPSS (Statistical Package for Social Sciences) and AMOS
(Analysis of Moment Structures) version 22.0 for Windows, and the statistical significance
was set at $p < 0.05$. In the first study, Exploratory Factor Analysis (EFA) and Confirmatory
Factor Analysis (CFA) were performed to assess the construct validity of the PSS. Multiple correlations among the items and the item totals were calculated to examine the scale's

internal validity, and the scale's reliability was tested using Cronbach's Alpha coefficient. In the second study, the relationship between psychosocial skills and negative emotional state was examined using Pearson correlation analysis.

3 RESULTS

3.1 *First study*

Determine what to measure. We decided to measure the children's psychosocial skills. The psychosocial skills in this context consisted of coping with stress, communication, social awareness, and problem-solving. Coping with stress refers to an individual's cognitive and behavioral efforts to manage stress (Carpenter, 1992). Communication is a basic skill that one learned to communicate effectively (Hollander et al., 2003). Social awareness is the ability to take the perspective of and empathize with others from diverse backgrounds and cultures, to understand social and ethical norms for behavior, and to recognize family, school, and community resources and supports (Collaborative for Academic, Social and Emotional Learning, 2015). Problem-solving is defined as a cognitive-affective-behavioral process through which an individual or group identifies or discovers an efficient way of coping with a problem encountered in everyday life (Yigiter, 2013). These skills are crucial for children in their daily lives.

Generate an item pool. We initially drafted a self-reporting scale to measure children's psychosocial skills that included four subscales that assessed children's stress coping, communication, social awareness, and problem-solving. Each subscale consisted of ten items. All items were developed using simple statements in the Indonesian language.

Determine the format for measurement. The Likert-type scale was selected to rate responses because it was easier to score and the respondents were familiar with the format. The items measured the children's agreement with statements describing them on a four-point scale, with responses ranging from not according to me (0), less suited to me (1), moderately according to me (2), and completely according to me (3). The choice of a four-point instead of a five-point scale was intended to force apathetic or ambivalent respondents to choose a final response category (Garland, 1991).

Review of the item pool by experts. The authors selected four experts from the areas of education and psychology to review the initial item pool. The educational experts of the study consisted of a classroom teacher, a PE teacher, a sports teacher, and a researcher in education. We asked the teachers and the educational researcher to review the scale and provide input on it. They checked the quality of each item regarding its content, clarity, and legibility, especially its suitability for children in the fourth- to sixth-grades of elementary school. They also reviewed the response options for their compatibility with the information obtained by the authors. The authors obtained feedback from these experts to refine and revise the scale's items. They suggested improving the item's statements by using easier words that could be understood by the children. The scale's statements were restructured to form simple sentences. The experts also suggested avoiding the use of educational terminology that would elude the children. For the response options, the experts suggested using the appropriate reading and comprehension level for children's responses to the statements, by considering their level of thinking skills. The expert from the specialty area of social psychology recommended revisions and corrections of the scale's items. This expert checked each item's compatibility with the concept of the psychosocial skills to be investigated and reviewed by the scale's response options. The expert judged that ten items (statements) for each subscale was sufficient for extracting information. This expert also revised statements that were unclear, ambiguous or lengthy. The expert agreed with the use of the four-point rating scale without a neutral option to ascertain a firm position on the children's attitudes through their responses to the items.

Include scale validation items. In this step, the authors selected items that had been suggested by experts according to their field of expertise. The authors selected ten statements to measure each skill, which summed to 40 items. The valid items in the scale's English version are shown in Table 3.

Table 3. The final statements in the psychosocial skills scale.

Item	Stress coping statements
1	I avoid contact when having problems with a friend.
2	I do something that is fun to solve a problem with a friend.
3	I avoid anything that makes me disappointed.
4	I think that every problem in the school will resolve itself automatically.
5	I do something to calm down when I face problems at school.
6	I avoid feeling disappointed, or I forget about problems at school.
7	I engage in exercise/sport.
8	According to me, any problem can be resolved well.
9	I engage in a hobby/interest that helps me feel relaxed and happy.
10	I pray diligently.

Item	Communication statements
1	I say "please" and "thank you" when I ask for something from someone.
2	The clothes which I wear make others feel comfortable.
3	I am not cursing/using abusive language in a public place.
4	My hair is clean and tidy.
5	I have a good body condition.
6	I look in their eyes while talking to someone.
7	My nails are cleanly and neatly trimmed.
8	I am angry and impatient when something is not as I would like.
9	I try not to criticize when others do something different from me.
10	I am grateful to those who helped or gave me a gift.

Item	Social awareness statements
1	I do not care about friends who tease or call my name.
2	I try to understand the feelings of a friend who is angry, upset, or sad.
3	I feel pity for the people affected by the disaster/accident.
4	I do things that please my parents, (such as helping at home) without being asked.
5	I speak to my parents when opinions are different.
6	I listen to older people without getting angry.
7	I make friends easily.
8	I invite others to participate in community activities.
9	I smile, wave or nod at others.
10	I participate in school activities (such as extracurricular sports, boy scouts, etc.).

Item	Problem-solving statements
1	I like to solve problems and make decisions.
2	I love to collaborate with groups to complete tasks.
3	I resolve problems quickly and easily.
4	I can learn quickly and easily.
5	I know the details of the task and do it right.
6	I am an intelligent person and can think in complicated situations.
7	I am more concerned about facing uncertain problems.
8	I try to sort the problems faced starting from the easiest to the most difficult.
9	I like to do something that can be done well.
10	I can make difficult decisions easily and be firm about them.

Administer the items to a development sample. We conducted a pilot study of the validated subscales using a sample with similar characteristics to the research sample. As explained in the description of the study's participants, 745 children from the fourth- to sixth grades of elementary schools in the Yogyakarta area comprised the development sample from the urban, suburban, and disaster areas (see Table 4).

Evaluation of the items. In this step, the authors evaluated the scale's items based on the data obtained from the pilot study of the development sample. Statistical analyses were

Table 4. The name of factors and distribution of items.

Scale	Factors	Item distribution
Stress coping	Reactivity to stress	7, 8, 9, 10
	Assess situation	1, 2, 3, 6
	Relaxation	4, 5
Communication	Verbal	3, 6, 8, 9
	Non-verbal	1, 2, 4, 5, 7, 10
Social awareness	Cognitive empathy	1, 5, 6
	Emotional empathy	2, 3, 4, 7, 8, 9, 10
Problem-solving	Decision-making process	1, 2, 3, 4, 5, 6, 7, 8, 9, 10

performed to assess the scale's construct and internal validity, and its internal consistency/ reliability. Construct validity was tested using Exploratory Factor Analysis (EFA) and Confirmatory Factor Analysis (CFA). The internal validity was examined by calculating multiple correlations of the scores on the individual items with the total score, and the scale's internal consistency/reliability was verified using Cronbach's Alpha coefficient. Before the EFA, the authors performed the required Kaiser-Meyer-Olkin Measure of Sampling Adequacy (KMO MSA). The following scores were obtained: stress coping = 0.728 (moderate), communication = 0.827 (good), social awareness = 0.874 (good), and problem-solving = 0.905 (very good). The scores on Bartlett's Test of Sphericity were 1,148.691 for coping with stress, 1,170.953 for communication, 1,557.175 for social awareness, and 1,840.836 for problem-solving, with 45 degrees of freedom and a probability of < 0.001, indicating significant results. Thus, the sample was deemed appropriate for further analysis. The next step of the EFA, the extraction of factors to view eigenvalues in the scree plot, showed that three components of stress coping, two components of communication, two components of social awareness, and one component of problem-solving had eigenvalues greater than one. Varimax rotation of the factors was used to maximize the relationship between the variables with multiple iterations or rounds. The varimax method was selected to rotate the initial extraction factor results and eventually obtain the results in one column where the values were as close as possible to zero. Item statements were disqualified if the rotated factor loading was less than 0.30 (< 0.30). The rotation factor results indicated that there were no items with a rotated factor loading less than 0.30. The distribution of the items and the names of each factor, based on the rotated factors, are presented in Table 4.

In the next steps, a CFA was performed to verify the model's goodness of fit. The fit indices were the Root Mean Square Error of Approximating (RMSEA), the Goodness of Fit Index (GFI), Adjusted Goodness of Fit Index (AGFI), and Comparative Fit Index (CFI), which indicated that the model had a good fit. This is shown in Table 5.

The internal validity of the items for each subscale were examined by calculating the correlations between each item's score and the total score on each subscale. Pearson's correlation coefficient was significant ($p < 0.01$ in the two-tailed test) between the items' scores and the total score on each subscale. To examine the instrument's internal consistency/reliability, the authors calculated Cronbach's Alpha coefficient for each subscale, which indicated adequate reliability: 0.727 for the coping with stress subscale; 0.699 for the communication subscale; 0.794 for the social awareness subscale; and 0.835 for the problem-solving subscale. Thus, the item statements on the scale had high internal consistency/reliability. Finally, the analysis of the scale's construct validity indicated a good model fit; a significant correlation between all the items and the total score showed good internal validity; and the internal consistency, as measured by Cronbach's Alpha, was acceptable for this sample.

Completion of the final version of the scale. Based on the results of the analysis of the scale's construct validity using factor analysis, internal validity, and consistency/reliability, the item statements in each scale significantly contributed to the indicators. Furthermore, the statistical results indicated the scale's structure contained four different subscales. Each subscale consisted of ten items; coping skills for stress (reactivity to stress, assessment of

Table 5. CFA Indexes of the scale of a good fit model.

No.	Scale	RMSEA	GFI	AGFI	CFI
1	Stress coping	0.055	0.973	0.954	0.936
2	Communication	0.051	0.973	0.957	0.942
3	Social awareness	0.054	0.973	0.956	0.952
4	Problem-solving	0.057	0.966	0.947	0.953

Table 6. Summary of Pearson correlation analysis.

Variables	Stress coping	Communication	Social awareness	Problem-solving
Depression	−0.045	−0.065*	−0.083**	−0.058*
Anxiety	0.026	0.021	0.021	0.011
Stress	0.035	0.032	0.025	−0.006

Note: *p < 0.05, **p < 0.01.

the situation, and relaxation factors), communication skills (verbal and non-verbal factors), social awareness skills (cognitive and emotional empathy factors), and problem-solving skills (decision-making process factor). The final version of the scale included general information about the respondent's name, date of birth, age, sex, school's name, grade), and instructions for completion (how to answer the questions and the four possible answers) for use in the subsequent study.

3.2 *Second study*

The Pearson's correlation analysis was performed to examine the relationship between a negative emotional state and psychosocial skills variables. There is a significant negative correlation between depression and communication ($r = -0.065$, $p = 0.032$), social awareness ($r = -0.083$, $p = 0.009$), and problem-solving ($r = -0.058$, $p = 0.049$) but not between depression and stress coping ($r = -0.045$, $p = 0.098$). There is not a significant correlation between anxiety and stress with psychosocial skills components. The conclusion is that the relationship between negative emotional state and psychosocial skills is fragile and there tend to be no connection between them. This can be seen in Table 6.

4 DISCUSSION

The results of this study provide empirical evidence that the PSS is a reliable and valid measure of children's psychosocial (coping with stress, communication, social awareness, and problem-solving) skills. The overall scale consists of four subscales for use with fourth- to sixth-grade children in elementary schools. This scale was developed using a sample of children from various areas of Yogyakarta, Indonesia. The development of the PSS involved education and psychology experts who provided direct input on the generation of the items. Efforts were made to ensure that the items were developmentally appropriate for the sample children in wording and content. Each of the four subscales were developed with the purpose of measuring a specific skill. The coping with stress subscale was developed to assess children's ability to deal with stressful problems in their school and daily activities. This subscale's purpose is consistent with the assumption that individuals cope with stress by using avoidance measures to reduce stressful problems (Aslam & Tariq, 2010). The communication skills subscale was developed to measure children's verbal and non-verbal communication skills. This scale's purpose was based on the premise that there are three levels of communication: logical (words), para-verbal (tone, volume, a rate of speech, and so on), and non-verbal (facial expression, position, movement, clothing, and so on) communication (Preja, 2013). The social awareness skills subscale measures children's cognitive and emotional empathy. This subscale is consistent with the definition of social awareness as being closely related to the ability to empathize to understand (cognitive) and feel emotions (emotional/affective) in response to others' situations (Cotton, 2001; Blair, 2005; Smith, 2006; Zhou & Ee, 2012). The problem-solving skills subscale was developed to assess the problem-solving ability of the children in their daily activities. Problem-solving skills involve the use of cognitive, affective, and behavioral processes to solve problems encountered in everyday life (Karademir & Tasçi, 2015; Yigiter, 2013; Thompson et al., 2013).

This study found that depression among the children had a significant and negative relationship with communication, social awareness, and problem-solving skills. A depressed state might have negatively influenced the psychosocial skills of communication, social awareness, and problem-solving. Therefore, a reduction in depression should be followed by increased communication, social awareness, and problem-solving skills of the children and vice versa. This finding is consistent with the notion that depression involves some contributing factors, such as genetics, environment, lifestyle, brain activity, psychology, and personality (Moghaddam et al., 2012). The current study's findings are similar to that of a study by Yasin and Dzulkifli (2010) which reported a significant negative relationship between social support and depression. Another study found that the ability/inability to solve a problem with a positive attitude was associated with the risk of depression (Becker-Weidman et al., 2010). However, a study by Tully et al. (2016) reported different findings, specifically a positive correlation between social awareness and depression by the level of cognitive empathy associated with elevated depression.

Anxiety and stress did not have a significant relationship with any of the psychosocial skills. It means that the children's anxiety state did not relate to or significantly alter their psychosocial skills. The current study's findings contradict some previous studies. A study by Aslam and Tariq (2010) found that resilient individuals were less vulnerable to anxiety, and another study concluded that the anxiety associated with differences in the communicative behavior of individuals involved physical changes and changes in speech and voice (Almeida et al., 2011). In the overall consideration, the relationship between stress and psychosocial skills found in the current studies also does not match with previous studies. Stress can be an early symptom of a medical problem among children, resulting in their loss of social interaction (Jellesma, 2013). The Study that have examined the relationship between stress and psychosocial skills reported an association between stress and coping strategies (Chou et al., 2011). Other studies have concluded that stress is a significant predictor of empathy among students (Park et al., 2015) and that the presence of stress and having old problems predicted excellent problem-solving skills (Karademir & Tasçi, 2015).

5 CONCLUSION

In the first study, the PSS (coping with stress, communication, social awareness, and problem-solving) was developed and validated. The scale was used to measure the psychosocial skills of fourth- to sixth-grade children in elementary schools in the Yogyakarta area. Despite using published guidelines and appropriate statistical analyses for the scale's development, this study has several limitations. First, our sample consists of fourth- to sixth-grade elementary school children from the Yogyakarta area of Indonesia; therefore, our results cannot be generalized to children who live in other geographical locations in Indonesia. Second, experts in fields specializing in psychosocial skills were not involved in this study. Third, in the preliminary examination of the scale's construct validity, the authors did not analyze the correlations between the PSS factors and the factors of other scales.

A fragile relationship between a negative emotional state and the psychosocial skills of children was reported in the second study. Depression, stress, and anxiety tended not to have a close relationship with stress coping, communication, social awareness, and problem-solving skills. It means that an increase or decrease in the negative emotional state may not affect the psychosocial skills of children and vice versa. Further research on intervention programs is needed to harmonize the two components so that the children will have the psychological and social strength to meet the challenges of everyday life.

ACKNOWLEDGMENTS

The authors would like to thank the experts and elementary school children who were involved in this study. This work was supported by JSPS KAKENHI Grant Number JP24500704.

REFERENCES

Almeida, A.A.F., Behlau, M., & Leite, J.R. (2011). Correlation between anxiety and communicative performance. *Revista da Sociedade Brasileira de Fonoaudiologia, 16*(4), 384–389. doi:10.1590/S1516-80342011000400004

Aslam, N. & Tariq, N. (2010). Trauma, depression, anxiety, and stress among individuals living in earthquake affected and unaffected areas. *Pakistan Journal of Psychological Research, 25*(2). 131–148.

Becker-Weidman, E.G., Jacobs, R. H., Reinecke, M. A., Silva, S. G., & March, J. (2010). Social problem-solving among adolescents treated for depression. *Behaviour Research Therapy, 48*(1), 11–18. doi:10.1016/j.brat.2009.08.006

Blair, R.J. (2005). Responding to the emotions of others: dissociating forms of empathy through the study of typical and psychiatric populations. *Consciousness and Cognition, 14*(4), 698–718. doi:10.1016/j.concog.2005.06.004

Carpenter, B.N. (1992). Personal coping: theory, research, and application. In L.J. Haas (Ed), *Handbook of Primary Care Psychology*, 1–13. Westport, CT: Praeger.

Cavojová, V., Sirota, M., & Belovicová. (2012). Slovak validation of the basic empathy scale in pre-adolescents. *Studia Psychologica, 54*(3), 195–208.

Chou, P., Chao, Y.Y., Yang, H., Yeh, G., & Lee, T.S. (2011). Relationships between stress coping and depressive symptoms among overseas university preparatory Chinese students: A cross-sectional study. *BMC Public Health, 11*(352) 1–7. doi:10.1186/1471-2458-11-352

Collaborative for Academic, Social, Emotional Learning. (2015). *Social and Emotional Learning Core Competencies*. Chicago, IL: Author.

Cotton, K. (2001). Developing empathy in children and youth. *School Improvement Research Series Close Up, 13*. Retrieved from http://www.nwrel.org/scpd/sirs/7/cu13.html.

DeVellis, R.F. (2003). *Scale development: theory and applications second edition*. Thousand Oaks, California: Sage Publication. Inc.

Erdogan, T. & Bayraktar, G. (2014). Effects of sports on communication skills: A research on teacher candidates. *Research on Humanities and Social Sciences, 4*(2), 68–74.

Garland, R. (1991). The mid-point on a rating scale: is it desirable? *Marketing Bulletin, 2*, 66–70.

Gorucu, A. (2016). The investigation of the effects of physical education lessons planned in accordance with cooperative learning approach on secondary school students' problem solving skills. *Educational Research and Reviews, 11*(10), 998–1007. doi:10.5897/ERR2016.2756

Henley, R. (2005). *Helping children overcome disaster trauma through post emergency psychosocial sports program*. Bern, Switzerland: Swiss Academy for Development.

Hollander, D.B., Wood, R.J., & Herbert, E.P. (2003). Protecting students against substance abuse behaviors: Integrating personal and social skills into physical education. *Journal of Physical Education, Recreation, & Dance, 74*(5), 45–48. doi:10.1080/07303084.2003.10608485

Jellesma, F.C. (2013). Stress and yoga in children. *Journal of Yoga and Physical Therapy, 3*, 1–3. doi:10.4172/2157-7595.1000136

Kadhiravan, S. & Kumar, K. (2012). Enhancing stress coping skills among college students. *Journal of Arts, Science & Commerce* III, *4*(1), 49–55.

Kar, N. (2009). Psychological impact of disaster on children: Review of assessment and interventions. *World Journal of Pediatrics, 5*, 5–11. doi:10.1007/s12519-009-0001-x

Karademir, T. & Taşçi, M. (2015). Examination of problem solving skills of volleyball trainers. *Journal of Sport and Social Sciences, 2*(1), 1–8.

Lovibond, S.H. & Lovibond, P.F. (1995). *Manual for the depression anxiety stress scales second edition*. Sydney, Australia: Psychology Foundation.

Moghaddam, J.B., Mehrdad, H., Salehian, M.H., & Shirmohammadzadeh, M. (2012). Effects of different exercise on reducing male students depression. *Annals of Biological Research, 3*(3), 1231–1235.

Nopembri, S., Saryono, Jatmika, H.M., & Sugiyama, Y. (2013). The opinions of physical education teachers on the psychosocial skills of elementary students in the Yogyakarta area. *The 7th Asian-South Pacific Association of Sports Psychology International Congress* (August 7–10, 2013).

Park, K.H., Kim, D. H., Kim, S. K., Yi, Y. H., Jeong, J. H., Chae, J., Hwang, J., & Roh, H. (2015). The relationships between empathy, stress and social support among medical students. *International Journal of Medical Education, 6*, 103–108. doi:10.5116/ijme.55e6.0d44

Preja, C.A. (2013). Verbal and non-verbal communication in sport culture. *Palestrica of the Third Millennium-Civilization and Sport, 14*(3), 239–243.

Smith, A. (2006). Cognitive empathy and emotional empathy in human behavior and evolution. *The Psychological Record, 56*(1), 3–12. doi:10.1007/BF03395534

Tasgin, Ö. (2011). Examining Problem Solving Skills of Physical Education and Sport Students from Several Factors. *Collegium Antropologicum, 35*(2), 325–328.

Thompson, D., Bhatt, R. & Watson, K. (2013). Physical activity problem-solving inventory for adolescents: Development and initial validation. *Pediatric Exercise Science, 25*(3), 448–467. doi:10.1123/pes.25.3.448

Tully, E., Ames, A.M., Garcia, S.E. & Donohue, M.R. (2016). Quadratic associations between empathy and depression and the moderating influence of dysregulation. *The Journal of Psychology: Interdisciplinary and Applied, 50*(1), 15–35. doi:10.1080/00223980.2014.992382

Wagner, E. F., Myers, M. G., & McIninch, J. L. (1999). Stress-coping and temptation-coping as predictors of adolescent substance use. *Addictive Behaviors, 24*(6), 769–779. doi:10.1016/S0306-4603(99)00058-1

Yasin, M.A.S.M. & Dzulkifli, M.A. (2010). The relationship between social support and psychological problems among students. *International Journal of Business and Social Science, 1*(3), 110–116.

Yigiter, K. (2013). The examining problem solving skills and preferences of Turkish university students in relation to sport and social activity. *Educational Research International, 1*(3), 34–40.

Zhou, M. & Ee, J. (2012). Development of the social emotional competence questionnaire. *The International Journal of Emotional Education, 4*(2), 27–42.

Developing assessment instruments of communication skills for vocational school students

S. Suranto
Universitas Negeri Yogyakarta, Indonesia

ABSTRACT: This research aims to develop assessment instruments which are feasible for assessing communication skills of vocational school students. The criteria of the instruments feasibility consists of validity, reliability, readability, effectiveness and practicality. The data collection techniques include focus group discussions and questionnaires. Descriptive data analysis technique was employed to reveal the quality of an instrument in terms of its practicality, readability, and effectiveness. Moreover, the content validity was applied using *Aiken V* formula to know the validity of the instruments. In addition, the reliability was tested using *Cronbach's Alpha* coefficient. The assessment instruments for assessing the communication skills which have been developed must accomodate the noble characters such as politeness and tolerance as indicators.

1   INTRODUCTION

The unemployment rate in Indonesia is still quite high. Data from 2015shows that the open unemployment rate in the country reached 8.49 percent or 9.44 million of the total population, in which high school graduates dominate the number (The Central Bureau of Statistics, 2015). Apart from limited job vacancies, unemployment also exists due to inadequate quality of human resource of educational institution graduates. There still exists the gap between competence demanded in the workplace and the competence of the workforce.

In reality, Indonesia's educational system has designed vocational schools which serve to equip learners or graduates with comprehensive modules and trainings specifically for employment and the professional world. It has been recognized that the graduates of vocational schools possess two-tiered competences. As Jafar (2011: 3) describes, "the quality of vocational education follows two standards which are in-school success standards and out-of-school success standards". The former assesses the learner's success in fulfilling the curricular demands that have been formulated to meet the requirements of the working world. The latter assesses the learner's performance based on national or international competence standards after they are involved in the real working world.

The competences that need to be owned by graduates include oral and written communication skills. According to a survey by Pennsylvania State University, oral and written communication skills of prospective workers hold a figure of 83.5 percent as the most required skill by the employers (Tubbs and Moss, 2001). However, the reality in Indonesia shows that the right format of the productive field-learning models that contain effective communication skills, designed to develop graduates' communication skills in the working world, is still questionable.

Communication skills serve as an important competence that determines an individual's performance in an organization. Curtis, Floyd, and Winsor (1988: 66) identify that 70 percent of job scope requires communication in many different ways. Thus, a worker's productivity is determined by their communication skills which include the ability to produce, send, and receive messages both verbally and non-verbally in oral and written communications.

Communication skills are related to the character values such as politeness and tolerance. Schein (1997) describes that character is the basis for communication.

On one hand, oral communication is the ability to speak (a speaking ability), such as being able to explain and present ideas clearly to an audience. These capabilities include the ability to adjust to the way a speaker talks, to use the right approach and style, and to understand the importance of non-verbal cues. On the other hand, written communication is the ability to write effectively, such as writing letters, Sort Message Service (SMS), and so forth. In oral communication, body language is important as Mulyana (2005: 159) states that "gestures, body movements, posture, head movements, facial expressions, and eye contact are behaviors that imply a potential message." Tubbs dan Moss (2001: 192) described that the quality of communication within an organization deals with the whole performance. Some of the required skills are as follows: (1) an active listener, (2) an effective presenter, (3) a quick thinker, and (4) a win-win negotiator. In addition, Permanasari (2014: 28) claims that communication skills comprise of a person's ability in using media, by which they can benefit from interacting with the media.

Communication skills that become the focus of this research are the theoretical concepts of communication processes and procedures that involve the skills of being communicators, designing messages, choosing channels, understanding the speakers, understanding the communication procedures and speaking in a socio-cultural setting. In addition, the ability of using social media that has become the main media of communication in the community needs to be considered. Wang, Jung Ki, and Kim (2017: 133) assert, "Mobile technology and social media exert a substantial impact on our society and daily lives". Similarly, Soffer (2013: 49) says that social media and internet have become media which replace the national newspapers." Another opinion is expressed by Manaf, Taibi, and Manan (2017: 15), "Mass media are said to report issues with a certain agenda, set to influence public opinion.".Also, Suter and Norwood (2017) stress that communication in living rooms and in public serves as "connection of private familial spheres to larger public discourses and structures; and inherent openness to critique, resistance, and transformation of the status quo."

In this research, the assessment instrument that is considered feasible to reveal information about communication skills is a test. Mardapi (2011: 2) asserts that a test is one of the strategies to estimate the level of a person's ability indirectly i.e. through a person's response to the stimulus or questions. The developed product is an assessment instrument of communication skills for vocational school students that are expected to meet teachers' needs. The implementation of the instrument is expected to be applicable in the school and acceptable by educational stakeholders. In this case, the developed instrument is a practice-based test instrument.

The practice-based test is an assessment that requires a response in the form of skills to perform an activity or behavior in accordance with the required competence. Practice-based tests are also called action tests intended to reveal students' motoric skills. Motoric domains are the domains that deal with the ability and skill or the ability to act after a person receives a particular learning experience. Meanwhile, psychomotor is related to the achievement of learning through skill as a result of the knowledge achievement. Psychomotor learning results are a continuation of cognitive learning outcomes and affective learning outcomes (tendency to behave or act). Labibah Lala (2016) says, "The learner's competence in the psychomotor domain involves the ability to perform an action. The assessment of skill competence is an assessment conducted by the teachers to measure the level of student's achievement in skill competencies covering aspects of imitation, manipulation, precision, articulation, and naturalization."

2 RESEARCH METHODS

This research expects to test assessment instruments of communication skills for students in vocational schools. The instrument adopts Mardapi's (2005) development model which include 10 steps: (1) formulating constructs, (2) formulating indicators, (3) making instrument blueprints, (4) determining standards or parameters; (5) writing instrument items, (6) performing validation process, (7) revising, (8) testing, (9) performing validity test, and (10) having produced feasible instruments.

The instrument feasibility criteria consists of three aspects, namely: (1) feasibility in terms of readability, effectiveness, and practicality; (2) validity; and (3) reliability. Readability, effectiveness and practicality are some criteria to determine whether or not the instruments are eligible to be used. The instruments were tested by 20 teachers who are the members of MGMP (*Musyawarah Guru Mata Pelajaran*) or Teacher Networks in a provincial level of Yogyakarta. The data analysis is to reveal the quality of the instrument viewed from the aspects of readability, practicality, and effectiveness performed using descriptive analysis techniques based on the assessment of the 20 MGMP teachers. The data analysis to determine the feasibility of the instrument viewed from the aspect of readability, practicality, and effectiveness is based on the mean score > 3.4–4.2 of the maximal score of 5 on the eligible classification, referring to the quantitative data conversion to the qualitative data on a scale of 5 using the rule which is a modification of the rule developed by Al-Rashid (1994: 27–29).

The validity test of instrument items uses a content validity based on the assessment of eight experts and analyzed with Aiken V formula, with validity index criteria $V_{count} > V_{table} = 0.75$. The Aiken V index is used to prove the validity of the content based on the expert panel's assessment of each instrument item, to the extent that it represents the constructs and indicators which are measured through the test instrument. The formula for Aiken's Index V is as follows.

$$V = \Sigma \frac{s}{N(c-1)}$$

Note:
V: Validity index
S: The score set by each rater is reduced to the lowest score of the desired score
c: The number of categories that can be selected by raters
n: The number of raters

Meanwhile, the reliability test aims at determining the suitability between the assessment results made by two assessors (raters), using the criteria of Cronbach's Alpha coefficient of at least 0.7.

3 RESEARCH FINDINGS AND DISCUSSIONS

3.1 *Research findings*

The data presented in this paper is the results of descriptive analysis on feasibility of the instruments as well as the results of the validity and reliability test.

3.1.1 *The Feasibility of instruments viewed from the aspect of readability, effectiveness, and practicality*

The feasibility test of the practice-based test instruments of communication skills in this research was achieved by sending the developed instruments along with the assessment questionnaires to the respondents consisting of 20 teachers. Then, they were asked to provide an assessment regarding the quality of the instrument as well as provide suggestions and opinions. The assessment on instrument readability is addressed by the aspects of instrument guidance clarity, the clarity of communication skills indicators, language used, and grammar. The assessment utilizes a multi-level scale score with maximum score of 5. The assessment of the language aspects consists of: 1) the use of standard Indonesian language, and 2) the formulation of communicative statements. The writing assessment includes the assessment of: 1) letter form, 2) letter size, and 3) punctuation.

The assessment of the instrument practicality is on the aspects of: (a) practicality in the sense that the instrument is easy to be carried out in the tests; (b) practicality in terms of ease to perform a work review after performing tests; (c) practicality in the sense that the instrument is easy to be used because the instruments are equipped with instructions. The assessment of the instrument effectiveness is on the aspects of: (a) the effectiveness of components which are developed with the theory of communication skills; (b) the effectiveness of the instrument indicators in measuring communication skills; (c) the effectiveness of the test items substance; and (d) the conformity between test items and learning materials.

Based on the procedure of quantitative to qualitative data conversion as presented in Table 1, the mean score of the eligibility of communication skill assessment instrument is = 3.95 from the maximum of 5. It is in the mean score range > 3.4–4.2 which is included in the classification of feasible or requires a minor revision. Thus, it can be concluded that based on the validity assessment of 20 respondents, communication skills assessment instruments developed in this study viewed from the aspects of readability, effectiveness, and practicality, are classified as feasible to use.

3.1.2 The validity and reliability of instruments

The validity test of the assessment instruments was carried out by performing the content validity through expert agreements. In this study, researchers technically asked the communication experts in the Vocational High School of Office Administration Expertise Program as a validator to assess each item of the instrument. Instrument items are categorized as having content validity if they can measure specific objectives that are relevant to the content related. Thus, the validators are asked to validate the instruments by comparing the contents of the test instruments with the teaching materials that have been taught.

The data of content validity assessment results for eight items of practical communication skills were obtained from eight experts as raters or validators. The test results of the instrument were analyzed using the Aiken formula. The practice test instrument items of communication skills are categorized valid if the experts believe that the instruments could measure the achieved capabilities defined in the validated items. The content validity test using Aiken formula refers to the index of validity, obtained by using Aiken's formula. Employing eight raters and a rating scale of 1 to 5, the instrument item is categorized as valid if Aiken V's validity index is greater than $V_{tabel} = 0.75$.

Table 1. The feasibility of the instrument viewed from the aspect of readability, effectiveness, and practicality.

No.	Aspects of assessment	Mean score
1	Instruments readability	3.91
2	Instruments effectiveness	3.90
3	Instruments practicality	4.06
Mean of the total score		3.95

Table 2. The results of instrument items validity test.

Items	V	V_{table}	Note
1	0.81	0.75	Valid
2	0.94	0.75	Valid
3	0.81	0.75	Valid
4	0.75	0.75	Valid
5	0.94	0.75	Valid
6	0.81	0.75	Valid
7	0.72	0.75	Not valid
8	0.75	0.75	Valid

The results of the analysis show that there is only one item of instrument out of 8 that is not valid, because it only has an Aiken's validity index value (V count 0,72 < V table 0,75), so that the instrument item is dropped. The eight items of the instrument include: item 1, telephone handling; item 2, presentation; item 3, interview; item 4, lobbying and negotiation; item 5, managing the meeting; item 6, communicating via mail; item 7, interpersonal communication; and item 8, searching for information on the internet.

The instrument reliability test was performed by looking at the score suitability between rater 1 and rater 2 using Cronbach's Alpha coefficient > 0,7. In this research, the assessment instrument was tried out to assess the communication skills of 32 students. In this practice test, there were two teachers acting as raters. The test results show that the coefficient value indicates $\alpha = 0.848$. With regard to this, the instruments developed in this study are classified as reliable.

3.2 *Discussions*

In this research, eight instruments were developed to assess communication skills. They consist of: (1) telephone handling, (2) presentations, (3) interviews, (4) lobbying and negotiation, (5) managing meetings, (6) communicating via mail (written communication), (7) interpersonal communication, and (8) searching for information on the internet. One important step in the development of instruments is the development of components or indicators. An indicator is anything that clearly and consistently explains the definition. Shavelson (2001: 7) says that designing indicators can be carried out through the following steps: conceptualizing potential indicator, refining poor indicator, and designing alternative indicators.

The elaboration of indicators are designed to provide accurate information about the various conditions and provide information on how the evaluated components produce a whole effect. Chamidi (2005: 14) states that in the simplest sense, the indicator is a symptom that address a particular issue or condition. To assess the effectiveness of an assessment instrument, it is necessary to review the required criteria of effectiveness. Some criteria of assessment instrument effectiveness presented by Kandak & Egen may be considered for adoption in assessing the effectiveness of the assessment instrument. Kandak & Egen (in Kaluge, 2004: 76) says that, "effective instrument in the real assessment must be valid, systematic, and practical." Mardapi (2006: 16) explains that the instrument developments may take a few steps. The first step is formulating variables based on synthesis of theories. The second step is determining construct, dimensions and indicator variables that have been explicitly stated in the formulation of variable constructs. The next step is making the instrument blueprints in the form of a specification table, which contains dimensions, indicators, item numbers and number of items for each dimension and indicator. It is followed by determining the standards or parameters within a continuum span, ranging from one pole to another opposite pole (from low to high, from negative to positive). The next step is writing instrument items which may take the form of statements or questions concerning the characteristics or circumstances, attitudes or perceptions. The sixth step is performing validation processes, both theoretical and empirical validations. Theoretical validation is obtained through an expert judgement in examining how far the dimension provides the exact description of the construct, how far the indicator becomes the exact description of the dimension, and how far the precise instrument blue prints can measure the indicator. The seventh step is performing a field trial which is part of the empirical validation process. The validity test may use either the internal criterion of the total score of the instrument as the criterion or the external criterion. If the content of the items is considered valid or feasible, the instruments become the final instruments which are used to measure the research variables.

The assessment instruments developed in this research are the instruments which are used to assess the students communication skills of Administrative Skills Competence in Vocational High Schools. One of the reasons of developing this assessment instrument is the increasing demand of communication skills in the working world. Tubbs and Moss (2001: 168) present the survey results that the most required skills by companies hiring fresh graduates are spoken and written communication skills. As pointed out by Kamaruzzaman (2016) in his research findings, speaking skills are indeed deemed important in the working world apart from physical, material, and mental preparation. In addition, Blizard (2012: 314) stresses four key skills in effective communication which include "understanding communication from another's perspective, listening, emotional intelligence, and conflict management."

The business world and industry require that communication skills are improved through learning to meet the demands of the working world. In order to obtain useful information as a consideration of improving communication skills, the development of a communication skills assessment instrument is viewed as a strategic effort. The availability of instrument kits will motivate school principals and teachers to assess students' communication skills, and in return, information on communication skills achievement can be used as a consideration in improving the quality of learning.

Preliminary studies taken to develop the instrument are performing theoretical and field studies in the form of observations and interviews on empirical conditions in the field. The information obtained from the preliminary study is used as the materials for developing the initial draft of the instrument development which include: (1) formulating the constructs, (2) formulating the indicators, (3) making the instrument blueprints. This is relevant to the instrument development proposed by Mardapi (2006: 16) covering 10 steps: (1) formulating constructs, (2) formulating indicators, (3) making instrument blueprints, (4) determining the standards or parameters (5) writing instrument items, (6) performing validation process, (7) revising, (8) testing, (9) conducting a validity test, and (10) having produced an eligible instrument.

The assessment instrument of the students' communication skills at Vocational High Schools developed in this research is a fairly simple assessment instrument, but it could gather sufficient information. Therefore, it becomes one of the alternatives that can be used by school principals or teachers to conduct an assessment on the quality of communication skills. This assessment instrument has been tested in which the results indicate that the instrument offers clarity of the manual usage, is quite practical and is effective.

The results of descriptive analysis on the feasibility of the instrument viewed from the aspect of readability has achieved the mean score of 3.91, the effectiveness aspect is 3.90, and the practicality aspects is 4.06. The mean of the total score is 3.95 from the maximum score of 5, included in the mean score range > 3.4–4.2 belonging to the classification of feasible. This indicates that the feasibility level of the communication skills assessment instrument is in a good category. Thus, the validation through the process of examination performed by the teacher has led to the formulation of instrument items which are fitted with the indicators. Mardapi (2005, 15–20) says that the validation process, both theoretical and empirical validation, is an important step. Theoretical validation is obtained through an expert judgement to examine how far the dimension provides the exact description of the construct, how far the indicator becomes the exact description of the dimension, and how far the precise instrument blueprints can measure the indicator.

The indicator system is designed to provide accurate information on various conditions and information on how the components being assessed may produce the whole effects. This concept is in line with Shavelson's (2001: 7) opinion stating that designing an indicator can be performed through the following steps: (1) conceptualizing potential indicators, (2) refining poor indicators, (3) designing alternative indicators of system options, and (4) evaluating the options and begin developing or refining an individual indicator.

The results of the quantitative test, which are analyzed using the items validity test employing the Aiken formula and inter-rater reliability test utilizing Cronbach's Alpha coefficient criterion, indicate that the instrument items being developed are valid items and the instruments are reliable. Validity is the supporting evidence and theory to the interpretation of test instruments assessment results which are in accordance with test objectives (Mardapi, 2011: 39). The type of validity used in this research is a content validity, which is estimated by testing on the feasibility or relevance of content through rational analysis performed by the competent panel (Azwar, 2016: 42). Meanwhile Satyadi and Kartowagiran (2014: 295) add that another purpose of the validity test is to examine the extent to which the content of instrument represents the aspects that are considered to be the conceptual framework of the materials being tested. In this research, the assessment instruments are in the form of a communication practical test validated by teachers and evaluation experts. Referring to the opinion of Allen & Yen (1979: 98), although the assessment is based on individual subjective judgment, some people who are competent in the field being measured are involved. Hence, the results can be accounted for.

The reliability test results indicate that the instrument of communication skills assessment developed in this research reaches the coefficient of $\alpha = 0.848$. Thus, it is concluded that this instrument is reliable. This level of reliability suggests that instruments can be trusted as a means for assessing communication skills. The coefficient used is the consistency between assessors (Mardapi, 2011: 86). In this study, two raters are employed which are intended to avoid the possibility of subjectivity and bias. As Gwet (2012: 8) stresses, "being in agreement with ourselves does not suggest that we will be in agreement with others." In addition, Azwar (2016: 88) argues that in order to minimize the effect of subjective judgment, the rating procedure is better performed by at least two people.

The research findings show that there is one out of eight instrument items which is invalid. The item is instrument number 7 that measures interpersonal communication skills. This finding indicates that the instrument item number 7 is not feasible and requires major revisions. Suranto Aw (2011: 12) describes interpersonal communication as a communication which has a great effect in influencing other people. This is because the people involved in the communication process usually meet directly and do not use media devices in delivering the message in which spatial distance does not really occur between them (face-to-face). Since they face each other, each can immediately know the response given, and reduce the level of dishonesty when the communication occurs. Hardjana (2003) argues that interpersonal communication will be more effective if the message is received and understood as intended by the sender of the message, and the message is followed up by a voluntary action of the recipient of the message, which can lead to improvement of the quality of interpersonal relationships.

The characteristics of communication skills instruments developed in this study differ from other assessment instrument models in several manners. Firstly, the communication skill assessment instruments are used as a data gathering tool in conducting communication skills practice-based tests implemented at the end of the semester as well as in competency tests in schools. Secondly, this assessment instrument consists of 7 components or 7 items of communication skills practice-based tests, namely (a) telephone handling, (b) presentations, (c) interviews, (d) lobbying and negotiation, (e) managing meetings, (f) communicating via mail (written communication), (g) interpersonal communication, and (h) searching for information on the internet. With regard to this, the users of the instruments can apply these seven components, or choose certain components that are deemed to have a higher priority. Lastly, this assessment instrument can be used to assess Vocational Practice-based Test both for grade X and XI, and XII and for semester final exam in Vocational High Schools offering Office Administration Expertise Programs.

Despite that, this research has a few limitations. First, the seven components of the communication skills assessment instrument developed in this study do not cover other potential communication skill components such as the ones used in different contexts of the environment and conditions. Second, the assessment instrument developed in this research was tested only using two teachers as raters. Therefore, the assessment process does not involve independent raters from outside of the school (independent appraisal). Since it only relies on the assessment of the internal party (internal appraisal), it is possible to reduce the level of objectivity in assessment results. To reduce this limitation, a crosscheck of the assessment results should be carried out between sources such as assessors and teachers.

4   CONCLUSIONS AND SUGGESTIONS

Based on the research findings and the discussions described above, this paper offers a few concluding points. Firstly, the components of the communication skills in this research consist of

a) telephone handling, (b) presentations, (c) interviews, (d) lobbying and negotiation, (e) managing meetings, (f) communicating via mail (written communication), (g) interpersonal communication, and (h) searching information from the internet. Secondly, the results of the test show that all developed instruments meet the validity and reliability requirements. The validity coefficient of Aiken V shows > 0.75 and the reliability coefficient is $\alpha = 0.848 > 0.7$. Thus, the instruments are feasible to be used by teachers to assess the communication skills of students taking Office Administration programs in Vocational High Schools. Lastly, the test results of readibility, practicality, and effectiveness of the instruments indicate that the instrument of communication skills assessment is included in the category of good or feasible to be used.

This research also offers a few suggestions for future studies. First, the developed product needs to be disseminated and implemented to assess students' communication skills. Assessments should be performed right after the the learning activities are completed (e.g. at the end of the semester), in order to obtain actual and accurate information of the assessed communication skills. Second, to improve the efficiency of the application of the assessment instruments, an assessor (rater) needs to compile data entry in Microsoft Office Excel. This will help to ease the data entry and analysis, as well as in making recommendations for assessment results.

REFERENCES

Aiken, L.R. 1985. "Three coefficient for analyzing the reliability and validity of ratings". *Educational and Psychological Measurement*. Retrievedon 16 April 2016 from Sagepub.com.

Allen, M.J. & Yen, W.M. 1979. *Introduction to measurement theory*. Monterey: Brooks/Cole Publishing Company.

Azwar, S. 2016. *Reliabilitas dan validitas* [relaibility and validit]. Yogyakarta: PustakaPelajar.

Borg, W.R. & Gall, M.D. 1983. *Educational research: An Introduction*. NewYork& London: Longman

Castanedal, JAF., 2014. Learning to Teach and Professional Identity: Images or Personal and Professional Recognition. *PROFILE*, 16(2): 1–10. doi 10.15446/profile.v16n2.38075.

Craig, R.T. 2016. *Communication Theory as a field*. Retrievedondownloaded on 5 June 2017 from http://www.elt.astate.edu/mhays/craig.

Crawford, D.C. 2006. "Suggestions to Assess Non-formal Education Programs" in *ProQuest Education Journals*.

Curtis, DB., Floyd, JJ., & Winsor, JL., 1998. *Komunikasi Bisnis dan Profesional* [Business Communication and Professional]. Bandung: RemajaRosdakarya.

Ebel, R.L. & Frisbie, D.A. 1986. *Essential of educational measurement*. New Jersey: Prentice- Hall, Inc.

Hardjana, A.M. 2003. *Komunikasi Intrapersonal & Interpersonal* [Intrapersonal & Interpersonal Communication]. Jakarta: Kanisius.

Gwet, K.L. 2012. *Handbook of inter-rater reliability*. Washington: Advanced Analysis, LCC.

Jafar, J. 2011. *Materi uji kompetensi 2008* [2008 Competency test material]. Retrievedon 13 August 2008 fromhttp://209.85.175.104/search?q=cache:GfP_vhf6W8YJ:download. ditpsmk.net /~t4 mu/01. MATERI_UJI_KOMPETENSI_2008.

Kaluge, A.H. 2004. *Pengembangan model penilaian proses belajar matematika yang komprehensif dan kontinu pada pembelajaran kooperatif di SMP* [Development of a comprehensive and continuous assessment model of mathematics instruction in Junior High School]. Unpblished Thesis, Universitas Negeri Surabaya, Indonesia

Kamaruzzaman. 2016. Kecakapan Komunikasi [Communication Ability]. *Journal of Conseling Gusjigang*. 2 (2): 1–10.

Manaf, A.M.A., Taibi, M., & Manan, K.A. 2017. Media Agenda and Public Agenda (Agenda Media and Public): A Study of Issues. *Malaysian Journal of Communication*. 33 (2): 13–26.

Mardapi, D. 2006. *Evaluasi pendidikan* [Evaluation pf Education]. A paper presented in The National Education Convention, at Jakarta State University.

Mardapi, D. 2011. *Pengukuran, penilaian, dan evaluasi pendidikan* [Measurement, Assessment and Evaluation of Education]. Yogyakarta: Nuha Medika.

Mizikaci, F. 2007. A systems approach to program evaluation model for quality in higher education. *ProQuest Education Journals*. 130 (125–140).

Permanasari, R. 2014. "Proses Komunikasi Interpersonal Berdasarkan Teori Penetrasi Sosial (Studi Deskriptif Kualitatif Proses Komunikasi Interpersonal antara Personal Trainer dengan

Pelanggan di Club House Casa Grande Fitness Center" (Interpersonal Communication Process based on Social Penetration Theory (Descriptive Qualitative Study). *Jurnal Ilmu Komunikasi.* http:// e-journal.uajy.ac.id/id/ eprint/6490. Retrieved on 12 October 2015.

Rasjid, H.A. 1994. Social Statistic. Bandung: Padjadjaran University.

Sanders, J.R. & Sullins, C.D. 2006. *Evaluation school programs an educator's guide. California:* Corwin Press.

Schein, E.H. 1997. Organizational Culture and Leadership. Second Edition. San Fransisco, LA: Jossey Bass A Willey Co.

Setyadi, H., & Kartowagiran, B., 2014. *Evaluasi kurikulum (The evaluation of curriculum).* A paper presented in The National Education Convention, at Universitas Semarang, Indonesia.

Suranto Aw. 2011. *Komunikasi interpersonal (Interpersonal Communication).* Yogyakarta: GrahaIlmu.

Suter, E.A. & Norwood, K.M., 2017. Critical Theorizing in Family Communication Studies. *Journal of Communication Theory.* Volume 27, Issue 2. 2017. Retrievedon 5 June 2017 from http://onlinelibrary. wiley.com/doi/10.1111/comt.12117/.

Tubbs, S.L. & Moss, S. 2001. *Human communication.* Bandung: RemajaRosdakarya.

Venesaar, U., Ling, H., Voolaid, K., 2011. "Evaluation of The Entrepreneurship Education Programme in University: a New Approach. " *AE Journal Vol. XIII No.30 June 2011.* http://www.amfiteatrueco-nomic.ro/home.ro.aspx.Retrieved on 12 March 2016.

Wang,Y., Jung Ki, E., & Kim, YH. 2017. Exploring the Perceptual and Behavioral Outcomes of Public Engagement on Mobile Phones and Social Media. *International Journal of Strategic Communication.* 11(2): 1–12.

Character Education for 21st Century Global Citizens – Retnowati et al. (Eds)
© *2019 Taylor & Francis Group, London, ISBN 978-1-138-09922-7*

The implementation of affective domain assessment in elementary school

H. Sujati
Universitas Negeri Yogyakarta, Indonesia

ABSTRACT: Affective domain assessment is considered to be more advanced compared to cognitive and psych motoric assessments. This study was aimed at describing the implementation of affective domain qualitatively in Percobaan IV Elementary School, Wates Kulonprogo, Yogyakarta Province, Indonesia. This research employed qualitative approach with teacher as the research subject. The data was collected through interview and observation. The number of research subjects was determined by snowball sampling method. The validity test of data was done by member check and triangulation technique. The collected data was analyzed using flow methods from Miles and Huberman. The results of this study showed that the teacher has set various attitudes in the design of learning at the planning stage. However, the teacher has not yet elaborated the attitudes that will be measured into indicators subsequently developed into an assessment instrument. In the framework of attitude assessment, teachers did not develop assessment instruments. They used the instruments which were already available in the teacher's book. At the appraisal level, teachers used only a single technique, i.e. observation. Teachers also have not involved the relevant parties so that the subjectivity of teachers in the assessment became high. At the level of data processing, teachers used the application program so that the results were less clear to describe the state of each individual.

1 INTRODUCTION

In this day and age, the imbalance in the students' learning achievement assessment has drawn attention by some educators to conduct a further study. The assessment is still dominated by cognitive and creativity domain while affective domain assessment is neglected. Popham (Depdiknas, 2008) states that affective domain is also considered as one aspect which determines the success of someone's learning. This idea is in line with Goleman (2007) who states that 80% of someone's success is affected by his or her emotional intelligence and only 20% of it is determined by the brain intelligence (IQ).

The importance of affective learning brings to a conclusion that every educator is required not only to educate learners, but also to have competence in conducting an assessment in affective domain. Government Regulation No. 32 of 2013states that the graduate standard competence includes attitude, knowledge, and skill. Therefore, a professional teacher is intended to be able to conduct an assessment in those three domains. Those ideas are also stated in the Regulation of the Minister of *Education and Culture* of the Republic of Indonesia No. 66 of 2013 about Assessment Education Standard which claims that students' learning assessment covers affection competence (attitude), knowledge, and skills which are conducted in an even way so that it could be used to determine the relative position of every learner towards the standard which has been agreed together.

Percobaan IV Elementary School of Wates, Kulon Progo is one of the elementary schools that is currently is adapting Curriculum 2013 of Indonesian National Education. If the Regulation of the Minister of *Education and Culture* of the Republic of Indonesia No. 104

of 2014 is properly used, this school should have implemented authentic assessment which also includes affective domain assessment inside. This assessment also covers spiritual and social attitude assessment. Based on some preliminary studies, there are a lot of teachers who feel doubtful in conducting affective domain assessment. They have not implemented the assessment as it should be. Thus, it needs a review on the implementation of affective domain assessment through a further study.

In this study, the research problem is formulated as follows: "How is the affective domain assessment implemented in Percobaan IV Elementary School of Wates, Kulonprogo?" This general question is then elaborated into three substitute questions, which are: How is the plan, implementation, and the data analysis of the affective domain assessment in Percobaan IV Elementary School of Wates, Kulonprogo?

Shermis and Vista (2001) state that affective behavior covers attitude, interest, motivation, opinion, personality, and appreciation. Anderson (1981) has the same idea and states that affective characteristic deals with emotions and feelings. Anderson (1981) mentions some affective characteristics, such as attitude, confidence, locus of control, and anxiety. Anderson (Gable 2013) says that affective characteristics contain some aspects such as attitude, self-efficacy, score, self-concept, and interest. Meanwhile, Curriculum 2013 focuses on affective domain and attitude.

Angelo & Cross (1993) define assessment as an approach designed to discover what improvement students have learned after making mistakes in the learning process or how well the learning process takes place. Meanwhile et al. (2001) define assessment as a set of procedures designed to provide information regarding to development, growth, and achievement of a process which is then compared to a certain standard. The data on assessment is obtained through formal observations, informal observations, tests and other measures. The use of assessment is also stated by Mardapi (2011) who says that assessment can be used to determine the achievement of learning outcomes.

According to Majid and Firdaus (2014) the assessment of the affective domain in learning is useful to measure the advancement of learners' attitudes. In this time, assessment of education is still dominated by the assessment of the cognitive domain, whereas the assessment of the affective domain is also significant. They tend to use more cognitive assessment because conducting affective assessment is more complex. Naga (1992) states that conducting affective assessment is more complex because the objects are not easy to identify. Those objects are ability, success, attitudes, interests or other. According to Anderson (1981), there are two methods that can be used to measure affective domain, those are observation method and self-report. The use of observational methods is based on the assumption that affective characteristics can be seen from the behaviors or activities displayed or psychological reactions. The self-reporting method assumes that one who knows one's affective state is only himself.

Mardapi (2011) states that the assessment of the affective domain can produce both quantitative and qualitative data. Quantitative data can be obtained using attitude scales and assessment scales, while qualitative data can be collected through observations, interviews, and anecdotal notes. According to Suparno (2002), the weakness of quantitative data lies in its inability to build moral awareness of the learners. This happens because the assessment results are not able to touch the moral intelligence of learners. On the other hand, qualitative data is used because it uses a verbal description of the behavior of learners. It can explain the advantages and disadvantages of certain aspects of behavior and how to improve them.

According to The Ministry of Education and Culture (2015), affective assessment includes three steps, namely planning, implementation, and processing. The steps of attitude assessment planning are as follows: (1) determining the attitude that will be developed in schools (2) formulating indicators in accordance with the competence of attitudes to be developed; (3) designing learning activities that can elicit predetermined attitudes; and (4) preparing assessment instruments. An observation on students' attitudes inside and outside the class will be a part of affective assessment. The results of attitudes observation will be documented after each of the themes is completed by the teacher.

608

2 RESEARCH METHOD

This study employs qualitative approach since it can produce descriptive data in the form of written or oral form of data obtained from an observable object (Moleong, 2009). The study was conducted in Percobaan IV Elementary School of Wates, Kulonprogo. The primary school was chosen as a research site because this elementary school is one of several elementary schools in Kulonprogo Regency that consistently implements the 2013 Curriculum since the implementation of the curriculum is proclaimed by the Ministry of Education and Culture.

In this study, there are some sources of information, such as the principal, the teachers, and the students. The number of information sources was determined by using the principle of snowball sampling. Semi-structured interviews is used to collect data. Source triangulation and member check are used to validate the data. The flow method from Miles and Huberman is used to analyze the data.

3 RESEARCH RESULTS

Percobaan IV Elementary School of Wates has determined several attitudes that would be developed in classroom learning. The attitudes developed by classroom teachers would include discipline, responsibility, honesty, confidence, self-confidence, cooperation, curiosity and tolerance. However, in the process of developing attitude measurement instrument, the teachers have not formulated indicators of the attitudes to be measured. In the lesson plan, teachers simply included attitudes that would be assessed without including indicators of achievement. They stated that they did not know how the indicators were formulated and did not have enough time to develop them. Therefore, teachers have never developed an attitude assessment instrument, since the instrument already exists in the teacher's book.

The results of field observation indicated that not all of the teachers have prepared an attitude observation sheet. Those who did not make an observation sheet argued that they could monitor students' attitudes using a diary. However, the observation sheet was not used maximally in the real practice. Those teachers said that they did not have enough time to use the observation sheet they had made. Instead, the teacher kept a record of the most prominent events of the day. Records were made towards any events that occurred inside and outside the classroom.

To provide final grades of attitudes, teachers carried out processing of qualitative data obtained from the field. The qualitative data was narrative descriptions of events related to the attitudes documented in the diary. Beside the classroom teachers, the data was also attained from other staffs, such as religion teachers and physical education teachers. The diary by the teacher was then transferred into a journal.

4 DISCUSSION

This study found that teachers have conducted an affective field assessment as required by the 2013 National Education Curriculum. The assessment is the implementation of a teacher's professional responsibilities as it is stated in The Ministry of Education and Culture Regulation No. 104 that teachers have the responsibility to assess the cognitive, affective, and psychomotor aspects. The implementation of the assessment of learning results affective domain by teachers was a manifestation of the implementation of professional duties of an educator as stated in Law of Indonesia No. 14 of 2005.

The assessment of attitude includes three levels, namely planning, implementation and processing. At the planning stage, the teacher has set the expected attitudinal goals in learning, such as curiosity, honesty, responsibility, discipline, and self-reliance. The decision to determine the various attitudes was in line with the Evaluation Guidelines issued by The Ministry of Education and Culture (2015). In the guidelines, it was stated that at the stage of attitude assessment planning, teachers need to first establish the specific attitudes that are expected to arise as a result of the strategies and teaching process applied by the teacher. The determination of these attitudes is very important in the process of character formation of learners.

Although the teachers have listed some students' attitudes that were expected to appear during the learning process, there was only one single technique which was used by the teacher to assess them. In assessing attitudes, teachers only used observation technique with the help of observation sheet instruments in the form of check lists and field notes. The use of single technique like this was certainly not suggested by the 2013 National Education Curriculum. Based on The Ministry of Education and Culture Regulation No. 104 of 2014, teachers should start using multiple techniques in assessing the competency of learners. Kunandar (2015) states that authentic assessment can only be applied when teachers use multiple techniques in the assessment. It is not only applicable in cognitive and psychomotor assessment, but also affective domain.

The teacher never elaborated the competency of the attitudes into competency indicators and did not know how to develop an affective appraisal instrument due to the limited time they had. Teachers tended to use assessment instruments that were already available in the teacher's book, even if the instrument was still very simple. The National Education Department (2008) has firmly affirmed that in order for teachers to measure affective aspects validly, teachers need to describe the competence of these attitudes into indicators. If the indicators have been determined, then the instrument specifications can be formulated.

The Ministry of Education and Culture (2010) states that indicators are the sign of achieving basic competencies characterized by behavioral changes, both in terms of knowledge, attitude, and skills. Therefore, each indicator formulated by the teacher should be a description of the basic competencies. This indicator becomes the basis for developing the learning design. Therefore, indicators illustrated the design of learning that must be pursued. The lack of formulation of the affective domain indicator could reflect how unclear the lesson was which would be learned. In addition, the attainment of certain attitudes became unclear.

The tendency of the teachers to use the instruments which were already available in the teacher's book gave a grim picture that the 2013 National Education Curriculum objectives would not be maximally achieved. The 2013 National Education Curriculum as a refinement of School-Based Curriculum has the spirit to restore teachers as creative educators who are able to generate new ideas. Creative teachers love to do new things and do unexpected things (Mulyasa 2002).

In this study, based on the attitude assessment, teachers rely on the use of field notes to write down some significant behaviours. According to Anderson (1981) there are two methods that can be used to measure the affective domain, namely the method of observation and self assessment. The use of observational methods is based on the assumption that affective characteristics can be seen from the behaviours or attitudes displayed and or psychological reactions.

In the attitude assessment, teachers have not utilized self-assessment techniques. Koesoema (2010) argues that attitude assessment should make the best use of self-assessment techniques. This technique provides an opportunity for learners to be honest and open in describing themselves. The one who most knows the state of the learner is himself. Therefore, in the attitude assessment the teachers should facilitate learners so that he is able to be honest in assessing himself. Honesty enables individuals to perfect themselves as human beings.

In addition to the attitude assessment within the classroom, teachers also carry out the recording of student attitudes outside the classroom as an affective assessment material. Teachers note outstanding students' attitudes outside the classroom in diaries, textbooks, and journals achieved from teachers, students and other school stakeholders. When applying affective attitudes, the teacher coordinates with the religion teacher and the physical education teacher. Classroom teachers collect diaries and journals prepared by religion teachers and physical teachers in religion and physical education learning.

In the assessment of the affective domain, the teacher provides an assessment in the form of qualitative descriptions. The descriptions were generated through the 2013 Curriculum application program created by one of the teachers of Percobaan IV Elementary School of Wates. The application automatically describes numbers (4, 3, 2, and 1) with their qualitative descriptions. Assessment in the form of qualitative description is in accordance with the opinion Mardapi (2011) which states that the assessment of affective areas can produce data both qualitative and qualitative.

5 CONCLUSION

Based on the results of research and discussion above, some conclusions were attained as follows:

1. In terms of affective assessment planning at Percobaan IV Elementary School of Wates, Kulonprogo, teachers have established attitudes that will be developed in classroom learning, such as responsibility, honesty, discipline, courtesy, confidence, tolerance, curiosity, cooperation, carefulness, persistence, and diligence. However, the teachers have not formulated the attitude indicators to be assessed in details. The attitudes are shown in the planning of learning activities tailored to the theme and the material to be studied. In the implementation stage of attitude assessment, teachers have not developed the instrument as well. They used the instruments available in the teacher's book.
2. In conducting an attitude assessment, teachers have not involved the parties which have concern in this matter. The assessment is still subjective which was attained from daily observations. Similarly, teachers have not used other techniques such as self-assessment, peer assessment, and so on.
3. The results of the assessment have been obtained in the form of qualitative narrative. The narrative is obtained based on the results of the process by using the application assessment program.

REFERENCES

Anderson, L.W. 1981. *Assessing affective characteristics in the schools*. Boston: Allyn and Bacon.
Angelo, T.A. & Cross, P.K. 1993. *Classroom assessment techniques: A handbook for college teachers*. San Fransisco: Jessey-Bass Inc.
Depdiknas. 2008. *Pengembangan perangkat penilaian afektif* [Development of an affective assessment tool]. Jakarta: Dirjen Pendidikan Dasar dan Menengah.
Gable, R.K. 2013. *Instrument development in the affective domain*. New York, NY: Springer Science.
Goleman. 2007. *Social Intelligence*. Jakarta: Gramedia.
Koesoema, D. 2010. *Pendidikan karakter* [Character building]. Jakarta: Grasindo.
Kunandar. 2015. *Penilaian autentik: penilaian hasil belajar peserta didik berdasarkan kurikulum 2013* [Authentic assessment: assessment of learners' learning outcomes based on the 2013 curriculum]. Jakarta: Raja Grasindo Persada.
Law of Indonesia Number 14 of 2005 about Teachers and Lecturers.
Majid, A. & Firdaus, A. 2014. *Penilaian Autentik Proses dan Hasil Belajar* [Authentic assessment of learning processes and results]. Bandung: Interes.
Mardapi, D. 2011. *Penilaian pendidikan karakter* [Character education assessment]. Yogyakarta: UNY Press.
Moleong, L.J. 2009. *Metode penelitian kualitatif* [Qualitative research methods]. Bandung: Remaja Rosda Karya.
Mulyasa, E. 2002. *Kurikulum berbasis kompetensi* [Competency-based curriculum]. Bandung: Remaja Rosda Karya.
Naga, D.S. 1992. *Pengantar teori skor* [Introduction to score theory]. Jakarta: Gunadarma
Shermis, M.D. & Di Vesta, F.J. 2001. *Classroom Assessment in Action*. New York, NY: Rowman & Littlefield Publishers, Inc.
Suparno, P. 2002. *Pendidikan budi pekerti di sekolah* [Education Character in School]. Yogyakarta: Kanisius.
The Ministry of Education and Culture. 2015. The Principles of Learning Outcomes Assessment. Jakarta: The General Director of Basic Education.
The Ministry of Education and Culture Regulation No. 64 of 2013 about the Curriculum Standard of Primary and Secondary Education.
The Ministry of Education and Culture Regulation No. 66 of 2013 about the Standard of Primary and Secondary Education.

Creating managing conducive school culture to character education

Character Education for 21st Century Global Citizens – Retnowati et al. (Eds)
© 2019 Taylor & Francis Group, London, ISBN 978-1-138-09922-7

Emotional analysis of student drawings of the human figure as an indicator of intelligence

D.A. Febriyanti
Universitas Diponegoro, Indonesia

ABSTRACT: The assessment of students' intelligence is an essential part of conducting effective learning strategies. According to projective theories, children attempt to express their unconscious emotional condition in ways which avoid conflict with their surroundings. One of these ways is projective drawing. Previous research has shown that children's drawings, especially children's drawings of the human figure, deliver information about their psychological functions. Nevertheless, there were still questions about the correlation between children's drawings and children's intelligence. This study aims to look at the correlation between children's drawings and intelligence scores from a different point of view. The basic assumption underpinning this study is that there is a correlation between the emotional aspects of children's drawings and their personality aspects according to the Wechsler Intelligence Scale for Children (WISC) test. Participants of this study consisted of 71students from elementary schools located in Semarang, Indonesia who had agreed to join this study. All participants completed the WISC test and Koppitz's Human Figure Drawing test (HFD). The results show that further research is needed to explore this correlation even though there are some aspects of WISC that have significant correlation with HFD scores.

1   INTRODUCTION

Various studies in the areas of psychology and child development have revealed the important meaning behind children's images beyond the process and production of the images they produce. These studies are more focused on the image content. The images produced by the child can give an idea of the intellectual condition (Imuta et al., 2013), self-assessment (Dey & Gosh, 2016), understanding of the surrounding environment (Sorin, 2005) and emotional state of the child (Skybo et al., 2007).

As one type of projective psychological technique, human figure drawings created by children can describe their various psychological conditions. A study in Turkey (Metin & Ustun, 2010) concluded that human figure drawings made in early childhood can illustrate the state of a child's sibling relationships at home. This study uses a human drawing test version of Kinetic Family Drawing. Projective drawing in children with cancer can also provide information and can be used as a screening tool to determine the level of distress they are experiencing (Hermann, 2015). The method of human figure drawing analysis used in this research was Draw-A-Person: Screening Procedure for Emotional Disturbances (DAP:SPED).

Although the results of Metin & Ustun's study had proven a correlation between the IQ scores obtained from the Goodenough–Harris version of human images while other research proves the opposite, different versions of human drawing tests provide the facility to see cognitive potential in general (Groth-Marnat, 1984; Cohen & Swerdlik, 2009). This facility raises the author's interest to further examine the projective drawing test as an assessment tool to ascertain a student's potential ability in the education process. The assessment of students' abilities is accompanied by intelligence tests.

The intelligence test chosen in this research was the Wechsler Intelligence Scale for Children (WISC). In addition to the wealth of intelligence aspects that can be expressed through this individual intelligence test, this test also provides the ability to reveal personality aspects, both in the early versions of the development of this test (Groth-Marnat, 1984) and in subsequent versions (Groth-Marnat, 2009) and even personality analysis relating to a psychiatric diagnosis (Gregory, 2011). The results of the study using WISC show that the assessment properties in this test can be used to reveal potential student risk behavior and a pattern of individual antisocial behavior (Isen, 2010) based on the composition of IQ verbal score, IQ performance score, and total IQ score. The personality aspect can also be expressed through the analysis of the subtest scores of the WISC, both through the score of each subtest (Groth-Marnat, 2009).

The study results of Miller (2006) show that human figure drawings created by elementary school students are related to their executive functions and achievement levels. Miller's study results are also supported by the study on drawing as an expression of the child's perspective of his school experiences (Einarsdottir et al., 2009). This research indicates that, through their drawing, children use their cognitive ability to create a perspective, reasoning and describing their school conditions. Therefore, in drawing activities, they use both cognitive abilities and their affective functions to express details of their schooling experience in their picture.

Based on previous research, the problem focused on in this research can be formulated as a question; is there any relationship between the personality aspects of elementary students obtained from human drawing tests and cognitive and personality aspects derived from the WISC test results?

2 LITERATURE REVIEW

2.1 *Human figure drawing*

The projective drawing test used in this study was a human drawing test developed by Koppitz (1968). This test was developed with the aim of understanding the occurrence of emotional problems in self-tests. The basic assumption behind this test is that the emotional and neurological conditions that cause problems in a child's behavior will emerge as indicators in their pictures.

Based on the scoring system compiled by Koppitz (1968), the procedure for conducting a human drawing test begins with instructions for each subject to draw on the drawing sheets that have been arranged in a certain format. Male subjects were instructed to draw human images of

Table 1. List of emotional indicators in human figure drawing (Koppitz, 1968).

No.	Emotional indicators	No.	Emotional indicators
1	Broken/sketchy lines	22	Hands hidden behind back or in pockets
2	Poor integration of parts of figure	23	Legs pressed together
3	Shading of the face or part of it	24	Genitals
4	Shading of the body and/or limbs	25	Monster or grotesque figure
5	Shading of the hands and/or neck	26	Three or more figures spontaneously drawn
6	Gross asymmetry of limbs	27	Figure cut off by edge of paper
7	Figure slanting by 15° or more	28	Baseline, grass, figure on edge of paper
8	Tiny figure, 2" or less in height	29	Sun or moon
9	Big figure, 9" or more in height	30	Clouds, rain, snow
10	Transparencies	31	Omission of eyes
11	Tiny head, 1/10th of total height of figure	32	Omission of nose
12	Large head, as large or larger than body	33	Omission of mouth
13	Vacant eyes, circles without pupils	34	Omission of body
14	Side glances of both eyes, both eyes turned to one side	35	Omission of arms
15	Crossed eyes, both eyes turned inward	36	Omission of legs
16	Teeth	37	Omission of feet
17	Short arms, not long enough to reach waistline	38	Omission of neck
18	Long arms, that could reach below knee-line		
19	Arms clinging to side of body		
20	Big hands, as big as face		
21	Hands cut off, arms without hands and fingers		

Note: Score 1 for each appearance of an emotional indicator; score 0 if no appearance of emotional indicator.

males, while female subjects were instructed to make a human image of a female. Subsequent subject image results are based on the appearance of emotional indicators in the image. Based on a review by Skybo et al. (2007) of the series of manuscripts compiled by Koppitz, the emotional indicators appearing in the children's drawings have been presented in Table 1.

2.2   Wechsler intelligence scale for children

The WISC test is an individualized intelligence test. This test was developed in 1932 by David Wechsler, as a critique of the Binet–Simon intelligence test that had been used extensively (Gregory, 2011). This scale was developed based on various theoretical reviews of various intelligence tests that existed at the time, such as Binet–Simon, Army Alpha and Army Beta. The main criticism of the WISC test is a concept of intelligence that is no longer based on the concept of mental age, but is based on the index score of the number of items successfully completed in each subtest.

The WISC intelligence test consists of 12 subtests. Each subtest reveals aspects of intelligence and aspects of personality (Groth-Marnat, 2009). Personality aspects is an emotional analysis in WISC. The 12 subtests and their respective aspects of intelligence and personality have been presented in Table 2.

Table 2.   Intelligence and personality aspects of WISC.

No.	Subtest	Intelligence aspects	Personality aspects
1	Information	Ability to obey instructions Ability to respond quickly Long-term memory	Curiosity Interest Culture
2	Comprehension	Logical thinking	Social reasoning Realistic thinking Attitude norms Emotional stability
3	Arithmetic	Concentration Auditory memory Numerical reasoning School learning	Motivation Interest Problem-solving belief
4	Similarities	Abstract reasoning Verbal reasoning	Diligence Strength of observation
5	Vocabulary	Verbal productivity Memory Verbal synthetic-analytic	Childhood experience Life around the world of education
6	Digit span	Concentration Short-term memory	Attention Anxiety
7	Picture completion	Concentration Visual awareness Visual arrangement Visual memory	Experience related to the environment
8	Picture arrangement	Logical thinking Nonverbal reasoning Strength of observation	Social comprehension Social sensitivity Interpersonal social sensitivity
9	Block design	Observational accuracy Logical reasoning Visual-motor coordination	Achievement motivation
10	Object assembly	Perceptual organization Visual-motor organization	Working diligence Productivity willingness
11	Symbol	Visual-motor dexterity Associative learning	Emotional stability Comprehensive task performance
12	Mazes	Visual-motor coordination Visual-motor organization	Working diligence

3 METHOD

3.1 *Participants*

This study was conducted at an elementary school which stated a willingness to be involved as a participant. The study participants were all elementary school students in the classroom designated by the school. All students designated by the school to become participants of this study completed a battery of WISC intelligence test procedures and human drawing tests. The school principle promised that 80 students could participate in this research, but only 71students participated in the WISC test.

3.2 *The WISC test*

The WISC test implementation procedures refer to the WISC manual. Considering the time allocation of data collection, which was determined by the school, all students took 11 WISC subtests, consisting of six verbal subtests and five performance subtests. These subtests were: information, comprehension, arithmetic, similarities, vocabulary, digit span, picture completion, picture arrangement, block design, object assembly, and symbol. A total of 12 students could not complete the WISC test because they has been picked up by their parents, so only 59 data results of the WISC test were collected.

3.3 *The human figure drawing test*

The procedure for the human figure drawing test began with instructions for each subject to draw on a sheet that has been arranged in a certain format. Subjects were asked to draw a complete human being. The complete picture was not limited to the body parts or figures. The drawing sheets used were legal size (21.59 cm × 27.94 cm). There was a page border surrounding the drawing area and a place to write down the subject's scores obtained from the drawing. Subsequent subject's image results are based on the appearance of emotional indicators in the image. A total of 20 of the 59 students who completed the WISC test did not attend the drawing session so only 39 students followed the complete data collection procedure for later analysis.

4 WITH RESULTS

An overview of the WISC scores and emotional indicator scores based on Koppitz's system of human figure production obtained by the participants in this study are presented in Table 3.

Table 3. Descriptive statistics for participant scores.

	Mean	Std. deviation	N
IQ total	96.69	10.563	39
Verbal IQ (VIQ)	95.77	12.121	39
Performance IQ (PIQ)	98.13	12.226	39
Discrepancy of VIQ & PIQ	12.10	8.711	39
Information	10.74	3.408	39
Comprehension	7.15	3.208	39
Arithmetic	9.64	2.987	39
Similarities	9.51	3.136	39
Vocabularies	7.59	5.604	39
Digit Span	9.62	3.049	39
Picture Completion	10.38	2.943	39
Picture Arrangement	9.21	2.922	39
Block Design	10.05	3.268	39
Object Assembly	8.62	3.274	39
Symbol	10.41	3.582	39

Table 4.   Correlation of IQ score of WISC and HFD score.

		HFD score
Full IQ	Pearson correlation	.080
	Sig. (2-tailed)	.629
	N	39
Verbal IQ (VIQ)	Pearson correlation	.132
	Sig. (2-tailed)	.423
	N	39
Performance IQ (PIQ)	Pearson correlation	−.016
	Sig. (2-tailed)	.921
	N	39
Discrepancy of VIQ & PIQ	Pearson correlation	−.055
	Sig. (2-tailed)	.741
	N	39

Table 5.   Correlation between WISC subtests and HFD score.

		HFD score
Information	Pearson correlation	.012
	Sig. (2-tailed)	.942
	N	39
Comprehension	Pearson correlation	.136
	Sig. (2-tailed)	.410
	N	39
Arithmetic	Pearson correlation	−.160
	Sig. (2-tailed)	.330
	N	39
Similarities	Pearson correlation	.386*
	Sig. (2-tailed)	.015
	N	39
Vocabulary	Pearson correlation	.391*
	Sig. (2-tailed)	.014
	N	39
Digit Span (DS)	Pearson correlation	.019
	Sig. (2-tailed)	.910
	N	39
Picture Completion (PC)	Pearson correlation	−.082
	Sig. (2-tailed)	.619
	N	39
Picture Arrangement (PA)	Pearson correlation	.292
	Sig. (2-tailed)	.071
	N	39
Block Design (BD)	Pearson correlation	.031
	Sig. (2-tailed)	.852
	N	39
Object Assembly (OA)	Pearson correlation	.115
	Sig. (2-tailed)	.487
	N	39
Symbol	Pearson correlation	−.340*
	Sig. (2-tailed)	.034
	N	39

* Correlation is significant at the 0.05 level (2-tailed).

Multiple correlation analysis using Pearson Product Moment shows a significant correlation between the personality aspects from the WISC test results and the emotional indicator score of the Koppitz system from the HFDs, as presented in Table 4.

Taking the significance level at $\alpha < 0.01$, it can be seen from the table that none of the IQ scores of the WISC test results significantly correlate with the HFD scores. The result of a statistical analysis of the correlation between the scores of each WISC subtest with the HFD score is presented in Table 5.

The result of the correlation calculation between the WISC subtests and the HFD score shows that there are three subtests which have a correlation with the HFD score at significance level $p < 0.05$. The three subtests are similarities, vocabularies, and symbols. The similarity subtest has a positive correlation with the HFD score with $\rho xy = 0.386$. The vocabulary subtest also has a positive correlation with the HFD score with $\rho xy = 0.391$. While the symbol subtest has a negative correlation with the HFD score with $\rho xy = 0.340$.

5 DISCUSSION

Discussion of the results of the data analysis obtained in this study needs to consider the limitations present in the study. The main limitation was the small amount of data as a result of the small number of study participants. One of the main conditions of research using survey methods is a large and representative sample (Cresswell, 2012). Only one school was willing to be a participant but with a limited time allocation so that the total amount of data collected was also very limited. Given the allocation of time allowed by the school, this implementation of WISC only included 11 subtests. This procedure can be used when it is not possible to conduct a complete set of 12 WISC subtests (Wechsler, 1993). The computation of the WISC score is sourced from the 11 subtests. Researchers did a proration to calculate verbal IQs and performance IQs. Although there is a slight difference in scores, results of previous research (Glass et al., 2008) show that a proration analysis of the verbal comprehension index and the performance reasoning index of WISC is highly correlated with the same index analysis of linearly performed tests.

The results of the data analysis show that, in general, the WISC IQ score, consisting of full-scale IQ, verbal IQ, and performance IQ, does not correlate with the total HFD score. The total HFD score is a sum of the number of emotional indicators that appear on the production of the human figure drawing created by the subject. The WISC IQ score is divided into three components; verbal IQ, performance IQ and full-scale IQ. A large discrepancy between verbal IQ and performance IQ is one source of clinical interpretation. A discrepancy of greater than 15 points is an indication of the subject having emotional or neurological problems (Groth-Marnat, 2009). Research in antisocial communities (Isen, 2010) showed discrepancies of up to six points between performance IQ and verbal IQ in adolescents which correlated to antisocial behavior. In the adult community, discrepancies of up to three points between performance IQ and verbal IQ already indicates a correlation with antisocial behavior. However, in the group of children, the discrepancy between the two scores is so enlarged that it can be negligible. The results of this research can explain the results of the statistical analysis which shows no relationship between the difference score between PIQ and VIQ and the HFD score.

The total IQ score of the subjects did not correlate with the HFD score, nor did the verbal IQ score, the IQ performance score, or the difference between the IQ verbal score and the IQ performance score. The results of this statistical analysis reinforce the results of previous studies which conclude that human figure drawings are not a valid measurement of child intelligence (Wilcock et al., 2011), especially in older children or adolescents (Ngernyam, 2014). This explanation is acceptable within the scope of the link between intelligence test results as a measure of cognitive ability and the production of human figure drawing.

The personality aspects of the WISC test that correlate with human figure drawings should be seen from three subtests that correlate significantly with the HFD score. The three subtests are the symbols subtest, the vocabulary subtest, and the similarities subtest. The vocabulary and similarities subtests are tests of verbal abilities. These two subtests have a positive relationship with the emotional indicator score of the HFD test. Research conducted by Gigi (2016) concluded that the production of human figure drawings is also influenced by cognitive style. People with a verbal cognitive style tend to add more detail in their human figure drawings. The emotional indicator in the Koppitz system consists of items with adding detail so as to enable people with a verbal cognitive style to get a bigger emotional indicator score. This linkage can also be explained through the theory that people with good verbal skills and understanding will develop a more critical attitude to their environmental situation (Groth-Marnat, 2009).

In slight contrast to the relationship between the vocabulary and similarities subtests and the HFD score, the symbol subtest has a negative relationship with the HFD score. The intelligence aspect of the symbol subtest contains elements of visual-motor ability as well as the personality aspect of emotional stability and complete comprehension of tasks (Groth-Marnat, 2009). Based on previous research (De Klerk et al., 2014), a comprehensive understanding of the tasks and duties is also needed to recognize the emotional expressions of others so as not to create situations that could trigger a conflict within the environment.

6 CONCLUSION

In general, there is no correlation between the personality aspects of the WISC intelligence test and the emotional indicator scores obtained from the Koppitz scoring system of the human image created by elementary school students in this study. There are only three subtests of the WISC with a personality aspect which correlates with the HFD score. Based on the limitations of this study, more research is needed to explore the benefits of the HFD test as a child psychological assessment tool.

REFERENCES

Cohen, R.J. & Swerdlik, M. (2009). *Psychological testing and measurement: Introduction to test and measurement* (7th ed.). New York, NY: McGraw-Hill.

Cresswell, J.W. (2012). *Educational research: Planning, conducting, and evaluating quantitative and qualitative research* (4th ed.). Boston, MA: Pearson.

De Klerk, H.M., Dada, S. & Alant, E. (2014). Children's identification of graphic symbols representing four basic emotions: Comparison of Afrikaans-speaking & Sepedi-speaking children. *Journal of Communication Disorders, 52*, 1–15.

Dey, A. & Gosh, P. (2016). Do human figure drawings of children and adolescents mirror their cognitive style and self-esteem? *The International Journal of Art and Design Education, 35*(1), 68–84.

Einarsdottir, J., Dockett, S., & Perry, B. (2009). Making meaning: Children's perspectives expressed through drawings. *Early Child Development & Care, 179*(2), 217–232.

Gigi, A. (2016). Human figure drawing (HFD) test is affected by cognitive style. *Clinical and Experimental Psychology, 2*(1), 1–3.

Glass, L.A., Ryan, J.J., Bartles, J.M., & Morris, J. (2008). Estimating WISC-IV indexes: Proration versus linear scaling. *Journal of Clinical Psychology, 64*(10), 1175–1180.

Gregory, R.J. (2011). *Psychological testing: History, principles, and applications*. Boston, MA: Allyn and Bacon.

Groth-Marnat, G. (1984). *Handbook of psychological assessment*. New York, NY: Van Nostrand Reinhold

Groth-Marnat, G. (2009). *Handbook of psychological assessment* (5th ed.). Hoboken, NJ: John Wiley & Sons.

Hermann, D.S. (2015). Distress screening in childhood cancer patients. *Journal of Infant, Child, and Adolescent Psychotherapy, 14*(2), 129–142.

Imuta, K., Scarf, D., Pharo, H., & Hayne, H. (2013). Drawing a close to the use of human figure drawings as a projective measure of intelligence. *PLOS One, 8*(3), 1–8.

Isen, J. (2010). A meta-analytic assessment of Wechsler's P > V sign in antisocial populations. *Clinical Psychology Review, 30*, 423–435.

Koppitz, E.M. (1968). *Psychological evaluation of children's human figure drawings*. New York, NY: Grune & Stratton.

Metin, O. & Ustun, E. (2010). Reflection of sibling relationships into the kinetic family drawings during the preschool period. *Procedia Social and Behavioral Sciences, 2*(2), 2440–2447.

Miller, J.M. (2006). *Relationship of human figure drawing with executive functioning and achievement* (Unpublished thesis, Rochester Institute of Technology, New York).

Neuman, W.L. (2014). *Social research methods: Qualitative and quantitative approaches* (7th ed.). London, UK: Pearson Education.

Ngernyam, N. (2014). Comparison of IQ derived from objective and projective human figure drawing. *International Journal of Child Development and Mental Health, 2*(1), 7–20.

Skybo, T., Ryan-Wenger, N. & Su, Y.H. (2007). Human figure drawings as a measure of children's emotional status: Critical review for practice. *Journal of Pediatric Nursing, 22*(1), 15–28.

Sorin, R. (2005). Changing images of childhood: Reconceptualising early childhood practice. *International Journal of Transitions in Childhood, 1*, 12–21.

Wechsler, D. (1993). Buku petunjuk wechsler intelligence scale for children [Wechsler intelligence scale for children]. Yogyakarta, Indonesia: Faculty of Psychology, Universitas Gadjah Mada.

Wilcock, E., Imuta, A.K., & Hayne, H. . (2011). Children's human figure drawings do not measure intellectual ability. *Journal of Experimental Child Psychology*, 110(3), 444–452.

Do demographics really affect lecturer performance?

S. Sukirno
Universitas Negeri Yogyakarta, Indonesia

ABSTRACT: As a developing country, Indonesia is facing problems regarding low lecturer performance. This study explored the effects of demographics on lecturer performance in higher-education institutions in Indonesia. A total of 347 usable questionnaires were obtained which was about a 46.3% rate of return. Descriptive statistics by cross-tabulation were employed to explore the demographic characteristics in detail, related to lecturer performance within universities. Regression analysis was employed to test the research hypothesis. This study shows that the effect of demographics varies when viewed from the level of awards and performance of lecturers The results are discussed and recommendations for the future are provided.

1   INTRODUCTION

In today's economic knowledge, the importance of education has been recognized worldwide. Within the education system of any country, teachers have a vital position, as the success of educational institutions is mostly dependent on the teachers who educate the most valued assets of the country, students; therefore, the teachers' performance is a fundamental concern for all educational institutions (Khan et al., 2012). The success of students in the classroom learning process cannot be separated from the role and competency of the teaching staff. Competency is an underlying characteristic of a person, relating to the effectiveness of an individual's performance in their job, or the basic characteristics of individuals who have a causal relationship or a cause and effect with the criteria referenced, that is effective or excellent or superior performance in the workplace or in certain situations (Hakim, 2015).

Based on a meta-analytic review of over 30 years of study, experience–performance and age–performance relationships are a critical concern for theory, research and practice. Interestingly, although work experience and employee age are often positively correlated, their respective effects on performance are different (Fu, 2009).

Aktas et al. (2013) have also proven that gender and experience affect lecturer performance. Nevertheless, Alexander (2004) found that teacher experience produces a negative coefficient and did not significantly distinguish teacher performance in Texas. Schools with more teachers with teaching experience of five or more years achieved better results than schools with more teachers with less than five years of teaching experience (Adeyemi, 2008). Similarly, people who have been in one or more positions with a company for longer than other employees are more productive (McEnrue, 1988).

Ackah and Heaton (2003) found that the careers of men and women do differ, with men receiving more internal promotions while women were more likely to seek career progression in another organization and were less successful in terms of earnings. Watson (2003) studied gender-based differences in the financial performance and business growth of small and medium-sized Australian enterprises and also found that female lecturers tend to perform better than male lecturers. In contrast, Gunbayi (2007) concluded that there was no significant difference in rewards based on gender, marital status, educational levels or seniority levels of the teachers from a sample of 204 teachers from urban high schools in city centers in Afyon and Usak in the west of Turkey.

Omolayo and Owolabi (2007) found no significant difference in reward system practices, Performance Development and Management (PDM), lecturer satisfaction, lecturer commitment and lecturer performance according to education level. In contrast, Peterson et al. (2003) explained that personal characteristics such as the number of years in education could be used to predict the level of employee commitment. Therefore it means that educational level has not yet been considered as an element in determining a reward system, as an indicator of lecturer performance, or as a measure of lecturer satisfaction and lecturer commitment.

Research by Mustafa and Othman (2010) in Indonesia found that, on average, there is a significant difference in teaching performance between male and female teachers. Male teachers had lower teaching performance compared to their female counterparts. Apparently, people working in public universities felt safer in terms of comfortable working conditions, stable income, and future career, so that they were more satisfied and performed better than people working in private universities. Surprisingly little empirical evidence is available regarding the predictive validity of demographics on lecturer performance in Indonesia. On the basis of the literature review, it is hypothesized that 'Demographics (i.e. gender, university status, age, experience, educational level) affect lecturer performance in Indonesia'.

2 METHODS

A total of 347 questionnaires were returned and could be further analyzed, which was about a 46% rate of response. Approximately 60% (210) of the samples were male and 40% (137) were female. This research found that 39% of the respondents work in public universities and 61% in private universities. Based on the lecturers' educational attainment level, the majority (70%) of lecturers held a master's degree and many (approximately 25%) had obtained a bachelor's degree. Only about 4% of lecturers held doctoral degrees and the rest (approximately 1%) of the lecturers held only a diploma degree or lower.

In order to collect the data relating to the three factors influencing job performance, a semi-open-ended questionnaire with self-rating was employed. Lecturer performance was measured using six items from the study of Smeenk et al. (2008). Respondents individually assessed their performance by assigning only one choice, from 1 = Very poor to 5 = Very good. The six items measure the lecturer performance load of more than 0.50 and are in in one factor.

Descriptive statistics by cross-tabulation was employed to explore the demographics characteristics in detail relating to lecturer performance in universities. The effect of age, gender, and entry qualifications on academic performance (measured as the final coursework mark obtained) was determined. In addition, a regression analysis was also conducted to investigate the different impacts of demographics on lecturer performance.

3 RESULTS AND DISCUSSION

Table 1 briefly describes the data relating to the response rate of this research. More respondents came from public universities than from private universities. The ratio was 49% of respondents from public universities and 45% from private universities.

The response rate and usable rate of this survey were, respectively, 55% and 46%. Working closely with the universities for three months, approximately 347 usable questionnaires (matched between raters) were returned and could be further analyzed, which was approximately a 46% rate of return. The researcher decided to exclude the unmatched questionnaires from the analysis (Dechawatanapaisal, 2005).

The data shows that 57% (199) of the respondents were male and 43% (148) were female. The number of peers in the 25-years-of-age group is small (2%), as is the number in the older-than-57 group (7%). The majority of peers were in the middle age ranges (15% for 25–33 years; 22% for 33–41 years; 36% for 41–49 years; 18% for 49–57 years).

Based on the data analysis presented in Table 1, demographics have a significant impact on lecturer performance, but only university status contributes significantly. Lecturers work-

Table 1. Impact of demographics on lecturer performance.

Variable	Unstandardized coefficients	Standardized coefficients	T	Sig.
(Constant)	3.860		18.533	0.000
Gender	−0.007	−0.006	−0.102	0.919
University status	−0.172	−0.143	−2.625	**0.009**
Age	−0.018	−0.038	−0.459	0.646
Education level	−0.004	−0.004	−0.074	0.941
Experience	0.038	0.117	1.408	0.160

R	R-squared	Adjusted R-squared
0.181	0.033	0.019
SEE	F	Sig.
0.572	2.314	0.044

ing in public universities tend to have a lower performance compared to lecturers working in private universities. The majority of demographic categories have no impact on lecturer performance. This finding is similar to a previous observation conducted by Ashraf et al. (2009).

Furthermore, this finding supports earlier research by Akiri and Ugborugbo (2008) that found no significant difference in the productivity of male and female teachers in secondary schools in Delta State, Nigeria, although the male teachers were generally more productive than their female counterparts. In terms of teaching experience, nevertheless, this study contradicts the study conducted by Adeyemi (2008), and Marfu'ah et al. (2017), which concluded that a teacher's teaching experience has a significant impact on their lecturing performance.

4 CONCLUSIONS AND RECOMMENDATIONS

Pertaining to the main research objective, it can be generally concluded that gender, age, experience, and level of education do not significantly affect lecturer performance in higher-education institutions in Indonesia. However, university status, perhaps surprisingly, does have a significant effect on lecturer performance. Following on from these research findings, public universities in Indonesia should be able to learn from their private counterparts in managing their lecturers' performance.

REFERENCES

Ackah, C. & Heaton, N. (2003). Human resource management careers: Different paths for men and women? *Career Development International*, 8(3), 134–142.

Adeyemi, T.O. (2008). Teachers' teaching experience and students' learning outcomes in secondary schools in Ondo State. *Nigeria Educational Research and Review*, 3(6), 204–212.

Akiri, A.A. & Ugborugbo, N.M. (2008). An examination of gender's influence on teachers' productivity in secondary schools. *Journal of Social Science*, 17(3), 185–191.

Aktas, M., Kurt, H., Aksu, O., and Ekici, G. (2013). Gender and experience as predictor of biology teachers' education process self-efficacy perception and perception of responsibility from student success. *International Journal on New Trends in Education and Their Implications*, 4(3), 37–47.

Ashraf, M.A., Ibrahim, Y. & Joarder, M.H.R. (2009). Quality education management at private universities in Bangladesh: An exploratory study. *Jurnal Pendidik dan Pendidikan*, 24, 17–32.

Dechawatanapaisal, D. (2005). *The effect of cognitive dissonance and human resource management on learning work behavior for performance improvement: An empirical investigation of Thai corporations* (Doctoral dissertation, Asian Institute of Technology, Bangkok, Thailand).

Fu, F.Q. (2009). Effects of salesperson experience, age, and goal setting on new product performance trajectory: A growth curve modeling approach. *Journal of Marketing Theory and Practice, 17*(1), 7–20.

Gunbayi, I. (2007). School climate and teachers, perceptions on climate factors: Research into nine urban schools. *The Turkish Online Journal of Educational Technology, 6*(3), 70–78.

Hair, J.F., Black, W. C., and Babin, B. J. (2006). *Multivariate data analysis.* Singapore: Pearson Prentice Hall.

Hakim, A. (2015). Contribution of teacher competence (pedagogical, personality, professional competence and social) on the performance of learning. *The International Journal of Engineering and Science (IJES), 4*(2), 1–12.

Anwar, K., ., Ishak, M. I., and Khan, S. (2012). Teachers' stress, performance & resources: The moderating effects of resources on stress & performance. *International Review of Social Sciences and Humanities, 2*(2), 21–29.

Marfu'ah, S., Djatmiko, I.W., & Khairudin, M. (2017) Learning Goals Achievement of a Teacher in Professional Development. *Jurnal Pendidikan Teknologi dan Kejuruan, 23*(3), 295–303.

McEnrue, M.P. (1988). Length of experience and the performance of managers in the establishment phase of their careers. *Academy of Management Journal, 31*(1), 175–185.

Mustafa, M.N. & Othman, N. (2010). The effect of work motivation on teacher's work performance in Pekanbaru Senior High Schools, Riau Province, Indonesia. *Sosiohumanika, 3*(2), 259–272.

Omolayo, B. & Owolabi, A.B. (2007). Monetary reward: A predictor of employees' commitment to medium scale organization in Nigeria. *Bangladesh e-Journal of Sociology, 4*(1), 42–48.

Peterson, D., Puia, K.G.M. & Suess, F.R. (2003). Yo tengo la camiseta (I have the shirt on): An exploration of job satisfaction and commitment among workers in Mexico. *Journal of Leadership & Organizational Studies, 10*(2), 73–88.

Smeenk, S., Teelken, C., Eisinga, R. & Doorewaard, H. (2008). An international comparison of the effects of HRM practices and organizational commitment on quality of job performances among European university employees. *Higher Education Policy, 21*(3), 323–344.

Watson, J. (2003). SME Performance: Does gender matter? In *Proceedings of the 16th Annual Conference of the Small Enterprise Association of Australia and New Zealand, 28 September – 1 October 2003, University of Ballarat, Victoria.*

Character Education for 21st Century Global Citizens – Retnowati et al. (Eds)
© 2019 Taylor & Francis Group, London, ISBN 978-1-138-09922-7

The meaning of cleverness: Elementary school students' perspectives

A. Listiara & D. Rusmawati
Universitas Diponegoro, Indonesia

ABSTRACT: In family life and school settings, the sayingling of a child as clever or not clever becomes one of the risk factors that can interfere with students' characters and learning experiences. This descriptive study was conducted to explore elementary school students' views about the meaning of cleverness. A total of 67 second-graders were involved as participants. This study utilized an open-ended questionnaire. Data analysis was achieved using a descriptive analysis method. The result revealed that the concept of clever children is not only related to academic or school activities, but also to moral-related activities, strategies for self-development, orientation towards social life, and the need to obtain help from parents, teachers or significant others who are deemed to be more competent. This phenomenon is discussed in this paper.

1   INTRODUCTION

The psychometric approach states that psychometric instruments that measure intelligence, and related constructs such as scholastic aptitude tests, school achievement tests, and specific abilities, are popular in European and American life (Neisser et al., 1996). Instruments such as intelligence tests usually measure a person's general intelligence. The measurement results show a certain Intelligence Quotient (IQ) score. The results of these measurements often lead people to make rudimentary judgments on the intellectual level of children, namely clever or smart, or not clever, even though the grade classification shown through IQ scores is Below average, Average, Above average, and Superior.

Many scholars argue that simply to value the concept of intelligence through intelligence test scores means ignoring many important aspects of mental ability (Neisser et al., 1996). Intelligence includes not only the ability to adapt to the environment (adaptation), but also the ability to modify the environment to suit oneself (shaping), and the ability to discover new environments that are more appropriate for one's skills, values, and desires, or choice (Sternberg & Grigorenko, 2004).

Because intelligence is adaptive and culture-specific (McDevitt & Ormrod, 2013), intelligence will manifest itself differently, depending on individual experience, according to the different social and cultural settings that surround it (Cocodia, 2014; Cole, as cited in Stevenson, 2010). The difference in the concept of intelligence between the Eastern and Western world still continues even today (Sternberg & Grigorenko, 2004). Culture provides a specific and unique information system and meaning in the environment in which the culture emerges (Matsumoto, 2007). This leads to a recommendation that the meaning of intelligence and measurement of one's intellectual abilities should take into account the actual and everyday circumstances encountered by the individual (Neisser et al., 1996; Stern, 1999). Culture also provides the basic guidelines for behavior (Matsumoto, 2007). This suggests a new understanding in which intelligence has an indigenous meaning and conception for a particular group of people sharing a common culture (Cocodia, 2014).

The parents of first-grade students argue that schooling a child means implementing the government's program, then hoping that the child becomes clever, knowledgeable, gets good grades, develops his/her talents, and becomes employable (Listiara & Ariati, 2011). The school is expected to develop the intelligence and talents of children so that their lives can be successful because it will help them to get a job in the future. It is quite possible that this parental expectation is conveyed to the child. However, it is important to note that parents' or teachers' praise for children's intelligence can undermine children's motivation and performance. It can be concluded that the children become more concerned with performance-oriented goals, leading to discrepancy (Ilhan-Beyaztas & Dawson, 2017; Mueller & Dweck, 1998).

Having intelligence enables us to adapt well to our everyday environment (Sternberg, as cited in Stevenson, 2010). Anthropologists, sociologists, and psychologists have long been interested in cognitive function or how non-Western societies think (Stern, 1999). Because every child is intelligent to a certain degree (McDevitt & Ormrod, 2013), we assume that individual, social, and cultural diversities will also affect how primary school students think about the concepts of cleverness. Accordingly, this research proposes to understand the meaning of cleverness based on students' perspectives. This is necessary because their understanding of the concept originates from their own experiences.

1.1 *Concepts and perceptions of intelligence: Research in some cultures*

Research in Turkey, with student teachers as participants, stated that being able to understand math is an indicator of intelligence (Ilhan-Beyaztas & Dawson, 2017). However, research on the Canadian Cree community's indigenous concepts of being smart or intelligent found that the nearest equivalent qualities are being wise, respectful, and attentive (Berry & Bennet, as cited in Stevenson, 2010). For another Canadian community, being smart means being able to behave appropriately, like adults, for instance, regularly getting work on time, trying to make others feel comfortable, and being able to avoid conflict. Being smart could also mean being physically active, for example, doing regular exercise or physical activity by learning hunting and navigation skills for men, or sewing for women, and recognizing the limits to the capacity to endure pain. Individuals who are not smart are the ones who demonstrate the potential to engage in behaviors or actions that can harm himself or others (Stern, 1999).

From the Confucian perspective, the image of a clever person is one who dedicates themselves to character cultivation so that they will be able to embody benevolence and act according to what is right. An intelligent person dedicates a great deal of their time to learning. They are able to enjoy learning and persist in life-long learning with a high degree of enthusiasm (Yang & Sternberg, 1997).

Skills and behaviors that are valued and supported in a particular society are very likely to differ from behavioral demands in other cultures (Stern, 1999). Smiling expressions that are often associated with positive traits and give comfort can lead to different reactions. A smiling face leads to an attribution of higher intelligence in 18 out of 44 cultures involved in a cross-cultural study of the linkage between smiles and perceptions of intelligence. Nevertheless, six other cultures showed different results, that is, smiling faces were perceived as significantly less intelligent (Krys et al., 2016). A study of the perception of intelligence among college students in the United States found that individual characteristics, such as age and facial expressions, affect perceptions of the individual's intelligence, compared with the use of glasses that are often assumed to show a person's intelligence (Borgen, 2015).

The phenomenon above shows that different cultural factors will place a different emphasis on the varieties of expression of one's intelligence (Stern, 1999). These studies are very useful for developing theories derived from the experiences of people from different cultures.

1.2 *How primary-grade students think*

Primary school students (grades 1, 2, and 3), aged seven to nine years old, begin to understand that there are many ways to know something, more than others. They also begin to understand that learning and remembering activities are influenced by the ability to think that they can establish control, that is, by making specific efforts. However, in accordance with their motor development, these children have not been able to learn effectively compared with older students (Snowman & McCown, 2012). They are also in transition from the pre-operational stage to the concrete operational stage Step by step, they will learn how to solve problems based on concrete experience (McDevitt & Ormrod, 2013).

The primary grade is children's first school experience and they will be eager to learn to read and write (Snowman & McCown, 2012). Classroom learning situations, social interaction with teachers, peers, and administrative personnel, and learning the school rules all provide useful experiences for their cognitive, moral, social, and emotional development. In the Holman Inuit community in Canada, children become clever by learning through observing and experiencing (Stern, 1999). This finding emphasizes that children's cognitive development arises through conversation and interaction with adults and peers whom they perceive to be more capable in the culture in which they live (Vygotsky, as cited in Woolfolk, 2007).

2 METHOD

This study was conducted using a descriptive method. The data was collected using instruments that were adapted to real conditions and cultures that were part of the participants' daily life. Data from this study will be analyzed descriptively because it is intended to provide information and a description of the meaning of cleverness from the perspective of elementary school students.

2.1 Participants

This research involved 67 second-graders (28 boys and 39 girls) from one public elementary school in Semarang city. The children ranged in age from seven to nine years old, their mean age was 7.9 years old (SD = 0.53). All participants lived with their families. A total of 56.71% of participants had extended family members also living in the family home.

2.2 Instrument

The data was collected using a questionnaire consisting of four open-ended questions relating to the children's understanding of cleverness (*pintar*) and what makes a clever student from their point of view. We posed questions such as 'how can you become clever?' or 'who helps you to be clever?' All of the questions were in the Indonesian language. The participants were encouraged to write more than one answer to each question.

We asked two senior educational psychology lecturers from the Faculty of Psychology of the Universitas Diponegoro to make a professional judgment regarding the face validity of the instrument.

2.3 Data analysis

The data was analyzed by means of categorization of the various responses of participants based on the research focus. We wanted to explore the meaning of cleverness, therefore we used the analytical method suggested in the indigenous psychology research of Anggoro and Widhiarso (2010). Data analysis is achieved by: 1) preparing the data to be analyzed by collecting the participants' answers; 2) re-reading all the answers that have a similar meaning so as to extract common themes; 3) selecting the theme groups that appear by repeated reading in accordance with the focus of research; 4) excluding answers that are inconsistent with the research focus; 5) in the final stages, selecting thematic groups and reorganizing them according to their frequency of occurrence.

3 RESULTS AND DISCUSSION

The data analysis gave several categorized themes that described the participants' understanding of: 1) the definition of clever children, 2) strategies to become a clever student, 3) goals to be achieved by being clever, and 4) people that help you to become clever.

3.1 The definition of clever children

It is interesting that when the second-graders described behaviors that are depicted as clever, they mentioned studying as a behavior that was perceived as clever. These studying behaviors included studying earnestly, studying at home and at school, studying group participation, and studying every day. It might not have been easy for the participants to describe the concept of cleverness because their cognitive development is still progressing towards concrete operational stages. In this study, we continued to help students by giving individual instruction to enable them to accomplish tasks with joy.

Two major themes emerged from the participants' varied answers. First, the theme of conducting academic-related activities, for instance studying, reading, attending school, writing, and doing homework. Second, the theme of performing moral-related activities, such as helping parents, taking parental advice, not cheating on examinations, not playing computer games, and helping friends. The descriptive scores related to this question are given in Table 1.

It is common for children at ages seven to nine years old to have difficulty in providing answers to questions about abstract concepts such as 'how can you become clever?' Even if their answers described studying hard, the meaning of studying hard is still an abstract concept. However, responses such as daily learning or studying earnestly, study group participation, studying at school, or studying at home, looked more concrete, as well as other participants' responses such as reading, writing, counting, and doing homework.

The activity of counting does not seem to dominate the participant's response. This phenomenon appears to differ from the characteristics of cleverness defined by the Turkish student teachers who prioritized the understanding of mathematics lessons (Ilhan-Beyaztas & Dawson, 2017). The theme of moral-related activities in participant responses about the meaning of cleverness is similar to the results of previous research which found that being clever also means taking care of oneself, not harming oneself or others, being friendly, and being responsive to change (Berry & Bennet, as cited in Stevenson, 2010; Stern, 1999; Yang & Sternberg, 1997).

3.2 Strategies to become clever

The participants were aware that there are many ways to be clever. The development of their cognitive abilities allowed them to choose their preferred means. Studying was the most frequent answer from the participants (92.54%). In addition, there are activities that are perceived differently from what students understand to learning activities to be smart

Table 1. Ideas about clever children (N = 67).

Theme	Response	fq	%
Academic-related activities	Studying	65	97.01
	Reading	13	19.40
	Attending school	6	8.96
	Writing	5	7.46
	Doing homework	3	4.48
	Listening to the teacher	2	2.99
	Seeking knowledge	2	2.99
	Painting	1	1.49
	Going up a class	1	1.49
	Counting	1	1.49
Moral-related activities	Helping parents	4	5.97
	Obeying parents' advice	2	2.99
	Not cheating	2	2.99
	Not playing computer games	2	2.99
	Helping friends	1	1.49
	Persistence	1	1.49
	Diligence	1	1.49

We categorized these into two major themes: academic strategies (such as studying, reading, attending school, writing, and counting) and non-academic or self-development strategies (obeying parents' advice, not playing computer games, helping parents, tidying up their own uniforms, dressing themselves, and taking a bath without any help). The distribution of participants' opinions about strategies to become clever is shown in Table 2.

Participants responded with similar answers that they will not play or spend much time playing games as moral-related activities (2.99%) in concepts of smart kids and as one of non-academic or self-development strategies (4.48%) in strategies to be clever. In this paper, 'playing' means playing electronic games such as on a Personal Computer (PC), laptop, or mobile phone.

Some Indonesians seem to spend time playing games on cell phones or laptops; there are many internet shops that children visit because they want to play games without their parents' knowledge. However, Montessori (as cited in Woolfolk, 2007) says that 'play is children's work'. The human brain develops with stimulation and play, for children, will actually provide stimulation at every stage of their development. By playing, children have the opportunity to practice, to have fun, to imagine, to learn the rules, and to cooperate.

As well as the role of teachers, parental involvement is especially important in providing the rules and support that help children to play properly. There are many challenging traditional games that can help children to improve themselves without having to compete through a computer screen or hold a cell phone all the time.

3.3 Reasons to be clever

Another interesting result can be seen from the participants' responses on the purpose of being clever. Four major themes emerged, including in-school goals, after-school goals, social life goals, and other. Responses were dominated by the goal of going up a class (34.33%). It is a real goal for most students that by studying and learning, he/she either wants to move up a class, graduate, or move on to the next level. Nevertheless, an unexpected response to this third question arose from the participants, that cleverness is needed so that they can achieve their goals, for the future, to be successful, to apply for jobs, to earn money, and to register for schools. The perception that clever means successful in many things has been found in research involving student teachers from Turkey (Ilhan-Beyaztas & Dawson, 2017). The responses also indicated that being clever is also needed in order to be liked by friends and can make parents happy (see Table 3).

It is important to consider that children do not think like adults (Piaget, as cited in Woolfolk, 2007). There were participants who perceived cleverness to be important for finding jobs, making money, and registering for schools. In this study, parents' jobs and educational

Table 2. Strategies to become clever (N = 67).

Theme	Response	fq	%
Academic strategies	Studying	62	92.54
	Loving to read	18	26.87
	Attending school	6	8.96
	Writing	4	5.97
	Counting	2	2.99
	Listening to the teacher	2	2.99
	Doing homework	2	2.99
	Drawing & coloring	1	1.49
	Taking courses	1	1.49
Non-academic/self-development strategies	Obeying parents' advice	3	4.48
	Not playing computer games	3	4.48
	Helping parents	1	1.49
	Tidying up uniform by him/herself	1	1.49
	Getting dressed by him/herself	1	1.49
	Taking a bath by him/herself	1	1.49

Table 3.   Goals for becoming clever (N = 67).

Theme	Response	fq	%
In-school goals	Going up a class	23	34.33
	Earning good test scores	9	13.43
	Not being stupid	6	8.96
	Having a good ranking	5	7.46
	Not cheating	3	4.48
After-school goals	Achieving success in the future	16	23.88
	Applying for jobs and making money	3	4.48
	Registering for or entering school	2	2.99
Social life goals	Being liked by friends	1	1.49
	Making parents happy	1	1.49
	Teaching younger siblings	1	1.49
Other	Requirement	1	1.49

background have not been considered. It is possible that ease of access to information and technology could be considered as an important factor for a child's cognitive development. These conditions will affect how well individuals will adapt, modify, and select their environments (Sternberg & Grigorenko, 2004).

In the Indonesian cultures known to some Western societies, people enjoy being involved in long and fruitless conversations (Lewis, 2006). In fact, children will easily become involved in conversations between adults, such as their parents and neighbors, when they are talking about the complexity of finding schools if the children are not qualified or their grades are considered bad. Children are also often exposed to information from television programs, newspapers, or other information media about the narrowness of job opportunities or the lack of professions through which it appears to be easy to make money.

3.4   *People who help you to be clever*

The last research question about people who help the children to be clever elicited five main response themes. The themes were parents and teachers (both of them), family, teachers, relatives, and others. Most of them answered that both parents and teachers (37.31%) were the most important figures. Responses that mentioned either parents or teachers were the next most frequent (29.85%). Other family members, such as brothers and sisters, as well as grandparents and relatives of their parents also contributed in helping the children. Besides the teachers, friends and students who rent rooms in their parents' homes were perceived to help the children become clever. The distribution of participants' opinions about people who help them to become clever is shown in Table 4.

Based on the above description, family members, relatives, peers, teachers, and people around the child, whether planned or through daily events, have shown behaviors and give external support to behaviors that should or should not be done if the child wants to be regarded as clever The phenomenon emphasizes the fact that social interaction affects cognitive development because social interaction creates cognitive structure and thinking processes (Palincsar, as cited in Woolfolk, 2007).

3.5   *Limitations of the study*

The number of participants involved is very likely to require enlargement to support conclusions about the meaning of cleverness in accordance with the Indonesian cultural background. We also recommend involving participants with higher levels of cognitive development. This research has not yet achieved the preparation of constructs of cleverness that can be developed as instruments to measure the indigenous characteristics of Indonesian students.

Table 4. People who help children to be clever (N = 67).

Theme	Response	fq	%
Parents & Teachers	Both of them	25	37.31
Family	Parents (only)	20	29.85
	Siblings	17	25.37
Teacher	Teachers (only)	20	29.85
Relatives	Uncles/aunts/grandparents/cousins	9	13.43
Others	Peers/boarding students	3	4.48

4 CONCLUSION

The definition of cleverness has its own meaning that children learn from their family, teachers, friends, and people around them. Social interaction with the environment will help children to build their knowledge. Thus, it is expected they will know the rules as to which of the behaviors should and should not be done in order to become a clever child. The children perceived that their parents and teachers were the main people involved in helping them to be clever. It is recommended that parents or families, teachers, and significant others should provide appropriate support and motivation to the children. This will help the children to enjoy learning and perform useful games in order to promote their good characters.

ACKNOWLEDGMENTS

We would like to express our gratitude to the Faculty of Psychology, Universitas Diponegoro, for the support of research funding for this study. We are also grateful to the participants, the parents, and the school. Thanks to Rheza and Ikko for designing the layout of the questionnaire, collecting the data, and transferring the participant responses.

REFERENCES

Anggoro, W.J. & Widhiarso, W. (2010). *Konstruksi dan identifikasi properti psikometris instrumen pengukuran kebahagiaan berbasis pendekatan indigenous psychology: Studi multitrait-multimethod* [The construction and identification of the psychometric property of happiness measurement instrument based on indigenous psychology approach: A study of multitrait-multimethod]. *Jurnal Psikologi, 37*(2), 176–188.

Borgen, A. (2015). The effect of eyeglasses on intelligence perceptions. *The Red River Psychology Journal, 1*, 1–14.

Cocodia, E.A. (2014). Cultural perceptions of human intelligence. *Journal of Intelligence, 2*(4), 180–196. doi:10.3390/jintelligence2040180.

Ilhan-Beyaztas, D. & Dawson, E. (2017). A cross-national study of student teachers' views about intelligence: Similarities and differences in England and Turkey. *Universal Journal of Educational Research, 5*(3), 510–516. doi:10.13189/ujer.2017.050324.

Krys, K., Vauclair, C.-M., Capaldi, C.A., Lun, V.M.-C., Bond, M.H., Domínguez-Espinosa, A., ... Yu, A.A. (2016). Be careful where you smile: Culture shapes judgments of intelligence and honesty of smiling individuals. *Journal of Nonverbal Behavior, 40*(2), 101–116. doi:10.1007/s10919-015-0226-4.

Lewis, R.D. (2006). *When cultures collide: Leading across cultures* (3rd ed.). Boston, MA: Nicholas Brealey.

Listiara, A. & Ariati, J. (2011). *Why schooling? Exploring parents' motives in schooling their children.* Poster presented at the 2nd International Conference of Indigenous and Cultural Psychology, AAICP & Psychology Department, Faculty of Medicine and Health Science, University of Udayana.

Matsumoto, D. (2007). Culture, context, and behavior. *Journal of Personality, 75*(6), 1285–1320.

McDevitt, T.M. & Ormrod, J.E. (2013). *Child development and education* (5th ed.). Upper Saddle River, NJ: Pearson Education.

Mueller, C.M. & Dweck, C.S. (1998). Praise for intelligence can undermine children's motivation and performance. *Journal of Personality and Social Psychology, 75*(1), 33–52.

Neisser, U., Boodoo, G., Bouchard, T.J., Boykin, A.W., Brody, N., Ceci, S.J., Halpern, D.F., Loehlin, J.C., Perloff, R., Sternberg, R.J., and Urbina, S. (1996). Intelligence: Knowns and unknowns. *American Psychologist, 51*(2), 77–101.

Snowman, J. & McCown, R. (2012). *Psychology applied to teaching* (13th ed.). Wadsworth, OH: Cengage Learning.

Stern, P.R. (1999). Learning to be smart: An exploration of the culture of intelligence in a Canadian Inuit community. *American Anthropologist, 101*(3), 502–514.

Sternberg, R.J. & Grigorenko, E.L. (2004). Intelligence and culture: How culture shapes what intelligence means, and the implications for a science of well-being. *Philosophical Transactions of the Royal Society B, 359*(1449), 1427–1434. doi:10.1098/rstb.2004.1514.

Stevenson, A. (2010). *Cultural issues in psychology: A student's handbook*. Hove, UK: Routledge.

Woolfolk, A. (2007). *Educational psychology* (10th ed.). Boston, MA: Pearson Education.

Yang, S.Y. & Sternberg, R.J. (1997). Conceptions of intelligence in ancient Chinese philosophy. *Journal of Theoretical and Philosophical Psychology, 17*(2), 101–118.

Challenges in developing character education at a "risk school" in Yogyakarta, Indonesia

A. Efianingrum
Universitas Negeri Yogyakarta, Indonesia

ABSTRACT: This study uses a qualitative research method to describe social practices, implementation of school rules, and challenges in developing character education at a "risk school" for students at risk of dropping out of education. It was conducted at a private senior high school in Yogyakarta, Indonesia (namely *SMA Gadjah Mada*), and its research subjects were school administrators, teachers, and students. The data collection techniques used were observation and interviews. The results show that the school faced difficulties in developing the character of its students. Students in the school have low academic achievement, poor discipline, and the potential for violence or bullying, and the characteristics of the students do not support the implementation of school rules. School acts as a field for the contestation of values. The challenges in developing character education can be categorized into two groups: structural challenges and cultural challenges. Structural challenges are related to school policies in developing school programs that are appropriate for students' characteristics, while cultural challenges are related to the development of the character values of students.

1 INTRODUCTION

Nowadays, there is an ongoing process of social and culture transformation taking place that is supported by information technology. Tapscott has stated that the advancement of technology and information creates a new generation, while Drucker argues that globalization produces unpredicted realities (cited in Tilaar, 2002). Societies face many risks, and a "risk society" is defined as one that is characterized by the possibility of physical, mental, and social risk caused by such changes (Beck, 1998).

There are various problems in society (Macfud, 2011), such as: 1) disorientation—undetermined direction and objectives in social life; 2) distrust—lack of trust among people and between generations; 3) disobedience and rebellion; 4) disintegration of society. The transformation of society and culture has relevant connections to the process of education.

Education has a primary role in radically changing people's mindsets, ways of life, and behaviors. It is thus a power for the future because it can be an effective means of creating change (Morin, 2005). Russell (1993) explains that the aim of education is to create better people who are good citizens. In the educational context, schools deal with a number of risks, both to themselves as organizations and to their students.

Some research has shown that incidents may be recorded and tallied within schools because they occurred there. Students (as a group) face both personal risks (such as being the victims or perpetrators of violence) and environmental risks (Chapin & Gleason, 2004). Bullying behaviors among students occur at all levels in schools, and the intensity of bullying and victimization has increased (Freeman, 2014; Fu et al., 2013). 'Risky' teenagers are identified as those who come from families with poor intimacy, who demonstrate failure in learning, and who have experienced bad treatment from others (Kingery et al., 1998; Nazim, 2013). Boys experience higher levels of victimization than girls for all forms of school bullying (Kouri-Kassabri et al., 2004). Bullying behavior is associated with substance abuse, self-harm, and the carrying of weapons (Smalley et al., 2017).

Bullying as a social phenomenon is a serious concern for students, parents, teachers, and school officials, and a prevention strategy is seen as a key component in addressing it (Hong & Garbarino, 2012). There are two main discourses around improving education in school. The first is related to the academic achievement discourse that emphasizes the reconstruction process. The second is the cultural discourse that emphasizes aspects of reconstruction (Suyata, 2000).

Another opinion states that two approaches are found in the process of school change. The first is a structural approach that focuses on changes in structural and bureaucratic aspects. The second is a cultural approach that focuses on the culture of excellence, such as changes of mind, words, behavior, and the 'heart' of each school. A cultural approach to improving the performance of a school may be more effective than structural approaches (Sastraprat-edja, 2001). This present article presents a description of a "risk school" for students at risk of dropping out of education, as well as of the structural and cultural challenges faced by schools in dealing with challenging students.

Discussion of character education is an important issue which needs to be developed, notably in risk schools. Lickona explains that character education is paramount in creating positive characters (cited in Darmiyati, 2011). The main aims of education are to ensure that new generations are intelligent and well-behaved. Good characters in students are created from a sustainable process grounded in both theory and practice. Character education varies in its practices, reflecting both assumptions and consequences, and its complex nature cannot easily be simplified (Koesoema, 2012).

Social practices in schools can be explained from Bourdieu's (1993) perspective through the concepts of "habitus" and "field". Bourdieu defines habitus as the system of durable, transposable dispositions and structures predisposed to function as determining structures. School rules are instruments for developing students' characters, but there is no simple approach to such development.

2 RESEARCH METHOD

This research used a qualitative approach to establish the meaning and intentions of the actions of the subjects. This method required the researcher to carry out a careful analysis in order to gain detailed qualitative data and a deep understanding of the subjects. It was conducted at a private senior high school in Yogyakarta, Indonesia (namely *SMA Gadjah Mada*). The selection of this school was based on its characteristics and risks in terms of both physical aspects of its environment that are not conducive and the non-physical environment (sociocultural).

The subjects of this research were school administrators, teachers, and students. The data collection techniques used were observation and in-depth interviews. Observation of the daily life at the school was performed both inside and outside the classroom, and in-depth interviews were conducted with teachers, students, and security staff. The data gathered was analyzed through the following steps: data reduction, data display, and conclusion drawing. The data was validated by using method triangulation, source triangulation, and research duration extension.

3 FINDINGS AND DISCUSSION

Data about the social practices in the school, implementation of school rules, and the challenges faced in developing character education at the participating school are described systematically in the following subsections.

3.1 *Social practice in the risk school*

Most of the students in this school were transferred from other schools. There is an opinion that schools in a city like Yogyakarta are excellent; however, that opinion may change because of cases of vandalism, gang fights, violent activity (*klithih*), and other kinds of violence using weapons that have taken place among students and other violence by using weapons that occur among the students.

The participating school accepts transfer students from other schools. There are also schools that are the destination for students who are not promoted through the grades or who are troublesome. The school has rules of conduct governing student manners and behavior in school. However, it is not easy to apply them in daily life because of characteristics of the students' habits that are different from regular school students, especially from the aspect of learning motivation and discipline.

In the observation, the learning process in the classroom did not go as expected. The number of students who arrived late, did not attend school, or who left early was reflected in low student achievement. In fact, it was often the teachers who waited for their students to arrive. Although only a few students attended classes, the teachers remained patient and eager to teach.

In extracurricular activities, the students were also less enthusiastic to participate, with only a small number of students attending and following the activities. Schools had invited students and tried to contact parents to motivate their children to study at school.

Although few in number, there are physical artifacts in this school that are expected to inspire a spirit of learning. However, the culture in this school showed that the dimensions of positive values and physical artifacts were not present in the everyday life of the school. Based on Bourdieu's (1993) scheme of habitus and field, social practices are not just determined by school systems, but are also a reflection of the habit and capital of students. Most of the students in this school have low levels of intellectual capital, but have strong social capital, such as social networking and solidarity among students.

3.2 *Implementation of school rules*

The school faces difficulties in implementing school rules. The characteristics of the students do not allow the implementation of school rules, because most students have different and particular habits. The background of most of the students is that they come from dysfunctional families, or that the parents pay less attention to their children than is usual.

This school was established and survives because it has a spirit of wanting to help educate the life of the nation, regardless of the status or background of its students. In this school, there is no complicated selection system for prospective students. The system of acceptance of new learners is simply through interviews with prospective students and parents. These aim to find out their motivation for study, their reasons for choosing the school, and the reason why they wish to move to the school.

These students usually have a history of low achievement at school, not because they are of low ability, but because their spirit and motivation to learn is poor. Most of the students at this school have family, economic, or academic problems, or experience the type of personal problems typical of students.

Life and social relationships in this school are generally different from regular schools. Student activities in classroom learning were observed to be less enthusiastically performed than in other schools. Students had low motivation to learn and achieve, and they rarely visited the school library. When school activities were over, students usually interacted with their groups in order to preserve school identity. The conditions within their families, being mostly dysfunctional, also resulted in after-school activities being poorly attended, but not all students went straight home. They had the means, space, and time to spend with friends and other former pupils, and they spent more time out of school.

Socio-historically, the existence of gangs in schools is also a medium for meetings of students and former pupils. The existence of these informal forums strengthens the solidarity of students and alumni, who are members of a gang known as the Gadjah Mada Hooligans.

Social relations among students and between students and teachers showed closeness and intimacy. Different student characteristics encouraged the teachers to choose counseling and mentoring strategies based on more cultural relationships. Students do not hesitate to tell stories and convey complaints to their teachers. A number of violent cases involving students of this school have created a stigma that suggests this school is an "outcast" school with violent students. The ever-present student violence could not be separated from the aggressive habits of students in their words and behavior. Interestingly, because of the solidarity between students at the school, student violence occurs more frequently between schools. Factors in the presence of student gangs in schools and a history of inter-school hostilities also suggest reasons why students in this school were often involved in student violence.

3.3 *Challenges in developing character*

In consequence of the environment described, the attempt to develop students' characters at the risk school faces some challenges. Suyata (2000) mentions two dimensions to such issues, namely the structural and the cultural. Emphasizing education merely in the process of reconstruction is no longer sufficient, because there is a belief that social and cultural systems are core issues as well as the key to success in education. The success of education often relies on intangible factors such as cultural values and beliefs. However, cultural factors are frequently ignored in attempts to improve education.

The challenges in developing character education can be categorized as structural and cultural.

3.3.1 *Structural challenges*

In the context of student character development, the first issue to be addressed relates to structural challenges relating to policies, programs, and school curriculums. Structural challenges apply to the development of school programs that are appropriate to students' characteristics and needs. The solutions to such problems should consider the students as the target subjects of policy, because their characteristics differ from students attending regular schools. In fact, their low motivation to achieve needs a specific and effective character development program, and the non-academic dimension should be a priority in deciding the programs followed. This process is simple to carry out as long as the whole school is included in formulating and implementing such programs. Under the facilities and supervision of the Education Board, the dynamics of programs can be developed and evaluated.

3.3.2 *Cultural challenges*

These challenges relate to issues such as cultural values, mindsets, and the habits of school society, and affect the development of character values in students. The strategy for dealing with such challenges is not simple, because it is connected to the transformation of values and habits. The school should remove negative cultures and initiate positive ones. The continuity of school programs with programs at home should be also be evaluated.

To meet these challenges, an alternative character education program should be developed by considering related aspects. Character education should be modified using multiple approaches, multiple methods, and multiple media that can meet the varied needs of students. The development of students' characters in this school requires cooperation with various parties through innovative programs. In the risk school, the teachers should approach students as their colleagues or dialog partners. However, without simultaneous support from both parents and school, such programs will end in failure.

4 CONCLUSION

Discussing schools has never been simple due to their different characteristics. Every school has different supporting aspects that may improve their quality, and there are various problems related to the input of students, infrastructure, and human resources. Schools such as the one investigated in this study face challenges in developing the character of their students that are unique in terms of student behavior and learning motivation.

Risk schools offer protection for students at risk of dropping out of education. Structural and cultural challenges should be overcome through proper character development programs. Thus, the Education Board should provide the headmaster and teachers of the risk school with proper programs for developing a positive school culture, and should fully support their implementation.

REFERENCES

Beck, J.D. (1998). Risk revisited. *Community Dentistry and Oral Epidemiology*, *26*, 220–225.
Bourdieu, P. (1993). *The field of cultural production: Essays on art and leisure.* New York, NY: Columbia University Press.
Chapin, J. & Gleason, D. (2004). Student perceptions of school violence: Could it happen? *Journal of Adolescent Research*, *19*(3), 360–376.
Darmiyati, Z. (2011). *Pendidikan karakter dalam perspektif teori dan praktek* [*Character education in theory and prespective*]. Yogyakarta, Indonesia: UNY Press.
Freeman, G.G. (2014). The implementation of character education and children's literature to teach bullying characteristics and prevention strategies to preschool children: An action research project. *Early Childhood Education Journal, 42*(5), 305–316.
Fu, Q., Land, K.C., & Lamb, V.L. et al. (2013). Bullying victimization, socioeconomic status and behavioral characteristics of 12th grades in United States, 1989 to 2009: Repetitive trends and persistent risk differentials. *Child Indicators Research*, *6*(1), 1–21.
Hong, J.S. & Garbarino, J. (2012). Risk and protective factors for homophobic bullying in schools: An application of the social-ecological framework. *Educational Psychology Review*, *24*(2), 271–285.
Kingery, P.M., Coggeshall, M.B., & Alford, A.A. et al. (1998). Violence at school: Recent evidence from four national surveys. *Psychology in the Schools, 35*(33), 247–258.
Koesoema, D.A. (2012). *Pendidikan karakter: Utuh dan menyeluruh* [*Character education: Whole and thorough*]. Yogyakarta, Indonesia: Kanisius.
Kouri-Kassabri, M. et al. (2004). The contributions of community, family and school variables to student victimization. *American Journal of Community Psychology*, *34*(3–4), 187–204.
Morin, M.L. (2005). Labour law and new forms of corporate organization. *International Labour Review*, *144*(1), 5–30.
Nazim, A.M. 2013. Ciri-ciri remaja berisiko: Kajian literature [Risk teens features: A literature review]. *Islamiyyat*, *35*(1), 111–119.
Russell, J.E.A. (1993). *Human resource management.* Singapore: McGraw Hill.
Sastrapratedja, M. (2001). Budaya Sekolah [School culture]. *Majalah Ilmiah Dinamika Pendidikan*, *2*(8), 1–17.
Smalley, B.K. Warren, J.C., & Barefoot, K.N. et al. (2017). Connection between experiences of bullying and risky behaviors in middle and high school students. *School Mental Health*, *9*(1), 87–96.
Suyata. (2000). *Refleksi sistem pendidikan nasional dan mencerdaskan kehidupan bangsa* [Reflections on the national education system and the introduction of life of the nation]. Paper presented at Pokja Sistem Pendidikan Nasional untuk Mencerdaskan Kehidupan Bangsa.
Tilaar, H.A.R. (2002). *Perubahan sosial dan pendidikan: Pengantar pedagogik transformatif untuk Indonesia* [*Social change and education: Introduction to transformative pedagogy for Indonesia*]. Jakarta, Indonesia: Grasindo.

Development of a physics test for 10th graders based on Marzano's higher-order thinking skills

E. Ermansah & E. Istiyono
Universitas Negeri Yogyakarta, Indonesia

ABSTRACT: Research was conducted to develop a Marzanoian Higher-Order Thinking Skills (HOTS) physics test for 10th-grade senior high school students. A blueprint was developed based on Marzano's HOTS consisting of 13 indicators, and these were used to craft the test items. The instruments consisted of two sets of tests, each containing 35 items. Seven anchor items were validated by an expert on measurement in physics education, a physics education expert, and practitioners. The instruments were tried out on 437 students from five schools in Sleman District, Indonesia. Polytomous data was analyzed using the Partial Credit Model (PCM). The results show that all of the items are 63 and its instrument are a proven fit with PCM, the reliability of the instruments is 0.77, and the item difficulty index ranges from –0.89 to 0.74, which means all items are in the "good" category. Thus, the Marzanoian HOTS physics test is eligible to be used to measure HOTS in physics in senior high school students.

1 INTRODUCTION

Science education in Indonesia can be described as being of low quality. This is indicated by the findings of both the Trends in International Mathematics and Science Study and the Program for International Students Assessment, which are seen as setting the parameters for the quality of science education (OECD, 2016). One of the common mistakes of assessment processes in Indonesia is that they measure Lower-Order Thinking Skills (LOTS) rather than Higher-Order Thinking Skills (HOTS).

Physics is one of the branches of science education that gives us knowledge about living with nature and improving it through technology. Physics is a core subject in science education in Indonesia, and is the subject of a national examination known as the UNBK *(Ujian Nasional Berbasis Komputer)*.

Assessment is the procedure by which information about student learning outcomes is obtained, and a test may be one of the methods used to perform such assessment. One of the frequently used types of testing is the multiple choice test, a format which offers a number of advantages over other options. Science education in Indonesia includes multiple choice testing of the physics curriculum. However, the physics tests used in senior high schools only measure LOTS. A new model of test should be devised to overcome this weakness by including HOTS. One of the alternatives is a reasoning-based multiple choice test.Such tests are often developed according to Bloom's taxonomy, but a new test based on Marzanoian principles now needs to be developed. Heong et al. (2011) put forward the view that Bloom's taxonomy has limitations in current educational environments. A newer taxonomy, such as that proposed by Marzano, can address these limitations. Marzano's approach accommodates many advanced factors that affect students' thinking, and provides us with theory drawn from real research to help students develop their thinking skills. Based on this, a new Marzanoian physics higher-order thinking skills (physics HOTS) test should be developed.

2 LITERATURE REVIEW

According to the curriculum regulations, every education provider should not only plan and process learning activities, but should also carry out assessment to improve their effectiveness and efficiency. Reynolds et al. (2009) define assessment as a procedure to measure students' knowledge and skills. The opinion of Newman and Wehlage (1993, p. 2) is that "HOTS requires students to manipulate information and ideas in ways that transform their meaning and implications, such as when students combine facts and ideas in order to synthesize, generalize, explain, hypothesize, or arrive at some conclusion or interpretation." Since this idea was proposed, it has been used as a basis for evaluation processes.

A multiple choice test is a preferred format because of its various advantages, such as the fact that one test can address most of the learning materials, answers can be corrected quickly and easily, and the answer to every question is either true or false. It is seen as an objective test method (Sudjana, 1990).

McLoughlin and Luca (2000) state that HOTS are the capacity to go beyond the information given, to adopt a critical stance, to evaluate, to have metacognitive awareness and problem-solving capacities. The characteristics of HOTS identified by Resnick (1987) are that they are non-algorithmic, complex, have multiple solutions, involve decision-making and variation in interpretation, reflect multiple criteria, and require effortful implementation. Conklin (2012) describes the characteristics of HOTS as follows: "higher order thinking skills encompass both critical thinking and creative thinking."

Johnson (2000) puts forward the view that thinking skills can be distinguished as critical thinking and creative thinking. Ennis (1987) defines critical thinking as reflective thinking that focuses on the pattern of decision-making about what should be believed and what must be done. Based on this definition, Ennis's critical thinking theory consists of 12 components: (1) formulating the problem, (2) analyzing the argument, (3) asking and answering questions, (4) assessing the credibility of information sources, (5) observing and assessing the results of observations, (6) making and assessing deductions, (7) making and assessing inductions, (8) evaluating, (9) defining and assessing definitions, (10) identifying assumptions, (11) deciding and implementing, and (12) interacting with others.

In senior high schools in Indonesia, the multiple choice tests used generally measure only lower-order thinking, especially in physics subjects, and often ignore the measurement of higher-order thinking (HOT).

A development that is required is a multiple choice test able to measure HOT. Halaydina and Downing (1989) proposed that a two-tier multiple choice test could be used not only to increase higher-order thinking skills, but also to see the reasoning used by participants to answer the questions posed. One example of two-tier multiple choice testing is reasoning-based multiple choice. A reasoning-based multiple choice test consists of questions with multiple answers which require reasoning to be applied in making the correct choice. Each answer and reasoning consists of five choices. Istiyono et al. (2014) propose that reasoning-based multiple choice questions feature the following assessment characteristics: a score of 1 if both the answer and reasoning are wrong, 2 if the answer is right but the reasoning is wrong, 3 if the answer is wrong but the reasoning is right, and 4 if both answer and reasoning are right.

Megawati (2012) suggests that sometimes teachers makes mistakes in developing tests. First, the test may not pass the reliability and validity test or the threshold and deception effectivity test. Second, the assessment time needed for each quality test is too long. Due to such limitations in developing such tests, there is a need to develop a Marzanoian test. Marzano's taxonomy can address the limitations of Bloom's taxonomy and thus allow many advanced factors that affect student thinking to be measured and provide theory from research to help students to develop thinking skills (Heong et al., 2011).

Thinking is one of the variety of skills that need to be developed through the education process. HOT is a very important skill for educational purposes (Zohar & Dori, 2003). HOTS can be learned and can be taught to students (Thomas et al., 2011). By training HOTS, a better rate of learning can be achieved and student weaknesses can be addressed (Heong et al., 2011).

Istiyono et al. (2013) state that a test providing HOTS measurement should be developed. Such a test could help students to train and improve their thinking skills. Therefore, it is important to develop a Marzanoian higher-order thinking skills test for physics that is valid and reliable.

3 METHOD

This research was developed from a quantitative approach by using a model developed by Oriondo and Antonio. The research was conducted between early November 2016 and the end of March 2017. The development of the test began with a construct test, validation, and development of the instrument, which was then used at the end of March 2017. The research was conducted in five senior high schools in Sleman, Yogyakarta, Indonesia. The subjects of this research were all of the students in the 10th grade of each school, and 430 students participated.

The stages in developing the test using the Oriondo and Antonio model were: (1) construct the test, (2) experiment with the test, and (3) refine the test. The construction steps consisted of (1) early analysis, (2) defining specific test purpose, (3) subject analysis, (4) writing the blueprint, (5) writing the test items following Marzano's HOTS indicators, (6) writing the scoring manual, (7) validation, and (8) correction of the items, test constructs, and scoring manual. The testing steps of the instrument consisted of (1) testing in a high school (*Sekolah Menengah Atas* or SMA) in Sleman District, (2) data analysis, and (3) item correction, and the test instrument was then refined in consequence.

3.1 *Data analysis*

Analysis of this research was carried out using the Partial Credit Model (PCM) to test HOTS in physics in senior high schools. This form of analysis was considered appropriate because PCM, as a part of the Rasch model, requires samples to conduct a calibration, as compared to polytomous calibration in two-Parameter Logistic (2-PL) or 3-PL models (Keeves & Masters, 1999). The response characteristics also follow the PCM.

Data analysis was carried out in consideration of: (1) content validation, (2) goodness of fit of instrument items, (3) reliability, (4) item curve characteristic (ICC), (5) threshold index, and (6) information function and SEM.

3.2 *Content validity*

The process of content validation involved four experts in assessment and two physicists. This reflected the material indicators and the written material. The validation consisted of four categories—Correlates; Does not really correlate; Does not correlate; Really does not correlate—and all of the validations consist of scores 1, 2, 3 and 4. The next step was to calculate the Aiken index for each validated item. This step of the data analysis of the validity of the test instrument was carried out using Aiken's formula (Azwar, 2015).

$$V = \frac{s}{(n(c-1))} \tag{1}$$

V= Aiken's index; I0 = Minimum scale; r = from $I_0 + 1$ to $I_0 + (c - 1)$; and s = $r - I_0$

The second step was to convert each questionnaire item to qualitative data with a range of Aiken's index between 1 and 0. The result would be valid if the Aiken's V index was between 0.037 and 1.00.

3.3 *Goodness of fit*

The analysis of data was intended to indicate the validity and reliability of every item. This analysis was intended to show the overall goodness of fit test or just as a sample. Adam and Khoo (1996) state that the goodness of fit analysis can be referred to the average of the INFIT mean of square (mean INFITMNSQ) and its standard deviation or can observe the value of INFIT t (mean INFIT t) and its standard deviation. The reliability output of estimates of item tests can be interpreted and categorized as indicated in Table 1 (Sumintono & Widhiarso, 2015).

The output threshold, b, is assessed for each item. The item is appropriate for the test if the threshold is between −2.0 and 2.0 (−2.0 ≤ b ≤ 2.0). The threshold is easy if it is −2 and difficult if it is 2.0. The higher b is, the higher the ability required to answer the test.

3.4 *Information function and standard error of measurement*

Based on the information function, SEM can define the test for the student with high, middle, or low ability (θ).

4 RESULTS

The research and development produced Marzanoian physics HOTS tests for senior high schools. Two sets of questions were developed, A and B, each containing 35 items. Every set tested topics such as elasticity, Hooke's law, static fluids, temperature and heat, and optical instruments, and every item addressed aspects such as cognition, analysis, and practical knowledge. The results of the item estimation using PCM are shown in Table 2.

Figure 1 shows the information function and SEM. Based on Figure 1, the test is suitable for students with physics HOT (θ) in the high category, that is, $-1.9 \le \theta \le 2.1$.

The research and development resulted in two types of test instruments of Marzano's higher-order thinking skills for physics in senior high school students: an A test and a B test. The results of the testing produced 63 valid and reliable items. It can therefore be said that the Marzanoian HOTS physics test is valid and reliable.

Table 1.　Interpretation of reliability value.

Reliability test	Interpretation of reliability
>0.94	Perfect
0.91–0.94	Really Good
0.81–0.90	Good
0.67–0.80	Good Enough
<0.67	Weak

Table 2.　Item estimation results.

		Estimates	
No	Description	Item	Testee
1	Mean and standard deviation	−0.02 ± 0.28	−0.29 ± 0.18
2	Reliability	0.83	0.87
3	Mean and standard deviation of INFIT MNSQ	1.00 ± 0.08	0.99 ± 0.17
4	Mean and standard deviation of OUTFIT MNSQ	0.96 ± 0.10	0.96 ± 0.12
5	Mean and standard deviation of INFIT t	0.21 ± 1.68	0.29 ± 1.31
6	Mean and standard deviation of OUTFIT t	−0.51 ± 1.28	−0.18 ± 0.43

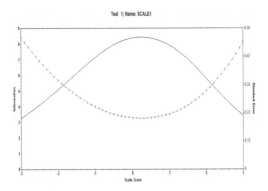

Test 1; Name: SCALE1

Figure 1.   The information function and SEM.

5   DISCUSSION

Initial analysis was conducted to establish the basic problems in the development of a Marzanoian HOTS physics test instrument for senior high schools. The analysis was conducted
by developing the test instrument for such testing.

5.1   *Content validity*

The characteristics of the test item parameters were reviewed for validation, reliability, and
index of difficulty. Based on the responses and assessments of four assessment experts, two
from physics education and two physicists, the value of the Aiken V index is 0.89 for Marzano's HOTS, so that, overall, the items that have been developed are valid according to the
experts' judgment and can be used in the next phase of the trial, the empirical trial test. This
is in accordance with the stages of interpretation developed by Kowsalya et al. in which the
items are valid if they are in the range of 0.67 to 1.00.

5.2   *Goodness of fit*

Based on the data analysis, the result of the trial is considered to fit according to the partial
credit model with polytomous data of four categories, and with the valid questions having
INFIT MNSQ in the range 0.83 to 1.11.
    The results of the analysis presented in Table 3 show that the difficulty level of the ques-
tion items in the Marzanoian HOTS physics test instrument is between −0.89 to 0.74. Based
on theory (Hambleton et al. 1985), it is said that a test is a fit if the difficulty index is in the
range −2.0 to 2.0 or can be expressed by the equation ($-2.0 \leq b \leq 2.0$). Therefore, it can be
concluded that all items developed are feasible or fit.

5.3   *Reliability*

Based on the analysis results obtained by estimates of reliability of the test instrument sets,
the summary of item estimates value is 0.83, and the summary of case estimates is 0.87.
Therefore, it can be concluded that the test instruments are reliable.
    Based on the analysis obtained, the information function and standard error of measure-
ment indicate that this test is suitable for physics students with high-level thinking ability (θ)
in the high category, that is, $-1.9 \leq \theta \leq 2.1$.
    This shows that if the ability of learners is at the high cognitive level then the test can be
completed by them. This is in accordance with the results of previous research (Istiyono,
2016). By training students' higher-order thinking skills, their learning performance can be
improved and their weaknesses can be reduced (Heong et al., 2011). Therefore, the use of
HOTS questions can be an indicator of educational quality improvement. Halaydina and

645

Table 3. Mean of difficulty index for categories 1, 2, 3, and 4.

No.	Aspect	Sub-aspect	Mean of difficulty (b)	Difficulty index (%)			
				Category 1	Category 2	Category 3	Category 4
1	Knowledge recall	Decision-making	−0.88	18.23	21.48	18.58	41.70
		Inductive reasoning	−0.61	16.30	21.84	20.96	40.90
2	Comprehension	Deductive reasoning	−0.24	15.65	17.25	24.30	42.80
		Constructing	−0.52	15.80	27.00	21.05	36.15
3	Analysis	Classifying	−0.01	13.90	19.60	31.45	35.05
		Analyzing error	−0.14	19.24	23.44	25.42	31.90
		Constructing support	−0.20	16.58	23.25	32.50	27.68
		Analyzing perspectives	0.07	16.58	17.63	30.22	35.57
		Investigation	0.08	24.83	17.57	20.20	37.40
4	Knowledge utilization	Problem-solving	0.09	23.53	17.90	24.47	34.10
		Experimental inquiry	0.87	18.47	24.03	35.90	21.60
		Invention	0.30	31.27	17.20	21.13	30.40
		Decision-making	0.89	27.97	25.83	39.63	6.57

Downing (1989) state that there are advantages to two-tier multiple choice questions, one of which is that they can be used to measure the cognitive abilities of students at a higher level.

6 CONCLUSION

The conclusions which can be drawn from this research and development of Marzanoian higher-order thinking skills physics test instruments are that an appropriate instrument consisting of package A and package B of 35 items, each with seven anchor items, has been developed. The physics subjects covered for the 10th grade of senior high school include elasticity, Hooke's law, static fluids, temperature and heat, and optical devices.

The Marzanoian physics higher-order thinking skills test instrument has feasibility characteristics based on the results of validation analysis from the subject experts, measurement experts and physics teachers. These produced an Aiken's V index of 0.895, which is categorized as "excellent".

Based on the data analysis, the result of the trial is considered fit using the partial credit model with polytomous data in four categories, and the entire validity of questions with INFIT MNSQ is in the range 0.83 to 1.11.

All items in the Marzanoian higher-order thinking skills physics test instrument are in the "good" criterion, with difficulty levels ranging from −2.0 to 2.0.

Based on the information function, the Marzanoian higher-order thinking skills physics test instrument can be very appropriately used to measure students' higher thinking ability in an ability range from −1.9 to 2.1.

REFERENCES

Adam, R.J. & Khoo, S.T. (1996). Quest: The interactive test analysis system version 2.1. *Victoria, Australia: The Australian Council for Educational Research.*

Azwar, S. (2015). *Sikap manusia teori dan penerapannya* [Human's behavior theory and its application]. *Yogyakarta, Indonesia: Pustaka Pelajar.*

Conklin, W. (2012). Higher order thinking skills to develop 21st century learners. *Huntington Beach, CA: Shell Education.*

Ennis, R.H. (1987). A taxonomy of critical thinking dispositions and abilities. In J. Baron & R. Sternberg (Eds.), Teaching thinking skills: Theory and practice. *New York, NY: W.H. Freeman.*

Halaydina, T.M. & Downing, S.M. (1989). A taxonomy of multiple choice item writing rules. *Applied Measurements in Education, 2*(1), 37–50.

Hambleton, R.K., Swaminathan, H., & Rogers, H.J. (1985). Item response theory: Principles and applications. Boston, United States of America: Kluwer.

Heong, M.Y., Othman, W., Yunos, J.M., Tee, T.K., bin Hassan, R. & Mohaffyza, M. (2011). The level of Marzano higher order thinking skills among technical education students. *International Journal of Social Science and Humanity, 1*(2), 121–125.

Istiyono, E. (2016). The application of GPCM on MMC test as a fair alternative assessment model in physics learning. In *Proceedings of International Conference on Research, Implementation and Education of Mathematics and Sciences (ICRIEMS), 16–17 May 2016, Yogyakarta, Indonesia* (pp. PE25–PE29). Yogyakarta, Indonesia: Faculty of Mathematics and Natural Science, Yogyakarta State University. Retrieved from http://seminar.uny.ac.id/icriems/sites/seminar.uny.ac.id.icriems/files/prosiding/PE-04.pdf.

Istiyono, E., Mardapi, D. & Suparno. (2013). *Pengembangan tes kemampuan berpikir tingkat tinggi fisika (PhysTHOTS) peserta didik SMA* [Developing higher order thinking skill test of physics (PhysTHOTS) for senior high school students]. *Jurnal Penelitian dan Evaluasi Pendidikan, 14*(1), 1–12.

Istiyono, E., Mardapi, D. & Suparno. (2014). Effectiveness of reasoned objective choice test to measure higher order thinking skills in physics implementing of curriculum 2013. In *Proceedings of International Conference on Educational Research and Evaluation (ICERE) 2014*. Yogyakarta, Indonesia: Graduate Program, Universitas Negeri Yogyakarta. Retrieved from http://eprints.uny.ac.id/24883/.

Johnson, E.B. (2000). Contextual teaching and learning. *Thousand Oaks, CA: Corwin Press.*

Keeves, J.P. & Masters, G.N. (1999). Advances in measurement in educational research and assessment. *Amsterdam, The Netherlands: Pergamon.*

Kowsalya, D.N., Lakshmi, H.V.,M & Suresh, K.P. (2012). Development and Validation of a Scale to assess Self-Concept in Mild Intellectually Disabled Children. *Int. j. Sci. Educ., vol. 2, no. 4.*

Marzano, R.J., & Guskey, T.R. (2001). Designing a New Taxonomy of Educational Objectives. *Michigan, United States of America: Corwin Press.*

McLoughlin, C. & Luca, J. (2000). Cognitive engagement and higher order thinking through computer conferencing: We know why but do we know how? In A. Herrmann & M.M. Kulski (Eds.), *Flexible Futures in Tertiary Teaching: Proceedings of the 9th Annual Teaching Learning Forum, 2000.* Perth, Australia: Curtin University of Technology. Retrieved from http://clt.curtin.edu.au/events/conferences/tlf/tlf2000/mcloughlin.html.

Megawati, E. (2012). *Pengembangan perangkat tes kimia dalam rangka pembentukan bank soal di Kabupaten Paser di Kalimantan Timur* [Developing a chemistry in the framework of establishment of item bank in Paser East Kalimantan] (Unpublished thesis, Universitas Negeri Yogyakarta, Indonesia).

Newman, F.M. & Wehlage, G.G. (1993). Five standards of authentic instruction. *Educational Leadership, 50*(7), 8–12.

OECD. (2016). Indonesia – OECD Data. *Paris, France: Organisation for Economic Co-operation and Development.* Retrieved from https://data.oecd.org/indonesia.htm.

Oriondo, L.L., & Antonio, D. (1998). Evaluating educational outcomes (test, measurement, and evaluation). *Quezon City, Philippine: REX Printing Company*..Resnick, L.B. (1987). Education and learning to think. *Washington, DC: National Academy Press.*

Reynolds, C.R., Livingston, R.B. & Willson, V. (2009). Measurement and assessment in education (2nd ed.). *Upper Saddle River, NJ: Pearson.*

Sudjana, N. (1990). *Teori-teori belajar untuk pengajaran* [Learning theories for teaching]. Jakarta, Indonesia: Lembaga Penerbit Fakultas Ekonomi UI.

Sumintono, B. & Widhiarso, W. (2015). *Aplikasi model rasch untuk penelitian ilmu-ilmu sosial* [Application of rasearch model for social sciences research]. Jakarta, Indonesia: TrimKom.

Thomas, A., & Thorne, G. (2011). High Order Thinking – It's HOT. Retrieved from http://cdl.org/resource-library/pdf/feb00PTHOT.pdf.

Zohar, A. & Dori, Y.J. (2003). Higher order thinking skills and low-achieving students: Are they mutually exclusive? *The Journal of Learning Sciences, 12*(2), 145–181.

Character Education for 21st Century Global Citizens – Retnowati et al. (Eds)
© 2019 Taylor & Francis Group, London, ISBN 978-1-138-09922-7

Author index